# THE CHEMISTRY PROBLEM SOLVER®

REGISTERED TRADEMARK

## A Complete Solution Guide to Any Textbook

**Staff of Research and Education Association**
**Dr. M. Fogiel, Director**

special chapter reviews by
**A. Lamont Tyler, Ph.D.**
Chairperson and Professor of Chemical Engineering
University of Utah
Salt Lake City, Utah

Research and Education Association
61 Ethel Road West
Piscataway, New Jersey 08854

# THE CHEMISTRY
# PROBLEM SOLVER®

## Year 2005 Printing

Copyright © 2001, 1998, 1996, 1993, 1977 by
Research & Education Association, Inc. All rights
reserved. No part of this book may be
reproduced in any form without permission of
the publisher.

Printed in the United States of America

Library of Congress Control Number 00-134401

International Standard Book Number 0-87891-509-5

K04

# WHAT THIS BOOK IS FOR

For as long as chemistry has been taught in schools, students have found this subject difficult to understand and learn. Despite the publication of hundreds of textbooks in this field, each one intended to provide an improvement over previous textbooks, students continue to remain perplexed, and the subject is often taken in class only to meet school/departmental requirements for a selected course of study.

In a study of the problem, REA found the following basic reasons underlying students' difficulties with chemistry taught in schools:

(a) No systematic rules of analysis have been developed which students may follow in a step-by-step manner to solve the usual problems encountered. This results from the fact that the numerous different conditions and principles which may be involved in a problem, lead to many possible different methods of solution. To prescribe a set of rules to be followed for each of the possible variations, would involve an enormous number of rules and steps to be searched through by students, and this task would perhaps be more burdensome than solving the problem directly with some accompanying trial and error to find the correct solution route.

(b) Textbooks currently available will usually explain a given principle in a few pages written by a professional who has an insight in the subject matter that is not shared by students. The explanations are often written in an abstract manner which leaves the students confused as to the application of the principle. The explanations given are not sufficiently detailed and extensive to make the student aware of the wide range of applications and different aspects of the principle being studied. The numerous possible variations of principles and their applications are usually not discussed, and it is left for the students to discover these for themselves while doing exercises. Accordingly, the average student is expected to rediscover that which has been long known and practiced, but not published or explained extensively.

(c) The examples usually following the explanation of a topic are too few in number and too simple to enable the student to obtain a thorough grasp of the principles involved. The explanations do not provide sufficient basis to enable a student to solve problems that may be subsequently assigned for homework or given on examinations.

The examples are presented in abbreviated form which leaves out much material between steps, and requires that students derive the omitted material themselves. As a result, students find the examples difficult to understand--contrary to the purpose of the examples.

Examples are, furthermore, often worded in a confusing manner. They do not state the problem and then present the solution. Instead, they pass through a general discussion, never revealing what is to be solved for.

Examples, also, do not always include diagrams/graphs, wherever appropriate, and students do not obtain the training to draw diagrams or graphs to simplify and organize their thinking.

(d) Students can learn the subject only by doing the exercises themselves and reviewing them in class, to obtain experience in applying the principles with their different ramifications.

In doing the exercises by themselves, students find that they are required to devote considerably more time to chemistry than to other subjects of comparable credits, because they are uncertain with regard to the selection and application of the theorems and principles involved. It is also necessary for students to discover those "tricks" not revealed in their texts (or review books) that make it possible to solve problems easily. Students must usually resort to methods of trial-and-error to discover these "tricks" and as a result they find that they may sometimes spend several hours to solve a single problem.

(e) When reviewing the exercises in classrooms, instructors

usually request students to take turns in writing solutions on the boards and explaining them to the class. Students often find it difficult to explain in a manner that holds the interest of the class, and enables the remaining students to follow the material written on the boards. The remaining students seated in the class are, furthermore, too occupied with copying the material from the boards, to listen to the oral explanations and concentrate on the methods of solution.

This book is intended to aid students in chemistry to overcome the difficulties described, by supplying detailed illustrations of the solution methods which are usually not apparent to students. The solution methods are illustrated by problems selected from those that are most often assigned for class work and given on examinations. The problems are arranged in order of complexity to enable students to learn and understand a particular topic by reviewing the problems in sequence. The problems are illustrated with detailed step-by-step explanations, to save the students the large amount of time that is often needed to fill in the gaps that are usually found between steps of illustrations in textbooks or review/outline books.

The staff of REA considers chemistry a subject that is best learned by allowing students to view the methods of analysis and solution techniques themselves. This approach to learning the subject matter is similar to that practiced in various scientific laboratories, particularly in the  medical fields.

In using this book, students may review and study the illustrated problems at their own pace; they are not limited to the time allowed for explaining problems on the board in class.

When students want to look up a particular type of problem and solution, they can readily locate it in the book by referring to the index which has been extensively prepared. It is also possible to locate a particular type of problem by glancing at just the material within the boxed portions. To facilitate rapid scanning of the problems, each problem has a heavy border around it. Furthermore, each problem is identified with a number immediately above the problem at the right-hand margin.

To obtain maximum benefit from the book, students should familiarize themselves with the section, "How To Use This Book," located in the front pages.

To meet the objectives of this book, staff members of REA have selected problems usually encountered in assignments and examinations, and have solved each problem meticulously to illustrate the steps which are usually difficult for students to comprehend. Gratitude for their patient work in this area is due to Barbara Dunkle, Sheldon Lipski, Victor Collymore, Roberto Dominjanni and the contributors who devoted short periods of time to this work. Barbara Dunkle and Sheldon Lipski deserve special praise for their exceptional outstanding efforts and competence.

Gratitude is also expressed to the many persons involved in the difficult task of typing the manuscript with its endless changes, and to the REA art staff who prepared the numerous detailed illustrations together with the layout and physical features of the book.

Finally, special thanks are due to Helen Kaufmann for her unique talents to render those difficult border-line decisions and constructive suggestions related to the design and organization of the book.

Max Fogiel, Ph.D.
Program Director

# HOW TO USE THIS BOOK

Chemistry students will find the *Chemistry Problem Solver®* to be an invaluable supplement to their textbooks. The book is subdivided into 32 chapters, each dealing with a separate topic. The subject matter is developed beginning with fundamental gas laws and extending through organic chemistry, biochemistry, and applied problems. Also included are sections on equilibrium, chemical kinetics, thermodynamics, and quantum chemistry.

Twenty-three of our 32 chapters begin with a special section entitled "Basic Attacks and Strategies for Solving Problems in this Chapter." This section explains the principles that are applicable to the topics in the chapter. By reviewing these principles, the student can acquire a good grasp of the underlying techniques and strategies through which problems related to the chapter can be solved.

# HOW TO LEARN AND UNDERSTAND A TOPIC THOROUGHLY

1. Refer to your class text and read the section pertaining to the topic. You should become acquainted with the principles discussed there. These principles, however, may not be clear to you at the time.

2. Locate the topic you are looking for by referring to the Table of Contents in the front of this book. After turning to the appropriate chapter, read the section "Basic Attacks and Strategies for Solving Problems in this Chapter." This section is a review of the important principles related to the chapter, and it will help you to understand further how and why problems in the chapter are solved in the manner shown.

3. Turn to the page where the topic begins and review the problems under each topic, in the order given. For each topic, the problems are arranged in order of complexity, from the simplest to the more difficult. Some problems may appear similar to others, but each problem has been selected to illustrate a different point or solution method.

To learn and understand a topic thoroughly and retain its contents, it will be generally necessary for students to review the problems several times. Repeated review is essential in order to gain experience in recognizing the principles that should be applied, and to select the best solution technique.

# HOW TO FIND A PARTICULAR PROBLEM

To locate one or more problems related to any particular subject matter, refer to the index. In using the index, be certain to note that the numbers given there refer to problem numbers, not to page numbers. This arrangement of the index is intended to facilitate finding a problem more rapidly, since two or more problems may appear on a page.

If a particular type of problem cannot be readily found, it is recommended that the student refer to the Table of Contents and then turn to the chapter which is applicable to the problem being sought. By scanning or glancing at the material that is boxed, it will generally be possible to find problems related to the one being sought, without consuming considerable time. After the problems have been located, the solutions can be reviewed and studied in detail.

For the purpose of locating problems rapidly, students should acquaint themselves with the organization of the book as found in the Table of Contents.

In preparing for an exam, it is useful to find the topics to be covered in the exam from the Table of Contents, and then review the problems under those topics several times. This should equip the student with what might be needed for the exam.

# CONTENTS

# UNITS CONVERSION FACTORS

This section includes a particularly useful and comprehensive table to aid students and teachers in converting between systems of units.

The problems and their solutions in this book use **SI (International System)** as well as English units. Both of these units are in extensive use throughout the world, and therefore students should develop a good facility to work with both sets of units until a single standard of units has been found acceptable internationally.

In working out or solving a problem in one system of units or the other, essentially only the numbers change. Also, the conversion from one unit system to another is easily achieved through the use of conversion factors that are given in the subsequent table. Accordingly, the units are one of the least important aspects of a problem. For these reasons, a student should not be concerned mainly with which units are used in any particular problem. Instead, a student should obtain from that problem and its solution an understanding of the underlying principles and solution techniques that are illustrated there.

| To convert | To | Multiply by | For the reverse, multiply by |
|---|---|---|---|
| acres | square feet | $4.356 \times 10^4$ | $2.296 \times 10^{-5}$ |
| acres | square meters | 4047 | $2.471 \times 10^{-4}$ |
| ampere-hours | coulombs | 3600 | $2.778 \times 10^{-4}$ |
| ampere-turns | gilberts | 1.257 | 0.7958 |
| ampere-turns per cm. | ampere-turns per inch | 2.54 | 0.3937 |
| angstrom units | inches | $3.937 \times 10^{-9}$ | $2.54 \times 10^8$ |
| angstrom units | meters | $10^{-10}$ | $10^{10}$ |
| atmospheres | feet of water | 33.90 | 0.02950 |
| atmospheres | inch of mercury at 0°C | 29.92 | $3.342 \times 10^{-2}$ |
| atmospheres | kilogram per square meter | $1.033 \times 10^4$ | $9.678 \times 10^{-5}$ |
| atmospheres | millimeter of mercury at 0°C | 760 | $1.316 \times 10^{-3}$ |
| atmospheres | pascals | $1.0133 \times 10^5$ | $0.9869 \times 10^{-5}$ |
| atmospheres | pounds per square inch | 14.70 | 0.06804 |
| bars | atmospheres | $9.870 \times 10^{-7}$ | 1.0133 |
| bars | dynes per square cm. | $10^6$ | $10^{-6}$ |
| bars | pascals | $10^5$ | $10^{-5}$ |
| bars | pounds per square inch | 14.504 | $6.8947 \times 10^{-2}$ |
| Btu | ergs | $1.0548 \times 10^{10}$ | $9.486 \times 10^{-11}$ |
| Btu | foot-pounds | 778.3 | $1.285 \times 10^{-3}$ |
| Btu | joules | 1054.8 | $9.480 \times 10^{-4}$ |
| Btu | kilogram-calories | 0.252 | 3.969 |
| calories, gram | Btu | $3.968 \times 10^{-3}$ | 252 |
| calories, gram | foot-pounds | 3.087 | 0.324 |
| calories, gram | joules | 4.185 | 0.2389 |
| Celsius | Fahrenheit | (°C × 9/5) + 32 = °F | (°F − 32) × 5/9 = °C |

| To convert | To | Multiply | For the reverse, multiply by |
|---|---|---|---|
| Celsius | kelvin | °C + 273.1 = K | K − 273.1 = °C |
| centimeters | angstrom units | $1 \times 10^8$ | $1 \times 10^{-8}$ |
| centimeters | feet | 0.03281 | 30.479 |
| centistokes | square meters per second | $1 \times 10^{-6}$ | $1 \times 10^6$ |
| circular mils | square centimeters | $5.067 \times 10^{-6}$ | $1.973 \times 10^5$ |
| circular mils | square mils | 0.7854 | 1.273 |
| cubic feet | gallons (liquid U.S.) | 7.481 | 0.1337 |
| cubic feet | liters | 28.32 | $3.531 \times 10^{-2}$ |
| cubic inches | cubic centimeters | 16.39 | $6.102 \times 10^{-2}$ |
| cubic inches | cubic feet | $5.787 \times 10^{-4}$ | 1728 |
| cubic inches | cubic meters | $1.639 \times 10^{-5}$ | $6.102 \times 10^4$ |
| cubic inches | gallons (liquid U.S.) | $4.329 \times 10^{-3}$ | 231 |
| cubic meters | cubic feet | 35.31 | $2.832 \times 10^{-2}$ |
| cubic meters | cubic yards | 1.308 | 0.7646 |
| curies | coulombs per minute | $1.1 \times 10^{12}$ | $0.91 \times 10^{-12}$ |
| cycles per second | hertz | 1 | 1 |
| degrees (angle) | mils | 17.45 | $5.73 \times 10^{-2}$ |
| degrees (angle) | radians | $1.745 \times 10^{-2}$ | 57.3 |
| dynes | pounds | $2.248 \times 10^{-6}$ | $4.448 \times 10^5$ |
| electron volts | joules | $1.602 \times 10^{-19}$ | $0.624 \times 10^{18}$ |
| ergs | foot-pounds | $7.376 \times 10^{-8}$ | $1.356 \times 10^7$ |
| ergs | joules | $10^{-7}$ | $10^7$ |
| ergs per second | watts | $10^{-7}$ | $10^7$ |
| ergs per square cm. | watts per square cm. | $10^{-3}$ | $10^3$ |
| Fahrenheit | kelvin | (°F + 459.67)/1.8 | 1.8K − 459.67 |
| Fahrenheit | Rankine | °F + 459.67 = °R | °R − 459.67 = °F |
| faradays | ampere-hours | 26.8 | $3.731 \times 10^{-2}$ |
| feet | centimeters | 30.48 | $3.281 \times 10^{-2}$ |
| feet | meters | 0.3048 | 3.281 |
| feet | mils | $1.2 \times 10^4$ | $8.333 \times 10^{-5}$ |
| fermis | meters | $10^{-15}$ | $10^{15}$ |
| foot candles | lux | 10.764 | 0.0929 |
| foot lamberts | candelas per square meter | 3.4263 | 0.2918 |
| foot-pounds | gram-centimeters | $1.383 \times 10^4$ | $1.235 \times 10^{-5}$ |
| foot-pounds | horsepower-hours | $5.05 \times 10^{-7}$ | $1.98 \times 10^6$ |
| foot-pounds | kilogram-meters | 0.1383 | 7.233 |
| foot-pounds | kilowatt-hours | $3.766 \times 10^{-7}$ | $2.655 \times 10^6$ |
| foot-pounds | ounce-inches | 192 | $5.208 \times 10^{-3}$ |
| gallons (liquid U.S.) | cubic meters | $3.785 \times 10^{-3}$ | 264.2 |
| gallons (liquid U.S.) | gallons (liquid British Imperial) | 0.8327 | 1.201 |
| gammas | teslas | $10^{-9}$ | $10^9$ |
| gausses | lines per square cm. | 1.0 | 1.0 |
| gausses | lines per square inch | 6.452 | 0.155 |
| gausses | teslas | $10^{-4}$ | $10^4$ |
| gausses | webers per square inch | $6.452 \times 10^{-8}$ | $1.55 \times 10^7$ |
| gilberts | amperes | 0.7958 | 1.257 |
| grads | radians | $1.571 \times 10^{-2}$ | 63.65 |
| grains | grams | 0.06480 | 15.432 |
| grains | pounds | $1/_{7000}$ | 7000 |
| grams | dynes | 980.7 | $1.02 \times 10^{-3}$ |
| grams | grains | 15.43 | $6.481 \times 10^{-2}$ |

| To convert | To | Multiply | For the reverse, multiply by |
|---|---|---|---|
| grams | ounces (avdp) | $3.527 \times 10^{-2}$ | 28.35 |
| grams | poundals | $7.093 \times 10^{-2}$ | 14.1 |
| hectares | acres | 2.471 | 0.4047 |
| horsepower | Btu per minute | 42.418 | $2.357 \times 10^{-2}$ |
| horsepower | foot-pounds per minute | $3.3 \times 10^4$ | $3.03 \times 10^{-5}$ |
| horsepower | foot-pounds per second | 550 | $1.182 \times 10^{-3}$ |
| horsepower | horsepower (metric) | 1.014 | 0.9863 |
| horsepower | kilowatts | 0.746 | 1.341 |
| inches | centimeters | 2.54 | 0.3937 |
| inches | feet | $8.333 \times 10^{-2}$ | 12 |
| inches | meters | $2.54 \times 10^{-2}$ | 39.37 |
| inches | miles | $1.578 \times 10^{-5}$ | $6.336 \times 10^4$ |
| inches | mils | $10^3$ | $10^{-3}$ |
| inches | yards | $2.778 \times 10^{-2}$ | 36 |
| joules | foot-pounds | 0.7376 | 1.356 |
| joules | watt-hours | $2.778 \times 10^{-4}$ | 3600 |
| kilograms | tons (long) | $9.842 \times 10^{-4}$ | 1016 |
| kilograms | tons (short) | $1.102 \times 10^{-3}$ | 907.2 |
| kilograms | pounds (avdp) | 2.205 | 0.4536 |
| kilometers | feet | 3281 | $3.408 \times 10^{-4}$ |
| kilometers | inches | $3.937 \times 10^4$ | $2.54 \times 10^{-5}$ |
| kilometers per hour | feet per minute | 54.68 | $1.829 \times 10^{-2}$ |
| kilowatt-hours | Btu | 3413 | $2.93 \times 10^{-4}$ |
| kilowatt-hours | foot-pounds | $2.655 \times 10^6$ | $3.766 \times 10^{-7}$ |
| kilowatt-hours | horsepower-hours | 1.341 | 0.7457 |
| kilowatt-hours | joules | $3.6 \times 10^6$ | $2.778 \times 10^{-7}$ |
| knots | feet per second | 1.688 | 0.5925 |
| knots | miles per hour | 1.1508 | 0.869 |
| lamberts | candles per square cm. | 0.3183 | 3.142 |
| lamberts | candles per square inch | 2.054 | 0.4869 |
| liters | cubic centimeters | $10^3$ | $10^{-3}$ |
| liters | cubic inches | 61.02 | $1.639 \times 10^{-2}$ |
| liters | gallons (liquid U.S.) | 0.2642 | 3.785 |
| liters | pints (liquid U.S.) | 2.113 | 0.4732 |
| lumens per square foot | foot-candles | 1 | 1 |
| lumens per square meter | foot-candles | 0.0929 | 10.764 |
| lux | foot-candles | 0.0929 | 10.764 |
| maxwells | kilolines | $10^{-3}$ | $10^3$ |
| maxwells | webers | $10^{-8}$ | $10^8$ |
| meters | feet | 3.28 | $30.48 \times 10^{-2}$ |
| meters | inches | 39.37 | $2.54 \times 10^{-2}$ |
| meters | miles | $6.214 \times 10^{-4}$ | 1609.35 |
| meters | yards | 1.094 | 0.9144 |
| miles (nautical) | feet | 6076.1 | $1.646 \times 10^{-4}$ |
| miles (nautical) | meters | 1852 | $5.4 \times 10^{-4}$ |
| miles (statute) | feet | 5280 | $1.894 \times 10^{-4}$ |
| miles (statute) | kilometers | 1.609 | 0.6214 |
| miles (statute) | miles (nautical) | 0.869 | 1.1508 |
| miles per hour | feet per second | 1.467 | 0.6818 |
| miles per hour | knots | 0.8684 | 1.152 |
| millimeters | microns | $10^3$ | $10^{-3}$ |

| To convert | To | Multiply | For the reverse, multiply by |
|---|---|---|---|
| mils | meters | $2.54 \times 10^{-5}$ | $3.94 \times 10^{4}$ |
| mils | minutes | 3.438 | 0.2909 |
| minutes (angle) | degrees | $1.666 \times 10^{-2}$ | 60 |
| minutes (angle) | radians | $2.909 \times 10^{-4}$ | 3484 |
| newtons | dynes | $10^{5}$ | $10^{-5}$ |
| newtons | kilograms | 0.1020 | 9.807 |
| newtons per sq. meter | pascals | 1 | 1 |
| newtons | pounds (avdp) | 0.2248 | 4.448 |
| oersteds | amperes per meter | $7.9577 \times 10$ | $1.257 \times 10^{-2}$ |
| ounces (fluid) | quarts | $3.125 \times 10^{-2}$ | 32 |
| ounces (avdp) | pounds | $6.25 \times 10^{-2}$ | 16 |
| pints | quarts (liquid U.S.) | 0.50 | 2 |
| poundals | dynes | $1.383 \times 10^{4}$ | $7.233 \times 10^{-5}$ |
| poundals | pounds (avdp) | $3.108 \times 10^{-2}$ | 32.17 |
| pounds | grams | 453.6 | $2.205 \times 10^{-3}$ |
| pounds (force) | newtons | 4.4482 | 0.2288 |
| pounds per square inch | dynes per square cm. | $6.8946 \times 10^{4}$ | $1.450 \times 10^{-5}$ |
| pounds per square inch | pascals | $6.895 \times 10^{3}$ | $1.45 \times 10^{-4}$ |
| quarts (U.S. liquid) | cubic centimeters | 946.4 | $1.057 \times 10^{-3}$ |
| radians | mils | $10^{3}$ | $10^{-3}$ |
| radians | minutes of arc | $3.438 \times 10^{3}$ | $2.909 \times 10^{-4}$ |
| radians | seconds of arc | $2.06265 \times 10^{5}$ | $4.848 \times 10^{-6}$ |
| revolutions per minute | radians per second | 0.1047 | 9.549 |
| roentgens | coulombs per kilogram | $2.58 \times 10^{-4}$ | $3.876 \times 10^{3}$ |
| slugs | kilograms | 1.459 | 0.6854 |
| slugs | pounds (avdp) | 32.174 | $3.108 \times 10^{-2}$ |
| square feet | square centimeters | 929.034 | $1.076 \times 10^{-3}$ |
| square feet | square inches | 144 | $6.944 \times 10^{-3}$ |
| square feet | square miles | $3.587 \times 10^{-8}$ | $27.88 \times 10^{6}$ |
| square inches | square centimeters | 6.452 | 0.155 |
| square kilometers | square miles | 0.3861 | 2.59 |
| stokes | square meter per second | $10^{-4}$ | $10^{-4}$ |
| tons (metric) | kilograms | $10^{3}$ | $10^{-3}$ |
| tons (short) | pounds | 2000 | $5 \times 10^{-4}$ |
| torrs | newtons per square meter | 133.32 | $7.5 \times 10^{-3}$ |
| watts | Btu per hour | 3.413 | 0.293 |
| watts | foot-pounds per minute | 44.26 | $2.26 \times 10^{-2}$ |
| watts | horsepower | $1.341 \times 10^{-3}$ | 746 |
| watt-seconds | joules | 1 | 1 |
| webers | maxwells | $10^{8}$ | $10^{-8}$ |
| webers per square meter | gausses | $10^{4}$ | $10^{-4}$ |

# CHAPTER 1

# UNITS OF MEASUREMENT

> **Basic Attacks and Strategies for Solving Problems in this Chapter. See pages 1 to 12 for step-by-step solutions to problems.**

Any measured or calculated quantity has two parts — a numerical value and a unit. Sometimes, when the unit is likely to be understood, it is not written. For example, if the United States Weather Service reports that the temperature in Phoenix is 108, it is understood to be 108°F. However, fewer errors in scientific and engineering calculations are made when the rule of always writing both the value and the unit for each quantity is strictly observed as illustrated below.

2 grams (2 g), 0.5 seconds (0.5 s), 6.28 meters (6.28 m), 98.6°F

There are only seven quantities — length, mass, time, thermodynamic (or absolute) temperature, electric current, amount of a substance, and luminous intensity — for which base units are defined and from which all other units can be derived. These quantities and their respective units and symbols are listed in the table below.

| Table 1 | | |
|---|---|---|
| PHYSICAL QUANTITY | SI UNIT | SYMBOL |
| Length | meter | m |
| Mass | kilogram | kg |
| Time | second | s |
| Amt. of substance | mole | mol |
| Thermodynamic temp. | kelvin | K |
| Electric current | ampere | A |
| Luminous intensity | candela | cd |

A unit of velocity is derived, for example, by dividing a unit of length by a unit of time to obtain miles per hour, feet per second, furlongs per fortnight, etc. In this chapter, only base units for length, mass, and absolute temperature are discussed, along with derived units of area (length squared), volume (length cubed), density (mass divided by volume), and the common Celsius (°C) and Fahrenheit (°F) temperature scales. If the rules illustrated with these units are learned and understood, they can be applied to any units that may be encountered.

Units can be treated like ordinary algebraic variables in equations when quantities are added, subtracted, multiplied, or divided. Two rules, borrowed from algebra, must be remembered:

1. The numerical values of two quantities can be added or subtracted **only** if the units are the same. This is true whether or not the units are the same quantity — e.g., the mixture of mass and length units in (b) or the two different length units in (c) below.

   a) 5 feet − 2 feet = 3 feet  $(5x − 2x = 3x)$
   b) 5 grams − 2 feet = ?    $(5y − 2x = ?)$
   c) 5 meters − 2 feet = ?    $(5z − 2x = ?)$

2. Numerical values and their associated units can **always** be multiplied or divided.
   a) 15 grams ÷ 5 cm$^3$ = 3 g/cm$^3$    $(15y ÷ 5x = 3y/x)$
   b) 7 cm × 5 cm = 35 cm$^2$ $(7y × 5y = 35y^2)$

## Conversion of Units

Most new scientific work is now performed and reported in the International System of Units (SI units) but, for practical reasons, many other systems of units are still in use and will remain so for many years. It is essential, therefore, that rules for converting from one system or set of units to another be understood. There is one simple rule that covers most cases for converting from one unit or set of units to another:

Multiply or divide by the appropriate conversion factor or factors **including the units** — i.e., be certain to multiply and/or divide the units of each conversion factor as well as the numerical values.

The exception to this rule occurs when the zero points of two units, such as exists with the Celsius and Fahrenheit temperature scales, are different. The conversion then requires a linear transformation — i.e., $y = mx + b$, where $m$ is the ratio of the magnitude of the unit $y$ to the unit $x$, and $b$ is the value of $y$ when $x$ is zero. Except for the example of temperature scales, such cases are rare, indeed. In virtually all other cases, the zero point is independent of the unit; a velocity of zero, for example, is zero in any velocity unit — ft/sec, miles/hr, m/sec, etc.

Conversion factors are tabulated in handbooks but, in simple cases, can be drawn from memory. Conversion factors for units formed by adding the common prefixes — centi (c), milli (m), kilo (k), etc. — are simply the factors associated with those prefixes — $10^{-2}$, $10^{-3}$, $10^3$, etc., — respectively. Let us convert, as an example, a length measurement from feet to inches. To do this we multiply or divide by the appropriate conversion factor including its units. In this case, most people know the conversion factor(s) from memory — 12 inches/ft or 1 ft/12 inches.

15 ft × 12 inches/ft = 180 inches                   1-1

Note that the numerical values ($15 \times 12 = 180$) and the units (ft × inches/ft = inches) are each multiplied separately to obtain the correct answer which includes both a value and a unit. The reciprocal conversion factor, 1 ft/12 inches, can also be used, but it is necessary, then, to divide rather than to multiply.

15 ft ÷ 1 ft/12 inches = (15 ÷ 1/12)[ft ÷ (ft/in)] = 180 in          1-2

Observe how careful multiplication or division of the units as well as the numerical values can help to avoid errors in the conversion of units. In this case, suppose that the conversion factor in the first equation, 12 in/ft, was used but division rather than multiplication was employed mistakenly.

15 ft ÷ 12 inches/ft = (15/12)[ft ÷ in/ft] = 1.25 ft$^2$/in          1-3

The unexpected unit in the answer signals that an error has been made.

Several successive multiplication and/or division steps can be used to convert one or more units. If careful multiplication and/or division of the units, as well as the numerical values is observed, the correct unit will result. If the expected unit is not obtained, the probability of an error in the calculation is signaled. Let us convert, as an example, the density of water,

62.4 lb/ft³, to the appropriate value in grams per cubic centimeter. The answer, in this case, is known in advance to be 1.0 g/cm³ or 1.0 g/cc. The needed conversion factors (454 g/lb, 2.54 cm/in, 12 in/ft) can be obtained from a handbook with a table of conversion factors.

$$62.4 \text{ lb/ft}^3 \times 454 \text{ g/lb} \div [12 \text{ in/ft} \times 2.54 \text{ cm/in}]^3 =$$

$$62.4 \times 454 \div (12 \times 2.54)^3 \, [\text{lb/ft}^3 \times \text{g/lb} \div (\text{in/ft} \times \text{cm/in})^3] = 1.0 \text{ g/cm}^3$$

$$1\text{-}4$$

Many tables of conversion factors contain a factor for conversion directly from centimeters to feet (30.48 cm/ft) and from cubic centimeters to cubic feet (28,317 cm³/ft³). These factors can also be used and the calculation is simplified as follows.

$$62.4 \text{ lb/ft}^3 \times 454 \text{ g/lb} \div (30.48 \text{ cm/ft})^3 = 1.0 \text{ g/cm}^3 \qquad 1\text{-}5$$

$$62.4 \text{ lb/ft}^3 \times 454 \text{ g/lb} \div 28{,}317 \text{ cm}^3/\text{ft}^3 = 1.0 \text{ g/cm}^3 \qquad 1\text{-}6$$

As shown, more than one procedure can lead to the correct answer, and some may result in fewer or simpler calculations than others. If careful attention is paid to the multiplication and division of units as well as numerical values, the chance of making an error is diminished.

Step-by-Step Solutions to
Problems in this Chapter,
"Units of Measurement"

## LENGTH

● **PROBLEM** 1

The Eiffel Tower is 984 feet high. Express this height in meters, in kilometers, in centimeters, and in millimeters.

Solution: A meter is equivalent to 39.370 inches. In this problem, the height of the tower in feet must be converted to inches and then the inches can be converted to meters. There are 12 inches in 1 foot. Therefore, feet can be converted to inches by using the factor 12 inches/1 foot.

984 feet × 12 inches/1 foot = 118 × 10² inches.

Once the height is found in inches, this can be converted to meters by the factor 1 meter/39.370 inches.

11808 inches × 1 meter/39.370 inches = 300 m.

Therefore, the height in meters is 300 m.

There are 1,000 meters in one kilometer. Meters can be converted to kilometers by using the factor 1 km/1000 m.

300 m × 1 km/1000 m = .300 km.

As such, there are .300 kilometers in 300 m.

There are 100 centimeters in 1 meter, thus meters can be converted to centimeters by multiplying by the factor 100 cm/1 m.

300 m × 100 cm/1 m = 300 × 10² cm.

There are 30,000 centimeters in 300 m.

There are 1,000 millimeters in 1 meter; therefore,

1

meters can be converted to millimeters by the factor 1000 mm./1 m.

$$300 \text{ m} \times 1,000 \text{ mm}/1 \text{ m} = 300 \times 10^3 \text{ mm.}$$

There are 300,000 millimeters in 300 meters.

The unaided eye can perceive objects which have a diameter of 0.1 mm. What is the diameter in inches?

Solution: From a standard table of conversion factors, one can find that 1 inch = 2.54 cm. Thus, cm can be converted to inches by multiplying by 1 inch/2.54 cm. Here, one is given the diameter in mm, which is .1 cm. Millimeters are converted to cm by multiplying the number of mm by .1 cm/1 mm. Solving for cm, you obtain:

$$0.1 \text{ mm} \times .1 \text{ cm}/1 \text{ mm} = .01 \text{ cm.}$$

Solving for inches:

$$.01 \text{ cm} \times \frac{1 \text{ inch}}{2.54 \text{ cm}} = 3.94 \times 10^{-3} \text{ inches.}$$

## AREA

● PROBLEM 3

One cubic millimeter of oil is spread on the surface of water so that the oil film has an area of 1 square meter. What is the thickness of the oil film in angstrom units?

Solution: Since one is asked to give the final thickness of the film in angstroms, it is useful to convert the other dimensions given to angstroms first. $1\overset{\circ}{A} = 10^{-10} \text{m} = 10^{-7} \text{ mm}$. Therefore, $1 \text{ mm} = 1\overset{\circ}{A}/10^{-7} = 10^7 \overset{\circ}{A}$.

Cubing both sides of this equation gives the number of cubic angstroms in 1 cubic millimeter.

$$(1 \text{ mm})^3 = (10^7 \overset{\circ}{A})^3$$

$$1 \text{ mm}^3 = 10^{21} \overset{\circ}{A}^3 = \text{volume of oil.}$$

The final area of the film is given as 1 m$^2$. One

knows that $1 m = 10^{10} \overset{o}{A}$; therefore, $1 m^2 = (10^{10} \overset{o}{A})^2 = 10^{20} \overset{o}{A}^2$.
The volume is equal to the area of the film multiplied
by the thickness. Thus, one can find the thickness of
the film by dividing the volume by the area.

$$thickness = \frac{10^{21} \overset{o}{A}^3}{10^{20} \overset{o}{A}^2} = 10 \overset{o}{A}.$$

● **PROBLEM** 4

How much area, in square meters, will one liter of paint
cover if it is brushed out to a uniform thickness of
100 microns?

Solution: Because one is asked to give the final area in
square meters, one should first convert the volume and
thickness to meter units. One liter is equal to 1,000 cc.
Since 1 m = 100 cm, one can convert centimeters to meter
units by cubing both sides of the equality:

$$1m^3 = (100cm)^3$$
$$1m^3 = 1.0 \times 10^6 cc$$
$$\frac{1m^3}{1.0 \times 10^6} = 1cc$$
$$10^{-6}m^3 = 1cc$$

Therefore, 1000 cc or 1 liter is equal to $10^{-6} m^3 \times 1,000$
or $10^{-3} m^3$. There are $10^6$ microns in 1 m. Thus, 1 micron =
$10^{-6}$ m and 100 microns = $10^{-4}$ m. The area of the film is
equal to the volume divided by the thickness.

$$Therefore, \; area = \frac{10^{-3} \; m^3}{10^{-4} \; m} = 10 \; m^2.$$

# VOLUME

● **PROBLEM** 5

Determine the number of cubic centimeters in one cubic inch.

Solution: One meter equals 39.37 inches and, since there
are 100 centimeters in 1 meter, there are 39.37 inches in
100 cm. Thus, 1 inch is equal to 100/39.37 cm.

$$1 \text{ inch} = \frac{100}{39.37} \text{ cm} = 2.54 \text{ cm}.$$

By cubing both sides of this equation, one can solve for the number of cubic centimeters in 1 cubic inch.

$$(1 \text{ inch})^3 = (2.54 \text{ cm})^3$$

$$1 \text{ inch}^3 = 16.4 \text{ cc}.$$

● **PROBLEM** 6

Calculate the number of liters in one cubic meter.

Solution: There are 1,000 milliliters (ml) or cubic centimeters (cc) in one liter. Thus, if one wishes to convert one cubic meter to liters, the cubic meter must be converted to cubic centimeters.

$$\begin{aligned}
1 \text{ meter} &= 100 \text{ centimeters} \\
(1 \text{ meter})^3 &= (100 \text{ centimeters})^3 \\
&= 1,000,000 \text{ centimeters}^3 \\
&= 1 \times 10^6 \text{ cubic centimeters}
\end{aligned}$$

Cubic centimeters can be converted to liters by multiing the number of cubic centimeters by the factor 1 liter/1,000 cubic centimeters.

$$1 \times 10^6 \text{ cubic centimeters} \times 1 \text{ liter}/1,000 \text{ cubic centimeters}$$

$$= 1,000 \text{ liters}.$$

There are 1,000 liters in one cubic meter.

● **PROBLEM** 7

What is the volume, in cubic centimeters, of a cube which is 150.0 mm along each edge?

Solution: There are 10 mm in 1 cm; therefore, millimeters can be converted to centimeters by multiplying the number of millimeters by 1 cm/10 mm.

length of edge in cm = 150 mm × 1 cm/10 mm = 15 cm.

The volume of a cube is equal to the length of the side cubed.

$$\text{volume} = (15 \text{ cm})^3 = 3375 \text{ cc}.$$

What volume (in cc) is occupied by a block of wood of dimensions 25.0 m × 10.0 cm × 300 mm. All edges are 90° to one another.

Solution: Since all of the edges are 90° to one another, one knows that the block is a rectangular solid. The volume of a rectangle is equal to the length times the width times the height. If one wishes to find the volume in cubic centimeters, the lengths of all of the sides must be first expressed in centimeters.

There are 100 cm in 1 m; thus, to convert meters to centimeters, the number of meters must be multiplied by 100 cm/1 m.

25.0 m × 100 cm/1 m = 2500 cm.

There are 10 mm in 1 cm; thus, to convert milli-meters to centimeters, multiply the number of milli-meters by 1 cm/10 mm.

300 mm × 1 cm/10 mm = 30 cm

Solving for the volume:

volume = 2500 cm × 10.0 cm × 30 cm = $7.50 \times 10^5$ cc.

What is the volume in liters of a rectangular tank which measures 2.0 m by 50 cm by 200 mm?

Solution: One liter is equal to 1000 cc; therefore, one should find the volume of the tank in cubic centimeters first and then convert to liters. This method is best for this problem because the sides of the tank are given in units which can quickly be converted to centimeters.

There are 100 cm in 1 m. Thus, 2 m is equal to 2.0 m × 100 cm/1 m or 200 cm. There are 10 mm in 1 cm, therefore 200 mm is equal to 200 mm × 1 cm/10 mm or 20 cm.

Solving for the volume of the tank in cubic centimeters:

volume = 200 cm × 50 cm × 20 cm = $2.0 \times 10^5$ cc.

To convert from cubic centimeters to liters multiply by 1 liter/1000 cc.

volume in liters = $2.0 \times 10^5$ cc × 1 liter/1000 cc

= 200 liters.

A rectangular box is 5.00 in. wide, 8.00 in. long, and 6.0 in. deep. Calculate the volume in both cubic centimeters and in liters.

Solution: The volume of a solid is found by multiplying the height times the length times the width.

volume = (6.0 in) × (8.0 in) × (5.0 in) = 240 in$^3$.

From a standard conversion table, one finds that 1 inch = 2.54 cm. One finds the volume of cubic inches in cubic centimeters by cubing both sides of this equality.

1 inch = 2.54 cm

(1 inch)$^3$ = (2.54cm)$^3$

1 inch$^3$ = 16.4 cc.

Thus, one can convert the volume of the rectangle from cubic inches to cubic centimeters by multiplying the number of cubic inches by the conversion factor, 16.4 cc/1 inch$^3$.

volume of rectangle = 240 in$^3$ × 16.4 cc/1 in$^3$

= 3936 cc.

There are 1000 cc in 1 liter. Therefore, to convert from cubic centimeters to liters, multiply the number of cubic centimeters by 1 liter/1000 cc.

volume in liters = 3936 cc × 1 liter/1000 cc

= 3.936 liters.

# MASS

A student made three successive weighings of an object as follows: 9.17 g, 9.15 g, and 9.20 g. What is the average weight of the object in milligrams?

Solution: The average of a set of weights is found by adding together all of the weights and then dividing by the number of weighings used.

$$\text{avg. weight} = \frac{(9.17 \text{ g} + 9.15 \text{ g} + 9.20 \text{ g})}{3} = \frac{27.52 \text{ g}}{3} = 9.17 \text{ g}.$$

Now that the average weight in grams has been determined, convert it to milligrams using the conversion factor of 1,000 mg/g.

$$9.17 \text{ g} \times \frac{1000 \text{ mg}}{g} = 9170 \text{ mg}.$$

● **PROBLEM** 12

A silver dollar weighs about 0.943 ounces. Express this weight in grams, in kilograms, and in milligrams.

Solution: One ounce is equal to 28.35 g; thus, to convert from ounces to grams, one multiplies the number of ounces by the conversion factor, 28.35 g/1 ounce.

no. of grams = 0.943 ounces × 28.35 g/1 ounce

= 26.73 g.

There are 1,000 g in 1 kg; therefore, to convert from grams to kilograms, one multiplies the number of grams by 1 kg/1,000 g.

no. of kg = 26.73 g × 1 kg/1000 g = .02673 kg.

There are 1,000 mg in one gram; thus, to convert from grams to milligrams, multiply the number of grams by the conversion factor, 1000 mg/1 g.

no. of mg = 26.73 g × 1000 mg/1 g = 26,730 mg.

● **PROBLEM** 13

It is estimated that $3 \times 10^5$ tons of sulfur dioxide, $SO_2$, enters the atmosphere daily owing to the burning of coal and petroleum products. Assuming an even distribution of the sulfur dioxide throughout the earth's atmosphere (which is not the case), calculate in parts per million by weight the concentration of $SO_2$ added daily to the atmosphere. The weight of the atmosphere is $4.5 \times 10^{15}$ tons. (On the average, about 40 days are required for the removal of the $SO_2$ by rain).

Solution: Here, one is asked to find the number of tons of $SO_2$ per $10^6$ tons, i.e. per million, of atmosphere. This is done by using the following ratio: Let x = no. of tons of $SO_2$ per $10^6$ tons of atmosphere.

$$\frac{3.0 \times 10^5 \text{ tons } SO_2}{4.5 \times 10^{15} \text{ tons atm}} = \frac{x}{10^6 \text{ tons atm}}$$

$$x = \frac{3.0 \times 10^5 \text{ tons } SO_2 \times 10^6 \text{ tons atm}}{4.5 \times 10^{15} \text{ tons atm}}$$

$$x = 6.67 \times 10^{-5} \text{ tons } SO_2 \text{ or } 6.67 \times 10^{-5} \text{ ppm } SO_2.$$

## DENSITY

● **PROBLEM** 14

The density of alcohol is 0.8 g/ml. What is the weight of 50 ml. of alcohol?

Solution:  Density is defined as weight per unit volume.

$$\text{density} = \frac{\text{weight}}{\text{volume}} = \frac{g}{ml}.$$

Thus, one can solve for the weight of the alcohol by multiplying the density by the volume.

weight = density × volume

weight = 0.8 g/ml × 50 ml = 40 g.

● **PROBLEM** 15

Calculate the density of a block of wood which weighs 750 kg and has the dimensions 25 cm × 0.10 m × 50.0 m.

Solution:  The density is a measure of weight per unit volume and is usually expressed in g/cc. Therefore, one must find the weight of this block in grams and the volume in cubic centimeters. The density is then found by dividing the weight by the volume.

1 kg = 1,000 g; therefore, 750 kg = 750 × 1,000 g

= $7.5 \times 10^5$ g.   To find the volume in cubic centimeters, all of the dimensions must be converted to centimeters first.

1 m = 100 cm; thus, .10 m = 10 cm and

50.0 m = 5,000 cm.

Volume = 25 cm × 10 cm × 5,000 cm = $1.25 \times 10^6$ cc.

Solving for the density:

$$\text{density} = \frac{\text{weight}}{\text{volume}} = \frac{7.5 \times 10^5 \text{ g}}{1.25 \times 10^6} = .60 \text{ g/cc.}$$

● **PROBLEM** 16

The density of concentrated sulfuric acid is 1.85 g/ml. What volume of the acid would weigh 74.0 g?

Solution: Density is defined as weight per unit volume.

$$\text{density} = \frac{\text{weight}}{\text{volume}} = \frac{\text{g}}{\text{ml}}.$$

Therefore: $\text{volume} = \frac{\text{weight}}{\text{density}}$ .

Solving for the volume:

$$\text{volume} = \frac{74.0 \text{ g}}{1.85 \text{ g/ml}} = 40.0 \text{ ml.}$$

● **PROBLEM** 17

If 2.02 g of hydrogen gas occupies 22.4 liters, calculate the density of hydrogen gas in grams per milliliter.

Solution: Density $= \frac{\text{weight}}{\text{volume}} = \frac{\text{g}}{\text{ml}}.$

One is given the weight as 2.02 g but the volume is given in liters. Therefore, before calculating the density in g/ml, one must convert liters to milliliters. 1 liter = 1,000 ml; therefore, 22.4 l = 22,400 ml. Solving for the density:

$$\text{density} = \frac{2.02 \text{ g}}{22,400 \text{ ml}} = 9.0 \times 10^{-5} \text{ g/ml.}$$

● **PROBLEM** 18

One kilogram of metallic osmium, the "heaviest" substance known, occupies a volume of 44.5 cm$^3$. Calculate the density of osmium in grams per cm$^3$.

Solution: One is told that one kilogram of osmium occupies 44.5 cm$^3$ and is then asked how many grams of osmium occupy one cm$^3$. To find the density in grams per cm$^3$, one kilogram must be first converted to grams after which this number of grams is divided by 44.5 cm$^3$, the volume that they occupy. There are 1000 grams in one kilogram. As such, kilograms are converted to grams by multiplying the number of kilograms present by the factor 1,000 g/1 kg.

1 kg × 1,000/1 kg = 1,000 g.

Therefore, 1,000 g occupy 44.5 cm$^3$. To find the number of grams present in 1 cm$^3$, 1,000 g is divided by 44.5 cm$^3$.

1000 g/44.5 cm$^3$ = 22.5 g/cm$^3$

The density of osmium is 22.5 g/cm$^3$.

# TEMPERATURE

● **PROBLEM** 19

Body temperature on the average is 98.6°F. What is this on (a) the Celsius scale and (b) the Kelvin scale?

Solution:     (a) One converts °F to °C by using the following equation.

°C = 5/9 (°F − 32)

°C = 5/9 (98.6 − 32)

  = 5/9 (66.6) = 37.00°C.

(b) °C can be converted to °K by adding 273.15 to the Celsius temperature.

°K = °C + 273.15

°K = 37.0 + 273.15 = 310.15°K.

● **PROBLEM** 20

Liquid helium boils at 4°K. What is the boiling temperature on the Fahrenheit scale?

Solution:  The temperature in Kelvin is the temperature in degrees Centigrade added to 273. In this problem, the boiling point is given in °K. Hence, the temperature should be converted to °C and, then, to Fahrenheit using the relation

°F = 9/5 °C + 32

The boiling point of helium can be converted to °C by subtracting 273 from the boiling point in °K.

°C = °K - 273

°C = 4°K - 273 = - 269°C.

After the temperature is converted to the Centigrade scale, the temperature on the Fahrenheit scale can be determined.

°F = 9/5 °C + 32

°F = 9/5 (- 269°C) + 32 = - 452°F

The boiling point of helium on the Fahrenheit scale is - 452°F.

● **PROBLEM** 21

The freezing point of silver is 960.8°C and the freezing point of gold is 1063.0°C. Convert these two readings to Kelvin (°K), Fahrenheit (°F), and Rankine (°R).

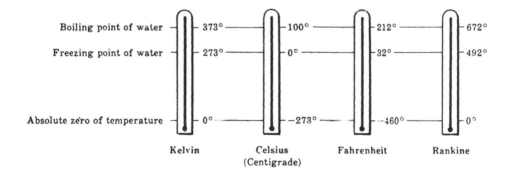

|  | Kelvin | Celsius (Centigrade) | Fahrenheit | Rankine |
|---|---|---|---|---|
| Boiling point of water | 373° | 100° | 212° | 672° |
| Freezing point of water | 273° | 0° | 32° | 492° |
| Absolute zero of temperature | 0° | -273° | -460° | 0° |

Solution:  Kelvin: Temperatures measured in Celsius (°C) are converted to °K by adding 273.15 to the original measurement.

freezing point of silver = 960.8°C + 273.15 = 1234°K

freezing point of gold =  1063.0°C + 273.15 = 1336.2°K

Fahrenheit: °C are converted to °F by using the equation °F = 9/5 (°C) + 32.

freezing point of silver 9/5 (960.8°C) + 32 = 1761°F

freezing point of gold = 9/5 (1063.0°C) + 32 = 1945°F

Rankine:  The Rankine scale is an absolute scale

used by engineers. Its unit is the Fahrenheit degree. Absolute zero is equal to zero degrees Rankine. Convert °K to °R by using the equation

$$°R = 9/5 \ (°K)$$

freezing point of silver = 9/5 (1234°K) = 2221°R

freezing point of gold = 9/5 (1336.2°K) = 2405°R.

# CHAPTER 2

# GASES

---

**Basic Attacks and Strategies for Solving Problems in this Chapter. See pages 13 to 60 for step-by-step solutions to problems.**

---

Gases, which represent one of the three important states of matter, differ remarkably from solids and liquids. Gases expand to uniformly fill a container and, unlike either solids or liquids, the volume is strongly dependent on the temperature and pressure. The pressure, volume, and temperature — called the state variables — are simply related for gases at low to moderate pressures by the ideal gas law or the ideal equation of state.

$$PV = nRT \qquad\qquad 2\text{-}1$$

The terms in this equation are defined as follows:

$P$ = absolute pressure ($P = 0$ is a perfect vacuum)

$V$ = volume

$n$ = number of moles of gas*

$T$ = absolute temperature ($T = 0$ where all molecular motion ceases) in kelvins

$R$ = universal gas constant; it has the same value for all gases

In this chapter, the principles governing the pressure exerted by a column of static fluid are illustrated, and the Laws of Boyle, Charles, Gay-Lussac, and Avogadro (all early scientists who studied gas properties) are presented. These laws are combined into the ideal gas law, and a basic procedure for using the ideal gas law to calculate gas properties is outlined. It should be

---

*A mole is the mass numerically equal in grams to the molecular weight. It contains Avogadro's number, 6.02 × 10²³, of molecules which is the identical amount for any species. A lb. mole is also defined; it is the mass numerically equal in lbs. to the molecular weight. Note that a lb. mole does not contain Avogadro's number of molecules.

remembered that ideal gases, i.e., gases which obey the ideal gas law or ideal equation of state exactly, do not exist. But the behavior of real gases is often so closely approximated by the ideal gas law that it is very useful. A useful rule of thumb is that at pressures below 1 atmosphere and temperatures above the critical temperature (the temperature above which a gas cannot be liquified by an increase in pressure), real gases obey the ideal gas law extremely well.

## Pressure

The pressure exerted by a column of fluid is equal to the weight of the fluid divided by the cross sectional area of the column. One simple equation, called the barometric equation, can be used to calculate this pressure.

$$P = \rho g h \text{ (barometric equation)} \qquad\qquad 2\text{-}2$$

where $\rho$ is the fluid density, $g$ is the gravitational constant, and $h$ is the height of the column. The principles or rules used to attack static pressure problems are as follows:

1. The pressure contribution of different static fluids in the same column, whether calculated from the barometric equation or obtained another way, can simply be added together to obtain the total pressure.

2. In a static fluid, all points at the same elevation have the same pressure.

In Problem 24, for example, the pressure at level $h$, is the same in every column. Therefore:

$$P_1 = P_2 + \rho g h_1 = P_3 + \rho g(h_3 + h_1) \qquad\qquad 2\text{-}3$$

Since the level in column 2 is the same as the level open to the atmosphere, $P_2 = P_{atm}$. From the values of $h_1$ and $h_3$, $P_1$ and $P_3$ are easily calculated and vice versa.

## Gas Laws

The Laws of Boyle, Charles, Gay-Lussac, and Avogadro were developed from experimental observations.

Boyle's Law:   The pressure of a quantity of a gas is inversely proportional to volume at constant temperature; $P \propto 1/V$ at constant $T$, $n$.

| Charles' Law: | The volume of a quantity of a gas is proportional to the absolute temperature at constant pressure; $V \propto T$ at constant $P$, $n$. |

| Gay-Lussac's Law: | The pressure of a quantity of a gas is proportional to the temperature at constant volume; $P \propto T$ at constant $V$, $n$. |

| Avogadro's Law: | At the same temperature and pressure, equal volumes of all gases contain the same number of molecules or moles; $V \propto n$ at constant $T$ and $P$. |

The four laws above can be combined into a single law — the ideal gas law or ideal equation of state.

| Boyle's Law: | $P = k_1/V$ | ($T$ = constant) |
| Charles' Law: | $V = k_2T$ | ($P$ = constant) |
| Gay-Lussac's Law: | $P = k_3T$ | ($V$ = constant) |
| Avogadro's Law: | $V = k_4n$ | ($P$ and $T$ constant) |
| Combined Law: | | |

$$\frac{PV}{nT} = \text{constant or } PV = nRT \qquad\qquad 2\text{-}4$$

The value of $R$, the universal gas constant, depends on the units of pressure, volume, and temperature and is given in Table 1 for several common sets of units.

| Table 1 - Value and Units of Universal Gas Constant ($R$) | |
| --- | --- |
| 82.057 (atm cc)/(mol°K) | 0.7302 (atm cubic feet)/(lb. mol $R$) |
| 0.082 (atm liter)/(mol °K) | 10.73 (psi cubic feet)/(lb. mol $R$) |
| 62.361 (torr liter)/(mol °K) | 1545 (lbf/ft² cubic feet)/(lb. mol $R$) |
| 8.31 (Pascal m³)/(mol°K) | or (ft lbf)/(lb. mol K) |
| or joule/(mol°K) | |

In solving ideal gas law problems, it is not necessary to remember the separate Laws of Boyle, Charles, Gay-Lussac, and Avogadro. They are all special cases of the ideal gas law which can be expressed in the following forms:

$$\frac{P_1 V_1}{n_1 T_1} = \frac{P_2 V_2}{n_2 T_2} \quad \text{or} \quad PV = nRT \qquad\qquad 2\text{-}5$$

The best procedure for attacking gas law problems is to tabulate, in one column, what is known and, in a second, what is to be calculated. Then select the form of the gas law that is the most convenient. Observe that the form on the left above permits variables which do not change (e.g., $T_1$ and $T_2$ in a constant temperature problem) to be canceled from both sides of the equation. Let us examine Problem 54 in the text from this perspective.

**Known**                 **To be calculated**

$P_1, V_1, T_1$                 $V_2$

$P_2, T_2$

$n_1 = n_2 = \text{constant}$

Remember to convert Celsius to Kelvin.

$$V_2 = \frac{P_1 V_1 n_2 T_2}{P_2 n_1 T_1} = \frac{(600 \text{ torr})(350 \text{ ml})(298K)}{(25 \text{ torr})(873K)} = 24,600 \text{ ml}$$

Note that the unit for each quantity is written, and the unit of the answer is calculated along with its numerical value as discussed in Chapter 1.

Let us illustrate this procedure again using Problem 56 in the text.

**Known**                          **To be calculated**

$P_{O_2}, V_{O_2}$                          $n = n_{O_2} + n_{N_2}$

$P_{N_2}, V_{N_2}$                          $n_{N_2} = P_{N_2} V_{N_2} / RT$

$T = \text{constant}$                   $n_{O_2} = P_{O_2} V_{O_2} / RT$

$\qquad\qquad\qquad\qquad\qquad V_f = V_{O_2} + V_{N_2} = 3 \text{ liters} + 2 \text{ liters} = 5 \text{ liters}$

$\qquad\qquad\qquad\qquad\qquad P_f$

From the ideal gas law:

$$P_f = n_f(RT) / V_f = \left(n_{O_2} + n_{N_2}\right)(RT) / V_f$$

$$= \left(P_{O_2} V_{O_2} / RT + P_{N_2} V_{N_2} / RT\right) RT / 5 \text{ liters}$$

$$= [(195 \text{ torr})(3 \text{ liters}) + (530 \text{ torr})(2 \text{ liters})]/5 \text{ liters} = 329 \text{ torr}$$

## PRESSURE

Given the setup in the figure, what would be the pressure of the gas (in atm) if $P_{atm}$ is 745 Torr and $P_{liq}$ is the equivalent of a mercury column 3.0 cm high?

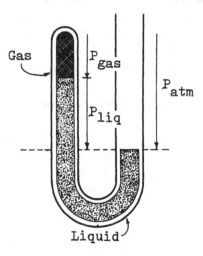

**Solution:**  A manometer is used to measure the pressure of a trapped sample of gas. If the right hand tube is open to the atmosphere, the pressure which is exerted in the right-hand surface is atmospheric pressure, $P_{atm}$. If the liquid level is the same in both arms of the tube, the pressures must be equal; otherwise, there would be a flow of liquid from one arm to the other. At the level indicated by the dashed line in the figure, the pressure in the left arm is equal to the pressure of the trapped gas, $P_{gas}$, plus the pressure of the column of liquid above the dashed line, $P_{liq}$. One can write

$$P_{gas} = P_{atm} - P_{liq}$$

Here, one is given that $P_{atm}$ is 745 Torr and that $P_{liq}$ is equivalent to a mercury column 3.0 cm high. One wishes to find $P_{gas}$ in atm, thus, one must convert 745 Torr to atm and find the $P_{liq}$. There are 760 Torr in 1 atm, which means

$$\text{no. of atm in } P_{atm} = \frac{745 \text{ Torr}}{760 \text{ Torr}/1 \text{ atm}} = .98 \text{ atm}$$

One atmosphere pressure supports 76 cm of mercury, thus 3 cm of mercury supports 3/76 atm. Therefore, $P_{liq}$ = 3 cm/76 cm/atm = .039 atm.

Solving for $P_{gas}$

$$P_{gas} = .980 \text{ atm} - .039 \text{ atm} = .941 \text{ atm}.$$

● **PROBLEM** 23

Consider the manometer, illustrated below, first constructed by Robert Boyle. When h = 40 mm, what is the
pressure of the gas trapped in the volume, $V_{gas}$. The
temperature is constant, and atmospheric pressure is
$P_{atm}$ = 1 atm.

<u>Solution</u>: We do not need to know any gas law to solve this problem. All we must realize is that the pressure exerted on the gas, $P_{total}$, is equal to the sum of the pressure exerted by the mercury, $P_{Hg}$, and the pressure exerted by the air, $P_{atm}$. Since 1 mm Hg = 1 torr and 1 atm = 760 torr,

$P_{Hg}$ = 40 mm Hg = 40 torr and $P_{atm}$ = 1 atm = 760 torr.

Then $P_{total} = P_{Hg} + P_{atm}$ = 40 torr + 760 torr = 800 torr.

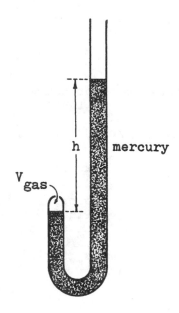

h    mercury

$V_{gas}$

● **PROBLEM** 24

Consider gases confined by a liquid, as shown in the diagram below. Find an expression for the pressures $P_1$, $P_2$, and $P_3$ in terms of the density of the liquid, $\rho$ (g/m$\ell$), the heights $h_1$ and $h_3$ (mm), and the barometric pressure $P_{atm}$ (mm Hg).

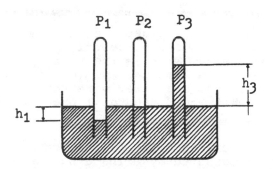

$P_1$    $P_2$    $P_3$

$h_3$

$h_1$

Solution: The device shown above is called a eudiometer. It is used to compare the pressures of several gases. The pressure of the confined gas is equal to the barometric pressure plus the pressure needed to depress the column of liquid (for $P_1$) or to the barometric pressure minus the pressure needed to support the column of liquid (for $P_3$). The pressure of the liquid column is given by

$$\text{pressure of liquid} = \text{height or depth of column (mm)} \times \rho \text{ (g/m}\ell)(\text{m}\ell/\text{cm}^3)(\text{cm}/10\text{mm})$$

15

Hence:

P₁ = barometric pressure + pressure needed to depress
                                     liquid

$$= P_{atm} + h_1 \times \rho \left( m\ell/cm^3 \right) cm/10mm$$

$$= P_{atm} + 0.1 \ h_1 \ (g/cm^2) \ \rho$$

P₂ = barometric pressure + pressure needed to depress
                                     liquid (or - pressure needed
                                     to elevate liquid)

$$= P_{atm} + 0 \times \rho \times ml/cm^3 \times cm/10mm$$

$$= P_{atm}$$

P₃ = barometric pressure - pressure needed to elevate
                                     liquid

$$= P_{atm} - h_3 \times \rho \times m\ell/cm^3 \times cm/10mm$$

$$= P_{atm} - 0.1 \ h_3 \ (g/cm^2) \rho$$

● **PROBLEM** 25

(a) A diver descends to a depth of 15.0 m in pure water
(density 1.00 g/cm³). The barometric pressure is 1.02
standard atmospheres. What is the total pressure on the
diver, expressed in atmospheres? (b) If, at the same
barometric pressure, the water were the Dead Sea
(1.20 g/cm³), what would the total pressure be?

Solution: Pressure is defined as force per unit area.
Atmospheric is measured by using a barometer (usually
a mercury barometer; see figure). It is constructed by
inverting a tube longer than 76 cm filled with mercury
into a dish of mercury. The atmosphere will support only
that height of mercury which exerts an equivalent
pressure; any excess mercury will fall into the reservoir
and leave a space with zero air pressure above it.

The pressure exerted on the diver from above is
equal to the sum of the pressure exerted by the sea
water and the pressure exerted by the atmosphere. One
standard atmosphere equals the pressure exerted by
exactly 76 cm (= exactly 760 mm) of mercury at 0°C

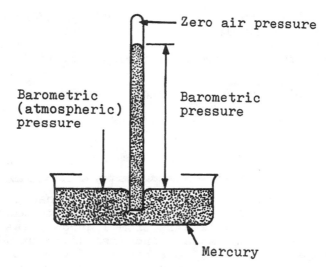

Mercury Barometer

(density Hg = 13.5951 g/cm$^3$) and at standard gravity,
980.665 cm/s$^2$. Thus, 1 standard atm equals 13.5951 g/cm$^3$
× 76 cm (exactly) × 980.665 cm/s$^2$= 1.01325 × 10$^6$ dynes/cm$^2$.
The pressure exerted by the water is found similarly:
pressure of water = density × height × standard gravity

$$= 1.00 \text{ g/cm}^3 \times 1.5 \times 10^3 \text{ cm} \times 980.665 \text{ cm/s}^2$$

$$= 1.47 \times 10^6 \text{ dynes/cm}^2.$$

However, the problem states that atmospheric pressure is 1.02 standard atmospheres and, therefore, equals
(1.02)(1.01325 × 10$^6$ dynes/cm$^2$) = 1.03 × 10$^6$ dynes/cm$^2$.
To this atmospheric pressure, the pressure of the water
is added to yield a total pressure of

$$(1.03 \times 10^6 \text{ dynes/cm}^2) + (1.47 \times 10^6 \text{ dynes/cm}^2)$$

$$= 2.50 \times 10^6 \text{ dynes/cm}^2.$$

This answer expressed in atmospheres gives

$$\frac{2.50 \times 10^6 \text{ dynes/cm}^2}{1.01 \times 10^6 \text{ dynes/cm}^2/\text{atm}} = 2.48 \text{ atm.}$$

(b) This part is very similar to part (a). The total
pressure exerted is the pressure of the water plus the
pressure of the atmosphere. The pressure of the atmosphere
from part (a) is 1.03 × 10$^6$ dynes/cm$^2$. The pressure of the
water must be calculated:

pressure of water = density × height × standard gravity

$$= 1.20 \text{ g/cm}^3 \times 1.5 \times 10^3 \text{ cm} \times 980.665 \text{ cm/s}^2$$

$$= 1.77 \times 10^6 \text{ dynes/cm}^2.$$

The total pressure is $1.77 \times 10^6$ dynes/cm² + 1.03 $\times 10^6$ dynes/cm² equals $2.80 \times 10^6$ dynes/cm².

This answer expressed in atmospheres is

$$\frac{2.80 \times 10^6 \text{ dynes/cm}^2}{1.01 \times 10^6 \text{ (dynes/cm}^2\text{)/atm}} = 2.77 \text{ atm.}$$

● **PROBLEM** 26

How many full strokes of a bicycle pump (chamber 4.0 cm diameter and 40.0 cm long) would you need to make in order to pump up an automobile tire from a gauge pressure of zero to 24 pounds per square inch (psi) (1.63 atm.)? Assume temperature stays constant at 25°C and atmospheric pressure is one atmosphere. Note, that gauge pressure measures only the excess over atmospheric pressure. A typical tire volume is about 25 liters.

Solution: One atmosphere equals 14.7 psi, therefore, the amount of pressure needed to fill the tire is 24 + 14.7 psi or 38.7 psi. Converting back to atm:

$\begin{matrix} \text{atm. contained in} \\ \text{inflated tire} \end{matrix}$ = 38.7 psi × 1 atm/14.7 psi = 2.63 atm.

When the tire is deflated the pressure is 1 atm and the volume is 25 ℓ. Each atm of pressure occupies 25 ℓ. Therefore, the volume of the tire when inflated is 25ℓ × 2.63 atm.

Volume of inflated tire = 25ℓ × 2.63 = 65.75ℓ

Since there is 25ℓ present in the tire before inflation, the volume that the pump must contribute is 65.75 - 25ℓ or 40.75ℓ.

The volume of air forced into the tire at each stroke of the pump is equal to the volume of the pump.

Volume of pump = $\pi r^2 h$ = $\pi \times (2 \text{ cm})^2 \times 40$ cm = 503 cm³/stroke

There are 1000 cm³ in 1 liter, therefore 40750 cm³ of air must be pumped into the tire. If 503 cm³ is pumped in per stroke the number of strokes necessary to fill the tire is $\dfrac{40750 \text{ cm}^3}{503 \text{ cm}^3\text{/stroke}}$ .

No. of strokes = $\dfrac{40750 \text{ cm}^3}{503 \text{ cm}^3\text{/stroke}}$ = 81 strokes.

# BOYLE'S LAW, CHARLES' LAW, LAW OF GAY-LUSSAC

● **PROBLEM** 27

100 ml. of gas are enclosed in a cylinder under a pressure of 760 Torr. What would the volume of the same gas be at a pressure of 1520 Torr?

*P*=Pressure

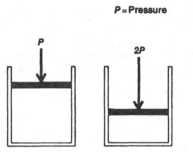

If the pressure on a gas is doubled (at a constant temperature), the volume is halved (Boyle's Law).

Solution:  Since this problem deals with the pressure and volume of a gas at a constant temperature, Boyle's law can be used. Boyle's law states that the volume, V, of a given mass of gas, at constant temperature, varies inversely with the pressure, P. It can be stated as

$$V = k \times 1/P,$$

where k is a constant.

Hence,   $k = PV$

For a particular system, at constant temperature, k is constant. Therefore, if either the pressure or the volume is changed, the other must adjust accordingly.

Here, $P = 760$ Torr  and  $V = 100$ ml    so

$k = 760 \times 100 = 76000$   Torr-ml.

If P is doubled to 1520 Torr, then

$k = 76000$ Torr-ml $= 1520$ Torr x V

$$V = \frac{76000 \text{ Torr-ml}}{1520 \text{ Torr}} = 50 \text{ ml}$$

Since k is a constant for a given system, another form of Boyle's Law can be expressed as

$$P_1 V_1 = P_2 V_2$$

This says that the pressure of the original system times the volume of the original system is equal to the new pressure times the new volume. Here,

$$760 \text{ Torr} \times 100 \text{ ml} = 1520 \text{ Torr} \times V_2$$

$$50 \text{ ml} = V_2 .$$

● **PROBLEM** 28

What pressure is required to compress 5 liters of gas at 1 atm. pressure to 1 liter at a constant temperature?

<u>Solution</u>:  In solving this problem, one uses Boyle's Law: The volume of a given mass of gas at constant temperature varies inversely with the pressure. This means that, for a given gas, the pressure and the volume are proportional; at a constant temperature, and their product equals a constant.

$$P \times V = K$$

where P is the pressure, V is the volume and K is a constant. From this one can propose the following equation

$$P_1 V_1 = P_2 V_2 ,$$

where $P_1$ is the original pressure, $V_1$ is the original volume, $P_2$ is the new pressure and $V_2$ is the new volume.

In this problem, one is asked to find the new pressure and is given the original pressure and volume and the new volume.

$$P_1 V_1 = P_2 V_2$$

$$1 \text{ atm} \times 5 \text{ liters} = P_2 \times 1 \text{ liter}$$

$$\frac{1 \text{ atm} \times 5 \text{ liters}}{1 \text{ liter}} = P_2$$

$$5 \text{ atm.} = P_2 .$$

$P_1 = 1$ atm.
$V_1 = 5$ liters
$P_2 = ?$

$V_2 = 1$ liter

● **PROBLEM** 29

A mass of gas is under a pressure 760 mm Hg and occupies a volume of 525 ml. If the pressure were doubled, what volume would the gas now occupy? Assume the temperature is constant.

<u>Solution</u>:  This question can be answered by using Boyle's Law. This law states that, under constant temperature, the pressure is inversely proportional to the volume of a gas. In other words, PV = K, where P = pressure, V = volume and K = constant. Therefore, for a given gas at constant temperature, $P_1 V_1 = P_2 V_2$, where $P_1$ and $V_1$ are

the initial pressure and volume, and $P_2$ and $V_2$ are the final values.

In this problem $P_1$ = 760 mm Hg, $V_1$ = .525 liter (525 ml) and $P_2$ = 1520 mmHg. The unknown, $V_2$, can be determined by substituting these values into $P_1V_1 = P_2V_2$.

(760)(.525) = 1520 ($V_2$)   Solving for $V_2$, you obtain $V_2$ = .2625 $\ell$ = 262.5 ml.

● **PROBLEM** 30

A gaseous sample of neon, maintained at constant temperature, occupies 500 ml at 2.00 atm. Calculate the volume when the pressure is changed to each of the following: (a) 4.00 atm; (b) 760 torr; (c) 1.8 × $10^{-3}$ torr.

Solution: We need a relationship between volume and pressure. Such a relationship is provided by Boyle's Law, which states that the product of pressure P and volume V of an ideal gas is a constant, k, or PV = k.

We must first determine k for the neon sample. We are given a value of P corresponding to a given value of V.

Therefore, k = PV = 2.00 atm × 500 ml = 2.00 atm × 0.500 $\ell$ = 1.00$\ell$-atm. Now that k is determined, we can obtain the value of V corresponding to the given values of P by using the formula V = k/P.

(a)   P = 4.00 atm:   $V = \dfrac{k}{P} = \dfrac{1.00\,\ell\text{-atm.}}{4.00\ \text{atm}} = 0.25\ell$

(b) P = 760 torr = 760 torr × $\dfrac{1\ \text{atm}}{760\ \text{torr}}$ = 1 atm :

$V = \dfrac{k}{P} = \dfrac{1.00\,\ell\text{-atm.}}{1\ \text{atm}} = 1.00\ell$

(c) P = 1.8 × $10^{-3}$ torr = 1.8 × $10^{-3}$ torr × $\dfrac{1\ \text{atm}}{760\ \text{torr}}$

= 2.3 × $10^{-6}$ atm:

$V = \dfrac{k}{P} = \dfrac{1.00\,\ell\text{-atm.}}{2.3 \times 10^{-6}\ \text{atm}} = 4.35 \times 10^5\ \ell.$

A sample of gaseous krypton, maintained at constant pressure, is found to have a volume of 10.5ℓ at 25°C. If the
system is heated to 250°C, what is the resulting volume?

Solution:  We need a relationship between volume and
temperature. Such a relationship is provided by Charles'
Law, which states that volume V and absolute temperature
T are proportional, or, as an equality, V = kT, where k
is a constant.

We can determine k for our system from the initial
volume and temperature. Thus,

$$k = \frac{V}{T} = \frac{10.5\ \ell}{25°C} = \frac{10.5\ \ell}{298.15°K} = 0.0352\ \ell \cdot °K^{-1}$$

The volume corresponding to a temperature of 250°C,
which is 523.15°K, is V = kT = 0.0352 ℓ·°K$^{-1}$ × 523.15°K
= 18.41 ℓ.

A certain gas occupies a volume of 100 ml at a temperature
of 20°C. What will its volume be at 10°C, if the pressure
remains constant?

Solution:  In a gaseous system, when the volume is changed
by increasing the temperature, keeping the pressure
constant, Charles' Law can be used to determine the new
volume. Charles' Law states that, at a constant pressure,
the volume of a given mass of gas is directly proportional
to the absolute temperature. Charles' Law may also be
written

$$\frac{V_1}{T_1} = \frac{V_2}{T_2}$$

where $V_1$ is the volume at the original temperature $T_1$ and
$V_2$ is the volume at the new temperature $T_2$.

To use Charles' Law, the temperature must be expressed on the absolute scale. The absolute temperature
is calculated by adding 273 to the temperature in degrees
Centigrade.  In this problem, the centigrade temperatures
are given and one must convert them to the absolute scale.

$$T_1 = 20°C + 273 = 293°K$$

$$T_2 = 10°C + 273 = 283°K$$

Using Charles' Law,

$V_1$ = 100 ml

$T_1$ = 293°K

$T_2$ = 283°K

$V_2$ = ?

$$\frac{V_1}{T_1} = \frac{V_2}{T_2} \qquad\qquad V_2 = \frac{V_1 T_2}{T_1}$$

$$V_2 = \frac{(100 \text{ ml})(283°K)}{293°K}$$

$V_2$ = 96.6 ml.

● **PROBLEM** 33

Assume that one cubic foot of air near a thermonuclear explosion is heated from 0°C to 546,000°C. To what volume does the air expand?

<u>Solution</u>: Charles' Law($V_1/T_1$ = k, where $V_1$ is the initial volume, $T_1$ the initial absolute temperature, and k is a constant)states that the volume is directly proportional to the temperature.

$$V \ \alpha \ T$$

$\alpha$ means 'is proportional to.'

The absolute temperature can be found by adding 273 to the temperature in °C.

$T_1$ = 0 + 273 = 273

$T_2$ = 546,000 + 273 = 546,273

Using the Charles' Law, one can set up the following ratio

$$\frac{V_1}{T_1} = \frac{V_2}{T_2} \ ,$$

where $V_1$ is the original volume, $V_2$ is the final volume, $T_1$ is the original temperature, and $T_2$ is the final temperature. Using this ratio, one can determine the final volume.

$$\frac{V_1}{T_1} = \frac{V_2}{T_2}$$

Substituting, one obtains

$$\frac{1 \text{ ft}^3}{273} = \frac{V_2}{546,273}$$

$$V_2 = \frac{546,273}{273} \ (1 \text{ ft}^3) = 2001 \text{ ft}^3.$$

● **PROBLEM** 34

The volume of a sample of gaseous argon maintained at constant pressure was studied as a function of temperature and the following data were obtained:

| Temperature, T (°K) | Volume, V (ℓ) |
|---|---|
| 250 | 0.005 |
| 300 | 0.006 |
| 350 | 0.007 |

Calculate and confirm the Charles' Law constant for this system. Determine the temperature corresponding to a volume of 22.4 ℓ.

**Solution:** Charles' law states that the volume and absolute temperature of an ideal gas are directly proportional, i.e. $V = kT$, k is a constant. To determine the Charles' law constant k, we can use any one of the three sets of temperature-volume measurement. Choosing the first of these, we obtain

$$k = \frac{V}{T} = \frac{0.005 \ \ell}{250°K} = 2.00 \times 10^{-5} \ \ell\text{-}°K^{-1}.$$ This value

is confirmed by showing that the other two sets of data give the same value for k:

$$k = \frac{V}{T} = \frac{0.006 \ \ell}{300°K} = 2.00 \times 10^{-5} \ \ell\text{-}°K^{-1}$$ and

$$k = \frac{V}{T} = \frac{0.007 \ \ell}{350°K} = 2.00 \times 10^{-5} \ \ell\text{-}°K^{-1}.$$

To determine the temperature corresponding to 22.4 ℓ, we solve Charles' law for T, obtaining

$$T = \frac{V}{k} = \frac{22.4 \ \ell}{2.00 \times 10^{-5} \ \ell\text{-}°K^{-1}} = 1.12 \times 10^6 \ °K$$

or about a million degrees Kelvin.

Consider the gas thermometer illustrated below. At 0°C,
the volume of the gas is 1.25 liters. Assuming that the
cross-sectional area of the graduated arm is 1 cm$^2$,
what is the distance (in cm.) from the 0°C reading and
a reading at 35°C?

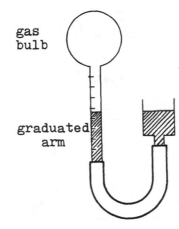

gas
bulb

graduated
arm

leveling bulb
(adjust height until the
liquid level in the leveling bulb is at the same
height as the liquid level
in the graduated arm)

Solution: Since volume, V, and absolute temperature, T,
are the only two variables being considered, we can
approach this problem by means of Charles' Law, V = kT,
where k is the Charles' law constant, to be determined
for our system.

We can determine k by using the initial values
of 0°C (0°C = 273.15°K) and 1.25 liters (= 1250 ml =
1250 cm$^3$). Then,

$$k = \frac{V}{T} = \frac{1250 \text{ cm}^3}{273.15°K} = 4.576 \text{ cm}^3 \text{-}°K^{-1}$$

The volume at 35°C (308.15°K) can now be determined:
V = kT = 4.576 cm$^3$-°K$^{-1}$ × 308.15°K = 1410.1 cm$^3$ = 1.410 ℓ.
The difference between this volume and the initial volume
is 1410.1 cm$^3$ - 1250 cm$^3$ = 160.1 cm$^3$. This volume of
liquid will be displaced in the graduated arm.

The difference in height of the liquid in the
graduated arm, can now be calculated from the volume
displaced and the cross-sectional area:

$$\text{height} = \frac{\text{volume}}{\text{area}} = \frac{160.1 \text{ cm}^3}{1 \text{ cm}^2} = 160.1 \text{ cm}.$$

The air in a tank has a pressure of 640 mm of Hg at 23°C. When placed in sunlight the temperature rose to 48°C. What was the pressure in the tank?

Solution: The law of Gay-Lussac deals with the relationship existing between pressure and the absolute temperature (°C + 273°), for a given mass of gas at constant volume. The relationship is expressed in the law of Gay-Lussac: volume constant, the pressure exerted by a given mass of gas varies directly with the absolute temperature. That is:

$P \alpha T$ (volume and mass of gas constant).

The variation that exists between pressure and temperature at different states can be expressed as

$$\frac{P_1}{T_1} = \frac{P_2}{T_2}$$

where $P_1$ = pressure of original state, $T_1$ = absolute temperature of original state, $P_2$ = pressure of final state, and $T_2$ = absolute temperature of final state.

Thus this problem is solved by substituting the given values into Gay-Lussac's Law.

$P_1$ = 640 mm Hg $\qquad$ $T_1$ = 23°C + 273° = 296°K

$P_2$ = ? $\qquad$ $T_2$ = 48°C + 273° = 321°K

Substituting and solving,

$$\frac{640 \text{ mm}}{296°K} = \frac{P_2}{321°K}$$

$$P_2 = 640 \text{ mm} \times \frac{321°K}{296°K}$$

$$= 694 \text{ mm of Hg.}$$

A sealed glass bulb contained helium at a pressure of 750 mm Hg and 27°C. The bulb was packed in dry ice at - 73°C. What was the resultant pressure of the helium?

Solution: The only parameters mentioned in this problem are pressure (P) and temperature (T). One is given the initial pressure and temperature and the final temperature. One is asked to determine the pressure at - 73°C. The

Law of Gay-Lussac relates temperature and pressure. It can be stated mathematically

$$\frac{P_1}{T_1} = \frac{P_2}{T_2}$$

where $P_1$ is the initial pressure, $T_1$ the initial absolute temperature, $P_2$ the final pressure and $T_2$ the final absolute temperature.

The temperature in °C is converted to °K by adding 273.

$T_1 = 27°C + 273 = 300°K$

$T_2 = -73°C + 273 = 200°K$

Solving for $P_2$

$$\frac{750 \text{ mm Hg}}{300°K} = \frac{P_2}{200°K}$$

$$P_2 = \frac{(750 \text{ mm Hg})(200°K)}{300°K} = 500 \text{ mm Hg}.$$

## COMBINED GAS LAWS

● PROBLEM 38

A gas occupies a volume of 1.0 liter at a temperature of 27°C and 500 Torr pressure. Calculate the volume of the gas if the temperature is changed to 60°C and the pressure to 700 Torr.

Solution: For a given mass of gas, the volume is inversely proportional to the pressure and directly proportional to the absolute temperature. This combined law can also be written:

$$\frac{P_1 V_1}{T_1} = \frac{P_2 V_2}{T_2}$$

where $P_1$ is the original pressure, $V_1$ is the original volume, $T_1$ is the original absolute temperature, $P_2$ is the new temperature, $V_2$ is the new volume, and $T_2$ is the new absolute temperature.

In this problem, one is given the original pressure, temperature, and volume and the new temperature and pressure. One is asked to calculate the new volume. The temperatures are given in °C; they must be converted to the absolute scale before using the combined law. This can be done by adding 273 to the temperature in °C.

Converting the temperature:

$T_1 = 27° + 273 = 300°K$

$T_2 = 60° + 273 = 333°K$

Using the combined law:

$$\frac{P_1 V_1}{T_1} = \frac{P_2 V_2}{T_2}$$

$P_1 = 500$ Torr          $P_2 = 700$ Torr

$V_1 = 1.0$ liter          $V_2 = ?$

$T_1 = 300°K$              $T_2 = 333°K$

$$\frac{(500 \text{ Torr})(1.0 \text{ liter})}{(300°K)} = \frac{(700 \text{ Torr}) (V_2)}{(333°K)}$$

$$V_2 = \frac{(500 \text{ Torr})(1.0 \text{ liter})(333°K)}{(300°K)(700 \text{ Torr})}$$

$V_2 = 0.79$ liter.

● **PROBLEM** 39

Calculate the pressure required to compress 2 liters of a gas at 700 mm pressure and 20°C into a container of 0.1 liter capacity at a temperature of - 150°C.

Solution: One is dealing with changing volumes, pressures and temperatures of a gas. Therefore, this problem can be solved using the combined gas law. It states that as the pressure increases, the volume decreases and that as the temperature increases, the volume increases. These factors are related by the equation

$$\frac{P_1 V_1}{T_1} = \frac{P_2 V_2}{T_2}$$

where $P_1$, $V_1$ and $T_1$ are the initial pressure, volume and temperature and $P_2$, $V_2$, and $T_2$ are the final values.

For any problem dealing with gases, the first step always involves converting all of the temperatures to the degree Kelvin scale by the equation

$°K = °C + 273$

For this question

$T_1 = 20°C = 20 + 273 = 293°K$

$T_2 = - 150°C = - 150 + 273 = 123°K.$

This seems to indicate that the pressure would decrease. But one is also told that the volume decreases, which would have the effect of increasing the pressure. Therefore, one cannot predict the final change in volume.

For the sake of clarity, set up a table as given below.

$P_1 = 700$ mm                    $P_2 = ?$

$V_1 = 2$ liters                    $V_2 = 0.1$ liter

$T_1 = 293°K$                    $T_2 = 123°K$

Since one is given 5 of the 6 values, it is possible to use the combined gas law equation to determine $P_2$

$$\frac{P_1V_1}{T_1} = \frac{P_2V_2}{T_2}$$

$$P_2 = \frac{T_2V_1P_1}{T_1V_2}$$

$$= \frac{123°K \ (2 \ liters)(700 \ mm)}{293°K \ (0.1 \ liter)}$$

$$= 5877 \ mm.$$

● PROBLEM 40

750 ml of gas at 300 torr pressure and 50°C is heated until the volume of gas is 2000 ml at a pressure of 700 torr. What is the final temperature of the gas?

Solution:  Here, one is given a gaseous system involving pressure, volume and temperature, where two of the variables are changed in going from the original system to the final system. This indicates that the combined gas law should be used. It can be stated: For a given mass of gas, the volume is inversely proportional to the pressure and directly proportional to the absolute temperature. This gas law can also be stated

$$\frac{P_1V_1}{T_1} = \frac{P_2V_2}{T_2}$$

where $P_1$ is the original pressure, $V_1$ is the original volume, $T_1$ is the original absolute temperature, $P_2$ is the final pressure, $V_2$ is the final volume, and $T_2$ is the final absolute temperature.

In this problem, you are given the original pressure, volume and temperature and the final pressure and volume. You are asked to find the final temperature. The temperature in °C must be converted to the absolute scale before using the combined law. This can be done by adding 273 to the temperature in °C.

Converting the temperature:

$T_1 = 50 + 273 = 323°K$

Using the combined law:

$$\frac{P_1 V_1}{T_1} = \frac{P_2 V_2}{T_2}$$

$P_1 = 300$ torr                $P_2 = 700$ torr

$V_1 = 750$ ml                $V_2 = 2000$ ml.

$T_1 = 323°K$                $T_2 = ?$

$$\frac{(300 \text{ torr})(750 \text{ ml})}{323°K} = \frac{(700 \text{ torr})(2000 \text{ ml})}{T_2}$$

$$T_2 = \frac{(700 \text{ torr})(2000 \text{ ml})(323°K)}{(300 \text{ torr})(750 \text{ ml})}$$

$T_2 = 2010°K$

Convert $T_2$ to centigrade by subtracting 273 from it.

$2010 - 273 = 1737°C.$

● **PROBLEM** 41

500 liters of a gas at 27°C and 700 torr would occupy what volume at STP?

Solution: STP (Standard Temperature and Pressure) is defined as being 0°C and 760 torr, thus, in this problem, one is asked to find the new volume of a gas when the temperature and pressure are changed. One refers to the combined gas law in such a case. This law can be stated: For a given mass of gas, the volume is inversely proportional to the pressure and directly proportional to the absolute temperature. Stated algebraically

$$\frac{PV}{T} = K$$

where P is the pressure, V is the volume, T is the absolute temperature, and K is a constant. This means that if two of the variables are changed, the third changes so that the relation PV/T = K remains true. This means that one can now state that

$$\frac{P_1V_1}{T_1} = \frac{P_2V_2}{T_2}$$

where $P_1$ is the original pressure, $V_1$ is the original volume, $T_1$ is the original temperature, $P_2$ is the new pressure, $V_2$ is the new volume, and $T_2$ is the new absolute temperature. In this problem, one is given the temperature on the Celsius scale. It must be converted to the absolute scale before using the combined gas law. This can be done by adding 273 to the temperature in °C.

$T_1 = 27 + 273 = 300°K$

$T_2 = 0 + 273 = 273°K$

One knows $P_1$, $V_1$, $T_1$, $P_2$, and $T_2$. One is asked to find $V_2$.

Using the Combined Gas Laws:

$$\frac{P_1V_1}{T_1} = \frac{P_2V_2}{T_2}$$

$P_1 = 700$ torr

$V_1 = 500$ liters

$T_1 = 300°K$

$P_2 = 760$ torr

$V_2 = ?$

$T_2 = 273°K$

$$\frac{700 \text{ torr} \times 500 \text{ liters}}{300°K} = \frac{760 \text{ torr} \times V_2}{273°K}$$

$$V_2 = \frac{700 \text{ torr} \times 500 \text{ liters} \times 273°K}{760 \text{ torr} \times 300°K}$$

$V_2 = 419$ liters.

● **PROBLEM** 42

A chemist has a certain amount of gas under a pressure of 33.3 atm; it occupies 30 ℓ at 273°C. For his research, however, the gas must be at standard conditions. Under standard conditions what will the volume of the gas be?

Solution: Standard conditions are defined to be 0°C and

1 atm. Hence, the gas is cooled and the pressure on it is decreased. The combined gas law relates pressure P, volume V and absolute temperature T.

$$k = \frac{PV}{T}$$

where k is a constant that is characteristic to the system. Hence,

$$\frac{P_I V_I}{T_I} = \frac{P_F V_F}{T_F}$$

where the subscript I indicates the initial values and the subscript F indicates the final states.

In this problem one is asked to solve for $V_F$. The temperatures in °C are converted to °K by adding 273.

$$T_I = 273°C + 273 = 546°K$$

$$T_F = \quad 0°C + 273 = 273°K$$

Solving for $V_F$:

$$\frac{(33.3 \text{ atm})(30 \text{ } \ell)}{(546°K)} = \frac{(1 \text{ atm})(V_F)}{(273°K)}$$

$$V_F = 499.5 \text{ } \ell.$$

● **PROBLEM** 43

On a hot day, the pressure in an automobile tire increases. Assuming that the air in a tire at 59°F increases in pressure from 28.0 lbs/in² (1.9 atm.) to 30.0 lbs/in² (2.04 atm.), (a) what is the temperature of the air in the tire, assuming no change in volume? (b) What will the pressure be if the temperature rises to 106°F?

Solution: (a) Because pressure, volume, and temperature are involved, one thinks of the combined gas law. This law can be stated: For a given mass of gas, the volume is inversely proportional to the pressure and directly proportional to the absolute temperature. This law can be stated algebraically as:

$$\frac{PV}{T} = K,$$

where P is the pressure, V is the volume, T is the absolute temperature, and K is a constant. This means that if two of the parameters are changed, the third will adjust itself so that PV/T = K is still true. The following equation is thus true:

$$\frac{P_1 V_1}{T_1} = \frac{P_2 V_2}{T_2}$$

where $P_1$ is the original temperature, $V_1$ is the original volume, $T_1$ is the original absolute temperature, $P_2$ is the new pressure, $V_2$ is the new volume, and $T_2$ is the new absolute temperature. When $V_1 = V_2$, as in this problem, it follows from the formula that pressure is directly proportional to absolute temperature. In this case, the law can be written

$$\frac{P_1}{T_1} = \frac{P_2}{T_2}$$

In this problem, the temperature is given in °F, and must be converted to °K before it is used in the combined gas law. This is done by first converting it to °C and then adding 273. °F are converted to °C by use of the following equation:

°C = 5/9 (°F - 32)

$T_1$ in °C = 5/9 (59° - 32°) = 15°C

$T_1$ = 15 + 273 = 288°K.

One is given $P_1$, $P_2$, and $T_1$ and asked to find $T_2$.

$$\frac{P_1}{T_1} = \frac{P_2}{T_2}$$

$P_1$ = 28.0 lbs/in$^2$

$T_1$ = 288°K

$P_2$ = 30.0 lbs/in$^2$

$T_2$ = ?

$$\frac{28.0 \text{ lbs/in}^2}{288°K} = \frac{30.0 \text{ lbs/in}^2}{T_2}$$

$$T_2 = \frac{30.0 \text{ lbs/in}^2 \times 288°K}{28.0 \text{ lbs/in}^2} = 309°K.$$

The absolute temperature can now be converted to °F by first subtracting 273°, and then using the equation for conversion from °C to °F.

°F = 9/5 °C + 32

$T_2$ in °C = 309° - 273 = 36°C

$T_2$ in F = 9/5 (36°) + 32 = 97°F.

(b) One can again use the shortened form of the combined gas law, here:

$$\frac{P_1}{T_1} = \frac{P_2}{T_2}$$

Here, one is given $T_1$, $T_2$, and $P_1$ and asked to find $P_2$. The temperatures must be converted to the absolute scale before use. The same method as used in part (a) will be applied.

$T_1$ in °C = 5/9 (59 - 32) = 15°C

$T_1$ in °K = 15 + 273 = 288°K

$T_2$ in °C = 5/9 (106 - 32) = 41°C

$T_2$ = 41 + 273 = 314°K

The $P_2$ can now be found

$$\frac{P_1}{T_1} = \frac{P_2}{T_2}$$

$P_1$ = 28.0 lbs/in²

$T_1$ = 288°K

$P_2$ = ?

$T_2$ = 314°K

$$\frac{28.0 \text{ lbs/in}^2}{288°K} = \frac{P_2}{314°K}$$

$$P_2 = \frac{314°K \times 28.0 \text{ lbs/in}^2}{288°K} = 30.5 \text{ lbs/in}^2.$$

● **PROBLEM 44**

When J.F. Piccard made a stratosphere flight in a balloon, the balloon seemed to be only half filled as it left the ground near Detroit. The gas temperature was about 27°C, the pressure 700 mm, and the volume of gas in the balloon 80,000 cubic feet (2.26 x 10⁶ liters). What was the gas volume at high altitude where the temperature was -3°C, and the pressure 400 mm?

Solution: The solution to this problem necessitates the use of the combined gas law. It is stated as follows: For a given mass of gas, the volume is inversely proportional to the pressure and directly proportional to the absolute temperature. The general formula for the combined gas laws may be written

$$\frac{P_1V_1}{T_1} = \frac{P_2V_2}{T_2} \qquad \text{or} \quad V_2 = V_1 \times \frac{P_1}{P_2} \times \frac{T_2}{T_1}$$

in which $P_1$ and $T_1$ are pressure and absolute temperature of the gas at volume $V_1$; and $P_2$ and $T_2$ are pressure and absolute temperature at volume $V_2$.

Thus, if

$P_1 = 700$ mm $\quad\quad$ and $\quad\quad$ $P_2 = 400$ mm

$V_1 = 80,000$ cu.ft. $\quad\quad\quad\quad$ $T_2 = 273° - 3° = 270°K$

$T_1 = 27° + 273° = 300°K$

then $V_2$ can be determined by substitution in the general formula above.

$$V_2 = (80,000)\left(\frac{700 \text{ mm}}{400 \text{ mm}}\right)\left(\frac{270°K}{300°K}\right) = 125,000 \text{ cu. ft.}$$
$$(3.54 \times 10^6 \text{ liters})$$

## AVOGADRO'S LAW-THE MOLE CONCEPT

● **PROBLEM** 45

How many moles are there in one atom?

<u>Solution</u>: A mole of atoms is defined as containing Avogadro's Number of atoms. Avogadro's number is $6.02 \times 10^{23}$. Therefore, the number of moles in one atom is equal to 1 atom divided by $6.02 \times 10^{23}$ atoms/mole.

$$\text{No. of moles} = \frac{1 \text{ atom}}{6.02 \times 10^{23} \text{ atoms/mole}}$$

$$= 1.66 \times 10^{-24} \text{ moles.}$$

● **PROBLEM** 46

During the course of World War I, $1.1 \times 10^8$ kg of poison gas was fired on Allied soldiers by German troops. If the gas is assumed to be phosgene ($COCl_2$), how many molecules of gas does this correspond to?

<u>Solution</u>: To solve this problem we must convert mass to number of moles and then multiply by Avogadro's number to obtain the corresponding number of molecules.

The number of moles of gas is given by

$$\text{moles} = \frac{\text{mass (grams)}}{\text{molecular weight of } COCl_2 \text{ (g/mole)}}$$

The mass of gas is $1.1 \times 10^8$ kg $= 1.1 \times 10^8$ kg $\times 10^3$ g/kg $= 1.1 \times 10^{11}$ g. The molecular weight of $COCl_2$ is obtained by adding the atomic weights (atm. wgt.) of its constituents. Thus,

$$\text{molecular weight } (COCl_2) = \text{atm wgt}(C) + \text{atm wgt}(O) \\ + 2 \text{ x atm wgt}(Cl)$$

$$= 12.0 \text{ g/mole} + 16.0 \text{ g/mole} \\ + 2 \times 35.5 \text{ g/mole}$$

$$= 99 \text{ g/mole}.$$

Hence, the number of moles of gas is

$$\text{moles} = \frac{\text{mass}}{\text{molecular weight}} = \frac{1.1 \times 10^{11} \text{ g}}{99 \text{ g/mole}}$$

$$\approx 1.1 \times 10^9 \text{ moles}.$$

Multiplying the number of moles by Avogadro's number, we obtain the number of molecules of gas:

$$\text{number of molecules} = \text{moles} \times \text{Avogadro's number}$$

$$= 1.1 \times 10^9 \text{ moles} \times 6 \times \\ 10^{23} \text{ molecules/mole}$$

$$= 6.6 \times 10^{32} \text{ molecules}.$$

● **PROBLEM** 47

For a single year, the motor vehicles in a large city produced a total of $9.1 \times 10^6$ kg of the poisonous gas carbon monoxide (CO). How many moles of CO does this correspond to?

Solution:    The number of moles of a substance is equal to the quotient of the mass (in grams) of that substance and its molecular weight (in g/mole), or

$$\text{moles} = \frac{\text{mass (g)}}{\text{molecular weight (g/mole)}}$$

The mass of CO is $9.1 \times 10^6$ kg $= 9.1 \times 10^6$ kg $\times 1000$ g/kg $= 9.1 \times 10^9$ g. The molecular weight of CO is the sum of the atomic weight of C and the atomic weight of O, or

molecular weight of (CO) = atomic weight (C) + atomic weight (O)

$$= 12 \text{ g/mole} + 16 \text{ g/mole}$$

$$= 28 \text{ g/mole}.$$

Hence,

$$\text{moles} = \frac{\text{mass (g)}}{\text{molecular weight (g/mole)}} = \frac{9.1 \times 10^9 \text{ g}}{28 \text{ g/mole}}$$

$$= 3.3 \times 10^8 \text{ moles}.$$

● **PROBLEM** 48

An automobile travelling at 10 miles per hour produces 0.33 lb of CO gas per mile. How many moles of CO are produced per mile?

Solution:  The number of moles of a substance is equal to the quotient of the mass (in grams) of that substance and its molecular weight (in g/mole), or

$$\text{moles} = \frac{\text{mass (g)}}{\text{molecular weight (g/mole)}}$$

The mass of CO is 0.33 lb = 0.33 lb × 454 g/lb = 150 g. The molecular weight of CO is the sum of the atomic weight of C and the atomic weight of O, or

$$\text{molecular weight (CO)} = \text{atomic weight (C)} + \text{atomic weight (O)}$$

$$= 12 \text{ g/mole} + 16 \text{ g/mole}$$

$$= 28 \text{ g/mole}.$$

Hence,

$$\text{moles} = \frac{\text{mass (g)}}{\text{molecular weight (g/mole)}} = \frac{150 \text{ g}}{28 \text{ g/mole}}$$

$$= 5.4 \text{ moles per mile}.$$

● **PROBLEM** 49

A flask containing $H_2$ at 0°C was sealed off at a pressure of 1 atm and the gas was found to weigh, 4512 g. Calculate the number of moles and the number of molecules of $H_2$ present.

Solution:  A mole is defined as $6.023 \times 10^{23}$ molecules of a substance. Hydrogen has a molecular weight of 1.008, therefore, 1 mole of $H_2$ weighs 2.016 grams. The number of moles present is therefore

$$\text{moles} = \frac{\text{grams}}{\text{M.W.}} = \frac{.4512\ g}{2.016\ g/\text{mole}} = .2238 \text{ mole of } H_2$$

To calculate the number of molecules, recall that 1 mole of any gas at STP has $6.023 \times 10^{23}$ molecules/mole. Therefore,

$$\text{no. of molecules} = 6.023 \times 10^{23}\ \frac{\text{molecules}}{\text{mole}} \times \text{no. of moles}$$

$$= 6.023 \times 10^{23}\ \frac{\text{molecules}}{\text{mole}} \times 0.2238 \text{ moles}$$

$$= 1.348 \times 10^{23} \text{ molecules of } H_2.$$

# THE IDEAL GAS LAW

Three researchers studied 1 mole of an ideal gas at 273°K in order to determine the value of the gas constant, R. The first researcher found that at a pressure of 1 atm the gas occupies 22.4 ℓ. The second researcher found that the gas occupies 22.4 ℓ at a pressure of 760 torr. Finally, the third researcher reported the product of pressure and volume as 542 cal. What value for R did each researcher determine?

Solution: This problem is an application of the ideal gas equation, PV = n RT, where P = pressure, V = volume, n = number of moles, R = gas constant, and T = absolute temperature. Specifically, we are trying to determine R from the relation R = PV/nT. All three researchers worked with one mole of gas (n = 1) at T = 273°K. Their results are as follows:

First researcher: P = 1 atm, V = 22.4 ℓ.

$$R = \frac{PV}{nT} = \frac{1 \text{ atm} \times 22.4\ \ell}{1 \text{ mole} \times 273°K} = 0.0821\ \frac{\ell\text{-atm}}{K\text{-mole}}$$

Second researcher: P = 760 torr = 760 torr $\times \frac{1 \text{ atm}}{760 \text{ torr}}$ = 1 atm, V = 22.4 ℓ.

$$R = \frac{PV}{nT} = \frac{1 \text{ atm} \times 22.4\ \ell}{1 \text{ mole} \times 273°K} = 0.0821\ \frac{\ell\text{-atm}}{K\text{-mole}}$$

Third researcher: $PV = 542$ cal $= nRT$
$= 1$ mole $(R)(273°K)$

$$R = \frac{PV}{nT} = \frac{542 \text{ cal}}{(1 \text{ mole})(273°K)} = 1.99 \text{ cal/mole }°K.$$

● **PROBLEM** 51

How many moles of hydrogen gas are present in a 50 liter steel cylinder if the pressure is 10 atmospheres and the temperature is 27°C? R = .082 liter-atm/mole °K.

Solution:  In this problem, one is asked to find the number of moles of hydrogen gas present where the volume, pressure and temperature are given. This would indicate that the Ideal Gas Law should be used because this law relates these quantities to each other. The Ideal Gas Law can be stated:

$$PV = nRT,$$

where P is the pressure, V is the volume, n is the number of moles, R is the gas constant (.082 liter-atm/mole °K), and T is the absolute temperature. Here, one is given the temperature in °C, which means it must be converted to the absolute scale. P, V, and R are also known. To convert a temperature in °C to the absolute scale, add 273 to the temperature in °C.

$$T = 27 + 273 = 300°K$$

Using the Ideal Gas Law:

$$PV = nRT \qquad \text{or} \qquad n = \frac{PV}{RT}$$

$P = 10$ atm

$V = 50$ liters

$R = .082$ liter-atm/mole °K

$T = 300°K$

$n =$ number of moles of $H_2$ present

$$n = \frac{(10 \text{ atm})(50 \text{ liters})}{(.082 \text{ liter-atm/mole }°K)(300°K)}$$

$$= 20 \text{ moles.}$$

The barometric pressure on the lunar surface is about $10^{-10}$ torr. At a temperature of 100°K, what volume of lunar atmosphere contains (a) $10^6$ molecules of gas, and (b) 1 millimole of gas?

Solution: This problem is an application of the ideal gas equation, PV = nRT, where P = pressure, V = volume, n = number of moles, R = gas constant, and T = absolute temperature. Solving for V, V = nRT/P. In the first part of this problem, we use the definition

$n = \dfrac{\text{number of molecules}}{\text{Avogadro's number}} = \dfrac{N}{A}$ . We must then substitute

into the formula V = nRT/P =(N/A) (RT/P)in order to obtain the volume corresponding to N molecules. In the second part of this problem we can use V = nRT/P directly, remembering that 1 millimole = $10^{-3}$ mole.

Hence, for the first part of the problem,

$$V = \frac{N}{A}\frac{RT}{P} = \frac{10^6 \text{ molecules}}{6.02 \times 10^{23} \text{ molecules/mole}}$$

$$\times \ \frac{0.0821 \ \ell \text{ -atm/°K - mole} \times 100\text{°K}}{1.316 \times 10^{-13} \text{ atm}}$$

$= 1.04 \times 10^{-4} \ \ell$, where we have used P = $10^{-10}$ torr =

$10^{-10}$ torr $\times \dfrac{1 \text{ atm}}{760 \text{ torr}} = 1.316 \times 10^{-13}$ atm.

For the second part of the problem,$V = \dfrac{nRT}{P}$

$$= \frac{10^{-3} \text{ mole} \times .0821 \ \ell \text{ -atm/°K - mole} \times 100\text{°K}}{1.316 \times 10^{-13} \text{ atm}}$$

$= 6.24 \times 10^{10} \ \ell.$

Describe the curve one would obtain by plotting pressure versus volume for an ideal gas in which the temperature and number of moles of gas are held constant.

Solution: This problem requires a plot of the ideal gas equation. This equation reads

$$PV = nRT,$$

where P = pressure of gas, V = volume of gas, n = number of moles of gas, R = gas constant, and T = absolute temperature of gas. Since R is a constant and n and T are held constant, the product nRT may be combined into a single constant, call it k. Then, PV = nRT = k, or

$$PV = k$$

which is the equation of a hyperbola. Plotting P on the abscissa axis and V on the ordinate axis, we obtain the following curve:

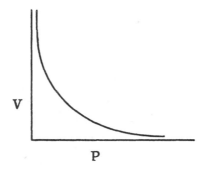

Note: Even if we did not know PV = k is the graph of a hyperbola, we could still obtain this graph. For example, if k is a constant, P and V must vary inversely with each other. In other words, if P is large, V must be small for k to remain constant. Likewise, if V is large, P must be small.

● **PROBLEM** 54

A sample of a gas exhibiting ideal behavior occupies a volume of 350 m$\ell$ at 25.0°C and 600 torr. Determine its volume at 600°C and 25.0 torr.

Solution: We can solve this problem by determining the number of moles, n, of gas from the ideal gas equation, PV = nRT (where P = pressure, V = volume, n = number of moles, R = gas constant, and T = absolute temperature) by using the first set of conditions and then substituting this value of n along with P and T from the second set of conditions into the ideal gas equation and solving for V. However, we can save useless calculation by denoting the first and second sets of conditions by subscripts "1" and "2", respectively, to obtain

$$P_1V_1 = nRT_1, \quad \text{and} \quad P_2V_2 = nRT_2,$$

where n is the same in both cases. Dividing the second of these equations by the first we obtain

$$\frac{P_2V_2}{P_1V_1} = \frac{nRT_2}{nRT_1} \quad \text{or} \quad \frac{P_2V_2}{P_1V_1} = \frac{T_2}{T_1},$$

where we have cancelled n and R (both constant). Solving for $V_2$, we obtain

$$V_2 = \frac{P_1V_1}{P_2} \times \frac{T_2}{T_1}$$

$$= \frac{600 \text{ torr} \times 350 \text{ m}\ell}{25.0 \text{ torr}} \times \frac{873.15^\circ K}{298.15^\circ K}$$

$$= 24.6 \times 10^3 \text{ m}\ell$$

$$= 24.6 \ \ell.$$

● PROBLEM 55

A sample of gas occupies 14.3 $\ell$ at 19°C and 790 mm Hg. Determine the number of moles of gas present. What volume will this same amount of gas occupy at 190°C and 79.0 mm Hg?

Solution: This problem involves an application of the ideal gas equation, PV = nRT, where P = pressure, V = volume, n = number of moles, R = gas constant, T = absolute temperature. We will solve this equation for n and then use this value of the number of moles, and the ideal gas equation, to obtain V at the new temperature and pressure.

For our gas, we initially have V = 14.3 $\ell$, T = 19°C = 292.15°K, P = 790 mm Hg x 1 atm/760 mm atm/760 mm Hg = 1.039 atm and R = 0.0821 $\ell$-atm/°K-mole. Solving the ideal gas equation for n, we obtain

$$n = \frac{PV}{RT} = \frac{1.039 \text{ atm} \times 14.3\ \ell}{0.0821\ \ell\text{-atm}/^\circ K\text{-mole} \times 292.15^\circ K}$$

$$= 0.6194 \text{ mole.}$$

Under the new conditions, we have T = 190°C = 463.15°K and P = 79.0 mm Hg = 79.0 mm Hg x 1 atm/760 mm Hg = 0.1039 atm. Solving the ideal gas equation for V, the new volume is

$$V = \frac{nRT}{P} = \frac{0.6194 \text{ mole} \times 0.0821\ \ell\text{-atm}/^\circ K\text{-mole} \times 463.15^\circ K}{0.1039 \text{ atm}}$$

$$= 226.68 \ \ell.$$

A 3-liter bulb containing oxygen at 195 torr is connected to a 2-liter bulb containing nitrogen at 530 torr by means of a closed valve. The valve is opened and the two gases are allowed to equilibrate at constant temperature, T. Calculate the equilibrium pressure.

<u>Solution</u>: We will solve for the final or equilibrium pressure, $P_f$, by using the ideal gas law,

$$P_f = \frac{nRT}{V_f}$$

where n is the number of moles of $O_2$ $\left(n_{O_2}\right)$ plus the number of moles of $N_2$ $\left(n_{N_2}\right)$, that is, the total number of moles of gas, R is the gas constant, T the absolute temperature, and $V_f$ the final volume (3 liters + 2 liters = 5 liters). We cannot solve this equation directly, since we do not know T.

The quantities $n_{O_2}$ and $n_{N_2}$ are calculated using the initial conditions and the ideal gas law,

$$n_a = \frac{P_a V_a}{RT}$$

where $P_a$ is the original partial pressure of gas a and $V_a$ is the volume initially occupied by gas a. Thus,

$$n_{O_2} = \frac{P_{O_2} V_{O_2}}{RT} = \frac{195 \text{ torr} \times 3 \text{ liters}}{RT} , \qquad \text{and}$$

$$n_{N_2} = \frac{P_{N_2} V_{N_2}}{RT} = \frac{530 \text{ torr} \times 2 \text{ liters}}{RT}$$

Multiplying both equations by RT and then adding,

$$n_{O_2} RT + n_{N_2} RT$$

$$= \left(\frac{195 \text{ torr} \times 3 \text{ liters}}{RT}\right) RT + \left(\frac{530 \text{ torr} \times 2 \text{ liters}}{RT}\right) RT$$

$$\left(n_{O_2} + n_{N_2}\right) RT = 195 \text{ torr} \times 3 \text{ liters} + 530 \text{ torr}$$

$$\times 2 \text{ liters}.$$

Dividing both sides by $V_f$, we obtain

$$\left(n_{O_2} + n_{N_2}\right) \frac{RT}{V_f} = \frac{195 \text{ torr} \times 3 \text{ liters} + 530 \text{ torr} \times 2 \text{ liters}}{V_f} .$$

But $n_{O_2} + n_{N_2} = n$, and $P_f = nRT/V_f$. Hence,

$$P_f = \frac{nRT}{V_f} = \left( n_{O_2} + n_{N_2} \right) \frac{RT}{V_f}$$

$$= \frac{195 \text{ torr} \times 3 \text{ liters} + 530 \text{ torr} \times 2 \text{ liters}}{V_f}$$

$$= \frac{195 \text{ torr} \times 3 \text{ liters} + 530 \text{ torr} \times 2 \text{ liters}}{5 \text{ liters}}$$

$$= 329 \text{ torr}.$$

● **PROBLEM** 57

Determine the molecular weight of a gas if 4.50 g of it occupies 4.0 liters at 950 torr Hg and 182°C.
R = .082 liter - atm/mole - °K.

Solution: One can find the molecular weight of this gas, once one knows the number of moles present in 4.50 g. The molecular weight is equal to 4.50 g divided by the number of moles present

$$MW = \frac{4.50 \text{ g}}{\text{no. of moles}}$$

The number of moles of the gas can be found by using the Ideal Gas Law. This law is stated

$$PV = nRT,$$

where P is the pressure, V is the volume, n is the number of moles, R is the gas constant (0.082 liter-atm/mole-°K) and T is the absolute temperature. One must convert the pressure in torr to atm to use the gas constant. There are 760 torr in 1 atm. Thus, torr are converted to atm by multiplying the number of torr by 1 atm/760 torr.

P (in atm.) = 950 torr × 1 atm/760 torr

= 1.25 atm.

To use the Ideal Gas Law, the temperature must be in °K. Here, it is given in °C. Temperature in °C can be converted to °K by adding 273 to the temperature in °C.

T = 182 + 273 = 455°K

One can now use the Ideal Gas Law to solve for n.

$$n = \frac{PV}{RT}$$

P = 1.25 atm.
V = 4.0 liters
R = 0.082 liter-atm/mole-°K
T = 455°K

$$n = \frac{1.25 \text{ atm} \times 4.0 \text{ liters}}{0.082 \frac{\text{liter-atm}}{\text{mole-°K}} \times 455°K}$$

= 0.13 moles

In 4.50 g of this gas, there are 0.13 moles. The molecular weight can now be found.

$$MW = \frac{4.50 \text{ g}}{0.13 \text{ moles}} = 34.7.$$

● **PROBLEM** 58

A research worker isolated an unknown compound from one of his reaction products. This compound was a volatile liquid with a boiling point of 65°C. A .134 g sample of the liquid was vaporized and the gas was collected. The temperature of the collecting chamber was kept at 98°C in order that liquefaction would not occur. After all the sample was vaporized, the volume of the gas was measured as .0532 liters at 98°C and 737 mm. Calculate the molecular weight.

Solution: We are dealing with the variables m = mass, V = volume, P = pressure, and T = temperature. Whenever this occurs, it indicates that one should employ the ideal gas equation, PV = nRT, where n = moles and R = universal gas constant. An ideal gas is one in which the gas molecules take up no space and do not attract one another. Real gases do not completely meet these conditions, but at values in the neighborhood of standard conditions ( 0°C and 1 atm), the difference in the real and ideal values is small enough such that the ideal gas law is accurate enough to carry out the calculations.

One further bit of information is required. Since M.W. (Molecular Weight) does not appear anywhere in the equation, a relation must be found between M.W. and one of the 4 variables. The relation is n = moles = mass/M.W. Substituting this into the ideal gas equation we obtain

$$PV = \frac{m}{M.W.} \quad RT$$

Solving for the molecular weight, we obtain:

$$MW = \frac{mRT}{PV}$$

$m$ = mass of the sample = .134 g

$R$ = gas constant = .082 liter - atm/mole - °K

$T$ = 98°C = 273 + 98 = 371°K

$V$ = 0.0532 ℓ

$P$ = 737 mm. For the formula, P must be given in atmospheres. To do this, we multiply 737 mm by the conversion factor of $\frac{1 \text{ atm}}{760 \text{ mm}}$. Thus,

$$737 \text{ mm} = 737 \text{ mm} \cdot \frac{1 \text{ atm}}{760 \text{ mm}} = \frac{737}{760} \text{ atm.}$$

$$MW = \frac{(0.134)(0.082)(371)}{(737/760)(0.0532)} = 79.018 \text{ g/mole.}$$

● **PROBLEM** 59

Assuming ideal gas behavior, what is the molecular weight of a gas with a density of 2.50 g/liter at 98°C and .974 atm?

Solution:  You must employ the ideal gas law to answer this question. This law, also called the equation of state for an ideal gas, relates pressure, volume, temperature, and moles to each other quantitatively. It states PV = nRT, where R is called the universal gas constant, P = pressure, V = volume, T = temperature in Kelvin, and n = number of moles. You can also write this equation as n/V = P/RT to determine moles per liter. By substituting,

$$\frac{.974 \text{ atm}}{(.0821 \text{ liter-atm/mole-°K})(371°K)} = .0320 \text{ moles/liter.}$$

Therefore, according to the above calculation, you know there are .0320 moles in one liter. However, the density is 2.50 g/liter. Hence, you can compare the two, and obtain the fact that .0320 moles weighs 2.50 g. From this, the molecule weight determination follows via a proportion. If 0.032 moles weighs 2.50 g, then 1 mole weighs x grams, which equals the molecular weight. In other words,

$\frac{.032}{2.50} = \frac{1}{x}$ . Solving for x, we obtain x = 78.1 g.

● **PROBLEM** 60

A cylinder contains oxygen at a pressure of 10 atm and a temperature of 300°K. The volume of the cylinder is 10 liters. What is the mass of the oxygen?

<u>Solution</u>:  In this problem,we are dealing with 4 variables M = mass, V = volume, P = pressure, and T = temperature. Three of them are given; we must calculate the fourth. These variables are governed by the ideal gas law.

An ideal gas is one in which the molecules have no attraction for one another and the molecules themselves occupy no space (a situation contrary to fact). Real gases do not completely satisfy either of these conditions, but under conditions close to STP (0°C and 1 atm), the real gases come very close to meeting these conditions. Therefore, the difference between the real and ideal values is small enough such that the ideal gas law equation can be used. It i·s:

$$PV = n\,RT$$

where n is the number of moles. Moles is defined as grams/molecular weight. R is a gas constant equal to .082 (liter)-(atm)/(°K)-(mole).

We are asked to find the mass of oxygen, yet there is no value for mass in the ideal gas equation. Therefore, we must find an equation that involves mass and moles. That equation is moles = grams/M.W. The molecular weight of $O_2$ = 32 g/ mole.  Therefore, substitute the mole equation into the ideal gas equation

$$PV = nRT = \frac{M}{M.W.}\,RT$$

The only unknown value is mass and the solution is arrived at by substituting in the appropriate values

$$(10\ \text{atm})(10\ \text{liter}) = \frac{m}{32}\,(.082)(300°K)$$

$$m = \frac{32 \times 10 \times 10}{.082 \times 300} = 130.08\ \text{grams}.$$

● **PROBLEM** 61

Metallic sodium violently liberates $H_2$ gas according to the reaction

$$2Na(s) + 2H_2O(\ell) \rightarrow 2NaOH + H_2(g)$$

If you collect the gas at 25°C under 758 mm Hg of pressure, and it occupies 2.24 liters, (a) find the number of moles of hydrogen produced, (b) find the number of moles of water decomposed and (c) find the percent error in (a) if the gaseous hydrogen had been collected dry, assuming the vapor pressure of water is 23.8 mm at 25°C. (R = .0821 $\ell$-atm/mole °K.)

<u>Solution</u>:  (a) You want to determine the number of moles of hydrogen gas liberated given its pressure, volume and temperature. Therefore, you make use of the equation of state, which indicates that PV = nRT, where P = pressure, V = volume, R = universal gas constant, T = temperature in Kelvin (Celsius plus \273°), and n = moles. Thus, to find n, substitute into this equation and solve for n.

In 1 atm, there is 760 mm of pressure. You use atmospheres as the units of pressure, since R is expressed in atmospheres. Hence,

$$P = 758 \text{ mm} = \frac{758}{760} \text{ atms} = .997 \text{ atm}.$$

We are given R in the problem statement and that T = 298°K and V = 2.24 ℓ. Substituting,

$$n = \frac{(.997)(2.24)}{(.0821)(298°K)} = .0913 \text{ mole } H_2.$$

Now that you know how many moles of $H_2$ are liberated, you also know how much water decomposed via the stoichiometry of the reaction.

In (b), therefore, you make use of the equation

$$2Na(s) + 2H_2O(\ell) \rightarrow 2NaOH + H_2(g),$$

which tells you that for every mole of $H_2(g)$ produced you have 2 moles of $H_2O$ decomposed. You calculated a release of .0913 moles of $H_2$. Therefore, 2 × .0913 = .183 mole of water must have decomposed. For (c) you must realize that if the hydrogen gas were wet (not dried), the pressure given is not the true pressure of the $H_2$. For in this pressure is included the vapor pressure of water. Thus, the actual pressure of $H_2$ is only 758 − 23.8 = 734.2, where 23.8 was the given vapor pressure of water at 25°C. The % error is, then

$$\frac{23.8}{734.2} \times 100 = 3.24 \text{ %}.$$

● **PROBLEM** 62

Compare the number of $H_2$ and $N_2$ molecules in two containers described as follows: (1) A 2-liter container of Hydrogen filled at 127°C and 5 atm. (2) A 5-liter container of nitrogen filled at 27°C and 3 atm.

<u>Solution</u>:  Avogadro's Law states that equal numbers of molecules are contained in equal volumes of different gases if the pressure and temperature are the same. Therefore, if the conditions were the same, then there would be equal numbers of molecules of $H_2$ and $N_2$.

For reasons of clarity, it is useful to set up a table like that shown below.

| gas | V | T | P |
|-----|-----|-----|-----|
| $H_2$ | 2 liters | $400°K$ | 5 atm |
| $N_2$ | 5 liters | $300°K$ | 3 atm |

By using the general gas equation

$$PV = nRT$$

where n = number of moles, and R = 0.082 atm-liter/°K-mole, one can find the number of moles of $H_2$ and $N_2$.

Thus, for $H_2$

$$n = \frac{PV}{RT} = \frac{(5 \text{ atm})(2\ell)}{(0.082 \text{ atm-}\ell/\text{mole-}°K)(400°K)} = 0.305 \text{ moles}$$

for $N_2$

$$n = \frac{PV}{RT} = \frac{(3 \text{ atm})(5 \ \ell)}{(0.082 \text{ atm-}\ell/\text{mole-}°K)(300°K)} = 0.610 \text{ moles.}$$

There are twice as many moles of $N_2$ as there are of $H_2$. Since the number of molecules is directly proportional to the number of moles, then there are twice as many molecules of $N_2$ as molecules of $H_2$.

● PROBLEM 63

A 0.100 $\ell$ container maintained at constant temperature contains $5.0 \times 10^{10}$ molecules of an ideal gas. How many molecules remain if the volume is changed to 0.005 $\ell$? What volume is occupied by 10,000 molecules at the initial temperature and pressure?

Solution: Since only the volume changed and no molecules were added to or withdrawn from the system, the number of molecules at a volume of 0.005 $\ell$ is the same as that at a volume of 0.100 $\ell$, or $5.0 \times 10^{10}$ molecules.

In the second part of the problem we require a relationship between volume and number of molecules, taking into account the fact that the number of molecules can change. We will obtain such a relationship by modifying the ideal gas equation, $PV = nRT$, where P = pressure, V = volume, n = number of moles, R = gas constant, and T = absolute temperature. Since the number of moles is equal to the number of molecules, N, divided by Avogradro's number, A, or n = N/A, we can write the ideal gas equation as $PV = nRT = N/A \ RT$, or

PVA = NRT.

The initial and final pressure and temperature are the same. Also, A and R are constants. Denoting the initial volume and number of molecules by $V_i$ and $N_i$, respectively, and the final volume and number of molecules by $V_f$ and $N_f$, respectively, we obtain

$$PV_iA = N_iRT \quad \text{and} \quad PV_fA = N_fRT.$$

Dividing the second of these by the first we obtain

$$\frac{PV_fA}{PV_iA} = \frac{N_fRT}{N_iRT} \qquad \text{or} \qquad \frac{V_f}{V_i} = \frac{N_f}{N_i} \quad ,$$

where we have cancelled all the constants. Solving for the final volume we obtain

$$V_f = V_i \times \frac{N_f}{N_i} = 0.100 \; \ell \times \frac{10{,}000}{5 \times 10^{10}} = 2 \times 10^{-8} \; \ell.$$

● **PROBLEM** 64

At 100°C and 720 mm Hg or 720 torr, what is the density of carbon dioxide, $CO_2$?

<u>Solution</u>: According to the kinetic theory of gases, the parameters of pressure, volume, and temperature of a gas are mutually dependent - that is, a change in one influences the others. In this problem, you want the density of $CO_2$. Density is defined as mass/volume. Thus, density =(mass of 1 mole $CO_2$)/(volume of 1 mole $CO_2$). The mass of one mole of carbon dioxide is 44 g. (This number represents the sum of the atomic weights of all atoms in the gas.) Consequently, this problem involves only the calculation of $CO_2$'s volume at 100°C and 720 torr.

At standard temperature and pressure, one mole of $CO_2$ or any gas occupies 22.4 liters. (Recall that standard temperature and pressure is 760 mm Hg and 0°C.) Because the temperature and pressure has been altered, the volume will no longer be 22.4 liters for one mole of $CO_2$.

According to the combined gas law, PV/T = constant.

Thus at standard temperature and pressure,

$$\frac{(760 \text{ mm Hg})(22.4 \text{ liters})}{273°} = \text{a constant.}$$

(You add 273° to the Celsius temperature, since T must be in Kelvin.) At 100°C and 720 mm Hg, you have

$$\frac{(720 \text{ mm Hg}) \text{ V}}{373°} = \text{a constant.}$$

According to this law, both expressions are equal, i.e.

$$\frac{(760 \text{ mm Hg})(22.4 \text{ liters})}{273°} = \frac{(720 \text{ mm Hg})(V)}{373°}$$

Solving for V in this equation, you obtain V = 32.3 liters.

The density is, therefore, 44/32.3 = 1.32 g/liter.

● **PROBLEM** 65

Using the ideal gas law, show that for an ideal gas of density $\rho$, $P/\rho T$ is a constant.

Solution: The ideal gas law reads PV = nRT, where P = pressure, V = volume, n = number of moles, R = gas constant, and T = absolute temperature. The number of moles is equal to the mass m of a sample of gas divided by the molecular weight MW of that gas: n = m/MW. Then PV = nRT = (m/MW)RT. Dividing both sides by the volume, we obtain $P = \left(\frac{m}{MW}\right)\frac{RT}{V}$ . But mass per unit volume is density, hence $\rho$ = m/V and this equation becomes P = ($\rho$/MW) RT or P/($\rho$T) = R/MW. Since R is a universal constant and MW, the molecular weight, is constant for a particular gas, P/($\rho$T) = R/MW = constant.

● **PROBLEM** 66

The density of a gas is measured as 0.222 g/ℓ at 20.0°C and 800 torr. What volume will 10.0 g of this gas occupy under standard conditions?

Solution: Since we know the density at a given temperature and pressure, we can apply the equation P/($\rho$T) = constant, which holds for ideal gases of density $\rho$ at pressure P and absolute temperature T. We must first determine the constant for our system and then use this value along with standard pressure (760 torr) and temperature (273.15°K) to determine the density of the gas under standard conditions. From this density, we obtain the volume that 10.0 g of the gas occupies under standard conditions.

The value of the constant is, as previously indicated, $P/(\rho T)$

$$= \frac{800 \text{ torr}}{0.222 \text{ g/}\ell \times 20.0°C} = \frac{800 \text{ torr}}{0.222 \text{ g/}\ell \times 293.15°K} = 12.28 \frac{\text{torr-}\ell}{\text{g} - °K}$$

Hence, for our gas,

$$\frac{P}{\rho T} = 12.28 \frac{\text{torr} - \ell}{\text{g} - °K} \quad \text{or} \quad \rho = \frac{P}{T} \times \frac{1}{12.28 \text{ torr} - \ell/ \text{ g} - °K} \cdot$$

Under standard conditions, the density of our gas is

then $\rho = \dfrac{P}{T} \times \dfrac{1}{12.28 \text{ torr} - \ell/\text{g} - °K}$

$$= \frac{760 \text{ torr}}{273.15 °K} \times \frac{1}{12.28 \text{ torr} - \ell/\text{g} - °K} = 0.2265 \text{ g/}\ell.$$

Density is defined as $\rho$ = mass/volume. Hence, volume = mass/$\rho$, and the volume occupied by 10.0 g of our gas under standard conditions is

$$\text{volume} = \frac{\text{mass}}{\rho} = \frac{10.0 \text{ g}}{.2265 \text{ g/}\ell} = 44.15 \, \ell.$$

● **PROBLEM** 67

At standard conditions, it is found that 28.0 g of carbon monoxide occupies 22.4 $\ell$. What is the density of carbon monoxide at 20°C and 600 torr?

Solution: The density of carbon monoxide under STP or standard conditions (pressure = 1 atm or 760 torr and temperature = 0°C or 273.15°K) is $\rho$ = mass/volume. 28 g of CO, carbon monoxide, represents one mole and a mole of any gas occupies 22.4 liters of volume under standard conditions. Thus, $\rho$ = 28g/22.4 $\ell$ = 1.25 g/$\ell$.

We can apply the equation $P/(\rho T)$ = constant, which holds for ideal gases of density $\rho$ at pressure P and absolute temperature T. We will first determine the constant for our system and then use this value to calculate the density at 20°C = 293.15°K and 600 torr.

The value of the constant for our system is

$$\text{constant} = \frac{P}{\rho T} = \frac{760 \text{ torr}}{1.25 \text{ g/}\ell \times 273.15 °K} = 2.225 \frac{\text{torr} - \ell}{\text{g} - °K} \cdot$$

Hence, for our system, $\dfrac{P}{\rho T} = 2.225 \dfrac{\text{torr} - \ell}{\text{g} - °K}$ or

$\rho = \dfrac{P}{T} \times \dfrac{1}{2.225 \text{ torr} - \ell/\text{g} - °K}$. At 20°C = 293.15°K and

600 torr, the density of CO is

$$\rho = \frac{P}{T} \times \frac{1}{2.225 \text{ torr } - \ell/g - ^\circ K}$$

$$= \frac{600 \text{ torr}}{293.15^\circ K} \times \frac{1}{2.225 \text{ torr } - \ell/g - ^\circ K} = .9198 \text{ g}/\ell.$$

● **PROBLEM** 68

One gram of an unknown gaseous compound of boron (B, atomic weight = 10.8 g/mole) and hydrogen (H, atomic weight = 1.0 g/mole) occupies 0.820 liter at 1.00 atm at 276 K. What is the molecular formula of this compound?

Solution: The general formula for the compound is $B_a H_b$, where a and b are to be determined. To find these, we must first find the molecular weight of the compound by using the ideal gas law,

$$PV = nRT,$$

where P = pressure, V = volume, n = number of moles, R = gas constant, and T = absolute temperature. n equals the mass, m, divided by the molecular weight MW, PV = $nRT = \frac{m}{MW} RT$, or MW = $\frac{mRT}{PV}$ .

$$MW = \frac{mRT}{PV} = \frac{1 \text{ g} \times 0.082 \text{ liter } - \text{ atm/mole}-^\circ K \times 276^\circ K}{1.00 \text{ atm} \times 0.820 \text{ liter}}$$

$$= 27.6 \text{ g/mole.}$$

To find a and b, we use the relation $MW_{B_a H_b}$ = a × atomic weight of B plus b × atomic weight of H, or 27.6 g/mole = a × 10.8 g/mole + b × 1.0 g/mole.

By trial and error, we find that a = 2 and b = 6 (2 × 10.8 g/mole + 6 × 1.0 g/mole = 27.6 g/mole), so that the formula of our compound is $B_2 H_6$ (diborane).

● **PROBLEM** 69

Of the several ways in which the supersonic transport (SST) acts as a pollutant, one is by the elimination of carbon dioxide ($CO_2$). During normal flight, $CO_2$ is released at a temperature of about 627°C and at a press-ure of 0.41 atm at an estimated rate of 6.6 × 10⁴ kg per hour. To what volume does this amount of gas correspond?

Solution:   This problem is an application of the ideal
gas law, PV = nRT, where P = pressure, V = volume, n =
number of moles, R = gas constant, and T = absolute temperature. Solving for V,

$$V = \frac{nRT}{P}$$

We know that R = 0.082(liter - atm/mole -$^{o}$K),   T =
627°C = 900°K, and P = 0.41 atm. n is obtained by dividing
the mass of $CO_2$ (6.6 × 10$^4$ kg = 6.6 × 10$^4$ kg × 10$^3$ g/kg =
6.6 × 10$^7$ g) by the molecular weight of $CO_2$. The molecular
weight of $CO_2$ is equal to

atomic mass$_C$ + 2 atomic mass$_O$  = 12 g/mole + 2(16 g/mole)
or 44 g/mole. Hence

$$n = \frac{mass\ CO_2}{molecular\ weight\ CO_2} = \frac{6.6 \times 10^7\ g}{44\ g/mole} = 1.5 \times 10^6\ moles.$$

Substituting the values of n, R, T, and P into the
equation for V gives

$$V = \frac{nRT}{P} = \frac{1.5 \times 10^6\ moles \times 0.082\ liter\text{-}atm/mole\text{-}^{o}K\ \times 900°K}{0.41\ atm}$$

$$= 2.7 \times 10^8\ liters.$$

● PROBLEM 70

KMnO$_4$ is added to a 350 ml solution of an approximately 3%
$H_2O_2$ solution. It is found that 3.94 cm$^3$ of $O_2$ gas is released. It is collected over $H_2O$ at a pressure (barometric)
of 0.941 atm and 28°C. Assuming that the pressure due to
water vapor is 0.0373 atm, what is the actual concentration
of the $H_2O_2$ solution?

Solution:   The concentration of a solution is given in
terms of molarity, i.e., moles/liter. This means that
the number of moles of $H_2O_2$ must be determined, since you
already know the volume.

To determine the number of moles of $H_2O_2$, you know
that, when oxidized, 1 mole of $H_2O_2$ yields 1 mole of $O_2$.
Thus, if you know the number of moles of $O_2$ produced, you
also know the moles of $H_2O_2$ originally present. To find
the number of moles of $O_2$, you must use the equation of
state. This tells us PV = nRT, where P = pressure, V =
volume, n = number of moles, R = universal gas constant,
(.082057  l-atm mol$^{-1}$ deg$^{-1}$), and T = temperature in
degrees Kelvin (degree Celsius plus 273°). Hence,

$$n_{O_2} = \frac{PV}{RT} \ .$$

All the values needed to find $n_{O_2}$ are known or can be calculated. The pressure of $O_2$ will not be .941 atm, since this figure also includes vapor pressure from water. Therefore, the actual pressure of $O_2$ is .941 - .0373. That is, the difference between barometric and vapor pressure. Substituting, you obtain

$$n_{O_2} = \frac{(.941 - .0373)(3.94 \times 10^{-3})}{(.082)(301)} = 1.44 \times 10^{-4}$$

moles of $O_2$. This means there were originally $1.44 \times 10^{-4}$ moles. Now, recalling the definition of molarity, which represents the concentration of a solution, you have

$$\frac{1.44 \times 10^{-4} \text{ moles}}{.350 \text{ liters}} = 4.12 \times 10^{-4} \text{ M}$$

which is the concentration of the $H_2O_2$ solution.

• PROBLEM 71

The van der Waal equation is a modification of the ideal gas equation. It reads

$$\left(P + \frac{an^2}{V^2}\right)(V - nb) = nRT$$

where P = pressure, V = volume, n = number of moles, R = gas constant, T = absolute temperature, and (a) and (b) are constants for a particular gas. The term $an^2/V^2$ corrects the pressure for intermolecular attraction and the term - nb corrects the volume for molecular volume. Using this equation, determine whether a gas becomes more or less ideal when: (a.) the gas is compressed at constant temperature; (b.) more gas is added at constant volume and temperature; and (c.) The temperature of the gas is raised at constant volume.

Solution: As the behavior of a gas approaches ideality, PV/nRT approaches 1. Hence, we must manipulate the van der Waal equation into a form in which it can be determined whether PV/nRT approaches 1 as the variables are changed. Thus,

$$\left(P + \frac{an^2}{V^2}\right)(V - nb) = nRT$$

$$PV - Pnb + \frac{an^2}{V^2} V - \frac{an^2}{V^2} nb = nRT$$

$$PV - Pnb + \frac{an^2}{V} - \frac{an^3 b}{V^2} = nRT$$

$$PV = nRT + Pnb - \frac{an^2}{V} + \frac{an^3 b}{V^2} \quad .$$

Dividing by nRT we obtain

$$\frac{PV}{nRT} = 1 + \frac{Pb}{RT} - \frac{an}{RTV} + \frac{an^2b}{RTV^2}$$

(a.) When the gas is compressed at constant temperature, volume decreases and pressure increases. Hence, the last three terms on the right side increase, and PV/nRT deviates from 1. The gas thus becomes less ideal.

(b.) If more gas is added at constant volume and temperature, both n and P increase, hence the last three terms on the right side increase and the gas becomes less ideal.

(c.) As the temperature of the gas is increased at constant volume, the pressure increases. The last two terms on the right side become smaller while the second term (Pb/RT) increases slightly. Hence, the gas comes closer to being ideal.

● **PROBLEM** 72

Using Van der Waal's equation, calculate the pressure exerted by 1 mole of carbon dioxide at 0°C in a volume of (a) 1.00 liter, (b) 0.05 liter.

Solution:  Real gases are more compressible than ideal gases because their molecules attract each other. The intermolecular attraction is provided for by adding to the observed pressure a term of $n^2a/V^2$ in the ideal gas law (PV = nRT) where n is the number of moles, V is the volume, and a is a van der Waal's constant. At very high pressures, real gases occupy larger volumes than ideal gases; this effect is provided for by subtracting an excluded volume nb from the observed volume V to give the actual volume in which the molecules move (b is a constant). Thus, van der Waal's full equation is:

$$\left(P + \frac{n^2a}{V^2}\right)(V - nb) = nRT$$

To solve this problem one must 1) convert the temperature from units of °C to °K by adding 273° to the °C temperature. This is done because     T is expressed in the absolute. (P = pressure and R = gas constant.) 2) Find the values for the constants (a) and (b) (these can be found in any reference text). For $CO_2$; a = 3.592 liter$^2$ atm/mole$^2$; b = 0.04267 liter/mole. 3) Substitute the known values into the van der Waal's equation and solve for pressure.

Known values:

n = 1   R = 0.082 $\frac{liter-atm}{mole-°K}$     T = 273°K

(a) V = 1.00 liter. Substituting,

$$\left[P + \frac{(1)^2(3.592)}{(1.00)^2}\right]\left[1.00 - (1)(0.04267)\right] = (1)(0.082)(273)$$

$$P = \frac{(0.082)(273°K)}{(1.00 - 0.04267)} - \frac{3.592}{(1.00)^2}$$

$$= 23.38 - 3.592$$

$$= 19.79 \text{ atm}$$

(b) V = 0.05 liter. Substitute

$$\left[P + \frac{(1)^2(3.592)}{(0.05)^2}\right](0.05 - (1)\ 0.04267) = (1)(0.082)(273\ K)$$

$$P = \frac{(0.082)(273°K)}{(0.05 - 0.04267)} - \frac{3.592}{(0.05)^2}$$

$$= 3054 - 1437$$

$$= 1617 \text{ atm.}$$

● PROBLEM 73

Using the data from the accompanying figure, calculate the pressure exerted by .250 moles of $CO_2$ in .275 liters at 100°C. Compare this value with that expected for an ideal gas.

*van der Waals constants*

| Gas | a, liter$^2$ atm/mol$^2$ | b, liter/mol |
|---|---|---|
| Helium | 0.0341 | 0.0237 |
| Argon | 1.35 | 0.0322 |
| Nitrogen | 1.39 | 0.0391 |
| Carbon dioxide | 3.59 | 0.0427 |
| Acetylene | 4.39 | 0.0514 |
| Carbon tetrachloride | 20.39 | 0.1383 |

Solution: To solve this problem you must understand the concept of an "ideal gas" and the formulas associated with it. Boyle's law states that PV = constant, if the temperature is fixed. In other words, if T is constant, then a fixed mass of gas occupies a volume inversely proportional to the pressure exerted on it. If a gas obeys this law, it is termed an "ideal gas." When a gas is ideal, PV = nRT, where P = pressure, V = volume, n = moles, R = universal gas constant, and T = temperature in Kelvin. When a gas is not ideal, it doesn't obey Boyle's Law;

and instead of the Ideal Gas law you use Van der Waal's equation,

$$\left(P + \frac{n^2a}{V^2}\right)(V - nb) = nRT, \quad \text{where, for } CO_2, \ a =$$

3.59 liter$^2$-atm/mole$^2$ and b = .0427 liter/mole. These values are called Van der Waal's constants. Substitute the values into these equations and solve for P. As such,

$$\left[P + \frac{(.250)^2(3.59)}{(.275)^2}\right][.275 - (.250)(.0427)]$$

$$= (.250)(.08206)(373).$$

Solving for P, you obtain P = 26.0 atm.

If you had considered the gas as ideal, then PV = nRT, and P = 27.8 atm.

● **PROBLEM** 74

Of the following two pairs, which member will more likely deviate from ideal gas behavior? (1) $N_2$ versus CO, (2) $CH_4$ versus $C_2H_6$.

Solution: Characteristics of ideal gases include: (1) Gases are composed of molecules such that the actual volume of the molecules is negligible compared with the empty space between them. (2) There are no attractive forces between molecules. (3) Molecules do not lose net kinetic energy in collisions. (4) The average kinetic energy is directly proportional to the absolute temperature.

From these assumptions came the ideal equation of state: PV = nRT, where P = pressure, V = volume, n = moles, R = universal gas constant, and T = temperature. No gas is ideal; they don't obey these assumptions absolutely. To reflect these limitations the ideal gas law can be written as:

$$\left(P + \frac{n^2a}{V^2}\right)(V - nb) = nRT.$$

The term V = nb can be thought of as representing the free volume minus the volume occupied by the gas molecules themselves. The magnitude of (b) is proportional to a gas molecule's size. Thus, the greater the size of a gas molecule, the greater the nb is, and, the greater the deviation, since the volume is assumed to equal exactly V in the Ideal Gas Law. V - nb is smaller than V. The greater nb, the smaller it becomes.

The term n/V, a concentration term, when squared, gives probability of collisions. (a) gives a measure of the cohesive force between molecules. Real molecules do have an attraction for each other. The greater the value of (a), the larger $n^2a/V^2$ becomes, and the larger $(P + n^2a/V^2)$ gets. The ideal gas law assumes a value of exactly P. Thus, as (a) increases, the more deviation.

To determine which gas deviates the most, look at the size of the gas molecules and their dipole moment. The dipole moment is an indication of unlike charges separated by a given amount of distance. Unlike charges on separate molecules will be attracted to each other. Thus, a higher dipole moment indicates greater cohesive attraction among molecules and greater deviation. Proceed as follows:

(1) $N_2$ versus CO. The CO gas has a net dipole moment, while $N_2$ does not. This is because O is more electromagnetic than C. In $N_2$ both of the atoms are the same and thus their electronegativities are equal and no net dipole moment exists. Thus, CO deviates to a greater extent.

(2) $CH_4$ versus $C_2H_6$. $C_2H_6$ is a much larger molecule and occupies more volume. Therefore, it is more likely to deviate from ideal gas behavior.

● **PROBLEM** 75

What do the terms critical temperature and pressure mean? If you want to liquefy a gas, what physical properties must you consider? If one gas has a higher critical temperature than another, what can be said about the relative forces of attraction between like molecules?

Solution: The critical temperature is the temperature above which a gas may not be liquefied regardless of the pressure. The critical pressure is the pressure necessary to liquefy a gas at its critical temperature.

Both temperature and pressure are factors in the condensation of a gas. Molecules of a particular compound have a lower kinetic energy in their liquid state than in their gaseous state.

kinetic energy = mass × (velocity)$^2$

Therefore, kinetic energy is proportional to the square of the velocities of the particles. In the gaseous state the velocities of the molecules are greater, when the temperature of a system is decreased the velocities of the molecules decrease and, hence, the kinetic energies decrease. If the pressure of the system is increased, the particles are forced closer together. At a certain point, when the kinetic energy will be low and/or the particles will be close enough together, the gas will liquefy. If a gas has a  high critical temperature, less energy is necessary to liquefy it. In the liquid state the molecules are much closer together than in the gas. When less energy is needed to force the molecules together, the inter-molecular forces of attraction are greater.

# CHAPTER 3

# GAS MIXTURES AND PHYSICAL PROPERTIES OF GASES

> **Basic Attacks and Strategies for Solving Problems in this Chapter. See pages 61 to 94 for step-by-step solutions to problems.**

## Mixtures of Ideal Gases

In mixtures of ideal gases there is negligible interaction between molcules — either between the same or different species. Therefore, each component in a mixture of ideal gases behaves as though it were present all by itself.

The pressure exerted by each component in a mixture of gases is called that component's partial pressure. Since the components in a gas mixture do not interact, the partial pressure of a species in a mixture is the total pressure that component would exert if it were the only one present in the system. The ideal gas law from Chapter 2 applies equally well for the individual components in gas mixtures if we replace the total pressure with the partial pressure and the total number of moles with the number of moles of the individual component.

$$P_i V = n_i RT \qquad\qquad 3\text{-}1$$

This equation is identical to Equation 2-1 of Chapter 2 except that $P_i$ is the partial pressure of species $i$, and $n_i$ is the number of moles of species $i$.

$P_i$ = partial (absolute) pressure of species $i$
$V$ = volume
$n_i$ = number of moles of species $i$
$T$ = absolute temperature
$R$ = universal gas constant

It should be obvious intuitively that the total number of moles in a system ($n$) will equal the sum of the numbers of moles of each of the individual components.

$$n = n_1 + n_2 + n_3 \ldots + n_m = \sum n_i \qquad \text{3-2}$$

It follows, therefore, that at constant temperature and volume, the total pressure of a system containing several gaseous species ($P$) is simply the sum of the partial pressures.

$$P = \sum P_i = (\sum n_i)RT/V = nRT/V \qquad \text{3-3}$$

This result (that each species exerts a partial pressure as though it were present alone, and the sum of the partial pressures equals the total pressure) is known as Dalton's Law of Partial Pressures. An important corollary of Dalton's Law can be obtained by multiplying and dividing Equation 3-1 by $n$, the total number of moles in the system.

$$P_i = (n_i/n)nRT/V = (n_i/n)P = N_i P \qquad \text{3-4}$$
$$\text{or}$$
$$P_i/P = n_i/n = N_i$$

The quantity $n_i/n$ (written $N_i$) is defined as the mole fraction of species $i$ and is simply the fraction of the total number of moles or molecules in the system that are of component $i$. Note that the mole fraction represents a fraction of molecules; since unlike molecules normally have different masses, the mole fraction and the mass fraction are not the same. This concept is explained in greater detail in Chapter 4.

In this chapter, problems illustrating principles governing ideal gas mixtures are presented. These problems are best attacked by applying the ideal gas law to each individual species. While Dalton's Law of Partial Pressures is exact only for mixtures of ideal gases, it is often an excellent approximation for real gases at pressures and temperatures where deviations from ideal gas behavior are not significant.

## Kinetic Theory of Gases / Graham's Law of Gaseous Diffusion

The kinetic theory of gases, first developed in the mid 1800s, explains many observed properties of gases. The basic postulate of the kinetic theory is that gas molecules move about at high velocities and collide elastically with other molecules and the vessel walls. Collisions with the walls of the container exert a force we observe as the gas pressure. An important observation is that the kinetic energy associated with the motion of a molecule is proportional to the absolute temperature and is the same for all species.

$$E_k = 1/2 \, m \, v^2 = 3/2 \, k \, T \qquad \qquad 3\text{-}5$$

where the terms in Equation 3-5 are defined as follows:

$E_k$ = kinetic energy of the gas molecule
$m$ = mass of the gas molecule
$v$ = velocity of the molecule
$k$ = Boltzmann constant ($1.38 \times 10^{-23} \, J/K$)
$T$ = absolute temperature

There is a distribution of gas velocities in any gas, and several different averages are defined in Problem 100. The important fact to remember is that the average energy is a function of temperature and not the mass of the molecule. If Equation 3-5 is multiplied by $N_A$, the number of molecules in a mole, the following equation is obtained:

$$E_k = 1/2 N_A m v^2 = 1/2 M v^2 = 3/2 \, RT \qquad \qquad 3\text{-}6$$

where ($N_A \times m$) is the molecular weight, $M$, of the gas, and ($N_A \times k$) is the universal gas constant, $R$.

Graham's Law of Gaseous Diffusion follows logically from Equation 3-6. The rate at which a gas molecule will diffuse is proportional to its velocity. Since all gases in a mixture are at the same temperature and have, therefore, the same average kinetic energy per molecule, the molecular velocity is inversely proportional to the square root of the molecular weight.

$$v \propto D \propto \frac{1}{\sqrt{M}} \qquad \qquad 3\text{-}7$$

In Equation 3-7, $D$ is the diffusivity; the equation provides a means for comparison of the diffusivities of different gaseous species. Graham's Law of Diffusion states, however, that the rate of diffusion is inversely proportional to the gas density, $\rho$. From the ideal gas law the density can be calculated as follows and is proportional to the molecular weight for a given temperature and pressure.

$$\rho = \frac{m}{V} = \frac{n \cdot m}{V}$$

$$PV = nRT \therefore \frac{n}{V} = \frac{P}{RT}$$

$$\frac{n \cdot M}{V} = \frac{PM}{RT} \qquad\qquad\qquad 3\text{-}8$$

It follows that, at constant temperature and pressure, the rate of diffusion is inversely proportional to the square root of the molecular weight of a gas.

$$r_1/r_2 = (\rho_2/\rho_1)^{1/2} = (M_2/M_1)^{1/2} \qquad\qquad 3\text{-}9$$

Problems in this chapter which involve the molecular diffusion of different gases all require a comparison of rates of diffusion or diffusivity. These problems are most efficiently attacked by observing the ratios in Equation 3-9 and determining which of the quantities can be calculated from those given.

## MOLE FRACTION

● PROBLEM 76

Calculate the mole fractions of ethyl alcohol, $C_2H_5OH$, and water in a solution made by dissolving 9.2 g of alcohol in 18 g of $H_2O$. M.W. of $H_2O$ = 18, M.W. of $C_2H_5OH$ = 46.

Solution: Mole fraction problems are similar to % composition problems. A mole fraction of a compound tells us what fraction of 1 mole of solution is due to that particular compound. Hence,

mole fraction of solute = $\dfrac{\text{moles of solute}}{\text{moles of solute + moles of solvent}}$

The solute is the substance being dissolved into or added to the solution. The solvent is the solution to which the solute is added.

The equation for finding mole fractions is:

$\dfrac{\text{moles A}}{\text{moles A + moles B}}$ = mole fraction A

Moles are defined as grams/molecular weight (MW). Therefore, first find the number of moles of each compound present and then use the above equation.

moles of $C_2H_5OH$ = $\dfrac{9.2 \text{ g}}{46.0 \text{ g/mole}}$ = .2 mole

moles of $H_2O$ = $\dfrac{18 \text{ g}}{18 \text{ g/mole}}$ = 1 mole

mole fraction of $C_2H_5OH$ = $\dfrac{.2}{1 + .2}$ = .167

mole fraction of $H_2O$ $= \dfrac{1}{1 + .2} = .833.$

Note, that the sum of the mole fractions is equal to 1.

● **PROBLEM** 77

Of the many compounds present in cigarette smoke, such as the carcinogen 3,4-benzo[a]pyrene, some of the more abundant are listed in the following table along with their mole fractions:

| Component | Mole Fraction |
|-----------|---------------|
| $H_2$ | 0.016 |
| $O_2$ | 0.12 |
| CO | 0.036 |
| $CO_2$ | 0.079 |

What is the mass of carbon monoxide in a 35.0 ml puff of smoke at standard temperature and pressure (STP)?

<u>Solution</u>: Assuming ideal gas behavior, we will calculate the number of moles of ideal gas in a 35.0 ml volume. Using the mole fraction of CO in smoke, we will then obtain the number of moles of CO in a 35.0 ml volume and finally convert this to a mass.

The molar volume of an ideal gas is 22.4 liter/mole when STP conditions exist, i.e. when temp. = 0°C and pressure = 1 atm. Hence, the number of moles of ideal gas in a 35.0 ml volume is obtained by dividing this volume by the molar volume of gas, or 35.0 ml/22.4 liter/mole = 0.035 liter/22.4 liter/mole $= 1.56 \times 10^{-3}$ mole.

The mole fraction is defined by the equation

$$\text{mole fraction CO} = \frac{\text{moles CO}}{\text{total number of moles}}$$

solving for the number of moles of CO,

moles CO = mole fraction CO × total number of moles
$= 0.036 \times 1.56 \times 10^{-3}$ mole
$\equiv 5.6 \times 10^{-5}$ mole.

This is converted to a mass by multiplying by the molecular weight of CO (28 g/mole). Hence,

mass of CO = moles of CO × molecular weight of CO
$= 5.6 \times 10^{-5}$ mole × 28 g/mole
$= 1.6 \times 10^{-3}$ g.

It has been estimated that each square meter of the earth's surface supports $1 \times 10^7$ g of air above it. If air is 20% oxygen ($O_2$, molecular weight = 32 g/mole) by weight, approximately how many moles of $O_2$ are there above each square meter of the earth?

Solution: This problem is solved by first calculating what weight of oxygen is present in $1 \times 10^7$ g of air and then dividing by the molecular weight of oxygen to convert this mass to moles.

Using the definition of weight percent,

$$\text{weight \% of } O_2 = \frac{\text{weight of } O_2}{\text{total weight}} \times 100\%$$

$$20\% = \frac{\text{weight of } O_2}{1 \times 10^7 \text{ g}} \times 100\%$$

$$20 = \frac{\text{weight of } O_2}{1 \times 10^7 \text{ g}} \times 100$$

Solving for the weight of $O_2$,

$$\text{weight of } O_2 = \frac{20}{100} \times 1 \times 10^7 \text{ g} = 2 \times 10^6 \text{ g}.$$

The number of moles of $O_2$ is equal to the weight of $O_2$ divided by the molecular weight, or

$$\text{moles of } O_2 = \frac{\text{weight of } O_2}{\text{molecular weight}} = \frac{2 \times 10^6 \text{ g}}{32 \text{ g/mole}}$$

$$\cong 6 \times 10^4 \text{ moles}.$$

Therefore, each square meter of the earth's surface supports $6 \times 10^4$ moles of $O_2$.

# DALTON'S LAW OF PARTIAL PRESSURES

A mixture of nitrogen and oxygen gas is collected by displacement of water at 30°C and 700 torr Hg pressure. If the partial pressure of nitrogen is 550 torr Hg, what is the partial pressure of oxygen? (Vapor pressure of $H_2O$ at 30°C = 32 torr Hg.)

Solution: Here, one uses Dalton's Law of partial pressures. This law can be stated: Each of the gases in a gaseous mixture behaves independently of the other gases and exerts its own pressure; the total pressure of the mixture being the sum of the partial pressures exerted by each gas present. Stated algebraically

$$P_{total} = p_1 + p_2 + \ldots p_n$$

where $P_{total}$ is the total pressure and $p_1$, $p_2$, $\ldots p_n$ are the partial pressures of the gases present. In this problem, one is told that oxygen and nitrogen are collected over $H_2O$, which means that there will also be water vapor present in the gaseous mixture. In this case, the equation for the total pressure can be written

$$P_{total} = P_{O_2} + P_{N_2} + P_{H_2O}$$

One is given $P_{total}$, $P_{N_2}$, and $P_{H_2O}$ and asked to find $P_{O_2}$. This can be done by using the law of partial pressures

$$P_{total} = P_{O_2} + P_{N_2} + P_{H_2O} \qquad\qquad P_{total} = 700 \text{ torr}$$

$$P_{O_2} = ?$$

$$700 \text{ torr} = P_{O_2} + 550 \text{ torr} + 32 \text{ torr} \qquad P_{N_2} = 550 \text{ torr}$$

$$P_{O_2} = (700 - 550 - 32)\text{torr} \qquad P_{H_2O} = 32 \text{ torr}$$

$$= 118 \text{ torr}$$

$$P_{O_2} = 118 \text{ torr}.$$

● **PROBLEM** 80

Methane is burnt in oxygen to produce carbon dioxide and water vapor according to the following reaction, $CH_{4\,(g)} + 2O_{2\,(g)} \rightarrow CO_{2\,(g)} + 2H_2O_{(g)}$. The final pressure of the gaseous products was 6.25 torr. Calculate the partial pressure of the water vapor.

Solution: The partial pressure of each of the components in a mixture of gaseous substances will be equal to the product of the mole fraction of that component in the gas and the total pressure of the system. Given the total vapor pressure of the gases, one can calculate the mole fraction of the water vapor.

Let $P_A$ = partial pressure of the water vapor, $N_T$ = total number of moles of gases produced, $N_A$ = number of moles of water vapor, and $P_T$ = total pressure. This law can now be expressed in these terms:

$$P_A = \frac{N_A}{N_T} P_T \quad .$$ From the stoichiometry of the

equation, the total number of moles of products is 3.
Out of these 2 moles are water vapor. Thus, substituting,

$$P_A = 2/3 \ (6.25 \ torr) = 4.16 \ torr \ of \ water \ vapor.$$

● **PROBLEM** 81

A mixture of gaseous oxygen and nitrogen is stored at
atmospheric pressure in a 3.7 ℓ iron container maintained
at constant temperature. After all the oxygen has reacted with the iron walls of the container to form solid
iron oxide of negligible volume, the pressure is measured
at 450 torr. Determine the final volume of nitrogen and
the initial and final partial pressures of nitrogen and
oxygen.

Solution:  The partial pressure of each component, which
is independent of any other component in a gaseous mixture
at a defined volume, is equal to the pressure each component would exert if it were the only gas in that volume.
The total pressure of a gaseous mixture is the sum of the
partial pressures of the components. These two definitions
are sufficient to solve this problem.

Once all the oxygen has reacted, no oxygen is present
in the gaseous phase, so that the final partial pressure
of oxygen is zero. By the second definition, final total
pressure = final partial pressure of $N_2$ + final partial
pressure of $O_2$

450 torr = final partial pressure of $N_2$ + 0 torr

or,  final partial pressure of $N_2$ = 450 torr.

But  since, after all the oxygen has reacted, only
nitrogen fills the entire volume, the final volume of
nitrogen is 3.7  ℓ. By the first definition, the initial
partial pressure of nitrogen is the same as the final
partial pressure of nitrogen, 450 torr. We now again
employ the second definition to determine the last remaining unknown quantity, the initial partial pressure
of oxygen. Proceeding, we obtain

initial total pressure = initial partial pressure of $O_2$
                         + initial partial pressure of $N_2$

1 atm = initial partial pressure of $O_2$ + 450 torr

or,  initial partial pressure of $O_2$ = 1 atm - 450 torr

$$= 760 \text{ torr} - 450 \text{ torr}$$

$$= 310 \text{ torr}.$$

Summarizing these results in tabular form we have:

|  | Initial | Final |
|---|---|---|
| partial pressure of $O_2$ | 310 torr | 0 torr |
| partial pressure of $N_2$ | 450 torr | 450 torr |
| total pressure | 760 torr | 450 torr |

● **PROBLEM** 82

A cylinder contains 40 g He, 56 g $N_2$, and 40 g Ar.
(a) What is the mole fraction of each gas in the
mixture? (b) If the total pressure of the mixture is
10 atm, what is the partial pressure of He?

Solution: (a) The mole fraction of a component in a
system is defined as the number of moles of that component divided by the sum of all the moles present in
the system. Here, one must first calculate the number
of moles present of each component, then the mole
fractions can be found. The number of moles of each
gas can be found by dividing the number of grams
present of the gas by the molecular weight. (MW He = 4,
MW $N_2$ = 28, MW Ar = 40.)

$$\text{no. of moles} = \frac{\text{no. of grams}}{\text{MW}}$$

$$\text{no. of moles of He} = \frac{40 \text{ g}}{4 \text{ g/mole}} = 10 \text{ moles}$$

$$\text{no. of moles of } N_2 = \frac{56 \text{ g}}{28 \text{ g/mole}} = 2 \text{ moles}$$

$$\text{no. of moles of Ar} = \frac{40 \text{ g}}{40 \text{ g/mole}} = 1 \text{ mole}$$

The total number of moles of gas in the system is
the sum of the number of moles of the three gases. Thus,
there are 13 moles of gas in the system. The mole
fraction can now be found for each gas by dividing the
number of moles of each gas by 13, the total number of
moles

$$\text{mole fraction} = \frac{\text{no. of moles}}{\text{total no. of moles in system}}$$

$$\text{mole fraction of He} = \frac{10 \text{ moles}}{13 \text{ moles}} = .77$$

$$\text{mole fraction of } N_2 = \frac{2 \text{ moles}}{13 \text{ moles}} = .15$$

mole fraction of Ar = $\frac{1 \text{ mole}}{13 \text{ moles}}$ = .08

(b) In a system where various gases are present, the partial pressure of each gas is proportional to the mole fraction of the gas. The relationship between the partial pressure of a particular gas and the total pressure is

partial pressure = total pressure × mole fraction

In this problem, one is given the total pressure of the system and one has found the mole fraction of He. One can now find the partial pressure of He

partial pressure of He = 10 atm × .77 = 7.7 atm.

● **PROBLEM** 83

What is the partial pressure of each gas in a mixture which contains 40 g. He, 56 g. $N_2$, and 16 g. $O_2$, if the total pressure of the mixture is 5 atmospheres.

Solution: In a mixture of gases, the partial pressure of each gas is proportional to its mole fraction. Here, to calculate the partial pressures of the various gases contained in this system, one must first calculate the mole fraction of each component. This is done by calculating the number of moles present of each component and dividing that by the total number of moles present in the system. To calculate the partial pressure of each component, the mole fraction must then be multiplied by the total pressure of the system.

(a) To calculate the number of moles present of each component, divide the number of grams present by the molecular weight of the element.

number of moles = $\frac{\text{number of grams present}}{\text{molecular weight}}$

Moles of He = $\frac{40 \text{ g}}{4 \text{ g/mole}}$ = 10 moles

Moles of $N_2$ = $\frac{56 \text{ g}}{28 \text{ g/mole}}$ = 2 moles

Moles of $O_2$ = $\frac{16 \text{ g}}{32 \text{ g/mole}}$ = 0.5 moles

(b) To calculate the total number of moles present, add the number of moles of all components together

total number of moles = moles He + moles $N_2$ + moles $O_2$

= 10 + 2 + 0.5

= 12.5 moles

(c) To calculate the mole fraction of each component, divide the number of moles present of each component by the total number of moles in the system.

mole fraction He = $\frac{10}{12.5}$ = .80

mole fraction $N_2$ = $\frac{2}{12.5}$ = .16

mole fraction $O_2$ = $\frac{0.5}{12.5}$ = .04

(d) To find the partial pressure of each component, multiply the mole fraction by the total pressure in the system.

partial pressure He = (.80) × 5 atm = 4 atm

partial pressure $N_2$ = .16 × 5 atm = 0.8 atm

partial pressure $O_2$ = .04 × 5 atm = 0.2 atm.

● **PROBLEM** 84

The composition of dry air by volume is 78.1% $N_2$, 20.9% $O_2$ and 1% other gases. Calculate the partial pressures, in atmospheres, in a tank of dry air compressed to 10.0 atmospheres.

Solution: A partial pressure is the individual pressure caused by one gas in a mixture of several gases. The total pressure, $P_{total}$, according to Dalton's Laws, is the sum of these individual partial pressure, $p_1$, $p_2$ and $p_3$. i.e. $P_{total} = p_1 + p_2 + p_3$.

The term dry air indicates that no water vapor is present.

The partial pressure is found by multiplying the percent of each gas in the volume by the total pressure.

partial pressure = proportion by volume × total pressure

$P_{N_2}$ = .781 × 10 atm = 7.81 atm

$P_{O_2}$ = .209 × 10 atm = 2.09 atm

$P_{other\ gases}$ = .010 × 10 atm = .1 atm.

The total pressure, as required, is 10 atm.

A sample of air held in a graduated cylinder over water has a volume of 88.3 ml at a temperature of 18.5°C and a pressure of 741 mm (see figure). What would the volume of the air be if it were dry and at the same temperature and pressure?

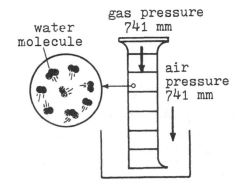

Solution: This problem is an application of Dalton's Law which states that the total pressure of a mixture, $P_{total}$, of gases is the sum of the individual partial pressures, $P_1$, $P_2$. Stated mathematically

$$P_{total} = P_1 + P_2 + P_3 \ldots$$

The mixture of gases in this problem is located in the cylinder and is made up of dry air and water vapor. At 18.5°C the vapor pressure of water is 16 mm. The pressures are related by the equation

$$P_{total} = P_{dry\ air} + P_{water\ vapor}$$

Using this equation, we can calculate $P_{dry\ air}$. Having obtained this value, the problem becomes a direct pressure-volume relationship. As the volume increases, the pressure decreases and vice versa. These factors are related by the equation

$$P_1 V_1 = P_2 V_2$$

where $P_1$ and $V_1$ represent pressure and volume in the presence of water vapor, and $P_2$ and $V_2$ represent the same quantities for dry air. By Dalton's Law:

$$P_{dry\ air} = P_{total} - P_{water\ vapor}$$

$$= 741 - 16$$

$$= 725 \text{ mm when water vapor is present.}$$

If there is no water vapor present and the dry air were to exert the entire pressure of 741 mm then, we predict that the dry air volume will be smaller. Next substitute the values into the pressure-volume equation and solve for unknown volume $V_2$

$$V_2 = \frac{P_1 V_1}{P_2}$$

$$= \frac{725(88.3)}{741}$$

$$= 86.4 \text{ ml}$$

Therefore, as predicted the volume has decreased.

● **PROBLEM** 86

What would be the final partial pressure of oxygen in the following experiment? A collapsed polyethylene bag of 30 liters capacity is partially blown up by the addition of 10 liters of nitrogen gas measured at 0.965 atm and 298°K. Subsequently, enough oxygen is pumped into the bag so that, at 298°K and external pressure of 0.990 atm, the bag contains a full 30 liters. (assume ideal behavior.)

Solution: One should first find the number of moles of gas that fill a volume of 30 liters at .990 atm. One then solves for the number of moles of $N_2$ that was pumped in and the number of moles of $O_2$ that are needed to fill the bag. The partial pressures of these gases are equal to the mole  fraction of the gas times the total pressure.

To solve for the partial pressure of $O_2$ one should find:

1) the total number of moles of gas that will fill 30 liters at .990 atm.

2) the number of moles of $N_2$ present

3) the partial pressure of $N_2$

4) the partial pressure of $O_2$.

Solving:

1) One uses the Ideal Gas Law to solve for the number of moles of gas needed to fill 30 liters at 298°K and .990 atm. The Ideal Gas Law can be stated as

$$PV = nRT \qquad \text{or} \qquad n = \frac{PV}{RT}$$

where n is the number of moles, P is the pressure, V is the volume, R is the gas constant (.082 liter-atm/mole-°K) and T is the absolute temperature. Solving for n in this system

$$P = .990 \text{ atm}$$
$$V = 30 \text{ liters}$$
$$R = .082 \text{ liter-atm/mole-}^{\circ}K$$
$$T = 298^{\circ}K$$

$$n = \frac{.990 \text{ atm} \times 30 \text{ liters}}{.082 \text{ liter-atm/mole-}^{\circ}K \times 298^{\circ}K}$$

$$= 1.22 \text{ moles}$$

2) One also uses the Ideal Gas Law to find the number of moles of $N_2$ already present.

$$P = .965 \text{ atm}$$
$$V = 10 \text{ liters}$$
$$R = .082 \text{ liter-atm/mole-}^{\circ}K$$
$$T = 298^{\circ}K$$

$$n = \frac{.965 \text{ atm} \times 10 \text{ liters}}{.082 \text{ liter-atm/mole-}^{\circ}K \times 298^{\circ}K}$$

$$n = 0.395 \text{ moles.}$$

3) The partial pressure is related to the total pressure by the following

$$P_x = P_o X$$

where $P_x$ is the partial pressure, $P_o$ is the total original pressure and X is the mole fraction of the gas. The mole fraction is defined as the number of moles of the gas present divided by the total number of moles in the system. Solving for the partial pressure of $N_2$, $P_{N_2}$,

$$P_N + P_o X \qquad\qquad P_{N_2} = ?$$

$$P_o = .990 \text{ atm}$$

$$X = .395/1.22$$

$$P_{N_2} = .990 \text{ atm} \times \frac{.395}{1.22}$$

$$P_{N_2} = .321 \text{ atm.}$$

4) The total pressure of the system is equal to the sum of the partial pressures. Thus, .990 atm is the sum of $P_{N_2}$ and $P_{O_2}$. One has already found $P_{N_2}$, therefore, one can now solve for $P_{O_2}$

$$P_{O_2} = P_{total} - P_{N_2} \qquad\qquad P_{total} = .990 \text{ atm}$$

$$P_{N_2} = .321 \text{ atm}$$

$$P_{O_2} = .990 \text{ atm} - .321 \text{ atm}$$

$$P_{O_2} = .669 \text{ atm}.$$

● **PROBLEM** 87

200 ml of oxygen is collected over water at 25°C and 750 torr. If the oxygen thus obtained is dried at a constant temperature of 25°C and 750 torr, what volume will it occupy? What volume will be occupied by the water vapor removed from the oxygen if it is maintained at 25°C and 750 torr? (The equilibrium vapor pressure of water at 25°C is 28.3 torr.)

Solution:  This problem will be approached by determining the partial pressures of oxygen and of water vapor in the initial mixture and then using these partial pressures to determine the separate volumes of each component.

The partial pressures exerted by the oxygen and the water vapor must add up to the total pressure. Since the partial pressure of water vapor is equal to its equilibrium vapor pressure, we can determine the partial pressure of oxygen:

total pressure = partial pressure of $O_2$ + partial pressure of water

750 torr = partial pressure of $O_2$ + 28.3 torr

or,    partial pressure of $O_2$ = 750 torr - 28.3 torr
= 721.7 torr.

By definition, this partial pressure is the pressure the oxygen would exert if it filled the entire 200 ml volume. We must determine the volume this amount of oxygen would fill at 750 torr. To do this we use the relationship $P_1 V_1 = P_2 V_2$, which is valid for constant temperature and constant number of moles of gas. Let $P_1 = 721.7$ torr, $V_1 = 200$ ml, and $P_2 = 750$ torr. Then,

$$V_2 = \frac{P_1}{P_2} \times V_1 = \frac{721.7 \text{ torr}}{750 \text{ torr}} \times 200 \text{ ml} = 192.45 \text{ ml}.$$

Hence, the dried oxygen would occupy a volume of 192.45 ml.

We follow a similar procedure in calculating the volume that the water vapor would occupy at 750 torr. The vapor pressure of water, 28.3 torr, is the pressure the water vapor would exert if it filled the entire 200 ml volume. Again, applying the relationship $P_1 V_1 = P_2 V_2$, remembering that the only reason we can do so is that we have constant temperature and constant number of moles of gas, we can calculate the volume that the water vapor occupies under a pressure of 750 torr. Let $P_1$ = 28.3 torr, $V_1$ = 200 m$\ell$, and $P_2$ = 750 torr. Then the volume at 750 torr is

$$V_2 = \frac{P_1}{P_2} \times V_1 = \frac{28.3 \text{ torr}}{750 \text{ torr}} \times 200 \text{ m}\ell = 7.55 \text{ m}\ell.$$

As a check, we ascertain that the individual volumes at 750 torr add up to the total volume at 750 torr:

Volume of $O_2$ + volume of water = 192.45 m$\ell$ + 7.55 m$\ell$

$$= 200 \text{ m}\ell.$$

● **PROBLEM** 88

Suppose 100 ml of oxygen were collected over water in the laboratory at a pressure of 700 torr and a temperature of 20°C. What would the volume of the dry oxygen gas be at STP?

Solution: STP means Standard Temperature and Pressure, which is 0°C and 760 torr. In this problem oxygen is gathered over water, therefore, Dalton's law of partial pressure (each of the gases in a gaseous mixture behaves independently of the other gases and exerts its own pressure, the total pressure of the mixture being the sum of the partial pressures exerted by each gas present; that is $P_{total}$ = $p_1$ + $p_2$ + $p_3$ ... $p_n$) is used to calculate the original pressure of the oxygen. There is both water vapor and oxygen gas present.

After you obtain the original pressure of the oxygen, you can use the combined gas law to calculate the final volume of the oxygen. The combined gas law stated that for a given mass of gas, the volume is inversely proportional to the pressure and directly proportional to the absolute temperature. It can be written as follows:

$$\frac{P_1 V_1}{T_1} = \frac{P_2 V_2}{T_2}$$

where $P_1$ is the original pressure, $V_1$ is the original volume, $T_1$ is the original absolute temperature, $P_2$ is the final temperature, $V_2$ is the final volume, and $T_2$ is the final absolute temperature. You are given the temperature in °C, so it must be converted to the absolute temperature by adding 273.

(a) Finding the original pressure of oxygen. The partial pressure of water is 17.5 torr.

$$P_{total} = 700 \text{ torr} = P_{O_2} + P_{H_2O}$$

$$P_{O_2} = 700 - 17.5$$

$$= 682.5 \text{ torr.}$$

(b) Converting the temperature to the absolute scale.

$$T_1 = 20 + 273 = 293°K$$

$$T_2 = 0 + 273 = 273°K$$

(c) Using the combined law

$$\frac{P_1V_1}{T_1} = \frac{P_2V_2}{T_2}$$

$P_1 = 682.5$ torr                    $P_2 = 760$ torr

$V_1 = 100$ ml                        $V_2 = ?$

$T_1 = 293°K$                         $T_2 = 273°K$

$$\frac{(682.5 \text{ torr})(100 \text{ ml})}{293°K} = \frac{(760 \text{ torr}) V_2}{273°K}$$

$$V_2 = \frac{(682.5 \text{ torr})(100 \text{ ml})(273°K)}{(760 \text{ torr})(293°K)} = 83.7 \text{ ml.}$$

● **PROBLEM** 89

If 40 liters of $N_2$ gas are collected at 22°C over water at a pressure of .957 atm, what is the volume of dry nitrogen at STP (Standard temperature and pressure)?

Solution:  The key to this problem is the recognition that $PV/T = k$, where P is the  pressure, V the volume, T the absolute temperature and k a constant. If you let the STP conditions (1-atm.and 0°C), be represented by $P_1V_1/T_1$ and the conditions of the gas collected over the water by $P_2V_2/T_2$,then

$$\frac{P_1V_1}{T_1} = \frac{P_2V_2}{T_2} ,$$

since both equal the same constant, k. Therefore, to answer the question, you have to substitute the known values into the equation and solve for the unknown, $V_1$.

However, there is one precaution. Over water, you have vapor pressure. At 22°C, water has a vapor pressure of .026 atm. As such, this makes the initial pressure of $N_2$ .957 - .026, or .931 atm. Add 273° to the Celsius temperature to obtain the temperature in Kelvin. Therefore, $T_2 = 22°C + 273° = 295°K$ and $T_1 = 0°C + 273 = 273°K$. **Therefore, by substitution**

$$\frac{(1)(V_1)}{273} = \frac{(.931)(40)}{295} \ .$$

Now, solving for $V_1$, the volume at STP, you obtain 34.4 liters.

● **PROBLEM** 90

A sample of **hydrogen** is collected in a bottle over water. By carefully raising and lowering the bottle, the height of the water outside is adjusted so that it is just even with the water level inside (see figure). When a sample of gas was collected the initial conditions were: volume = 425 ml, pressure = 753 mm and the temperature of the water (and thus, the gas also) = 34°C. Calculate the volume of the hydrogen if it were dry and at a pressure of 760 mm and a temperature of 0°C (STP).

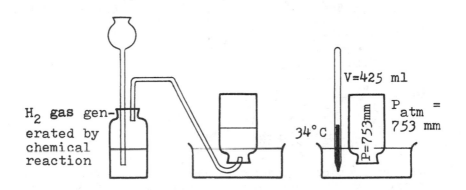

Solution: First, apply Dalton's Law and then make use of the Combined Gas Law to obtain the desired answer.

Dalton's Law states that the total pressure of a mixture of gases, $P_{total}$, is equal to the sum of the individual partial pressures, $P_{H_2}$ and $P_{H_2O}$. In this example the individual gases are hydrogen and water vapor. Therefore, the total pressure is the sum of these two gases. At 35°C, the partial pressure of $H_2O$ vapor is 40 mm,

$$P_{total} = P_{H_2} + P_{H_2O}$$

$$P_{H_2} = P_{total} - P_{H_2O}$$

$$= 753 - 40 = 713 \text{ mm}$$

This means that if no water vapor was present the $H_2$ gas would fill the container at 713 mm of pressure.

Since pressure is indirectly proportional to the volume of $H_2$ gas, the effect of changing the pressure of

713 mm to STP pressure (760 mm) is to decrease the volume.

In addition, however, the temperature of the $H_2$ gas is directly proportional to the volume of the gas, so that the effect of changing the temperature of (34°C + 273° =) 307°K to STP temperature (273°K) is also to decrease the volume of the gas. Therefore, we can predict that the net effect of changing the pressure and temperature of the gas to STP conditions is to decrease the volume.

Using the combined gas law equation to solve for the dry $H_2$ gas volume, $V_2$,

$$\frac{P_1 V_1}{T_1} = \frac{P_2 V_2}{T_2}$$

where     $P_1$ = 713 mm          $P_2$ = 760 mm

           $V_1$ = 425 ml          $V_2$ = volume at STP

           $T_1$ = 307°K          $T_2$ = 273°K

Thus, substituting,

$$V_2 = \frac{P_1 V_1 T_2}{P_2 T_1}$$

$$= \frac{(713 \text{ mm})(425 \text{ ml})(273°K)}{(760 \text{ mm})(307°K)} = 355 \text{ ml}.$$

We see that this answer is in total agreement with our prediction that the volume at STP, $V_2$ = 355 ml, is less than the volume at the initial conditions, $V_1$ = 425 ml.

● **PROBLEM** 91

A 20 g chunk of Dry Ice ($CO_2$) is placed in an "empty" 0.75 liter wine bottle and tightly corked. What would be the final pressure in the bottle after all the $CO_2$ has evaporated and the temperature has reached 25°C?

Solution:   The final pressure in the bottle will equal the original pressure plus the pressure contributed by the $CO_2$. Since the bottle was originally open in atmospheric pressure, the original pressure is 1 atm. The pressure contributed by the $CO_2$ is found by using the Ideal Gas Law. This law is stated

$$PV = nRT \qquad \text{or} \qquad P = \frac{nRT}{V}$$

where P is the pressure of the gas, n is the number of moles, R is the gas constant, 0.082 liter-atm/mole -°K, V is the volume, and T is the absolute temperature. One is given R and V in the problem. n is found by dividing the number of grams of $CO_2$ present by the molecular weight (MW of $CO_2$ = 44 ).

$$\text{no. of moles} = \frac{20 \text{ g}}{44 \text{ g/mole}} = .45 \text{ moles}$$

One can convert °C to °K by adding 273° to the temperature in °C.

$$°K = °C + 273°$$

$$T = 25 + 273° = 298°K$$

Solving for $P_{CO_2}$:

$$P_{CO_2} = \frac{.45 \text{ moles} \times .082 \text{ liter - atm/ mole - °K} \times 298°K}{.75 \text{ liter}}$$

$$= 14.81 \text{ atm.}$$

$$P_{total} = P_{CO_2} + P_{original}$$

$$P_{total} = 14.81 \text{ atm} + 1 \text{ atm} = 15.81 \text{ atm.}$$

● **PROBLEM** 92

The volume of hydrogen evolved during the course of a reaction is measured by the displacement of water as shown in the diagram below. Hydrogen is evolved in flask A and displaces water from flask B into beaker C. If, during a particular run of this experiment in which atmospheric pressure is 765 torr and the water temperature is 293.15°K, 65.0 ml of water is displaced, how much water would be displaced at 760 torr and 298.15°K? (The equilibrium vapor pressure of water at 293.15°K is 17.5 torr and at 298.15°K is 23.8 torr.)

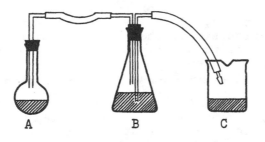

A        B        C

Solution: We can convert from the volume under non-standard conditions in the problem to the volume under standard conditions by means of the ideal gas equation, PV = nRT (where P = pressure, V = volume, n = number of moles, R = gas constant, and T = absolute temperature). For our nonstandard conditions we may write

$$P_nV_n = nRT_n$$

and for final conditions (760 torr and 298.15°K) we may write

$$P_sV_s = nRT_s.$$

Note that n and R are the same in both cases. Dividing the second equation by the first we obtain

$$\frac{P_sV_s}{P_nV_n} = \frac{nRT_s}{nRT_n} = \frac{T_s}{T_n} \quad \text{or} \quad V_s = \frac{P_nV_n}{V_s} \times \frac{T_s}{T_n} .$$

The pressure experienced by the hydrogen under non-standard conditions is equal to the pressure exerted by the atmosphere (765 torr) plus that from water vapor pressure at 20°C (17.5 torr), or $P_n$ = 765 torr + 17.5 torr = 782.5 torr. Under the final conditions, the pressure experienced by hydrogen is equal to the sum of the atmospheric pressure (760 torr) and the vapor pressure of water at 298.15°K (23.8 torr), or

$$P_s = 760 \text{ torr} + 23.8 \text{ torr} = 783.8 \text{ torr.}$$

We can now calculate the volume of hydrogen under the final conditions. Thus,

$$V_s = \frac{P_nV_n}{P_s} \times \frac{T_s}{T_n} = \frac{765 \text{ torr} \times 65.0 \text{ ml}}{783.8 \text{ torr}} \times \frac{298.15°K}{293.15°K}$$

$$= 64.5 \text{ ml.}$$

● **PROBLEM 93**

A technician is working with dry ice (solid $CO_2$) in a closed 6000 liter fume hood maintained at a constant pressure of 1.00 atm and a constant temperature of 27°C. If he has not been conditioned to tolerate $CO_2$, he will succumb to $CO_2$ poisoning when the concentration rises to 10%, by volume, of the atmosphere. If the ventilation stops, what minimum weight of dry ice must sublimate to constitute a hazard? Assume that there was no $CO_2$ present initially.

Solution: To solve this problem we will convert the concentration, 10% by volume, to a partial pressure and then determine what mass of $CO_2$ gives rise to this pressure.

At the point where $CO_2$ is harmful, it occupies 10% by volume of the total atmosphere. Its partial pressure p is therefore 10% of the total pressure, or p = 10% × 1 atm = 0.10 atm.

The ideal gas law reads

$$pV = nRT = \frac{m}{MW} RT$$

where V is the volume, n the number of moles, R the gas constant, T the absolute temperature, m the mass, and MW the molecular weight, and we have used n = m/MW. Solving for m

$$m = \frac{pV(MW)}{RT}$$

Now p = 0.10 atm, V = 6000 liters, MW = 44 g/mole of $CO_2$, R = 0.082 liter-atm/mole-degree, and T = 27°C = 300°K. Hence,

$$m = \frac{pV(MW)}{RT} = \frac{0.10 \text{ atm} \times 6000 \text{ liters} \times 44 \text{ g/mole}}{0.082 \text{ liter-atm/mole-deg} \times 300°K}$$

$$\overset{\sim}{=} 1100 \text{ g.}$$

Thus, 1100 g, or 1.10 kg, must sublimate before there is sufficient $CO_2$ in the atmosphere to be harmful.

# GRAHAM'S LAW OF GASEOUS DIFFUSION

● PROBLEM 94

Under standard temperature and pressure conditions, compare the relative rates at which inert gases, Ar, He, and Kr diffuse through a common orifice.

Solution: This problem involves the application of Graham's law of diffusion. It states that the relative rates at which gases will diffuse will be inversely proportional to the square roots of their respective densities or molecular weights. That is, rate $\propto 1/\sqrt{mass}$. Thus, to compare the rates of diffusion of Ar, He, and Kr, look up their weights in the Periodic Table of Elements and substitute this value into $1/\sqrt{M} \propto \mu_m$, where M = mass of that element and $\mu_m$ = rate. Therefore, $\mu_{Ar} : \mu_{He} : \mu_{Kr}$

$$= \frac{1}{\sqrt{M}_{Ar}} : \frac{1}{\sqrt{M}_{He}} : \frac{1}{\sqrt{M}_{Kr}}$$

$$= \frac{1}{\sqrt{39.95}} : \frac{1}{\sqrt{4.003}} : \frac{1}{\sqrt{83.80}}$$

$$= .1582 : .4998 : .1092.$$

Two gases, HBr and $CH_4$, have molecular weights 81 and 16, respectively. The HBr effuses through a certain small opening at the rate of 4 ml/sec. At what rate will the $CH_4$ effuse through the same opening?

**Solution**: The comparative rates or speeds of effusion of gases are inversely proportional to the square roots of their molecular weights. This is written

$$\frac{rate_1}{rate_2} = \frac{\sqrt{MW_2}}{\sqrt{MW_1}}$$

For this case $\dfrac{rate_{HBr}}{rate_{CH_4}} = \dfrac{\sqrt{MW_{CH_4}}}{\sqrt{MW_{HBr}}}$

One is given the $rate_{HBr}$, $MW_{CH_4}$ and $MW_{HBr}$ and asked to find $rate_{CH_4}$.

Solving for $rate_{CH_4}$:

$$\frac{rate_{HBr}}{rate_{CH_4}} = \frac{\sqrt{MW_{CH_4}}}{\sqrt{MW_{HBr}}}$$

$rate_{HBr} = 4$ ml/sec.

$rate_{CH_4} = ?$

$$\frac{4 \text{ ml/sec}}{rate_{CH_4}} = \frac{\sqrt{16}}{\sqrt{81}}$$

$MW_{CH_4} = 16$

$MW_{HBr} = 81$

$$rate_{CH_4} = \frac{4 \text{ ml/sec} \times \sqrt{81}}{\sqrt{16}}$$

$$= \frac{4 \text{ ml/sec} \times 9}{4} = 9 \text{ ml/sec.}$$

The time required for a volume of gas, X, to effuse through a small hole was 112.2 sec. The time required for the same volume of oxygen was 84.7 sec. Calculate the molecular weight of gas X.

**Solution**: This problem involves the rate of effusion through a small hole and the method to solve it involves a modified statement of Graham's Law. High density gaes effuse more slowly than low density gases. The time required for effusion is inversely proportional to the rate of effusion. It is mathematically stated as

$$\frac{r_1}{r_2} = \frac{t_2}{t_1} = \frac{\sqrt{M.W._2}}{\sqrt{M.W._1}} \; ,$$

where, $r_1$, $r_2$ are the rates of the two gases; $t_2$, $t_1$ are the times; and $M.W._2$ and $M.W._1$, = molecular weights.

Knowing 3 of the 4 values, one can use this equation to find the 4th value. We can write

$$\frac{t_{O_2}}{t_X} = \frac{\sqrt{M.W._{O_2}}}{\sqrt{M.W._X}}$$

$$\frac{84.7 \text{ sec}}{112.2 \text{ sec}} = \frac{\sqrt{32}}{\sqrt{X}}$$

$M.W._X = 56.2$ amu.

Molecular Weight is expressed in terms of a.m.u. (atomic mass units).

At standard conditions, 1 liter of oxygen gas weighs almost 1.44 g, whereas 1 liter of hydrogen weighs only .09 g. Which gas diffuses faster? Calculate how much faster.

Solution:  This question deals with diffusion of gases, which is governed by Graham's law. A gas that has a high density diffuses more slowly than one with a lower density. The rates of diffusion of 2 gases are inversely proportional to the square roots of their densities, as shown by the equation

$$\frac{r_1}{r_2} = \frac{\sqrt{d_2}}{\sqrt{d_1}} \quad , \quad \text{where } r_1, r_2 = \text{rates of diffusion of}$$

the gases, and $d_1$, $d_2$ are their respective densities.

Therefore, by Graham's law, we know that $H_2$ diffuses faster because of its lower density. ($H_2$'s density $= \frac{M}{V} = \frac{.09 \text{ g}}{1 \text{ } \ell} = .09$ g/$\ell$ while $O_2$'s density $= \frac{1.44}{1} = 1.44$ g/liter.)

To find out how much faster $H_2$ diffuses, plug the necessary factors into the equation

$$\frac{r_{H_2}}{r_{O_2}} = \frac{\sqrt{d_{O_2}}}{\sqrt{d_{H_2}}} \quad . \qquad \text{Rewriting,}$$

$$r_{H_2} = r_{O_2} \frac{\sqrt{d_{O_2}}}{\sqrt{d_{H_2}}} = r_{O_2} \frac{\sqrt{1.44}}{\sqrt{.09}}$$

$$r_{H_2} = 4r_{O_2}$$

Thus, the calculated comparison is in agreement with the predicted comparison that $H_2$ diffuses more quickly. To be precise, the rate of diffusion of $H_2$ is 4 times that of the rate of diffusion of $O_2$.

● **PROBLEM** 98

The relative rates of diffusion of $NH_3(g)$ and HX (g) can be determined experimentally by simultaneously injecting $NH_3(g)$ and HX(g), respectively, into the opposite ends of a glass tube and noting where a deposit of $NH_4X(s)$ is formed. Given an 1m tube, how far from the $NH_3$ injection end would you expect the $NH_4X(s)$ to be formed when HX is a) HF, b) HCl?

<u>Solution</u>: The rate of diffusion of a gas is observed to be inversely proportional to the square root of its molecular weight. This is called Graham's law.

$$\text{Rate} = \frac{\text{constant}}{\sqrt{MW}}$$

For HF:    According to Graham's Law:

the speed of $NH_3$ is proportional to $\frac{1}{\sqrt{17}} = .243$
(MW of $NH_3 = 17$)

the speed of HF is proportional to $\frac{1}{\sqrt{20}} = .224$
(MW of HF = 20)

Thus, for an arbitrary amount of time, let us say 1 min., $NH_3$ moves .243 cm and HF moves .224 cm. Because the gases are ejected from either end of the tube, after 1 minute the gases move (.243 + .224) cm or .467 cm/min. closer together. When this distance is equal to 100 cm the solid forms. One can determine the time needed by dividing 100 cm by .467 cm/min

time after which solid forms $= \frac{100 \text{ cm}}{.467 \text{ cm/min}} = 214.13$ min.

Thus, after 214.13 min. the HF meets the $NH_3$.

The distance the $NH_3$ would travel is found by multiplying the distance it travels in 1 minute, .243 cm, by the time elapsed till the $NH_4F$ deposits.

distance $NH_3$ travels $= .243$ cm/min $\times$ 214.13 min

                    = 52.03 cm.

   b) One solves for HCl in a similar manner

   MW of HCl = 36.5

   the speed of HCl is proportional to $\dfrac{1}{\sqrt{36.5}}$ = .165

      As previously found, the speed of NH₃ is proportional
to $\dfrac{1}{\sqrt{17}}$ = .243.

        Speed is a measure of distance covered over a
unit time. Let us designate the speeds here as
cm./min. Thus, HCl moves .165 cm/min and NH₃ moves
.243 cm/min. The solid forms after the sum of the
distances travelled by the two gases is 100 cm. In
1 min. the gases travel a distance of (.165 + .243)cm
or .408 cm. The time it takes for the gases to travel
100 cm. is found by dividing 100 cm. by .408 cm/min.

   time after which the solid will deposit =

      = $\dfrac{100 \text{ cm}}{.408 \text{ cm/min}}$ = 245.10 min.

The distance the NH₃ travels is then found by multiplying 245.10 min by the distance the gas travels in 1
minute.

   distance travelled by NH₃ = .243 cm/min × 245.10 min

                        = 59.56 cm.

● PROBLEM 99

> Two balloons at the same temperature of equal volume
> and porosity are each filled to a pressure of 4 atmos-
> pheres, one with 16 kg of oxygen, the other with 1 kg
> of hydrogen. The oxygen balloon leaks to a pressure
> of ½ atmosphere (atm) in 1 hour. How long will it take
> for the hydrogen balloon to reach a pressure of ½ atm?

<u>Solution</u>:  This problem deals with effusion, the
escape of molecules through a hole, and is therefore
an application of Graham's Law: Rate of effusion of
a gas is inversely proportional to the square root of
its density

   $\dfrac{\text{rate}_1}{\text{rate}_2} = \dfrac{\sqrt{d_2}}{\sqrt{d_1}}$

      From this we know that large molecules effuse less
rapidly than small molecules, since they move more slowly.

This allows us to predict that hydrogen effuses more rapidly than oxygen. By applying Graham's Law, we can determine how much faster.

$$\frac{rate_{hydrogen}}{rate_{oxygen}} = \frac{\sqrt{16 \text{ kg per unit volume}}}{\sqrt{1 \text{ kg per unit volume}}} = 4.$$

There, hydrogen effuses 4 times more quickly, which is in agreement with our prediction.

Since it takes 1 hour for oxygen to leak to a pressure of ½ atm, it requires ¼hr. = 15 min for hydrogen to decrease to the same pressure inside the balloon.

# KINETIC THEORY OF GASES

• PROBLEM 100

Calculate the most probable speed $v_p$, the arithmetic mean speed $\bar{v}$, and the root mean square speed $v_{rms}$ for hydrogen molecules at 0°C.

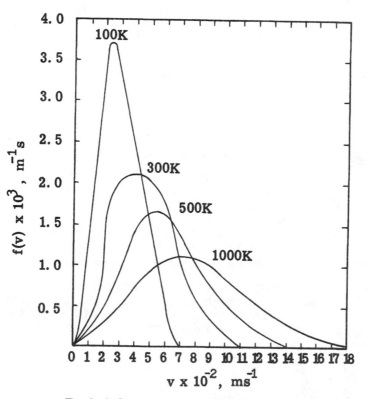

Probability density of various speeds v for oxygen at 100, 300, 500, and 1000 K.

Solution: Before beginning this problem, we must first understand the basic differences in the three types of speeds. The arithmetic mean speed $\bar{\nu}$ is obtained by summing all the speeds and dividing by the total number of molecules N.

$$\bar{\nu} = \frac{1}{N} \sum_{i=1}^{N} \nu_i$$

The symbol $\sum \nu_i$ indicates the sum of the speeds $\nu_1$, $\nu_2$, $\nu_3$, ... $\nu_N$ of all the N individual molecules. The equation for mean speed is

$$\bar{\nu} = \left(\frac{8kT}{\pi m}\right)^{\frac{1}{2}} = \left(\frac{8RT}{\pi M}\right)^{\frac{1}{2}}$$

where k is Boltzmann's constant, R is the gas constant, T is the absolute temperature, m is the weight of 1 molecule, and M is the molecular weight.

The most probable speed is obtained from the plot of $f(\nu)$ versus $\nu$, where $f(\nu)$ is the fraction of molecules with speed $\nu$. The equation of this curve

$$f(\nu) = \left(\frac{m}{2\pi kT}\right)^{3/2} e^{-m\nu^2/2kT} 4\pi\nu^2$$

is differentiated with respect to $\nu$ and set equal to zero $\left(\frac{df(\nu)}{d\nu} = 0\right)$. The shape of the curve of $f(\nu)$ versus $\nu$ is indicated in the accompanying figure.

$$\frac{df(\nu)}{d\nu} = \left(\frac{m}{2\pi kT}\right)^{3/2} e^{-m\nu^2/2kT} \left(8\pi\nu + 4\pi\nu^2 \left(\frac{-m\nu}{kT}\right)\right)$$

This derivative vanishes at $\nu = \nu_p$, called the most probable speed, where

$$\nu_p = \left(\frac{2kT}{m}\right)^{\frac{1}{2}} = \left(\frac{2RT}{M}\right)^{\frac{1}{2}}$$

The root-mean-square speed, $\nu_{rms}$, is defined by

$$\nu_{rms} = \left(\frac{1}{N} \sum_{i=1}^{N} \nu_i^2\right)^{\frac{1}{2}}$$

The symbol $\sum \nu_i^2$ indicates the sum of the squares of the velocities, $\nu_1$, $\nu_2$, $\nu_3$, ..., $\nu_n$ of all the N individual molecules.

Since the velocity distribution is continuous, the root-mean-square velocity is obtained by multiplying each velocity squared by the probability of that velocity, integrating over all velocities, and taking the square root:

$$v_{rms} = \left[ \int_0^\infty v^2 f(v) \ dv \right]^{\frac{1}{2}}$$

Substituting for $f(v)$ and simplifying one obtains

$$v_{rms} = \left( \frac{3kT}{m} \right)^{\frac{1}{2}} = \left( \frac{3RT}{M} \right)^{\frac{1}{2}}$$

At any given temperature, these velocities are inversely proportional to the square root of the molecular weight. Lighter molecules move more rapidly so that their average kinetic energies are exactly equal to those for the heavier molecules.

To solve this problem for hydrogen at 0°C one must 1) find the molecular weight of hydrogen in kg, 2) convert 0°C into the absolute temperature scale, 3) substitute these values into the equations for $v_p$, $\bar{v}$, and $v_{rms}$.

The molecular weight of hydrogen is 2.016 g/mole, which is $2.016 \times 10^{-3}$ kg/mole. The temperature in degrees Kelvin is 273°K. Other pertinent information R = 8.314 J/K-mole, $\pi$ = 3.142. Thus,

$$v_p = \left( \frac{2RT}{M} \right)^{\frac{1}{2}}$$

$$= [(2)(8.314 \ \text{J/K-mol})(273°K)/(2.016 \times 10^{-3} \ \text{Kg/mol})]^{\frac{1}{2}}$$

$$= 1.50 \times 10^3 \ \text{m/s}$$

$$\bar{v} = \left( \frac{8RT}{\pi M} \right)^{\frac{1}{2}}$$

$$= [(8)(8.314 \ \text{J/K-mol})(273°K)/(3.142)(2.016 \times 10^{-3} \text{Kg/mol})]^{\frac{1}{2}}$$

$$= 1.69 \times 10^3 \ \text{m/s}.$$

$$v_{rms} = \left( \frac{3RT}{M} \right)^{\frac{1}{2}}$$

$$= [(3)(8.314 \ \text{J/K-mol})(273°K)/(2.016 \times 10^{-3} \ \text{Kg/mol})]^{\frac{1}{2}}$$

$$= 1.84 \times 10^3 \ \text{m/s}.$$

The relative ratios of $v_p$ : $\bar{v}$ : $v_{rms}$ are 1 : 1.13 : 1.23.

Given that the average speed of an $O_2$ molecule is 1,700 km/h at 0°C, what would you expect the average speed of a $CO_2$ molecule to be at the same temperature?

Solution: The speed of a molecule is related to the mass and temperature by the equation

$$\tfrac{1}{2} mv^2 = 3/2 \ kT$$

where m is the mass, v is the speed, k is Boltzmann's constant, and T is the absolute temperature. Since k and T are constant 3/2 kT is constant for this system. Thus

$$\tfrac{1}{2} m_{O_2} v^2_{O_2} = \tfrac{1}{2} m_{CO_2} v^2_{CO_2}$$

where $m_{O_2}$ is the mass of $O_2$, $v_{O_2}$ is the speed of $O_2$, $m_{CO_2}$ is the mass of $CO_2$, and $v_{CO_2}$ is the speed of $CO_2$. The masses are equal to the molecular weights here. Solving for $v_{CO_2}$:

$$v_{CO_2} = \sqrt{\frac{\tfrac{1}{2} m_{O_2} v^2_{O_2}}{\tfrac{1}{2} m_{CO_2}}}$$

$$m_{O_2} = 32$$

$$m_{CO_2} = 44$$

$$v_{O_2} = 1700 \text{ km/h.}$$

$$v_{CO_2} = \sqrt{\frac{\tfrac{1}{2}(32)(1700)^2}{\tfrac{1}{2}(44)}}$$

$$= 1700 \ \sqrt{\frac{32}{44}} = 1700 \times (.85)$$

$$= 1450 \text{ km/h.}$$

A chemist possesses 1 $cm^3$ of $O_2$ gas and 1 $cm^3$ of $N_2$ gas both at Standard Temperature and Pressure (STP). Compare these gases with respect to (a) number of molecules, and (b) average speed of molecules.

Solution: This problem requires the use of the kinetic theory of gases and Avogadro's principle.

Avogadro proposed the principle that equal volumes of gases at the same temperature and pressure contain equal numbers of molecules. Both gases have equal volumes, 1 $cm^3$, and are at the same temperature and pressure (STP). According to Avogadro's Principle, both have the same

number of molecules. The average speed of these molecules
can be found from kinetic theory of gases. The rate of
diffusion of a gas is directly proportional to the
average speed of the molecules. According to Graham's
law, there is an inverse proportionality between diffusion
rate and the square root of the molecular weight (mass).
In other words,

$$\frac{\text{rate of diffusion of } N_2}{\text{rate of diffusion of } O_2} = \frac{V_{N_2}}{V_{O_2}} = \sqrt{\frac{M_{O_2}}{M_{N_2}}} \text{,}$$

where V = velocity and M = mass, $M_{N_2}$ = 28, $M_{O_2}$ = 32. In
this case, mass = molecular weight. Thus,

$$\frac{V_{N_2}}{V_{O_2}} = \sqrt{\frac{32}{28}} = 1.07 \text{.}$$

Thus, if the average speed of $O_2$ is one and the average
speed of $N_2$ is 1.07.

● **PROBLEM** 103

Calculate (a) the number of collisions per second per
molecule, (b) the number of collisions per cubic meter
per second, and (c) the number of moles of collisions
per liter per second of oxygen at a temperature of 25°C
and pressure of 1 atm. The molecular diameter of oxygen
is 3.61 × 10⁻¹⁰ m.

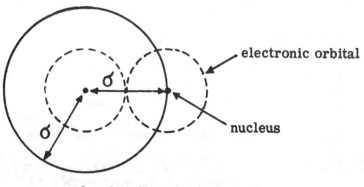

molecular diameter

Figure A

Solution: Molecules of real gas attract one another at
large distances and repel one another at very short
distances. Assuming that molecules are rigid, noninter-
acting spheres with diameter, σ, one can drive an
approximate equation for the number of collisions per
second. It is also assumed that all the molecules travel
with the same speed, the arithmetic mean velocity $\bar{v}$.

If two identical molecules just touch each other, then the distance separating their centers is the molecular diameter $\sigma$. Thus, a moving molecule collides with other molecules whose centers come within a distance of $\sigma$. (Hence, the effective collision radius is equal to $\sigma$, see figure a.) The quantity $\pi\sigma^2$, is called the collision cross section for the rigid spherical molecule. A molecule moving with a constant velocity of $\bar{v}$ will sweep out $(\pi\sigma^2\bar{v})$ (meters)$^3$/sec and strike $\pi\sigma^2\bar{v}n$ molecules per second, where n is the concentration of molecules. In these calculations, we have assumed several things some of which are not completely accurate. Actually the number of collisions per second is off by a factor of $\sqrt{2}$. Thus, the number of collisions per molecule per second is:

$$z = \sqrt{2}\ \pi\sigma^2\bar{v}n$$

Since there are n molecules per unit volume, there will be $\sqrt{2}\ \pi\sigma^2\bar{v}n^2$ collisions per unit time. The number of collisions per unit time per unit volume is given by

$$Z = \frac{\sqrt{2}}{2}\ \pi\sigma^2\bar{v}n^2$$

The factor of one-half comes from the fact that the number of collisions is one-half the number of total collisions. This is true because the number of total collisions is equal to the sum of all of the collisions of all of the particles, but in each collision two molecules are involved which means that in the total number each collision is really counted twice.

To solve the problem we must: 1) find the arithmetic mean speed, 2) find the concentration of molecules, n, using the ideal gas law, 3) substitute the values in the equations for $z = \sqrt{2}\ \pi\sigma^2\bar{v}n$ and $Z = \frac{\sqrt{2}}{2}\ \pi\sigma^2\bar{v}n^2$. The information known:

$\sigma = 3.61 \times 10^{-10}$m $\qquad$ P = 1 atm

T = 25°C + 273°= 298°K ; M = 32 $\times$ 10$^{-3}$ Kg/mole = molecular weight of oxygen

R = 0.082 $\frac{\text{liter - atm}}{\text{°K - mole}}$ = 8.312 J/K-mole.

1) $\bar{v} = \left(\frac{8RT}{\pi M}\right)^{\frac{1}{2}}$

$\qquad = [(8)(8.314\ \text{J/K-mole})(298)/\pi(32 \times 10^{-3}\ \text{Kg/mol})]^{\frac{1}{2}}$

$\qquad = 444$ m/s.

2) $n = \frac{N}{V} = \frac{\text{number of molecules}}{\text{m}^3}$ = concentration

According to the ideal gas law, PV = $\bar{N}$RT, where $\bar{N}$ = moles or $\frac{\bar{N}}{V} = \frac{P}{RT}$ . Avogardo's number, $N_a$ = 6.02 $\times$

$10^{23}\ \dfrac{\text{molecules}}{\text{mole}}$ , so that $\dfrac{\bar{N}\ Na}{V} = \dfrac{PNa}{RT}$ . But $\bar{N}\ Na = N$ so that

$$\dfrac{N}{V} = n = \dfrac{PNa}{RT} = \dfrac{(1\ \text{atm})\left[6.022\times10^{23}\dfrac{\text{molecules}}{\text{mole}}\right]\left[10^{3}\ \dfrac{\text{liter}}{m^3}\right]}{\left[0.082\ \dfrac{\text{liter atm}}{°K\text{-mole}}\right](298\ °K)}$$

$$= 2.46 \times 10^{25}\ \dfrac{\text{molecules}}{m^3}$$

3) $z = \sqrt{2}\ \pi\sigma^2\bar{\nu}n$

$$= (1.414)(3.14)(3.6\times10^{-10}\ m)^2(444\ m/s)$$
$$\left(2.46\times10^{25}\ \dfrac{\text{molecules}}{m^3}\right)$$

$$= 6.32 \times 10\ \ \dfrac{\text{molecules}}{s}$$

This is the number of collisions per molecule per second.

$$Z = \dfrac{\sqrt{2}}{2}\ \pi\sigma^2\bar{\nu}n^2$$

$$= \dfrac{z\ n}{2} = \dfrac{\left[6.32 \times 10^9\ \dfrac{\text{molecules}}{s}\right]\left[2.46 \times 10^{25}\ \dfrac{\text{molecules}}{m^3}\right]}{2}$$

$$= 7.77 \times 10^{34}\ \dfrac{\text{collisions}}{m^3\ s}$$

To answer part c of the problem, we use the conversion factor $\dfrac{10^{-3}\ m^3/\text{liter}}{6.02 \times 10^{23}\ \text{molecules/mole}}$ . Thus

$$Z = \dfrac{(7.77 \times 10^{34}\ \text{collisions}/m^3\text{-s})(10^{-3}\ m^3/\ell)}{(6.022 \times 10^{23}\ \text{molecules/mole})}$$

$$= 1.29 \times 10^8\ \text{moles/liter-s},$$

which is the number of moles of collisions per liter per second.

● **PROBLEM** 104

Calculate the mean free path for oxygen at 25°C at
(a) 1 atm pressure and (b) $10^{-3}$ torr.

Solution: The mean free path is the average distance traversed by a molecule between collisions. In a second, a molecule will, on the average, traverse $\nu$ meters and collide Z times (where $Z = \sqrt{2}\ \pi\sigma^2\bar{\nu}n$). Thus, the mean free path is

$$\ell = \frac{v}{Z} = \frac{1}{\sqrt{2}\ \pi\sigma^2 n}$$

Substituting $n = P/kT$ from the ideal gas law, we have

$$\ell = \frac{kT}{\sqrt{2}\ \pi\sigma^2 P}$$

To solve this problem, we must 1) find n, the concentration, 2) substitute the known values into the equation for the mean free path

a) $P = 1$ atm          $\sigma = 3.61 \times 10^{-10}$ m

$T = 25°C + 273 = 298°K$   $R = 0.082$ liter-atm/K-mole

$$n = \frac{PN_A}{RT} = \frac{(1\ \text{atm})\left[6.022\times10^{23}\ \frac{\text{molecules}}{\text{mole}}\right]\left[10^3\ \frac{\text{liter}}{\text{m}^3}\right]}{\left[0.082\ \frac{\text{liter-atm}}{\text{K-mole}}\right](298°K)}$$

$$n = 2.46 \times 10^{25}\ \frac{\text{molecules}}{\text{m}^3}$$

Thus, $\ell = \dfrac{1}{\sqrt{2}\ \pi\sigma^2 n}$

$$= [(1.414)(3.14)(3.61\times10^{-10}\ \text{m})^2(2.46\times10^{25})]^{-1}$$

$$= 7.02 \times 10^{-8}\ \text{m}$$

b) $P = 10^{-3}$ torr $= \dfrac{10^{-3}\ \text{torr}}{760\ \frac{\text{torr}}{\text{atm}}} = 1.3157 \times 10^{-6}$ atm.

$T = 298°K$          $\sigma = 3.61 \times 10^{-10}$ m

Here, it is necessary to convert torr units to units of atmospheres.

$$n = \frac{(1.3157\times10^{-6}\ \text{atm})\left[6.022\times10^{23}\ \frac{\text{molecules}}{\text{mole}}\right]\left[10^3\ \frac{\text{liter}}{\text{m}^3}\right]}{\left[.082\ \frac{\text{liter-atm}}{\text{K-mole}}\right](298°K)}$$

$$= 3.24 \times 10^{19}\ \text{molecules/m}^3.$$

Thus, $\ell = \dfrac{1}{\sqrt{2}\ \pi\sigma^2 n}$

$$= [(1.414)(3.14)(3.61\times10^{-10}\ \text{m})^2(3.24\times10^{19}\ /\text{m}^3)]^{-1}$$

$$= 0.053\ \text{m} = 5.3\ \text{cm}.$$

In water vapor at 25°C, what is the average speed of a water molecule in meters/sec? Recall that 1 Joule = 1 Kg - $m^2$/$sec^2$. The Boltzmann constant, k = 1.3806 × $10^{-23}$ J/deg.

**Solution:** Recall, that for all bodies in motion, kinetic energy (K.E.) = $\frac{1}{2}$ $M\bar{V}^2$, where M = mass of body and $\bar{V}$ = average velocity of the molecules. Thus, if K.E. and the mass are known, then $\bar{V}$ can be solved. To find K.E., note that for gases average K.E. is proportional to temperature. This can be written as K.E. = 3/2 kT, where k is the Boltzman constant 1.3806 × $10^{-23}$ J/deg and T = temperature in Kelvin (Celsius plus 273°). You are given the temperature of the system (25°C or 298°K). Thus, K.E. = 3/2(1.38 × $10^{-23}$)(298) = 6.17 × $10^{-21}$ J/molecule.

Since K.E. = $\frac{1}{2}MV^2$, 6.17 × $10^{-21}$ = $\frac{1}{2}$ $MV^2$. If you knew the mass of one water molecule, you could substitute it into this expression and find V. The mass of the $H_2O$ molecule can be found by using the molecular weight (M.W.) and Avogadro's number (6.02 × $10^{23}$/mole). $H_2O$ has an M.W. of 18 grams/mole. Thus, the mass of one molecule =

$$\frac{18 \text{ g/mole}}{6.02 \times 10^{23} \text{ molecule/mole}} = 2.99 \times 10^{-23} \text{ grams/molecule.}$$

Substituting and solving for V,

$$V = \sqrt{\frac{2(K.E.)}{M}} = \sqrt{\frac{2\left[6.17 \times 10^{-21} \dfrac{kg - m2}{sec^2 - molecule}\right]}{\left[2.99 \times 10^{-23} \dfrac{g}{molecule}\right]\left[.001 \dfrac{kg}{g}\right]}}$$

V = 642 m/sec.

(Note: .001 kg/g is a conversion factor.)

The root mean square (rms) speed of hydrogen ($H_2$) at a fixed temperature, T, is 1600 m/sec. What is the rms speed of oxygen ($O_2$) at the same temperature?

**Solution:** To solve this problem we make use of the following equation for the rms speed, $v_{rms}$,

$$v_{rms} = \sqrt{\frac{3 RT}{M}},$$

where R is the gas constant, T the absolute temperature, and M the molecular weight. For oxygen and hydrogen we have

$$v_{rms(O_2)} = \sqrt{\frac{3 RT}{M_{O_2}}} \quad \text{and} \quad v_{rms(H_2)} = \sqrt{\frac{3 RT}{M_{H_2}}}$$

Dividing the first of these equations by the second gives

$$\frac{v_{rms(O_2)}}{v_{rms(H_2)}} = \frac{\sqrt{\dfrac{3\ RT}{M_{O_2}}}}{\sqrt{\dfrac{3\ RT}{M_{H_2}}}} = \sqrt{\frac{3\ RT/M_{O_2}}{3\ RT/M_{H_2}}}$$

or, since R is a constant and T is the same for both gases,

$$\frac{v_{rms(O_2)}}{v_{rms(H_2)}} = \sqrt{\frac{M_{H_2}}{M_{O_2}}}$$

Solving for $v_{rms(O_2)}$,

$$v_{rms(O_2)} = \sqrt{\frac{M_{H_2}}{M_{O_2}}} \times v_{rms(H_2)}$$

$$= \sqrt{\frac{2.0\ g/mole}{32.0\ g/mole}} \times 1600\ m/sec = 400\ m/sec.$$

● **PROBLEM** 107

Graham's law states that the rate at which gas molecules escape through a small orifice (rate of effusion) is inversely proportional to the square root of the density of the gas. Derive Graham's law from the following assumptions: (a) temperature is directly proportional to the average kinetic energy of the molecules; (b) the rate of effusion is directly proportional to the root mean square speed of the molecules;(c) the density of a gas at constant temperature and pressure is directly proportional to the molecular mass.

Solution: Consider the assumptions individually. The first states that temperature, T, is proportional to the average kinetic energy $\frac{1}{2}\ m\bar{u}^2$, where m = molecular mass, u = speed, and the bar over $u^2$ indicates the average. Hence

$$T = k_1 \times \frac{1}{2}\ \overline{mu}^2$$

where $k_1$ is a constant.

The second assumption states that the rate of effusion is directly proportional to the root mean square speed of the molecules, $\sqrt{\bar{u}^2}$, or

$$rate = k_2\sqrt{\overline{u^2}}$$

where $k_2$ is a constant.

The third assumption states that the density, p, is directly proportional to the molecular mass. or

$$p = k_3 \, m$$

where $k_3$ is a constant. **Thus,**

$$T = k_1 \times \tfrac{1}{2} \, m\bar{u}^2 = k_1 \times \tfrac{1}{2} \times \frac{p}{k_3} \times \bar{u}^2.$$

Taking the square root of both sides we obtain

$T^{\frac{1}{2}} = \left(\dfrac{k_1}{2k_3}\right)^{\frac{1}{2}} p^{\frac{1}{2}} \sqrt{\bar{u}^2}$ . Multiplying by $\left(\dfrac{2k_3}{k_1}\right)^{\frac{1}{2}}$ this becomes

$\left(\dfrac{2k_3}{k_1}\right)^{\frac{1}{2}} T^{\frac{1}{2}} = p^{\frac{1}{2}} \sqrt{\bar{u}^2}$. At constant temperature, the expression

$\left(\dfrac{2k_3}{k_1}\right)^{\frac{1}{2}} T^{\frac{1}{2}}$ is a constant, say $k_4$. Then, $k_4 = p^{\frac{1}{2}} \sqrt{\bar{u}^2} =$

$p^{\frac{1}{2}} \times \dfrac{\text{rate}}{k_2}$ . Multiplying by $\dfrac{k_2}{p^{\frac{1}{2}}}$ and letting $k_2 \times k_4 = k_5$,

a constant, we obtain: rate $= \dfrac{k_2 \times k_4}{p^{\frac{1}{2}}} = \dfrac{k_5}{p^{\frac{1}{2}}}$ , which states

that the rate of effusion is inversely proportional to the square root of the density, which is Graham's law.

# CHAPTER 4

# AVOGADRO'S HYPOTHESIS; CHEMICAL COMPOUNDS AND FORMULAS

> **Basic Attacks and Strategies for Solving Problems in this Chapter. See pages 95 to 143 for step-by-step solutions to problems.**

## Avogadro's Law

Avogadro's hypothesis, first introduced by the Italian Renaissance physicist Amedeo Avogadro, states that for two gases at the same temperature and pressure, equal volumes will contain equal numbers of molecules. This number will be the same for all gases.

Avogadro's hypothesis has been so extensively verified that it is now widely known as Avogadro's Law. It follows, from the ideal gas law, that a mole of any gas will contain the same number of molecules. This number, $6.02 \times 10^{23}$, is called Avogadro's number and applies to all matter, not just to gases. One mole of any species will contain Avogadro's number of molecules. One mole of any species is the amount of that species whose mass in grams is numerically equal to the molecular or atomic weight.

In order to calculate the number of molecules in a given mass of material, it is necessary to first compute the number of moles. To calculate the number of moles of a substance, divide the mass in grams by the molecular weight as shown in Equation 4-1.

$$n \text{ (moles)} = m \text{ (grams)}/M\text{(grams/mole)} \qquad 4\text{-}1$$

The number of molecules is simply the number of moles times Avogadro's number (the molecules per mole for any substance).

Number of molecules $= n \times N_A$ 
4-2

## Definitions of Atomic Weight, Molecular Weight, and Equivalent Weight

Calculations involving chemical reactions require the understanding of the basic ideas of atomic weight, molecular weight, equivalent weight, and moles. You cannot proceed until you master these concepts and they become so familiar that their use is automatic. The definitions of these terms follow.

Atomic Weight: The atomic weight of an atom is the mass of the atom compared to Carbon–12 which is defined as 12 amu or 12 atomic mass units. Atomic weights are tabulated for all elements. Unless reference is made to a specific isotope, the tabulated values of atomic weights are average values for the mixture of isotopes that occur naturally.

Molecular Weight: The molecular weight is the sum of the atomic weights of all the atoms which form the molecule.

Equivalent Weight: The equivalent weight is the molecular weight divided by the valence — i.e., the number of electrons/molecule which enter the reaction in question. Equivalent weight depends on the reaction and a substance may have different equivalent weights for different reactions.

Mole: The mass of a substance (in grams) numerically equal to the molecular weight. Sometimes lb moles, kg moles, etc., are defined as the mass (in lb, kg, etc., respectively) numerically equal to the molecular weight. These quantities are not moles, but can be used for convenience in calculations employing those units. Note, however, that Avogadro's number of molecules constitutes only a (gram) mole.

The key to determining a chemical formula from weight percent data of the elemental constituents is the reduction of the amount of each element to moles. Then, from the molar ratio, the empirical or simplest formula will usually become obvious. When amounts of gases are expressed as volumes, the number of moles can be determined from the ideal gas law as discussed in Chapters 2 and 3.

● PROBLEM 108

How many mercury atoms would there be in a 100 g piece of swordfish said to contain 0.1 ppm (part per million by weight) of mercury?

<u>Solution</u>: One can find the number of atoms present by multiplying the number of moles of mercury by the number of atoms in 1 mole. The number of atoms in one mole is equal to Avogadro's Number or $6.02 \times 10^{23}$ atoms/mole. To solve for the number of mercury atoms present find:

(1) the amount of mercury present

(2) the number of moles of mercury

(3) the number of atoms of mercury.

Solving:

(1) Because there is 0.1 ppm of mercury in the swordfish, one multiplies $0.1 \times 10^{-6}$ by the amount of fish present.

amount of mercury = $0.1 \times 10^{-6} \times 100$ g = $1.0 \times 10^{-5}$ g.

(2) The number of moles is found by dividing the weight of the mercury present by the weight of one mole, the molecular weight (MW = 200.6)

no. of moles = $\dfrac{1.0 \times 10^{-5} \text{ g}}{200.6 \text{ g/mole}}$ = $4.99 \times 10^{-8}$ moles.

(3) The number of atoms is found by multiplying the number of moles by the number of atoms in one mole, $6.02 \times 10^{23}$ atoms.

no. of atoms = $4.99 \times 10^{-8}$ moles $\times 6.02 \times 10^{23}$ atoms/mole

$= 3.00 \times 10^{16}$ atoms.

If the dot under a question mark has a mass of $1 \times 10^{-6}$ g, and you assume it is carbon, how many atoms are required to make such a dot?

Solution: Two facts must be known to answer this question. You must determine the number of moles in the carbon dot. You must also remember the number of atoms in a mole of any substance, $6.02 \times 10^{23}$ atoms/mole (Avogadro's number).

A mole is defined as the weight in grams of a substance divided by the atomic weight (or molecular weight).

The atomic weight of carbon is 12 g/mole. Therefore, in the dot you have

$\dfrac{1 \times 10^{-6} \text{ g}}{12 \text{ g/mole}}$ moles of carbon. Therefore, the number

of atoms in such a dot is the number of moles × Avogadro's number,

$\left( \dfrac{1 \times 10^{-6} \text{ g}}{12 \text{ g/mole}} \right) 6.02 \times 10^{23}$ atoms/mole = $5 \times 10^{16}$ atoms.

What is the approximate number of molecules in a drop of water which weighs 0.09 g?

Solution: The number of molecules in a mole is defined to be $6.02 \times 10^{23}$ molecules. Thus, to find the number of molecules in a drop of water, one must know the number of moles making up the drop. This is done by dividing the weight of the drop by the molecular weight of $H_2O$. (MW of $H_2O$ = 18)

number of moles = $\dfrac{0.09 \text{ g}}{18 \text{ g/mole}}$ = .005 moles.

The number of molecules present is now found by multiplying the number of moles by Avogadro' number $(6.02 \times 10^{23})$.

no. of molecules = .005 moles $\times 6.02 \times 10^{23}$ molecules/mole

$= 3.01 \times 10^{21}$ molecules.

What is the difference between the number of carbon atoms in 1.00 g of C-12 isotope (atomic mass = 12.000 g/mole) and 1.00 g of C-13 isotope (atomic mass = 13.003 g/mole)?

Solution: The difference in the number of carbon atoms in each sample is equal to the difference in the number of moles times Avogadro's number $6.02 \times 10^{23}$. Hence, we must begin by calculating the number of moles of C-12 and of C-13 in 1.00 g samples of each.

The number of moles is equal to the mass divided by the atomic weight. Therefore,

$$\text{moles C-12} = \frac{\text{mass C-12}}{\text{atomic mass C-12}} = \frac{1.00 \text{ g}}{12.000 \text{ g/mole}} = 0.083 \text{ mole}$$

$$\text{and moles C-13} = \frac{\text{mass C-13}}{\text{atomic mass C-13}}$$

$$= \frac{1.00 \text{ g}}{13.003 \text{ g/mole}} = 0.077 \text{ mole}.$$

The difference in the number of moles between the two samples is (moles C-12) - (moles C-13) = 0.083 - 0.077 = 0.006 mole. Multiplying by Avogadro's number gives the difference in the number of carbon atoms in the two samples:

$$\text{number of carbon atoms} = 0.006 \text{ mole} \times 6.02 \times 10^{23} \frac{\text{molecules}}{\text{mole}}$$

$$= 3.61 \times 10^{21} \text{ molecules}.$$

During a moon landing, one of the experiments performed was the measurement of the intensity of solar wind. As a collector, an aluminum strip of about 3000 $cm^2$ area was used. It was found that in 100 min, a mass of $3.0 \times 10^{-10}$ g of H atoms was collected (by the sticking of H atoms to the strip). What was the intensity of the solar wind (in numbers of atoms per $cm^2$ per second)?

Solution: The intensity of the solar wind is the number of H atoms striking the unit area in unit time,

$$\text{intensity} = \frac{\text{number of atoms}}{\text{area} \times \text{time}}.$$

We must determine the number of atoms. This is accomplished by converting mass to moles and then multiplying by Avogadro's number.

The number of moles is given by

$$\text{moles} = \frac{\text{mass of H atoms}}{\text{atomic weight of H}} = \frac{3.0 \times 10^{-10} \text{ g}}{1.0 \text{ g/mole}}$$

$$= 3.0 \times 10^{-10} \text{ mole.}$$

Since there are an Avogadro's number of atoms in one mole, the number of atoms in $3.0 \times 10^{-10}$ mole is

$$\text{number of atoms} = \text{moles} \times \text{Avogadro's number}$$

$$= 3.0 \times 10^{-10} \text{ mole} \times 6 \times 10^{23} \text{atoms/mole}$$

$$= 18 \times 10^{13} \text{ atoms.}$$

The intensity of the solar wind is then

$$\text{intensity} = \frac{\text{number of atoms}}{\text{area} \times \text{time}} = \frac{18 \times 10^{13} \text{ atoms}}{3000 \text{ cm}^2 \times 100 \text{ min}}$$

$$= \frac{18 \times 10^{13} \text{ atoms}}{3000 \text{ cm}^2 \times 100 \text{ min} \times 60 \text{ sec/min}}$$

$$= 1 \times 10^7 \text{ atoms/cm}^2\text{-sec.}$$

● **PROBLEM** 113

The most abundant element in sea water is chlorine (Cl, atomic weight = 35.5 g/mole), which is present as chloride ion, $Cl^-$, in a concentration of 19 g $Cl^-$ per 1 kg of sea water. If the volume of the earth's oceans is $1.4 \times 10^{21}$ liters, how many moles of $Cl^-$ are present in the oceans? Assume that the density of sea water is 1.0 g/cm$^3$.

Solution: The total number of moles of $Cl^-$ is equal to the total mass of $Cl^-$ divided by its atomic weight. Hence we must find the total mass of $Cl^-$. This will be done by determining the total mass of the oceans and using this value in conjunction with the concentration of $Cl^-$ to find the total mass of $Cl^-$.

The volume of the oceans is $1.4 \times 10^{21}$ liters = $1.4 \times 10^{21}$ liters $\times$ 1000 cm$^3$/liter = $1.4 \times 10^{24}$ cm$^3$. Multiplying this volume by the density of sea water gives the total mass of the oceans, or $1.4 \times 10^{24}$ cm$^3$ $\times$ 1.0 g/cm$^3$ = $1.4 \times 10^{24}$ g. Expressing this total mass as kilograms, we obtain

$$1.4 \times 10^{24} \text{ g} = 1.4 \times 10^{24} \text{ g} \times 10^{-3} \text{ kg/g} = 1.4 \times 10^{21} \text{ kg.}$$

For every kg of sea water, there are 19 g of $Cl^-$. Hence, for $1.4 \times 10^{21}$ kg there are $1.4 \times 10^{21}$ kg $\times$ 19 g $Cl^-$/kg of sea water = $2.7 \times 10^{22}$ g $Cl^-$.

Therefore, the number of moles of $Cl^-$ is

$$\text{moles } Cl^- = \frac{\text{mass } Cl^-}{\text{atomic weight } Cl^-}$$

$$= \frac{2.7 \times 10^{22} \text{ g}}{35.5 \text{ g/mole}} = 7.6 \times 10^{20} \text{ moles.}$$

● **PROBLEM** 114

An experiment to measure Avogadro's number involves the collection of a beam of alpha particles to which electrons are added to give neutral helium. Suppose that for a particular run it takes a number of electrons equivalent to $6.40 \times 10^{-5}$ coulomb/sec to neutralize the beam, and that after 24 hours of collection $1.14 \times 10^{-4}$ g of neutral helium are collected. From this data, determine Avogadro's number. The charge on one electron is $1.602 \times 10^{-19}$ coulombs.

<u>Solution</u>: Remembering that it takes two electrons to neutralize one alpha particle, i.e., one positively charged He atom, we will calculate the total charge required during the 24 hour run and, from this, the number of electrons. The number of alpha particles neutralized, or the number of helium atoms produced, is then equal to one-half the number of electrons. Using the mass of helium collected and the atomic mass of helium (4.0026 g/mole), we will calculate the corresponding number of moles. Dividing the number of atoms by the number of moles will give the number of atoms per mole, or Avogadro's number.

24 hours is equivalent to 24 hours × 60 min/hr × 60 sec/min = 86,400 sec. Thus, in 24 hours it took

$6.40 \times 10^{-5}$ coulomb/sec × 86,400 sec = 5.530 coulomb

to neutralize the beam. Since the charge on one electron is $1.602 \times 10^{-19}$ coulomb, dividing this number into the total number of coulombs gives the equivalent number of electrons, or

$$\text{number of electrons} = \frac{\text{coulombs}}{\text{coulombs per electron}}$$

$$= \frac{5.530 \text{ coulomb}}{1.602 \times 10^{-19} \text{ coulomb/electron}}$$

$$= 3.452 \times 10^{19} \text{ electrons.}$$

It takes two electrons to neutralize one alpha particle, hence the number of alpha particles neutralized by this number of electrons is number of alpha particles = $\dfrac{\text{number of electrons}}{2}$

$$= \dfrac{3.452 \times 10^{19}}{2} = 1.726 \times 10^{19} \text{ alpha particles.}$$

But one alpha particle produces one helium atom, so that the beam produced $1.726 \times 10^{19}$ helium atoms.

The mass of helium corresponding to this number of helium atoms produced by the beam is given as $1.14 \times 10^{-4}$ g. Dividing by the atomic mass of helium, we convert this mass to moles:

$$\text{moles helium} = \dfrac{\text{mass helium}}{\text{atomic mass helium}} = \dfrac{1.14 \times 10^{-4} \text{ g}}{4.0026 \text{ g/mole}}$$

$$= 2.84 \times 10^{-5} \text{ mole.}$$

Hence, $1.726 \times 10^{19}$ helium atoms correspond to $2.84 \times 10^{-5}$ mole of helium. Dividing these two numbers gives Avogadro's number:

$$\text{Avogadro's number} = \dfrac{\text{number of atoms}}{\text{moles}}$$

$$= \dfrac{1.726 \times 10^{19}}{2.84 \times 10^{-5} \text{ moles}}$$

$$= 6.08 \times 10^{23} \text{ atoms/mole.}$$

Comparing this with the currently accepted value of $6.02 \times 10^{23}$ atoms/mole, we see that the estimate calculated is in error by

$$\dfrac{6.08 \times 10^{23} - 6.02 \times 10^{23}}{6.02 \times 10^{23}} \times 100 \text{ \%}$$

$$= \dfrac{6.08 - 6.02}{6.02} \times 100 \text{ \%} = \dfrac{0.06}{6.02} \times 100 \text{ \%} = 1 \text{ \%.}$$

● **PROBLEM 115**

A chemist wants to calculate Avogadro's number by the inspection of a solid cube of AgCl. The density of the cube is 5.56 g/cm³. The spacing between the $Ag^+$ and $Cl^-$ ions in the cube is $2.773 \times 10^{-8}$ cm from their centers. From these data, perform this calculation.

Solution: To solve this problem, the following quantities must be determined: the volume of one mole of AgCl, cubic edge length of one mole, the number of ions on an edge, and the total number of ions. The total number of ions divided by 2 will yield Avogadro's number. From the atomic weights of Ag and Cl (107.868 amu and 35.453 amu) one mole of AgCl weighs 143.321 grams. Because density = mass/volume, the volume of one mole =

$$= \frac{143.321 \text{ g/mole}}{5.56 \text{ g/cm}^3} = 25.78 \text{ cm}^3/\text{mole}.$$

The volume of a cube = $(e)^3$, where "e" is the length of the cube edge. Therefore,

$$e = \sqrt[3]{25.78 \text{ cm}^3/\text{moles}} = 2.954 \text{ cm}/\sqrt[3]{\text{mole}}.$$

Because the spacing between ions is $2.77 \times 10^{-8}$ cm, the number of ions along the edge is

$$\frac{2.954 \text{ cm}/\sqrt[3]{\text{mole}}}{2.773 \times 10^{-8} \text{ cm}} = 1.065 \times 10^8 \text{ ions}/\sqrt[3]{\text{mole}}$$

The total number of ions in 1 mole of AgCl must be

$$(1.065 \times 10^8/\sqrt[3]{\text{mole}})^3 = 1.209 \times 10^{24}/\text{mole}.$$

This number is not Avogadro's number. Because you have two ions per formula unit, there is one Avogadro's number of $Ag^+$ and one Avogadro's number of $Cl^-$. It follows, therefore, that you must divide by two. As such,

$$\frac{1.209 \times 10^{24}/\text{mole}}{2} = 6.04 \times 10^{23}/\text{mole} = \text{Avogadro's number.}$$

# ATOMIC AND MOLECULAR WEIGHTS

● PROBLEM 116

1.3625 g of an unknown metal X reacts with oxygen to form 1.4158 g of the oxide $X_2O$. What is the atomic mass of X?

Solution:  This problem will be solved by calculating the mass and, from this, the number of moles of oxygen in the compound. The corresponding number of moles of metal is twice this amount. The molecular weight is the number of moles of metal divided by the weight of metal that was used.

The mass of oxygen in the compound is

mass O = mass oxide - mass X

= 1.4158 g - 1.3625 g = 0.0533 g.

The corresponding number of moles is

moles O = mass O ÷ atomic mass O

$$= 0.0533 \text{ g} \div 15.9994 \text{ g/mole} = 0.0033 \text{ mole}.$$

Since, in the oxide, there are two moles of X per mole of O, the number of moles of X present is 2 × 0.0033 mole = 0.0066 mole X. Hence, 1.3625 g of X corresponds to 0.0066 mole X. The atomic mass of X is then:

$$\text{atomic mass X} = \frac{\text{mass X}}{\text{moles X}} = \frac{1.3625 \text{ g}}{0.0066 \text{ mole}} = 206.4394 \text{ g/mole}.$$

● **PROBLEM** 117

How many moles are contained in 196 g of $H_2SO_4$?

Solution: To calculate the number of moles of a given compound from a certain number of grams of that compound one should use the following relation:

$$\text{number of moles} = \frac{\text{number of grams}}{\text{molecular weight}}$$

When determining the molecular weight of a compound one must first know the atomic weights of the elements within the compound. One then determines the molecular weight by adding together the weights of the individual elements making up the compound. The molecular weights of the elements contained in $H_2SO_4$ are MW of H = 1, MW of S = 32, and MW of O = 16. If there is more than one atom of a particular element, the molecular weight of the element must be multiplied by the number of atoms present in the compound.

Calculating the molecular weight of $H_2SO_4$:

| H | 2 | 1.0 | = | 2 |
|---|---|-----|---|----|
| S | 1 | 32 | = | 32 |
| O | 4 | 16 | = | 64 |
| | | Total | | 98 |

The molecular weight of $H_2SO_4$ is thus 98 g. The number of moles of $H_2SO_4$ in 196 g can now be calculated.

$$\text{number of moles} = \frac{196 \text{ g}}{98 \text{ g/mole}} = 2.0 \text{ moles}.$$

● **PROBLEM** 118

How many moles are present in 100 g quantities of each of the following? (a) $CaCO_3$, (b) $H_2O$, (c) HCl, (d) $Al_2(SO_4)_3$?

<u>Solution</u>: One can find the number of moles of a specific compound in a certain number of grams of that compound by dividing the number of grams present by the molecular weight. The molecular weight is defined as the weight of one mole or Avogadro's Number ($6.02 \times 10^{23}$) of particles.

$$\text{number of moles} = \frac{\text{weight in grams of sample}}{\text{molecular weight}}$$

One calculates the molecular weight of a compound by adding together the molecular weights of the elements present. When calculating the molecular weight, one must take into account the number of atoms of each element in the compound. This is done by multiplying the molecular weight of the element by the number of atoms present of the particular element. This method will be used in the following examples. Once the molecular weight is determined, the number of moles present in 100 g of the compound can be found by using the equation

$$\text{number of moles in 100 g} = \frac{100 \text{ g}}{\text{molecular weight}}$$

(a) $CaCO_3$

There is one atom of Ca present in $CaCO_3$. Thus, the molecular weight of Ca, 40, is multiplied by one. There is one atom of C. Thus the molecular weight, 12, of C is multiplied by one. Because there are 3 atoms of O present, the molecular weight of O, 16, is multiplied by 3.

| | | |
|---|---|---|
| 1 atom of Ca | $1 \times 40$ = | 40 |
| 1 atom of C | $1 \times 12$ = | 12 |
| 3 atoms of O | $3 \times 16$ = | 48 |
| molecular weight of $CaCO_3$ | = | 100 |

The number of moles of $CaCO_3$ in 100 g may now be found.

$$\text{number of moles in 100 g} = \frac{100 \text{ g}}{100 \text{ g/mole}} = 1 \text{ mole.}$$

(b) $H_2O$

(1) Calculation of molecular weight. The MW of H is 1 and the MW of O is 16.

| | | |
|---|---|---|
| 2 atoms of H | $2 \times 1$ = | 2 |
| 1 atom of O | $1 \times 16$ = | 16 |
| molecular weight of $H_2O$ | = | 18 |

(2) Calculation of number of moles in 100 g of $H_2O$.

$$\text{number of moles in 100 g} = \frac{100 \text{ g}}{18 \text{ g/mole}} = 5.55 \text{ moles.}$$

(c) HCl

MW of H = 1, MW of Cl = 35.5.

```
1 atom of H          1 ×  1   =  1
1 atom of Cl         1 × 35.5 = 35.5
molecular weight of HCl        = 36.5
```

$$\text{number of moles of HCl in 100 g} = \frac{100 \text{ g}}{36.5 \text{ g/mole}} = 2.74 \text{ moles}$$

(d) $Al_2(SO_4)_3$

MW of Al = 27    MW of S = 32    MW of O = 16

```
 2 atoms of Al      2 × 27  =   54
 3 atoms of S       3 × 32  =   96
12 atoms of O      12 × 16  =  192
molecular weight of Al₂(SO₄)₃=  342
```

$$\text{number of moles of } Al_2(SO_4)_3 \text{ in 100 g} = \frac{100 \text{ g}}{342 \text{ g/mole}}$$

$$= .292 \text{ moles.}$$

● **PROBLEM** 119

What is the molecular weight of a substance, each molecule of which contains 9 carbon atoms and 13 hydrogen atoms and $2.33 \times 10^{-23}$ g of other components?

Solution:  The molecular weight of a compound is the sum of the weights of the components of the compound. It is the weight of one mole of the substance, thus this compound weighs the sum of the weight of 9 moles of C, 13 moles H and $(6.02 \times 10^{23}) \times (2.33 \times 10^{-23}$ g). Because each molecule of the third substance weighs $2.33 \times 10^{-23}$ g, one mole of it weighs $(6.02 \times 10^{23}) \times (2.33 \times 10^{-23}$ g). There are $6.02 \times 10^{23}$ molecules of this other substance in one mole of the compound (Avogadro's number).

molecular weight = (9 × MW of C) + (13 × MW of H) +

$$(2.33 \times 10^{-23} \text{g})(6.02 \times 10^{23} \text{ /mole})$$

= (9 × 12.01 g/mole) + (13 × 1.00 g/mole)+

(14.03 g/mole)

= 108.09 g/mole + 13.0 g/mole +

14.03 g/mole

= 135.12 g/mole.

Determine the relative abundance of each isotope in naturally occurring gallium from the following data: At. wt. Ga = 69.72. Masses if isotopes $^{69}$Ga = 68.926, $^{72}$Ga = 70.925.

Solution: The relative abundance of the various isotopes of gallium can be found by using the following equation:

atomic weight of Ga = % of $^{69}$Ga × at. wt. of $^{69}$Ga +

% of $^{71}$Ga × at wt. of $^{71}$Ga

It is given that Ga consists only of the two isotopes $^{69}$Ga and $^{71}$Ga. Thus, if one lets x = fraction of $^{69}$Ga, then 1 - x = fraction of $^{71}$Ga. Using the above equation one can solve for x.

69.72   = (x × 68.926) + (1 - x) × 70.925

69.72   = 68.926x + 70.925 - 70.925x

- 1.205   = - 1.999x

x = 1.2/2.0 = 0.60.

This means that the $^{69}$Ga makes up 60 % of Ga. $^{71}$Ga makes up 1 - .60 or 40% of Ga.

One method for determining the molecular weight of large, biologically important molecules, is by measuring the density by standard procedures and determining the average volume occupied by a single molecule by X-ray crystallographic analysis. If a biochemist measures the density of a sample of deoxyribonucleic acid (DNA) as 1.1 g/cm$^3$ and X-ray analysis of the same sample estimates the volume of a single DNA molecule as 0.91 × 10$^{-15}$ cm$^3$, what is the molecular weight of this type of DNA?

Solution: The number density (molecules/cm$^3$) is the reciprocal of the volume of a single molecule and is related to the density (g/cm$^3$) by the following formula:

number density = $\dfrac{1}{\text{volume per molecule}}$

= $\dfrac{\text{density}}{\text{molecular weight}}$ × Avogadro's number

The validity of this formula is readily seen by considering the dimensions of the quantities involved.

Dividing the density ($g/cm^3$) by the molecular weight (g/mole) gives a quantity with units of

$$\frac{g/cm^3}{g/mole} = mole/cm^3 \text{ (molar density).}$$

Multiplying this number by **Avogadro's** number (molecules/mole) gives a quantity with dimensions of

$$\frac{mole}{cm^3} \times \frac{molecules}{mole} = molecules/cm^3 \text{ (number density).}$$

Solving for the molecular weight in the above expression we obtain

molecular weight = density × Avogadro's number × volume

per molecule

$$= 1.1 \ g/cm^3 \times 6 \times 10^{23} \ molecules/mole \times$$

$$\times \ 0.91 \times 10^{-15} \ cm^3/molecule$$

$$= 6 \times 10^8 \ g/mole.$$

● **PROBLEM** 122

When metal M is heated in halogen $X_2$, a compound $MX_n$ is formed. In a given experiment, 1.00 g of titanium reacts with chlorine to give 3.22 g of compound. What is the corresponding value of n?

<u>Solution:</u> In the compound $MX_n$, there is one mole of M and n moles of X. In the compound here, one is given that 1.00 g of it is Ti (titanium) and that the compound weighs 3.22 g. This means that 3.22 g - 1.00 g or 2.22 g of the compound is Cl. To find the simplest formula for the compound, one must first determine the number of moles of each component present. This is done by dividing the weight present by the molecular weight

$$\text{no. of moles} = \frac{\text{no. of grams}}{MW}$$

For Ti: (MW = 47.9)

$$\text{no. of moles} = \frac{1.00 \ g}{47.9 \ g/mole} = 2.09 \times 10^{-2} \ moles.$$

For Cl: (MW = 35.5)

$$\text{no. of moles} = \frac{2.22 \ g}{35.5 \ g/mole} = 6.25 \times 10^{-2} \ moles.$$

Since the simplest formula for the compound is $TiCl_n$, n can be found using the following ratio.

$$\frac{1\ Ti}{n} = \frac{2.09 \times 10^{-2}\ \text{moles Ti}}{6.25 \times 10^{-2}\ \text{moles Cl}}$$

$$n = \frac{1\ Ti \times 6.25 \times 10^{-2}\ Cl}{2.09 \times 10^{-2}\ Ti} = 2.99\ Cl$$

The formula for the compound is therefore $TiCl_3$. 2.99 Cl is rounded off to the nearest whole number, 3.

● **PROBLEM** 123

If the density of ethylene is 1.25 g/liter at S.T.P. and the ratio of carbon to hydrogen atoms is 1 : 2, what is molecular weight and formula of ethylene?

**Solution:** This problem is solved once you know that at S.T.P. (Standard Temperature and Pressure, 0°C, and 1 atm) one mole of any gas occupies 22.4 liters. Assume that one mole of ethylene gas is present. Density = mass/volume. As such the mass of the gas = (1.25 g/ℓ)(22.4 ℓ) = 28.0 g. Therefore one mole of ethylene weighs 28 g.

From the ratio given in the question you know the molecular formula can be represented as $(CH_2)_x$. To obtain the actual molecular formula, look for a compound that has a molecular weight of 28 g yet maintains the carbon : hydrogen ratio of 1 : 2. By looking at the periodic table for atomic weights and through some arithmetic, you will find that the only formula that meets these requirements is $(CH_2)_2$ or $C_2H_4$. This formula can also be found by dividing the weight of 1 $CH_2$ into 28 g. This solves for x in the expression $(CH_2)_x$. MW of $CH_2$ = 14.

$$\text{no. of } CH_2 = \frac{28\ g}{14\ g/CH_2} = 2\ CH_2 = C_2H_4 .$$

● **PROBLEM** 124

Phosphorus (atomic weight = 30.97) combines with another element such that 1 g of phosphorus requires 0.7764 g of the other element. If the atomic ratio of phosphorus to the other element is 4 : 3, what is the atomic weight of the unknown element?

**Solution:** First determine how many moles of phosphorus reacted. From this, one can calculate the number of moles of the unknown element that reacted using the atomic ratio. This ratio represents the relative mole amounts. If the number of moles of the unknown element is determined, the atomic weigth can be calculated because the number of moles is equal to number of grams/atomic weight. (At. wt. of P = 30.97).

One can now determine the atomic weight of the unknown element. The number of moles of phosphorus that reacted

$$= \frac{1.0 \text{ g}}{30.97 \text{ g/mole}} = 0.03229 \text{ moles.}$$

Let x = the number of moles of the unknown element. The atomic ratio of phosphorus to the unknown element is 4 : 3. Thus,

$$\frac{4 \text{ atoms}}{3 \text{ atoms}} = \frac{.03229 \text{ moles}}{x} \quad . \quad \text{Solving,}$$

$$x = 0.02422 \text{ moles unknown element.}$$

Therefore, the atomic weight of the unknown element is equal to

$$\frac{0.7764 \text{ g}}{0.02422 \text{ moles}} = 32.06 \text{ g/mole.}$$

● **PROBLEM** 125

1.0g of scandium (Sc) combines with oxygen (O) to form 1.5338 g of oxide. Assuming the oxide contains two atoms of scandium for every three atoms of oxygen, calculate the atomic weight of Sc. Oxygen has an atomic weight of 15.9994 amu.

Solution: This problem can be solved once a proportion is set up.

The oxide weighs 1.5338 g and the Sc, 1.0 g. Therefore, the weight of the oxygen that reacted must be 1.5338 - 1 = .5338 g.

The problem stated that 3 oxygen atoms reacted for every 2 Sc atoms. Therefore, the total atomic weight of the oxygen in the compound is 3 × 15.9994.

Let y be the atomic weight of Sc, then 2y is the total weight involved in the formation of the oxide. Therefore, there is .5338 g of oxygen whose atoms (3 of them) have an atomic weight of 3 × 15.9994, reacting with 1 g of Sc whose atoms weigh 2y. This can be represented as

$$\frac{.5338 \text{ g}}{3(15.9994)} = \frac{1.00 \text{ g}}{2y}$$

Solving for y, results in 44.96, which is the atomic
weight of scandium.

● **PROBLEM** 126

A given sample of pure compound contains 9.81 g of zinc,
$1.8 \times 10^{23}$ atoms of chromium and 0.60 mole of oxygen atoms.
What is the simplest formula?

<u>Solution</u>:  The simplest formula for this compound is found
from the simplest ratio of moles Zn : moles Cr : moles O.
One is given the number of moles of O but must find the
number of moles of Zn and Cr. One is given that there is
9.81 g of Zn. One can find the number of moles by dividing
9.81 g by the MW of Zn. (MW of Zn = 65.4).

$$\text{no. of moles} = \frac{9.81 \text{ g}}{65.4 \text{ g/mole}} = 0.15 \text{ moles}$$

One is given that there are $1.8 \times 10^{23}$ atoms of Cr
present. The number of moles can be found by dividing the
number of atoms by the number of atoms in one mole, $6.02 \times 10^{23}$.

$$\text{no. of moles} = \frac{1.8 \times 10^{23} \text{ atoms}}{6.02 \times 10^{23} \text{ atoms/mole}} = 0.30 \text{ moles}$$

The ratio of Zn : Cr : O is .15 : .30 : .60 or
1 : 2 : 4. The simplest formula is $ZnCr_2O_4$.

● **PROBLEM** 127

Two different compounds of elements A and B were found to
have the following composition: first compound, 1.188 g of
A combined with 0.711 g of B; second compound, 0.396 g of
A combined with 0.474 g of B. (a) Show that these data are
in accord with the law of multiple proportions. (b) If the
formula for the first compound is $AB_2$, what is the formula
for the second?

<u>Solution</u>:  (a) The law of multiple proportions can be
stated: When two elements combine to form more than one
compound, the different weights of one that combine with a
fixed weight of the other are in the ratio of small whole
numbers. This means that if one solves for the expected
amount of B that is used in forming the second compound
from the ratio of A : B in experiment one, the experimental
amount should be a multiple of the calculated value. This
is seen more clearly after looking at the data.

In experiment 1, A combines with B in a ratio of
1.188 g A : 0.711 g B or 1 : .598. In experiment 2, A

combined with B in a ratio of 0.396 A : 0.479 B or
1 : 1.20. The law of multiple proportions states that
.598 should be a small multiple of 1.20. 1.20/.598 = 2,
thus the law is supported.

(b) If the formula for the first compound is $AB_2$,
one knows that the proportion of the number of moles of
A to B is 1 : 2. Thus 1 unit volume of A weighs 1.188 g
and 2 units of B weigh 0.711 g. Therefore, 1 unit of B
weighs 0.711 g/2 = 0.356 g. Using this data, one solves
for the number of units of A in the second compound by
dividing the weight of A in compound 2 by 1.188 g.

no. of units of A in compound 2 = $\dfrac{.396 \text{ g}}{1.188 \text{ g/unit}}$ = .311 units

no. of units of B in compound 2 = $\dfrac{.474 \text{ g}}{.356 \text{ g/unit}}$ = 1.331 units

The ratio of A to B in compound 2 can now be found.
Let x = no. of B atoms

$$\frac{A}{B} = \frac{.311}{1.331} = \frac{1}{x}$$

$$x = \frac{1.331}{.331} = 4.0$$

Therefore, the second compound is $AB_4$. This result
could also have been obtained by using the data from part
(a). It was determined that twice as much B is present in
the second compound. Thus, if the first compound is $AB_2$,
then the second compound must be $AB_4$.

● **PROBLEM** 128

When 10.00 g of phosphorus was reacted with oxygen, it
produced 17.77 g of a phosphorus oxide. This phosphorus
oxide was found to have a molecular weight of approximately
220 in the vapor state. Determine its molecular formula.

Solution: The molecular formula of a substance indicates
the relative number of atoms in a molecule of the substance.
Therefore, to solve this problem, you must first calculate
the ratios of the gram-atoms to each other, the empirical
formula, and then extrapolate to the molecular formula via
the molecular weight.

The number of gram-atoms of phosphorus (P) is

$$\frac{\text{wt. in grams of P}}{\text{atom weight}} = \frac{10}{30.97} = 0.323 \text{ gram-atoms P.}$$

For oxygen we have $\dfrac{7.77 \text{ g}}{16.00}$ = .484 **gram-atoms**

The weight in grams of oxygen is 7.77 because the final product weighs 17.77 g and the phosphorus weighs 10.00 g. Since, the only other element is oxygen, its weight must be the difference.

The ratio of the gram-atoms of P and O is respectively 1 : 1.5 or 2 : 3. Therefore, the empirical formula of the oxide is $P_2O_3$.

To calculate the molecular formula, we must use the stated molecular weight of 220. We must look for a formula that totals to this molecular weight AND maintains the 2 : 3 ratio of P : O as expressed in the empirical formula. With some arithmetic, we find that the only formula that meets these two requirements is $P_4O_6$. 4 : 6 is the same as 2 : 3. The atomic weight of P and O is respectively 30.97 and 16. We have four P atoms for a total of 123.88 and we have 6 O atoms for a total of 96. Now add: 123.88 + 96 = 219.88, which is approximately 220.

Another method for determining the molecular formula is to divide the molecular weight of the molecule, 220, by the weight of 1 $P_2O_3$, 110.

$$\text{no. of } P_2O_3 = \frac{220 \text{ g}}{110 \text{ g/mole of } P_2O_3} = 2 \text{ moles of } P_2O_3.$$

The formula is, therefore $2 \times P_2O_3$ or $P_4O_6$.

● **PROBLEM** 129

A chemist reacts metal "B" with sulfur and obtains a compound of metal and sulfur. Assuming metal "B" weighed 2.435 g (MW = 121.75 amu), and the compound weighs 3.397 g, what is the simplest or empirical formula of the compound? The atomic weight of sulfur is 32.06 g/mole.

Solution:  To calculate the simplest formula, one must know the mole ratio of the elements that reacted.

A mole is defined as the weight in grams of a substance divided by its atomic weight. The number of moles of "B"

$$= \frac{2.435 \text{ g}}{121.75 \text{ g/mole}} = .0200 \text{ moles of "B".}$$

The number of grams of sulfur that reacted must be 3.397 - 2.435 = .962 g.

The increase in weight can only be derived from the addition of sulfur.

The number of moles of sulfur is

$$\frac{0.962 \text{ g}}{32.06 \text{ g/mole}} = .0300 \text{ moles of sulfur.}$$

The ratio of "B" to sulfur moles is .0200 B/.0300 S = 2 : 3. Therefore, the simplest formula must be $B_2S_3$.

● **PROBLEM** 130

It has been determined experimentally that two elements, A and B react chemically to produce a compound or compounds. Experimental data obtained on combining proportions of the elements are:

|  | Grams of A | Grams of B | Grams of Compound |
|---|---|---|---|
| Experiment 1 | 6.08 | 4.00 | 10.08 |
| Experiment 2 | 18.24 | 12.00 | 30.24 |
| Experiment 3 | 3.04 | 2.00 | 5.04 |

(a) Which two laws of chemical change are illustrated by the above data? (b) If 80 g of element B combines with 355 g of a third element C, what weight of A will combine with 71 g of element C? (c) If element B is oxygen, what is the equivalent weight of element C?

Solution: (a) If one adds the weight of A to the weight of B and obtains the weight of the compound formed, the Law of Conservation of Mass is illustrated. This law states that there is no detectable gain or loss of mass in a chemical change. Using the data from the experiments described, you find the following:

For experiment 1      6.08 g A + 4.00 g B = 10.08 g of compound

6.08 + 4.00 = 10.08

For experiment 2      18.24 g A + 12.00 g B
= 30.24 g of compound

18.24 + 12.00 = 30.24

For experiment 3      3.04 g A + 2.00 g B = 5.04 g compound

3.04 + 2.00 = 5.04

From these calculations one can see that the Law of Conservation of Mass is shown.

Another important law of chemistry is the Law of Definite Proportions. This law is stated: When elements combine to form a given compound, they do so in a fixed and invariable ratio by weight. One can check to see if this law is adhered to by calculating the ratio of the weight of A to the weight of B in the three experiments. If all of these ratios are equal, the Law of Definite Proportions is shown.

$$\text{Experiment 1} \quad \frac{6.08 \text{ g A}}{4.00 \text{ g B}} = 1.5$$

$$\text{Experiment 2} \quad \frac{18.24 \text{ g A}}{12.00 \text{ g B}} = 1.5$$

$$\text{Experiment 3} \quad \frac{3.04 \text{ g A}}{2.00 \text{ g B}} = 1.5$$

The Law of Definite Proportions is illustrated here.

(b) From the Law of Definite Proportions, one can find the number of grams of B that will combine with 71 g of C. After this weight is found, one can find the number of grams of A that will react with 71 g of C by finding the amount of A that reacts with that amount of B. It is assumed that the amount of A that reacts with 71 g of C is equal to the amount of A that will react with the amount of B that reacts with 71 g of C. The amount of A that will react with this amount of B can be found by remembering, from the previous section of this problem, that A reacts with B in a ratio of 1.5.

(1) Finding the amount B that would react with 71 g of C.

One is told that 80 g of B reacts with 355 g of C. By the Law of Definite Proportions, a ratio can be set up to calculate the number of grams of B that will react with 71 g of C.
Let x = the number of grams of B that will react with 71 g of C.

$$\frac{80 \text{ g B}}{355 \text{ g C}} = \frac{x \text{ g B}}{71 \text{ g C}}$$

$$x = \frac{71 \times 80}{355} = 16.$$

16 grams of B will react with 71 g of C.

(2) It is assumed that the same amount of A that will react with 16 g of B will react with 71 g of C. Therefore, using the fact that the ratio of the amount of A that reacts to the amount of B is equal to 1.5 (this fact was obtained in part (1)), one can calculate the amount of A that will react with 71 g of C.
Let x = the number of grams of A that will react with 16 g of B.

$$\frac{x \text{ g A}}{16 \text{ g B}} = 1.5$$

$$x = 16 \times 1.5 = 24 \text{ g.}$$

24 g of A will react with 16 g of B or 71 g of C.

(c) In finding the equivalent weight of C when B is taken to be oxygen, the Law of Definite Proportions is used again. The equivalent weight of oxygen is 8. Knowing that 16 g of B react with 71 g of C, one can set up the following ratio x = weight of C if the weight of B is taken to be 8.

$$\frac{71 \text{ g C}}{16 \text{ g B}} = \frac{x \text{g C}}{8 \text{g B}}$$

$$x = \frac{8 \times 71}{16} = 35.5 \text{ g.}$$

The equivalent weight of C when B is taken to be oxygen is 35.5 g.

# EQUIVALENT WEIGHTS

● PROBLEM 131

If the atomic weight of oxygen was 50, what would its equivalent weight be?

Solution:  The relationship of the atomic weight and the equivalent weight is

$$\frac{\text{Atomic weight}}{\text{Equivalent weight}} = \text{valence number}$$

The valence number is a measure of the number of atoms of hydrogen that will combine with one atom of the element. Two hydrogen atoms combine with one oxygen. The valence number of oxygen is therefore 2. One can now solve for the equivalent weight of oxygen when the atomic weight is taken as 50.

$$\frac{50}{\text{equivalent weight}} = 2$$

$$\text{equivalent weight} = \frac{50}{2} = 25.$$

● PROBLEM 132

A compound of vanadium and oxygen is analyzed and found to contain 56.0 % vanadium. What is the equivalent weight of vanadium in this compound?

Solution:     To find the equivalent weight of vanadium one can use the Law of Definite Proportions. This law states that when elements combine to form a given compound, they do so in a fixed and invariable ratio by weight. This means that the ratio of the weight of vanadium to the weight of the oxygen that reacts is equal to the ratio of the equivalent weight of the vanadium to the equivalent weight of the oxygen.

$$\frac{\text{weight of V}}{\text{weight of O}} = \frac{\text{equivalent weight of V}}{\text{equivalent weight of O}}$$

Solving for the equivalent weight of vanadium in this compound, one assumes he has 100 g of the compound for calculations. Because 56 % of this compound is vanadium, it means that 56 g of it is vanadium. This indicates that (100 % - 56 %) = 44 % of the compound is oxygen and that in 100 g of the compound there are 44 g of oxygen.

Here one will assume that the equivalent weight of oxygen is its atomic number, 8. Solving for the equivalent weight of vanadium:

$$\frac{56 \text{ g}}{44 \text{ g}} = \frac{\text{equivalent weight of V}}{8}$$

$$\text{equivalent weight of V} = \frac{8 \times 56 \text{ g}}{44 \text{ g}} = 10.2$$

The equivalent weight of vanadium in this compound is 10.2.

● **PROBLEM** 133

2.0 g of molybdenum (Mo) combines with oxygen to form 3.0 g of a molybdenum oxide. Calculate the equivalent weight of Mo in this compound.

Solution: The equivalent weight of a compound can be determined by use of the Law of Definite Proportions. This law states that when elements combine to form a given compound, they do so in a fixed and invariable ratio by weight. This means that the following ratio is maintained:

$$\frac{\text{weight of Mo in reaction}}{\text{weight of O in reaction}} = \frac{\text{equivalent weight of Mo}}{\text{equivalent weight of O}}$$

The equivalent weight of oxygen is equal to 8. The weight of oxygen in the reaction can be determined by using the Law of Conservation of Mass. This law states that there is no detectable gain or loss of mass in a chemical change. The problem states that 2.0 g of Mo is added to an unknown amount of O to form 3.0 g of a molybdenum oxide. Using the Law of Conservation of Mass the following relation is true:

weight of compound = weight of O + weight of Mo

3.0 g = weight of O + 2.0 g

1.0 g = weight of O

One can now use the Law of Definite Proportions to solve for the equivalent weight of Mo.

$$\frac{\text{weight of Mo}}{\text{weight of O}} = \frac{\text{equivalent weight of Mo}}{\text{equivalent weight of O}}$$

$$\frac{2.0 \text{ g}}{1.0 \text{ g}} = \frac{\text{equivalent weight of Mo}}{8}$$

$$\text{equivalent weight of Mo} = 8 \times \frac{2.0 \text{ g}}{1.0 \text{ g}} = 16.$$

● **PROBLEM** 134

It was found that a magnesium oxide contained .833 g of oxygen and 1.266 g of magnesium. Calculate the gram-equivalent weight of magnesium.

<u>Solution</u>: You begin this problem by establishing what is meant by the term gram-equivalent weight.

Gram-equivalent weight may be defined as the number of grams of an element that will involve a gain or loss of N electrons, i.e., the Avogadro's number ($6.02 \times 10^{23}$) of electrons, when the element enters into chemical combination with another element. In this problem, you know that oxygen is present. Oxygen has an oxidation state of - 2. In a reaction, one mole of oxygen will gain two moles of electrons. The molecular weight of oxygen is 16 g/mole. Its equivalent weight becomes 16/2 = 8.00 g, since 2 moles of electrons will be gained and, by definition, equivalent weight is the amount of a substance that will gain or lose one mole of electrons.

The gram-equivalent weight of magnesium is that amount of the element that combines with 8.00 g of oxygen. You are told that 1.266 g Mg combines with .833 g $O_2$. If you let x = grams of Mg that will combine with 8.00 of oxygen, you can set up the following proportion:

$$\frac{1.266 \text{ g Mg}}{.833 \text{ g } O_2} = \frac{x \text{ g Mg}}{8.00 \text{ g } O_2}$$

Solving for x, you obtain x = 12.16 g of Mg, which is its gram-equivalent weight.

● **PROBLEM** 135

A chemist forms magnesium oxide by burning magnesium in oxygen. The oxide obtained weighed 1.2096 grams. It was formed from .7296 g of magnesium. Determine the mass equivalent of magnesium in this reaction.

Solution: An equivalent is defined as that mass of oxidizing or reducing agent that picks up or releases the Avogadro number of electrons in a particular reaction. One equivalent of any reducing agent reacts with one equivalent of any oxidizing agent. In this problem, the key is to determine the number of equivalents of oxygen involved. Once this is known, you also know the number of equivalents of magnesium.

Since the oxide weighed 1.2096 g and the magnesium weighed .7296 g, the mass of the combined oxygen must be 1.2096 - .7296 = .4800 g. Before the oxygen reacted, its oxidation state was zero. After the reaction, however, it was - 2. As such, each oxygen atom gained 2 electrons. Therefore, the Avogadro number of electrons will be taken up by one half of a mole of O atoms.

It follows, therefore, that there are 8.000 g per equivalent for oxygen, since 1 mole of oxygen atoms weighs 16 grams. It was found, however, that there were .4800 g of oxygen. As such

$$\frac{.4800 \text{ g}}{8.00 \text{ g/equiv}} = .0600 \text{ equiv of oxygen.}$$

This means that magnesium also has .06 equiv. 0.7296 g of Mg participated in the reaction. Therefore, the grams per equivalent of Mg =

$$\frac{.7296 \text{ g Mg}}{.060 \text{ equiv}} = 12.16 \text{ g/equiv.}$$

● **PROBLEM 136**

For the oxidation of VO by $Fe_2O_3$ to form $V_2O_5$ and FeO, what is the weight of one equivalent of VO and of $Fe_2O_3$?

Solution: The equation for this reaction is

$$2VO + 3Fe_2O_3 \rightarrow V_2O_5 + 6FeO$$

Here V is a reducing agent and Fe is an oxidizing agent. One equivalent of an oxidizing agent is defined as the mass of the substance that picks up the Avogadro number of electrons. One equivalent of a reducing agent is defined as that mass of the substance that releases the Avogadro number of electrons. The oxidation state of O is always - 2, thus the oxidation state of V in VO is + 2.

The oxidation state of V in $V_2O_5$ is + 5 because 5 O contribute (5 × - 2) or - 10, thus 2V must be + 10 and V is + 5. The half-reaction for V is

$$V^{+2} \rightarrow V^{+5} + 3e^-$$

This means that there are 3 equivalents per mole of

VO. One finds the weight of one equivalent by dividing the molecular weight by 3. (MW of VO = 66.94).

$$\text{weight of 1 equiv of V} = \frac{66.94 \text{ g/mole}}{3 \text{ equiv/mole}} = 22.31 \text{ g}$$

One uses a similar method for Fe. The oxidation state of O in $Fe_2O_3$ is $(3 \times -2)$ or $-6$, this means the 2Fe must be $+6$ and Fe must be $+6/2$ or $+3$. The oxidation state of O in FeO is $-2$, thus the oxidation state of Fe is $+2$.

The half-reaction for the Fe is then

$$Fe^{+3} + 1e^- \rightarrow Fe^{+2}$$

In $Fe_2O_3$ there are 2 moles of Fe, therefore the half-reaction becomes

$$2Fe^{+3} + 2e^- \rightarrow 2Fe^{+2}$$

There are thus 2 equiv per mole of $Fe_2O_3$. The weight of one equivalent is equal to the weight of one mole $Fe_2O_3$ divided by 2. (MW of $Fe_2O_3$ = 159.70)

$$\text{wt of 1 equiv} = \frac{159.7 \text{ g/mole}}{2 \text{ equiv/mole}} = 79.85 \text{ g/equiv.}$$

● **PROBLEM** 137

In acting as a reducing agent a piece of metal M, weighing 16.00 g, gives up $2.25 \times 10^{23}$ electrons. What is the weight of one equivalent of the metal?

Solution:  One equivalent of a reducing agent is defined as that mass of the substance that releases the Avogadro number of electrons. Avogadro's number is $6.02 \times 10^{23}$, thus one can find the number of equivalents in 16.00 g of the metal by dividing $2.25 \times 10^{23}$ by $6.02 \times 10^{23}$.

$$\text{no. of equiv} = \frac{2.25 \times 10^{23} \text{ electrons}}{6.02 \times 10^{23} \text{ electrons/equiv}} = .374 \text{ equiv.}$$

Thus, .374 equiv weigh 16.00 g, one can find the weight of one equivalent by dividing 16.0 g by 0.374 equiv.

$$\text{weight of 1 equiv} = \frac{16.0 \text{ g}}{0.374 \text{ equiv}} = 42.78 \text{ g/equiv.}$$

# CHEMICAL COMPOSITION-WEIGHT AND VOLUME PERCENT

● PROBLEM 138

What is the simplest formula of a compound that is composed of 72.4 % iron and 27.6 % oxygen by weight?

Solution: For purposes of calculation, let us assume that there is 100 g of this compound present. This means that there are 72.4 g of Fe and 27.6 g O. The simplest formula for this compound is $Fe_n O_m$, where n is the number of moles of Fe present and m is the number of moles of O. One finds the number of moles by dividing the number of grams by the molecular weight.

$$\text{number of moles} = \frac{\text{number of grams}}{MW}$$

For Fe: n = number of moles present. MW = 55.8.

$$n = \frac{72.4 \text{ g}}{55.8 \text{ g/mole}} = 1.30 \text{ moles}$$

For O: m = number of moles present. MW = 16.0.

$$m = \frac{27.6 \text{ g}}{16.0 \text{ g/mole}} = 1.73 \text{ moles.}$$

One solves for the simplest formula by finding the ratio of Fe : O.

$$\frac{Fe}{O} = \frac{1.30}{1.73} = .75 = \frac{3}{4}$$

Therefore, n = 3 and m = 4.

The simplest formula is $Fe_3O_4$.

● PROBLEM 139

An unknown compound consists of 82.98 % potassium and 17.02 % oxygen. What is the empirical formula of the compound?

Solution: The empirical formula of any compound is the

ratio of the atoms that make up the compound by weight. It is the simplest formula of a material that can be derived solely from its components. Therefore, we must determine the ratio of gram-atoms of potassium (K) to the number of gram-atoms of oxygen (O).

The number of gram-atoms of a substance equals the weight of the substance in grams divided by the weight per gram-atom of the substance. In other words,

$$\text{number of gram-atoms} = \frac{\text{weight in grams}}{\text{weight per gram-atom}}$$

In this problem, we are given the percentages of the elements that make up the compound. These percentages are, in reality, the weight in grams, since in the definition of weight, we imply percentage. The weight per gram atom is the atomic weight of the element which can be found in the periodic table of elements.

Therefore, the number of gram-atoms for potassium is

$$\frac{82.98 \text{ g} \quad \text{(wt of K)}}{39.10 \quad \text{(atomic weight)}} = 2.120 \text{ moles.}$$

For oxygen, the number of gram-atoms is

$$\frac{17.02 \text{ g (wt of O)}}{16.00 \text{ (atomic wt)}} = 1.062 \text{ moles.}$$

Recall, the empirical formula is the ratio of the elements by weight. Consequently, the ratio of potassium to oxygen is 2 : 1, since the gram - atom ratios are respectively 2.120 : 1.062. Therefore, the empirical formula is $K_2O$.

● **PROBLEM** 140

A certain hydrate analyzes as follows: 29.7 % copper, 15.0 % sulfur, 2.8 % hydrogen, and 52.5 % oxygen. Determine the empirical formula of this hydrate from these percentages.

<u>Solution</u>:  A hydrate is the chemical combination of water with another compound. For example, copper sulfate combines with water to form the hydrate of the composition $CuSO_4 \cdot 5H_2O$

$$CuSO_4 + 5H_2O \rightarrow CuSO_4 \cdot 5H_2O$$

One can find the empirical formula for a compound, when given the weight percents of the various elements making up the compound. This is done by finding the number of moles of each element in 100 g of the compound. The weight of each element is equal to the percent weight.

The number of moles is equal to the weight divided by the molecular weight of the element. In 100 g of this compound, there are 29.7 g Cu, 15.0 g S, 2.8 g H, and 52.5 g O. One can now determine the number of moles of each element present.

$$\text{number of moles} = \frac{\text{number of grams}}{\text{MW}}$$

MW of Cu = 63.5, MW of S = 32, MW of H = 1, MW of O = 16.

$$\text{number of moles of Cu} = \frac{29.7 \text{ g}}{63.5 \text{ g/mole}} = 0.47 \text{ moles}$$

$$\text{number of moles of S} = \frac{15.0 \text{ g}}{32.0 \text{ g/mole}} = 0.47 \text{ moles}$$

$$\text{number of moles of H} = \frac{2.8 \text{ g}}{1 \text{ g/mole}} = 2.8 \text{ moles}$$

$$\text{number of moles of O} = \frac{52.5 \text{ g}}{16 \text{ g/mole}} = 3.28 \text{ moles.}$$

To determine the empirical formula for this hydrate one must look at the ratio of Cu : S : H : O.

The ratio of the number of moles of these elements is

Cu : S : H : O

.47 : .47 : 2.8 : 3.28

To find the empirical formula, these numbers should be made into integers. This is done by making the lowest number equal to 1 and solving for the other three.

$$\frac{.47}{1} = \frac{.47}{x} \qquad x = \frac{.47}{.47} = 1$$

$$\frac{.47}{1} = \frac{2.8}{x} \qquad x = \frac{2.8}{.47} = 6$$

$$\frac{.47}{1} = \frac{3.28}{x} \qquad x = \frac{3.28}{.47} = 7$$

The ratio now becomes

Cu : S : H : O

1 : 1 : 6 : 7.

The empirical formula for the compound is $CuSH_6O_7$ or $CuSO_4 \cdot 3H_2O$.

● PROBLEM 141

A chemist finds that an unknown compound contains

50.05 % S and 49.95 % O by weight. Calculate its simplest formula.

Solution: To calculate a compound's simplest formula, you need the relative number of moles of atoms in the compound. The percentages by weight of the elements allow for this calculation.

If you had 100 g of the unknown compound, it would consist of 50.05 g of S and 49.95 g of O. A mole is defined as weight in grams/atomic weight.

Therefore, the number of moles of sulfur

$$(S) = \frac{50.05 \text{ g}}{32.06 \text{ g/mole}} = 1.561 \text{ moles.}$$

The number of moles of oxygen =

$$\frac{49.96 \text{ g}}{15.999 \text{ g/mole}} = 3.122 \text{ moles.}$$

You see that the ratio of S to O is

$$\frac{1.561}{3.122} = \frac{1}{2} \; ; \quad 1 : 2.$$

Therefore, the simplest formula can be expressed as $SO_2$.

● **PROBLEM** 142

A sample of the poisonous compound nicotine extracted from cigarette smoke was found to contain 74.0 % by weight of carbon (C, atomic weight = 12.0 g/mole), 8.65 % by weight of hydrogen (H, atomic weight = 1.01 g/mole), and 17.3 % by weight of nitrogen (N, atomic weight = 14.0 g/mole). What is the empirical formula of nicotine?

Solution: The empirical formula states the relative ratio of the atoms of the various elements in any substance; the actual number of atoms (molecular formula) is not implied. Problems of this sort are solved by assuming a sample of some convenient mass and then calculating the number of moles of each atom in this sample. Once the number of moles has been calculated for each component, these numbers are divided by the greatest common factor in order to obtain the proportions in which the various components appear in the molecule. There is no error in assuming a sample of definite mass, since the size of a sample of molecules does not affect the composition of the molecules.

Assume a sample weighing 100 g. Then the masses of C, H, and N in this sample are:

mass C = 74.0 % × 100 g = 74.0 g

mass H = 8.65 % × 100 g =  8.65 g

mass N = 17.3 % × 100 g = 17.3 g

We convert the mass of each element to the corresponding number of moles by dividing by the atomic mass of that element. Thus we obtain the following number of moles of each element:

moles C = 74.0 g/12.0 g/mole = 6.2 moles

moles H = 8.65 g/1.01 g/mole = 8.6 moles

moles N = 17.3 g/14.0 g/mole = 1.2 moles

The greatest common factor is 1.2 moles. Dividing the number of moles of each element by 1.2 moles, we obtain

$$\frac{\text{moles C}}{1.2 \text{ moles}} = \frac{6.2 \text{ moles}}{1.2 \text{ moles}} \approx 5.$$

$$\frac{\text{moles H}}{1.2 \text{ moles}} = \frac{8.6 \text{ moles}}{1.2 \text{ moles}} \approx 7.$$

$$\frac{\text{moles N}}{1.2 \text{ moles}} = \frac{1.2 \text{ moles}}{1.2 \text{ moles}} = 1.$$

Hence, C, H, and N appear in the proportions 5, 7, and 1, respectively. The empirical formula of nicotine is thus $C_5H_7N$.

● **PROBLEM** 143

A compound subjected to analysis was found to have the following composition by weight: 69.96 % carbon (atomic weight = 12.0 g/mole), 7.83 % hydrogen (atomic weight = 1.01 g/mole), and 22.21 % oxygen (atomic weight = 16.0 g/mole). If the molecular weight of this compound is 360 g/mole, what is its molecular formula?

Solution:  Molecular formula may be defined as the formula stating the actual number of each type atom in a particular compound. Problems of this sort are solved by assuming a sample of some convenient mass and then calculating the number of moles of each atom in this sample. Once the number of moles has been calculated for each component, these numbers are divided by the greatest common factor in order to obtain the proportions in which the various components appear in the molecule (empirical formula). From this, we check to see if the elements, in those proportions, give the actual molecular weight. If they do, we have the actual molecular formula. If they do not, we multiply by a factor until their weights do give the molecular weight. There is no error in assuming a sample of definite mass, since the size of a sample of molecules does not affect the composition of the molecules.

Assume a sample weighing 100 g. Then the masses of C, H, and O in the sample are:

mass C = 69.96 % × 100 g = 69.96 g

mass H = 7.83 % × 100 g = 7.83 g

mass O = 22.21 % × 100 g = 22.21 g

We convert the mass of each element to the corresponding number of moles by dividing by the atomic weight of that element. Thus, we obtain the following number of moles of each element:

moles C = 69.96 g/12.0 g/mole = 5.83 moles

moles H = 7.83 g/1.01 g/mole = 7.75 moles

moles O = 22.21 g/16.0 g/mole = 1.39 moles

The greatest common factor is 1.39 moles. Dividing the number of moles of each element by 1.39 moles, we obtain

$$\frac{\text{moles C}}{1.39 \text{ moles}} = \frac{5.83 \text{ moles}}{1.39 \text{ moles}} \cong 4.2$$

$$\frac{\text{moles H}}{1.39 \text{ moles}} = \frac{7.75 \text{ moles}}{1.39 \text{ moles}} \cong 5.6$$

$$\frac{\text{moles O}}{1.39 \text{ moles}} = \frac{1.39 \text{ moles}}{1.39 \text{ moles}} = 1.$$

Multiplying these numbers by 5 in order to get whole numbers we obtain:

Proportion of C atoms = 4.2 × 5 = 21

Proportion of H atoms = 5.6 × 5 = 28

Proportion of O atoms = 1 × 5 = 5.

Thus, the molecular formula of this compound is $C_{21}H_{28}O_5$. As a check, we determine the molecular weight of this compound to be

molecular weight = (21 × atomic weight of C) + (28 ×

atomic weight of H) + (5 × atomic

weight of O)

= (21 × 12.0 g/mole) + (28 × 1.01 g/mole)

+ (5 × 16.0 g/mole)

$$\cong 360 \text{ g/mole},$$

which is the experimentally determined molecular weight.

The most common constituent of gasoline is iso-octane. It is a hydrocarbon, composed by weight of 84.12 % carbon, and 15.88 % hydrogen. Given that it contains $5.27 \times 10^{21}$ molecules per gram, what is its molecular formula?

**Solution:** The molecular formula for iso-octane is $C_n H_m$, where n is the number of moles of C and m is the number of moles of H present in 1 mole of iso-octane. One can find n and m from the molecular weight of the compound. It is given that there are $5.27 \times 10^{21}$ molecules/g present in one gram of the compound. One knows that there are $6.02 \times 10^{23}$ molecules per mole. Thus,

$$MW = \frac{6.02 \times 10^{23} \text{ molecules/mole}}{5.27 \times 10^{21} \text{ molecules/g}} = 114.23 \text{ g/mole}$$

The C in this compound weighs 84.12 % of 114.23 g and the H weighs 15.88 % of 114.23 g.

weight of C = .8412 × 114.23 g/mole = 96.09 g/mole

weight of H = .1588 × 114.23 g/mole = 18.14 g/mole.

The number of moles of carbon present, n, is equal to 96.09 g divided by 12.01 g/mole, its atomic weight.

$$n = \frac{96.09 \text{ g}}{12.01 \text{ g/mole}} = 8.00 \text{ moles}$$

The number of moles of H, m, can be calculated in a similar manner.

$$m = \frac{18.14 \text{ g}}{1 \text{ g/mole}} = 18.14 \text{ mole},$$

where 1 g/mole is the atomic weight of hydrogen.

The formula is $C_8 H_{18}$, 18.14 is rounded off to the nearest whole number.

Calculate the percentage composition of aluminum sulfate, $Al_2(SO_4)_3$.

<u>Solution</u>: Because the formula of a compound is constant, the percentage composition of each element present can be calculated by using the parts by weight of each element in one molecular weight of the compound. The molecular weight of the compound can be calculated by adding together the weights of the various elements contained in the compound.

There are 2 atoms of aluminum present, and the molecular weight of aluminum is 27, so the contribution of the aluminum to the compound's total molecular weight is found by mulyipltying 27 by 2.

2 atoms of aluminum weigh 2 × 27 = 54

The weight contributions of the other elements present can be calculated in the same way.

The molecular weight of sulfur is 32, and there are 3 sulfur atoms present in the compound.

3 atoms of sulfur weigh  3 × 32 = 96

The molecular weight of oxygen is 16, and there are 12 oxygen atoms present in the compound.

12 atoms of oxygen weigh  12 × 16 = 192

The weights of the three elements contained in the compound are added together to find the total molecular weight of the compound.

| | |
|---|---|
| 2 atoms of aluminum | 54 |
| 3 atoms of sulfur | 96 |
| 12 atoms of oxygen | 192 |
| molecular weight of $Al_2(SO_4)_3$ | 342 |

The percentage composition of each element present can now be calculated by considering what fraction by weight each element is of the total compound.

The weight of two atoms of aluminum is 54 so the fraction by weight of aluminum in the compound can be found by dividing 54 by 342, the molecular weight of the compound. The percentage is then found by multiplying this fraction by 100.

percentage composition of aluminum

$$= \frac{\text{weight of aluminum in compound}}{\text{molecular weight compound}} \times 100$$

percentage composition of aluminum $= \frac{54}{342} \times 100 = 15.8$ %.

The same method can be applied to the sulfur and oxygen.

percentage composition of sulfur = $\frac{96}{342}$ × 100 = 28. 1%

percentage composition of oxygen = $\frac{192}{342}$ × 100 = 56.1 %

If the percent compositions of all the elements in a compound are added together they will equal 100.0.

| | |
|---|---|
| percent composition of aluminum | 15.8 |
| percent composition of sulfur | 28.1 |
| percent composition of oxygen | 56.1 |
| | 100.0 |

● **PROBLEM** 146

Two thirds of the atoms in a molecule of water ($H_2O$) are hydrogen. What percentage of the weight of a water molecule is the weight of the two hydrogen atoms? The atomic weight of hydrogen is 1.008 g/mole and of oxygen is 16.00 g/mole.

Solution: The most direct way to solve this composition problem is to consider the total weight of one molecule of water:

$$mass_{H_2O} = 2\ mass_H + mass_O$$

$$= 2(1.008\ g/mole) + 16.00\ g/mole$$

$$= 18.016\ g/mole.$$

The mass of two hydrogen atoms is 2 × 1.008 = 2.016 g/mole. Hence, the percentage mass of hydrogen in water is:

$$\%\ mass = \frac{2\ mass_H}{mass\ H_2O} \times 100\ \% = \frac{2.016\ g/mole}{18.016\ g/mole} \times 100\ \%$$

$$= 11.19$$

● **PROBLEM** 147

Calculate the percent composition by weight of ether $(C_2H_5)_2O$.

Solution: To understand this problem, one must first realize that the subscript 2 outside the brackets means that all elements inside the brackets are multiplied by 2. Carrying this procedure out, one obtains the formula

$C_4H_{10}O$ for ether.

Percent composition is found by dividing the individual weights of the components by the molecular weight. Therefore, first find molecular weight by using the table below.

| Element | Number of Atoms | Atomic Weight | Total Atomic Weight |
|---------|-----------------|---------------|---------------------|
| C       | 4               | 12            | 48                  |
| H       | 10              | 1             | 10                  |
| O       | 1               | 16            | 16                  |

M.W. = 48 + 10 + 16 = 74 g/mole.

Percent composition is then calculated from the formula

$$\% \text{ composition} = \frac{(\text{total atomic weight})}{\text{M.W.}} \times 100$$

$$\% \text{ C} = \frac{48}{74} \times 100 = 64.86 \%$$

$$\% \text{ H} = \frac{10}{74} \times 100 = 13.51 \%$$

$$\% \text{ O} = \frac{16}{74} \times 100 = 21.63 \%.$$

To double check, make sure that the total percent equals 100.

● **PROBLEM** 148

Calculate the percent weight of each element in magnesium chloride, $MgCl_2$.

Solution: As can be seen from the formula, for each molecule of Mg there are 2 molecules of Cl that must combine to form the compound.

The way to solve a percent composition problem is first to find the total molecular weight. Then divide the weight of each component by the total molecular weight to find the percentages.

First set up a table as shown below.

| Element | Number of Atoms or Moles | Atomic Weight | Total Atomic Weight |
|---------|--------------------------|---------------|---------------------|
| Mg      | 1                        | 24.3          | 24.3                |
| Cl      | 2                        | 35.45         | 70.9                |

M.W. of $MgCl_2$ = 24.3 + 70.9 = 95.2

To find percent composition use the formula

$$\% \text{ composition} = \frac{\text{(total atomic weight)}}{\text{total molecular weight}} \times 100$$

$$\% \text{ Mg} = \frac{24.3}{95.2} \times 100 = 25.5 \%$$

$$\% \text{ Cl} = \frac{70.9}{95.2} \times 100 = 74.5 \%.$$

To double check that the answer is correct, see if the sum of the percentages is 100 %.

Using the Periodic Table of Elements, find the following for sodium dihydrogen phosphate, $NaH_2PO_4$: (a) formula weight, (b) percent composition of oxygen, (c) weight in grams of 2.7 moles, and (d) percentage composition of oxygen in 2.7 moles.

Solution: This problem encompasses work in chemical stoichiometry. With this in mind, you proceed as follows:

(a) The formula weight = molecular weight, which is the sum of atomic masses of all the atoms in the substance. Na = 22.98, H = 1.008, O = 15.9999, and P = 30.97. Thus, formula weight = Na + 2H + P + 4O = 22.98 + 2(1.0080) + 30.97 + 4(15.9999) = 119.9.

(b) Percentage composition of oxygen is

$$\frac{\text{total weight of oxygen in compound}}{\text{total weight of compound}}$$

$$= \frac{4 \text{ oxygen atoms} \times 15.9994 \text{ mass/oxygen atom}}{119.9}$$

$$= .5334 \text{ g of oxygen or } 53.34 \% \text{ by weight in } NaH_2PO_4.$$

(c) Mole = $\frac{\text{mass of substance}}{\text{molecular weight}}$. You are given that there are 2.7 moles and you calculated the molecular weight. Thus, the weight in grams of 2.7 moles = (119.9 g/mole) (2.7 moles) = 324 g of $NaH_2PO_4$.

(d) Following a procedure similar to the one in part (b),

$$\text{percent composition} = \frac{\text{no. of moles of O} \times \text{MW of O}}{\text{weight of comp/no. of moles}} \times 100$$

$$= \frac{4 \text{ moles} \times 15.9 \text{ g/mole}}{324 \text{ g/2.7 moles}} \times 100$$

$$= 53.34 \text{ \%}.$$

The percent composition of any element in any compound does not change when the amount of the compound present is changed.

● **PROBLEM** 150

Calculate the weight of iron in 350 pounds(158.9 kgs) of $Fe_2O_3$. First, calculate the percent of iron in the compound.

Solution:  The percent composition of iron in the compound $Fe_2O_3$ can be calculated by first finding the molecular weight of the compound. After the molecular weight of the compound is found, the percent of this weight that is made up by the iron is found by dividing the weight of the iron by the molecular weight of the compound and then multiplying this fraction by 100.

The molecular weight of iron is 55.8, so that two atoms of iron weigh $2 \times 55.8 = 111.6$.

The molecular weight of oxygen is 16.0, so that 3 atoms of oxygen weigh $3 \times 16.0 = 48.0$

The molecular weight of $Fe_2O_3$ is found by adding the weights of the iron and oxygen together.

| | |
|---|---|
| weight of iron | 111.6 |
| weight of oxygen | 48.0 |
| molecular weight of $Fe_2O_3$ | 159.6 |

The percent composition of iron =

$$\frac{\text{weight of iron in compound}}{\text{molecular weight of compound}} \times 100$$

percent composition of iron $= \frac{111.6}{159.6} \times 100 = 69.9$

Since the percent composition by weight of iron in the compound is known, the weight of iron in 350 pounds of this compound can now be found. Because the percent composition of iron in this compound is 69.9, the percent of iron in 350 pounds of $Fe_2O_3$ is also 69.9.

Weight of iron in 350 lb of $Fe_2O_3 = 350 \times 0.699 = 245$ lb.

(111.07 kg)

What is the elemental percent composition (by weight) of a mixture that contains 20.0 g of $KAl(SO_4)_2$ and 60.0 g of $K_2SO_4$?

Solution: The elemental percent composition by weight is equal to the weight of the element present divided by the total weight of the mixture multiplied by 100.

$$\text{percent composition} = \frac{\text{wt of element}}{\text{total wt}} \times 100$$

The weight of the total mixture is equal to the sum of the weights of the $KAl(SO_4)_2$ and the $K_2SO_4$.

total weight = 20.0 g + 60.0 g = 80.0 g

The weights of the various elements present is found by dividing the molecular weight of the element multiplied by the number of atoms of that particular atom present in the molecule by the molecular weight of the compound, and then multiplying the quotient by the number of grams of the compound present.

Solving for the weights of the elements in $KAl(SO_4)_2$: MW = 258.12.

For K = MW = 39.1

$$\text{weight of K} = \frac{39.1 \text{ g/mole} \times 1}{258.12 \text{ g/mole}} \times 20.0 \text{ g} = 3.03 \text{ g}.$$

For Al: MW = 26.98

$$\text{weight of Al} = \frac{26.98 \text{ g/mole} \times 1}{258.12 \text{ g/mole}} \times 20.0 \text{ g} = 2.09 \text{ g}.$$

For S: MW = 32.06

$$\text{weight of S} = \frac{32.06 \text{ g/mole} \times 2 \text{ moles}}{258.12 \text{ g/mole}} \times 20.0 \text{ g} = 4.97 \text{ g}.$$

For O: MW = 16

$$\text{weight of O} = \frac{16 \text{ g/mole} \times 8 \text{ moles}}{258.12 \text{ g/mole}} \times 20.0 \text{ g} = 9.92 \text{ g}.$$

Solving for the weight of the elements in $K_2SO_4$: MW = 174.26

For K: MW = 39.1

$$\text{weight of K} = \frac{39.1 \text{ g/mole} \times 2 \text{ moles}}{174.26 \text{ g/mole}} \times 60.0 \text{ g} = 26.93 \text{ g}$$

For S; MW = 32.06

weight of S = $\dfrac{32.06 \text{ g/mole} \times 1 \text{ mole}}{174.26 \text{ g/mole}}$ × 60.0 g = 11.04 g

For O: MW = 16

weight of O = $\dfrac{16 \text{ g/mole} \times 4 \text{ moles}}{174.26 \text{ g/mole}}$ × 60.0 g = 22.04 g.

One can find the total weights of the various elements by taking the sum of their weights from the two compounds.

Total weight of K = 3.03 g + 26.93 g = 29.96 g

Total weight of Al = 2.09 g

Total weight of S = 4.97 g + 11.04 g = 16.01 g

Total weight of O = 9.92 g + 22.04 g = 31.96 g.

One can now determine the elemental percent composition of the mixture.

$$\% = \dfrac{\text{weight of element}}{\text{total weight of mixture}} \times 100$$

% of K = $\dfrac{29.96 \text{ g}}{80.0 \text{ g}}$ × 100 = 37.45 %

% of Al = $\dfrac{2.09 \text{ g}}{80.0 \text{ g}}$ × 100 = 2.61 %

% of S = $\dfrac{16.01 \text{ g}}{80.0 \text{ g}}$ × 100 = 20.01 %

% of O = $\dfrac{31.96 \text{ g}}{80.0 \text{ g}}$ × 100 = 39.95 %.

● **PROBLEM** 152

When a piece of magnesium ribbon weighing 0.32 g is burned in oxygen, the resultant oxide weighs 0.53 g. What is the percentage composition of the oxide?

Solution: To find the percent composition of magnesium and oxygen in the oxide, one must first know the weight of the magnesium and of the oxygen making up the compound. The percent composition of magnesium is equal to the weight of the magnesium divided by the weight of the compound multiplied by 100. The percent composition of the oxygen is found in the same manner by substituting the weight of oxygen for the weight of the magnesium. One is told (in this problem) that 0.32 g of magnesium is used, and that the weight of the oxide formed is 0.53 g. The weight of oxygen in the oxide is the difference in the weight of the oxide and the magnesium.

weight of oxygen = 0.53 g - 0.32 g = 0.21 g.

The oxygen in the oxide weighs 0.21 g. The percent composition of the oxide can now be determined.

percent Mg in oxide = $\dfrac{0.32 \text{ g}}{0.53 \text{ g}} \times 100 = 60$ %

percent O in oxide = $\dfrac{0.21}{0.53} \times 100 = 40.$ %.

The oxide is therefore 60 % Mg and 40 % O.

● **PROBLEM** 153

A 0.240 g sample of a compound of oxygen and element X was found by analysis to contain 0.096 g of X and 0.144 g of oxygen. (a) Calculate the percentage composition by weight. (b) Calculate from the above data, three possible atomic weights for X relative to oxygen (at. wt. = 16). (c) What additional information is needed to calculate the true atomic weight of X?

<u>Solution</u>:  (a) The percent composition by weight of each element in the compound is found by dividing the weight of that element by the weight of the compound and then multiplying the quotient by 100. For O, therefore, you obtain the following percentage:

$\dfrac{.144 \text{ g}}{0.240 \text{ g}} \times 100 = 60$ % O

For X:

$\dfrac{.096 \text{ g}}{0.240 \text{ g}} \times 100 = 40$ % X

(b) If you assume that the compound is XO, you can solve for the molecular weight of X by using the following ratio. Let x = MW of X.

$\dfrac{.144 \text{ g O}}{16.0 \text{ g/mole O}} = \dfrac{.096 \text{ g}}{X \text{ g/mole x}}$

MW X = 10.7

If the compound is taken to be $X_2O$, there are twice as many X atoms present as O atoms. In solving for the molecular weight of X in this case, use the following ratio

$\dfrac{.144 \text{ g}}{\text{weight of 1 mole of O}} = \dfrac{.096}{\text{weight of 2 moles of X}}$

Let x = MW of X

$$\frac{.144 \text{ g}}{16.0} = \frac{.096 \text{ g}}{x}$$

$$x = \frac{16.0 \times .096 \text{ g}}{.144 \text{ g}}$$

$$x = 5.35.$$

Assume also that the formula for the compound might be $XO_2$. Then, the following ratio should be used: Let $x$ = MW of X.

$$\frac{.144 \text{ g}}{\text{weight of 2 moles of O}} = \frac{.096}{\text{weight of 1 mole of X}}$$

$$\frac{.144 \text{ g}}{32.0} = \frac{.096 \text{ g}}{x}$$

$$x = \frac{.096 \text{ g} \times 32.0}{.144 \text{ g}} = 21.4.$$

(c) To solve for the true atomic weight of X, you must know the actual number of X atoms present per each O atom present.

● **PROBLEM** 154

A certain solution contains 5 % $FeSO_4$. How many pounds of Fe could be obtained from 1 ton of this solution?

<u>Solution</u>: If a solution contains 5 % $FeSO_4$, this means that 5 % of the total weight of the solution is $FeSO_4$. In this problem, 5 % of a ton is $FeSO_4$. To determine the weight of Fe in this amount of $FeSO_4$, one must calculate the weight percent of Fe in $FeSO_4$. This is done by dividing the molecular weight of Fe by the molecular weight of $FeSO_4$ and multiplying the quotient by 100.

To solve this problem:

(1) Determine the percent weight of $FeSO_4$

(2) Determine the weight of 5 % of a ton

(3) Determine how much of this weight is Fe.

(1) The molecular weight of a compound is determined by adding together the weight contributed by the elements of which it is composed. MW of Fe = 55.8, MW of S = 32, MW of O = 16.

Thus, for FeSO₄,

```
1 atom of Fe     1 × 55.8 = 55.8
1 atom of S      1 × 32   = 32
4 atoms of O     4 × 16   = 64
molecular weight of FeSO₄   151.8
```

$$\text{percent weight of Fe in FeSO}_4 = \frac{\text{weight of Fe}}{\text{weight of FeSO}_4} \times 100$$

$$= \frac{55.8}{151.8} \times 100 = 37 \text{ \%.}$$

37 % of FeSO₄ by weight is Fe.

(2) Determining 5 % of a ton. 1 ton = 2000 lbs.

5 % of a ton = .05 × 2000 lbs = 100 lbs.

Thus, 5 % of a ton is 100 lbs.

(3) Determining the weight of Fe in 100 lbs. 37 % of this is Fe.

weight of Fe = .37 × 100 lbs = 37 lbs(16.798 kgs)

● **PROBLEM** 155

Of the total number of atoms in the universe, approximately 93 % are hydrogen (H, atomic weight = 1.0 g/mole) and 7 % are helium (He, atomic weight = 4.0 g/mole). What percentage of the universe by weight is hydrogen?

Solution:  The percentage by weight is equal to the total weight of H atoms divided by the total weight of the universe. Suppose that there are N atoms in the universe. Then the number of hydrogen atoms is 93 % × N = 0.93 N and the number of helium atoms is 7 % × N = 0.07 N. The mass of hydrogen atoms is

mass H = number H × mass H = 0.93 N × 1.0 g/mole

and the mass of helium atoms is

mass He = number He × mass He = 0.07 N × 4.0 g/mole.

Then, on a per mole basis,

$$\text{\% g/mole} = \frac{\text{g/mole of H}}{\text{total g/mole}} \times 100 \text{ \%}$$

$$= \frac{0.93 \text{ N} \times 1.0 \text{ g/mole}}{0.93 \text{ N} \times 1.0 \text{ g/mole} + 0.07 \text{ N} \times 4.0 \text{ g/mole}} \times 100 \text{ \%}$$

On a weight basis,

$$\text{\% H by weight} = \frac{0.93 \text{ N} \times 1.0 \text{ g/mole}}{0.93 \text{ N} \times 1.0 \text{ g/mole} + 0.07 \text{ N} \times 4.0 \text{ g/mole}}$$

$$\times \frac{1 \text{ mole}}{1 \text{ mole}} \times 100 \text{ \%}$$

$$= \frac{0.93 \text{ N g}}{0.93 \text{ N g} + 0.28 \text{ N g}} \times 100 \text{ \%}$$

$$= \frac{0.93 \text{ N g}}{1.21 \text{ N g}} \times 100 \text{ \%} = \frac{0.93 \text{ g}}{1.21 \text{ g}} \times 100 \text{ \%} = 77 \text{ \%}.$$

● **PROBLEM** 156

The density of a 25.0 % sugar solution is 1.208 g/ml. What weight of sugar would be contained in 1.00 liter of this solution?

Solution: When a solution is said to be 25.0 % sugar, it means that 25.0 % of the weight of the solution is made up by the sugar. Thus one can determine the weight of the sugar in this solution by multiplying the total weight of the solution by .25. Here one is not given the weight of the solution but the density and the volume. The density is the weight of 1 ml of the solution, thus the weight of 1.0 liter or 1000 ml is equal to the volume 1000 ml times the density (1.208 g/ml).

weight of solution = 1000 ml × 1.208 g/ml = 1208 g.

The weight of the sugar in the solution can now be found.

weight of sugar = .25 × weight of solution

= .25 × 1208 g = 302 g.

● **PROBLEM** 157

A lunar surface probe analyzed a sample of soil and found that 58 % of the atoms it contained were oxygen (O, atomic weight = 16 g/mole), 18 % were silicon (Si, atomic weight = 28 g/mole), 9 % were aluminum (Al, atomic weight = 27 g/mole), and 15 % consisted of other elements with an average atomic weight of 30 g/mole. Determine the percent oxygen by weight in this sample.

Solution: The percent oxygen by weight is equal to the mass of oxygen divided by the total mass, multiplied by 100 %. Hence, in order to determine the mass of O and the total mass, we must determine the mass of each element.

For each element, this is given by

mass of element = % of atoms of element × atomic weight ×

number of moles in sample.

Consider a sample containing one mole of atoms. We are justified in choosing a sample of definite size since no matter what the size of the sample, the percent composition is the same. Applying the above expression for the mass of each element, we obtain the following:

mass O        = 58 % × 16 g/mole × 1 mole = 9.28 g

mass Si       = 18 % × 28 g/mole × 1 mole = 5.04 g

mass Al       =  9 % × 27 g/mole × 1 mole = 2.43 g

mass others = 15 % × 30 g/mole × 1 mole = 4.5 g.

The mass of O in our 1 mole sample is thus 9.28 g and the total mass of our sample is 9.28 g + 5.04 g + 2.43 g + 4.5 g = 21.25 g. Hence, the percent by weight of oxygen in the sample is

$$\frac{9.28 \text{ g}}{21.25 \text{ g}} \times 100 \text{ \%} \cong 44. \text{ \%}.$$

● **PROBLEM** 158

What is the weight of 1.0 liter of carbon monoxide (CO) at STP?

Solution: At STP (Standard Temperature and Pressure, 0°C and 760 torr), a mole of any gas has a volume of 22.4 liters. This means that the gram molecular weight of a gas is contained in 22.4 liters. In this problem, one is looking for the weight of 1 liter of CO. The molecular weight of CO is 28 g. Because there are 28 g of CO in 22.4 liters of gas, 28 g must be divided by 22.4 to find the weight of one liter.

weight of one liter $= \dfrac{\text{gram molecular weight}}{22.4}$

weight of 1 liter of CO $= \dfrac{28 \text{ g}}{22.4} = 1.25$ g.

● **PROBLEM** 159

Mammalian hemoglobin contains about 0.33 % iron (Fe, atomic weight = 56 g/mole) by weight. If the molecular

weight of hemoglobin is 68,000 g/mole, how many iron
atoms are there in each molecule of hemoglobin?

Solution: To solve this problem, we consider a one mole
sample of hemoglobin. This assumption introduces no error,
since the size of the hemoglobin sample does not affect
the number of Fe atoms per molecule of hemoglobin. We
then calculate the mass of Fe atoms in one mole of
hemoglobin and divide by the atomic weight of Fe to
obtain the number of moles of Fe per mole of hemoglobin.

Since the molecular weight of hemoglobin is
68,000 g/mole, the weight of one mole of hemoglobin is

68,000 g/mole × 1 mole = 68,000 g.

The weight of Fe in one mole of hemoglobin is then

weight Fe = % Fe by weight × weight of 1 mole of
hemoglobin

= 0.33 % × 68,000 g = 224.4 g.

The number of moles of Fe corresponding to this
weight is found by dividing this weight of Fe by the
atomic weight of Fe to obtain

$$\text{moles Fe} = \frac{\text{weight Fe}}{\text{atomic weight Fe}} = \frac{224.4 \text{ g}}{56 \text{ g/mole}} = 4 \text{ moles.}$$

Hence, there are 4 moles of Fe per mole of
hemoglobin. To convert from number of moles to number
of atoms or molecules, we multiply by Avogadro's number:
hence, 4 moles of Fe correspond to 4 moles × 6 × $10^{23}$
atoms/mole = 24 × $10^{23}$ atoms of Fe and 1 mole of hemoglobin
corresponds to 1 mole × 6 × $10^{23}$ molecules/mole = 6 ×
$10^{23}$ molecules of hemoglobin. Thus, the ratio of number
of atoms of Fe to molecules of hemoglobin is

$$\frac{24 \times 10^{23} \text{ atoms of Fe}}{6 \times 10^{23} \text{ molecules of hemoglobin}} = 4 \text{ atoms Fe/molecule hemoglobin}$$

There are 4 atoms of Fe in every molecule of
hemoglobin.

● **PROBLEM** 160

The average bromine content of sea water is 0.0064%.
(a) How much sea water, in cubic feet, would be required to
obtain one pound of bromine? (b) What volume of chlorine
gas, measured at STP, would be required to liberate
the bromine from one ton of salt water? One cubic foot
of sea water weighs about 63 pounds (28.35 kg)

Solution: (a) The weight of bromine in one cubic foot of sea water is found by multiplying 63 pounds (the weight of 1 cu.ft. of sea water) by 0.000064. This is because bromine composes 0.0064 % of the weight of sea water.

weight of Br per 1 cubic ft (liter)= .000064 x 63 lb
(1.01 kg/liter)
$$= 4.032 \times 10^{-3} \text{ lb}$$
$$(6.46 \times 10^{-5} \text{ kg})$$

One can find the number of cubic feet of sea water necessary to extract 1 lb of Br by dividing 1 lb by the number of pounds of Br in one cubic foot.

$$\text{no. of cubic ft} = \frac{1 \text{ lb}}{4.032 \times 10^{-3} \text{ lb/cubic ft}}$$

= 248.02 cubic feet (7022.93 liters)

(b) Since chlorine is more active than bromine, the latter may be liberated from its salt by treatment with chlorine.

$$2Br^- + Cl_2 \rightarrow 2Cl^- + Br_2$$

One mole of $Cl_2$ will liberate 2 moles of $Br^-$. Therefore, to find the volume of $Cl_2$ gas necessary to liberate the bromine in one ton of sea water, one must first calculate the amount of $Br^-$ present in 1 ton of sea water. One is given that 1 cubic foot of sea water weighs 63 pounds. Thus the number of cubic feet of sea water in 1 ton can be found. (1 ton = 2000 lbs).

$$\text{no. of cubic feet} = \frac{2000 \text{ lbs}}{63 \text{ lbs/cubic feet}} = 31.75 \text{ cu.ft.}$$
(899 liters)

In the previous section one found that each cubic foot contains $4.032 \times 10^{-3}$ lb of $Br^-$, thus one finds the amount of $Br^-$ in 31.75 cu. ft. by multiplying the number of cubic feet by the weight of one cubic foot.

weight of $Br^-$ in 1 ton sea water =

$$= 31.75 \text{ cu.ft.} \times 4.032 \times 10^{-3} \text{ lb/cu.ft.}$$

= .128 lb (0.058 kg)

One finds the number of moles present by dividing .128 lb by the molecular weight in pounds. (MW of $Br^-$ = 80 gr/mole.)

There are 454 g in 1 lb, therefore grams are converted to pounds by multiplying the number of grams by 1 lb/454 g.

$$\text{MW of } Br^- \text{ in lb} = 80 \text{ g/mole} \times \frac{1 \text{ lb}}{454 \text{ g}} = .1762 \text{ lb/mole}$$

One can now find the number of moles of $Br^-$ present in one ton.

$$\text{no. of moles} = \frac{\text{weight of Br}^-}{\text{MW in lbs}}$$

$$\text{no. of moles} = \frac{.128 \text{ lbs}}{.1762 \text{ lbs/mole}} = .73 \text{ moles.}$$

From the equation one knows that ½ as many moles of $Cl_2$ are needed as $Br^-$. Therefore, the amount of $Cl_2$ used is equal to ½ of .73 moles or .365 moles. The volume of one mole of gas at STP (Standard Temperature and Pressure, 0°C and 1 atm) is defined to be 22.4 liters. Therefore, one can find the volume of $Cl_2$ gas required for the reaction by multiplying the number of moles of gas present by 22.4 liters.

volume of $Cl_2$ = .365 × 22.4 liters = 8.18 liters.

● **PROBLEM** 161

What mass of calcium (Ca, atomic mass = 40.08 g/mole) must be combined with 1.00 g of phosphorus (P, atomic mass = 30.97 g/mole) to form the compound $Ca_3P_2$?

<u>Solution</u>:  From the formula of the compound we are trying to form, $Ca_3P_2$, we see that the ratio of moles of calcium to moles of phosphorus must be 3/2. By calculating the number of moles of P in 1.00 g, we can determine the required number of moles of Ca and then convert this to a mass.

The number of moles of P in 1.00 g is

$$\text{moles P} = \frac{\text{mass P}}{\text{atomic mass P}} = \frac{1.00 \text{ g}}{30.97 \text{ g/mole}} = 0.0322 \text{ mole.}$$

Then, using the ratio $\frac{\text{moles Ca}}{\text{moles P}} = \frac{3}{2}$ or moles Ca = $\frac{3}{2}$ moles P, the number of moles of Ca required to combine with 0.0322 mole P in a 3/2 ratio is

moles Ca = $\frac{3}{2}$ moles P = $\frac{3}{2}$ × 0.0332 mole = 0.0483 mole.

To convert this to a mass we multiply by the atomic mass, obtaining

mass Ca = moles Ca × atomic mass Ca

= 0.0483 mole × 40.08 g/mole = 1.94 g Ca.

● **PROBLEM** 162

Aluminum and oxygen react to form $Al_2O_3$. This oxide has a density = 3.97 g/ml and by chemical analysis is 47.1 weight-percent oxygen. The atomic mass of oxygen is 15.9999, what is the atomic mass of aluminum?

Solution: You can answer this question by setting up a proportion between the relative number of atoms in the oxide and the percentages by weight of the atoms in the compound. From $Al_2O_3$, you see there must be 2 atoms of Al for every 3 atoms of O. You are told that the atomic mass of oxygen is 15.9999. There exist only 2 elements in $Al_2O_3$. Thus, the weight-percent of the aluminum is 100 - 47.1 = 52.9. You have,

$$\frac{2 \text{ Al}}{3 \text{ O}} = \frac{\text{weight-percent Al}}{\text{weight-percent O}} = \frac{2 \text{ Al}}{3(15.9999)} = \frac{52.9}{47.1}$$

Solving for Al, which is the atomic mass, you obtain

$$Al = \frac{3}{2} (15.9999) \left(\frac{52.9}{47.1}\right) = 26.9 \text{ g/mole.}$$

● **PROBLEM** 163

On being heated in air, a mixture of FeO and $Fe_3O_4$ picks up oxygen so as to be converted completely to $Fe_2O_3$. If the observed weight gain is 5.00 percent of the initial weight, what must have been the composition of the initial mixture?

Solution: One should first determine the weight percent increase when the mixture is 100 % FeO or 100 % $Fe_3O_4$. 5.0 % increase will be some mixture in between. 5 % will be equal to the sum of the products of the percent weight gained by each compound and the fractions of the mixture that each compound contributes.

The reaction of FeO and O to form $Fe_2O_3$ is

$$2FeO + \frac{1}{2} O_2 \rightarrow Fe_2O_3.$$

The percent weight increase for the reaction is found by dividing the weight of $Fe_2O_3$ by the weight of 2FeO. Weight of 2FeO is 144, weight of $Fe_2O_3$ is 160.

$$\frac{\text{weight of } Fe_2O_3}{\text{weight of 2FeO}} = \frac{160}{144} = 1.1111$$

The percent weight increase is equal to (1.1111 - 1.0) X 100 = 11.11 %. This means that for each $Fe_2O_3$ formed by FeO there is an 11.11 % weight increase. One can solve for the weight increase for $Fe_3O_4$ by a similar method. The reaction is

$$2Fe_3O_4 + \frac{1}{2} O_2 \rightarrow 3Fe_2O_3$$

The weight of $2Fe_3O_4$ is 464, the weight of $3Fe_2O_3$ is 480.

$$\frac{\text{weight of } 3Fe_2O_3}{\text{weight of } 2Fe_3O_4} = \frac{480}{464} = 1.0345$$

percent weight increase $= (1.0345 - 1.0) \times 100 = 3.45$ %. Thus when $Fe_2O_3$ is formed from $Fe_3O_4$ there is a 3.45 % weight increase. The final mixture must have a 5 % weight increase. Let x = fraction of mixture that is FeO.

Mixture weight gain = (fraction FeO) × (wt gain percent

for FeO) + (fraction $Fe_3O_4$) ×

(wt. gain percent for $Fe_3O_4$)

$$5 = (x)(11.11) + (1 - x)(3.45)$$

$$5 = 11.11x + 3.45 - 3.45 x$$

$$1.55 = 7.66 x$$

$$\frac{1.55}{7.66} = x$$

$$.2024 = x$$

$$.7976 = 1 - x$$

The initial mixture is, therefore, 20.24 % FeO and 79.76 % $Fe_3O_4$.

● **PROBLEM** 164

Ethanol ($C_2H_5OH$, molecular weight = 46 g/mole) unlike most ingested substances, is absorbed directly by the stomach lining. If 44 g of pure ethanol (4 oz of whiskey or 5.5 oz of a martini) is consumed, the resulting concentration of ethanol in the blood is 0.080 g ethanol/100 ml blood. What percent of the ingested ethanol is in the blood? Assume that the total blood volume of an adult is 7.0 liters.

Solution:  We must first calculate the total mass of alcohol in the blood and then determine to what percent of the ingested mass of alcohol this corresponds.

The total mass of ethanol in the blood is equal to the product of the concentration of ethanol and the total blood volume, or 0.080 g ethanol/100 ml blood × 7.0 liters blood = 0.080 g ethanol/100 ml blood × 7000 ml blood = 5.6 g ethanol. Note: 7 liters = 7000 ml, since 1 ℓ = 1000 ml. This corresponds to

$$\frac{5.6 \text{ g}}{44 \text{ g}} \times 100 \text{ % } = 13 \text{ %} \text{ of the ingested ethanol.}$$

It is known that, when exposed to air, beryllium does not corrode but barium does. One explanation is that beryllium (Be) forms a tightly protective oxide coat whereas barium (Ba) does not. The density of BeO = 3.01 g/cc and BaO = 5.72 g/cc, find what happens to the volume per atom when the metals become oxides. The density of Be = 1.86 g/cc and of Ba = 3.598 g/cc.

Solution: To solve this problem, first calculate the volume per atom of Be and Ba, and then compare it with the volume per atom of their oxides. From this comparison, expansion or shrinkage can be determined. The volume per atom of any element can be found by knowing the atomic weight, the number of atoms per mole and the density of the element.

Thus, there are Avogadro's number or $6.02 \times 10^{23}$ atoms per mole. You are given the densities of the substances involved in this problem. Thus, by substitution you find volume of Be

$$= \frac{MW}{(6.02 \times 10^{23})(\text{density})} = \frac{9.01 \text{ g/mole}}{6.02 \times 10^{23} \text{ a/mole} \times 1.86 \text{ g/cc}}$$

$$= 8.05 \times 10^{-24} \text{ cc/atom}$$

From similar computations, you find that

BeO = $1.38 \times 10^{-23}$ cc/atom, Ba = $6.34 \times 10^{-23}$ cc/atom, and BaO = $4.45 \times 10^{-23}$.

One can find the comparative size of BeO to Be by subtracting the volume per atom of Be from that of BeO. This difference is then divided by the volume of Be and multiplied by 100. This product gives the percentage increase in size of the initial and final atoms when Be is oxidized. A similar process is used for Ba.

For Be:

$$\frac{1.38 \times 10^{-23} \text{ cc} - 8.05 \times 10^{-24} \text{ cc}}{8.05 \times 10^{-24}} \times 100 = 71 \text{ \%}$$

For Ba:

$$\frac{4.45 \times 10^{-23} \text{ cc} - 6.34 \times 10^{-23}}{6.34 \times 10^{-23}} \times 100 = -30 \text{ \%}.$$

From a comparison of the atomic volumes of Be and BeO, you see that Be expands by 71 % when it forms the oxide. From a comparison of Ba and BaO, you see that Ba shrinks by 30 %, when it forms the oxide.

# CHAPTER 5

# STOICHIOMETRY/WEIGHT AND VOLUME CALCULATIONS

> **Basic Attacks and Strategies for Solving Problems in this Chapter. See pages 145 to 189 for step-by-step solutions to problems.**

This chapter addresses problems of balancing chemical equations and calculating amounts, such as mass, volume, moles, etc., of species which enter chemical reactions. These kinds of calculations are called stoichiometric calculations. The key idea, as in Chapter 4, is to understand that chemical reactions are always written representing the number of molecules (or atoms for the elements) or moles of species which undergo reaction. Atomic or molecular ratios equal mole ratios. Conversion from moles to mass or, for gases, from moles to volume (see Chapters 3 and 4) must be understood to solve these problems.

Chemical reactions are represented by equations showing reactants and products for a particular reaction. The key to balancing chemical reaction equations lies in understanding that elements are conserved in chemical reactions (the exception is nuclear reactions or reactions involving the atomic nucleus. These very special kinds of reactions are discussed in Chapter 19). The same number of moles (or atoms) of each element must appear in both reactants and products. For simple reactions, which include all those in this text, the molecular ratios of reactants are small integers. These molecular ratios are the coefficients in chemical equations and are called the stoichiometric coefficients; they are usually small integers and are usually reduced to the smallest possible integers by dividing each coefficient by common factors.

For example, ammonia ($NH_3$) can be combined with oxygen ($O_2$) to form nitric, oxide (NO) and water ($H_2O$). Write the balanced equation for this reaction (see Problem 173).

$$a\ NH_3 + b\ O_2 \rightarrow c\ NO + d\ H_2O \qquad\qquad 5\text{-}1$$

Conserving elements leads to the following requirements.

For N    $a = c$
For H    $3a = 2d$
For O    $2b = c + d$

One additional relationship is needed to solve these equations for the four stoichiometric coefficients, $a, b, c,$ and $d$. This relationship is arbitrary and is usually selected so that $a, b, c,$ and $d$ will be small integers. Simply pick a value of one coefficient — for example let $a = 1$. It follows that $c = a = 1$, $d = 3/2a = 3/2$, and $b = (c + d)/2 = 5/4$. If $a, b, c,$ and $d$ are multiplied by 4 (this is equivalent to initially selecting $a = 4$), the coefficients become

$a = 4$
$b = 5$
$c = 4$
$d = 6$

or    $4NH_3\ 5O_2 \rightarrow 4NO + 6H_2O \qquad\qquad 5\text{-}2$

This is the same equation that appears in Problem 173.

In order to solve wt./wt. or vol./vol. or vol./wt. problems, it is critical to first change weights and/or volumes to moles.

The other key solution to problems in this chapter is the method of limiting reactants. Simply stated, when one of the reactants is used up, the reaction stops. Again, examples best illustrate the technique. Suppose a limited amount of ammonia (for example, 100 g) is burned in the atmosphere (21% $O^2$) which is, compared to 100 g, unlimited in supply. When 100 g of $NH_3$ is consumed, the reaction stops even though more oxygen is available.

Compute the weight of NO and $H_2O$ produced when 100 g of ammonia is burned in the atmosphere. Note that the molecular weight of ammonia is 17 $(14 + 3 \times 1)$. Therefore,

$100/17 = 5.88$ moles of $NH_3$ are consumed.

From Equation 5-2 on the previous page, 4/4(5.88) = 5.88 moles of NO are produced and 6/4(5.88) = 8.82 moles of $H_2O$ are produced.

5.88 moles NO = 176.4 g (the molecular weight of NO = 30)
8.82 moles $H_2O$ = 158.8 g (the molecular weight of $H_2O$ = 18)

Now compute the volume (at standard conditions) of $NH_3$, air, NO, and $H_2O$ in this reaction. From Chapter 3, one mole of any gas at standard temperature and pressure occupies 22.4 liters.

For $NH_3$ and NO:    5.88 moles (22.4 liters/mole) = 131.7 liters
For air (21% $O_2$):    5/4(5.88)/.21 moles (22.4 moles/liter) = 784 liters
For $H_2O$:    8.82 moles (22.4 liters/mole) = 197.6 liters

## BALANCING EQUATIONS

● PROBLEM 166

---
Balance the equations: (a) $Ag_2O \rightarrow Ag + O_2$;

(b) $Zn + HCl \rightarrow ZnCl_2 + H_2$; (c) $NaOH + H_2SO_4 \rightarrow Na_2SO_4 + H_2O$.

---

Solution: When balancing chemical equations, one must make sure that there are the same number of atoms of each element on both the left and right side of the arrow. For example, $H_2 + O_2 \rightarrow H_2O$ is not a balanced equation because there are 2 O's on the left side and only one on the right. $2H_2 + O_2 \rightarrow 2H_2O$ is the balanced equation for water because there are the same number of H and O atoms on each side of the equation.

(a) $Ag_2O \rightarrow Ag + O_2$ is not a balanced equation because there are 2 Ag on the left and only one on the right, and because there is only one O on the left and two O on the right. To balance this equation one must first multiply the left side by 2 to have 2 O's on each side.

$$2Ag_2O \rightarrow Ag + O_2$$

There are now 4 Ag on the left and only one on the right, thus the Ag, on the right must be multiplied by 4.

$$2Ag_2O \rightarrow 4Ag + O_2$$

The equation is now balanced.

(b) $Zn + HCl \rightarrow ZnCl_2 + H_2$

In this equation, there are 2 H and 2 Cl on the right and only one of each on the left, therefore, the equation can be balanced by multiplying the HCl on the left by 2.

$$Zn + 2HCl \rightarrow ZnCl_2 + H_2$$

Because there are the same number of Zn, Cl, and H on both sides of the equation, it is balanced.

(c) NaOH + H$_2$SO$_4$ → Na$_2$SO$_4$ + H$_2$O

Here, there are 1 Na, five O, 3 H and 1 S on the left and 2 Na, 1 S, five O, and 2 H on the right. To balance this equation, one can first adjust the Na by multiplying the NaOH by 2.

2NaOH + H$_2$SO$_4$ → Na$_2$SO$_4$ + H$_2$O

There are now 2 Na, six O, 4 H, and 1 S on the left and 2 Na, five O, 2 H, and 1 S on the right. Because there are two more H and one more O on the left than on the right, you can balance this equation by multiplying the H$_2$O by 2.

2NaOH + H$_2$SO$_4$ → Na$_2$SO$_4$ + 2H$_2$O

The equation is now balanced.

● **PROBLEM** 167

Balance the following by filling in missing species and proper coefficient: (a) NaOH + _____ → NaHSO$_4$ + HOH; (b) PCl$_3$ + __HOH → _____ + 3HCl; (c) CH$_4$ + ____ → CCl$_4$ + 4HCl.

Solution: To balance chemical equations you must remember that ALL atoms (and charges) must be accounted for. The use of coefficients in front of compounds is a means to this end. Thus,

(a)  NaOH + _____ → NaHSO$_4$ + HOH

On the right side of the equation, you have 1 Na, 3 H's, 5 O's, and 1 S. This same number of elements must appear on the left side. However, on the left side, there exists only 1 Na, 1 O, and 1 H. You are missing 2 H's, 1 S, and 4 O's. The missing species is H$_2$SO$_4$, sulfuric acid. You could have anticipated this since a strong base (NaOH) reacting with a strong acid yields a salt (NaHSO$_4$) and water. The point is, however, that H$_2$SO$_4$ balances the equation by supplying all the missing atoms.

(b)  PCl$_3$ + 3HOH → _____ + 3HCl.

Here, the left side has 1 P, 3 Cl's, 6 H's, and 3 O's. The right has 3 H's and 3 Cl's. You are missing 1 P, 3 O's and 3 hydrogens. Therefore, P(OH)$_3$ is formed.

(c)  CH$_4$ + _____ → CCl$_4$ + 4HCl

Here, there are 1 C, 8 Cl's, and 4 H's on the right and 1 C and 4 H's on the left. The missing compound, therefore, contains 8 Cl's and thus it is 4 Cl$_2$. One knows that it is

4 $Cl_2$ rather than $Cl_8$ or $8Cl$ because elemental chlorine gas is a diatomic or 2 atom molecule.

# CALCULATIONS USING CHEMICAL ARITHMETIC

● PROBLEM 168

Verify that the following data confirm the law of equivalent proportions: Nitrogen and oxygen react with hydrogen to form ammonia and water, respectively. 4.66 g of nitrogen is required for every gram of hydrogen in ammonia, and 8 g of oxygen for every gram of hydrogen in water. Nitrogen plus oxygen yields NO. Here, 14 g of nitrogen is required for every 16 g of oxygen.

Solution:  To verify, you must show that when two elements (nitrogen and oxygen) combine with a third element (hydrogen), they will do so in a simple multiple of the proportions in which they combine with each other. Thus, the nitrogen to oxygen ratio in NO must be a ratio of small integers with the nitrogen and oxygen ratio in $H_3N$ (ammonia) and $H_2O$ (water).

For ammonia and water, $\frac{N}{O} = \frac{4.66 \text{ g}}{8.00 \text{ g}} = .582$

For NO, $\frac{N}{O} = \frac{14 \text{ g}}{16 \text{ g}} = .875.$

If .582 and .875 are a ratio of small integers to each other, you verify the law of equivalent proportions. Therefore,

$$\frac{.582}{.875} = .665 \approx \frac{2}{3} ,$$

concluding that they are a ratio.

● PROBLEM 169

In a chemical reaction requiring two atoms of phosphorus for five atoms of oxygen, how many grams of oxygen are required by 3.10 g of phosphorus?

Solution:  Because the relationship between the phosphorus and oxygen is given in atoms, the relationship also holds for moles. There must be 2 moles of phosphorus for every 5 moles of oxygen. This is true because there is a set number of atoms in any one mole. Therefore, one must first find the number of moles of phosphorus present. From this, one can find the number of moles of oxygen present. From the number of moles of oxygen, one can find the weight by multiplying by the molecular weight.

The number of moles of phosphorus present is found by dividing the number of grams by the molecular weight (MW = 31).

$$\text{no. of moles} = \frac{3.10 \text{ g}}{31 \text{ g/mole}} = .10 \text{ moles}$$

Because there must be five moles of oxygen for every two moles of phosphorus, the following ratio can be used to determine the number of moles of oxygen. Let x = number of moles of O.

$$\frac{2 \text{ moles P}}{5 \text{ moles O}} = \frac{.10 \text{ moles P}}{x}$$

$$x = \frac{5 \text{ moles O} \times .10 \text{ moles P}}{2 \text{ moles P}} = .25 \text{ moles O}$$

The weight of the oxygen is then found by multiplying the number of moles by the molecular weight (MW = 16).

no. of grams of O = 16 g/mole × .25 mole = 4.00 g O.

● **PROBLEM** 170

Two atoms of scandium are to combine with three atoms of oxygen. If you start with 1 gram of scandium, how much oxygen is required? Scandium has an atomic weight of 44.96 g/mole. The at. wt. of oxygen is 15.999 g/mole.

Solution: The key to solving this problem is determining the number of moles that will be reacting. After this, one is able to calculate the mass by multiplying the atomic weight by the number of moles.

Two atoms of scandium react with 3 atoms of oxygen, which is equivalent to saying 2 moles of scandium react with 3 moles of oxygen. A mole is defined as weight in grams of the substance divided by the atomic weight or molecular weight.

If you start with 1 gram of scandium, then the number of moles of scandium is

$$\frac{1.00 \text{ g}}{44.96 \text{ g/mole}} = .0222 \text{ moles.}$$

Since the number of moles of oxygen must be in a 3 : 2 ratio to scandium, multiply 3/2 by .0222 to determine the number of moles of oxygen that will react.

$.0222 \left(\frac{3}{2}\right) = .0333$ moles of oxygen atoms. Recalling the definition of a mole, the mass of the oxygen that reacts is

.0333 (15.999) = .533 g of oxygen.

A metal has an atomic weight of 24. When it reacts with
a non-metal of atomic weight 80, it does so in a ratio
of 1 atom to 2 atoms, respectively. With this information,
how many grams of non-metal will combine with 33.3 g of
metal. If 1 g of metal is reacted with 5 g of non-metal,
find the amount of product produced.

<u>Solution</u>:  To answer this problem, write out the reaction
between the metal and non-metal, so that the relative
number of moles that react can be determined. You can
calculate the number of grams of material that react or
are produced. You are told that 1 atom of metal reacts
with 2 atoms of non-metal. Let X = non-metal and M =
metal. The compound is $MX_2$. The reaction is $M + 2X \rightarrow MX_2$.
Determine the number of moles that react. You have 33 g
of M with an atomic weight of 24.

Therefore, the number of moles $= \frac{33 \text{ g}}{24 \text{ g/mole}} = 1.375$ moles.

The above reaction states that for every 1 mole of M,
2 moles of X must be present. This means, therefore, that
$2 \times 1.375 = 2.75$ moles of X must be present. The non-metal
has an atomic weight of 80. Thus, recalling the definition

a mole, $2.75$ mole $= \frac{\text{grams}}{80 \text{ g/mole}}$ . Solving for grams of X,
you obtain $2.75 (80) = 220$ grams.

Let us consider the reaction with 1 g of M with
5 g of X to produce an unknown amount of $MX_2$. The solution
is similar to the other, except that here you consider the
concept of a limiting reagent. The amount of $MX_2$ produced
from a combination will depend on the substance that exists
in the smallest quantity. Thus, to solve this problem you
compute the number of moles of M and X present. The smaller
number, (based on reaction equation) is the one you employ
in calculating the number of moles of $MX_2$ that will be
generated. You have, therefore,

$$M_{\text{moles}} = \frac{1}{24} = .04166 \text{ moles M}$$

$$X_{\text{moles}} = \frac{5}{80} = .0625 \text{ moles X.}$$

Using 0.0625 moles X, only .03125 moles of $MX_2$ will be
produced, since the equation informs you that 1 mole of
$MX_2$ is produced for every two moles of X. For M, it is a
1 : 1 ratio, so that .04166 moles of $MX_2$ would be generated.
Therefore, X is the limiting reagent. The atomic weight of
$MX_2$ is 184. Thus, the amount produced is

184 g/mole (.03125 moles) = 5.75 g.

# WEIGHT-WEIGHT PROBLEMS

Upon the addition of heat to an unknown amount of $KClO_3$, .96 g of oxygen was liberated. How much $KClO_3$ was present?

Solution:   The key to answering this question is to write a balanced equation that illustrates this chemical reaction. From this, you can employ the mole concept to determine the weights of the substances involved.

Given that oxygen is liberated and that oxygen gas exists as $O_2$ and not $O$, you can write the balanced equation

$$2KClO_3 \rightarrow 2KCl + 3O_2.$$

All atoms are accounted for. The coefficients indicate the relative number of moles that react. For example, every two moles of $KClO_3$ yield 3 moles of oxygen. A mole is defined as weight in grams of a substance divided by its atomic or molecular weight. Therefore, you have

$\frac{.96 \text{ g}}{32 \text{ g/mole}}$ = .030 moles of oxygen. To calculate the weight of $KClO_3$, you look at the balanced equation.

You find that the number of moles of $KClO_3$ is 2/3 the number of moles of oxygen. As such, the number of

moles of $KClO_3$ = .030 $\left( \frac{2 \text{ moles } KClO_3}{3 \text{ moles } O_2} \right)$ = .020 moles of $KClO_3$. Recalling the definition of a mole, the number of grams of $KClO_3$ (MW of $KClO_3$ = 122.55) is

.020 × (122.55 g/mole) = 2.5 g.

Given the balanced equation

$$4NH_3(g) + 5O_2(g) \rightarrow 4NO(g) + 6H_2O(g),$$

how many grams of $NH_3$ will be required to react with 80 g of $O_2$?

Solution:   In this problem, one is asked to find the number of grams of $NH_3$ required to react with 80 g of $O_2$. From the equation, one can see that 4 moles of $NH_3$ react with 5 moles of $O_2$. This means that one must first determine the number of moles of $O_2$ in 80 g. One can then use the following ratio to determine the number of moles of $NH_3$ that will react with this number. This ratio holds because the ratio of $NH_3$ to $O_2$ is 4 : 5.

$$\frac{4}{5} = \frac{\text{number of moles of } NH_3}{\text{number of moles of } O_2}$$

After one knows the number of moles of $NH_3$ required, one can determine the weight of the $NH_3$ by multiplying the number of moles by the molecular weight.

To solve this problem:

(1) calculate the number of $O_2$ in 80 g

(2) calculate the number of moles of $NH_3$ that react with 80 g $O_2$

(3) find the number of grams of $NH_3$.

Solving:

(1) The number of moles of $O_2$ in 80 g is found by dividing 80 g by the molecular weight of $O_2$. The molecular weight of O is 16, which means the molecular weight of $O_2$ is twice that amount, or 32.

$$\text{number of moles} = \frac{\text{number of grams present}}{\text{molecular weight}}$$

$$\text{number of moles} = \frac{80 \text{ g}}{32 \text{ g/mole}} = 2.5 \text{ moles}$$

(2) The number of moles of $NH_3$ reacting with a certain number of moles of $O_2$ is in the ratio 4 : 5, as shown previously.

$$\frac{4}{5} = \frac{\text{number of moles of } NH_3}{\text{number of moles of } O_2}$$

$$\frac{4}{5} = \frac{\text{number of moles of } NH_3}{2.5 \text{ moles}}$$

$$\text{number of moles of } NH_3 = \frac{4 \times 2.5 \text{ moles}}{5} = 2 \text{ moles}$$

2 moles of $NH_3$ are necessary to react with 2.5 moles or 80 g of $O_2$.

(3) The number of grams of $NH_3$ in 2.0 moles is found by multiplying 2.0 moles by the molecular weight of $NH_3$ MW of $NH_3$ = 17.

$$\text{number of grams} = \text{number of moles} \times \text{molecular weight}$$

$$\text{number of grams} = 2.0 \text{ moles} \times 17 \text{ g/mole} = 34 \text{ g}.$$

● **PROBLEM** 174

In the commercial preparation of hydrogen chloride gas, what weight of HCl in grams may be obtained by heating 234 g. of NaCl with excess $H_2SO_4$?

The balanced equation for the reaction is

$$2NaCl + H_2SO_4 \rightarrow Na_2SO_4 + 2HCl$$

Molecular weights: NaCl = 58.5, HCl = 36.5.

**Solution**: This problem can be solved by using either the mole method or the proportion method.

**Mole method**: According to the equation, 2 moles of NaCl produce 2 moles of HCl, thus 1 mole of HCl is obtained for every mole of NaCl used.

To use this method, one must first find the number of moles of NaCl reacted because the number of moles of NaCl reacted equals the number of moles of HCl formed.

The number of moles of NaCl reacted can be found by dividing the number of grams used by the molecular weight.

$$\text{number of moles} = \frac{\text{number of grams present}}{\text{molecular weight of NaCl}}$$

$$\text{number of moles} = \frac{234 \text{ g}}{58.5 \text{ g/mole}} = 4 \text{ moles NaCl}$$

Since 4 moles of NaCl are reacted, one knows that 4 moles of HCl will be formed. Thus, the weight of the HCl formed is equal to 4 times the molecular weight of HCl.

$$\text{weight of HCl formed} = \text{number of moles} \times \text{molecular weight}$$

$$\text{weight of HCl formed} = 4 \text{ moles} \times 36.5 \text{ g/mol} = 146 \text{ g HCl}$$

**Proportion method**: An alternate method of solution to this problem is also possible. In this method, the molecular weights (multiplied by the proper coefficients) are placed below the formula in the equation and the amounts of substances (given and unknown) are placed above.

```
    234 g                                    X

    2 NaCl    +   H₂SO₄  →  Na₂SO₄  +  2 HCl
    2×58.5 g                            2×36.5 gr
```

X is the weight of HCl produced. Solving for X

$$\frac{234 \text{ g}}{117 \text{ gr}} = \frac{X}{73 \text{ g}}$$

$$X = \frac{(234 \text{ g}) \times (73 \text{ g})}{117 \text{ g}} = 146 \text{ g.}$$

A chemist decides to react 2 g of VO (vanadium oxide) with 5.75 g of $Fe_2O_3$ to produce $V_2O_5$ and FeO. How many grams of $V_2O_5$ can be obtained?

Solution: It is extremely important to write out a balanced equation that illustrates this chemical reaction. The balanced equation for this reaction is:

$$2VO + 3Fe_2O_3 \rightarrow 6FeO + V_2O_5$$

The coefficients tell us the relative number of moles of each reactant and the relative number of moles of each product. For example, 2 moles of VO will react for every three moles of $Fe_2O_3$. To find the amount of $V_2O_5$, therefore, you must first determine how many moles of $V_2O_5$ will be generated. To do this you must know the **number of moles of the reactants.**

$$\text{moles of VO} = \frac{\text{weight in grams}}{\text{molecular weight}} = \frac{2.00 \text{ g}}{66.94 \text{ g/mole}}$$

$$= .0299 \text{ moles}$$

$$\text{moles of } Fe_2O_3 = \frac{5.75 \text{ g}}{159.69 \text{ g/mole}} = .0360 \text{ moles}$$

According to the equation, the number of moles of VO must be 2/3 the number of moles of $Fe_2O_3$. However, 2/3 of .0360 = .0240, but you only have .0299 moles. This means that VO is in excess. In other words, for the amount of $Fe_2O_3$ present, there exists more VO than will react. As such, the products produced will depend on the number of moles of $Fe_2O_3$, and not on VO. $Fe_2O_3$ is called the limiting reagent. In general, then, if you have two quantities that are known, and want to compute a third, you determine the limiting reagent. The unknown quantity will be dependent upon it only.

As stated earlier, the number of moles of $Fe_2O_3$ was .0360. According to the reaction, for every 3 moles of $Fe_2O_3$, one mole of $V_2O_5$ is generated. Therefore, the number of moles of $V_2O_5$ = 1/3 (.0360) = .0120.

Recalling the definition of the mole, you have for $V_2O_5$ : .0120 moles = no. of grams/molecular weight. The molecular weight of $V_2O_5$ = 181.9.

Therefore the number of grams of $V_2O_5$ obtained is

.0120 moles (181.9 g/mole) = 2.18 g.

Determine the weights of $CO_2$ and $H_2O$ produced on burning 104 g. of $C_2H_2$. Molecular weights are $CO_2 = 44$ and $H_2O = 18$. The equation for the reaction is

$$2C_2H_2 + 5O_2 \rightarrow 4CO_2 + 2H_2O$$

Solution:  Two methods can be used to solve this problem. One is called the mole method and the other the proportion method.

Mole method: According to the equation, 2 moles of $C_2H_2$ react with 5 moles of $O_2$ to produce 4 moles of $CO_2$ and 2 moles of $H_2O$. Thus, in this problem, the first thing one has to determine is how many moles of $C_2H_2$ are contained in 104 g.

The molecular weight of $C_2H_2$ is 26, by dividing the amount of $C_2H_2$ present by the molecular weight of the compound, the number of moles present is found.

$$\text{number of moles of } C_2H_2 = \frac{\text{number of grams of } C_2H_2}{\text{molecular weight of } C_2H_2}$$

$$\text{number of moles of } C_2H_2 \text{ present} = \frac{104 \text{ g}}{26 \text{ g/mole}} = 4 \text{ moles.}$$

In the equation 2 moles of $C_2H_2$ are burned to form 4 moles of $CO_2$ and 2 moles of $H_2O$. Here there are 4 moles of $C_2H_2$ present, which is twice the amount in the empirical equation. Therefore, twice as much $CO_2$ and $H_2O$ will be formed. 8 moles of $CO_2$ and 4 moles of $H_2O$ will be formed.

$$2 \times (2C_2H_2 + 5O_2 \rightarrow 4CO_2 + 2H_2O)$$

$$= 4C_2H_2 + 10O_2 \rightarrow 8CO_2 + 4H_2O$$

To find the weight of 8 moles of $CO_2$, the molecular weight of $CO_2$ is multiplied by 8.

$$\text{weight of } CO_2 = \text{number of moles present} \times \text{molecular weight of } CO_2.$$

$$\text{weight of } CO_2 = 8 \text{ moles} \times 44 \text{ g/mole} = 352 \text{ g.}$$

The same procedure can be followed for $H_2O$.

$$\text{weight of } H_2O = \text{number of moles present} \times \text{molecular weight of } H_2O$$

$$\text{weight of } H_2O = 4 \text{ moles} \times 18 \text{ g/mole} = 72 \text{ g.}$$

Proportion method:

An alternate method of solution to this problem is the proportion method in which molecular weights (multiplied by the proper coefficients) are placed below the formula in the equation and the amounts of substances (given and unknown) are placed above. Here one has for $CO_2$:

$$\begin{array}{c} 104 \text{ g} \qquad\qquad\qquad X \\ 2\ C_2H_2\ +\ \ 5O_2\ \rightarrow\ 4\ CO_2\ +\ 2\ H_2O \\ 2\times26 \text{ g} \qquad\qquad\quad 4\times44 \text{ g} \end{array}$$

This becomes $\dfrac{104 \text{ g}}{52 \text{ g}} = \dfrac{X}{176 \text{ g}}$

Solving for X one has

$$X = \frac{(104 \text{ g})(176 \text{ g})}{(52 \text{ g})} = 352 \text{ gr } CO_2$$

A similar method can be applied to the $H_2O$

$$\begin{array}{c} 104 \text{ g} \qquad\qquad\qquad X \\ 2\ C_2H_2\ +\ 5O_2\ \rightarrow\ 4CO_2\ +\ 2\ H_2O \\ 2\times26 \text{ g} \qquad\qquad\qquad 2\times18 \text{ g} \end{array}$$

$$\frac{104 \text{ g}}{52 \text{ g}} = \frac{X}{36 \text{ g}}$$

$$X = \frac{(104 \text{ g})(36 \text{ g})}{(52 \text{ g})} = 72 \text{ g } H_2O.$$

● **PROBLEM** 177

Silver bromide, AgBr, used in photography, may be prepared from $AgNO_3$ and NaBr. Calculate the weight of each required for producing 93.3 lb of AgBr. (1 lb = 454 g.)

Solution: The reaction for the production of AgBr is written

$$AgNO_3 + NaBr \rightarrow AgBr + NaNO_3$$

This means that one mole of each $AgNO_3$ and NaBr are needed to form one mole of AgBr. In this problem, to determine the number of pounds of $AgNO_3$ and NaBr used to form 93.9 lbs of AgBr, one must first determine the number of moles of AgBr in 93.9 lbs. There will be one mole of each NaBr and $AgNO_3$ for each mole of AgBr formed. Once the number of required moles of NaBr and $AgNO_3$ are found, the weights of these compounds can be determined by multiplying the number of moles by the molecular weight. To solve this problem:

(1) solve for the number of moles of AgBr in 93.9 lbs

(2) determine the weights of NaBr and $AgNO_3$ used.

Solving:

(1) Molecular weights are given in grams, therefore, the grams should be converted to pounds for use in this problem. MW of AgBr = 188 g/mole.

There are 454 grams in one pound, thus grams can be converted to pounds by using the conversion factor 1 lb/454 grams.

MW of AgBr in lbs = 188 × 1 lb/454 g = .41 lbs.

The number of moles in 93.9 lbs can be found by dividing the 93.9 lbs by the molecular weight in pounds.

$$\text{no. of moles of AgBr} = \frac{93.9 \text{ lbs}}{.41 \text{ lbs/mole}} = 229 \text{ moles}$$

Therefore, 229 moles of each NaBr and $AgNO_3$ are needed to produce 93.9 lbs of AgBr.

(2) The weight of NaBr and $AgNO_3$ used is equal to the number of moles times the molecular weight. In this problem, one wishes to find the weight in pounds, not grams, thus the molecular weights must be converted to pounds before the conversion factor 1 lb/454 g can be used. MW of NaBr = 103, MW of $AgNO_3$ = 170.

MW of NaBr in lbs = 103 g/mole × 1 lb/454 g = .23 lbs/mole

MW of $AgNO_3$ in lbs = 170 g/mole × 1 lb/454 g

$$= .37 \text{ lbs/mole.}$$

Since it has already been calculated that 229 moles of each of these compounds are needed, one can calculate the weight needed of each by multiplying the number of moles by the molecular weight.

weight of NaBr = .23 lbs/mole × 229 moles = 52.7 lbs.

weight of $AgNO_3$ = .37 lbs/mole × 229 moles = 84.7 lbs.

● **PROBLEM** 178

What weight of sulfur must combine with aluminum to form 600 lbs (272.4 kg) of aluminum sulfide?

Solution: In this problem one wants to find out how much reactant (sulfur) was needed to produce a given amount of product (aluminum sulfide). The balanced equation for this reaction is:

$$2Al + 3S \rightarrow AL_2S_3$$

The first method for solving this problem is the pro-portion method. This procedure involves the ratios of weights and molecular weights. Set up the balanced e-quation showing the weights and molecular weights.

$$
\begin{array}{ccccc}
X \text{ lbs} & & & 600 \text{ lbs} \\
2 \text{ Al} & + & 3S & \rightarrow & AL_2S_3 \\
& & 3(32) \text{ lb} & & 150 \text{ lb}
\end{array}
$$

M.W. of S is 32. M.W. of $Al_2S_3$ is calculated below.

| Atom | Number of Atoms | Atomic Weight | Total Atomic Weight |
|------|-----------------|---------------|---------------------|
| Al | 2 | 27 | 54 |
| S | 3 | 32 | 96 |

M.W. of $Al_2S_3$ = 150.

Use the proportion equation,

$$\frac{weight_{reactant}}{moles \times M.W._{reactant}} = \frac{weight_{product}}{moles \times M.W._{product}}$$

Solve for $weight_{reactant}$.

$$weight_{reactant} = \frac{(3)(\mathbf{32})(600)}{150}$$

$$= 384 \text{ lbs. } (174.3 \text{ kg) sulfur}$$

The second method for solving the problem is by the mole method. We see that 1 mole of $Al_2S_3$ requires 3 moles of S. Therefore, if one can calculate the actual number of moles present of 1 substance one can obtain the number of moles of the other substance by setting up a ratio. Then, knowing the number of moles, one can calculate the weight by using the equation moles = weight/M.W.

One is given that 600 lbs. of $Al_2S_3$ are present. Therefore, the number of moles is 600/150 = 4 pound moles of $Al_2S_3$. Setting up the ratio,

$$\frac{S}{Al_2S_3} = \frac{3}{1} = \frac{x}{4}$$

it is seen that x = 12 lb. moles of S.

Next, solve the mole equation for weight of S

$$weight = moles \times M.W. = 12 \times 32 = 384 \text{ lbs.}$$
$$= 384 \text{ lbs. } (174.3 \text{ kg})$$

Heating of $NaNO_3$ decomposed it to $NaNO_2$ and $O_2$. How much $NaNO_3$ would you have to decompose to produce 1.50 g of $O_2$?

<u>Solution</u>:  The equation for this reaction is:

$$2NaNO_3 \rightarrow 2NaNO_2 + O_2$$

This means that for every mole of $O_2$ produced, 2 moles of $NaNO_3$ must be decomposed. One is given that 1.50 g of $O_2$ is formed, thus one should determine the number of moles. The number of moles of $NaNO_3$ needed will be twice this amount. One can solve for the weight by multiplying the number of moles by the molecular weight of the compound.

The number of moles of $O_2$ can be found by dividing 1.50 g by the molecular weight of $O_2$ (MW = 32.0).

$$\text{no. of moles} = \frac{1.50 \text{ g}}{32.0 \text{ g/mole}} = 4.69 \times 10^{-2} \text{ moles.}$$

The number of moles of $NaNO_3$ needed can be found by multiplying the number of moles of $O_2$ by 2.

$$\text{no. of moles} = 2 \times 4.69 \times 10^{-2} \text{ moles} = 9.38 \times 10^{-2} \text{ moles.}$$

One finds the weight of $NaNO_3$ needed by multiplying the number of moles by the molecular weight (MW = 85).

$$\text{weight of } NaNO_3 = 85 \text{ g/mole} \times 9.38 \times 10^{-2} \text{ moles} = 7.97 \text{ g.}$$

Baking powder consists of a mixture of cream of tartar (potassium hydrogen tartrate, $KHC_4H_4O_6$, molecular weight = 188 g/mole) and baking soda (sodium bicarbonate, $NaHCO_3$, molecular weight = 84 g/mole). These two components react according to the equation

$$KHC_4H_4O_6 + NaHCO_3 \rightarrow KNaC_4H_4O_6 + H_2O + CO_2.$$

How much baking soda must be added to 8.0 g of cream of tartar for both materials to **react completely**?

<u>Solution</u>:    From the equation, we know that one mole of $NaHCO_3$ reacts with one mole of $KHC_4H_4O_6$. Hence, if we convert 8.0 g of $KHC_4H_4O_6$ to moles, we know how many moles of $NaHCO_3$ must be added. Finally, all we need to do is to convert moles of $NaHCO_3$ to grams of $NaHCO_3$.

In order to convert from grams to moles, we use the relationship moles = mass/molecular weight. The number of moles of $KHC_4H_4O_6$ in 8.0 g is

$$moles = \frac{mass}{molecular\ weight} = \frac{8.0\ g}{188\ g/mole} = 4.3 \times 10^{-2}\ mole.$$

Hence, we must add $4.3 \times 10^{-2}$ mole of $NaHCO_3$. Using the relationship mass = moles × molecular weight, we find that $4.3 \times 10^{-2}$ mole of $NaHCO_3$ corresponds to

$$4.3 \times 10^{-2}\ moles \times 84\ g/mole = 3.6\ g.$$

Hence, 3.6 g of baking soda must be added.

● **PROBLEM** 181

Some solid CaO in a test tube picks up water vapor from the surroundings to change completely to $Ca(OH)_2(s)$. An observed total initial weight (CaO + test tube) of 10.860 g goes eventually to 11.149 g. What is the weight of the test tube?

**Solution:** The equation for the reaction is

$$CaO + H_2O \rightarrow Ca(OH)_2$$

This means that one mole of $H_2O$ reacts with one mole of CaO. The difference in the weights of the test tubes is the weight of the $H_2O$ that the CaO absorbed.

weight of $H_2O$ = 11.149 - 10.860 = .289 g.

One should now solve for the number of moles of $H_2O$ because the number of moles of water equals the number of moles of CaO present. From this one can find the weight of the CaO and the test tube. The number of moles equals the number of grams divided by the molecular weight (MW of $H_2O$ = 18.0).

$$number\ of\ moles\ of\ H_2O = \frac{.289\ g}{18.0\ g/mole} = .0161\ moles$$

Therefore, .0161 moles of CaO were originally present in the test tube. One finds the number of grams by multiplying by the molecular weight of CaO (MW of CaO = 56.08).

$$number\ of\ grams\ of\ CaO = .0161\ moles \times 56.08\ g/mole$$

$$= .900\ g.$$

The weight of the test tube is equal to .900 g subtracted from the original weight of the test tube and material.

weight of test tube = 10.860 - .900 = 9.960 g.

"Hard" water contains small amounts of the salts calcium bicarbonate ($Ca(HCO_3)_2$) and calcium sulfate ($CaSO_4$, molecular weight = 136 g/mole). These react with soap before it has a chance to lather, which is responsible for its cleansing ability. $Ca(HCO_3)_2$ is removed by boiling to form insoluble $CaCO_3$. $CaSO_4$ is removed by reaction with washing soda ($Na_2CO_3$, molecular weight = 106 g/mole) according to the following equation:

$$CaSO_4 + Na_2CO_3 \rightarrow CaCO_3 + Na_2SO_4.$$

If the rivers surrounding New York City have a $CaSO_4$ concentration of $1.8 \times 10^{-3}$ g/liter, how much $Na_2CO_3$ is required to "soften" (remove $CaSO_4$) the water consumed by the city in one day (about $6.8 \times 10^9$ liters)?

**Solution**: We must determine the amount of $CaSO_4$ present in $6.8 \times 10^9$ liters and, from this, the amount of $Na_2CO_3$ required to remove it.

The number of moles per liter, or molarity, of $CaSO_4$ corresponding to $1.8 \times 10^{-3}$ g/liter is obtained by dividing this concentration by the molecular weight of $CaSO_4$. Multiplying by $6.8 \times 10^9$ liters gives the number of moles of $CaSO_4$ that must be removed. Hence,

$$\text{moles } CaSO_4 = \frac{\text{concentration (g/liter)}}{\text{molecular weight of } CaSO_4} \times 6.8 \times 10^9 \text{ liters}$$

$$= \frac{1.8 \times 10^{-3} \text{ g/liter}}{136 \text{ g/mole}} \times 6.8 \times 10^9 \text{ liters}$$

$$= 9.0 \times 10^4 \text{ moles.}$$

From the equation for the reaction between $CaSO_4$ and $Na_2CO_3$, we see that one mole of $CaSO_4$ reacts with one mole of $NaCO_3$. Hence, $9.0 \times 10^4$ moles of $NaCO_3$ are required to remove all the $CaSO_4$. To convert this to mass, we multiply by the molecular weight of $NaCO_3$ and obtain

$$\text{mass } Na_2CO_3 = \text{moles } Na_2CO_3 \times \text{molecular weight } Na_2CO_3$$

$$= 9.0 \times 10^4 \text{ moles} \times 106 \text{ g/mole}$$

$$= 9.5 \times 10^6 \text{ g} = 9.5 \times 10^6 \text{ g} \times 1 \text{ kg/1000 g}$$

$$= 9.5 \times 10^3 \text{ kg,}$$

which is about 10 tons.

How many pounds of air (which is 23.19% $O_2$ and 75.46% $N_2$ by weight) would be needed to burn a pound of gasoline by a reaction whereby $C_8H_{18}$ reacts with $O_2$ to form $CO_2$ and $H_2O$?

Solution:  The equation for the reaction is

$$C_8H_{18} + 12\tfrac{1}{2} O_2 \rightarrow 8 CO_2 + 9 H_2O$$

From the equation, one knows that 12.5 moles of $O_2$ are needed to burn 1 mole of gasoline. To solve for the number of pounds of air necessary to burn 1 pound of gasoline:

(1) find the number of moles in 1 pound of $C_8H_{18}$

(2) determine the number of moles of $O_2$ needed

(3) solve for the number of moles of $O_2$ in 1 pound of the air

(4) calculate the number of pounds of air needed.

Solving:

(1) The molecular weight of $C_8H_{18}$ is 114.23 g/mole. There are 453.50 g in 1 lb, thus one can convert from grams to pounds by multiplying the number of grams by 1 lb/453.50 g.

$$\text{MW of } C_8H_{18} \text{ in lbs} = 114.23 \text{ g/mole} \times \frac{1 \text{ lb}}{453.50 \text{ g}}$$

$$= 2.52 \times 10^{-1} \text{ lbs/mole.}$$

One can find the number of moles in 1 lb by dividing 1 lb by the molecular weight of $C_8H_{18}$ in lbs.

$$\text{no. of moles} = \frac{1 \text{ lb}}{2.52 \times 10^{-1} \text{ lbs/mole}} = 3.97 \text{ moles}$$

(2) One needs 12.5 times as much $O_2$ as $C_8H_{18}$.

$$\text{no. of moles of } O_2 \text{ needed} = 12.5 \times \text{no. of moles } C_8H_{18}$$

$$= 12.5 \times 3.97 \text{ moles} = 49.63 \text{ moles}$$

(3) In 1 lb of the air, there is .2319 lb $O_2$ and .7546 lb $N_2$. Thus, to find the number of moles of $O_2$ in one pound of the air, one must divide .2319 lb by the molecular weight of $O_2$ in pounds (MW of $O_2$ = 32 g/mole).

$$\text{MW in lbs of } O_2 = 32 \text{ g/mole} \times \frac{1 \text{ lb}}{453.5 \text{ g}}$$

$$= 7.06 \times 10^{-2} \text{ lb/mole}$$

$$\text{no. of moles} = \frac{.2319 \text{ lb}}{7.06 \times 10^{-2} \text{ lb/mole}} = 3.29 \text{ moles.}$$

(4) There are 3.29 moles of $O_2$ in one lb and 49.63 moles of $O_2$ are needed to burn 1 lb of gas.

$$\text{no. of lbs of air needed} = \frac{49.63 \text{ moles}}{3.29 \text{ moles/lb}} = 15.10 \text{ lbs.}$$

● **PROBLEM** 184

A lunar module used Aerozine 50 as fuel and nitrogen tetroxide ($N_2O_4$, molecular weight = 92.0 g/mole) as oxidizer. Aerozine 50 consists of 50 % by weight of hydrazine ($N_2H_4$, molecular weight = 32.0 g/mole) and 50 % by weight of unsymmetrical dimethylhydrazine (($CH_3$)$_2N_2H_2$, molecular weight = 60.0 g/mole). The chief exhaust product was water ($H_2O$, molecular weight = 18.0 g/mole). Two of the reactions that led to the formation of water are the following:

$$2N_2H_4 + N_2O_4 \rightarrow 3N_2 + 4H_2O$$

$$(CH_3)_2N_2H_2 + 2N_2O_4 \rightarrow 2CO_2 + 3N_2 + 4H_2O.$$

If we assume that these reactions were the only ones in which water was formed, how much water was produced by the ascent of the lunar module if 2200 **kg** of Aerozine 50 were consumed in the process?

Solution:  Aerozine 50 consists of $N_2H_4$ and $(CH_3)_2N_2H_2$. From the first reaction, we see that 2 moles of $H_2O$ are produced per mole of $N_2H_4$ consumed, and, from the second reaction, we see that 4 moles of $H_2O$ are produced per mole of $(CH_3)_2N_2H_2$ consumed. Thus, if we determine the number of moles of $N_2H_4$ and the number of moles of $(CH_3)_2N_2H_4$ in 2200 kg of Aerozine 50, we can calculate the number of moles of water, and, from this, the mass of water produced, since moles = grams (mass)/molecular weight.

Since $N_2H_4$ and $(CH_3)_2N_2H_2$ each form 50% of Aerozine 50 by weight, the mass of each component in 2200 kg of Aerozine 50 is

mass $N_2H_4$ = 50 % × 2200 kg = 1100 kg = $1.1 \times 10^6$ g.

mass $(CH_3)_2N_2H_2$ = 50 % × 2200 kg = 1100 kg = $1.1 \times 10^6$ g.

(there are **1000g** per kg.)

To convert mass to moles, we divide by the molecular weight. Hence, 2200 kg of Aerozine 50 contains

moles $N_2H_4$ = $1.1 \times 10^6$ g/32.0 g/mole = $3.4 \times 10^4$ moles $N_2H_4$

moles $(CH_3)_2N_2H_2$ = $1.1 \times 10^6$ g/60.0 g/mole

$$= 1.8 \times 10^4 \text{ moles } (CH_3)_2N_2H_2.$$

$3.4 \times 10^4$ moles of $N_2H_4$ produces $2 \times 3.4 \times 10^4$ = $6.8 \times 10^4$ moles of $H_2O$ and $1.8 \times 10^4$ moles of $(CH_3)_2N_2H_2$ produces $4 \times 1.8 \times 10^4$ = $7.2 \times 10^4$ moles of $H_2O$.

The total number of moles of $H_2O$ produced is $6.8 \times 10^4$ + $7.2 \times 10^4$ = $1.4 \times 10^5$ moles. To convert this to mass, we multiply by the molecular weight of water. Hence,

$1.4 \times 10^5$ moles $\times$ 18.0 g/mole = $2.5 \times 10^6$ g

$$= 2.5 \times 10^3 \text{ kg of water were produced.}$$

● **PROBLEM** 185

It has been found that the following sequence can be used to prepare sodium sulfate, $Na_2SO_4$:

$$S(s) + O_2(g) \rightarrow SO_2(g)$$

$$2SO_2(g) + O_2(g) \rightarrow 2SO_3(g)$$

$$SO_3(g) + H_2O(\ell) \rightarrow H_2SO_4(\ell)$$

$$2NaOH + H_2SO_4 \rightarrow Na_2SO_4 + 2H_2O$$

If you performed this sequence of reactions, how many moles of $Na_2SO_4$ could possibly be produced if you start with 1 mole of sulfur? How many moles are possible with 4.5 g of water?

Solution: If you had the general equation,

$$aA + bB \rightarrow cC + dD,$$

a moles of A react with b moles of B to produce c moles of C and d moles of D. Thus, if you want to know how many grams of D will be produced, and you know how much A you have, calculate the number of moles of A present. From this, you can determine how many moles of D can be generated, since you know that a moles of A will produce d moles of D. With this in mind, you can proceed to answer these questions.

From the first equation, you see that if you start with 1 mole of sulfur (S), 1 mole of $SO_2$ can be generated (since all the coefficients are one, although 1 is not written). Now that 1 mole of $SO_2$ is generated, you proceed to the second equation. It states that for every 2 moles of $SO_2$, 2 moles of $SO_3$ are generated. Thus, in keeping with this relative ratio, 1 mole of $SO_3$ can be generated from one mole of $SO_2$. In the third reaction you again have 1 mole of $SO_3$ yielding 1 mole of $H_2SO_4$. At this point, you have 1 mole of $H_2SO_4$ (from 1 mole of starting sulfur). The last equation shows 1 mole of $H_2SO_4$ producing 1 mole of $Na_2SO_4$. Therefore, if you were to start with 1 mole of S, you would obtain 1 mole of $Na_2SO_4$.

Now let us consider $H_2O$. Water does not enter the sequence until the third equation. This means you start with

$$SO_3(g) + H_2O(\ell) \rightarrow H_2SO_4(\ell).$$

If you have 4.5 g of $H_2O$ (MW = 18 g/m), you possess 4.5 g/18 g/mole = .25 moles $H_2O$. From the reaction, 1 mole of $H_2O$ yields 1 mole of $H_2SO_4$. Thus, .25 moles of $H_2SO_4$ are generated from .25 moles of water. You have the same 1 : 1 ratio.

The last equation is also a 1:1 ratio. Therefore, if you start with 4.5 grams of $H_2O$, .25 moles of $Na_2SO_4$ will be produced.

● **PROBLEM** 186

When 4.90 g of $KClO_3$ was heated, it showed a weight loss of 0.384 g. Find the percent of the original $KClO_3$ that had decomposed.

Solution:  To solve this problem, you need to determine how many moles of $KClO_3$ decomposed. Once this is determined, calculate the initial number of moles of $KClO_3$ and divide to obtain a percentage. To do this, a balanced equation that illustrates this reaction must be written. Such an equation is

$$2KClO_3(s) \rightarrow 2KCL(s) + 3O_2(g)$$

The weight loss of .384 g must be the amount of $O_2$ liberated. The number of moles of $O_2$ is this weight divided by its molecular weight, or

$$\frac{.384 \text{ g}}{32 \text{ g/mole}} = .0120 \text{ moles of } O_2.$$

Going back to the original equation, you find that the number of moles of $KClO_3$ is 2/3 that of $O_2$. Therefore, the number of moles of $KClO_3$ that reacted is

$$(.0120 \text{ mol of } O_2)\left(\frac{2 \text{ mol } KClO_3}{3 \text{ mol } O_2}\right) = .0080 \text{ moles of } KClO_3.$$

The number of moles of $KClO_3$ that you started with, however, is

$$\frac{4.90 \text{ g}}{122.6 \text{ g/mole}} = .0400 \text{ moles, where } 122.6 \text{ g/mole is}$$

the molecular weight of $KClO_3$. Therefore, the percentage decomposition is

$$\frac{.0080}{.0400} \times 100 = 20\%.$$

● PROBLEM 187

An impure sample of aluminum sulfate, $Al_2(SO_4)_3$, is analyzed by forming a precipitate of insoluble barium sulfate, $BaSO_4$, by reacting aluminum sulfate with an excess of $BaCl_2$ (to insure complete precipitation). After washing and drying, 2.000 g of $BaSO_4$ was obtained. If the original sample weighed 1.000 g, what was the per cent of aluminum sulfate in the sample?

Solution: The problem is solved by calculating the amount of $BaSO_4$ that would form if the sample were pure, and comparing this with the amount actually formed.

$Al_2(SO_4)_3$ reacts with $BaCl_2$ according to the equation

$$Al_2(SO_4)_3 + 3BaCl_2 \xrightarrow{\rightarrow} 3BaSO_4 + 2AlCl_3.$$

The number of moles of $Al_2(SO_4)_3$, assuming the sample is pure, is equal to the mass of the sample (1.000 g) divided by the molecular weight of $Al_2(SO_4)_3$, (342.14 g/mole),

$$\frac{1.000 \text{ g}}{342.14 \text{ g/mole}} = 0.00292 \text{ mole } Al_2(SO_4)_3.$$

Since 3 moles of $BaSO_4$ are formed for every mole of $Al_2(SO_4)_3$ reacted, as indicated by the coefficients in the reaction, 1.000 g of pure $Al_2(SO_4)_3$ would produce 3 × .00292 mole = 0.00876 mole $BaSO_4$.

The actual number of moles of $BaSO_4$ formed is equal to the mass of $BaSO_4$ obtained after washing and drying (2.000 g) divided by the molecular weight of $BaSO_4$, (233.40 g/mole),

$$\frac{2.000 \text{ g}}{233.40 \text{ g/mole}} = 0.00857 \text{ mole } BaSO_4 \text{ formed.}$$

The purity of the sample is then given by the ratio of the number of moles $BaSO_4$ formed divided by the number of moles of $BaSO_4$ that would have formed from a pure sample, times 100 %:

$$\text{purity} = \frac{0.00857 \text{ mole}}{0.00876 \text{ mole}} \times 100 \text{ \%} = 97.8 \text{ \%}.$$

The original sample, therefore, contained 97.8 % $Al_2(SO_4)_3$ and 100 % - 97.8 % = 2.2 % impurities.

● **PROBLEM** 188

A silicious rock contains the mineral ZnS. To analyze for Zn, a sample of the rock is pulverized and treated with HCl to dissolve the ZnS (silicious matter is insoluable). Zinc is precipitated from solution by the addition of potassium ferrocyanide $K_4Fe(CN)_6$. After filtering, the precipitate is dried and weighed. The reactions which occur are

$$ZnS + 2HCl \rightarrow ZnCl_2 + H_2S$$

$$2ZnCl_2 + K_4Fe(CN)_6 \rightarrow Zn_2Fe(CN)_6 + 4 KCl$$

If a 2 gram sample of rock yields 0.969 gram of $Zn_2Fe(CN)_6$, what is the percentage of Zn in the sample? Atomic weight Zn = 65.4, molecular weight $Zn_2Fe(CN)_6$ = 342.6.

Solution: In this problem, one wants to find the percent of Zn in the original rock sample. One is told that the rock contains some ZnS but not 100 %. One must find the weight of Zn in $Zn_2Fe(CN)_6$ for which we are given the total amount present. Since no Zn is lost during the course of the reaction the final and initial weight of Zn is the same. Then, knowing the weight of Zn in the rock, we can calculate the % of Zn in the rock by dividing the Zn weight by the total weight of the rock.

The first step is to find the % of Zn in $Zn_2Fe(CN)_6$. Use the formula:

$$\text{\% Zn} = \frac{(\text{atomic weight Zn})(\text{ number of Zn atoms})}{\text{M.W. of } Zn_2Fe(CN)_6} \times 100$$

$$= \frac{65.4 \ (2)}{342.6} \times 100$$

= 38.2 % or, as a fraction = 0.382.

To find the weight of any part of a compound, multiply the decimal fraction by the total weight of compound. In this case, weight of Zn = (0.382)(0.969 g) = 0.360 g. The weight of Zn is constant; thus % of Zn in rock is calculated by

$$100 \times \frac{\text{weight of Zn}}{\text{total weight of rock}} = \frac{.360}{2} \times 100 = 18 \text{ \% Zn.}$$

● **PROBLEM** 189

Clay contains 30 % $Al_2O_3$, 55 % $SiO_2$, and 15 % $H_2O$. What weight of limestone is required per ton of clay to carry out the following chemical change?

$$6CaCO_3 + Al_2O_3 + SiO_2 \rightarrow 3CaO \cdot Al_2O_3 + 3CaO \cdot SiO_2 + 6CO_2.$$
(limestone)

<u>Solution</u>: From the coefficients in the reaction equation, one sees that for every mole of $Al_2O_3$ and $SiO_2$ present, 6 moles of $CaCO_3$ (limestone) are required for the reaction to occur. As such, to determine how much limestone is necessary to react with 1 ton of clay, compute the number of moles of $Al_2O_3$ and $SiO_2$ present. From this figure, the number of moles of $CaCO_3$ can be found. Because a mole is defined as grams divided by molecular weight (MW), the weight (grams) can be found once the molecular weight is calculated.

It is given that 1 ton or 2000 lb of clay will be used. Because clay contains 30 % $Al_2O_3$ and 55 % $SiO_2$, .30 × 2000 = 600 lbs of $Al_2O_3$ and .55 × 2000 = 1100 lbs of $SiO_2$ are present in one ton of clay. To calculate the mole amounts, one must convert lbs to grams. This can be done by using the conversion factor 454 g/lb. Thus, 600 lbs × 454 g/lb = $2724 \times 10^2$ grams of $Al_2O_3$ and 1100 × 454 g/lb = $4994 \times 10^2$ grams of $SiO_2$.

The MW of $Al_2O_3$ = 102 grams/mole. The MW of $SiO_2$ = 60 grams/mole. Therefore,

$$\text{moles of } Al_2O_3 = \frac{2724 \times 10^2 \text{ g}}{102 \text{ g/mole}} = 2.67 \times 10^3 \text{ moles}$$

and moles of $SiO_2 = \frac{4994 \times 10^2 \text{ g}}{60 \text{ g/mole}} = 8.32 \times 10^3 \text{ moles}$

Notice, in one ton of clay, the mole amounts of $SiO_2$ and $Al_2O_3$ are not equal. The mole amount of limestone required will be six times the mole amount of $Al_2O_3$, $2.67 \times 10^°$, and not $SiO_2$. The reason stems from the fact that $Al_2O_3$ is the limiting reagent. The amount of any reagent required (or product produced) depends only on the limiting reagent. $SiO_2$, with a mole of $8.3 \times 10^3$, is in excess; the amount of limestone required will not depend on it.

Consequently, moles of limestone = 6(moles of $Al_2O_3$) = 6($2.67 \times 10^3$) = $1.60 \times 10^4$ moles. The molecular weight of $CaCO_3$ (limestone) = 100 g/mole. Grams of $CaCO_3$ required per ton of clay = (moles of $CaCO_3$)( MW of $CaCO_3$)

$$= (1.60 \times 10^4)(100) = 1.60 \times 10^6.$$

● PROBLEM 190

A chemist has a mixture of $KClO_3$, $KHCO_3$, $K_2CO_3$, and $KCl$. She heats 1,000 g of this mixture and notices that the following gases evolve: 18 g of water ($H_2O$), 132 g of $CO_2$, and 40 g of $O_2$ according to the following reactions:

$$2KClO_3 \rightarrow 2KCl + 3O_2$$

$$2KHCO_3 \rightarrow K_2O + H_2O + 2CO_2$$

$$K_2CO_3 \rightarrow K_2O + CO_2$$

The KCl is inert under these conditions. Assuming complete decomposition, determine the composition of the original mixture.

Solution:  The solution of this problem involves the use of the mole concept and the ability to employ it using chemical (balanced) equations. You need to determine the number of moles of the gases generated from the masses given and their molecular weights. A mole = mass in grams/molecular weight. Once this is known, you can calculate the number of moles of substances in the mixture that had to exist to produce the given quantities. You proceed as follows:

The only source of the 18 g of $H_2O$ is from

$$2KHCO_3 \rightarrow K_2O + H_2O + 2CO_2.$$

The molecular weight of $H_2O$ = 18. Thus, you have 18 g/18 g/mole = 1 mole of $H_2O$. But the equation states that for every mole of $H_2O$, you originally had 2 moles of $KHCO_3$. Therefore, the mixture must have had 2 moles of $KHCO_3$. The molecular weight of $KHCO_3$ is 100.1 g. Therefore, the weight in grams of it was 2(100.1) = 200.21 g.

The $O_2$ is generated from only the reaction

$$2KClO_3 \rightarrow 2KCl + 3O_2.$$

You have 40 g of $O_2$ evolved.  Molecular weight of $O_2$ = 32. Thus you have 40/32 = 1.25 moles of $O_2$.  According to the equation, for every 3 moles of $O_2$ produced, there existed 2 moles of $KClO_3$.  Thus, you have 2/3(1.25) or .833 moles of $KClO_3$.  Molecular weight of $KClO_3$ = 122.6.  Therefore, the number of grams = (122.6)(.833) = 102.1 g.

The $CO_2$ gas has two sources:

$$2KHCO_3 \rightarrow K_2O + H_2O + 2CO_2 \quad \text{and} \quad K_2CO_3 \rightarrow K_2O + CO_2.$$

From the water evolved, you already know that you have 200.21 g of $KHCO_3$. Its molecular weight is 100.1g. Thus, **the** number of moles of it is, $200.2/100.1 = 2$. From the equation, however, you see that for every 2 moles of $KHCO_3$, you obtain 2 moles of $CO_2$. Maintaining this one to one ratio, 2 moles of $CO_2$ must be generated. The molecular weight of $CO_2$ is 44. Thus, from $KHCO_3$, $2(44) = 88$ g of $CO_2$ produced. The total number of grams $CO_2$ evolved was given as 132 g. This means, therefore, that the other source of $CO_2$, $K_2CO_3$, must give off $132 - 88 = 44$ grams of $CO_2$. You have

$$K_2CO_3 \rightarrow K_2O + CO_2$$

This equation shows a 1 : 1 mole ratio between $K_2CO_3$ and $CO_2$. You have 44 grams of $CO_2$ released or 1 mole, since the molecular weight of $CO_2$ is 44. Therefore, you must have 1 mole of $K_2CO_3$ in the mixture. The molecular weight of $K_2CO_3$ is 138.21. Thus, the number of grams is $(1)(138.21)$ for $K_2CO_3 = 138.21$ g.

In summary, you have 200.14 g of $KHCO_3$, 102.12 g of $KClO_3$ and 138.21 g of $K_2CO_3$. The total mass of these substances = 440.47 g. The original mixture was 1000 g. Thus,

$1000 - 440.47 = 559.53$ g is the mass of the inert $KCl$.

# REACTIONS WITH LIMITING REAGENTS

● PROBLEM 191

Chromic oxide ($Cr_2O_3$) may be reduced with hydrogen according to the equation

$$Cr_2O_3 + 3H_2 \rightarrow 2Cr + 3H_2O$$

(a) What weight of hydrogen would be required to reduce 7.6 g of $Cr_2O_3$? (b) For each mole of metallic chromium prepared, how many moles of hydrogen will be required? (c) What weight of metallic chromium can be prepared from one ton of $Cr_2O_3$? 1 lb = 454 g.

Solution: (a) From the equation for the reaction, one knows that it takes three moles of $H_2$ to reduce one mole of $Cr_2O_3$. Thus, in solving this problem one should first determine the number of moles of $Cr_2O_3$ in 7.6 g, then using the ratio

$$\frac{3}{1} = \frac{\text{number of moles of } H_2}{\text{number of moles of } Cr_2O_3}$$

One can find the number of moles of $H_2$ necessary to reduce 7.6 g of $Cr_2O_3$. After finding the number of moles of $H_2$ needed, one can obtain the weight by multiplying the number of moles by the molecular weight of $H_2$.

(1) Solving for the number of moles of $Cr_2O_3$ in 7.6 g. This is done by dividing 7.6 g by the molecular weight of $Cr_2O_3$(MW = 152).

$$\text{no. of moles} = \frac{7.6 \text{ g}}{152 \text{ g/mole}} = .05 \text{ moles}$$

(2) determining the number of moles of $H_2$ necessary. The ratio

$$\frac{3}{1} = \frac{\text{number of moles of } H_2}{\text{number of moles of } Cr_2O_3}$$

will be used. This ratio was made using the stoichiometric coefficients of the equation for the reaction

$$\frac{3}{1} = \frac{\text{number of moles of } H_2}{.05 \text{ moles}}$$

$$\text{no. of moles of H} = \frac{.05 \text{ moles} \times 3}{1} = .15 \text{ moles}$$

(3) solving for the weight of $H_2$. (MW = 2.) The weight of a compound is found by multiplying the number of moles present by the molecular weight.

$$\text{weight of } H_2 = .15 \text{ moles} \times 2 \text{ g/mole} = .30 \text{ g.}$$

(b) From the equation for the reaction, it is seen that 3 moles of $H_2$ are needed to form 2 moles of Cr. This means that for every mole of Cr formed 3/2 this amount of $H_2$ is needed. Thus, 1.5 mole of $H_2$ is necessary to form one mole of Cr.

(c) Using the equation for the reaction, one is told that for every mole of $Cr_2O_3$ reduced 2 moles of Cr are formed. Thus, one must determine the number of moles in 1 ton of $Cr_2O_3$; there will be twice as many moles of Cr formed. After one knows the number of moles of Cr formed, one can determine its weight by multiplying the number of moles by the molecular weight of Cr.

(1) Determining the number of moles of $Cr_2O_3$ in 1 ton of the compound (1 ton = 2000 lbs). The number of moles is found by dividing the weight of the $Cr_2O_3$ present by its molecular weight. Because the molecular weight is given in grams per mole, it must be converted to pounds per mole

before using it to determine the number of moles present. There are 454 g in one pound, thus grams can be converted to pounds by multiplying the number of grams by the conversion factor 1 lb/454 g. (MW of $Cr_2O_3$ = 152.)

MW of $Cr_2O_3$ in lbs = 152 g/mole × 1 lb/454 g

= .33 lbs/mole.

(2) determining the number of moles of $Cr_2O_3$ in 1 ton. The number of moles present can be found by dividing 1 ton (2000 lbs) by the molecular weight in pounds.

$$\text{number of moles} = \frac{2000 \text{ lbs}}{.33 \text{ lbs/mole}} = 6060 \text{ moles}$$

Thus, there are twice this many moles of Cr produced.

number of moles of Cr = 2 × 6060 moles = 12120 moles

(3) finding the weight of Cr formed. One knows that 12,120 moles of Cr are produced. To find the weight of this quantity, the number of moles is multiplied by the molecular weight. To find the weight in pounds, one must first convert the molecular weight from grams to pounds. This is done by multiplying the molecular weight by the conversion factor 1 lb/454 g. This is used because there are 454 g in 1 lb (MW of Cr = 52).

MW of Cr in pounds = 52 g/mole × 1 lb/454 g

= 0.11 lb/mole.

The weight of the Cr formed is now found by multiplying this molecular weight by the number of moles present.

weight = 0.11 lb/mole × 12120 moles = 1333 lbs.

● **PROBLEM** 192

When 10.0 g of silicon dust, Si, is exploded with 100.0 g of oxygen, $O_2$, forming silicon dioxide, $SiO_2$, how many grams of $O_2$ remain uncombined? The reaction equation is

$$Si + O_2 \rightarrow SiO_2.$$

Solution: From the equation, it can be seen that Si and $O_2$ react in a 1 : 1 ratio. This means that 1 mole of Si will react with 1 mole of $O_2$. To determine the amount of $O_2$ left unreacted after the reaction is performed, calculate the number of moles of $O_2$ and Si present and then subtract the number of moles of Si from the number of moles of $O_2$. (MW of Si = 28, MW of $O_2$ = 32.)

$$\text{number of moles} = \frac{\text{number of grams}}{\text{MW}}$$

$$\text{number of moles of } O_2 = \frac{100.0 \text{ g}}{32 \text{ g/mole}} = 3.12 \text{ moles}$$

$$\text{number of moles of Si} = \frac{10.0 \text{ g}}{28 \text{ g/mole}} = 0.357 \text{ moles}$$

number of moles of excess $O_2$ = 3.12 - 0.357 = 2.763 moles

weight of excess $O_2$ = number of moles × MW

$$= 2.763 \text{ moles} \times 32 \text{ g/mole} = 88.5 \text{ g.}$$

● **PROBLEM** 193

How many moles of $Al_2O_3$ can be formed when a mixture of 0.36 moles of aluminum and 0.36 moles of oxygen is ignited? Which substance and how much of it is in excess of that required?

$$4Al + 3O_2 \rightarrow 2Al_2O_3$$

Solution:  In this reaction 4 moles of Al and 3 moles of $O_2$ form 2 moles of $Al_2O_3$. One is told that 0.36 moles of Al and of $O_2$ are available for the reaction. One can see from the equation that a greater number of moles of Al are needed for the reaction than $O_2$. Thus, one should assume that all 0.36 moles of Al will be used, but not all 0.36 moles of the $O_2$. Since 4 moles of Al are needed for every 3 moles of $O_2$, the following ratio holds:

$$\frac{4}{3} = \frac{\text{number of moles of Al}}{\text{number of moles of } O_2}$$

If 0.36 moles of Al are used, one can solve for the number of moles of $O_2$ that are needed to react with them.

$$\frac{4}{3} = \frac{0.36 \text{ moles}}{\text{number of moles of } O_2}$$

$$\text{number of moles of } O_2 = \frac{3 \times 0.36}{4} = 0.27 \text{ moles}$$

Since only 0.27 moles of $O_2$ are needed, and there are 0.36 moles present, there is an excess of 0.09 moles of $O_2$.

From the reaction, one knows that there are 2 moles of $Al_2O_3$ formed for every 4 moles of Al reacted. Therefore, a ratio can be set up to determine the number of moles of $Al_2O_3$ formed from 0.36 moles of Al.

$$\frac{4}{2} = \frac{\text{number of moles of Al}}{\text{number of moles of Al}_2\text{O}_3}$$

If 0.36 moles of Al are reacted, one can determine the number of moles of $Al_2O_3$ formed.

$$\frac{4}{2} = \frac{0.36 \text{ moles}}{\text{number of moles of Al}_2\text{O}_3}$$

$$\text{number of moles of Al}_2\text{O}_3 = \frac{0.36 \times 2}{4} = 0.18 \text{ moles}$$

Note: One does not determine the moles of $Al_2O_3$ from $O_2$, since the latter is in excess.

● **PROBLEM** 194

Suppose the change $HC_2O_4^- + Cl_2 \rightarrow CO_3^{2-} + Cl^-$ is to be carried out in basic solution. Starting with 0.10 mole of $OH^-$, 0.10 mole of $HC_2O_4^-$, and 0.05 mole of $Cl_2$, how many moles of $Cl^-$ would be expected to be in the final solution?

Solution:  The equation for this reaction is

$$HC_2O_4^- + Cl_2 + 5OH^- \rightarrow 2Cl^- + 2CO_3^{2-} + 3H_2O$$

From this equation one knows that 1 mole of $HC_2O_4^-$, 1 mole of $Cl_2$, and 5 moles of $OH^-$ are needed to form 2 moles of $Cl^-$.

One needs 5 times as much $OH^-$ as $HC_2O_4^-$ or $Cl_2$. This means that if one wishes to react 0.10 moles of $HC_2O_4^-$ or $Cl_2$ there must be 0.50 moles of $OH^-$ present. There are 0.10 moles of $OH^-$ present, thus one needs only .02 moles of each $HC_2O_4^-$ and $Cl_2$. Some of these reactants will remain unchanged. From the equation, one notes that there are 2 moles of $Cl^-$ formed for each $Cl_2$ reacted. If .02 moles of $Cl_2$ are used then .04 moles of $Cl^-$ will be formed. $OH^-$ is called the limiting reagent.

● **PROBLEM** 195

What is the maximum weight of $SO_3$ that could be made from 25.0 g of $SO_2$ and 6.00 g of $O_2$ by the following reaction?

$$2SO_2 + O_2 \rightarrow 2SO_3$$

Solution:  From the reaction, one knows that for every 2 moles of $SO_3$ formed, 2 moles of $SO_2$ and 1 mole of $O_2$ must react. Thus, to find the amount of $SO_3$ that can be formed, one must first know the number of moles of $SO_2$ and $O_2$ present. The number of moles is found by dividing the number of grams present by the molecular weight.

$$\text{number of moles} = \frac{\text{number of grams}}{\text{MW}}$$

For $O_2$:   MW = 32

$$\text{no. of moles} = \frac{6.0 \text{ g}}{32.0 \text{ g/mole}} = 1.88 \times 10^{-1} \text{ moles}$$

For $SO_2$:   MW = 64.

$$\text{no. of moles} = \frac{25.0 \text{ g}}{64.0 \text{ g/mole}} = 3.91 \times 10^{-1} \text{ moles.}$$

Because 2 moles of $SO_2$ are needed to react with 1 mole of $O_2$, $3.76 \times 10^{-1}$ moles of $SO_2$ will react with $1.88 \times 10^{-1}$ moles of $O_2$. This means that $3.91 \times 10^{-1} - 3.76 \times 10^{-1}$ moles or $.15 \times 10^{-1}$ moles of $SO_2$ will remain unreacted. In this case, $O_2$ is called the limiting reagent because it determines the number of moles of $SO_3$ formed. There will be twice as many moles of $SO_3$ formed as there are $O_2$ reacting.

$$\text{no. of moles of } SO_3 \text{ formed} = 2 \times 1.88 \times 10^{-1} \text{ moles}$$

$$= 3.76 \times 10^{-1} \text{ moles.}$$

The weight is found by multiplying the number of moles formed by the molecular weight (MW of $SO_3$ = 80).

$$\text{weight of } SO_3 = 3.76 \times 10^{-1} \text{ moles} \times 80 \text{ g/mole} = 30.1 \text{ g.}$$

● **PROBLEM** 196

A chemist reacts ferric sulfate with barium chloride and obtains barium sulfate and ferric chloride. He writes the following balanced equation to express this reaction:

$$Fe_2(SO_4)_3 + 3BaCl_2 \rightarrow 3BaSO_4\downarrow + 2FeCl_3$$

(A) How much $BaCl_2$ should be used to react with 10 grams of $Fe_2(SO_4)_3$? (B) How much $Fe_2(SO_4)_3$ will be necessary to produce 100 g of $BaSO_4$? (C) From a mixture of 50 g of $Fe_2(SO_4)_3$ and 100 g of $BaCl_2$, how much $FeCl_3$ can be produced?

Solution:   To answer these questions, you must understand the mole concept, and how it is used to calculate the amount of material required in a chemical reaction.

A mole is defined as the number of grams of a substance divided by its molecular weight. In other words,

$$a \; mole = \frac{amount \; in \; grams \; of \; substance}{molecular \; weight \; of \; substance}$$

In the given equation:

$$Fe_2(SO_4)_3 + 3BaCl_2 \rightarrow 3BaSO_4\downarrow + 2FeCl_3,$$

the numbers before each compound are termed coefficients (the coefficient of $Fe_2(SO_4)_3$ is one, and by convention is not written). The equation is balanced. This means that all of the elements are equal in number on both sides of the equation. For example, we have a total of 12 oxygen atoms on the left and 12 on the right. Before doing any problem involving a chemical equation, we must always balance it.

All the atoms must be accounted for. The coefficients serve this purpose. After balancing, they also tell you the relative mole amounts that will react. In other words, in this reaction 1 mole of $Fe_2(SO_4)_3$ reacts with 3 moles of $BaCl_2$. For the reaction to occur, we must have a mole ratio of 1 : 3 between $Fe_2(SO_4)_3$ and $BaCl_2$. If such a condition exists, then the equation tells us that 3 moles of $BaSO_4$ will be produced per mole of $Fe_2(SO_4)_3$. In addition, 2 moles of $FeCl_3$ will also be produced. Therefore, the coefficients tell the relative number of moles of each substance that must be either present or produced.

With this information, we can now answer the questions.

(A) We have 10 grams of $Fe_2(SO_4)_3$ and want the number of grams of $BaCl_2$ that are required to **react** with it. The molecular weight of $Fe_2(SO_4)_3$ is 399.88 g/mole. This number is obtained by adding up all the atomic weights of the individual atoms in the formula. Recalling the definition of a mole, we have 10/399.88 moles of $Fe_2(SO_4)_3$.

The equation tells us that 3 moles of $BaCl_2$ react **with one** mole of $Fe_2(SO_4)_3$. Therefore, we must have 3 times the number of moles of $Fe_2(SO_4)_3$ or

$$3 \times \frac{10}{399.88} = \frac{30}{399.88} .$$ The number of moles of $BaCl_2$

required is, therefore, $\frac{30}{399.88}$. The molecular weight of $BaCl_2$ is 208.24 g/mole. Recall, mole = number of grams/molecular weight. The number of grams of $BaCl_2$ required is, therefore,

$$\frac{30}{399.88} \; moles \times 208.24 \; g/mole = 15.62 \; grams$$

(B) We want to produce 100 grams of $BaSO_4$. The molecular weight of $BaSO_4$ is 233.40 g/mole. Therefore, we want to produce 100/233.40 moles of $BaSO_4$. How much $Fe_2(SO_4)_3$ should be used? Again, we must go back to the equation and look at the coefficients to determine the mole requirements. We see that 3 moles of $BaSO_4$ are produced for every mole of $Fe_2(SO_4)_3$. Therefore, the required number of moles of $Fe_2(SO_4)_3$ is 1/3 the number of moles of $BaSO_4$ or

$$\frac{1}{3} \times \frac{100}{233.40} = \frac{100}{700.20}$$ . The molecular weight of

$Fe_2(SO_4)_3$ is 399.88 g/mole. Consequently, the number of grams required of $Fe_2(SO_4)_3$ is

$$\frac{100}{700.20} \text{ mole} \times 399.88 \text{ g/mole} = 57.11 \text{ grams}$$

(C) In this question, we are working with a mixture of two substances to produce a third. The first thing is to compute the moles:

$$\text{Moles } Fe_2(SO_4)_3 = \frac{50.00 \text{ g}}{399.88 \text{ g/mole}} = .125 \text{ moles}$$

$$\text{Moles } BaCl_2 \quad = \frac{100 \text{ g}}{208.24 \text{ g/mole}} = .48 \text{ moles}$$

Because we are dealing with two substances to yield one, we must also consider their mole ratios as well as with the product $FeCl_3$. The equation calls for three moles of $BaCl_2$ to react with one mole of $Fe_2(SO_4)_3$. We have .125 moles of $Fe_2(SO_4)_3$. 3 times this amount is .375 moles. However, you have .48 moles of $BaCl_2$. This means, therefore, that we have .48 − .375 or .105 moles of $BaCl_2$ that will not react. In other words, $BaCl_2$ is in excess. We always take the limiting reagent, which is $Fe_2(SO_4)_3$. The amount of $FeCl_3$ present will reflect the number of moles of $Fe_2(SO_4)_3$, and not $BaCl_2$.

Because 2 moles of $FeCl_3$ will be produced for every mole of $Fe_2(SO_4)_3$, .125 × 2 = .25 moles of $FeCl_3$ will be produced. The molecular weight of $FeCl_3$ is 162.20. Therefore, the maximum weight of $FeCl_3$ produced is .25(162.20) = 40.56 grams.

● **PROBLEM** 197

Through several successive reactions, a chemist uses carbon, CaO, HCl and $H_2O$ to produce $C_6H_4Cl_2$. Assuming an efficiency of 65 %, how much $C_6H_4Cl_2$ can be produced from 500 grams of carbon? Assume that 1/3 of the carbon is lost as 3 moles CO.

Solution: In solving this problem, you must account for all carbon atoms and employ the mole concept. You need not be concerned with the actual sequence of reactions nor the roles of CaO and $H_2O$.

Dichlorobenzene, $C_6H_4Cl_2$, consists of 6 carbon atoms.

You can determine that there were originally 9 moles of carbon present since 3 moles of CO are produced and the carbon present in the CO represents 1/3 of the original amount of carbon present.

A mole is defined as weight in grams/molecular weight. The molecular weight of carbon is 12. You started with 500 grams. Therefore, the number of moles of carbon is 500/12. It is stated above, however, that for every 9 moles of C, 1 mole of $C_6H_4Cl_2$ was produced. Therefore, the number of moles of $C_6H_4Cl_2$ is 1/9 of the moles of carbon that you started with. Namely, the number of moles of $C_6H_4Cl_2$ is $1/9 \times 500/12$. The problem calls for an efficiency of 65 %. Therefore, we must multiply this number of moles by

$$\frac{65}{100} \quad \text{or,} \quad \frac{1}{9} \times \frac{500}{12} \times \frac{65}{100} \, .$$

The molecular weight of $C_6H_4Cl_2$ is 147. Recalling the definition of a mole, the weight of $C_6H_4Cl_2$ produced is

$$147 \times \frac{1}{9} \times \frac{500}{12} \times \frac{65}{100} = 442 \quad \text{grams.}$$

# VOLUME-VOLUME PROBLEMS

● **PROBLEM** 198

Calculate the volume of $O_2$ necessary to burn 50 liters of CO completely. The balanced reaction is:

$$2CO + O_2 \rightarrow 2CO_2$$

Also, calculate the volume of $CO_2$ formed.

Solution: This is a volume-volume problem and the technique shown applies only to gases. It is assumed that all the gases must be at the same temperature and pressure. No molecular weights or molecular volumes are needed. One solves this problem by setting up a ratio between the actual volumes present and the mole volumes of the reacting compounds as shown in the equation below.

$$\frac{\text{volume CO present}}{\text{mole volume CO}} = \frac{\text{volume } O_2 \text{ needed}}{\text{mole volume } O_2}$$

The mole ratio for this reaction is 2 : 1, therefore, the mole volume ratio is 2 : 1. Substitute, to obtain

$$\frac{50 \text{ liters}}{2 \text{ liters}} = \frac{\text{volume } O_2}{1 \text{ liter}}$$

volume $O_2$ = 25 liters.

To find the volume of $CO_2$ produced, set up a similar proportion between CO and $CO_2$ or $O_2$ and $CO_2$. The mole ratio for CO and $CO_2$ is 1 : 1 and for $O_2$ and $CO_2$ is 1 : 2. Therefore use the equation

$$\frac{\text{volume reactant}}{\text{mole volume reactant}} = \frac{\text{volume } CO_2}{\text{mole volume } CO_2}$$

The values for CO as the reactant are:

$$\frac{50}{2} = \frac{\text{volume } CO_2}{2}$$

volume $CO_2$ = 50 liters.

The values of $O_2$ as the reactant are:

$$\frac{25}{1} = \frac{\text{volume } CO_2}{2}$$

volume $CO_2$ = 50 liters.

• **PROBLEM** 199

In the Ostwald process for the commercial preparation of nitric acid, ammonia gas is burned in oxygen in the presence of a Pt catalyst. The balanced equation is:

$$4NH_3 + 5O_2 \xrightarrow{\text{Pt}} 4NO + 6H_2O$$

What volume of $O_2$ and what volume of NO is formed in the combustion of 500 liters of $NH_3$. All gases are under the same conditions of temperature and pressure.

Solution: In order to solve this volume-volume problem set up a ratio between the volumes involved. No molecular weights are needed in volume-volume problems. First, one wants to find out how much $O_2$ is required to react with 500 liters of $NH_3$. As can be seen from the stoichiometry of the reaction, the ratio of the mole volumes (i.e. how much $NH_3$ reacts with $O_2$) is 4 : 5. Using the equation

$$\frac{\text{volume } NH_3 \text{ present}}{\text{mole volume } NH_3} = \frac{\text{volume } O_2 \text{ required}}{\text{mole volume } O_2}$$

$$\frac{500 \text{ liters}}{4 \text{ liters}} = \frac{\text{volume } O_2}{5 \text{ liters}}$$

volume $O_2$ = 625 liters.

To find the amount of NO produced, set up a similar equation. In this case, 4 mole volumes of $NH_3$ produce 4 mole volumes of NO as indicated by the stoichiometry of the reaction

$$\frac{\text{volume } NH_3 \text{ present}}{\text{mole volume } NH_3} = \frac{\text{volume NO produced}}{\text{mole volume NO}}$$

$$\frac{500 \text{ liters}}{4 \text{ liters}} = \frac{\text{volume NO produced}}{4 \text{ liters}}$$

volume NO produced = 500 liters.

It was mentioned in the problem that a Pt catalyst was used. A catalyst is a substance which speeds up the rate at which a reaction occurs by lowering the amount of energy needed to start a reaction (i.e. lowering the activation energy). Catalysts have no effect on the final concentrations of the compounds involved in the reaction. Therefore, the presence of a catalyst does not change the values calculated in this question.

● PROBLEM 200

Calculate the volume of oxygen necessary to burn completely 100 cubic feet (1 cubic foot = 28.316 liters) of butane gas according to the equation

$$2C_4H_{10} + 13O_2 \rightarrow 8CO_2 + 10H_2O$$

Solution: Since volumes are concerned, the procedure for solving this problem is to set up a ratio between the volumes present and the mole requirements. For this reaction, 2 moles of butane react with 13 moles of $O_2$.

When two gases are under the same temperature and pressure conditions, a mole of either gas will occupy the same volume. Therefore, since 13/2 times as many moles of $O_2$ are required, the volume of $O_2$ must be 13/2 times that of methane.

Given the volume of methane is 1000 cu. ft., the volume of $O_2 = \frac{13}{2} (1000) = 6500$ cubic feet.($1.84 \times 10^5$ liters)

## WEIGHT-VOLUME PROBLEMS

● PROBLEM 201

Glucose-1-phosphate, essential to the metabolism of carbohydrates in humans, has a molecular weight of

260 g/mole and a density of about 1.5 g/cm$^3$. What is the volume occupied by one molecule of glucose-1-phosphate?

Solution: In general, volume = mass/density. Hence, in order to determine the volume, we must determine the mass of one molecule of glucose-1-phosphate.

One mole of glucose-1-phosphate weighs 260 g (this is the meaning of a molecular weight of 260 g/mole). Since there is an Avogadro's number of molecules in a mole ($6.02 \times 10^{23}$ molecules/mole), the mass of one molecule of glucose-1-phosphate is given by

$$\text{mass} = \frac{\text{molecular weight}}{\text{Avogadro's number}} = \frac{260 \text{ g/mole}}{6 \times 10^{23} \text{ molecules/mole}}$$

$$= 4.3 \times 10^{-22} \text{ g/molecule.}$$

Hence, one molecule of glucose-1-phosphate weighs $4.3 \times 10^{-22}$ g and has a volume of

$$\text{volume} = \frac{\text{mass}}{\text{density}} = \frac{4.3 \times 10^{-22} \text{ g}}{1.5 \text{ g/cm}^3} \cong 2.9 \times 10^{-22} \text{ cm}^3.$$

● **PROBLEM** 202

What is the mass of 1 liter of carbon monoxide (CO) at standard temperature and pressure (STP).

Solution: At STP, 1 mole of any gas occupies a volume of 22.4 ℓ. The weight of any substance divided by its molecular weight (MW) is equal to the number of moles of that substance. Thus, the weight of 1 mole of any substance weighs its molecular weight (in grams).

The molecular weight of CO = 28 g/mole. Thus, 22.4 liters of CO at STP weigh 28 g. The mass of one liter of CO can then be found from the following proportion

$$\frac{28 \text{ g}}{22.4 \text{ liters}} = \frac{X \text{ grams of CO}}{1 \text{ liter}}$$

Solving for X, X = 1.25 g.

● **PROBLEM** 203

What is the weight of 1,000 cubic feet of air at STP?

<u>Solution</u>: To solve this problem, one must first recognize that air is a mixture of 80% $N_2$ and 20% $O_2$. One should also know that 22.4 cubic feet of any gas at STP weighs its ounce molecular weight. For example, in this problem,

22.4 ft.$^3$ of $O_2$ would weigh 32 oz.

22.4 ft.$^3$ of $N_2$ would weigh 28 oz.

Using this information, if one knows what 22.4 ft.$^3$ of air weighs, then one can determine how much 1000 ft. of air weighs.

Since air is a mixture, one can find its ounce molecular weight by multiplying the per cent composition of each gas times each gas' ounce molecular weight and adding these two values:

for $N_2$ : $(0.8)(MW\ N_2) = (0.8)(28\ oz) = 22.4\ oz$

for $O_2$ : $(0.2)(MW\ O_2) = (0.2)(32\ oz) = 6.4\ oz.$

The ounce molecular weight for air is $22.4 + 6.4 = 28.8$ oz., and has a volume of 22.4 ft.$^3$

If 22.4 ft.$^3$ of air has a weight of 28.8 oz then, the weight of 1000 ft.$^3$ can be found through the following proportion:

$$\frac{22.4\ ft.^3}{28.8\ oz.} = \frac{1000\ ft.^3}{unknown\ weight}$$

Solving for the unknown weight of air

$$weight\ of\ air = \frac{(1000\ ft.^3)(28.8\ oz.)}{(22.4\ ft.^3)} = 1.29 \times 10^3\ oz.$$

● **PROBLEM** 204

Chlorine may be prepared by the action of $KClO_3$ on HCl, and the reaction may be represented by the equation:

$$KClO_3 + 6HCl \rightarrow KCl + 3Cl_2 + 3H_2O$$

Calculate the weight of $KClO_3$ which would be required to produce 1.0 liter of $Cl_2$ gas at STP. R = .082 liter-atm/mole-°K.

<u>Solution</u>: From the equation one can see that 1 mole of $KClO_3$ reacts to form 3 moles of $Cl_2$. If one can find the number of moles of $KClO_3$, which will react to form 1 liter of $Cl_2$, then one can find its weight by multiplying the number of moles by the molecular weight of $KClO_3$. STP

(Standard Temperature and Pressure) is defined as 0°C and 1 atm.

One can find the number of moles of $Cl_2$ in 1 liter by using the Ideal Gas Law

$$n = \frac{PV}{RT} ,$$

where P is the pressure, V is the volume, R is the gas constant (0.082 liter-atm/mole-°K), T is the absolute temperature and n is the number of moles. Here, T is given in °C; it can be converted to °K by adding 273 to it. You have, then,

$$T = 0 + 273 = 273°K.$$

One can now solve for the number of moles of $Cl_2$ produced.

$$n = \frac{PV}{RT}$$

$$P = 1 \text{ atm}$$
$$V = 1 \text{ liter}$$
$$R = 0.082 \frac{\text{liter-atm}}{\text{mole-}°K}$$
$$T = 273$$

$$n = \frac{1 \text{ atm} \times 1 \text{ liter}}{0.082 \frac{\text{liter-atm}}{\text{mole-}°K} \times 273°K} = .045 \text{ moles.}$$

From the equation for the reaction, the following ratio is determined.

$$\frac{3}{1} = \frac{\text{number of moles of } Cl_2}{\text{number of moles of } KClO_3}$$

One knows that 1 liter contains .045 moles $Cl_2$, and if you substitute this value into the above ratio, one can find the number of moles of $KClO_3$ reacted.

$$\frac{3}{1} = \frac{.045 \text{ moles}}{\text{number of moles of } KClO_3}$$

$$\text{number of moles of } KClO_3 = \frac{.045 \text{ moles} \times 1}{3} = .015 \text{ moles}$$

The weight of .015 moles of $KClO_3$ can be found by multiplying .015 moles by the molecular weight of $KClO_3$. (MW of $KClO_3$ = 122.5.)

weight of $KClO_3$ = 122.5 g/mole × .015 moles = 1.84 g.

● **PROBLEM** 205

Lithium oxide ($Li_2O$, molecular weight = 30 g/mole) reacts with water ($H_2O$, molecular weight = 18 g/mole, density = 1.0 g/cm$^3$) to produce lithium hydroxide (LiOH) according to the following reaction:

$$Li_2O + H_2O \rightarrow 2LiOH.$$

What mass of $Li_2O$ is required to completely react with 24 liters of $H_2O$?

Solution:  One mole of $Li_2O$ reacts with exactly one mole of $H_2O$, as indicated by the coefficients in the reaction. Hence, if we determine the number of moles of $H_2O$ in 24 liters, we then know the required number of moles of $Li_2O$, and, from this, can calculate the mass of $Li_2O$ that we need.

The number of moles of water is moles = mass/molecular weight. But mass = density × volume, so that the number of moles of water is

$$moles = \frac{mass}{molecular\ weight} = \frac{density \times volume}{molecular\ weight}$$

$$= \frac{1.0\ g/cm^3 \times 24\ liters}{18\ g/mole} = \frac{1.0\ g/cm^3 \times 24000\ cm^3}{18\ g/mole}$$

$$= 1333.33\ moles$$

Thus, we require 1333.33 moles of $Li_2O$ to completely react with 24 liters of $H_2O$ (= 1333.33 moles). Multiplying this by the molecular weight of $Li_2O$, we determine that

$$1333.33\ moles \times 30\ g/mole = 40,000\ g = 40\ kg$$

of $Li_2O$ is needed.

● **PROBLEM** 206

What volume of hydrogen at STP is produced as sulfuric acid acts on 120 g. of metallic calcium. Equation for the reaction is

$$Ca + H_2SO_4 \rightarrow CaSO_4 + H_2$$

Solution:  This problem may be solved by using either the mole method or the proportion method.

Mole method:  In using the mole method, one looks at this equation and sees that for every mole of calcium acted on by $H_2SO_4$, one mole of hydrogen is produced.

This means that to find how much hyrdogen is produced, one must first find out how much calcium is present. There will be the same number of moles of hydrogen produced as there are calcium reacted. After one knows how many moles of hydrogen are produced, one can calculate the volume.

To calculate the number of moles of calcium present, one must divide the amount present by the molecular weight (molecular weight of Ca = 40).

$$\text{number of moles} = \frac{\text{number of grams present}}{\text{molecular weight}}$$

$$\text{number of moles of Ca} = \frac{120 \text{ g}}{40 \text{ g/mole}} = 3.00 \text{ moles}$$

Therefore, 3.00 moles of hydrogen gas are produced. At STP (Standard Temperature and Pressure), the volume of one mole of any gas occupies 22.4 liters. Thus, when 3 moles of gas are generated, as in this problem it occupies 3 × 22.4 liters = 67.2 liters.

Proportion method: In the proportion method, the molecular weights (multiplied by the proper coefficients) are placed below the formula in the equation and the a-mounts of substances (given and unknown) are placed above. In this case, because one is trying to find the volume of hydrogen and not its weight, the volume of one mole will be placed below the equation as shown.

```
120 g                                X( Vol.)

Ca    +    H₂SO₄    →    CaSO₄    +    H₂
40.0 g                               22.4 liters
```

$$\frac{120 \text{ g}}{40.0 \text{ g}} = \frac{X}{22.4 \text{ liters}}$$

X = unknown volume of $H_2$ produced.

Solving for X:

$$X = \frac{(22.4 \text{ liters})(120 \text{ g})}{40.0 \text{ g}} = 67.2 \text{ liters.}$$

● PROBLEM 207

The executioner in charge of the lethal gas chamber at the state penitentiary adds excess dilute $H_2SO_4$ to 196 g (about ½ lb) of NaCN. What volume of HCN gas is formed at STP?

Solution:  The equation for this reaction is

$$H_2SO_4 + 2NaCN \rightarrow 2HCN + Na_2SO_4$$

This means that for each NaCN reacted, one HCN is formed. After one knows the number of moles of HCN formed, one can find the volume that the gas occupies by multiplying the number of moles by 22.4 liters (volume of one mole of gas at STP). STP (Standard Temperature and Pressure) is defined as 273°K and 1 atm.

The number of moles of NaCN reacted can be found by dividing 196 g by the molecular weight of NaCN. (MW of NaCN = 49.)

$$\text{number of moles} = \frac{196 \text{ g}}{49 \text{ g.mole}} = 4.0 \text{ moles}$$

From the stoichiometry of the reaction, we see there will be 4.0 moles of HCN formed.

The volume of the gas can now be found.

volume = 4.0 moles × 22.4 liters/mole = 89.6 liters.

● PROBLEM 208

A chemist decides to prepare some chlorine gas by the following reaction:

$$MnO_2 + 4HCl \rightarrow MnCl_2 + 2H_2O + Cl_2 \uparrow$$

If he uses 100 g of $MnO_2$, what is the maximum volume of chlorine gas that can be obtained at standard temperature and pressure (STP)?

Solution: The solution of this problem entails a combination of stoichiometry and gas law theory.

The first thing to determine is how many moles of chlorine will be produced. The volume can be determined using the fact that each mole, at STP, occupies 22.4 ℓ.

Since the equation is balanced, you must look at the coefficients. The equation indicates that for every mole of $MnO_2$, 1 mole of chlorine gas is produced. A mole is defined as being equal to weights in grams of material/molecular weight of material.

The molecular weight of $MnO_2$ is 86.9. It follows, then, that the number of moles of $MnO_2$ is 100/86.9. This will be equated to the number of moles of chlorine gas produced.

At STP one mole of gas occupies 22.4 liters. However, we do not have 1 mole, but 100/86.9 moles. To calculate the new volume, you multiply the STP volume by the number of moles of chlorine gas produced. In other words, the volume of the chlorine gas is

$$22.4 \times \frac{100}{86.9} = 25.8 \text{ liters.}$$

What volume of ammonia at STP can be obtained when steam is passed over 4000 g of calcium cyanamide? The balanced reaction is

$$CaCN_2 + 3H_2O \rightarrow 2NH_3 + CaCO_3$$

(Molecular weight of $CaCN_2$ = 80, MW of $NH_3$ = 17.)

Solution: From the stoichiometry of the equation 1 mole of $CaCN_2$ produces 2 moles of $NH_3$. Thus, if one calculates the number of moles of $CaCN_2$, then one knows the number of moles of $NH_3$. To calculate the number of moles of $CaCN_2$ divide its weight by its molecular weight, thus, the number of moles of $CaCN_2$ equals

$$\frac{4000 \text{ g } CaCN_2}{80 \text{ g/mole}} = 50 \text{ moles.}$$

Therefore, the number of moles of $NH_3$ produced equals 100 moles.

At STP, one mole of any gas occupies 22.4 liters. Hence, the total volume of $NH_3$ produced at STP is

$$V_{NH_3} = (22.4 \text{ liter/mole})(100 \text{ mole } NH_3) = 2240 \text{ liter.}$$

Iron (III) oxide is reacted with carbon monoxide in a blast furnace to produce iron. The balanced reaction is:

$$Fe_2O_3 + 3CO \rightarrow 2Fe + 3CO_2$$

What volume of CO at STP is required to completely use up 31.94 kg of iron oxide? (MW of $Fe_2O_3$ = 159.7, MW of CO = 28.)

Solution: From the stoichiometry of the above reaction, one sees that 1 mole of $Fe_2O_3$ requires 3 moles of CO to react. Thus, three times as many moles of CO will react with a given number of moles of $Fe_2O_3$. Once the number of moles of CO is known, one can use the fact that 1 mole of any gas at STP has a volume of 22.4 ℓ to determine the volume of CO required.

The number of moles of $Fe_2O_3$ available for the reaction is the weight of $Fe_2O_3$ in grams divided by its molecular weight:

$$\text{moles } Fe_2O_3 = \frac{\text{weight } Fe_2O_3}{\text{MW } Fe_2O_3} = \frac{31940 \text{ g}}{159.7 \text{ g/mole}} = 200 \text{ moles.}$$

The number of moles of CO needed to completely react is 3(200 moles) = 600 moles.

The volume at STP of 600 moles of CO is the number of moles of CO times the volume of 1 mole:

volume CO at STP = (moles CO)(22.4 $\ell$/mole)

$$= (600 \text{ moles})(22.4 \text{ }\ell/\text{mole}) = 13,440 \text{ }\ell.$$

● **PROBLEM** 211

How many liters of phosphine ($PH_3$) gas at STP could be made from 30 g of calcium by use of the following sequence of reactions:

$$3Ca + 2P \rightarrow Ca_3P_2$$

$$Ca_3P_2 + 6HCl \rightarrow 2PH_3 + 3CaCl_2$$

(Molecular weights: Ca = 40, $PH_3$ = 34.)

Solution: From the stoichiometry of these two equations 3 moles of Ca will yield 2 moles of $PH_3$ gas. Thus, if one knows the number of moles of Ca, then one can determine the number of moles of $PH_3$ produced. The number of moles of Ca given is its weight divided by its molecular weight. Therefore,

$$\text{number of moles Ca} = \frac{30 \text{ g Ca}}{40 \text{ g/mole}} = 0.75 \text{ moles}$$

From 0.75 moles of Ca, one produces 2/3 as much or 2/3(0.75) = 0.50 moles of $PH_3$ gas. At STP, one mole of any gas occupies 22.4 liters, hence, the total volume of $PH_3$ produced at STP

$$= (22.4 \text{ }\ell/\text{mole})(0.50 \text{ moles } PH_3) = 11.2 \text{ }\ell.$$

● **PROBLEM** 212

Nitroglycerin ($C_3H_5(NO_3)_3$) explodes according to the following reaction:

$$4C_3H_5(NO_3)_3(\ell) \rightarrow 12CO_2(g) + 6N_2(g) + O_2(g) + 10H_2O(g),$$

producing only gaseous products. What is the total volume of gaseous products produced at standard temperature and pressure (STP) when 454 g of nitroglycerin explodes? The molecular weight of nitroglycerin is 227 g/mole.

Solution: This problem is an application of the ideal

gas equation, PV = nRT, where P = pressure, V = volume,
n = number of moles, R = gas constant, and T = absolute
temperature. Solving for V,

$$V = \frac{nRT}{P}$$

STP is, by definition, 0°C(= 273°K) and 1 atm pressure.
Hence, T = 273°K and P = 1 atm. Also, R = 0.082 liter-
atm/mole-deg. We must find the number of moles, n, of
gaseous products. The number of moles of nitroglycerin
we started with is equal to its mass divided by the
molecular weight, or 454 g/227 g/mole = 2 moles of
nitroglycerin. Dividing the equation for the reaction of
nitroglycerin by 2 (so that $4C_3H_5(NO_3)_3$ becomes
$2C_3H_5(NO_3)_3$), we obtain:

$2C_3H_5(NO_3)_3(\ell) \rightarrow 6CO_2(g) + 3N_2(g) + \frac{1}{2}O_2(g) + 5H_2O(g)$.

Thus, our 2 moles of nitroglycerin will produce a total of
6 + 3 + ½ + 5 = 14.5 moles of gaseous products, so that
n = 14.5 moles. Substituting the values of n, R, T, and
P into the equation for V gives:

$$V = \frac{nRT}{P} = \frac{14.5 \text{ moles} \times 0.082 \text{ liter-atm/mole-deg} \times 273°K}{1 \text{ atm}}$$

$$= 325 \text{ liters}.$$

● **PROBLEM** 213

A chemist performs the following reaction:

$2KClO_3(s) \rightarrow 2KCl(s) + 3O_2(g)$.

He collects the $O_2$ gas by water displacement at 20°C.
He observes a pressure of 753 mm Hg. Assuming the
pressure of water vapor is 17.5 mm Hg at 20°C and he
started with 1.28 g of potassium chlorate ($KClO_3$), what
volume of gas is produced? (R = .0821 ℓ-atm/mole °K.)

Solution:  You are asked to find the volume of gas
produced and can do so using the equation of state,
PV = nRT, where P = pressure, V = volume, n =
number of moles, R = universal gas constant and T =
temperature in Kelvin (Celsius plus 273°).

     In the problem you are told the pressure at which
the gas is collected, which must be modified due to the
presence of water vapor and the temperature. If you knew
n, you could substitute for the values in the equation of
state and solve for V.

     Your procedure will be to find the number of moles of
gas produced. This can be determined from the reaction e-

quation given and the use of stoichiometry. If you knew
how many moles of $KClO_3$ you started with, you would know
the number of moles of $O_2$ produced, since 3 moles of $O_2$
are generated for every 2 moles of $KClO_3$. You can determine how many moles of $KClO_3$ you started with. You are
told that the chemist has 1.28 g. Since the molecular
weight of $KClO_3$ is 122.55 g, you have

$\frac{1.28}{122.55}$ = $1.045 \times 10^{-2}$ moles of $KClO_3$. Therefore,

from reaction there are $\frac{3}{2}$ $(1.045 \times 10^{-2})$ = $1.57 \times 10^{-2}$

moles of $O_2$ gas produced.

Now you go back to PV = nRT. Recall that the gas
was collected by water displacement. Thus, it is saturated
by water vapor. Hence, the pressure of $O_2$ is only
753 mm Hg - 17.5 mm Hg, or 735.5 mm where 17.5 mm Hg is
the water vapor pressure. Since 1 atm = 760 mm,

(735.5 mm) $\left(\frac{1 \text{ atm}}{760 \text{ mm}}\right)$ = .968 atm.

Now, substitute and solve for the volume V of $O_2$
produced. Rewriting and substituting,

$V = \frac{nRT}{P} = \frac{(.0157)(.0821)(293^{\circ}K)}{.968}$ = .390 liters or 390 ml.

# CHAPTER 6

# SOLIDS

> **Basic Attacks and Strategies for Solving Problems in this Chapter. See pages 191 to 211 for step-by-step solutions to problems.**

Many solids form crystals or molecular structures in which the constituent atoms or molecules are in a repeating three-dimensional array. Metals, simple salts, and semiconducting solids like silicon are all crystalline. The most common non-crystalline or amorphous solid is glass. In this solid, the molecules are in a random orientation with respect to each other. Glasses do not have a sharp melting point, rather they have a melting range, thus leading some to classify glass as a supercooled or extremely viscous liquid.

A complete description of the possible crystalline structures of solids is beyond the scope of this text, but several common cubic crystalline structures, simple cubic, face centered cubic (fcc), and body centered cubic (bcc), are illustrated in the figures associated with Problems 218 and 219. A unit cell is the smallest repeating unit of the lattice.

A key solution in calculating the stoichiometric properties of unit cells is to ascertain how many cells share a particular atom or molecule. For cubic lattices, a corner species is shared by eight cells, an edge species by four cells, a face centered species by two cells, and a body centered species by only one cell.

In crystals, the atoms or molecules remain in a fixed orientation with respect to each other. The distances between atoms can be determined from x-ray diffraction patterns because the distances are comparable in magnitude to the wave lengths of x-rays. Hence x-ray diffraction patterns are determined by the regular spacings between the atoms or molecules.

The principles of stoichiometry discussed in Chapters 3 and 4 can be used, along with information on the size of unit cells, to calculate the densities, volumes, and many other properties of crystalline solids. For example, the theoretical density of a crystalline solid can be obtained by:

$$\rho = \frac{\text{mass of unit cell}}{\text{volume of unit cell}} \qquad \qquad 5\text{-}3$$

This is the maximum value of a pure crystal; because of imperfections in the crystalline lattice, the measure density is usually slightly less than that calculated by this procedure.

## Phase Diagrams

Phase diagrams are also discussed in this chapter. Phase diagrams are simply graphs, usually on pressure and temperature coordinates, showing the boundaries between phases. An example for a simple single-component phase diagram (e.g., water) is shown below and illustrates some very familiar as well as some lesser known properties of pure compounds.

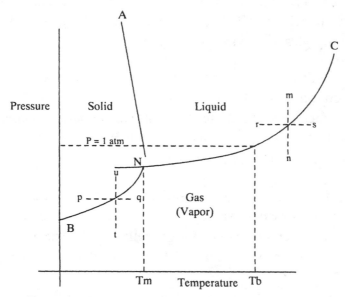

**Figure 1** — Phase diagram for single component system

The scales on the axes of the sample phase diagram shown above have deliberately been left off so that the diagram applies to any single-component system which has only one solid phase*. If the diagram above were to apply for water, then the normal boiling point, $T_b$ (the temperature where the liquid and gas phases are in equilibrium at 1 atm. total pressure), would be 100°C. The melting point, $T_m$, is 0°C, and at a pressure of one atmosphere, water is a liquid between 0°C and 100°C. At temperatures and pressures represented by points along the lines separating the phases, both phases can exist in equilibrium with each other.

The line *BN* represents the temperatures and pressures where the solid and gas phases are in equilibrium. Similarly, the line *NA* represents the points where the liquid and solid phases are in equilibrium; for water and many other species, it slants slightly to the left of vertical as shown in this figure illustrating the effect of pressure on lowering the melting point. Note that in a similar figure in Problem 231, the solid-liquid equilibrium boundary between the solid and liquid phases has a positive slope; either can occur. The line *NC* represents the points where the liquid and gas phases are in equilibrium and this line ends at point *C*, the critical point. Above the critical point (218 atm and 375°C for water), the liquid and gas phases are indistinguishable. In this part of the phase diagram the material is sometimes called a supercritical fluid since it is not quite correct to refer to it as either a liquid or gas. There is normally no corresponding end point at *A* where the solid and liquid phases become indistinguishable (except, perhaps, at extreme pressures).

Note that at point *N* (0.0098°C and 4.58 torr for water) the solid, liquid, and gas phases meet. At this point, called the triple point, all three phases can exist in equilibrium. The triple point is unique; that is, there is only one point where this can occur, and both the temperature and pressure must be exactly equal to those represented by the coordinates of the triple point. On the other hand, there are many values of temperature and pressure where both liquid and gas or liquid and solid or even gas and solid phases exist in equilibrium.

If the temperature is raised at constant pressure along the line *rs*, the material will change from a liquid to a gas. The same thing will happen if the pressure is lowered at constant temperature along the line *mn*. Similarly, the solid phase will sublime (change to a gas) if the appropriate state variable (temperature/pressure) changes along the lines *pq* or *ut*.

*Many single-component systems have more than one solid phase. These are typically different crystalline forms of the component. Sulfur, for example, has a monoclinic and orthorhombic solid crystalline phase. In a phase diagram for such a material, areas of the diagram exist which represent the temperatures and pressures where each of the solid phases is stable.

## CRYSTAL STRUCTURE

● **PROBLEM** 214

Distinguish between crystalline and amorphous solid substances, using some specific examples. To what extent is the distinction useful?

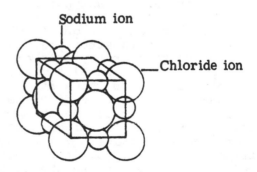

Sodium ion

Chloride ion

<u>Solution</u>: Crystalline substances can be generally thought of as species composed of structural units with specific geometric patterns. The accompanying drawing of sodium chloride would be an example of such a pattern. The important point is that there exists a regularity in the arrangement of structural units. Structures with regularity generally show a sharp and characteristic melting point, which is the case with crystalline substances.

Amorphous substances, however, tend to be shapeless and without definite order. That is, you have a randomness. For example, glassy or glasslike materials such as Plexiglas and silicate glasses. In substances with a general lack of order, the melting points vary over a range or temperature interval. For amorphous substances, this is

exactly what you find. It would, however, be <u>incorrect</u>
to state categorically that amorphous substances are
without ANY order. For they do tend to have short range
order even though they do contain long-range randomness.

Explain briefly the following terms; (a) isotropic and
anisotropic solids, (b) general and directional properties
of crystalline substances, (c) a plane, an axis, and a
center of symmetry and (d) polymorphism and allotropism.

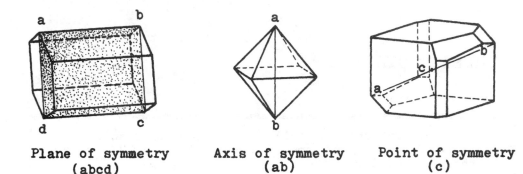

**Plane of symmetry**          **Axis of symmetry**          **Point of symmetry**
      **(abcd)**                    **(ab)**                        **(c)**

<u>Solution</u>:   These properties all pertain to the solid state.

      (a) An isotropic solid is one in which there are no
directional differences in structure. Isotropic solids
usually do not form regular crystals,and examples of an
isotropic amorphic solids are Plexiglass and glass. An
anisotropic solid is one in which its properties depend,
in general, upon the particular direction along which the
measurement is made. Crystalline substances, for example,
are anisotropic. Examples of anisotropic crystals are
cubic and tetrahedral lattices. (b) By general properties
of crystalline substances, you mean those such as density,
specific heat, melting point and chemical composition.
These are called "general" due to the fact that none of
them involve any special direction.Directional properties
refer to those such as refraction and absorption of light,
cohesion, expansion and electrical and thermal conductivity,
all of which are measured in a special direction. (c) Plane
of symmetry is said to exist in a crystal if it can be
divided by an imaginary plane passing through the center,
such that the two equal parts formed are an exact mirror
image of each other. An axis of symmetry is said to exist
in a crystal if you can draw a line through the crystal's
center and when you proceed to rotate the crystal about
this line through 360° the same appearance of itself is
produced more than once. A crystal will have a center of
symmetry, if every crystal face has a twin equidistant on
the opposite side of this center. The accompanying figures
illustrate each of these. (d) Polymorphism refers to the

existence of substance having a definite composition yet occurring in more than one crystalline form. When the substance is an element rather a compound, this phenomenon is termed allotropy.

It is known that both metallic and ionic crystals are good conductors of electricity, yet many of their other properties differ significantly. Explain why.

Solution:  The similarities and differences of the two will be seen after an investigation of their space lattices, i.e., those patterns of points describing the atomic or molecular arrangements of points in a unit of a crystal. For something to conduct electricity, it must have charges. In an ionic crystal, the lattice points are occupied by positive or negatively charged ions, each of which occupies a position exactly equivalent to every other species of like charge. The electric conductivity of ionic crystals is low but increases with Temp. because more electrons are excited into the conduction band of the crystal. Ionic crystals are held together by electrostatic forces between the ions.

In a metallic crystal, you have discrete atoms, not ions, at the lattice points. It seems that the species should not conduct electricity. It is important to realize, however, the valence or outer electrons of the metallic crystal are distributed over the crystal as a whole rather than being localized on each atom. The mobile electrons, therefore, which are able to move in an applied electric field, account for the electrical conductivity.

This mobility of electrons also explains why a metallic crystal is generally strong, lustrous and malleable. In the ionic crystal, you have strong electrostatic forces between the ions, which accounts for the high melting points. However, these crystals also tend to be hard and brittle because the crystal consists of parallel sheets of positively and negatively charged ions, lateral displacement may bring ions of like charge into the vicinity of each other, resulting in electrostatic repulsion of like charges. This, then, helps to explain why they possess a facile crystal fracture.

The properties of diamond and graphite differ vastly. Explain these differences on the basis of their fundamentally different space lattices.

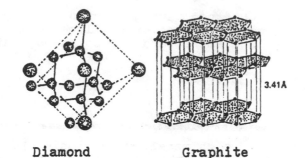

**Diamond Structure**

**Graphite Structure**

Solution: The arrangement (location) of all atoms in a unit of a crystal is termed a space lattice, i.e., those patterns of atoms describing the atomic or molecular arrangements. This problem calls for a comparison of such lattices in a diamond and graphite. The accompanying figures show the space lattices in question.

The diamond lattice surrounds each carbon atom by 4 others in a tetrahedral configuration. There exist covalent bonds between the carbon atoms, which means they share electrons. These covalent forces are responsible for the hard nature and high melting point of the substance. On the other hand, graphite has covalent bonding in two-dimensional, planar, hexagonal rings. As the figure suggests, the sheets of these atoms are held together by weak forces of attraction. This means, that the layers can slide easily over each other. This, then, accounts for the lubricating properties and flakiness associated with this material.

# LATTICE STRUCTURES AND UNIT CELLS

● **PROBLEM** 218

Excluding hexagonal unit cells, when counting the number of points inside a cell, a point on an edge is ¼ inside the cell, and a lattice point at a corner is 1/8 inside the cell. Justify these fractions. Calculate, also, the the net number of lattice points in the following unit cells: simple cubic, body-centered cubic, face-centered cubic, and tetragonal.

Solution: To solve this problem properly you need to know the definition of a unit cell and lattice point. You need, also, to know the actual structures of the unit cells given. You proceed as follows: A unit cell is that small fraction of a space lattice, which sets the pattern for the whole lattice. In other words, it is the smallest portion of the space lattice (which is just

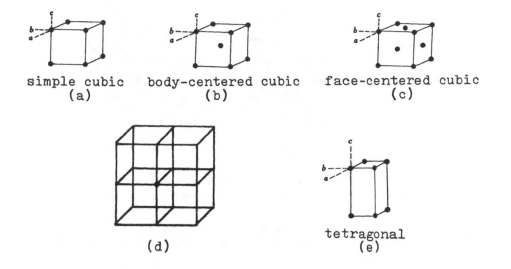

simple cubic
(a)

body-centered cubic
(b)

face-centered cubic
(c)

(d)

tetragonal
(e)

the pattern of points which describes the arrangement of atoms or molecules in the crystal), which, when moved repeatedly a distance equal to its own dimensions along the various directions, generates the whole space lattice equivalent to the original lattice. The lattice is, of course, just points that denote atoms or molecules; thus, the term lattice point. The accompanying illustrations give you the structures in question. From this, you can obtain justification for the fractions and find the net number of lattice points. For example, take the simple cubic structure. In a space lattice, each atom is shared with three other unit structures as seen in figure d.

Note; This point is shared by 8 unit cells, if you place another 4 cubes on the side. Thus, each point is only 1/8 in the cell. This same sort of procedure and reasoning can be used to justify all fractions in all unit cells given. Calculations now become easy. Simple cubic: Have 8 lattice points. Each is only 1/8 inside cell. Thus, you have net of (1/8)(8) = 1.

Face-centered cubic: 8 lattice points at corners, each only 1/8 inside cell. Have a point on each face. Each point is only ½ inside cell. Have six points. Thus, net = (8)(1/8) + (6)(½) = 4.

Body-centered: Again have (8)(1/8) from lattice points at corners. 1 point in center, which is entirely in cell. Thus net becomes (8)(1/8) + (1)(1) = 2.

Tetragonal: Only have a net of (8)(1/8) from lattice points at corners.

● PROBLEM 219

Refer to the **accompanying** figure and determine the number of unit particles (atoms) in a unit cell of each of the three type lattices.

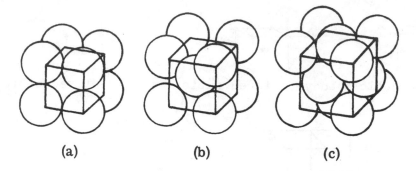

(a)                    (b)                    (c)

Unit cells in cubic lattices:  (a) simple cubic lattice;
(b) body-centered lattice;  (c) face-centered lattice.

Solution:  The following rules may be used to determine
the number of unit particles (atoms) associated with each
type of cubic unit cell:

1) An atom at a corner contributes 1/8 of its volume
to each of eight adjacent cubes.

2) An atom on an edge of a cube contributes 1/4 of
its volume to each of 4 adjacent cubes.

3) An atom in the face of a cube contributes 1/2
of its volume to each of 2 adjacent cubes.

4) An atom completely within a cube contributes all
of its volume to the unit cell.

The number of unit particles in each of the three
types of lattices can be found as follows:

Simple cubic:  In the simple cubic lattice all
eight units are of the same variety - they are all at
corners, thus rule 1 is used.

Total atoms in cell = 8 corners × 1/8 atom/corner = 1 atom

Therefore the area of unit cube is equivalent to 1 atom.

Body-centered: In the body-centered lattice there are
nine atoms contributing to the unit cube. There are 8
corners and one atom completely within the cube, thus,
rules 1 and 4 are used.
Total atoms in cell = 8 corners × 1/8 atom/corner

+ 1 atom in center = 2 atoms.

Face-centered:  In the face-centered lattice there
are 14 atoms contributing to the unit cube. There are
8 corners and 6 faces, thus, rules 1 and 3 will be used.

Total atoms in a unit cell = 8 corners × 1/8 atom/corner

+ 6 faces × 1/2 atom/corner

= 4 atoms.

What fraction of the total space in a body-centered cubic unit cell is unoccupied? Assume that the central atom touches each of the eight corner atoms of the cube.

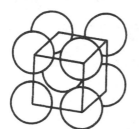

body-centered lattice

Figure 1

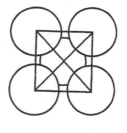

Figure 2

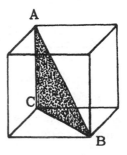

Solution: To determine the percent of the unit cube that is unoccupied by the atoms making up the lattice, one must subtract the volume taken up by the atoms from the volume of the cube. This volume is then divided by the volume of the cube and multiplied by 100 to find the percent of unoccupied space.

1) Determining the volume of the atoms: One must first determine the number of atoms contributing to the unit cube. In this lattice there are 8 corner atoms and 1 atom in the center. Corner atoms contribute 1/8 of their volume to the cube. The atom in the center contributes its entire volume.

no. of atoms in unit cube = (1/8 atom/corner × 8 corners)

+ 1 atom (in center) = 2 atoms.

Thus, the volume in the unit cube taken up by the atoms is equal to the volume of two atoms. The radius of these atoms is taken to be 1. The volume of a sphere is $4/3 \pi r^3$.

volume of 2 atoms = 2 × 4/3 $\pi(1)^3$ = 8.38

2) Volume of the cube: The corner atoms are assumed to be touching the central atom. The diagonal of the cube can be visualized as shown in Figure 2.

Because $\overline{AB}$ is shown to be 4r or 4 the side of the cube $\overline{AC}$ can be found using the Pythagorean Theorem. Once $\overline{AC}$ is known the volume of the cube can be found. From geometry, it is known that

$$\overline{CB} = \overline{AC} \times \sqrt{2}$$

$$\overline{AB}^2 = \overline{AC}^2 + \overline{CB}^2 \qquad\qquad \overline{AB} = 4$$

$$4^2 = \overline{AC}^2 + 2(\overline{AC})^2 \qquad\qquad CB = \overline{AC} \times \sqrt{2}$$

$$4^2 = 3\overline{AC}^2$$

$$\frac{4}{\sqrt{3}} = \overline{AC}$$

The volume of the cube is equal to the length of the side cubed.

$$\text{volume of cube} = \left(\frac{4}{\sqrt{3}}\right)^3 = 12.32$$

3) The space in the cube is equal to the volume of the spheres subtracted from the volume of the cube.

vol. of space = vol. of cube - vol. of spheres

vol. of space = 12.32 - 8.38 = 3.94

$$\text{percentage of cube taken up by space} = \frac{3.94}{12.32} \times 100 = 32 \%.$$

● **PROBLEM** 221

The spherical atomic nucleus of potassium has a radius of about $4 \times 10^{-13}$ cm.; $6.02 \times 10^{23}$ of these nuclei weigh about 39 g. Calculate the approximate density of the nucleus of potassium in tons per cubic centimeter. Volume of a sphere = $4/3\ \pi r^3$, where r is the radius.

Solution: The density is defined as the mass of a substance divided by its volume. To calculate the density one must know the weight and volume of the substance. Here one is given the weight in grams but the problem asks for the density in tons, thus one must convert grams to tons. The volume of the nuclei is not given, but can be found by using the radius of a nucleus, the formula for the volume of a sphere and the number of nuclei present.

1) Converting grams to tons

There are 454 g in one pound and 2000 lb in one ton, therefore, grams can be converted to tons by multiplying the number of grams by the following conversion factor 1 lb/454 g × 1 ton/2000 lb

number of tons = 39 g × 1 lb/454 g × 1 ton/2000 lb

$$= 4.3 \times 10^{-5} \text{ tons.}$$

2) Calculating volume of nuclei

The volume of a sphere is equal to $4/3\ \pi r^3$ where r is the radius of the sphere. One is given that the radius of a potassium nuclei is $4 \times 10^{-13}$ cm.

$$\text{volume of 1 nuclei} = 4/3\ \pi\ (4 \times 10^{-13}\ \text{cm})^3$$

$$= 4/3\ (3.14)(64 \times 10^{-39}\ \text{cm}^3)$$

$$= 2.6 \times 10^{-37}\ \text{cm}^3$$

The total volume of the nuclei is equal to the number of nuclei times the volume of one nucleus

total volume = no. of nuclei × volume of one nucleus

$$\text{total volume of potassium nuclei} = 6.02 \times 10^{23} \times 2.6 \times 10^{-37}\ \text{cm}^3$$

$$= 1.6 \times 10^{-13}\ \text{cm}^3$$

3) The weight in tons and the volume in cubic centimeters is now known. The density in tons per $\text{cm}^3$ can be calculated

$$\text{density} = \frac{\text{no. of tons (= mass)}}{\text{no. of cm}^3\ (= \text{volume})}$$

$$\text{density} = \frac{4.3 \times 10^{-5}\ \text{tons}}{1.6 \times 10^{-13}\ \text{cm}^3} = 2.7 \times 10^8\ \text{tons/cm}^3.$$

● **PROBLEM** 222

The density of KF is $2.48$ g/cm$^3$. The solid is made up of a cubic array of alternate $K^+$ and $F^-$ ions at a spacing of $2.665 \times 10^{-8}$ cm. between centers. From these data, calculate the apparent value of the Avogadro number.

Solution: Avogadro's number is the number of particles in one mole of a substance. Here, one is given the dimensions of a cubic lattice made of ½ $K^+$ ions and ½ $F^-$ ions. Due to the fact that KF crystallizes like NaCl, it can be seen that there are eight ions or 4 formula units of KF in the cube. To calculate Avogadro's number from the data, one should find:

(1) the volume of one cube

2) the weight of one cube using the density

3) the number of cubes in one mole of KF

4) Avogadro's Number by multiplying the number of cubes by 4, the number of formula units (KF) per cube.

Solving for Avogadro's Number:

1) The volume of the cube is found by cubing the length of the edge. From the crystallization of KF, it is seen that the length of the edge of cube is twice the spacing between ion centers. This is due to the fact that three ions make up an edge, and, as such, 2 spacings between the ions exist.

Therefore,

length of edge = $2.665 \times 10^{-8}$ cm $\times$ 2 = $5.33 \times 10^{-8}$ cm

volume of cube = $(5.33 \times 10^{-8})^3$ = $1.51 \times 10^{-22}$ cm$^3$.

2) Using the density, one knows that 1 cm$^3$ weighs 2.48 g. Therefore, the weight of one cube is equal to the density times the volume of a cube.

weight of one cube = $2.48$ g/cm$^3$ $\times$ $1.51 \times 10^{-22}$ cm$^3$

$$= 3.76 \times 10^{-22} \text{ g/cube}$$

3) One mole of KF weighs the sum of the weights of one mole of K$^+$ and one mole of F$^-$. (MW of K = 39.10, MW of F = 19.00.)

MW of KF = 39.10 + 19.00 = 58.10 g/mole

The number of cubes in one mole is then found by dividing 58.10 g/mole by the weight of one cube.

no. of cubes in one mole = $\dfrac{58.10 \text{ g/mole}}{3.76 \times 10^{-22} \text{ g/cube}}$

$$= 1.55 \times 10^{23} \text{ cubes/mole.}$$

4) One finds the number of KF particles in one mole (Avogadro's Number) by multiplying $1.55 \times 10^{23}$ cubes/mole by 4 formula units/cube, because there are 4 KF formula units in each cube.

Avogadro's Number = $4 \dfrac{\text{formula units}}{\text{cube}} \times 1.55 \times 10^{23} \dfrac{\text{cubes}}{\text{mole}}$

$$= 6.2 \times 10^{23} \text{ formula units/cube.}$$

● **PROBLEM** 223

Iron may crystallize in the face-centered cubic system. If the radius of an Fe atom is 1.26 Å, (a) Determine the length of the unit cell. (b) Calculate the density of Fe if its atomic weight is 55.85.

Solution: Picture the face of the cube with an iron

face-centered lattice

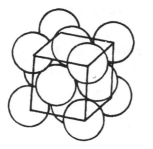 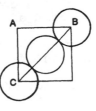

atom at each corner and one in the center of each face as shown in the figure.

The diagonal BC will be equal to the length L 4 radii of Fe atoms. Hence,

$$BC = 4 \times 1.26 \text{ Å} = 5.04 \text{ Å}$$

Using the Pythagorean Theorem, one knows that the squares of the two sides of a right triangle added together equal the square of the hypotenuse. Thus,

$$\overline{AB}^2 + \overline{AC}^2 = \overline{BC}^2$$

Because the sides of a square are equal – $\overline{AB} = \overline{AC}$. From this it can be seen that $2\,\overline{AB}^2 = \overline{BC}^2$. $\overline{AB}$ is defined as the length of the unit cell. Using

$$2\,\overline{AB}^2 = \overline{BC}^2,$$

it can be seen that

$$\overline{AB}^2 = \frac{\overline{BC}^2}{2} \qquad \text{and } AB = \sqrt{\frac{\overline{BC}^2}{2}} = \frac{BC}{\sqrt{2}}$$

Since we have already found BC, this equation can be solved. (BC = 5.04 Å.)

$$AB = \frac{\overline{BC}}{\sqrt{2}} = \frac{5.04 \text{ Å}}{1.42} = 3.55 \text{ Å}$$

The length of the unit cell is 3.55 Å.

(b) The density is defined as the mass in grams of a unit cell in one unit volume. In this problem we are given the mass of one mole of Fe and we have calculated the length of one side of the unit cube. To find the density one must calculate the number of atoms in one unit cell, the weight of one atom of Fe and the volume of one unit cube first.

The following rules may be used to determine the number of unit particles (atoms) associated with the face-centered lattice:

1) An atom at a corner contributes 1/8 of its volume to each of 8 adjacent cubes.

2) An atom in the face of a cube contributes 1/2 of its volume to each of 2 adjacent cubes.

In this lattice we have 8 corner atoms and 6 face atoms. To calculate the total number of atoms contributing to the unit cube one can use the above rules.

Total number of atoms = 8 corners × 1/8 atom/corner

+ 6 faces × 1/2 atom/face

= 4 atoms.

The unit cube therefore contains the equivalent of 4 atoms.

To find the weight of 1 atom one divides the molecular weight by Avogadro's Number $(6.02 \times 10^{23})$, because there are Avogadro's Number of atoms contained in a mole of atoms.

$$\text{weight of 1 atom} = \frac{55.85 \text{ g}}{6.02 \times 10^{23}} = 9.28 \times 10^{-23} \text{ g}$$

Since there are 4 atoms in this unit cube this figure is multiplied by 4.

$$\text{weight of 4 atoms} = \frac{4 \times (55.85 \text{ g})}{6.02 \times 10^{23}} = \text{weight of unit cube}$$

The volume of the cube is found by cubing the length of a side of the square because the volume is equal to the length × width × height of a rectangular solid. In a cube all three of these quantities are equal.

The volume of the cube = $(3.55 \times 10^{-8} \text{ cm})^3$.

$(1 \overset{\circ}{A} = 10^{-8} \text{ cm})$. We now have the quantities necessary to find the density. cc = cubic centimeters = $\text{cm}^3$.

$$\text{density} = \frac{\text{weight of unit cube}}{\text{volume of unit cube in cc}}$$

$$= \frac{\frac{4 \times (55.85 \text{ g})}{6.02 \times 10^{23}}}{(3.55 \times 10^{-8} \text{ cm})^3}$$

$$= 8.30 \text{ g/cc}$$

The density of Fe is therefore 8.30 grams per cubic centimeter.

Metallic gold crystallizes in the face-centered cubic
lattice. The length of the cubic unit cell, a, is
4.070 Å. (a) What is the closest distance between gold
atoms? (b) What is the density of gold?

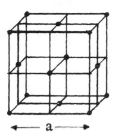

face-centered cubic

Solution:   An ordered array of atoms, ions, or molecules
is called a lattice. Every lattice is a three-dimensional
stacking of identical building blocks called unit cells.
The most symmetrical crystals have cubic lattices. There
are three kinds of elementary cubic lattices: the simple
cubic, the face-centered cubic, and the body-centered
cubic. The length of the cube edge is designated by the
symbol a. In a face-centered lattice, an atom would be
located at each corner of the unit cell and at the center
of each of the six faces.

(a) In the gold crystal above, the closest distance
from one corner atom to another corner atom is a,
4.070 Å. The distance from any corner atom to an atom
at the center of a face is one-half the diagonal of that
face. Using the fact that the diagonal of any square is
$\sqrt{2}$ × length of one side, the diagonal is then $a\sqrt{2}$.  Half
the diagonal is ½ $a\sqrt{2}$  or

½ (4.070 Å) $\sqrt{2}$ = 2.878 Å

This distance is the closest distance between atoms,
since this distance is shorter than the distance between
adjacent corners.

(b) The density is computed by first counting the
number of gold atoms that can occupy one unit cell. This
number is 1/8 times the number of occupied corners in
the unit cell plus ½ times the number of occupied face-
centers. Since a cube has 8 corners and 6 faces, the
number is

1/8 · 8 + ½ · 6 = 4

The volume of the cubic unit cell is $a^3$. The

density if the mass of 4 gold atoms (4 atomic Wt/Avogadro's #) divided by the volume of the unit cell, $a^3$. Thus, the formula for the density of a gold crystal lattice is

$$\text{Density} = \frac{(4 \text{ atoms})(\text{Atomic Wt. of gold})}{a^3 \text{ (Avogadro's number)}}$$

Substituting into the equation the Atomic Weight of gold, 197.0 g/mole, Avogadro's number, $6.023 \times 10^{23}$ atoms/mole, and converting 4.070 Å to $4.070 \times 10^{-8}$ cm, then

$$\text{Density} = \frac{(4 \text{ atoms})\ (197.0 \text{ g/mole})}{(4.070 \times 10^{-8} \text{ cm})^3 (6.023 \times 10^{23} \text{ atoms/mole})}$$

$$= 19.4 \text{ g/cm}^3.$$

● **PROBLEM** 225

Krypton crystallizes in the face-centered cubic system, with the edge of the unit cell 5.59 Å. Calculate the density of solid krypton. Assume 1 Å = $10^{-8}$ cm.

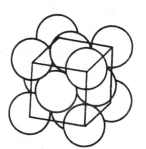

face-centered lattice

Solution:  To determine the density of krypton in g/cc, one can use the given information in the following way:

1) determine the number of atoms in one cube

2) determine the number of cubes in one cubic centimeter

3) determine the weight of one cube

4) determine the weight of one cubic centimeter (i.e. the density).

Determining the density:

1) In a face centered cube, there are 14 atoms contributing to the volume. 8 are corner atoms and 6 are

face atoms. Corner atoms contribute 1/8 of their volume to the cube and face atoms contribute 1/2 of their volume. Thus, the number of atoms making up the cube can be found.

no. of atoms = (1/8 atom/corner × 8 corners)

+ (1/2 atom/face × 6 faces) = 4 atoms.

2) 1 $\overset{\circ}{A}$ = $10^{-8}$ cm. Thus, the side of the cube is equal to 5.59 $\overset{\circ}{A}$ or 5.59 × $10^{-8}$ cm. The volume of a cube is equal to the length of the side cubed.

volume of one cube = (5.59 × $10^{-8}$ cm)$^3$ = 1.75 × $10^{-22}$ cc

The number of cubes in 1 cc is found by dividing 1 cc by the volume of one cube.

no. of cubes in 1 cc = $\dfrac{1 \text{ cc}}{1.75 \times 10^{-22} \text{ cc}}$ = 5.71 × $10^{21}$

3) The weight of one mole of krypton is 84. 6.02 × $10^{23}$ atoms (Avogadro's number) of it must weigh 84 g. Therefore, the weight of one atom is found by dividing 84 g by 6.02 × $10^{23}$ atoms. There are 4 atoms in one cube, thus the weight of one cube is equal to the weight of one atom times 4.

weight of 1 atom = $\dfrac{84 \text{ g}}{6.02 \times 10^{23} \text{ atoms}}$

weight of 4 atoms = 4 atoms × $\dfrac{84 \text{ g}}{6.02 \times 10^{23} \text{ atoms}}$

= 5.58 × $10^{-22}$ g.

4) The density is now found by multiplying the weight of 1 cube by the number of cubes in 1 cc.

density = no. of cubes in 1 cc × weight of 1 cube

density = 5.71 × $10^{21}$ /cc × 5.58 × $10^{-22}$ g = 3.19 g/cc.

● **PROBLEM** 226

It is known that AgCl has the same structure as NaCl. X-ray measurements of AgCl show a unit-cell edge length of .55491 nm. The density of the crystal is found to be 5.561 g/cm$^3$. Assuming $10^{-7}$ cm/nm, find the percentage of the sites that would appear empty. Assume that lattice vacancies are the only defects.

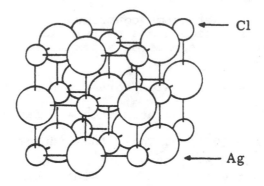

The Structure of Silver Chloride

<u>Solution</u>:  The key to solving this problem is to compute the theoretical density of the unit cell from its X-ray determined cubic edge length. Then, compare this value with the actual density to find the percent of the sites that would appear to be empty.

$$\text{Theoretical density} = \frac{\text{mass of 1 unit cell}}{\text{volume of 1 unit cell}}$$

Each unit cell possesses 4 ions of each $Ag^+$ and $Cl^-$.  This is determined from the fact that each corner ion contributes 1/8 of its volume to each of 8 adjacent unit cells, each face ion contributes ½ of its volume to the two adjacent cells, each edge ion contributes ¼ of its volume to each of 4 adjacent cells and each enclosed ion contributes its entire volume. As seen in the accompanying figure, there are 8 corner ions, 12 edge ions, 6 face ions and 1 totally enclosed ion. The molecular weight of AgCl is 143.32. Thus, the mass of 1 mole of unit cells is 4 times 143.32 g/mole of unit cells or 573.28 g/mole of unit cells. Since the unit cell is a cube,  volume = (cubic-edge length)$^3$ . Cubic edge length = .55491 nm or .55491 × $10^{-7}$ cm. Thus, volume becomes equal to $(.55491 × 10^{-7})^3 = 1.708 × 10^{-22} cm^3$.  As such, theoretical density =

$$\frac{\frac{573.28 \text{ g/mole of unit cells}}{6.022 × 10^{23} \text{ unit cells/mole}}}{1.708 × 10^{-22} \text{ cm}^3} = 5.571 \text{ g/cm}^3$$

The given density is 5.561 g/cm$^3$. Thus, the actual figure is

$$\frac{5.571 - 5.561}{5.571} × 100\% = .2\%$$      too small. This is

the percent of lattice that is vacant.

● **PROBLEM** 227

Find the ionic radius of I$^-$, given that the unit-cell edge length of LiI is measured to be .6240 nm. Assume

that the large, negative ions are in actual contact
with the diagonal.

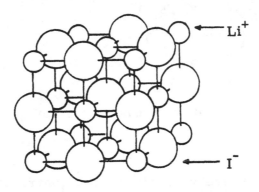

The Structure of Lithium Iodide

Solution:    To solve this problem, determine the length
of the face diagonal in LiI space lattice, which is the
pattern of the arrangement of atoms or molecules in a
crystal.Because LiI has the same form of the cubic unit
cell as NaCl, its space lattice will possess 3 $I^-$ ions
on the face diagonal. Thus.if one knows the length of the
face diagonal, one can divide  by 4 to deduce the radius
of $I^-$.   The face diagonal can be found (from the
Pythagorean Theorem) by multiplying ($\sqrt{2}$) times (cubic-
edge length). Thus, the face diagonal = ($\sqrt{2}$)
(.6240) = .8823 nm. However, face diagonal = 4 $I^-$ radii.

Thus, $I^-$ radius = $\frac{.8823}{4}$ = .2206 nm.

● PROBLEM 228

Consider a unit cell of sodium chloride in the accompanying
figure. (a) What fraction of a $Cl^-$  at each corner of the
cube is within the unit cell? (b) What fraction of a $Cl^-$
appears in each face? (c) What fraction of each $Na^+$ is within the unit cell? (Note that one complete $Na^+$ is in the
center.) (d) Add all the fractions for each ion within the
unit cell. What is the ratio of $Na^+$ to $Cl^-$?

Solution:    The following rules may be used to determine the
number of unit particles (atoms) associated with each type of
cubic unit cell:

     1) An atom at a corner contributes 1/8 of its volume
to each of 8 adjacent cubes.

     2) An atom on an edge of a cube contributes 1/4 of its
volume to each of 4 adjacent cubes.

     3) An atom in the face of a cube contributes 1/2 of

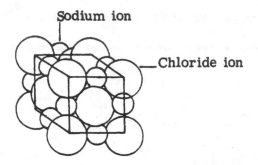

Crystal latice of sodium chloride

its volume to each of 2 adjacent cubes.

4) An atom completely within a cube contributes all of its volume to the unit cell.

(a) According to rule 1 an atom at a corner contributes 1/8 of its volume to the unit cell.

(b) From rule 3 one can see that an atom on the face of the unit cell contributes 1/2 of its volume to the cell.

(c) From rule 2 one can see that atoms on the edge of the unit cell contribute 1/4 of their volume to the unit cell. Therefore the sodium ions on the edge of the cell contribute 1/4 of their volume and the one in the center of the cell contributes its entire volume.

(d) There are 14 $Cl^-$ ions contributing to the cell. 8 are at corners and 6 are on faces - using parts a + b of this problem one can calculate the number of atoms that the $Cl^-$ ions contribute to the unit cell.

total number of $Cl^-$ ions = 8 corners × 1/8 atom/corner

+ 6 faces × 1/2 atom/face

= 4 atoms of $Cl^-$

The $Na^+$ ions can be dealt with in a similar manner. There are 13 $Na^+$ ions, 12 on edges and 1 in the center - using part c of this problem one can find the number of $Na^+$ ions contributing to the unit cell.

total number of $Na^+$ ions = 12 edges × 1/4 atom/edge

+ 1 atom in center

= 4 atoms of $Na^+$

There are 4 atoms of $Cl^-$ and 4 atoms of $Na^+$ contributing to the unit cell. The ratio will therefore be - $Na^+$ : $Cl^-$ is 4 : 4 or 1 : 1.

$Na^+$ and $Cl^-$ contribute equally to the unit cell.

Find the distance between a plane passing through the centers of one close-packed layer of spheres and another plane passing through the centers of an adjacent close-packed layer. You may assume that all spheres are identical with a radius of .200 nm.

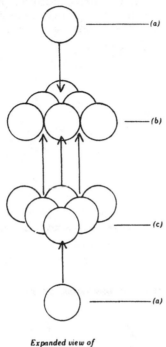

Expanded view of
stacking layers

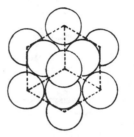

Edge view of stacking layers
showing face-centered
cubic unit cell

View perpendicular to
stacking layers showing
face-centered cubic
unit cell

Solution: The small fraction of a space lattice, which is the pattern of points that describes the arrangement of atoms or molecules in a crystal, is a unit cell. The unit cells contain only points that locate atomic or molecular centers. Since atoms are space-filling entities, their structures can be described as resulting from the packing together of representative spheres. The most efficient packing together of equal spheres is called closest packing. In this problem, you are asked for a distance between two planes, which is just the spacing between the layers of spheres. A face-centered cube, a type of unit cell, is a close packed structure. From the accompanying figure, you can see that the body diagonal of a face-centered cube is perpendicular to the close-packed stacking layers. Thus, if you stand it on a corner, the close packed layers will be parallel to the floor. To see this more clearly, you can use models. From this, you can relate the body diagonal to the spacing between layers. Namely, body diagonal = 3 times the spacing. Thus, you need to find the length of the body diagonal.

Body diagonal = $\sqrt{3}$ (edge length).

Edge length  = $1/\sqrt{2}$ (face diagonal).

The face diagonal = 4 times the radii of the spheres, as seen from the space lattice. You are told the radius of a sphere is .200 nm. Thus, face diagonal = 4(.200 nm) = .800 nm. Thus, edge length = $(1/\sqrt{2})(.800)$ = .5657 nm and body diagonal = $(\sqrt{3})(.5657)$ = .9798 nm. Recall, body diagonal is 3 times the spacing, which means the distance between the planes is $\cdot\dfrac{.9798 \text{ nm}}{3}$ = .327 nm.

# PHASE DIAGRAMS

● **PROBLEM** 230

The diagram below is an example of a phase diagram for a pure substance. To what phases do the regions A, B, and C correspond?

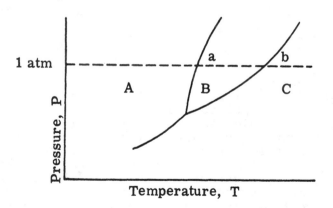

Solution:   Following the 1 atm constant pressure line from left to right, we are proceeding from low values of the temperature to high values. Therefore, we will intersect the three regions in the order solid-liquid-vapor. The regions A, B, C hence correspond to the solid, liquid, and vapor phases, respectively. Point a denotes the normal freezing (melting) point of the substance and the point b denotes the normal boiling point.

Draw a labelled phase diagram for a substance Z which has the following properties; normal boiling point = 220°C, normal freezing point 80°C, and triple point 60°C and .20 atm. Predict the freezing and boiling, if the pressure were .80 atm?

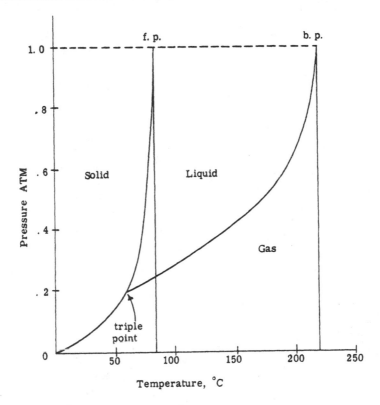

Temperature, °C

Solution: To draw this diagram you want to understand all the terms involved. The relation between solid, liquid, and gaseous states as a function of the given temperature and pressure can be summarized on a graph known as a phase diagram. From the given experimental observations, you can draw the diagram. The lines which separate the states in a diagram represent an equilibrium between the phases. The intersection of the three lines is called the triple point, where all three phases are in equilibrium with each other. By normal boiling and melting points you mean those readings taken at 1 atm. Thus, the phase diagram can be written as shown in the accompanying figure.

From the diagram you see that if the atm was .80, the b.p. and f.p. would drop, respectively, to 215°C and 85°C.

# CHAPTER 7

# PROPERTIES OF LIQUIDS

Basic Attacks and Strategies for Solving Problems in this Chapter. See pages 213 to 250 for step-by-step solutions to problems.

Liquids differ from solids in that the molecular structure is random, and the liquid will normally assume the shape of the vessel or container in which it is placed. Liquids are characterized by the following properties:

**Density:** The mass per unit volume occupied by a liquid is an important property used to characterize a liquid. Liquids are nearly incompressible and require extreme pressures to make a significant difference in the density. Also, only small differences in the density occur with moderate changes in the temperature; hence the density of a liquid is frequently considered to be a constant, except when very precise values are required. Water has a density of 1 gram per cubic centimeter; common hydrocarbons, such as those in gasoline, have densities of approximately 0.7 g/cc and float on water, while chloroform ($CHCl_3$), a common organic solvent, has a density of 1.5 g/cc and sinks in water.

**Viscosity:** The viscosity is the internal resistance of a liquid to shear stresses. Although not discussed in the problems of this text, it is an important characteristic that reflects the ability of a liquid to flow. Low viscosity liquids like water and gasoline flow and are pumped easily. Higher viscosity liquids like molasses flow very slowly (hence the cliché "slower than cold molasses"). Supercooled liquids, like glass, are so viscous that they do not flow at all and are usually considered solids.

**Vapor Pressure:** Liquids are in equilibrium with the gas or vapor above the liquid. The pressure of the gas above a pure liquid is called the vapor pressure or sometimes, for emphasis, the pure component vapor pressure. The vapor pressure is a strong function of the temperature.

**Surface Tension:** Liquids are characterized by attractive forces between the molecules. The strength of these attractive forces is characterized by the surface tension. At the surface of a liquid, the attractive forces tend to pull the surface atoms or molecules back into the liquid. This force can be measured as a force per unit length ($N/m$) of the liquid film.

The density of irregularly shaped solids is frequently measured by submerging them in a liquid to determine their volume. The solid volume is equal to the volume of liquid displaced. Since the determination of the mass is usually straightforward (it can be weighed on a balance), the density of irregular solids can be easily calculated once the volume is known.

## Freezing Point Depression and Boiling Point Elevation

The freezing point is lowered and the boiling point raised for liquids when a solute is dissolved. The freezing point depression and boiling point elevation are proportional to the number of moles dissolved per unit mass of solvent. The change in freezing or boiling point is calculated from the simple equation

$$\Delta T_f = k_f (m) \qquad\qquad 7\text{-}1$$

$$\Delta T_b = k_b (m) \qquad\qquad 7\text{-}2$$

where $k_f$ and $k_b$ are the freezing and boiling point constants respectively, and m is the molality (moles of solute/1000 g solvent) of the solution. The freezing and boiling point constants are determined experimentally and are useful because they are a characteristic of the solvent and usually not a characteristic of the solute. Freezing point depressions are frequently used to determined the molecular weights of solutes since the mass of solute and freezing point depression are easily measured.

Equations 7-1 and 7-2 can be written with the molecular weight shown explicitly as

$$\Delta T_f = k_f \frac{\text{(mass of solute/1000 g solvent)}}{\text{M.W.}} \qquad \text{7-3}$$

$$\Delta t_b = k_b \frac{\text{(mass of solute/1000 g solvent)}}{\text{M.W.}} \qquad \text{7-4}$$

The key to solving the problems involving freezing point depression and boiling point elevation is the use of these equations to solve for the unknown quantity — $\Delta T$, $k_f$ or $k_b$, or the molecular weight, (M.W.).

## Vapor Pressure–Raoult's Law

The vapor pressure is a characteristic of a liquid and is a strong function of temperature. It is precisely defined as the equilibrium gas (or vapor) pressure above a pure liquid at a specified temperature with no other species present. Except at extreme pressures, it is an excellent approximation to assume that the vapor pressure is independent of the total pressure, i.e., unaffected by the presence of other species in the gas phase.

There are empirical equations that permit the vapor pressure to be calculated, but these equations contain approximations and, for precise work, measurements should be used.

Raoult's Law predicts the partial pressure above a liquid which contains a mixture of compounds. It is correct only for ideal liquid mixtures, i.e., liquid mixtures where the attractive forces are the same between like and unlike molecules. It is a good approximation for mixutures of isomers (such as normal pentane, isopentane, and ncopentane) or adjacent members of homologous series (such as benzene, toluene, and xylene).

Raoult's Law is written as follows:

$$p_i = x_i P^{\circ}_i \qquad \text{7-5}$$

where $p$ is the partial pressure in the gas phase, $x$ is the mole fraction in the liquid phase, and $P^{\circ}$ is the pure component vapor pressure (or just vapor pressure). The subscript $i$ indicates that this equation can be applied to each species in the solution, i.e., $i = 1, 2, 3, \ldots$. If two of the terms in Equation 7-5 are known, the third can be easily calculated. If Raoult's Law holds for one species in a solution, it will hold for all the species. Departures from Raoult's Law are common in nature, but are not discussed in this text.

One key to working the problems involving Raoult's Law is to observe that the total pressure is the sum of the partial pressures. From the ideal gas law, the mole fractions in the vapor phase are simply the respective partial pressures divided by the total pressure (sum of the partial pressures). As well, the mole fractions of the various solutes must add up to 1.

$$YI = P_i/P \qquad\qquad\qquad 7\text{-}6$$

where $Y$ is the vapor phase mole fraction and $P$ is the total pressure.

## Clausius Clapeyron Equation of Vaporization

The Clausius Clapeyron equation is useful in calculating changes in vapor pressure with temperature. It contains several approximations and is derived by integrating the Clapeyron equation, which has a firm basis in theory.

$$\frac{dP^\circ}{dT} = \frac{\Delta H}{T\Delta V} \qquad\qquad\qquad 7\text{-}7$$

where $\Delta H$ and $\Delta V$ are the enthalpy and volume changes associated with vaporization of the liquid. If it is assumed that $\Delta V$ is simply the vapor phase volume ($\Delta V = RT/P$), and that $\Delta H$ is a constant, Equation 7-7 can be integrated to obtain the the Clausius Clapeyron equation.

$$ln\,(P^\circ_2/P^\circ_1) = -\Delta H^{vap}/R\,(1/T_2 - 1/T_1) \qquad\qquad 7\text{-}8$$

The assumptions are valid over limited temperature ranges for many species.

Note: Several problems (see Problems 263, 267, 268) in this text show the Clausius Clapeyron equation written as

$$log\,(P^\circ_2/P^\circ_1) = \Delta H/19.15\,(1/T_1 - 1/T_2) \qquad\qquad 7\text{-}9$$

This form of the equation may be confusing. The factor, 19.15, is the product of the gas constant, $R$, in joules/mole $^\circ$K (8.31) and the conversion factor from natural to common logarithms ($log_e$ to $log_{10}$), i.e., $8.31 \times 2.303 = 19.15$. To use this form of the equation, $\Delta H$ must be expressed in joules/mole and base

10 logarithms must be used.

It is possible, from Equation 7-7, to calculate changes in vapor pressure from the enthalpy of vaporization or to calculate the enthalpy (or heat) of vaporization from measurements of the vapor pressure at more than one temperature.

Osmotic pressure ($\pi$), for very dilute solutions, can be represented by the equation

$$\pi = nRT/V \qquad\qquad 7\text{-}10$$

where $n$ is the number of moles of solute in a volume, $V$, of solution. It is important to note that this equation does not apply except for very dilute solutions. With Equation 7-9 it is possible to determine the molecular weight of a species by measuring the osmotic pressure produced by a known mass of solute.

Surface tension is usually determined by measuring the force required to pull a ring of known diameter from a solution or by measuring the height that a liquid "climbs" the wall of a wetted vessel.

If a ring is pulled from a liquid, two liquid films (inside and outside) form between the ring and the liquid. The force required to pull the ring from solution divided by the length of surface ($2\pi D$) is the surface tension. When surface tension is measured by determining the height that the liquid climbs the wall of a wetted vessel, the same principle is used to determine the surface tension. The equation that relates surface tension to the height of the liquid is

$$\gamma = 1/2\; r \cdot h\rho g \qquad\qquad 7\text{-}11$$

where $\gamma$ is the surface tension, $\rho$ the density of the liquid, $r$ the radius of the tube containing the liquid, $h$ the height that the liquid film climbs above the free surface of the liquid, and $g$ is the constant downward acceleration of gravity.

Step-by-Step Solutions to
Problems in this Chapter,
"Properties of Liquids"

## DENSITY

● **PROBLEM** 232

What volume of a block of wood (density = 0.80 g/cm$^3$),
which weighs 1.0 kg, will be above the water surface
when the block is afloat? (Note: A floating object
displaces its own weight of water.)

Solution: Since a floating object displaces its own
weight in water, 1 kg of water is displaced by this
block of wood. One can find the volume of the block of
wood above the water by solving for the volume of the
block and subtracting the volume of 1 kg of water from
it. One uses the density to solve for the volume.

$$\text{density} = \frac{\text{weight}}{\text{volume}} = \frac{g}{cm^3}$$

Therefore:

$$\text{volume} = \frac{\text{weight}}{\text{density}}$$

Solving for the volume of the wood: 1 kg = 1000 g.

$$\text{volume} = \frac{1000 \text{ g}}{.80 \text{ g/cm}^3} = 1.25 \times 10^3 \text{ cm}^3$$

Solving for the volume of the water displaced:

By definition the density of water is 1.0 g/cm$^3$.

$$\text{volume} = \frac{1000 \text{ g}}{1.0 \text{ g/cm}^3} = 1.00 \times 10^3 \text{ cm}^3$$

volume of wood above water = volume of wood - volume of water

volume of wood above water = 1250 cm$^3$ - 1000 cm$^3$

$$= 250 \text{ cm}^3$$

● **PROBLEM** 233

A chemist dropped a 200 g object into a tank of water. It displaced 60 ml. of the water when it sunk to the bottom of the tank. In a similar experiment, it displaced only 50 g of an oil into which it was dropped. Calculate the density of the object and the oil.

Solution:  The density ($\rho$) of a substance is defined as its mass divided by its volume.

$$\rho = \frac{\text{mass}}{\text{volume}}$$

Thus, to solve for the densities of the object and the oil, one must first calculate their respective masses and volumes. The mass of the object is 200 g, but the volume is not given. An object dropped in any liquid displaces a volume of liquid equal to the volume of the object. The object displaces 60 ml of water; therefore, the volume of the object is 60 ml. Solving for the density of the object:

$$\rho = \frac{\text{mass}}{\text{volume}} = \frac{200 \text{ g}}{60 \text{ ml}} = 3.33 \text{ g/ml}.$$

One is given that the object displaces 50 g of the oil, and from the water experiment, it is known that the volume of the object is 60 ml. Since the object displaces the same volume of liquid as it occupies, 60 ml of the oil weighs 50 g. Solving for the density of the oil:

$$\rho = \frac{\text{mass}}{\text{volume}} = \frac{50 \text{ g}}{60 \text{ ml}} = .833 \text{ g/ml}.$$

● **PROBLEM** 234

Assuming that the density of water is .9971 g/cm$^3$ at 25°C and that of ice at 0° is .917 g/cm$^3$, what percent of a water jug at 25°C should be left empty so that, if the water freezes, it will just fill the jug?

Solution:  Density = $\frac{\text{mass}}{\text{volume}}$ .

When water freezes, the mass is constant, but its density decreases. This means, therefore, that an increase in volume occurs. To solve this problem, you must determine the change in volume of water from 25° to 0°. Because you are given the densities of the substance at each temperature, you can say for 1 mole of $H_2O$ (MW = 18)

$$\frac{mass}{volume} = \frac{18 \text{ g/mole}}{y} = \frac{0.997 g}{cm^3} \text{ for water at 25°C and}$$

$$\frac{mass}{volume} = \frac{18 \text{ g/mole}}{x} = \frac{0.917 g}{cm^3} \text{ for ice at 0°C.}$$

Rewriting, you have .997 y = 18 and .917 x = 18. Both expressions equal 18, which means .997 y = .917 x, or

$$\frac{x}{y} = \frac{.997}{.917} = 1.087.$$

This means, then, that at 0°C the volume is 1.087 times greater than the original at 25°C. If you let B equal the fraction of water to be put in the jug, then you can say 1.087 B = 1, or B = .920. In other words, only 92% of the jug should be filled at 25° to obtain a completely filled jug at 0°. This means that 8 % is empty space.

● **PROBLEM** 235

The molecular diameter of an $N_2$ molecule, as deduced from the Van der Waals b parameter, is $3.15 \times 10^{-8}$ cm. The density of liquid nitrogen is 0.8081 $g/cm^3$. On a hard-sphere model, what fraction of the liquid volume appears to be empty space?

<u>Solution</u>: To find the fraction of the liquid volume that appears to be empty space, one must subtract the volume of spheres from the total volume of the liquid. One can find the volume of the spheres that constitute one mole of $N_2$ liquid by multiplying the volume of one sphere by $6.02 \times 10^{23}$, the number of spheres in one mole. The volume of a sphere is equal to $4/3 \pi r^3$ where r is the radius of the sphere. The diameter of the $N_2$ sphere is given as $3.15 \times 10^{-8}$ cm, thus the radius is half of length or $1.58 \times 10^{-8}$ cm. Solving for the volume of 1 sphere:

volume of 1 sphere = $4/3 \pi (1.58 \times 10^{-8} \text{ cm})^3$

$$= 1.652 \times 10^{-23} \text{ cm}^3/\text{sphere}$$

Volume of 1 mole of spheres

$$= (6.02 \times 10^{23} \text{ spheres/mole}) \times (1.652 \times 10^{-23} \text{ cm}^3/\text{sphere})$$

$$= 9.945 \text{ cm}^3/\text{mole}$$

One can find the total volume of 1 mole of liquid $N_2$ by dividing the molecular weight by the density. (MW = 28)

215

$$\text{total volume} = \frac{28 \text{ g/mole}}{0.8081 \text{ g/cm}^3} = 34.65 \text{ cm}^3/\text{mole}$$

The volume of the empty space is equal to the volume of the spheres subtracted from the total volume.

volume of empty space per mole = 34.65 cm$^3$ - 9.945 cm$^3$

$$= 24.70 \text{ cm}^3$$

The percent of the volume taken up by empty space is equal to the volume of the space divided by the total volume and multiplied by 100.

$$\text{\% of empty space} = \frac{24.70 \text{ cm}^3}{34.65 \text{ cm}^3} \times 100$$

$$= 71.30 \text{ \%.}$$

# FREEZING POINT DEPRESSION AND BOILING POINT ELEVATION

● **PROBLEM** 236

The freezing point constant of toluene is 3.33°C per mole per 1000 g. Calculate the freezing point of a solution prepared by dissolving 0.4 mole of solute in 500 g of toluene. The freezing point of toluene is - 95.0°C.

Solution:   The freezing point constant is defined as the number of degrees the freezing point will be lowered per 1000 g of solvent per mole of solute present. The freezing point depression is related to this constant by the following equation.

freezing pt depression = molality of solute × freezing pt
constant

The molality is defined as the number of moles per 1000 g of solvent. Here, one is given that 0.4 moles of solute are added to 500 g of solvent, therefore there will be 0.8 moles in 1000 g.

$$\frac{0.4 \text{ moles}}{500 \text{ g}} = \frac{0.8 \text{ moles}}{1000 \text{ g}}$$

The molality of the solute is thus 0.8 m. One can now find the freezing point depression. The freezing point constant for toluene is 3.33°.

freezing point depression = molality × 3.33°

$$= 0.8 \times 3.33° = 2.66°$$

The freezing point of toluene is thus lowered by 2.66°.

freezing point of solution = (- 95°C) - 2.66° = - 97.66°C.

● PROBLEM 237

By how much will 50 grams of water have its freezing point depressed if you add 30 grams (molecular weight 80) of glucose to it?

Solution:    The addition of any substance to water will alter its boiling or freezing point. To determine the amount of change, you must know the concentration of solute (the substance dissolved) in the solvent (water). This information is required because the freezing point depression, $\Delta T_f$, equals the molal freezing point constant, $k_f$, times the molality (m).

$$\Delta T_f = k_f \ (m).$$

The concept of molality refers to the number of moles of solute per 1 kilogram of solvent.

The solute, glucose, weighs 30 grams. Therefore, the number of moles of glucose is 30 g/180 g/mole = 1/6 moles. You have 50 grams of water. However, molality refers to moles per thousand grams. As such, a conversion is required; namely 50/1000. Hence, the molality is

$$\frac{\frac{1}{6} \ \text{moles}}{\frac{50}{1000} \ \text{g}} = 3.33 \ m$$

Recall that the amount of depression of the freezing point is defined as $\Delta T_f = \left(k_f \ m\right)$. $k_f$ is given for water as - 1.86 deg mole$^{-1}$. You calculated m. The temperature depression is thus $\Delta T_f = k_f \times m = - 1.86° \times 3.33 \ m = - 6.2°$.

● PROBLEM 238

What is the freezing point of a solution of 92 g of alcohol ($C_2H_5OH$) and 500 g of $H_2O$?

Solution:    The freezing point is dependent on the number of solute particles. One mole of a substance dissolved in 1000 g of water lowers the freezing point 1.86°. One uses the following equation to find the freezing point depression:

freezing pt. depression = molality of solute × 1.86°.

Now, one must find the molality of the alcohol. The molality is defined as the number of moles in 1000 g of

$H_2O$. In this solution, there are 92 g of alcohol present. The number of moles is found by dividing 92 g by the molecular weight of the alcohol. MW of $C_2H_5OH$ = 46.

$$\text{no. of moles} = \frac{92.0 \text{ g}}{46 \text{ g/mole}} = 2 \text{ moles}$$

There are 2 moles in 500 g of water.

In 1000 g of $H_2O$, there would be twice this amount or 4 moles. Therefore, the molality of the alcohol is 4 m.

One can now find the freezing point depression.

$$\text{freezing pt. depression} = \text{molality} \times 1.86°$$

$$= 4 \times 1.86° = 7.44°.$$

The freezing point of $H_2O$ is 0°C.

freezing pt of solution = freezing pt of $H_2O$ −
freezing pt of depression

$$= 0° - 7.44° = - 7.44°.$$

● **PROBLEM** 239

Calculate the composition (molality) of an alcohol - water mixture which will not freeze above a temperature of - 10°C (+14°F). (MW of alcohol = 46; Freezing point constant for water ($K_f$) = 1.86°.)

<u>Solution</u>: For dilute solutions, the number of degrees that the freezing point is lowered by adding a solute to a solvent is equal to the molality of the solute times the freezing point constant ($K_f$) of the solvent.

freezing point depression = Molality × $K_f$

Because the solvent in this case is water, the $K_f$ for water is used in solving the above equation. The freezing point depression is found by subtracting the new freezing point of the water from the original one. The original freezing point of water is 0°C.

freezing pt. depression = 0 - (- 10) = 10°C.

Since the freezing point depression and the freezing point constant are known, one can now solve for the molality.

$$\text{molality} = \frac{\text{freezing point depression}}{\text{freezing point constant}}$$

$$\text{molality} = \frac{10°}{1.86°} = 5.4 \text{ molal}$$

Molality is defined as the number of moles of solute present per kilogram of solvent. Here, the solute is the alcohol and the solvent is water. The number of grams of the alcohol present in 1000 g of water can be found by multiplying the molality by the molecular weight of the alcohol.

46 g/mole × 5.4 moles/kg of $H_2O$ = 250 g/kg $H_2O$

Therefore, if 250 g of this alcohol are added to 1000 g of water, the freezing point of the water will be lowered by 10°C.

● **PROBLEM** 240

The molal freezing point constant for a certain liquid is 0.500°C. 26.4 g of a solute dissolved in 250 g of this liquid yields a solution which has a freezing point 0.125° below that of the pure liquid. Calculate the molecular weight of this solute.

Solution:    A mole of a substance in 1000 g of $H_2O$ gives a definite and known lowering of the freezing point. By determining the freezing point of a solution of known concentration, one can calculate the molecular weight of the dissolved substance. A general formula may be developed for this kind of calculation:

$$\text{molecular weight} = \frac{\text{grams solute}}{\text{kg of solvent}} \times \frac{K_f}{\Delta T_f}$$

where $K_f$ is the freezing point constant and $\Delta T_f$ is change in the freezing point. Here, 26.4 g of a solute is dissolved in 250 g of liquid. One can find the number of grams of solute using the following ratio: (Note: there are 1000 g in 1 kg.) Let X = number of g of solute in 1000 g

$$\frac{26.4 \text{ g}}{250 \text{ g}} = \frac{X}{1000 \text{ g}}$$

$$X = \frac{1000 \text{ g} \times 26.4 \text{ g}}{250 \text{ g}} = 105.6 \text{ g}$$

One can now solve for the molecular weight.

$$\text{molecular weight} = \frac{105.6 \text{ g}}{1 \text{ kg}} \times \frac{0.500°C}{0.125°}$$

$$= 422.4 \text{ g}$$

Liquid naphthalene normally freezes at 80.2°C. When 1 mole of solute is dissolved in 1000 g of naphthalene, the freezing point of the solution is 73.2°C. When 6.0 g of sulfur is dissolved in 250 g of naphthalene, the freezing point is 79.5°C. What is the molecular weight of sulfur?

Solution: In order to determine the molecular weight of sulfur, we must determine how many moles of sulfur corresponds to 6.0 g.

When 1 mole of solute is added to 1000 g of naphthalene, the normal freezing point is lowered by 7°C, from 80.2°C to 73.2°C. This corresponds to a concentration of 1 molal (1 mole solute/1000 g solvent = 1 molal). The ratio of sulfur to naphthalene in the sulfur-naphthalene solution is 6.0 g sulfur/250 g naphthalene = 0.024. Thus, adding 6.0 g of sulfur to 250 g of naphthalene is equivalent to adding 24 g of sulfur to 1000 g of naphthalene (because 24 g sulfur/1000 g naphthalene is also 0.024).

The observed freezing point depression in the sulfur-naphthalene solution is 0.7°C (from 80.2°C to 79.5°C). This is the freezing point depression one would obtain by adding 24 g of sulfur to 1000 g of naphthalene. But this is one-tenth the lowering one would obtain by adding 1 mole of sulfur (the lowering would then be 7.0°C), hence 24 g of sulfur must correspond to one-tenth of a mole.

Thus, the apparent molecular weight of sulfur is

$$\frac{24 \text{ g}}{0.1 \text{ mole}} = 240 \text{ g/mole.}$$

A chemist wishes to determine the molecular weight and molecular formula of fructose (a sugar). He places .946 g of it in 150 g of $H_2O$ (water) and finds that the freezing point of water is depressed to $-0.0651$°C. Determine the molecular weight and formula of fructose, assuming that the simplest formula of fructose is $(CH_2)O$.

Solution: To answer this question, you must know the quantitative relationship for the depression of the freezing point, $\Delta T_f$. This relation can be expressed as $\Delta T_f = k_f \times m$, where $\Delta T_f$ = the actual depression, $k_f$ = molal freezing point constant ($-1.86$ deg mol$^{-1}$ for water) and m = molality of the solution. Molality is defined as moles of solute per 1 kg of solvent.

In this problem, water is the solvent and the fructose compound is the solute. As such, you can express the freezing point relationship as

$$\Delta T_f = k_f \times \frac{\dfrac{\text{grams solute}}{\text{molecular weight}}}{\text{grams solvent}} \times 1000, \qquad \text{where}$$

the numerator is the number of moles of solute. You possess all of the unknowns, except the molecular weight, which can be calculated. In other words, solve for the molecular weight by substitution. Hence.

$$-0.0651°C = -1.86°C/\text{mole} \times \frac{\dfrac{.946\ g}{\text{mol. wt.}}}{150\ g} \times 1000\ g/\text{kg}$$

mol wt. = 180 g/mole.

To determine the molecular formula, you must figure out what formula has a molecule weight of 180 g, and yet is still a multiple of $(CH_2)O$.

Thus, one determines the weight of 1 $(CH_2)O$ and divides this into 180 g. (MW of $(CH_2)O$ = 30.)

$$\text{no. of } (CH_2)O \text{ in fructose} = \frac{180\ g}{30\ g/(CH_2)O} = 6\ (CH_2)O$$

Therefore, the formula for fructose is 6 × $(CH_2)O$ or $C_6H_{12}O_6$.

● **PROBLEM** 243

A chemist discovered a new compound and wished to determine its molecular weight. He placed 2.00 grams of the unknown compound in 10 grams of water and found that the normal freezing point of the water was depressed to - 1.22°C. What is the molecular weight of the new compound?

Solution:  To answer this question, you must know the quantitative relationship for the depression of the freezing point, $\Delta T_f$. This relationship can be expressed as $\Delta T_f = k_f \times m$, where $\Delta T_f$ is the actual  depression, $k_f$ = molal freezing point constant (- 1.86 deg mole$^{-1}$ for water), m = molality of the solution. Molality is defined as the number of moles of solute per 1 kilogram of solvent. In this problem, water is the solvent and the unknown compound is the solute. You can express the freezing point relationship as

$$\Delta T_f = k_f \times \frac{\text{g.solute/mol.wt.}}{\text{g.solvent}} \times 1000,$$

where the numerator is the number of moles of solute. You have all of the unknowns, except the molecular weight. Solving for the molecular weight of the unknown:

$$- 1.22° = - 1.86° \times \frac{2/mol.wt.}{10} \times 1000.$$

mol. wt. = 305 g/mole.

Calculate the approximate freezing point of a solution of 162 g of HBr in 500 g $H_2O$, assuming that the acid is 90% ionized.

<u>Solution:</u>    The freezing point of a solution is a colligative property, which means that it depends upon the number of solute particles present in the solution. The depression in the freezing point is related to the number of particles by the equation:

freezing pt. depression = $K_f$ × molality of solute.

where $K_f$ is the freezing point constant. The freezing point constant is defined as the number of degrees the freezing point is lowered when 1 mole of solute is added to 1 kg of solvent. $K_f$ for $H_2O$ is 1.86°. One now must determine the molality of the solution. The molality of unionized HBr is found by first determining the number of moles present. (MW of HBr = 81.) The number of moles present equals the weight divided by the molecular weight.

$$no. \text{ of moles} = \frac{162 \text{ g}}{81 \text{ g/mole}} = 2 \text{ moles}$$

One can now find the molality by dividing the number of moles by the number of kilograms of solvent (1000 g = 1 kg)

$$molality = \frac{2 \text{ moles}}{0.5 \text{ kg}} = 4 \text{ m}$$

The molality of the unionized HBr is 4.0 m, but one is given that the compound is 90% ionized. This means that the total number of particles in the solution will increase. One must therefore calculate the effective molality. HBr ionizes as shown:

HBR $\overset{\leftarrow}{\rightarrow}$ $H^+$ + $Br^-$

One HBr forms, one $H^+$ and one $Br^-$. Because the HBR is 90% ionized, only 10% of the original HBr is left and .90 × molality of $H^+$ and $Br^-$ ions are formed. The effective molality of the solution is the sum of the concentrations of the three components.

| HBr | .10 × 4m = .4 m |
| $H^+$ | .90 × 4m = 3.6 m |

Br⁻         .90 × 4m = <u>3.6 m</u>

effective molality = 7.6 m

One can now find the amount that the freezing point has lowered.

freezing pt. depression = $K_f$ × eff. molality

freezing pt. depression = 1.86° × 7.36 = 13.7°

The freezing point of water is lowered by 13.7°. It was originally 0°C, it is now - 13.7°C.

● **PROBLEM** 245

What percent of the AB particles are dissociated by water if the freezing point of a(.0100 m) AB solution is -0.0193°C? The freezing point lowering constant of water is - 1.86°C/mole.

<u>Solution</u>:   To answer this question,you must determine the number of particles dissociated and divide that number by the initial number of particles available for dissociation. To accomplish this, you must be able to relate quantitatively the freezing point depression with the concentration of particles in solution. From the equation, $\Delta T_f = k_f\, m$, where $\Delta T_f$ = the depression of the freezing point, $k_f$ = the freezing point depression constant, and m = molality of the solution, one can determine the concentration of the particles. From the data given, the total concentration of particles,

$$m = \frac{\Delta T_f}{k_f} = \frac{-\ .0193°C}{-\ 1.86°C/m} = .0104\ m$$

With this in mind, let x be the concentration of AB particles that dissociate. As such, you have x molal for $A^+$ and x molal for $B^-$, since AB → $A^+$ + $B^-$. You are told that you start with a molal solution of.0100. Therefore, the number of particles undissociated is .01 - x. As such, the total concentration of particles is x + x + (.0100 - x), which is equal to .0100 + x. Before, however, you said it was  .0104 m from the equation $\Delta T_f = k_f\, m$. It follows then, that you can equate the two. Thus, .0104 m = (.0100 + x)m. Solving for x, you obtain, x = .0004 m.

This number, x = .0004 m, represents the molal concentrations of each $A^+$ and $B^-$. Therefore, it must be the number of particles that dissociated. You started with .0100 m  concentration of particles, Therefore, the percent that dissociated is

$$\frac{.0004}{.0100} = .04 = 4\%.$$

What is the approximate boiling point at standard pressure of a solution prepared by dissolving 234 g of NaCl in 500 g of $H_2O$?

Solution:   The boiling point of a solution is a colligative property. This means that it depends on the ratio of number of solute particles to number of solvent particles or molality. Molality is defined as the number of moles of solute divided by the number of kilograms of solvent.

When water is the solvent the boiling point constant, $K_b$, is 0.52°. This means that when one mole of solute is in 1 kg of $H_2O$ the boiling point of water is raised 0.52°.

boiling pt elevation = $K_b$ × molality of solute

One can find the boiling point elevation once the molality of the NaCl is found. To determine the molality one must first determine the number of moles present. This is done by dividing the number of grams present by the molecular weight. (MW of NaCl = 58.5.)

no. of moles = $\dfrac{234 \text{ g}}{58.5}$ = 4.0 moles

There are 4.0 moles of NaCl in 500 g, or 0.5 kg, of $H_2O$. The molality can now be found.

molality = $\dfrac{\text{moles of solute}}{\text{kg of solvent}}$

molality = $\dfrac{4.0 \text{ moles}}{0.5 \text{ kg}}$ = 8.0 m

NaCl is an electrolyte which means that when it is placed in $H_2O$ it dissociates into its ions.

NaCl $\overset{\leftarrow}{\rightarrow}$ $Na^+$ + $Cl^-$

This means that for every NaCl present, there are two ions formed. There are thus twice as many moles of particles as molecules of NaCl present. The effective molality is, therefore, twice the true molality.

effective molality = 2 × 8.0 m = 16.0 m

One can now solve for the boiling point elevation.

boiling point elevation = 0.52° × 16.0 m = 8.32°

The boiling point of the water is raised 8.32°. It was originally 100°C, therefore the new boiling point is 100°C + 8.32° = 108.32°C.

Ethanol boils at 78.5°C. If 10 g of sucrose ($C_{12}H_{22}O_{11}$) is dissolved in 150 g of ethanol, at what temperature will the solution boil? Assume $K_b$ = 1.20°C/M for the alcohol.

Solution: When a non-volatile solute, such as sucrose, is dissolved in a solvent, such as ethanol, it will raise the boiling point of the solvent. The boiling point elevation can be found by using the equation

$$\Delta T_b = K_b \eta,$$

where $\Delta T_b$ is the boiling point elevation, $K_b$ is the elevation constant, and $\eta$ is the molality of the solution. You want to determine $\Delta T_b$ and are given $K_b$. Thus, you need to determine the molality of the solution. Molality is defined as the number of moles of solute per 1 kg of solvent, i.e. moles solute/1 kg solvent.

Solving for molarity of sucrose: (MW of $C_{12}H_{22}O_{11}$ = 342.3).

$$\text{Moles of solute} = \frac{\text{grams solute}}{\text{molecular wt. solute}}$$

$$= \frac{10 \text{ g}}{342.3 \text{ g/mole}} = 2.921 \times 10^{-2} \text{ moles.}$$

You have 150 g of solvent, ethanol. But molality is per 1 kg, so that you must multiply 150 g by 1 kg/1000 g.

$$\text{Molality} = \frac{.02921 \text{ mole}}{\frac{150}{1000} \text{ kg}} = .195 \text{ M}$$

Thus, the elevation of the boiling point is

$$\Delta T_b = K_b \eta = (1.20°C/M)(.195 \text{ M}) = .23°C.$$

Thus, the boiling point of the solution is

78.5°C + .23°C = 78.73°C.

Give the approximate boiling points at sea level for the following: (a) 2 molal HBr; (b) Suspension of 100 g of powdered glass in one liter of water; (c) $1.2 \times 10^{24}$ sucrose molecules/liter; (d) 0.5 molal $BaCl_2$.

<u>Solution</u>:    The boiling point is a colligative property and, therefore, depends upon the number of particles present in 1 kg of solvent. The boiling point constant for water is 0.52°C. This means that for each mole of particles dissolved in 1 kg of $H_2O$ the boiling point will be elevated by 0.52°C.

boiling pt. elevation = 0.52°C × molality of solute.

The molality is defined as the number of moles in one kg. of solvent.

(a) 2 molal HBr. HBr is a strong acid and will thus ionize completely when diluted with $H_2O$ to form $H^+$ and $Br^-$ ions.

$$HBr \rightleftharpoons H^+ + Br^-$$

Therefore, 2 particles will be formed by each HBr that ionizes. The effective molality of the solution is then twice the molality of the HBr.

effective molality = 2 × 2 molal = 4 molal.

One can now find the boiling point elevation.

boiling pt. elevation = 0.52° × 4 molal = 2.1°

The normal boiling point of $H_2O$ is 100°C the new boiling point is 100° + 2.1° or 102.1°C.

(b) Suspension of 100 g of powdered glass. These particles will not dissolve to their molecular components in $H_2O$, thus the boiling point of the water will not be changed.

(c) $1.2 \times 10^{24}$ sucrose molecules/liter. One liter of $H_2O$ weighs 1 kg, therefore, $1.2 \times 10^{24}$ sucrose molecules are dissolved in 1 kg of $H_2O$. The number of moles present is equal to the number of molecules present divided by Avogadro's number, the number of particles in one mole.

$$\text{no. of moles} = \frac{1.2 \times 10^{24} \text{ molecules}}{6.02 \times 10^{23} \text{ molecules/mole}} = 2 \text{ moles}$$

Hence, there are 2 moles of sucrose in 1 kg of water, thus the solution is 2 molal. Sucrose does not ionize in water, therefore, the true molality is equal to the effective molality. Solving for the boiling point elevation:

boiling pt. elevation = 0.52° × 2 molal = 1.04°

The boiling point of this solution is 100° + 1.04° or 101.04°.

(d) 0.5 molal $BaCl_2$. $BaCl_2$ is a strong electrolyte and will completely ionize in $H_2O$.

$$BaCl_2 \rightleftharpoons Ba^{++} + 2Cl^-$$

3 ions will be formed for each $BaCl_2$ present. The effective molality will therefore be 3 times the molality of unionized $BaCl_2$.

effective molality = 3 × 0.5 molal = 1.5 molal

One can now solve for the boiling point elevation.

boiling pt. elevation = 1.5 molal × 0.52° = 0.78°

The boiling point of $H_2O$, 100°, will be raised 0.78° to be 100.78°C.

● **PROBLEM** 249

The normal boiling point of benzene is 80.10°C. When 1 mole of a solute is dissolved in 1000 g of benzene, the boiling point of the resulting solution is 82.73°C. When 1.2 g of elemental sulfur is dissolved in 50 g of benzene, the boiling point of the solution is 80.36°C. What is the molecular weight of sulfur?

<u>Solution</u>:   In order to determine the molecular weight of sulfur, we must determine how many moles of sulfur correspond to 1.2 g.

When 1 mole of solute is added to 1000 g of benzene, the normal boiling point is raised by 2.63°C, from 80.10°C to 82.73°C. This corresponds to a concentration of 1 molal (1 mole solute/1000 g solvent = 1 molal). The ratio of sulfur to benzene in the sulfur-benzene solution is 1.2 g sulfur/50 g benzene = 0.024. Thus, adding 1.2 g of sulfur to 50 g of benzene is equivalent to adding 24 g of sulfur to 1000 g of benzene (because 24 g sulfur/1000 g benzene is also 0.024).

The observed boiling point elevation in the sulfur-benzene solution is 0.26°C (from 80.10°C to 80.36°C). This is the boiling point elevation one would obtain by adding 24 g of sulfur to 1000 g of benzene. But this is about one-tenth the rise one would obtain by adding 1 mole of sulfur (the rise would then be 2.63°C), hence 24 g of sulfur must correspond to about one-tenth of a mole.

Thus, the apparent molecular weight of sulfur is

$$\frac{24 \text{ g}}{0.1 \text{ mole}} = 240 \text{ g/mole}.$$

● **PROBLEM** 250

It is known that at 4.6 mm Hg, the boiling point of water is lowered to 0°C. But at 4.6 mm Hg, the freezing point is still 0°C. Explain (a) why the freezing point was not altered substantially by a change in pressure, and (b) the paradox of water freezing and boiling at the same temperature.

Solution: To solve this problem, you want to consider what would alter the temperature of any phase change and see if the condition exists in this case. The temperature of any phase transition is affected by pressure only because of the accompanying volume change. Since the difference in density between solid and liquid is usually small, as compared to liquid and gas, a pressure change will not alter to any significant degree the freezing point, since the volume change will be small. Thus, the boiling point was altered, but not the freezing point. The paradox of both the boiling point and freezing point being at the same point can be explained by a discussion of the triple point. At this point, some molecules are going from liquid to solid and others from solid to liquid, and similarly for the gas-liquid and liquid-solid transformations. An equilibrium is established when an equal number of molecules are going each way in all three processes. Thus, at the triple point, you have a set of conditions in which all three states may exist in equilibrium with each other. The freezing and boiling points, when at the same temperature, suggest such an equilibrium. Thus, the existence of one temperature for both boiling and freezing point can occur at a given pressure.

● **PROBLEM** 251

A G.T.O. has a 22 gal. cooling system. Suppose you fill it with a 50-50 solution by volume of $(CH_2OH)_2$, ethylene glycol, and water. At what temperature would freezing become a problem? Assume the specific gravity of ethylene glycol is 1.115 and the freezing point depression constant of water is 1.86 C°/mole. You might have placed in methanol $(CH_3OH)$ instead of $(CH_2OH)_2$. If the current cost of ethylene glycol is 12 cents/lb and the cost of methanol is 8 cents/lb, how much money would you save by using $CH_3OH$? And yet, ethylene glycol is the more desirable antifreeze. Why? density = .79 g/ml for $CH_3OH$ and 3.785 liter= 1 gallon

Solution: To determine the freezing point of the water-ethylene glycol solution, you employ the following equation

$$\Delta T_f = \frac{1000g \ K_f \ (N_2)}{M_1 \ (N_1 + N_2)}$$

where $\Delta T_f$ is the change in the freezing point of the solvent, water, $K_f$ the freezing point depression constant of the solvent, $M_1$ is the molecular weight of the solvent, $N_1$ the number of moles of solvent and $N_2$ the number of moles of solute.

You are given $K_f$ for water and the molecular weight

is 18; hence, you can solve for $\Delta T_f$ after determining $N_1$ and $N_2$.

You can calculate the number of moles of water and of ethylene glycol present by using their densities and volumes. Use the following steps:

1) Convert 22 gal (the volume of the tank) to liters. volume = 22 gal x 3.785 $\ell$/gal = 83.3 $\ell$

2) You are told that the mixture is 50-50 by volume or one half water and one half ethylene glycol. Hence there are 83.3/2 $\ell$ or 41.68 $\ell$ of each.

3) Use the density of each compound to find the number of grams of each present. (density of $H_2O$ = 1.0 and 1000 $cm^3$ = 1 $\ell$)

$$\text{mass of } H_2O = 1.0 \text{ g/cm}^3 \times 1000 \text{ cm}^3/\ell \times 41.68 \text{ } \ell$$

$$= 41.68 \times 10^3 \text{ g } H_2O$$

$$\text{mass of } (CH_2OH)_2 = 1.115 \text{ g/cm}^3 \times 1000 \text{ cm}^3/\ell \times 41.68 \text{ } \ell$$

$$= 46.47 \times 10^3 \text{ g of } (CH_2OH)_2$$

4) The number of moles of each is found by dividing its mass by its molecular weight. (MW of $H_2O$ = 18, MW of $(CH_2OH)_2$ = 62).

$$\text{moles of } H_2O = \frac{41.68 \times 10^3 \text{ g}}{18 \text{ g/mole}} = 2.32 \times 10^3 \text{ moles}$$

$$\text{moles of } (CH_2OH)_2 = \frac{46.47 \times 10^3 \text{ g}}{62 \text{ g/mole}} = 7.50 \times 10^2 \text{ moles}$$

You can now solve for $\Delta T_f$ by substituting into the above equation

$$\Delta T_f = \frac{1000 \text{ g } (1.86 \text{ °C/mole})(7.50 \times 10^2 \text{ mole})}{(18 \text{ g/mole}) (2.32 \times 10^3 + 7.50 \times 10^2 \text{ moles})}$$

$$= \frac{1.40 \times 10^6}{5.53 \times 10^4} = 25.2 \text{ °C}$$

The normal freezing point of water is 0°C, after adding the ethylene glycol, the freezing point becomes 0°-25.2°C or -25.2°C.

Determine the amount of money it is possible to save by using a methanol-water mixture by calculating how much 41.68 $\ell$ (half of the volume of the tank) of methanol costs and subtracting this value from the cost of 41.68 $\ell$ of ethylene glycol. To calculate the cost of each

1) 1 lb = 454 g. Since the costs are given in cents/lb, determine the weight in pounds of 41.68 $\ell$ of each. Earlier the weight of 41.68 $\ell$ of ethylene glycol was found to be 46.47 x $10^3$g. Converting this to pounds,

$$\text{weight of } (CH_2OH)_2 = 46.47 \times 10^3 g \times \frac{1 \text{ lb}}{454 \text{ g}} = 102.4 \text{ lb}$$

$$\text{weight of } (CH_3OH) = \text{density} \times \text{volume}$$

$$= .79 \text{ g/ml} \times 10^3 \text{ ml}/\ell \times 41.68 \, \ell$$

$$= 33.07 \times 10^3 \text{ g}$$

Converting to pounds

$$\text{weight of } CH_3OH = 33.07 \times 10^3 \text{ g} \times \frac{1 \text{ lb}}{454 \text{ g}} = 72.8 \text{ lb.}$$

The cost of each is found by multiplying the price per pound by the weight of each.

$$\text{cost of } (CH_2OH)_2 = 102.4 \text{ lb} \times \$.12/\text{lb} = \$12.29$$

$$\text{cost of } (CH_3OH) = 72.8 \text{ lb} \times \$.08/\text{lb} = \$5.82$$

The difference in cost is $12.29 - 5.82 = \$6.47$. Hence, you save $6.47 by using methanol.

You have found that a 50-50 mixture of water and ethylene glycol protects the car to a temperature of -25.2°C. The most desirable antifreeze will protect the car at the lowest temperature, hence, to determine whether ehtylene glycol is more desirable than methanol, calculate the lowest temperature at which the car will be protected by methanol. This is accomplished by using the above equation for freezing point depression.

$$\Delta T_f = \frac{1000 \text{ g } (K_f) (N_2)}{M_1 (N_2 + N_1)}$$

Previously, one has determined $K_f$, $N_1$ and $M_1$ to be 1.86°C/mole, $2.32 \times 10^3$ moles and 18 g/mole, respectively. Thus, to solve for $\Delta T_f$ determine $N_2$ first. As shown, the weight of the methanol is $33.07 \times 10^3$ g. Find $N_2$ by dividing this weight by the molecular weight of methanol, 32.

$$\text{No. of moles of methanol } (N_2) = \frac{33.07 \times 10^3 g}{32 \text{ g/mole}}$$

$$= 1.03 \times 10^3 \text{ moles.}$$

Solving for $\Delta T_f$:

$$\Delta T_f = \frac{1000 \text{ g } (1.86 \text{ °C/mole})(1.03 \times 10^3 \text{ moles})}{(18 \text{ g/mole})(1.03 \times 10^3 + 2.32 \times 10^3 \text{ moles})}$$

$$= 31.8°C.$$

The methanol will protect the car at temperatures as low as 0°C-31.8°C or -31.8°C. This is fine while the car is not

in motion and is cold, but when the car is driven the engine
becomes very warm.  Methanol has a boiling point of 60°C
and will evaporate while the engine is operating, whereas
the ethylene glycol has a much higher boiling point and
will not evaporate.  Thus, if methanol is used as an antifreeze it will have to be replaced after each use of the
car while the ethylene glycol will last all winter.  In the
end it is far more expensive and troublesome to use methanol.

# RAOULT'S LAW AND VAPOR PRESSURE

● **PROBLEM** 252

You have two 1-liter containers connected to each other by
a valve which is closed. In one container, you have liquid
water in equilibrium with water vapor at 25°C. The other
container contains a vacuum. Suddenly, you open the valve.
Discuss the changes that take place, assuming temperature
is constant with regard to (a) the vapor pressure, (b) the
concentration of the water molecules in the vapor, (**c**) the
number of molecules in the vapor state.

Solution:  The vapor pressure is the pressure exerted by
the gas molecules when they are in equilibrium with the
liquid. When the valve is opened some of the gas molecules
will move to the empty container. At this point the pressure will be less than the equilibrium pressure because the
concentration of the gas molecules will be lowered. Very
quickly, though, the equilibrium will be attained again by
the action of more liquid molecules vaporizing. Therefore,
the vapor pressure and the concentration of the gaseous
molecules of the system remains essentially unchanged.

Since the concentration of the gaseous molecules
remains unchanged when the volume of the system is doubled,
the number of molecules must also be doubled. This is true
because concentration is an expression of the number of
molecules per unit volume.

● **PROBLEM** 253

A chemist decides to find the vapor pressure of water by
the gas saturation method.  100 liters of $N_2$ gas is passed
through 65.44 g of water.  After passage of the gas,
63.13 g remained. The temperature of the $H_2O$ (water) is 25°C.
Find the vapor pressure of water at this temperature.

Solution: In the gas saturation method, a dry, unreactive gas
such as nitrogen or air is bubbled through a specific amount

of liquid maintained at constant temperature. After the gas has been bubbled away, the loss in weight of the liquid is determined. The weight loss is the number of grams of liquid in the vapor state. There exists an equation that relates the volume, pressure, weight loss, and molecular weight of the liquid.

$$P = \frac{g}{MV} RT,$$ where P = vapor pressure, g = grams of

vapor, M = molecular weight of liquid, R = universal gas constant ($.0821 \frac{\text{liter-atm}}{\text{mole °K}}$, V = volume, and T = temperature in Kelvin (Celsius plus 273°)

The pressure will be expressed in mm, so that you use the conversion factor of $\frac{760 \text{ mm}}{\text{atm}}$. g = 65.44 − 63.13 = 2.31 grams or the weight of the liquid in the vapor state.

The molecular weight of water = 18.02 g/mole. Thus,

$$P = \frac{gRT}{MV} = \frac{(2.31 \text{ grams})(.0821 \frac{\text{liter atm}}{\text{mole °K}})(298°K) \frac{760 \text{ mm}}{1 \text{ atm}}}{(18.02 \text{ g/mole})(200 \text{ liters})}$$

$$= 23.8 \text{ mm} = \text{vapor pressure of } H_2O.$$

● **PROBLEM** 254

When a swimmer leaves cold water on a warm, breezy day, he experiences a cooling effect. Why?

Solution: The equilibrium vapor pressure, is defined as the pressure exerted by the gas of that substance in equilibrium with the liquid state of the same substance. When the swimmer leaves the cold water, he is, coated with water. As the water evaporates a layer of gaseous water molecules form around the swimmer's body.

This vapor attains an equilibrium with the atmosphere which hinders further vaporization of the water. However, when the breeze comes, it blows away the vapor above the water, and thereby keeps the partial pressure of the vapor low above the skin. This, then, causes increased evaporation so that the partial pressure can be reestablished. This is true because the concentration and vapor pressure of the gaseous water molecules must be below a certain point for evaporation to occur. Once this pont has been reached no more liquid molecules of water can go into the gaseous state.

For the water to evaporate, it must go from a liquid to a gaseous phase, a process which requires heat. The water removes heat from the swimmer's body. Thus, the swimmer feels a cooling effect.

The vapor pressures of pure benzene and toluene at 60°C are 385 and 139 Torr, respectively. Calculate (a) the partial pressures of benzene and toluene, (b) the total vapor pressure of the solution, and (c) the mole fraction of toluene in the vapor above a solution with 0.60 mole fraction toluene.

Solution: The vapor pressure of benzene over solutions of benzene and toluene is directly proportional to the mole fraction of benzene in the solution. The vapor pressure of pure benzene is the proportionality constant. This is analogous to the vapor pressure of toluene. This is known as Raoult's law. It may be written as

$$P_1 = X_1 \ P_1^\circ$$

$$P_2 = X_2 \ P_2^\circ$$

where 1 and 2 refer to components 1 and 2, $P_1$ and $P_2$ represent the partial vapor pressure above the solution, $P_1^\circ$ and $P_2^\circ$ are the vapor pressures of pure components, and $X_1$ and $X_2$ are their mole fractions. Solutions are called ideal if they obey Raoult's law.

The mole fraction of a component in the vapor is equal to its pressure fraction in the vapor. The total vapor pressure is the sum of the vapor's component partial pressures.

To solve this problem one must

1) calculate the partial pressures of benzene and toluene using Raoult's law

2) find the total vapor pressure of the solution by adding the partial pressures

3) find the mole fraction of toluene in the vapor.

One knows the mole fraction of toluene in the solution is 0.60 and, thus, one also knows the mole fraction of benzene is (1-0.60) or 0.40. Using Raoult's law:

$$P_{benzene}^\circ = 385 \ Torr \qquad P_{Toluene}^\circ = 139$$

a) $P_{benzene} = (0.40)(385 \ Torr) = 154.0 \ Torr$

$P_{toluene} = (0.60)(139 \ Torr) = 83.4 \ Torr$

b) $P_{total} = 154.0 + 83.4 = 237.4 \ Torr$

c) The mole fraction of toluene in the vapor =

$$X_{toluene, \ vap} = \frac{P_{toluene}}{P_{toluene} + P_{benzene}} = \frac{83.4}{237.4} = 0.351.$$

At 90°C, benzene has a vapor pressure of 1022 Torr, and toluene has a vapor pressure of 406 Torr. From this information, calculate the composition of the benzene-toluene solution that will boil at 1 atm pressure and 90°C, assuming that the solution is ideal.

Solution: A solution boils when the sum of the partial pressures of the components becomes equal to the applied pressure (i.e. total pressure). To solve this problem, one must realize that the applied pressure is atmospheric pressure, 760 Torr.

One can determine the partial pressure of benzene ($P_a^o$) and of toluene ($P_b^o$) by using Raoult's law. Raoult's law states that the partial pressure p of a gas is equal to its vapor pressure p° times its mole fraction X in the solution.

$$p = p^o X$$

Dalton's law of partial pressure states that the sum of the partial pressures of a system is equal to the toal pressure of the system. For this particular benzene-toluene solution to boil the total pressure of the system must equal atmospheric pressure 760 Torr. If one lets the mole fraction of benzene equal $X_a$, then the mole fraction toluene is equal to $1 - X_a$. One determines the partial pressures by substituting into Raoult's law

$$P_a = P_a^o (X_a)$$

$$P_b = P_b^o (1-X_a).$$

Then, substituting into Dalton's law:

$$760 \text{ Torr} = P_a + P_b$$

$$760 \text{ Torr} = 1022 X_a + 406 (1-X_a)$$

$$= 1022 X_a + 406 - 406 X_a$$
$$X_a = 0.574$$

The mole fraction of benzene in the liquid is 0.574 and the mole fraction of toluene is $(1-0.574)=0.426$.

A chemist dissolves 300 g of urea in 1000 g of water. Urea is $NH_2CONH_2$. Assuming the solution obeys Raoult's law, determine the following: a) The vapor pressure of the solvent at 0° and 100°C and b) The boiling and freezing point of the solution. The vapor pressure of pure water is 4.6 mm and 760 mm at 0°C and 100°C, respectively and the $K_f = 1.86$ C /mole and $K_b = .52$ C /mole.

<u>Solution</u>:  To solve (a), you must employ Raoult's law, which states $P_{H_2O} = P°_{H_2O} X_{H_2O}$, where $P_{H_2O}$ = the partial pressure of water (the solvent), $P°_{H_2O}$ = the vapor pressure of water when pure, and $X_{H_2O}$ is the mole fraction of $H_2O$ in the solution.  The mole fraction is equal to $\dfrac{N_{H_2O}}{N_{solute} + N_{H_2O}}$, where $N_{H_2O}$ = moles of $H_2O$, and $N_{solute}$ = moles of solute. You are told the $P°_{H_2O}$ for $H_2O$ at the temperatures in question.  Thus, to find the vapor pressures of the solution, you must calculate $X_{H_2O}$ and substitute into Raoult's law.

To calculate the concentration of urea, remember moles = $\dfrac{grams}{molecular\ weight}$.  Since you have 300 grams of urea and the molecular weight of urea ia 60.06 g, you have $\dfrac{300}{60.06}$ or 4.995 moles of urea.  Similarly, for water (MW=18), you have $\dfrac{1000\ g}{18\ g/mole}$ = 55.55 moles of water.  Thus, the mole fraction of water = $\dfrac{55.55}{4.995 + 55.55}$ = .917.  Therefore, at 0°,

$\left( P_{H_2O} = P°_{H_2O} X_{H_2O} = (4.6\ mm)(.917) = 4.2\ mm \right)$ at 760°,

$\left( P_{H_2O} = P°_{H_2O} X_{H_2O} = (760)(.917) = 697\ mm. \right)$ To solve (b), you use the equations $\Delta T_f = K_f m$, where $\Delta T_f$ = freezing point depression, $K_f$ = molal freezing pt. depression constant, and m = molality and $\Delta T_b = K_b m$, where $\Delta T_b$ = boiling point elevation, $K_b$ = molal boiling point depression constant.  If you find $\Delta T_f$, you can calculate the freezing point, since the normal freezing point is decreased by this amount to give you the new freezing point.  Boiling point works in the same way, except that you add to the normal boiling point.  To calculate $\Delta T_f$ and $\Delta T_b$, you must know molality.  Molality = moles solute per 1 kilogram solvent.  You have 4.995 moles of urea and 1000 g of water or solvent, so that $\dfrac{4.995\ mole}{1\ kg} \approx 5$ molal solution,

Thus, $\Delta T_f = (1.86)(5) = 9.3°C$ and $\Delta T_b = (.52)(5) = 2.6°C$. Therefore, the freezing point is now 0.0°C − 9.3°C = −9.3°C and the boiling point is = 100°C + 2.6°C = 102.6°C.

● **PROBLEM** 258

A solution of 20.0 g of a non-volatile solute in 100 g of benzene at 30°C has a vapor pressure 13.4 torr lower than the vapor pressure of pure benzene. What is the mole fraction of solute? Vapor pressure of benzene at 30°C = 121.8 torr.

Solution:  At constant temperature, the lowering of the vapor pressure by a non-volatile solute is proportional to the concentration of the solute in the solution (the mole fraction). This is called Raoult's Law. The mole fraction of a solute is defined as the number of moles of solute divided by the sum of the number of moles of solute and of solvent.

$$\text{mole fraction of solute} = \frac{\text{moles solute}}{\text{moles solute + moles solvent}}$$

Raoult's Law is used to solve this problem. It can be stated:

$$P = P^0 X_2,$$

where $P^0$ is the vapor pressure of the pure solvent, P is the vapor pressure of the solution, and $X_2$ is the mole fraction of the solute.

Using Raoult's Law for this solution:

13.4 torr = (121.8 torr)($X_2$)

0.111 = $X_2$

The mole fraction of the solute is 0.111.

● PROBLEM 259

The vapor pressure of benzene at 75°C is 640 torr. A solution of 3.68 g of a solute in 53.0 g benzene has a vapor pressure of 615 torr. Calculate the molecular weight of the solute. (MW of benzene = 78.0.)

Solution:    At constant temperature, the lowering of the vapor pressure by a non-volatile solute is proportional to the concentration of the solute in the solution.

Hence,    $P^0 - P = P^0 X_2,$

where $P^0$ is the original pressure, P is the final pressure and $X_2$ is the mole fraction of the solute. Here, one solves for the mole fraction of the solute, and from that, one can determine the molecular weight.

$X_2$ = mole fraction

$P^0 - P = P^0 X_2$        $P^0$ = 640 torr

P = 615 torr

$$640 \text{ torr} - 615 \text{ torr} = (640 \text{ torr}) \cdot X_2$$
$$25 \text{ torr} = (640 \text{ torr}) \cdot X_2$$
$$\frac{25 \text{ torr}}{640 \text{ torr}} = X_2$$

mole fraction = .039

Mole fraction is defined as the number of moles of each component divided by the sum of the number of moles in the solution.

mole fraction of solute =

$$= \frac{\text{no. of moles of solute}}{\text{no. of moles of solute} + \text{no. of moles of benzene}}$$

To find the number of moles of solute, one must first know the number of moles of benzene present. This is found by dividing the number of grams of benzene by the molecular weight of benzene. (MW = 78.)

$$\text{no. of moles of benzene} = \frac{\text{no. of grams}}{MW}$$

$$\text{no. of moles} = \frac{53.0 \text{ g}}{78.0 \text{ g/mole}} = .679 \text{ moles}$$

One can now solve for the number of moles of solute. Let X = moles of solute

$$\frac{\text{mole fraction}}{\text{of solute}} = \frac{\text{moles solute}}{\text{moles benzene} + \text{moles solute}}$$

$$.039 = \frac{X}{.679 + X}$$

$$.039 \, (.679 + X) = X$$

$$.265 + .039X = X$$

$$.265 = .961 \, X$$

$$.275 = X$$

Thus there are .275 moles of solute present. One is told that there are 3.68 g of this solute, therefore, there are .275 moles in 3.68 g. The molecular weight is found by dividing 3.68 g by 0.275 moles.

$$\text{molecular weight} = \frac{3.68 \text{ g}}{0.275 \text{ mole}} = 133.8 \text{ g/mole}.$$

● **PROBLEM** 260

Water at 30°C has a vapor pressure of 31.82 mm Hg. When 25.0 g of ethylene glycol is added to 1000 g of water, the vapor pressure is lowered to 31.59 mm Hg. Determine the molecular weight of ethylene glycol.

<u>Solution</u>:  This problem is an application of Raoult's law.

Let $p^0$ denote the vapor pressure of pure water, $p$ the vapor pressure of the ethylene glycol - water solution, $n_w$ the number of moles of water in the solution, and $n_e$ the number of moles of ethylene glycol. Then from Raoult's law,

$$p = p^0 \, \frac{n_w}{n_e + n_w} \; .$$

Multiplying by minus one and adding $p^0$ to both sides of this equation gives

$$p^0 - p = p^0 - p^0 \, \frac{n_w}{n_e + n_w}$$

$$= p^0 \left( 1 - \frac{n_w}{n_e + n_w} \right)$$

$$= p^0 \left( \frac{n_e + n_w}{n_e + n_w} - \frac{n_w}{n_e + n_w} \right) ,$$

or, $\qquad p^0 - p = p^0 \, \dfrac{n_e}{n_e + n_w}$

Since the number of moles is equal to the number of grams divided by the molecular weight, this equation becomes

$$p^0 - p = p^0 \, \frac{g_e/M_e}{g_e/M_e + g_w/M_w} ,$$

where $g_e$ denotes the number of grams of ethylene glycol, $g_w$ the number of grams of water, $M_e$ the molecular weight of ethylene glycol (which we are trying to determine), and $M_w$ the molecular weight of water.

For this problem, $p^0 = 31.82$ mm, $p = 31.59$ mm, $g_e = 25.0$ g, $g_w = 1000$ g, and the molecular weight of water, $M_w = 18$. Substituting into the last equation, we obtain

$$31.82 \text{ mm} - 31.59 \text{ mm} = 31.82 \text{ mm} \left( \frac{25.0 \text{ g}/M_e}{1000 \text{ g}/18 + 25.0 \text{ g}/M_e} \right) ,$$

or, $\quad 0.23 \text{ mm} = 31.82 \text{ mm} \left( \dfrac{25.0 \text{ g}/M_e}{1000 \text{ g}/18 + 25.0 \text{ g}/M_e} \right) .$

Solving for $M_e$,

$$\frac{0.23 \text{ mm}}{31.82 \text{ mm}} \left( \frac{1000 \text{ g}}{18} + \frac{25.0 \text{ g}}{M_e} \right) = \frac{25.0 \text{ g}}{M_e} ,$$

$$\frac{0.23 \text{ mm}}{31.82 \text{ mm}} \times \frac{1000 \text{ g}}{18} = \frac{25.0 \text{ g}}{M_e} - \frac{0.23 \text{ mm}}{31.82 \text{ mm}} \times \frac{25.0 \text{ g}}{M_e}$$

$$= \frac{1}{M_e} \times 25.0 \text{ g} \left( 1 - \frac{0.23 \text{ mm}}{31.82 \text{ mm}} \right) ,$$

or, $\quad M_e = \dfrac{31.82 \text{ mm}}{0.23 \text{ mm}} \times \dfrac{18}{1000 \text{ g}} \times 25.0 \text{ g} \left( 1 - \dfrac{0.23 \text{ mm}}{31.82 \text{ mm}} \right) = 62.$

Thus, the molecular weight of ethylene glycol is 62.

# CLAUSIUS-CLAPEYRON EQUATION OF VAPORIZATION

● **PROBLEM** 261

If the vapor pressure of ethyl alcohol, $C_2H_5OH$, is 0.132 atm at 34.9°C, and 0.526 atm at 63.5°C what do you predict it will be at 19.0°C?

Solution: Equilibrium vapor pressure is the pressure exerted by a vapor when the vapor is in equilibrium with its liquid. The magnitude of the equilibrium vapor pressure depends (1) on the nature of the liquid and (2) on its temperature. The vapor pressure is related to temperature by the following equation

$$\log p = \frac{- \Delta H}{19.15 \text{ T}} + C$$

where p is the vapor pressure, T is the absolute temperature, $\Delta H$ is the heat required to transform one mole of liquid to the ideal gas state, and C is a constant dependent on the liquid and units used for expressing pressure. One is given p and T for two trials, thus $\Delta H$ for ethyl alcohol can be found.

For $\quad$ p = 0.132 atm

$\qquad$ T = 34.9°C + 273 = 307.9°K

a) $\qquad \log 0.132 = \dfrac{- \Delta H}{19.15 \ (307.9°K)} + C$

For $\quad$ p = 0.526 atm

$\qquad$ T = 63.5°C + 273 = 336.5°K

b) $\qquad \log 0.526 \text{ atm} = \dfrac{- \Delta H}{19.15 \ (336.5°K)} + C$

One can solve for $\Delta H$ by subtracting equation a from equation b.

$\log 0.526 = - \Delta H/19.15(336.5°K) + C$

$- \ [\log 0.132 = - \Delta H/19.15(307.9°K) + C]$

_____

$\log 0.526 - \log 0.132 = -\Delta H/19.15(336.5°K) + \Delta H/19.15(307.9°K)$

$- \ .279 - (- .879) = - \Delta H/6444 + \Delta H/5896$

$(6444)(5896)(.60) = (- \Delta H/6444 + \Delta H/5896)(6444)(5896)$

$2.28 \times 10^7 = - 5896\Delta H + 6444\Delta H$

$2.28 \times 10^7 = 548\Delta H$

$\dfrac{2.28 \times 10^7}{548} = \Delta H$

$4.16 \times 10^4 \ \text{J/mole} = \Delta H$

After solving for $\Delta H$ one can find the vapor pressure at 19.0°C by a similar method.

For $\qquad p = 0.132 \text{ atm}$

$\qquad T = 34.9°C + 273 = 307.9°K$

$\qquad \Delta H = 4.16 \times 10^4 \ \text{J/mole}$

c) $\qquad \log 0.132 = \dfrac{- \ 4.16 \times 10^4 \ \text{J/mole}}{19.15 \ (307.9°K)} + C$

$\qquad \log 0.132 = - \ 7.055 + C$

For $\qquad p = ?$

$\qquad T = 19.0°C + 273 = 292°K$

$\qquad \Delta H = 4.16 \times 10^4 \ \text{J/mole}$

d) $\qquad \log p = \dfrac{- \ 4.16 \times 10^4 \ \text{J/mole}}{(19.15)(292°K)} + C = - \ 7.439 + C$

Subtracting equation c from equation d:

$\log p = - \ 7.439 + C$

$- \ [\log 0.132 = - \ 7.055 + C]$

_____

$\log p - \log 0.132 = - \ .384$

$\log p = - \ .384 + \log 0.132$

$\log p = - \ .384 + (- .879)$

log p = - 1.263

p = .0545 atm.

If the vapor pressure of methyl alcohol, $CH_3OH$, is 0.0526 atm at 5.0°C and 0.132 atm at 21.2°C, what do you predict the normal boiling point will be?

Solution: The normal boiling point of a liquid is the temperature at which the vapor pressure of the liquid is equal to the prevailing atmospheric pressure. Atmospheric pressure is defined as 1 atm. The pressure is related to the temperature by the equation

$$\ln \ p = \frac{-\ \Delta H}{RT} + C$$

where p is the vapor pressure, $\Delta H$ is the heat required to transform one mole of liquid to the ideal-gas state; R is the gas constant, 8.314 J/mole °K; T is the absolute temperature; and C is a constant dependent on the liquid. One can solve for $\Delta H$ by using the values obtained for p at 5°C and 21.2°C.

For    p = 0.0526 atm

T = 5.0°C + 273 = 278°K

R = 8.314 J/mole °K

$$\ln (0.0526) = \frac{-\ \Delta H}{(8.314 \ J/mole \ °K)(278°K)} + C$$

a)    $- 2.945 = -\ \Delta H/2.311 \times 10^3$ J/mole + C

For    p = 0.132 atm

T = 21.2°C + 273 = 294.2°K

R = 8.314 J/mole °K

$$\ln (0.132) = \frac{-\ \Delta H}{(8.314 \ J/mole \ °K)(294.2°K)} + C$$

b)    $- 2.025 = -\ \Delta H/2.446 \times 10^3$ J/mole + C

One can solve for $\Delta H$ by subtracting equation a from equation b.

$- 2.025 = -\ \Delta H/2.446 \times 10^3$ J/mole + C

$- [- 2.945 = -\ \Delta H/2.311 \times 10^3$ J/mole + C]
_____

$.920 = -\ \Delta H/2.446 \times 10^3$ J/mole $+ \Delta H/2.311 \times 10^3$ J/mole

$(2.446 \times 10^3 \text{ J/mole})(2.311 \times 10^3 \text{ J/mole})(.920) =$

$(-\Delta H/2446 \times 10^3 \text{ J/mole} + \Delta H/2.311 \times 10^3 \text{ J/mole})(2.446 \times 10^3 \text{ J/mole})$

$\times (2.311 \times 10^3 \text{ J/mole})$

$5.200 \times 10^6 \text{ J}^2/\text{mole}^2 = -(2.311 \times 10^3 \text{ J/mole}) \Delta H$

$+ (2.446 \times 10^3 \text{ J/mole}) \Delta H$

$5.200 \times 10^6 \text{ J}^2/\text{mole}^2 = (.135 \times 10^3 \text{ J/mole}) \Delta H$

$\dfrac{5.200 \times 10^6 \text{ J}^2/\text{mole}^2}{.135 \times 10^3 \text{ J/mole}} = \Delta H$

$3.85 \times 10^4 \text{ J/mole} = \Delta H$

Using a similar method the normal boiling point can be found now that one knows $\Delta H$.

For     p = 0.0526 atm

     T = 5.0°C + 273 = 278°K

     R = 8.314 J/mole °K

     $\Delta H = 3.85 \times 10^4 \text{ J/mole}$

     $\ln 0.0526 = \dfrac{- 3.85 \times 10^4 \text{ J/mole}}{(8.314 \text{ J/mole °K})(278°K)} + C$

c)     $- 2.945 = - 16.657 + C$

For     p = 1.0 atm

     T = ?

     R = 8.314 J/mole °K

     $\Delta H = 3.85 \times 10^4 \text{ J/mole}$

     $\ln 1.0 = \dfrac{- 3.85 \times 10^4 \text{ J/mole}}{(8.314 \text{ J/mole °K})T} + C$

d)     $0 = - \dfrac{4.63 \times 10^3 °K}{T} + C$

Subtracting equation d from equation c one can obtain T.

$- 2.945 = - 16.657 + C$

$- \left[ 0 = - \dfrac{4.63 \times 10^3 °K}{T} + C \right]$

$\overline{\phantom{xxxxxxxxxxxxxxxxxxxxxxxxxxxxxxxxxxxxxxx}}$

$- 2.945 = - 16.657 + \dfrac{4.63 \times 10^3 °K}{T}$

$13.712 = \dfrac{4.63 \times 10^3 \ °K}{T}$

$$T = \frac{4.63 \times 10^3 \ ^\circ K}{13.712} = 337.7^\circ K$$

Boiling point in $^\circ C = 337.7^\circ K - 273 = 64.7^\circ C$.

● PROBLEM 263

Find the vapor pressure of $CCl_4$ (carbon tetrachloride) at $38^\circ C$, if $H^1 = 32,000$ J/mole and at $23^\circ C$, the vapor pressure is .132 atm.

Solution: To solve this problem, employ the Clausius-Clapeyron equation, which permits evaluation of vapor pressure in terms of $\Delta H^1$ (the heat required to transform one mole of liquid to the ideal-gas state), the original vapor pressure and the two temperatures. The equation states:

$$\log \frac{p_1}{p_2} = \frac{\Delta H^1}{19.15} \left( \frac{1}{T_2} - \frac{1}{T_1} \right) ,$$

where $p_1$ and $p_2$ are vapor pressures and $T_1$ and $T_2$ are the temperatures in Kelvin. Let $p_2$ be the vapor pressure of $CCl_4$ at $38^\circ C$. Substitute the known values for solution,

$$\log \frac{.132}{p_2} = \frac{32,000}{19.15} \left( \frac{1}{311} - \frac{1}{296} \right).$$

Solve for $p_2$, to obtain $p_2 = .250$ atm.

● PROBLEM 264

Two substances A and B have the same 0.132-atm vapor pressure at $15^\circ C$, but A has a heat of vaporization of 25,000 J/mole, whereas B, 35,000 J/mole. What will be the respective normal boiling points?

Solution: The vapor pressure, heat of vaporization and absolute temperature are related in the Clausius-Clapeyron equation. This equation is written:

$$\ln \frac{p_1}{p_2} = \frac{\Delta H}{R} \left( \frac{1}{T_2} - \frac{1}{T_1} \right)$$

where $p_1$ is the initial pressure, $p_2$ is the final pressure, $\Delta H$ is the heat of vaporization when $T_2$ is the boiling point, R is the gas constant, 8.314 J/mole $^\circ K$, $T_1$ is the initial temperature and $T_2$ is the final temperature. To solve for the boiling point of A, note that at the boiling point of a liquid the vapor pressure is equal to the atmospheric pressure. Atmospheric pressure is 1 atm.

$p_1 = 0.132$ atm $\qquad T_1 = 15^\circ C + 273 = 288^\circ K$

$p_2 = 1$ atm $\qquad\qquad T_2 = ?$

$$\Delta H = 25,000 \text{ J/mole} \qquad R = 8.314 \text{ J/mole } ^{\circ}K$$

$$\ln \frac{0.132}{1} = \frac{25,000 \text{ J/mole}}{8.314 \text{ J/mole } ^{\circ}K} \left( \frac{1}{T_2} - \frac{1}{288^{\circ}K} \right)$$

$$- 2.025 = 3.00 \times 10^3 \, ^{\circ}K \, (1/T_2 - 3.47 \times 10^{-3}/^{\circ}K)$$

$$- 6.75 \times 10^{-4}/^{\circ}K = 1/T - 3.47 \times 10^{-3}/^{\circ}K$$

$$2.795 \times 10^{-3}/^{\circ}K = 1/T$$

$$357.78 = T$$

The boiling point of A in $^{\circ}C$ is

$$357.78 - 273 = 84.78^{\circ}C$$

Solving for the boiling point of B:

$p_1 = 0.132$ atm $\qquad T_1 = 15^{\circ}C + 273 = 288^{\circ}K$

$p_2 = 1.0$ atm $\qquad T_2 = ?$

$\Delta H = 35,000$ J/mole

$R = 8.314$ J/mole $^{\circ}K$

$$\ln \frac{0.132}{1.0} = \frac{35,000 \text{ J/mole}}{8.314 \text{ J/mole } ^{\circ}K} \left( \frac{1}{T_2} - \frac{1}{288^{\circ}K} \right)$$

$$- 2.025 = 4.21 \times 10^3 \, ^{\circ}K \, (1/T_2 - 3.47 \times 10^{-3}/^{\circ}K)$$

$$- 4.81 \times 10^{-4}/^{\circ}K = 1/T_2 - 3.47 \times 10^{-3}/^{\circ}K$$

$$2.989 \times 10^{-3}/^{\circ}K = 1/T_2$$

$$334.56^{\circ}K = T_2$$

The boiling point of B is equal to $334.56^{\circ}K$ or $334.56 - 273 = 61.56^{\circ}C$.

● **PROBLEM** 265

At a 4000-m altitude the atmospheric pressure is about 0.605 atm. What boiling point would you expect for water under these conditions?

Solution: The pressure and temperature of a liquid that vaporizes are related in the Clausius-Clapeyron equation. This equation is

$$\ln \frac{P_1}{P_2} = \frac{\Delta H}{R} \left( \frac{1}{T_2} - \frac{1}{T_1} \right)$$

where $P_1$ is the initial pressure, $P_2$ is the final pressure, $\Delta H$ is the heat of vaporization, R is the gas constant, 8.314 J/mole $^{\circ}K$, $T_2$ is the boiling point at $P_2$ and $T_1$ is the boiling point at $P_1$. The heat of varporization of $H_2O$ is

40,600 J/mole.  One can assume for purposes of calculation that $P_1$ is 1 atm.  The boiling point of $H_2O$ at 1 atm is 373°K.  Solving for the boiling point at 0.605 atm:

$$P_1 = 1 \text{ atm} \qquad\qquad T_1 = 373°K$$

$$P_2 = 0.605 \text{ atm} \qquad T_2 = ?$$

$$\Delta H = 40,600 \text{ J/mole}$$

$$R = 8.314 \text{ J/mole °K}$$

$$\ln \frac{1}{0.605} = \frac{40,600 \text{ J/mole}}{8.314 \text{ J/mole °K}} \left( \frac{1}{T_2} - \frac{1}{373°K} \right)$$

$$.5025 = 4.88 \times 10^3 \text{ °K} \, (1/T_2 - 2.681 \times 10^{-3}/°K)$$

$$1.0297 \times 10^{-4}/°K = 1/T_2 - 2.681 \times 10^{-3}/°K$$

$$2.784 \times 10^{-3}/°K = 1/T_2$$

$$359.20°K = T_2$$

The boiling point in °C is 359.20°K−273 or 86.2°C.

● **PROBLEM** 266

Liquid nitrogen is an excellent bath for keeping temperatures around 77°K, its normal boiling point.  What pressure would you need to maintain over the liquid nitrogen if you wanted to set the bath temperature at 85°K?  Heat of vaporization is about 5560 J/mole.

Solution:   One uses the Clausius-Clapeyron equation to solve for the final pressure.  This equation is written

$$\ln \frac{P_1}{P_2} = \frac{\Delta H}{R} \left( \frac{1}{T_2} - \frac{1}{T_1} \right)$$

where $P_1$ is the initial pressure, $P_2$ is the final pressure at 85°K, $\Delta H$ is the heat of vaporization, R is the gas constant, 8.314 J/mole-°K, $T_1$ is the initial temperature and $T_2$ is the final temperature.  The pressure of a liquid at its boiling point is equal to the atmospheric pressure, which is 1 atm.  Solving for the final pressure:

$$P_1 = 1 \text{ atm} \qquad\qquad T_1 = 77°K$$

$$P_2 = ? \qquad\qquad\qquad T_2 = 85°K$$

$$\Delta H = 5560 \text{ J/mole}$$

$$R = 8.314 \text{ J/mole °K}$$

$$\ln \frac{1}{P_2} = \frac{5560 \text{ J/mole}}{8.314 \text{ J/mole } °K} \left( \frac{1}{85°K} - \frac{1}{77°K} \right)$$

$$-\ln P_2 = 6.687 \times 10^2 °K (1.176 \times 10^{-2} °K - 1.298 \times 10^{-2} °K)$$

$$-\ln P_2 = 6.687 \times 10^2 °K (-1.22 \times 10^{-3}/°K)$$

$$-\ln P_2 = {}^-8.1581 \times 10^{-1}$$

$$P_2 = 2.26 \text{ atm.}$$

● **PROBLEM** 267

If the $\Delta H'$ of water is 40,600 J/mole at 100°C, what is the boiling point of water at a pressure of .750 atm.

Solution: The boiling point is defined as the temperature at which the vapor pressure of a liquid equals that of the surrounding atmosphere. The normal or standard boiling point is that temperature where the vapor pressure of a gas e-quals one atmosphere. For water, this is 100°C.

   In this problem, therefore, you have two pressures for water, one temperature, and $\Delta H'$, which equals the heat required to transform one mole of liquid to the ideal-gas state. These parameters are used to solve for the non-standard boiling point in the Clausius-Clapeyron equation, which states

$$\log \frac{p_1}{p_2} = \frac{\Delta H'}{19.15} \left[ \frac{1}{T_2} - \frac{1}{T_1} \right] \quad ,$$

where $p_1$ and $p_2$ are vapor pressures and $T_1$ and $T_2$ are their respective temperatures in degrees Kelvin. If you let $T_2$ be the boiling point of water at .750 atm, sub-stituting into the equation:

$$\log \frac{1.00}{.750} = \frac{40,600}{19.15} \left[ \frac{1}{T_2} - \frac{1}{373} \right]$$

   Solving for $T_2$, you obtain 365°K or 92°C, which is the boiling point of water at .750 atm.

● **PROBLEM** 268

If the vapor pressure of $CCl_4$ (carbon tetrachloride) is .132 atm at 23°C and .526 atm at 58°C, what is the $\Delta H'$ in this temperature range?

Solution: Molecules in a vapor exert a pressure that is characteristic of its liquid state. The pressure exerted by this vapor, when in equilibrium with the liquid, is called the equilibrium vapor pressure. The equilibrium vapor pressure can be expressed quantitatively as

$$\log p = \frac{-\Delta H'}{19.15\ T} + C$$

where p = vapor pressure, T = temperature in degrees Kelvin, $\Delta H'$ = heat required to transform one mole of liquid to the ideal-gas state, and C = a constant that is characteristic of the liquid. C can be eliminated from the expression by denoting that at two temperature ranges, $T_1$ and $T_2$, you have:

$$\log p_1 - \log p_2 = \frac{-\Delta H'}{19.15\ T_1} + \frac{\Delta H'}{19.15\ T_2} \qquad \text{or}$$

$$\log \frac{p_1}{p_2} = \frac{\Delta H'}{19.15} \left( \frac{1}{T_2} - \frac{1}{T_1} \right) .$$

In this problem, you are asked to find $\Delta H'$.

Let $p_1$ = .132 atm at $T_1$ = 296°K and $p_2$ = .526 atm at $T_2$ = 331°K, then substitute these values in the above equation. Therefore,

$$\log \frac{.132}{.526} = \frac{\Delta H'}{19.15} \left( \frac{1}{331} - \frac{1}{296} \right)$$

Solving for $\Delta H'$, you obtain

$\Delta H'$ = 32,000 J/mole.

## OSMOTIC PRESSURE

● PROBLEM 269

A sugar solution was prepared by dissolving 9.0 g of sugar in 500 g of water. At 27°C, the osmotic pressure was measured as 2.46 atm. Determine the molecular weight of the sugar.

Solution: The molecular weight of the sugar is found by determining the concentration, C, of sugar from the equation for osmotic pressure,

$\pi$ = CRT,

where $\pi$ is the osmotic pressure, R = universal gas constant = 0.08206 liter-atm/mole-°K , and T is the absolute temperature.

The osmotic pressure is measured as $\pi$ = 2.46 atm and the absolute temperature is T = 27°C + 273 = 300°K, hence

$\pi$ = CRT,

2.46 atm = C × 0.08206 liter-atm/mole-°K × 300°K,

or   $C = \dfrac{2.46 \text{ atm}}{0.08206 \text{ liter-atm/mole} - °K \times 300°K}$

= 0.10 mole/liter.

If we assume that the volume occupied by the sugar molecules in so small a concentration can be neglected, then in 1 liter of solution there is approximately 1 liter of water, or 1000 g of water, and

C = 0.10 mole/1000 g.

Therefore, there is 0.10 mole of sugar dissolved in 1000 g of water.

9.0 g of sugar dissolved in 500 g of water is equivalent to 18.0 g of sugar dissolved in 1000 g of water (9.0 g/500 g = 18.0 g/1000 g). But since C = 0.10 mole/1000 g, the 18.0 g of sugar must correspond to 0.10 mole of sugar. Therefore, the molecular weight of the sugar is

18 g/0.1 mole = 180 g/mole.

● **PROBLEM** 270

A chemist dissolves 10 g of an unknown protein in a liter of water at 25°C. The osmotic pressure is found to be 9.25 mmHg. What is the protein's molecular weight. Based upon the number of moles in 10 g of protein, what would the freezing point depression and boiling point elevation be? Assume R = Universal Gas Constant = .0821 liter-atm/mole°K, $k_f$ = 1.86°C/m, and $k_b$ = .52°C/m.

Solution:   The osmotic pressure ($\pi$) can be related to the molar concentration (C), Universal Gas Constant (R) and teperature (T) in degrees Kelvin of a solution via the formula $\pi$ = CRT. After solving for C one can determine the molecular weight of the protein .Recall, C = moles/liter = N/V, where N = moles and V = volume in liters. Hence, $\pi$ = N/V RT. But N = moles = No. of grams/molecular wt = g/M.W. Rewriting,

$$\pi = \left(\dfrac{\frac{g}{M.W.}}{V}\right) RT \quad \text{or} \quad M.W. = \dfrac{gRT}{\pi V}.$$

Substituting for the known values,

$$M.W. = \dfrac{(10 \text{ g})(.0821)(298°K)}{9.25 \text{ mmHg (1)}} \times \dfrac{760 \text{ mm}}{\text{atm}} = 20,100 \dfrac{g}{mole}.$$

Note, you assumed a volume of 1 liter and multiplied by the conversion factor 760 mm/atm, since R is in atm. and $\pi$ in mmHg. Now that you know that the molecular weight is 20,100, the number of moles in the 10 g of protein is 10/20,100 = $4.98 \times 10^{-4}$ moles. To find the freezing point depression,

$\Delta T_f$, and boiling point elevation, $\Delta T_b$, you must find the molality, since $\Delta T_b = k_b m$ and $\Delta T_f = k_f m$, where $k_f$ = molal freezing point depression constant, $k_b$ = molal boiling point elevation constant and m = molality. Molality = moles of solute per 1 kg solvent. Thus, molality = .000498 mole/1 kg, since in 1 liter of water you have 1000 ml, which weighs 1 kg. Therefore,

$$\Delta T_f = (1.86^\circ C)(.000498) = 9.3 \times 10^{-4} \, ^\circ C \qquad \text{and}$$

$$\Delta T_b = .52^\circ C(.000498) = 2.59 \times 10^{-4} \, ^\circ C.$$

## SURFACE TENSION

● PROBLEM 271

What would be the surface tension of a liquid (density $0.876 \ g/cm^3$) which rises 2.0 cm in a capillary where water at 20°C rises 4.5 cm?

Solution:   Surface tension refers specifically to the force within a liquid that acts parallel to the surface and tends to stretch the surface out.  The equation that relates surface tension to the density of a liquid is

$$\gamma = 1/2 \ rh\rho g$$

where $\gamma$ is the surface tension, r is the radius of the tube, h is the height the liquid rises, $\rho$ is the density of the liquid and g is the downward acceleration of gravity.

This equation can be rewritten

$$rg = \frac{2\gamma}{h\rho}$$

Because r and g are constant for a given capillary tube

$$\frac{2\gamma \ H_2O}{h_{H_2O} \ \rho_{H_2O}} = \frac{2\gamma \ liq}{h_{liq} \ \rho_{liq}}$$

One can find $\gamma \ H_2O$ and $\rho_{H_2O}$ from standard tables.

$$\gamma H_2O = 72.62 \ dyn/cm, \ \rho_{H_2O} = 1 \ g/cm^3.$$

Solving for $_{liq}$:

$$\frac{2 \ (72.62 \ dyn/cm)}{(4.5 \ cm)(1 \ g/cm^3)} = \frac{2\gamma \ liq}{(2.0 \ cm)(0.876 \ g/cm^3)}$$

$$\gamma_{liq} = \frac{2(72.62 \text{ dyn/cm})(2.0 \text{ cm})(0.876 \text{ g/cm}^3)}{2(4.5 \text{ cm})(1 \text{ g/cm}^3)}$$

$$\gamma_{liq} = 28.27 \text{ dyn/cm}$$

Given that the surface tension of water is 72.62 dyn/cm at 20°C, how high should water rise in a capillary that is 1.0 mm in diameter?

<u>Solution</u>: The surface tension is related to the radius of the capillary tube by the equation:

$$\gamma = \tfrac{1}{2} rh\rho g$$

where $\gamma$ is the surface tension, r is the radius of the capillary tube, h is the height the water rises, $\rho$ is the density and g is the downward force of gravity. The density of $H_2O$ is 1 g/cm³, the force of gravity is 980 cm/s² and the surface tension is given as 72.62 dyn/cm. The radius is found by dividing the diameter of the tube by 2. (1 mm = $10^{-1}$ cm). Solving for h:

$$72.62 \frac{\text{dyn}}{\text{cm}} = 72.62 \frac{\text{g}}{\text{sec}^2}$$

$$72.62 \text{ g/sec}^2 = \tfrac{1}{2}(10^{-1}/2 \text{ cm})(h)(1 \text{ g/cm}^3)(980 \text{ cm/sec}^2)$$

$$h = \frac{72.62 \text{ g/sec}^2 (2)(2)}{(10^{-1} \text{ cm})(1 \text{ g/cm}^3)(980 \text{ cm/sec}^2)}$$

$$h = 2.96 \text{ cm.}$$

# CHAPTER 8

# SOLUTION CHEMISTRY

> **Basic Attacks and Strategies for Solving Problems in this Chapter. See pages 251 to 275 for step-by-step solutions to problems.**

It is important to understand the different ways in which the concentration of a solvent in a solution can be expressed. The different expressions of concentrations used in this chapter follow.

    a) Formality:   the number of gram formula weights per liter of solution (abbreviated F)

    b) Molarity:   the number of gram molecular weights per liter of solution (abbreviated M)

    c) Normality:   the number of gram equivalent weights per liter of solution (abbreviated N)

On occasion, each of these measures may be referred to as 1000 g of solvent rather than one liter of solution. They are then called weight-formality, weight-molarity, and weight-normality respectively. The weight-molarity is the only one of the three in common usage and is also called molality. Therefore, a fourth definition should be added to the list.

    d) Molality:   the number of gram molecular weights per 1000 g of solvent (abbreviated m)

The differences in (a), (b), and (c) are sufficiently subtle to justify additional explanation. The "formula weight" may differ from the "molecular weight" if hydration is present in the solute but, of course, loses its identity when dissolved in an aqueous solution. For example, the formula weight of

$Mg(NO_3)_2 \cdot 6H_2O$ is 256.43 but, in an aqueous solution, the water of hydration loses its identity and only magnesium ($Mg^{+2}$) and nitrate ($NO_3^-$) ions can be traced to the solute. Formula weights and molecular weights also differ for complicated molecules, such as polymers, where the formula is known but the number of monomer units in an individual molecule, and hence the size of the molecule, are uncertain.

Normality is related to the reactivity of a species. In the example given above, if 76.9 g of $Mg(NO_3)_2 \cdot 6H_2O$ is dissolved in water to form one liter of solution, the formality and molarity are both 0.3 (76.9/256.43). The salt completely ionizes to form magnesium ($Mg^{+2}$) ions and nitrate ($NO_3^-$) ions. The concentration of magnesium ions ($Mg^{+2}$) is 0.3M, and of nitrate ions ($NO_3^-$) is 0.6M. However, the normality of both the magnesium ions ($Mg^{+2}$) and the nitrate ions ($NO_3^-$) is 0.6N. For magnesium, the normality is twice the molarity because each ion has a charge of +2. While it is generally true, for ions, that the normality is equal to the molarity times the charge on the ion, it is important to examine the reaction context before specifying the normality.

In calculating the volumes of solutions required to mix a new solution of intermediate concentration or to neutralize species from separate solutions, it is important to calculate the amount (i.e., moles) of solute in each solution. In the problems in this text, non-ideal volume effects are, in general, neglected, and you may assume the volumes are additive. The procedure is best illustrated by an example.

Determine the volume of 0.1N $H_2SO_4$ and 0.1M $H_2SO_4$ needed to neutralize 100 ml of 0.5M NaOH. It is important to understand the meaning of molarity and normality.

In 100 ml of 0.5M NaOH, there exists 0.05 moles of $OH^-$ ions. Therefore, 0.05 moles of $H^+$ ions will be required for neutralization. In 0.1M $H_2SO_4$, there are 0.2 moles of $H^+$ per liter available for reaction, while in 0.1 N $H_2SO_4$, there are only 0.1 moles of $H^+$ per liter available for reaction. For the 0.1N solution, 500 ml (0.5 liters) will contain 0.05 moles of $H^+$, while for the 0.1M (0.2N) solution, 250 ml (0.25 liters) will contain the required 0.05 moles of $H^+$.

## DENSITY AND FORMALITY

In over 90% of cases, the concentration of ethanol ($C_2H_5OH$, density = 0.80 g/ml) in blood necessary to produce intoxication is 0.0030 g/ml. A concentration of 0.0070 g/ml is fatal. What volume of 80 proof (40% ethanol by volume) Scotch whiskey must an intoxicated person consume before the concentration of ethanol in his blood reaches a fatal level? Assume that all the alcohol goes directly to the blood and that the blood volume of a person is 7.0 liters.

<u>Solution</u>:   The difference between a fatal concentration of ethanol and an intoxicating concentration of ethanol is 0.0070 g/ml - 0.0030 g/ml = 0.0040 g/ml. We must determine the total volume of ethanol in the blood which corresponds to a concentration of 0.0040 g/ml and then calculate the volume of Scotch whiskey which will provide this concentration.

The total mass of ethanol in blood needed to raise the concentration from an intoxicating to a fatal level is equal to concentration of ethanol × volume of blood = 0.0040 g/ml × 7 liters = 0.0040 g/ml × 7000 ml = 28 g of ethanol. Dividing this mass by the density, we obtain the corresponding volume of ethanol, or 28 g/0.80 g/ml = 35 ml of ethanol.

The amount of Scotch whiskey that must be consumed must provide 35 ml of ethanol. But the Scotch whiskey is only 40% ethanol, or 0.40 ml ethanol/ml Scotch. Let v denote the volume of Scotch in ml. Then

$$\left(\begin{array}{c} \text{ratio of} \\ \text{ethanol to Scotch} \end{array}\right) \times \text{volume of Scotch} = \text{volume of ethanol}$$

0.40 ml ethanol/ml Scotch × v ml Scotch = 35 ml ethanol

or,  $v = \dfrac{35 \text{ ml ethanol}}{0.40 \text{ ml ethanol/ml Scotch}}$  88 ml Scotch.

Thus, under our assumptions, an intoxicated person must drink 88 ml of 80 proof Scotch whiskey (about 3 ounces) before ethanol reaches a fatal level in his blood.

● **PROBLEM 274**

What are the mole fractions of solute and solvent in a solution prepared by dissolving 98 g $H_2SO_4$ (M.W. 98) in 162 g $H_2O$ (M.W. 18)?

<u>Solution</u>:    The mole fraction of solute is defined as the moles of solute divided by the sum of the number of moles of the solute and the number of moles of the solvent.

$$\text{Mole fraction of solute} = \frac{\text{moles of solute}}{\text{moles of solute} + \text{moles of solvent}}$$

The mole fraction of the solvent is defined similarly.

$$\text{Mole fraction of solvent} = \frac{\text{moles of solvent}}{\text{moles of solute} + \text{moles of solvent}}$$

Here, the solute is $H_2SO_4$ and the solvent is $H_2O$. One is given the amount of $H_2SO_4$ and $H_2O$ in grams. Therefore, these quantities must be converted to moles.  This can be done by dividing the number of grams available by the molecular weight.

$$\text{No. of moles} = \frac{\text{no. of grams}}{MW}$$

$$\text{No. of moles of } H_2SO_4 = \frac{98 \text{ g}}{98 \text{ g/mole}} = 1 \text{ mole}$$

$$\text{No. of moles of } H_2O = \frac{162 \text{ g}}{18 \text{ g/mole}} = 9 \text{ moles}$$

Now that the number of moles of both solvent and solute are known, the mole fraction can be found.

$$\text{Mole fraction of } H_2SO_4 = \frac{1 \text{ mole}}{1 \text{ mole} + 9 \text{ moles}} = 0.1$$

$$\text{Mole fraction of } H_2O = \frac{9 \text{ moles}}{1 \text{ mole} + 9 \text{ moles}} = 0.9.$$

A wine has an acetic acid ($CH_3COOH$, 60 g/formula weight) content of 0.66% by weight. If the density of the wine is 1.11 g/ml, what is the formality of the acid?

Solution:     This problem involves the correct interpretation of percent by weight. 0.66% by weight means $0.66 \times 10^{-2}$ g acid per 100 g wine or $6.6 \times 10^{-2}$ g acid per 1000 g wine. To convert $6.6 \times 10^{-2}$ g acid to formula weight we divide by 60 g/formula weight and obtain $6.6 \times 10^{-2}$ g/60 g/formula weight = $1.1 \times 10^{-3}$ formula weight. To convert 1000 g wine to volume, we divide by the density of the wine and obtain 1000 g/1.11 g/ml = 900 ml = 0.90 liter. The formality is then

$$\text{formality} = \frac{\text{formula weights}}{\text{volume (liters)}} = \frac{1.10 \times 10^{-3} \text{ formula weight}}{0.90 \text{ liter}}$$

$$= 1.2 \times 10^{-3} \text{ formal} = 1.2 \times 10^{-3} \text{ F.}$$

## MOLALITY

2.3 g of ethanol ($C_2H_5OH$, molecular weight = 46 g/mole) is added to 500 g of water. Determine the molality of the resulting solution.

Solution:    This problem is a calculation of the molality of an aqueous solution. Molality is equal to the number of moles of solute per kilogram of solvent. In this case the solvent is water and the solute is ethanol.

Since 1 kg of water corresponds to twice the amount of water given in the problem it is desirable to calculate the amount of ethanol that would be added to 1 kg of water and still maintain the same concentration of 2.3 g ethanol per 500 g of water. For 2 × 500 g = 1000 g = 1 kg of water, we require 2 × 2.3 g = 4.6 g of ethanol.

We now know that our solution of 2.3 g ethanol in 500 g water corresponds to 4.6 g ethanol per 1 kg water, and all that remains to be done is to calculate the number of moles of ethanol in 4.6 g. We do this by dividing by the molecular weight to obtain 4.6 g/46 g/mole = 0.1 mole ethanol.

The concentration of the solution is then 0.1 mole ethanol/1 kg water = 0.1 molal.

What is the molality of a solution in which 49 g of $H_2SO_4$ (MW 98) is dissolved in 250 grams of water?

<u>Solution</u>:    Molality is defined as the number of moles of solute per kilogram of solvent.

$$\text{Molality} = \frac{\text{moles of solute}}{\text{no. of kg. of solvent}}$$

Here, the solute is $H_2SO_4$ and the solvent is water. In this problem, one is given the number of grams of solute and the number of grams of solvent. One must calculate the number of moles of solute and the number of kilograms of solvent.

The number of moles of solute ($H_2SO_4$) is found by dividing the number of grams available by the molecular weight.

$$\text{no. of moles} = \frac{\text{no. of grams}}{\text{MW}}$$

$$\text{no. of moles of } H_2SO_4 = \frac{49 \text{ g}}{98 \text{ g/mole}} = 0.5 \text{ moles}$$

Grams can be converted to kilograms by multiplying the number of grams by the conversion factor 1 kg/1000 gm. For the water, 250 × 1 kg/1000 gm = .250 kg. The molality can now be found.

$$\text{Molality} = \frac{\text{no. of moles of } H_2SO_4}{\text{no. of kg. of } H_2O}$$

$$\text{Molality} = \frac{0.5 \text{ moles}}{.250 \text{ kg}} = 2.0 \text{ moles/kg.}$$

Determine the mass of water to which 293 g of NaCl (formula weight = 58.5 g/mole) is added to obtain a 0.25 molal solution.

<u>Solution</u>:    To solve this problem we use the relationship

$$\text{molality} = \frac{\text{number of moles of solute}}{\text{kg of solvent}} \qquad \text{or,}$$

$$\text{molality} = \frac{\text{moles of NaCl}}{\text{kg of water}} \; .$$

The number of moles of NaCl in 293 g is determined by dividing 293 g by the formula weight of NaCl, or moles NaCl = mass NaCl/formula weight NaCl = 293 g/58.5 g/mole $\stackrel{\sim}{=}$ 5.0 moles NaCl. Then, solving for the kg of water,

$$\text{kg of water} = \frac{\text{moles of NaCl}}{\text{molality}} = \frac{5.0 \text{ moles NaCl}}{0.25 \text{ molal}} = 20 \text{ kg H}_2\text{O.}$$

## MOLARITY

● **PROBLEM** 279

Calculate the molarity of a solution containing 10.0 grams of sulfuric acid in 500 ml of solution. (MW of $H_2SO_4$ = 98.1.)

Solution:   The molarity of a compound in a solution is defined as the number of moles of the compound in one liter of the solution. In this problem, one is told that there are 10.0 grams of $H_2SO_4$ present. One should first calculate the number of moles that 10.0 g represents. This can be done by dividing 10.0 g by the molecular weight of $H_2SO_4$.

$$\text{number of moles} = \frac{\text{amount present in grams}}{\text{molecular weight}}$$

$$\text{number of moles of } H_2SO_4 = \frac{10.0 \text{ g}}{98.1 \text{ g/mole}} = 0.102 \text{ moles}$$

Since molarity is defined as the number of moles in one liter of solution, and since, one is told that there is 0.102 moles in 500 ml (½ of a liter), one should multiply the number of moles present by 2. This determines the number of moles in $H_2SO_4$ present in 1000 ml.

Number of moles in 1000 ml = 2 × 0.102 = 0.204.

Because molarity is defined as the number of moles in 1 liter, the molarity (M) here is 0.204 M.

● **PROBLEM** 280

Calculate the weight in grams of sulfuric acid in 2.00 liters of 0.100 molar solution. (MW of $H_2SO_4$ = 98.1.)

Solution:   Molarity is defined as the number of moles of a compound in a solution. In this problem, one is trying to find the amount of $H_2SO_4$ in 2.00 liters of a solution that is 0.100 molar in $H_2SO_4$. From the definition of molarity, one can see that in one liter of this solution there is 0.100 moles of $H_2SO_4$. This means that in 2.00 liters of the same solution, there is twice that amount.

number of moles of $H_2SO_4$ in 2 liters = 2 liters × 0.100 moles/liter = 0.200 moles

Since one knows that the molecular weight of $H_2SO_4$ is 98.1 g/mole and that there are 0.200 moles present in this solution, one can find the number of grams present by multiplying the MW by the number of moles present.

number of grams = MW × number of moles present

number of grams of $H_2SO_4$ = 98.1 g/mole × 0.200 moles

= 19.6 g.

● **PROBLEM** 281

Hydrogen peroxide solution for hair bleaching is usually prepared by mixing 5.0 g of hydrogen peroxide ($H_2O_2$, molecular weight = 34 g/mole) per 100 ml of solution. What is the molarity of this solution?

Solution:    Before employing the definition

$$molarity = \frac{number\ of\ moles}{volume\ (liters)}$$

we must convert 5.0 g of $H_2O_2$ to the corresponding number of moles. To do this, we use the formula

moles = mass/molecular weight.

Then,

$$moles = \frac{mass}{molecular\ weight}$$

$$= \frac{5.0\ g}{34\ g/mole} = 0.15\ moles\ H_2O_2$$

Converting 100 ml to liters, 100 ml = 100 ml × $\frac{1\ liter}{1000\ ml}$ = 0.10 liter. The molarity is then

$$molarity = \frac{number\ of\ moles}{volume\ (liters)} = \frac{0.15\ moles}{0.10\ liters}$$

= 1.5 moles/liter = 1.5 molar = 1.5 M.

Each liter of human blood serum contains about 3.4 g of sodium ions ($Na^+$, ionic weight = 23 g/mole). What is the molarity of $Na^+$ in human blood serum?

**Solution:** We are given that the concentration of $Na^+$ is 3.4 g/liter. To convert this concentration to molarity, we need to convert 3.4 g to the corresponding number of moles. This is accomplished by dividing 3.4 g by the ionic weight of $Na^+$ (which is essentially the atomic weight, since the missing electron does not detract much from the weight) to obtain 3.4 g/23 g/mole = 0.15 mole. The concentration is thus 0.15 mole/liter = 0.15 molar = 0.15 M.

● **PROBLEM** 283

A $10^6$ liter tank of seawater contains 16,600 kg of chlorine ($Cl^-$), 9200 kg of sodium ($Na^+$) and 1180 kg of magnesium ($Mg^{++}$). Calculate the molarity of each. Is all the charge accounted for?

**Solution:** You are given the volume and weight of the materials in seawater and asked to calculate the molarity. Since molarity = no. of moles/liters, you need to calculate the number of moles of $Cl^-$, $Mg^{++}$ and $Na^+$ present to solve this problem.

$$A\ mole = \frac{no.\ of\ grams}{atomic\ weight}.\ For\ Cl^-,\ then, the$$

$$no.\ of\ moles = \frac{1.66 \times 10^7\ g}{35.45\ g/moles} = 4.68 \times 10^5\ moles$$

$$Therefore,\ its\ molarity\ is\ \frac{4.68 \times 10^5}{10^6} = .468\ M\ Cl^-$$

$$For\ Mg^{++},\ the\ no.\ of\ moles = \frac{1.18 \times 10^6\ g}{24.3\ m/g}$$

$$Its\ molarity\ is,\ thus,\ \frac{\frac{1.18 \times 10^6\ g}{24.3\ m/g}}{10^6\ liters} = .048\ M$$

$$For\ Na^+,\ you\ have\ \frac{9.200 \times 10^6\ g}{23\ m/g} = no.\ of\ moles\ of\ Na^+.$$

$$Its\ molarity = \frac{\frac{9.200 \times 10^6\ g}{23\ m/g}}{10^6} = .400\ M.$$

The total positive charge is .400 + 2(.048) = .496 M. The concentration of $Mg^{++}$ must be doubled (in total positive charge) since each ion has 2 charges instead of one, as in $Na^+$. The total negative charge stems only from $Cl^-$, which must equal .468 M. Thus, 0.496 moles/liter - .468 moles/liter= .028 moles/liter negative charge is unaccounted for, since all opposite charges must cancel each other out.

● **PROBLEM** 284

What mass of calcium bromide $CaBr_2$, is needed to prepare 150 ml of a 3.5 M solution? (M.W. of $CaBr_2$ = 200.618 g/mole)

Solution: M is the molarity of the solution and is defined as the number of moles of solute per liter. The solute is the substance being added to solution. A 1 M solution contains 1 mole of solute per liter (1000 ml) of solution.

We are asked to calculate the mass, however, no term for mass appears in the molarity equation. Therefore, a connection must be found between mass and another variable. The connection is the mole equation (moles = mass/M.W.). After substitution, the molarity equation reads

$$M = \frac{\dfrac{grams(mass)}{M.W.}}{liters}$$

grams = liters (M.W.)(M)

= .15ℓ (200.618 g/mole)(3.5 moles/ℓ)

= 105.32 g of $CaBr_2$.

● **PROBLEM** 285

In the United States, alcohol is rated according to "proof", which is usually defined as twice the percent by volume of pure ethanol in solution, measured at 60°F. What is the molarity of ethanol in a "92 proof" solution of ethanol in water. The density of ethanol ($C_2H_5OH$, molecular weight = 46 g/mole) is 0.80 g/cm$^3$ at 60°F; the density of water ($H_2O$, molecular weight = 18 g/mole) is 1.0 gm/cm$^3$ at 60°F(15.5°C).

Solution: We will solve this problem by determining the number of moles of ethanol in one liter of solution.

A solution which is "92 proof" corresponds to 92/2 or 46% ethanol by volume. Consider one liter of solution. The volume of ethanol in this liter is 46% × 1 liter = 0.46 liter. Converting this to cm$^3$, 0.46 liter = 0.46 liter × 1000 cm$^3$/liter = 460 cm$^3$. To convert this to the corresponding mass we multiply by the density (mass = volume × density), obtaining 460 cm$^3$ × 0.80 g/cm$^3$ = 460 × 0.80 g of ethanol. The number of moles in our one liter solution is obtained by dividing the number of grams by the molecular weight (moles = mass/molecular weight), or, moles = 460 × 0.80 g/46 g/mole = 8.0 moles.

There are 8.0 moles of ethanol in one liter of solution, and the concentration is therefore

8.0 moles/1 liter = 8.0 molar.

● **PROBLEM** 286

A student has 50.00 mg crystal of Ba(OH)$_2$·8H$_2$O (M.W. = 315) and wants to make a solution of .12 M OH$^-$. How much water must the student add to obtain such a solution?

Solution: Molarity is defined as moles/liter. You are given the desired molarity, .12 M, and can calculate the number of moles of OH$^-$ that would have to be present for this molarity. With these two known values, the volume (liters) can be obtained. You have 50 mg of Ba(OH)$_2$·8H$_2$O or (1 g/1000mg) 50 mg/315 g/mole = 1.59 × 10$^{-4}$ moles of Ba(OH)$_2$·8H$_2$O.

Now, Ba(OH)$_2$·8H$_2$O contains 2 moles of OH$^-$, which means that you have 3.18 × 10$^{-4}$ moles of OH$^-$ present in this crystal. Recalling, the stated definition of molarity, liters (volume) = moles/molarity. Therefore, the volume (liter) of water to be added =

$$= \frac{3.18 \times 10^{-4} \text{ moles}}{.120 \text{ moles/liter}} = 2.65 \times 10^{-3} \text{ liters}$$

or 2.65 ml total solution volume.

● **PROBLEM** 287

Determine the molarity of a 40.0% solution of HCl which has a density of 1.20 g/ml.

Solution: The molarity is defined as the number of moles of a compound present in 1 liter of solution. Here one is told that the density of this solution is 1.20 g/ml. This means that one ml of the solution weighs 1.20 g. The solution is 40.0% HCl, thus 40.0% of the 1.20 g is made up by HCl and 60% by H$_2$O. Here one finds the molarity of the HCl by: 1) determining the total weight of 1 liter of the solution, 2) calculating the weight of HCl present, and 3) finding the number of moles of HCl present in 1 liter (molarity).

Solving: 1) If 1 ml of this solution weight 1.20 g, 1000 ml (1 liter) is equal to 1000 × 1.20 g.

weight of solution = density × 1000 ml

$$= 1.20 \text{ g/ml} \times 1000 \text{ ml} = 1200 \text{ g}$$

2) 40.0% of this weight is taken up by HCl.

weight of HCl = weight of solution × .40

$$= 1200 \text{ g} \times .40 = 480 \text{ g}.$$

3) The molecular weight of HCl is 36.5 g/mole, thus the number of moles of HCl present in this 1 liter of solution is equal to the weight of the HCl present divided by its molecular weight.

$$\text{no. of moles of HCl in 1 liter} = \frac{\text{weight of HCl}}{MW}$$

$$= \frac{480 \text{ g}}{36.5 \text{ g/mole}} = 13.15 \text{ moles}$$

The solution is therefore 13.15 moles/liter in HCl or 13.15 M.

● **PROBLEM** 288

If 25 ml of .11 M sucrose is added to 10 ml of .52 M sucrose, what would be the resulting molarity?

Solution: In this problem one mixes two solutions of different concentrations and wishes to find the resulting molarity. The answer is obtained by first calculating the total number of moles present and then dividing this by the total volume.

Molarity is defined as moles ÷ volume. Therefore,

moles = Molarity × Volume

For the .11 M solution,

no. of moles = .11 moles/liter × .025 liter

$$= .00275 \text{ moles}$$

For the .52 M solution,

no. of moles = .52 moles/liter × .010 liter

$$= .00520 \text{ moles}$$

total no. of moles = .00275 + .00520 = .00795

total no. of liters = .025 + .010 = .035

$$\text{Final Molarity} = \frac{.00795}{.035} = .23 \text{ M}.$$

A chemist wants to dilute 50 ml of 3.50 M $H_2SO_4$ to 2.00 M $H_2SO_4$. To what volume must it be diluted?

Solution: Molarity is defined as the number of moles of solute per liter of solution. In other words,

$$\text{molarity} = \frac{\text{no. of moles of solute}}{\text{liters of solution}}.$$

In this problem you have a 50 ml solution of 3.50 M. In a liter there are 1000 ml. Thus, the number of liters in this solution is .05. Substituting, you have

$$3.50 \text{ M} = \frac{\text{no. of moles of } H_2SO_4}{.05 \text{ }\ell}$$

Solving for the number of moles, you obtain .175 moles of $H_2SO_4$.

When you dilute this mixture, you will still have .175 moles of the solute, $H_2SO_4$. You are only increasing the volume of the solvent, water. Therefore, in the diluted 2.0 M solution you wish to have

$$2.00 = \frac{.175}{\text{liters of solvent}}.$$

Solving for liters of solvent, you obtain .0875 liters or 87.5 ml. Thus, to dilute to a molarity of 2, the total volume must be 87.5 ml.

What will be the final concentration of a solution of 50 ml of .5M $H_2SO_4$ added to 75 ml of .25M $H_2SO_4$ (the final volume is 125 ml)?

Solution: Two facts must be known to answer this question: (1) M = Molarity = the number of moles of solute per liter of solution and (2) that the number of moles of $H_2SO_4$ in the final solution is the sum of the number of moles of $H_2SO_4$ in each solution.

$$\text{molarity} = \frac{\text{no. of moles}}{\text{liters of solution}}$$

One can rearrange this equation to determine the number of moles present in each solution.

no. of moles = molarity × liters of solution

The total number of moles in the final solution is equal to the sum of the number of moles of each of the two initial solutions.

$$\text{total number of moles} = (0.05\,\ell)(.5 \text{ moles}/\ell) +$$

$$(.075\,\ell)(.25 \text{ moles}/\ell)$$

$$= 0.025 \text{ moles} + .019 \text{ moles}$$

$$= .044 \text{ moles}.$$

To obtain the final concentration, you return to the definition of molarity, supplying the known values. Thus, molarity (final concentration) =

$$\frac{0.044 \text{ moles}}{.125 \text{ liter}} = .35 \text{ M } H_2SO_4$$

(It was given that the volume of the final solution equaled 125 ml = .125 liters.)

● **PROBLEM** 291

What will be the final concentration of $I^-$ in a solution made by mixing 35.0 ml of 0.100 M $NaHSO_3$ with 10.0 ml of 0.100 M $KIO_3$? Assume volumes are additive.

Solution:  The equation for this reaction is

(1)    $5HSO_3^- + 2IO_3^- \rightarrow I_2 + 5SO_4^{-2} + 3H^+ + H_2O$

$I^-$ is formed in the following reaction:

(2)    $H_2O + I_2 + HSO_3^- \rightarrow 2I^- + SO_4^= + 3H^+$

One must first calculate the concentration of $I_2$ formed in reaction 1 before the amount of $I^-$ can be found. There are 5 moles of $HSO_3^-$ and 2 moles of $IO_3^-$ needed to form 1 mole of $I_2$. Molarity is defined as moles per liter. Thus the solution of $NaHSO_3$ contains

$$35 \text{ ml} \times \frac{0.100 \text{ moles}}{1000 \text{ ml}} = 3.50 \times 10^{-3} \text{ moles}$$

There are    $10.0 \text{ ml} \times \frac{.100 \text{ moles}}{1000 \text{ ml}} = 1.00 \times 10^{-3} \text{ moles}$

of $KIO_3$ present. In any given reaction one of the reactants is called the limiting reagent. This means that when the reaction proceeds, all of the limiting reagent is used but there are other reactants left in their original state. The limiting reagent limits the amount of product formed. Two moles of $IO_3^-$ are needed to form 1 mole of $I_2$. Thus, if all of the $KIO_3$ present reacts

$$\frac{1.00 \times 10^{-3} \text{ moles}}{2} \quad \text{or } 5.00 \times 10^{-4} \text{ moles of } I_2$$

will be formed. 5 moles of $NaHSO_3$ are needed to form 1 mole of $I_2$, if all of the $NaHSO_3$ present reacts,

$$\frac{3.50 \times 10^{-3} \text{ moles}}{5} \quad \text{or } 7.0 \times 10^{-4} \text{ moles of } I_2$$

will be formed. Therefore, when all of the $KIO_3$ reacts, only $5.0 \times 10^{-4}$ moles of $I_2$ are formed and when all of the $NaHSO_3$ is reacted, $7.0 \times 10^{-4}$ moles of $I_2$ are formed. Thus, $KIO_3$ is the limiting reagent and only $5.00 \times 10^{-4}$ moles of $I_2$ are formed. When $5.00 \times 10^{-4}$ moles of $I_2$ are formed, $5 \times 5.0 \times 10^{-4}$ moles, or $2.5 \times 10^{-3}$ moles of $NaHSO_3$ are used. This leaves $(3.5 \times 10^{-3} - 2.5 \times 10^{-3})$ moles or $1.0 \times 10^{-3}$ moles of $NaHSO_3$ left unreacted to react to form the $I^-$ in equation 2. $H_2O$, $HSO_3^-$ and $I_2$ react to form $I^-$. The number of moles present of each of these must be known to find the limiting reagent. This is done in a manner similar to that used for reaction 1. From equation 1, one knows that there is 1 mole of $H_2O$ formed for each $I_2$ formed. Thus, there are $5.0 \times 10^{-4}$ moles of $H_2O$ present.

From equation 2 one notices that one mole of each $H_2O$, $HSO_3^-$, and $I_2$ are needed to form 2 moles of $I^-$. There are $5.0 \times 10^{-4}$ moles of $H_2O$, $1.0 \times 10^{-3}$ moles of $HSO_3^-$, and $5.0 \times 10^{-4}$ moles of $I_2$ present. Thus, the $I_2$ is the limiting reagent because there is the least of it present to react. Therefore, $2 \times 5.0 \times 10^{-4}$ moles or $1.0 \times 10^{-3}$ moles of $I^-$ formed. If the volumes are additive, the final volume present is 45 ml. Thus, there are $1.0 \times 10^{-3}$ moles of $I^-$ present in the final volume of 45 ml of solution. Solving for the concentration

$$\frac{1.0 \times 10^{-3} \text{ moles}}{.045 \text{ liter}} = .002 \text{ M.}$$

● **PROBLEM** 292

What is the mole fraction of $H_2SO_4$ in a 7.0 molar solution of $H_2SO_4$ which has a density of 1.39 g/ml?

Solution: The mole fraction of $H_2SO_4$ is equal to the number of moles of $H_2SO_4$ divided by the sum of the number of moles of $H_2SO_4$ and of $H_2O$.

$$\text{mole fraction of } H_2SO_4 = \frac{\text{moles of } H_2SO_4}{\text{moles of } H_2SO_4 + \text{moles of } H_2O}$$

Since the solution is 7.0 molar in $H_2SO_4$, you have 7 moles of $H_2SO_4$ per liter of solution. If one knows how much 7 moles of $H_2SO_4$ weighs, and how much one liter of the solution weighs, the weight of water can be determined by taking the difference between the quantities. The weight of one liter of the solution can be calculated by multiplying the density by the conversion factor 1000 ml/1 liter.

weight of 1 liter of the solution

$$= 1.39 \text{ g/ml} \times 1000 \text{ ml/liter} = 1390 \text{ g/}\ell.$$

weight of the water in a one liter solution

$$= \text{weight of the total solution}$$
$$- \text{weight of 7 moles of } H_2SO_4$$

7 moles of $H_2SO_4$ weigh 7 times the molecular weight of $H_2SO_4$.

weight of 7 moles of $H_2SO_4 = 7 \times 98 \text{ g} = 686 \text{ g.}$

Hence, weight of the water = 1390 g - 686 g = 704 g.

The number of moles of $H_2O$ is found by dividing its weight by the molecular weight of $H_2O$.

no. of moles of $H_2O = \dfrac{704 \text{ g}}{18 \text{ g/mol}} = 39$ moles

The number of moles of both components of the system is now known, therefore, the mole fraction of $H_2SO_4$ can be obtained.

mole fraction $H_2SO_4 = \dfrac{\text{moles of } H_2SO_4}{\text{moles of } H_2SO_4 + \text{moles of } H_2O}$

mole fraction of $H_2SO_4 = \dfrac{7}{7 + 39} = 0.15.$

● **PROBLEM** 293

You have 100-proof (50 percent alcohol by volume) bonded Scotch whisky. Calculate its molarity, mole fraction, and molality. If the temperature were to drop to - 10°C, could you still drink the Scotch? Assume density = .79 g/ml for ethyl alcohol ($C_2H_5OH$) and $K_f = 1.86°C/m$ for water.

Solution: To find molarity, mole fraction, and molality, you need to know how many moles of alcohol are present in the whiskey. You can obtain this from a calculation using the density.

To find whether you can still drink the whisky  at

- 10°C, requires you to know its freezing point, since you cannot drink something if it is frozen. You can determine the freezing point of the Scotch by determining the freezing point depression of the water containing the alcohol. This is found from $\Delta T_f = K_f \, m$, where $\Delta T_f =$ the depression, $K_f =$ freezing point depression constant, and $m =$ molality.

You now proceed as follows:

You are told that 100 proof means 50 percent alcohol by volume. This means that 1 liter of components possess 500 $cm^3$ of ethyl alcohol (a liter = 1000 $cm^3$). Density = mass/volume. You know the density and volume, so that mass of alcohol = density × volume = .79(500) = 395 g.

Molecular wt. of the alcohol = 46.07 g. Thus, you have 395/46.07 = 8.57 moles of alcohol. Hence,

$$\text{molarity} = \frac{\text{moles of solute}}{\text{liters of solution}} = \frac{8.57}{1} = 8.57 \text{ M.}$$

$$\text{Molality} = \frac{\text{moles of solute}}{\text{kilograms of solvent}} \; .$$

To find the number of grams of solvent, you need to use the density again. In 1 liter of Scotch Whisky, you have 500 $cm^3$ of water. Density of water = 1 $g/cm^3$.

Thus, mass of water = 500 $cm^3$(1 $g/cm^3$) = 500 grams. Thus,

$$\text{Molality} = \frac{8.57}{\frac{500}{1000}} = 17.14 \text{ m}$$

You divided 500 by 1000 since molality is moles per 1000 g of solvent.

Mole fraction: (MW of $H_2O$ = 18.) moles of $H_2O$ = 500 g/18.0 g/mole.

Mole fraction of ethyl alcohol =

$$\frac{\text{moles ethyl alcohol}}{\text{moles ethyl alcohol + moles } H_2O} = \frac{8.57}{8.57 + \frac{500}{18}} = .2357.$$

Now, you answer whether the scotch can be consumed at - 10°C. You already calculated the molality. Thus, $\Delta T_f = k_f \, m = (1.86)(17.14) = 31.88°C$. Thus, the freezing point of whisky is 0°C - 31.88°C = - 31.88°C. The temperature is only - 10°C, however. This means whisky is still a liquid and can be consumed.

# NORMALITY

● **PROBLEM** 294

Calculate the normality of a solution containing 2.45 g
of sulfuric acid in 2.00 liters of solution. (MW of
$H_2SO_4$ = 98.1.)

Solution:    Normality is defined by the following equation.

$$Normality = \frac{grams\ of\ solute}{equivalent\ weight \times liters\ of\ solution}$$

In this problem, one is given the grams of solute
($H_2SO_4$) present and the number of liters of solution it
is dissolved in. One equivalent of a substance is the
weight in which the acid contains one gram atom of replaceable hydrogen. This means that when the acid is
dissolved in a solution, and it ionizes, that in one
equivalent weight, one hydrogen atom  is released. The
equivalent weight for acids is defined as:

$$equivalent\ weight = \frac{MW}{no.\ of\ replaceable\ H}$$

When $H_2SO_4$ is dissolved in a solution, there are
two replaceable H as shown in the following equation:

$$H_2SO_4 \rightleftharpoons 2H^+ + SO_4^=$$

This means that in calculating the equivalent weight
for $H_2SO_4$ the molecular weight is divided by 2.

$$equivalent\ weight\ of\ H_2SO_4 = \frac{98.1\ g}{2\ equivalents}$$

$$= 49\ g/equiv$$

The normality can now be calculated.

$$Normality = \frac{grams\ of\ solute}{equiv.\ wt \times liters\ of\ soln}$$

$$Normality = \frac{2.45\ g}{49\ g/equiv \times 2.0\ liters}$$

$$= 0.025\ equiv/liter.$$

How many grams of sulfuric acid are contained in 3.00 liters of 0.500 N solution? (MW of $H_2SO_4$ = 98.1.)

Solution:  The number of grams of sulfuric acid contained in this solution can be determined by using the definition of normality.

$$\text{Normality} = \frac{\text{grams of solute}}{\text{equivalent weight} \times \text{liters of solution}}$$

Here, one is given the normality (N) and the number of liters of solution. The equivalent weight for an acid is the molecular weight divided by the number of replaceable hydrogens,

$$\text{equivalent weight} = \frac{MW}{\text{no. of replaceable H}}$$

The number of replaceable hydrogens is determined by the number of hydrogens that will ionize when the acid is placed in solution. This number is 2 for $H_2SO_4$, as shown in the following equation:

$$H_2SO_4 \stackrel{\leftarrow}{\rightarrow} 2H^+ + SO_4^=$$

Once the number of replaceable hydrogen is known, the equivalent weight can be found.

$$\text{equivalent weight} = \frac{MW}{\text{no. of replaceable H}}$$

$$\text{equivalent weight of } H_2SO_4 = \frac{98.1 \text{ g}}{2 \text{ equiv}} = 49 \text{ g/equiv.}$$

At this point the number of grams of solute can be determined.

$$\text{grams of solute} = \text{Normality} \times \text{equiv weight} \times \text{liters of solution}$$

$$\text{grams of } H_2SO_4 = 0.500 \text{ equiv/}\ell \times 49 \text{ g/equiv} \times 3.0 \text{ liter}$$

$$= 73.5 \text{ g.}$$

Calculate the normality of a solution 40 ml of which is required to neutralize 0.56 g of KOH.

Solution:  KOH ionizes as shown in this equation.

$$KOH \stackrel{\leftarrow}{\rightarrow} K^+ + OH^-$$

This means that for each KOH molecule ionized, one $OH^-$ ion is formed. Thus to neutralize the KOH one must have 1 $H^+$ ion for each $OH^-$ ion. Thus, one must first determine the number of moles of KOH present in 0.56 g. This is done by dividing 0.56 g by the molecular weight of KOH (MW of KOH = 56)

number of moles of KOH = number of moles of $OH^-$

$$\text{number of moles of KOH} = \frac{0.56 \text{ g}}{MW} = \frac{0.56 \text{ g}}{56 \text{ g/mole}} = .01 \text{ mole}$$

Therefore, in the 40 ml of the solution there must be .01 moles of $H^+$ ions. The normality of an acid is defined as the number of equivalents of $H^+$ in 1 liter of solution. An equivalent may be defined as the weight of acid or base that produces 1 mole of $H^+$ or $OH^-$ ions. In this problem, equivalents = moles. There are 1000 ml in 1 liter, thus in 40 ml there are

$$40 \text{ ml} \times \frac{1 \text{ liter}}{1000 \text{ ml}}$$

$$\text{number of liters} = 40 \text{ ml} \times \frac{1 \text{ liter}}{1000 \text{ ml}} = .04 \text{ liters}.$$

The normality can now be found.

$$\text{normality} = \frac{\text{number of equivalents}}{\text{number of liters}}$$

$$= \frac{.01 \text{ equivalents}}{.04 \text{ liters}} = .25 \text{ N}.$$

● PROBLEM 297

If 20 ml of 0.5 N salt solution is diluted to 1 liter, what is the new concentration?

Solution:  When considering normality one must always keep in mind that it is a concentration defined as the number of equivalents per liter. Since the number of equivalents does not change during dilution, equivalents before dilution = equivalents after dilution, or in other words,

$N_1V_1 = N_2V_2$           1 liter = 1000 ml

where $N_1$ is the normality of the initial solution, $V_1$ the initial volume, $N_2$ the final normality and $V_2$ is final volume.

Solving for the final normality:

$(0.5)(0.020\ell) = N_2\ (1\ell)$

$$N_2 = \frac{(0.5)(0.020\ \ell)}{1\ \ell}$$

$$= 0.01\ N$$

A sulfuric acid solution has a density of 1.8 g/ml and is 90% $H_2SO_4$ by weight. What weight of $H_2SO_4$ is present in 1000 ml of the solution? What is the molarity of the solution? the normality?

Solution:   1) Here one is asked to find the weight of $H_2SO_4$ in 1000 ml of the solution and told that the solution weighs 1.8 g/ml and that 90% of this weight is made up by $H_2SO_4$. The total weight of the solution is 1000 ml times the weight of one ml.

Hence, weight of solution = 1000 ml × 1.8 g/ml = 1800 g. $H_2SO_4$ makes up 90% of this weight.

weight of $H_2SO_4$ = .90 × weight of solution

$$= .90 \times 1800\ g = 1620\ g.$$

2) The molarity is defined as the number of moles in one liter of solution. One has already found that 1620 g of $H_2SO_4$ is present in 1 liter of the solution. Thus, to calculate molarity, one should determine the number of moles present in 1620 g. This is done by dividing 1620 g by the molecular weight of $H_2SO_4$. (MW of $H_2SO_4$ = 98.1.)

moles of $H_2SO_4$ present = $\dfrac{1620\ g}{98.1\ g/mole}$ = 16.5 moles

Since there is 1 liter of solution, the molarity of $H_2SO_4$ is 16.5 M.

3) The normality is defined as the number of moles of ionizable hydrogens per liter of solution. From the following equation one can see that there are 2 ionizable hydrogens for each molecule of $H_2SO_4$.

$$H_2SO_4 \rightleftharpoons 2H^+ + SO_4^=$$

Therefore if there are 16.5 moles of $H_2SO_4$ in one liter of the solution the normality is twice this amount.

normality of an acid = no. of ionizable H × molarity

$$= 2 \times 16.5 = 33.0\ N.$$

A solution is prepared by dissolving 464 g NaOH in water and then diluting to one liter. The density of the resulting solution is 1.37 g/ml. Express the concentration of NaOH as (a) percentage by weight, (b) molarity, (c) normality, (d) molality, (e) mole fraction.

Solution:    (a) The percentage by weight of NaOH in this solution is found by dividing the weight of NaOH present, 464 g, by the weight of the solution and multiplying by 100. The weight of the solution is found by using the density of the solution. The density, 1.37 g/ml, tells the weight of 1 ml of solution, namely, 1.37 g. In one liter, there are 1000 ml, thus the weight of the solution is 1000 times the weight of one ml.

weight of 1 liter = 1000 ml × 1.37 g/ml = 1370 g.

The percentage by weight of the NaOH in the solution can now be found.

$$\text{percentage of NaOH} = \frac{\text{weight of NaOH}}{\text{weight of solution}} \times 100$$

$$= \frac{464 \text{ g}}{1370 \text{ g}} \times 100 = 33.9 \text{ \%}.$$

(b) The molarity is defined as the number of moles in one liter of solution. The molarity in this case can be found by determining the number of moles in 464 g of NaOH, which is the amount of NaOH in one liter. The number of moles can be found by dividing 464 g by the molecular weight of NaOH. (MW of NaOH = 40.)

$$\text{no. of moles} = \frac{464 \text{ g}}{\text{MW}} = \frac{464 \text{ g}}{40 \text{ g/mole}} = 11.6 \text{ moles}.$$

The molarity of this solution is thus 11.6 M.

(c) The normality of a basic solution is the number of moles of ionizable $OH^-$ ions in one liter of solution. There is one ionizable $OH^-$ ion in each NaOH as shown by the equation:

$$NaOH \rightleftharpoons Na^+ + OH^-$$

Therefore, there are the same number of $OH^-$ ions in the solution as NaOH molecules dissolved. Thus, the molarity equals the normality.

normality = molarity × 1 ionizable $OH^-$

normality = 11.6 N

(d) The molality is defined as the number of moles present in 1 kg of solvent. One has already found that the solution weighs 1370 g or 1.37 kg and that there are 11.6 moles of NaOH present. Therefore the molality can now be found.

$$molality = \frac{\text{no. of moles present}}{\text{no. of kg present}}$$

$$= \frac{11.6 \text{ moles}}{1.37 \text{ kg} - .339(1.370 \text{ kg})} = 12.8 \text{ m}$$

(e) The mole fraction is equal to the number of moles of each component divided by the total number of moles in the system. The components in this system are $H_2O$ and NaOH. One already has found that there are 11.6 moles of NaOH present, but not the number of moles of $H_2O$. This can be found by determining the weight of the water and dividing it by its molecular weight. This solution weighs 1370 g, the NaOH weighs 464 g, thus the weight of the water is equal to the difference of these two figures,

weight of $H_2O$ = weight of solution - weight of NaOH

$$= 1370 \text{ g} - 464 \text{ g} = 906 \text{ g}.$$

One can now find the number of moles of $H_2O$ present. (MW of $H_2O$ = 18.)

$$\text{no. of moles} = \frac{906 \text{ g}}{18 \text{ g/mole}} = 50.3 \text{ moles}$$

The total number of moles in the system is the sum of the number of moles of $H_2O$ and of the NaOH.

no. of moles in system = moles $H_2O$ + moles NaOH

$$= 50.3 + 11.6 = 61.9 \text{ moles}$$

One can now find the mole fractions.

$$mole\ fraction = \frac{\text{no. of moles of each component}}{\text{total no. of moles in system}}$$

$$\text{mole fraction of } H_2O = \frac{50.3}{61.9} = .81$$

$$\text{mole fraction of NaOH} = \frac{11.6}{61.9} = .19.$$

# NEUTRALIZATION

● **PROBLEM** 300

A mixture consisting only of KOH and Ca(OH)$_2$ is neutralized by acid. If it takes exactly 0.100 equivalents to neutralize 4.221 g of the mixture, what must have been its initial composition by weight?

Solution:    One equivalent of an acid is the mass of acid required to furnish one mole of $H_3O^+$; one equivalent of a base is the mass of base required to furnish one mole of $OH^-$ or to accept one mole of $H_3O^+$. Here one must find the mixture of KOH and Ca(OH)$_2$ that contains 0.100 equivalent of base. There is one equivalent for each $OH^-$ in a molecule of base. Thus, there is 1 equivalent of base for each KOH. There are two for each Ca(OH)$_2$. Since one is given that there are 0.100 equivalent  in 4.221 g of the mixture, one should solve for the equivalent per gram for KOH and Ca(OH)$_2$.

for KOH: MW = 56.1    equiv = 1

$$\frac{equiv}{g} = \frac{1}{56.1} = 1.78 \times 10^{-2} \text{ equiv/g}$$

For Ca(OH)$_2$ MW = 74.06    equiv = 2

$$\frac{equiv}{g} = \frac{2}{74.06} = 2.70 \times 10^{-2} \text{ equiv/g.}$$

There must be 0.100 equiv  present. Let

x = number of grams of KOH

4.221 - x = number of grams of Ca(OH)$_2$

$$0.100 \text{equiv.} = \left(1.78 \times 10^{-2} \frac{equiv}{g}\right)(x) + \left(2.70 \times 10^{-2} \frac{equiv}{g}\right)x$$

$$(4.221 - x)$$

$$0.100 \text{equiv.} = 0.0178x \frac{equiv}{g} + .114 \text{ equiv} - .0270x \frac{equiv}{g}$$

$$0.100 \text{equiv.} = - .0092x \frac{equiv}{g} + .114 \text{ equiv}$$

$$-.014 \text{equiv.} = - .0092x \frac{equiv}{g}$$

$$\frac{- \ .014 \ \text{equiv}}{- \ .0092 \ \text{equiv/g}} = x$$

$$1.52 \ g = x$$

$$2.701 \ g = 4.221 - x$$

The original mixture contains 1.52 g KOH and 2.701 g $Ca(OH)_2$.

● **PROBLEM** 301

Calculate the volume of 0.3 N base necessary to neutralize 3 liters of 0.01 N nitric acid.

Solution: For neutralization to occur, there must be the same number of hydrogen ions as there are hydroxide ions. This is shown by the following equation

$$H^+ + OH^- \rightleftharpoons H_2O$$

The $H^+$ ions come from the acid, the $OH^-$ ions from the base. The number of $H^+$ and $OH^-$ ions are equal to the number of equivalents.

Normality is defined as the number of equivalents of acid or base per liter of solution.

$$\text{Normality} = \frac{\text{equivalents}}{\text{liter}}$$

An equivalent is the number of grams of the acid or base multiplied by the number of replaceable hydrogens or hydroxides divided by the molecular weight of the acid or base.

$$\text{equivalent} = \frac{\text{grams of solute} \times \text{no. of replaceable H or OH}}{\text{MW of solute}}$$

The number of replaceable hydrogens or hydrodixes is defined as the number which ionize when the compound is placed in solution. For nitric acid, there is one replaceable hydrogen

$$HNO_3 \rightleftharpoons H^+ + NO_3^-$$

In brief, for neutralization there must be the same equivalent amount of base as there is acid. Because normality is defined as

$$\text{Normality} = \frac{\text{equivalents}}{\text{liters}}$$

equivalents are equal to the normality times the volume.

Thus, the normality of the base times its volume equals the normality of the acid times its volume.

$$N_{base} \, V_{base} = N_{acid} \, V_{acid'}$$

where N is the normality and V is the volume.

Here,     0.3 equiv/liter × $V_{base}$ = 0.01 equiv/liter × 3 liters

$$V_{base} = \frac{0.01 \times 3 \text{ liters}}{0.3}$$

$$= 0.1 \text{ liter or } 100 \text{ M.}$$

● **PROBLEM** 302

What weight of $Ca(OH)_2$, calcium hydroxide is needed to neutralize 28.0 g of HCl. What weight of phosphoric acid would this weight of $Ca(OH)_2$ neutralize?

Solution:   When we have neutralization problems, i.e., those where an acid reacts with a base, we must consider the meaning of equivalents. Equivalent weight may be defined as the molecular weight of a substance (grams/mole) divided by the number of protons ($H^+$) or hydroxyl ($OH^-$) ions available for reaction. In other words,

$$\text{Equivalent weight} = \frac{\text{molecular weight}}{\text{number of } H^+ \text{ or } OH^-}.$$

In a neutralization reaction, the number of equivalents of acid equals the number of equivalents of base.

The number of equivalents of acid and base are equal, thus, if we can calculate the number of equivalents of 1 reactant, we automatically know the number of equivalents present of the other reactant. To find this quantity use the equation

$$\text{no. of equivalents} = \frac{\text{grams}}{\text{grams/equivalent}}$$

The equivalent weight of HCl is

$$\frac{M.W.}{\text{no. of } H^+(OH^-)} = \frac{36.5}{1} = 36.5 \quad \text{and } Ca(OH)_2 \text{ is } \frac{74.1}{2} = $$

37 g/equiv.

$$\text{no. of equivalents HCl present} = \frac{28 \text{ g}}{36.5 \text{ g/equiv}} = .767$$

Therefore, there are also .767 equivalents of $Ca(OH)_2$. To find the grams of $Ca(OH)_2$ needed use no. of equiv = $\dfrac{\text{grams Ca(OH)}_2}{\text{grams/equiv}}$

$.767 = \dfrac{\text{grams}}{37}$

grams Ca(OH)$_2$ = 28.4 need to neutralize 28 g of HCl.

The second half of the problem is answered in exactly the same way. We know then that there must also be .767 equiv of H$_3$PO$_4$ (phosphoric acid) present due to the fact .767 equiv of base Ca(OH)$_2$ is present. Recall, equivalents of an acid must equal the equivalents of a base for neutralization to occur. The M.W. of H$_3$PO$_4$ = 98.0 g, which yields a value of $\dfrac{98}{3}$ = 32.7 grams per equiv. Therefore,

No. of equiv = $\dfrac{\text{grams H}_3\text{PO}_4}{\text{grams/equiv}}$

$.767 = \dfrac{\text{grams}}{32.7}$

grams H$_3$PO$_4$ = 25.1 g needed to neutralize 28.4 g of Ca(OH)$_2$.

# CHAPTER 9

# EQUILIBRIUM

Basic Attacks and Strategies for Solving Problems in this Chapter. See pages 277 to 303 for step-by-step solutions to problems.

Most physical and chemical phenomena encountered in chemistry (e.g., reactions, solubility, ionization, and dissociation) proceed to some equilibrium state where concentrations and other properties do not change with time. The equilibrium state, which might well appear to be a state where nothing is happening, is a dynamic state where reactions or other changes occur in opposite directions at equal rates.

The key to working all equilibrium problems lies in the definition of the equilibrium constant and conversion of this definition to an equation with only one unknown variable. For the general reaction in Equation 9-1

$$bB + dD \leftrightarrow rR + sS \qquad \qquad 9\text{-}1$$

(which can represent a classical reaction, ionization, solubility, dissociation, or any other equilibrium phenomenon) the equilibrium constant is defined as follows:

$$K_{eq} = (a_R^{\,r} a_S^{\,s})/(a_B^{\,b} a_D^{\,d}) \qquad \qquad 9\text{-}2$$

where $a_I^{\,i}$ is the activity of the species, $I$, raised to the stoichiometric coefficient $i$. For gases at low pressures, the activity is equal to the partial pressure in atmospheres; for liquids and solutes in dilute solution, the activity is equal to the concentration in moles/liter. The activity of pure species is one.

In order to solve for the equilibrium composition, it is necessary to convert the activity (the partial pressure or concentration) of each of the species to an expression in a single variable. This can almost always be done by referring to the stoichiometry of the reaction and writing the concentration of each species as a function of the fractional or absolute conversion of one species. An example will best illustrate the technique.

Consider the reaction in Problem 314:

$$H_2(g) + I_2(g) \leftrightarrow 2HI(g) \qquad\qquad 9\text{-}3$$

Since all species in this reaction are gases at moderate temperatures, the equilibrium constant is

$$K_{eq} = p_{HI}^2/(p_{H_2} p_{I_2}) = n_{HI}^2/(n_{H_2} n_{I_2}) (P/n_T)^0 \qquad\qquad 9\text{-}4$$

where $p$ represents the partial pressure in atm, $P$ the total pressure, and $n$ the number of moles. (Note that for gases at low pressures, $p_i = (n_i/n_T)P$). In order to solve Equation 9-4 for the equilibrium composition, $n_{HI}$, $n_{H2}$, $n_{I2}$, and $n_T$ must be written in terms of a single variable. A convenient variable is $x$, the fraction of the initial $H_2$ consumed at equilibrium. From Equation 9-3, the stoichiometric equation representing the reaction, the amounts of each species can be written in terms of the variable $x$.

$$
\begin{aligned}
n_{H_2} &= 1.0(1-x) \\
n_{I_2} &= 1.0(1-x) \\
n_{HI} &= 1.0(2x) \\
n_T &= 2
\end{aligned}
$$

Equation 9-4 can now be written in terms of a single variable, x, and solved as follows;

$$K = (2x)^2/(1-x)^2 \qquad\qquad 9\text{-}5$$

$$K^{1/2} = 2x/(1-x) \qquad\qquad 9\text{-}6$$

$$x = K^{1/2}/(2 + K^{1/2}) \qquad\qquad 9\text{-}7$$

The equilibrium constant varies with temperature (to be discussed in Chapter 15), but is independent of pressure. However, the equilibrium composition will be changed by changes in the pressure if the number of moles changes during the reaction. In Equation 9-4, the zero exponent on pressure is a consequence of the fact that there is no change in the number of moles in the reactants and products. If the number of moles changes, the exponent on total pressure will be non-zero and the composition will change as pressure changes.

The Principle of LeChatelier provides a way to determine quickly the qualitative effect of temperature and pressure changes (or changes in other important parameters) on equilibrium composition. LaChatelier's principle, simply stated, says that *"a system will respond to offset any imposed change."* Therefore, if the pressure is increased, the equilibrium composition will shift in a way that will lower the pressure. In gas phase reactions, this is equivalent to shifting in a way to reduce the total number of moles. If the moles of products of a reaction exceed the moles of the reactants, the equilibrium composition will shift toward the reactants and vice versa. If there is no change in the number of moles in a reaction (as in the HI synthesis represented by Equation 9-4), changes in the pressure will not affect the equilibrium composition (hence the zero exponent on total pressure).

Similarly, if the temperature is increased, the equilibrium will shift in a way to reduce the temperature. This means, for increases in temperature, the equilibrium composition will shift toward the reactants for exothermic reactions and toward the products for endothermic reactions. The reverse is true for decreases in temperature (Exothermic reactions produce heat, while endothermic reactions consume heat).

## THE EQUILIBRIUM CONSTANT

● **PROBLEM** 303

Determine the equilibrium constant for the reaction $H_2 + I_2 \rightleftharpoons 2HI$ if the equilibrium concentrations are: $H_2$, 0.9 moles/liter; $I_2$, 0.4 mole/liter; HI, 0.6 mole/liter.

<u>Solution</u>: The equilibrium constant (Keq) is defined as the product of the concentrations of the products divided by the product of the concentrations of the reactants. These concentrations are brought to the power of the stoichiometric coefficient of that component. For example for the reaction $2A + B \rightleftharpoons 3C$, the equilibrium constant can be expressed:

$$Keq = \frac{[C]^3}{[A]^2 [B]}$$

where [] indicates concentration.

One can now express the equilibrium constant for the reaction $H_2 + I_2 \rightleftharpoons 2HI$ as

$$Keq = \frac{[HI]^2}{[H_2][I_2]}$$

One is given the concentrations of the components for the system, thus one can solve for the equilibrium constant.

$$Keq = \frac{[HI]^2}{[H_2][I_2]}$$

$[H_2]$ = 0.9 mole/liter

$[I_2]$ = 0.4 mole/liter

$[HI]$ = 0.6 mole/liter

$$Keq = \frac{[0.6 \text{ mole/liter}]^2}{[0.9 \text{ moles/liter}][0.4 \text{ moles/liter}]}$$

$$= \frac{0.36}{0.36} = 1.0$$

● PROBLEM 304

One of the two most important uses of ammonia is as a reagent in the first step of the Osfwald process, a synthetic route for the production of nitric acid. This first step proceeds according to the equation

$$4NH_3(g) + 5O_2(g) \rightleftarrows 4NO(g) + 6H_2O(g) .$$

What is the expression for the equilibrium constant of this reaction?

Solution: This problem is an exercise in writing the equilibrium constant of a reaction. In general, for a reaction in which re-actants A,B,C,... go to products W,X,Y,... according to the equation

$$aA + bB + cC + \ldots \rightleftarrows wW + xX + yY + \ldots ,$$

where a,b,c,w,x,y,... are the stoichiometric coefficients, the equilibrium constant is given by

$$K = \frac{[W]^w [X]^x [Y]^y \ldots}{[A]^a [B]^b [C]^c \ldots} .$$

Hence, for the reaction

$$4NH_3 + 5 O_2 \rightleftarrows 4NO + 6H_2O ,$$

the equilibrium constant is given by

$$K = \frac{[NO]^4 [H_2O]^6}{[NH_3]^4 [O_2]^5} .$$

● PROBLEM 305

Given the reaction $A + B \rightleftarrows C + D$, find the equilibrium constant for this reaction if .7 moles of C are formed when 1 mole of A and 1 mole of B are initially present. For this same equilibrium, find the equilibrium composition when 1 mole each of A, B, C, and D are initial-ly present.

Solution: In general, an equilibrium constant measures the ratio of the concentrations of products to reactants, each raised to the power of their respective coefficients in the chemical equation. Thus, for this reaction, the equilibrium constant, K, is equal to

$$\frac{[C][D]}{[A][B]} .$$

You are asked to find K in the first part of this problem. Therefore, you must evaluate this expression. To do this, use stoichiometry. From the equation for the chemical reaction, you see that all the quantities are in equimolar amounts; 1:1:1:1 . It is given that .7 moles of C is produced. This means, therefore, that .7 mole of D must also be produced. If the equimolar quantities are to be maintained, then, .7 mole of each A and B must have been consumed. If you started with 1 mole of each, then .3 mole must be left. These mole amounts are the concentrations if you assume they are in a given amount of volume. Thus, substituting these values in

$$K = \frac{[C][D]}{[A][B]} \ ,$$

you obtain

$$K = \frac{(.7)^2}{(.3)^2} = 5.44 \ .$$

Therefore, you have calculated the equilibrium constant of this reaction to be 5.44.

The second part of this problem asks you to use this same equilibrium for a reaction that starts with 1 mole each of A, B, C, and D. To find its composition, determine the concentration of each species at equilibrium. This can be found from

$$K = \frac{[C][D]}{[A][B]} = 5.44 \ .$$

To find the composition, assume X moles/liter (concentration) react. Thus, at equilibrium, both C and D have a final concentration of 1 + x, since you started with 1 mole/liter and had x moles/liter of each produced. This means, therefore, that A's and B's initial concentration must be reduced by x moles/liter to 1 - x. Substituting,

$$5.44 = \frac{(1+x)^2}{(1-x)^2} \ .$$

Solving,

$$x = .40 \text{ moles/liter} \ .$$

Thus, the concentration of both A and B = 1-.4 = .6M and of both C and D = 1 + .4 = 1.4M .

● **PROBLEM** 306

At a certain temperature, $K_{eq}$ for the reaction $3C_2H_2 \rightleftarrows C_6H_6$ is 4. If the equilibrium concentration of $C_2H_2$ is 0.5 mole/liter, what is the concentration of $C_6H_6$ ?

**Solution:** The equilibrium constant (Keq) for this reaction is stated:

$$Keq = \frac{[C_6H_6]}{[C_2H_2]^3}$$

where [ ] indicate concentration. The $[C_2H_2]$ is brought to the third power because three moles of it react. Equilibrium is defined as the point where no more product is formed and no more reactant is dissipated; thus their concentrations remain constant. Here, one is given Keq and $[C_2H_2]$ and asked to find $[C_6H_6]$ . This can be done by substituting the given into the equation for the equilibrium constant.

$$K_{eq} = \frac{[C_6H_6]}{[C_2H_2]^3} \qquad K_{eq} = 4$$
$$[C_2H_2] = 0.5 \text{ moles/liter}$$

$$4 = \frac{[C_6H_6]}{(0.5)^3}$$

$$[C_6H_6] = (0.5)^3 \times 4 = 0.5 \text{ moles/liter .}$$

The following reaction

$$2H_2S(g) \rightleftarrows 2H_2(g) + S_2(g)$$

was allowed to proceed to equilibrium. The contents of the two-liter reaction vessel were then subjected to analysis and found to contain 1.0 mole $H_2S$, 0.20 mole $H_2$, and 0.80 mole $S_2$ . What is the equilibrium constant K for this reaction?

Solution: This problem involves substitution into the equilibrium constant expression for this reaction,

$$K = \frac{[H_2]^2 [S_2]}{[H_2S]^2} \quad .$$

The equilibrium concentration of the reactant and products are $[H_2S] = 1.0$ mole/2 liters $= 0.50M$, $[H_2] = 0.20$ mole/2 liters $= 0.10M$, and $[S_2] = 0.80$ mole/2 liters $= 0.40M$ , Hence, the value of the equilibrium constant is

$$K = \frac{[H_2]^2 [S_2]}{[H_2S]^2} = \frac{(0.10)^2(0.40)}{(0.50)^2} = 0.016$$

for this reaction.

You have the reaction $A_2 + B_2 \rightleftarrows 2AB$ with AB initially at 5 liters, $27°C$, and 25 atm. K = 50. Find (a) the initial concentration of AB , (b) the $A_2$ and $B_2$ concentrations at equilibrium and (c) the partial pressures of $A_2$ and $B_2$ and AB at equilibrium.

Solution: Use the equation of state, which indicates PV = NRT, where P = pressure, V = volume, N = moles, R = universal gas constant (.0821 liter-atm/mole-K), and T = temperature in degrees kelvin (Celsius plus $273°$). You can use this equation to find N, the moles of AB initially present.

(a) $PV = NRT$ or $N = \frac{PV}{RT} = \frac{(25)(5)}{(.0821)(300)} = 5.08$ moles .

Thus, the initial concentration = moles/liters = $5.08/5 = 1.02$.

(b) $K = 50$ (given). By definition, $K = \frac{[AB]^2}{[A_2][B_2]}$ . At equilibrium,

let x = moles/liter of AB that are dissociated. If its initial concentration is 1.02M, then the concentration is (1.02-x)M at equilibrium. From the reaction, 1 mole each of $A_2$ and $B_2$ is produced by every 2 moles of AB which react. Thus, if x moles/liter of AB dissociate, $\frac{1}{2}$x moles/liter of both $A_2$ and $B_2$ are produced. Recalling,

$K = \frac{[AB]^2}{[A_2][B_2]} = 50$, you now substitute to obtain $\frac{(1.02-x)^2}{(.5x)^2} = 50$.

Solving for x, x = .225 mole/liter. Therefore, at equilibrium, $[A_2] = [B_2] = x/2 = .1125$ mole/liter and $[AB] = 1.02 - .225 = .795$ mole/liter.

(c) The partial pressures of each will be $P = \frac{N}{V} RT$ from the equation of state, where $\frac{N}{V}$ is the concentration at equilibrium. Thus,

$P_{A_2} = .1125(.0821)(300) = 2.77$ atm, $P_{A_2} = P_{B_2}$ , since these concentrations are the same at equilibrium,

$P_{AB} = .795(.0821)(300) = 19.5$ atm.

# EQUILIBRIUM CALCULATIONS

● **PROBLEM** 309

At 986° C, you have the following equilibrium:

$$CO_2(g) + H_2(g) \rightleftharpoons CO(g) + H_2O(g) .$$

Initially, 49.3 mole percent $CO_2$ is mixed with 50.7 mole percent $H_2$ . At equilibrium, you find 21.4 mole percent $CO_2$ , 22.8 mole percent $H_2$ , and 27.9 mole percent of CO and $H_2O$ . Find K. If you start with a mole percent ratio of 60:40, $CO_2$ to $H_2$ , find the equilibrium concentrations of both reactants and products.

Solution: An equilibrium constant $K_{eq}$ measures the ratio of the concentrations of products to reactants, each raised to the power of their respective coefficients in the chemical equation. Thus, $K_{eq}$ for this reaction =

$$\frac{[CO][H_2O]}{[CO_2][H_2]} .$$

If you assume each substance occupies the same volume in liters, the concentration can be expressed in moles because concentration = moles/liter , i.e., liters cancel out of the equilibrium constant expression. Thus, to find $K$ , you need to find the number of moles of each of the products and reactants and then to substitute into the equilibrium expression. You are told the final product mole percents. The reactants, then, at equilibrium, have mole percents that equal their initial amounts minus the amount that decomposed to produce the products.

Thus, at equilibrium, $[CO_2]$ = 49.3 - 21.4 = 27.9, which was given. Similarily $[H_2]$ = 50.7 - 22.8 = 27.9, which was given. Thus, by substitution into

$$K = \frac{[CO][H_2O]}{[CO_2][H_2]} \quad , \text{ you obtain } \quad K = \frac{(.279)(.279)}{(.214)(.228)} = 1.60 \; ,$$

which is the equilibrium constant for this reaction.

The second part follows. You begin with a 60:40 ratio of $CO_2:H_2$ , which means that initially you have .600 moles $CO_2$ and .400 moles $H_2$ . To find the equilibrium concentrations, let $x$ = moles of $CO$ formed. Thus, $x$ = moles $H_2O$ formed since coefficients tell us they are formed in equimolar amounts. The fact that moles of a product form, means that $x$ moles of a reactant must have decomposed. Thus, at equilibrium, you have 0.600-x moles of $CO_2$ and 0.400-x moles of $H_2$ . Recalling,

$$K = \frac{[CO][H_2O]}{[CO_2][H_2]} \quad ,$$

you can now substitute these values to give $K = \dfrac{x^2}{(0.6-x)(0.4-x)}$ .

From previous part, $K$ = 1.60, therefore,

$$1.60 = \frac{x^2}{(0.6-x)(0.4-x)} \quad .$$

Solving, $x$ = 0.267 . Thus, $[CO]$ = $[H_2O]$ = $x$ = 0.267M, $[CO_2]$ = 0.6-x = 0.333M and $[H_2]$ = 0.4-x = 0.133M .

● **PROBLEM** 310

A chemist mixes nitric oxide and bromine to form nitrosyl bromide at 298° K, according to the equation $2NO_{(g)} + Br_{2(g)} \rightleftarrows 2NOBr_{(g)}$ . Assuming $K$ = 100, what is the quantity of nitrosyl bromide formed, if the reactants are at an initial pressure of 1 atm? $R$ = 0.0821 liter-atm./mole° K .

**Solution:** You are given the equilibrium constant for this reaction and asked to calculate the quantity of nitrosyl bromide produced. The first step is to write out the equilibrium expression and equate it with the given value. For the general reaction,

$$xA + yB \rightarrow zC \; , \quad K \text{ is defined } \frac{[C]^2}{[B]^y[A]^x} \quad ,$$

where the brackets represent concentrations. For this reaction,

$$K = \frac{[NOBr]^2}{[NO]^2[Br_2]} = 100 \ .$$

To find out how much NOBr is produced, you would have to know how many moles of NO and $Br_2$ were reacted. Once this is known, you can find the number of grams produced. You know that the equilibrium expression is based on concentration of reactants and products. Concentration is expressed in moles per liter. This means that if the volume of the NOBr and its concentration is known, you can find moles, since concentration X volume (liters) = moles. Let us represent the concentration as

$$\frac{N}{V} = \frac{moles}{Volume} \ .$$

Thus, the equilibrium expression becomes

$$K = \frac{(N_{NOBr}/V)^2}{(N_{NO}/V)^2 (N_{Br_2}/V)} \ .$$

Let x = moles of NOBr formed. Then, x moles of NO and x/2 moles of $Br_2$ are consumed, since the coefficients of the reaction show a 2:2:1 ratio among $NOBr:NO:Br_2$. The equilibrium expression becomes

$$100 = \frac{N_{NOBr}^2 \ V}{N_{NO}^2 \ N_{Br_2}} = \frac{x^2 V}{(2-x)^2(1-.5x)} \ .$$

If x moles of NOBr form, and you started with 2 moles of NO, then, at equilibrium, you have left 2-x moles of NO. You started with only 1 mole of $Br_2$ and ½x moles of it form NOBr; thus you have 1-.5x moles left. Therefore, you need to determine only the volume to find the quantity NOBr formed. V can be found from the equation of state, PV = NRT , where P = pressure, V = volume, N = moles, R = universal gas constant, and T = temperature in kelvin (celsius plus 273°). You are told that the reactants are under a pressure of 1 atm. at 298°K . N = 3, since the coefficients inform you that a relative sum of 3 moles of reagents exist. You know R. Thus,

$$V = \frac{NRT}{P} = \frac{(3)(.0821)(298)}{1} = 73.4 \ \text{liters} \ .$$

Now that V is known, the equilibrium expression becomes

$$\frac{x^2(73.4)}{(2-x)^2(1-0.5x)} = 100 \ .$$

Solving for x, you obtain x = .923 moles = moles of NOBr formed. Molecular weight = 110. Grams produced = .923 × 110 = 101.53g.

● **PROBLEM** 311

Carbon dioxide is reduced by hydrogen according to the equation

$$CO_2(g) + H_2(g) \rightleftarrows CO(g) + H_2O(g) \ .$$

One mole of $CO_2$ is reacted with one mole of $H_2$ at 825K and it is found that, at equilibrium, 0.27 atm of CO is present. Determine the equilibrium constant, K, at 825K .

Solution: This problem makes use 6f the fact that since there are equal numbers of moles of gaseous products and reactants, the partial pressure of each component is equal to the number of moles present at equilibrium.

For each mole of $CO_2$ and each mole of $H_2$ reacted, one mole of CO and one mole of $H_2O$ are produced. Thus, if at equilibrium there is 0.27 atm = 0.27 mole of CO, then there is 0.27 mole of $H_2O$ , 1 - 0.27 = 0.73 mole of $CO_2$ , and 1 - 0.27 = 0.73 mole of $H_2$ . If we let v denote the volume (in liters) in which the reaction takes place, then:

$$[CO] = [H_2O] = 0.27 \text{ mole}/v \ , \quad [CO_2] = [H_2] = 0.73 \text{ mole}/v \ .$$

Substituting these values into the expression for the equilibrium constant gives

$$K = \frac{[CO][H_2O]}{[CO_2][H_2]} = \frac{(0.27/v)(0.27/v)}{(0.73/v)(0.73/v)} = \frac{(0.27)^2}{(0.73)^2} = 0.137 \ .$$

● PROBLEM 312

Two moles of gaseous $NH_3$ are introduced into a 1.0-liter vessel and allowed to undergo partial decomposition at high temperature according to the reaction

$$2NH_3(g) \rightleftharpoons N_2(g) + 3H_2(g) \ .$$

At equilibrium, 1.0 mole of $NH_3(g)$ remains. What is the value of the equilibrium constant?

Solution: This problem involves substitution into the expression for the equilibrium constant for this reaction,

$$K = \frac{[N_2][H_2]^3}{[NH_3]^2}$$

Since 1.0 mole of $NH_3$ remains, 2.0 - 1.0 = 1.0 mole of $NH_3$ was consumed. Also, since one mole of $N_2$ and three moles of $H_2$ are formed per 2 moles $NH_3$ consumed, at equilibrium, there are 3/2 moles of $H_2$ and 1/2 mole of $N_2$ . The equilibrium concentrations are therefore $[NH_3]$ = 1.0 mole/1.0 liter = 1.0m, $[N_2]$ = (1/2 mole)/(1.0 liter) = 0.5M, and $[H_2]$ = (3/2 moles)/(1.0 liter) = 1.5M.

Substituting these into the expression for K gives

$$K = \frac{[N_2][H_2]^3}{[NH_3]^2} = \frac{(0.5)(1.5)^3}{(1)^2}$$

$$= 1.6875 \quad .$$

Four moles of $PCl_5$ are placed in a 2-liter flask. When the following equilibrium is established

$$PCl_5 \rightleftarrows PCl_3 + Cl_2 \quad ,$$

the flask is found to contain 0.8 mole of $Cl_2$. What is the equilibrium constant?

Solution: The equilibrium constant (Keq) for this reaction is:

$$Keq = \frac{[PCl_3][Cl_2]}{[PCl_5]}$$

where [ ] indicate concentrations. One is given that, at equilibrium, there are 0.8 mole of $Cl_2$ present. From the equation for the reaction, it follows that there is 0.8 mole of $PCl_3$ present also. This means that 0.8 moles of $PCl_5$ has reacted. Originally 4 moles of $PCl_5$ were present, thus, at equilibrium, 4.0 - 0.8 moles or 3.2 moles are present. These components are present in a two-liter flask. Therefore, the concentrations will be expressed in moles/2 liters. One can now solve for the equilibrium constant.

$$Keq = \frac{[PCl_3][Cl_2]}{[PCl_5]} \qquad \begin{array}{l} [PCl_3] = 0.8 \text{ moles/2liters} \\ [Cl_2] = 0.8 \text{ moles/2liters} \\ [PCl_5] = 3.2 \text{ moles/2liters} \end{array}$$

$$Keq = \frac{(0.8/2)(0.8/2)}{(3.2/2)} = \frac{(0.4M)(0.4M)}{(1.6M)}$$

$$= \frac{0.16M^2}{1.6M} = 0.1M \quad .$$

For the formation of hydrogen iodide,

$$H_2(g) + I_2(g) \rightleftarrows 2HI(g) \quad ,$$

the value of the equilibrium constant at 700K is 54.7. If one mole of $H_2$ and one mole of $I_2$ are the only materials initially present, what will be the equilibrium concentrations of $H_2$, $I_2$, and $HI$.

Solution: This problem is an application of the expression for the

equilibrium constant
$$K = \frac{[HI]^2}{[H_2][I_2]} \quad ,$$
except that the concentrations are unknown. However, as we shall see, we do not need to know the concentrations.

Let $x$ be the number of moles of $H_2$ that disappeared at equilibrium. Since one mole of $I_2$ is consumed and two moles of HI are formed per mole of $H_2$ consumed, then, at equilibrium, $x$ moles of $I_2$ will have reacted, $2x$ moles of HI will have formed, and $1-x$ moles of $H_2$ and $1-x$ moles of $I_2$ will remain. If we let the volume of the vessel in which the reaction takes place be $v$, then $[H_2] = (1-x)/v$ , $[I_2] = (1-x)/v$ , and $[HI] = 2x/v$ . The expression for the equilibrium constant is then

$$K = 54.7 = \frac{[HI]^2}{[H_2][I_2]} = \frac{(2x/v)^2}{[(1-x)/v][(1-x)/v)]} = \frac{4x^2}{(1-x)^2} \; \frac{v^2}{v^2}$$

$$= \frac{4x^2}{1-2x+x^2} \quad ,$$

or
$$54.7 - 2(54.7)x + 54.7x^2 = 4x^2$$
$$50.7x^2 - 109.4x + 54.7 = 0 \; .$$

Using the quadratic formula,

$$x = \frac{109.4 \pm \sqrt{(-109.4)^2 - 4(50.7)(54.7)}}{2(50.7)}$$

or
$$x = 0.79 \; , \; x = 1.37 \; .$$

Since $x$, the number of moles of $H_2$ or $I_2$ that reacted, cannot be greater than one, the answer $x = 1.37$ must be a nonphysical entity, and we retain only the answer $x = 0.79$. Hence, at equilibrium,

$$[H_2] = 1 - x = 1 - 0.79 = 0.21 \text{ mole}$$
$$[I_2] = 1 - x = 1 - 0.79 = 0.21 \text{ mole}$$
$$[HI] = 2x = 2(0.79) = 1.58 \text{ moles.}$$

● **PROBLEM** 315

For the reaction
$$CO_2(g) + H_2(g) \rightleftarrows CO(g) + H_2O(g),$$
the value of the equilibrium constant at $825°K$ is 0.137. If 5.0 moles of $CO_2$ , 5.0 moles of $H_2$, 1.0 mole of CO , and 1.0 mole of $H_2O$ are initially present, what is the composition of the equilibrium mixture?

Solution: This problem is an application of the expression for the equilibrium constant.

From the stoichiometry of the reaction, one mole of CO and one mole of $H_2O$ are produced for one mole of $CO_2$ and one mole of $H_2O$ that are reacted. Hence, if x moles of CO are produced at equilibrium, then x moles of $H_2O$ are produced, x moles of $CO_2$ are consumed, and x moles of $H_2$ are consumed. Therefore, at equilibrium, there are 1+x moles of CO, 1+x moles of $H_2O$, 5-x moles of CO, and 5-x moles of $H_2$ . If we let v denote the volume of the reaction vessel, the equilibrium concentrations are

$$[CO] = [H_2O] = (1+x)/v, \quad \text{and} \quad [CO_2] = [H_2] = (5-x)/v .$$

Substituting these values into the expression for the equilibrium constant gives

$$K = 0.137 = \frac{[CO][H_2O]}{[CO_2][H_2]} = \frac{[(1+x)/v][(1+x)/v]}{[(5-x)/v][(5-x)/v]} = \frac{(1+x)^2 v^2}{(5-x)^2 v^2} = \frac{1+2x+x^2}{25-10x+x^2} ,$$

or

$$0.137(25) - 0.137(10)x + 0.137x^2 = 1+2x+x^2$$

$$0.863x^2 + 3.370x - 2.425 = 0 .$$

Using the quadratic equation,

$$x = \frac{-3.370 \pm \sqrt{(3.370)^2 - 4(0.863)(-2.425)}}{2(0.863)}$$

or

$$x = 0.62 , \quad x = -4.52 .$$

The second of these is nonphysical and is therefore discarded.

The equilibrium concentrations are then

$$[CO] = 1+x = 1+0.62 = 1.62 \text{ moles}$$

$$[H_2O] = 1+x = 1+0.62 = 1.62 \text{ moles}$$

$$[CO_2] = 5-x = 5-0.62 = 4.38 \text{ moles}$$

$$[H_2] = 5-x = 5-0.62 = 4.38 \text{ moles}.$$

● PROBLEM 316

One mole of $H_2$ is to be reacted with one mole of $I_2$ . Assuming the equilibrium constant is 45.9, what will be the final concentrations of the chemical components in a 1 liter box at $490°$C ?

Solution: The first thing to do with this type of problem is to write a balanced equation for the reaction. A reaction between $H_2$ and $I_2$ is one which produces HI , As such, you have $H_2(g) + I_2(g) \rightarrow 2HI(g)$ . You are given the equilibrium constant of this reaction. To use this information, you must know the meaning of such a constant. Given the general reaction: $xA + yB \rightarrow zC$ , the equilibrium constant, K , is defined as

$$\frac{[C]^z}{[B]^y [A]^x} ,$$

where the brackets serve to indicate concentrations. For this problem,

$$K = \frac{[HI]^2}{[H_2][I_2]} \; .$$

You are asked to find the concentrations. Let $x$ = the no. of moles of $H_2$ that react. It follows, then, that $x$ is also equal to the number of moles of $I_2$ that react and $2x$ is equal to the number of moles of $HI$ that form. Substituting you obtain

$$\frac{[HI]^2}{[H_2][I_2]} = 45.9 \; .$$

You started with one mole of each $H_2$ and $I_2$. Therefore, at equilibrium you must have $1-x$ moles of each species. Since the volume of the box is one liter, this means that the concentration of each substance is equal to the number of moles of each substance. You can now substitute these values in the equilibrium equation.

$$\frac{(2x)^2}{(1-x)(1-x)} = 45.9 \; .$$

Solving for $x$ using the quadratic equation, you obtain $x = .772$. Therefore, at equilibrium you have

$$[H_2] = 1 - x = .228 \text{ mol/liter}$$
$$[I_2] = 1 - x = .228 \text{ mol/liter}$$
$$[HI] = 2x = 1.554 \text{ mol/liter} \; .$$

● **PROBLEM** 317

For the reaction

$$2HI(g) \; \rightleftarrows \; H_2(g) + I_2(g) \; ,$$

the value of the equilibrium constant at 700K is 0.0183. If 3.0 moles of HI are placed in a 5-liter vessel and allowed to decompose according to the above equation, what percentage of the original HI would remain undissociated at equilibrium?

Solution: This problem is an application of the equilibrium expression

$$K = \frac{[H_2][I_2]}{[HI]^2} \; .$$

Since two moles of HI are involved in the production of one mole of $H_2$ and one mole of $I_2$, if $x$ moles of $H_2$ (and therefore $x$ moles of $I_2$) are present at equilibrium, then $2x$ moles of HI have been consumed and $3-2x$ moles of HI remain. Therefore, at equilibrium, $[H_2] = [I_2] = x$ moles/ 5 liters and $[HI] =$

$(3-2x)$ moles/5 liters.

Substituting these into the expression for $K$ gives

$$K = 0.0183 = \frac{[H_2][I_2]}{[HI]^2} = \frac{(x/5)(x/5)}{[(3-2x)/5]^2} = \frac{x^2}{(3-2x)^2} \times \frac{5^2}{5^2}$$

$$= \frac{x^2}{(3-2x)^2} = \frac{x^2}{9-12x + 4x^2} \quad ,$$

or

$$9(0.0183) - 12(0.0183)x + 4(0.0183)x^2 = x^2$$
$$0.9268x^2 + 0.2196x - 0.1647 = 0 \quad .$$

Using the quadratic formula,

$$x = \frac{-0.2196 \pm \sqrt{(0.2196)^2 - 4(0.9268)(-0.1647)}}{2(0.9268)}$$

or

$$x = -0.56, \quad x = 0.32 \quad .$$

Since negative moles are a nonphysical entity, the first answer is discarded and the second retained.

Thus, at equilibrium,

$$[H_2] = x = 0.32 \text{ mole}$$
$$[I_2] = x = 0.32 \text{ mole}$$
$$[HI] = 3-2x = 3-2(0.32) = 2.36 \text{ moles.}$$
$$\% = \frac{.64}{3.00} = .213 \times 100\% = 21.3\%$$

● **PROBLEM** 318

For the reaction $I_2(g) \rightleftarrows 2I(g)$, $K = 3.76 \times 10^{-5}$ at $1000^{\circ}K$. Suppose you inject 1 mole of $I_2$ into a 2.00 liter-box at $1000^{\circ}K$. What will be the final equilibrium concentrations of $I_2$ and of $I$?

**Solution:** Final equilibrium concentrations can be determined from the equilibrium constant expression. For this reaction, K, the equilibrium constant, equals

$$\frac{[I]^2}{[I_2]} \quad .$$

You are given this value as $3.76 \times 10^{-5}$ at $1000^{\circ}K$. Equating, therefore, you obtain .

$$3.76 \times 10^{-5} = \frac{[I]^2}{[I_2]} \quad .$$

You inject 1 mole of $I_2$ in a 2 liter box, which means the initial concentration is 1 mole/2 liter = 0.5M. Let $x$ = the amount of $I_2$ that decomposes. Then, at equilibrium $[I_2] = .5 - x$, i.e., the initial amount minus the decomposed amount. Whatever decomposes yields I. For every mole of $I_2$ that decomposes, two moles of I are generated. This can be seen from the coefficients in the checmical equation. Thus, at equilibrium, $[I] = 2x$. Substituting these values in the equilibrium constant expression, you obtain

$$\frac{(2x)^2}{(.5-x)} = 3.76 \times 10^{-5} \quad .$$

Solving for $x$ you obtain $x = 2.17 \times 10^{-5}$, using the quadratic equation. Thus, the concentrations become

$$I = 2x = 4.34 \times 10^{-5} M$$

and

$$I_2 = .5 - x = .498M .$$

Given the equilibria $H_2S + H_2O \rightleftarrows H_3O^+ + HS^-$ and $HS^- + H_2O \rightleftarrows H_3O^+ + S^{-2}$, find the final concentration of $S^{-2}$, if the final concentrations of $H_3O^+$ and $H_2S$ are 0.3 M and 0.1 M, respectively.

$k_1 = 6.3 \times 10^{-8}$ for $H_2S + H_2O \rightleftarrows H_3O^+ + HS^-$ and

$k_2 = 1 \times 10^{-14}$ for $HS^- + H_2O \rightleftarrows H_3O^+ + S^{-2}$.

<u>Solution</u>: This problem can be solved by writing the equilibrium constant expressions for the equilibria. These expressions give the ratio of the concentrations of products to reactants, each raised to the power of its coefficient in the equilibrium equation.

Therefore, for $H_2S + H_2O \rightleftarrows H_3O^+ + HS^-$, we can write

$$k_1 = 6.\smile \times 10^{-8} = \frac{[H_3O^+][HS^-]}{[H_2S]}$$

For $HS^- + H_2O \rightleftarrows H_3O^+ + S^{-2}$,

$$k_2 = 1 \times 10^{-14} = \frac{[H_3O^+][S^{-2}]}{[HS^-]}$$

Note: water concentration is not included in the equilibrium expression since its concentration is assumed to be constant.

One is not given any information concerning $[HS^-]$. $[HS^-]$ is common to both $k_1$ and $k_2$, so that if one solves one equation for $[HS^-]$ and substitutes it into the other equation, $[HS^-]$ is eliminated. Proceed as follows:

If one solves for $[HS^-]$ in $k_1$, one obtains

$$[HS^-] = \frac{k_1[H_2S]}{[H_3O^+]} = \frac{6.3 \times 10^{-8} [H_2S]}{[H_3O^+]}$$

Substituting this into $k_2$, one obtains

$$k_2 = 1 \times 10^{-14} = \frac{[H_3O^+][S^{-2}]}{[HS^-]} = \frac{[H_3O^+][S^{-2}]}{\dfrac{6.3 \times 10^{-8}[H_2S]}{[H_3O^+]}}$$

Rewriting in terms of $[S^{-2}]$,

$$[S^{-2}] = \frac{(1 \times 10^{-14})(6.3 \times 10^{-8})[H_2S]}{[H_3O^+]^2}$$

One is given that $[H_2S]$ and $[H_3O^+]$ equal 0.1 M and 0.3 M, respectively.

Therefore, one can substitute these values into this equation to solve for $[S^{-2}]$.

$$[S^{-2}] = \frac{(1 \times 10^{-14})(6.3 \times 10^{-8})(0.1)}{(0.3)^2} = 7 \times 10^{-22} \text{ M.}$$

● PROBLEM 320

At $1000°K$, $K = 2.37 \times 10^{-3}$ for the reaction, $N_2(g) + 3H_2(g) \rightleftarrows 2NH_3(g)$. If you inject one mole of each $N_2$ and $H_2$ in a one-liter box at $1000°K$, what per cent of the $H_2$ will be converted to $NH_3$ at equilibrium?

Solution: To answer this question, determine the concentration of $H_2$ at equilibrium. Once this is known, subtract it from the initial concentration. The difference yields the amount that reacted to produce $NH_3$. By multiplying the quotient of the difference divided by the original amount of $H_2$ by 100, you obtain the per cent.

To find $[H_2]$ at equilibrium, employ the quilibrium constant expression. It states that K, the equilibrium constant, is equal to the concentration ratio of the products to reactants, each raised to the power of its coefficient in the chemical reaction. For this reaction, then,

$$K = 2.37 \times 10^{-3} = \frac{[NH_3]^2}{[H_2]^3[N_2]} \ .$$

To find $[H_2]$, proceed as follows. Let x = amount of $N_2$ that dissociates. The initial concentrations of all species are 1M, since molarity = moles/liter, and 1 mole of each is placed in a one-liter box. If the $N_2$'s initial concentration is one and x moles/liter dissociates, then, at equilibrium, there is $(1 - x)$ moles/liter left of $N_2$. In other words, at equilibrium, $[N_2] = 1 - x$. Now, from the chemical equation, it is seen that for every mole of $N_2$ that reacts, 3 moles of $H_2$ are necessary. Thus, when x moles/liter of $N_2$ dissociate, 3x moles/liter of $H_2$ are required. The initial concentration is 1, so that, at equilibrium

$$[H_2] = (1 - 3x) \text{ moles/liter.}$$

Notice, also, that for every mole/liter of nitrogen that dissociates, 2(mole/liter) of ammonia is obtained. Substituting these values into the equilibrium constant expression, one obtains

$$2.37 \times 10^{-3} = \frac{[NH_3]^2}{[N_2][H_2]^3} = \frac{(2x)^2}{(1-x)(1-3x)^3} \ .$$

Solving for x, one obtains x = .0217 moles/liter of $N_2$ that dissociate. Thus, $[H_2] = 1 - 3x = .935$ moles/liter at equilibrium. The initial concentration was 1 mole/liter. The difference is the amount that dissociated, i.e., the $H_2$. The difference = 1 - .935 = .065. Thus, the percent that dissociated equals

$$\frac{.065M}{1M} \times 100 = 6.5\% \ .$$

● PROBLEM 321

For $PCl_5 \rightleftarrows PCl_3 + Cl_2$, K = 33.3 at $760°K$. Suppose 1.00g of $PCl_5$ is injected into a 500-ml evacuated flask at $760°K$ and it is allowed to come into equilibrium. What percent of the $PCl_5$ will decompose? M.W. of $PCl_5$ = 208.235.

Solution: To find the percent of decomposition of $PCl_5$, you need to know its initial concentration and final concentration (i.e., equilibrium concentration).

Initially, there is 1.00g of $PCl_5$ in the 500ml flask. Concentration = moles/liter. There are 1000ml in 1 liter, so that 500ml is .5 liters. The molecular weight of $PCl_5$ = 208.235 grams/mole. A mole is defined as

$$\frac{grams \ (mass)}{molecular \ wt.}$$

Thus, number of moles of

$$PCl_5 = \frac{1g}{208.235g/mole} = .0048 \ moles.$$

Therefore, the initial concentration of

$$PCl_5 = \frac{.0048 \ moles}{.5 \ liter} = .0096M.$$

To find its concentration at equilibrium, use the equilibrium constant expression, which states that K, the equilibrium constant, is equal to the ratio of the concentrations of products to reactants, each raised to the power of its coefficient in the equation. Here, you have

$$K = \frac{[Cl_2][PCl_3]}{[PCl_5]} \ .$$

But, you are told K = 33.3. You can equate to obtain

$$33.3 = \frac{[Cl_2][PCl_3]}{[PCl_5]} \ .$$

Let x = concentration of $[Cl_2]$ at equilibrium. If this is the case, then x = $[PCl_3]$ also, since from the chemical equation it is shown that they are formed in equimolar amounts. If x moles/liter of each $Cl_2$ and $PCl_3$ form, and the only source is $PCl_5$, then the $[PCl_5]$ at equilibrium is the initial concentration minus

$$x \ \frac{moles}{liter} = .0096 - x \ .$$

Substituting these values into the equilibrium constant expression, you obtain

$$33.3 = \frac{(x)(x)}{.0096 - x} \ .$$

Solving for x, using the quadratic equation, you obtain

$$x = 9.597 \times 10^{-3} = [PCl_3] = [Cl_2]$$

at equilibrium. $[PCl_5]$ = .0096 - .009597 = $3 \times 10^{-6}$. This means percent dissociation =

$$100 \times \frac{\text{(Initial - final) concentration}}{\text{initial concentration}} =$$

$$100 \times \frac{.0096 - 3 \times 10^{-6}}{.0096} = 99.9 \text{ \% dissociated.}$$

● **PROBLEM** 322

At 273°K and 1 atm, 1 liter of $N_2O_4$ decomposes to $NO_2$ according to the equation

$$N_2O_4 \rightleftarrows 2NO_2.$$

To what extent has decomposition proceeded when the original volume is 25 percent less than the existing volume?

Solution: Volume is proportional to the number of moles present when Pressure (P) and Temperature (T) are held constant. This can be seen from the equation of state,

$$PV = nRT \quad \text{or} \quad V = \frac{nRT}{P},$$

where R = universal gas constant, n = moles, and V = volume. When P and T are constant, RT/P is a constant, and V varies directly with n. Thus, in this problem you can discuss volumes in terms of fractions decomposed.

Let a be the volume fraction of the original $N_2O_4$ that decomposes. Since moles are proportional to volumes and 2 moles of $NO_2$ are produced for every mole of $N_2O_4$, this fraction, a, results in the production of 2a of $NO_2$ and 1 - a of $N_2O_4$. The total final volume fraction is 1 - a + 2a = (1 + a); that is, the final volume = (1 + a) times the original volume. The original volume is 1 liter of $N_2O_4$. The existing volume is

$$(1 + a)(1 \text{ liter}) = (1 + a) \text{ liter.}$$

But 1 = .75 (1 + a). Solving, a = .333. Thus, you have 33% decomposition.

● **PROBLEM** 323

At 395 K, chlorine and carbon monoxide react to form phosgene, $COCl_2$, according to the equation

$$CO \text{ (g)} + Cl_2 \text{ (g)} \rightleftarrows COCl_2 \text{ (g).}$$

The equilibrium partial pressures are $P_{Cl_2} = 0.128$ atm, $P_{CO} = 0.116$ atm, and $P_{COCl_2} = 0.334$ atm. Determine the equilibrium constant $K_p$ for the dissociation of phosgene

and the degree of dissociation at 395 K under a pressure
of 1 atm.

Solution: After obtaining an expression for $K_p$ for the
dissociation of phosgene, the degree of dissociation under
1 atm of total pressure will be obtained by combining $K_p$
with Dalton's law of partial pressures.

The dissociation of phosgene may be written as

$$COCl_2 \text{ (g)} \rightleftarrows CO \text{ (g)} + Cl_2 \text{ (g)}.$$

By definition, $K_p$ is the product of the partial pressures
of the products divided by the product of the partial
pressure of the reactants. Hence,

$$K_p = \frac{P_{CO}\, P_{Cl_2}}{P_{COCl_2}} = \frac{(0.116 \text{ atm})(0.128 \text{ atm})}{(.334 \text{ atm})}$$

$$= 0.0444 \text{ atm.}$$

$K_p$ for the dissociation of phosgene is thus 0.0444 atm.

Let $\alpha$ denote the fraction of the original number of
moles of phosgene that decomposed. Then, $1 - \alpha$ is the
fraction of the original number of moles of phosgene remaining. From the stoichiometry of the dissociation reaction, one mole of CO and one mole of $Cl_2$ are formed for
every mole of phosgene that decomposes. Thus, $\alpha$ moles of
phosgene decomposes to $\alpha$ moles of CO and $\alpha$ moles of $Cl_2$.

From Dalton's law of partial pressures,

$$P_{COCl_2} = X_{COCl_2}\, P_T$$

$$P_{CO} = X_{CO}\, P_T \quad \text{and} \quad P_{Cl_2} = X_{Cl_2}\, P_T$$

where $X_{COCl_2}$ is the mole fraction of phosgene, $X_{CO}$ is the
mole fraction of CO, $X_{Cl_2}$ is the mole fraction of $Cl_2$, and
$P_T$ is the total pressure. Mole fraction may be defined as
the number of moles of that particular substance divided by
the total number of moles present. Since the total number
of moles present after $\alpha$ moles of phosgene decomposes is

$(1 - \alpha)$ (from remaining phosgene)+ $\alpha$ (from CO)+ $\alpha$ (from $Cl_2$)

$$= 1 + \alpha$$

we have $\quad P_{COCl_2} = X_{COCl_2}\, P_T = \frac{1 - \alpha}{1 + \alpha}\, P_T$

$$P_{CO} = X_{CO}\, P_T = \frac{\alpha}{1 + \alpha}\, P_T$$

and
$$P_{Cl_2} = X_{Cl_2} P_T = \frac{\alpha}{1 + \alpha} P_T.$$

Substituting these into the expression for $K_p$, we obtain

$$K_p = \frac{P_{CO} P_{Cl_2}}{P_{COCl_2}} = \frac{\left(\frac{\alpha}{1 + \alpha} P_T\right)\left(\frac{\alpha}{1 + \alpha} P_T\right)}{\frac{1 - \alpha}{1 + \alpha} PT} = \frac{\alpha^2}{(1 + \alpha)(1 - \alpha)} P_T$$

$$= \frac{\alpha^2}{(1 - \alpha^2)} P_T ,$$

Now, the total pressure is $p_T = 1$ atm and $K_p$ has been determined as $K_p = 0.0444$ atm. Hence,

$$K_p = \frac{\alpha^2}{1 - \alpha^2} P_T ,$$

$$0.0444 \text{ atm} = \frac{\alpha^2}{1 - \alpha^2} \times 1 \text{ atm},$$

$$0.0444 - 0.0444 \ \alpha^2 = \alpha^2,$$

or, $\alpha = \left(\frac{0.0444}{1.0444}\right)^{\frac{1}{2}} = 0.206.$

The degree of dissociation of phosgene is equal to the fraction, $\alpha$, of original moles that have dissociated. Hence, the degree of dissociation of phosgene at 395 K under a pressure of 1 atm is 0.206.

● **PROBLEM** 324

If 1.588 g of nitrogen tetroxide gives a total pressure of 1 atm when partially dissociated in a 500-cm$^3$ glass vessel at 25°, what is the degree of dissociation, $\alpha$? What is the value of $K_p$? The equation for this reaction is $N_2O_4 \rightarrow 2NO_2$.

Solution: The determination of the density of a partially dissociated gas provides one method for measuring the extent to which the gas dissociates. When a gas dissociates, more molecules are produced, and at constant temperature and pressure the volume increases. The density at constant pressure then decreases, and the difference between the density of the undissociated gas and that of the partially dissociated gas is directly related to the degree of dissociation.

If $\alpha$ represents the fraction of gas dissociated, then $1 - \alpha$ represents the fraction of gas undissociated. Assuming one starts with 1 mole of gas the number of moles of gaseous products in the balanced chemical equation is $1 + \sum v_i$ (where $\sum v_i$ is the summation of all the coefficients; positive coefficients for the products and negative coefficients for the reactants). Therefore, the number of moles of gas present at equilibrium is

$$(1 - \alpha) + (1 + \sum v_i)\alpha = 1 + \alpha \sum v_i$$

The density of an ideal gas at constant pressure and temperature is inversely proportional to the number of moles for a given weight and the ratio of the density $\rho_1$ of the undissociated gas to the density $\rho_2$ of the partially dissociated gas is given by the expression

$$\frac{\rho_1}{\rho_2} = 1 + \alpha \sum v_i$$

$$\alpha = \frac{\rho_1 - \rho_2}{\rho_2 \sum vi}$$

If there is no dissociation, then $\alpha = 0$ and $\rho_1 = \rho_2$, if dissociation is complete, then $\alpha = 1$, $\rho_2 \sum v_i = \rho_1 - \rho_2$ and $\rho_1 = (1 + \sum v_i)\rho_2$.

Molecular weights are proportional at constant temperature and pressure to the gas densities. Therefore, molecular weights can be substituted for the densities:

$$\alpha = \frac{M_1 - M_2}{M_2 \sum v_i}$$

where $M_1$ is the molecular weight of the undissociated gas, and $M_2$ is the average molecular weight of the gases when the gas is partially dissociated.

The dissociation of nitrogen tetroxide proceeds by the following equation:

$$N_2O_4 = 2NO_2 \qquad \text{and} \qquad K_p = \frac{P_{NO_2}^2}{P_{N_2O_4}}$$

The degree of dissociation is $\alpha$; $(1 - \alpha)$ is proportional to the number of moles of undissociated $N_2O_4$; $2\alpha$ is proportional to the number of moles of $NO_2$; and $(1 - \alpha) + 2\alpha$ or $1 + \alpha$ is proportional to the total number of moles.

If the total pressure of $N_2O_4$ plus $NO_2$ is $P$, the partial pressures are:

$$P_{N_2O_4} = \frac{1 - \alpha}{1 + \alpha} P \qquad \text{and} \qquad P_{NO_2} = \frac{2\alpha}{1 + \alpha} P$$

Then,  $K_p = \dfrac{\left(\dfrac{2\alpha}{1+\alpha}P\right)^2}{\left(\dfrac{1-\alpha}{1+\alpha}P\right)} = \dfrac{4\,\alpha^2\,P}{1-\alpha^2}$

$M_1 = 92.02$ g/mole.

Using the Combined gas Law,

$PV = n\ RT$

and setting n, the number of moles of substance, equal to the mass of substance, m, divided by its molecular weight, $M_2$, one obtains

$PV = \dfrac{m}{M_2}\ RT$

Rearranging and solving for $M_2$,

$M_2 = \dfrac{RT}{P} \times \dfrac{m}{V}$

$M_2 = \dfrac{RT}{P}\dfrac{m}{V} = \dfrac{(0.082\ \text{liter atm}\,^\circ K^{-1}\ \text{mole}^{-1})(298^\circ K)(1.588\ g)}{(1\ \text{atm})(0.500\ \text{liter})}$

$= 77.68$ g/mole

$\alpha = \dfrac{M_1 - M_2}{M_2\ \sum \nu_i}$

where $\sum \nu_i$ for the reaction equals $\big[$(coefficient of $NO_2$) − (coefficient of $N_2O_4$) = (2 − 1) = 1$\big]$ and $M_1 = 92.02$ and $M_2 = 77.68$.

$\alpha = \dfrac{92.02 - 77.68}{77.68} = 0.1846$

$K_p = \dfrac{4\ \alpha^2\ P}{1 - \alpha^2} = \dfrac{4(0.1846)^2(1\ \text{atm})}{1 - (0.1846)^2} = 0.141.$

# THE SHIFTING OF EQUILIBRIUM-LE CHATELIER'S PRINCIPLE

● PROBLEM 325

A solute of formula AB is slightly dissociated into $A^+$ and $B^-$. In this system, there is a dynamic equilibrium such that $A^+ + B^- \rightleftarrows AB$. Explain what happens if more acid is introduced into this system.

Solution: An acid is a species which, when added to a
solvent (such as $H_2O$), dissociates into protons ($H^+$) and
anions. In this particular case, the proton is represented
as $A^+$. When more acid is added to this general solvent
system, more $A^+$ is introduced. The increased concentration
of $A^+$ places a stress on the equilibrium and the result
is a shift in this equilibrium. According to Le Châtelier's
principle, an equilibrium system will readjust to reduce
a stress if one is applied. Thus, the equilibrium $A^+$ +

$B^- \rightleftarrows AB$ will readjust to relieve the stress of the increased $A^+$ concentration. The stress is relieved by the
reaction of $A^+$ with $B^-$ to produce more AB. The concentration
of $B^-$ will decrease as compared to its concentration prior
to the addition of the acid. Also, the concentration of
the product AB will increase with the addition of the acid.

● **PROBLEM** 326

At $986^\circ C$, $K = 1.60$ for the reaction, $H_2(g) + CO_2(g) \rightleftarrows H_2O(g) + CO(g)$.
If you inject one mole each of $H_2$, $CO_2$, $H_2O$, and $CO$ simultaneously in a
20-liter box at time $t = 0$ and allow them to equilibrate at $986^\circ C$, what
will be the final concentrations of all the species? What would happen to
these concentrations if additional $H_2$ was injected and a new equilibrium
was established?

Solution: Final concentrations of the species can be found by using the
equilibrium constant expression. This expression equates K, the equilibrium constant, to the concentration ratio of products to reactants,
each raised to the power of its coefficient in the chemical equation.
Thus, for this reaction, you can say

$$K = \frac{[H_2O][CO]}{[H_2][CO_2]} = 1.60 \quad .$$

Initially, there was 1 mole of each component in the 20 liter container.
Since, concentration = moles/liter, all had an initial concentration of

$$\frac{1 \text{ mole}}{20 \text{ liter}} = .05M \quad .$$

At equilibrium, let $x$ = the number of moles/liter of $H_2$ that have reacted. Thus, its concentration, at equilibrium, becomes $.05 - x$. If $x$
moles/liter of $H_2$ react, the same number of moles/liter of $CO_2$ must
react also, since they react in equimolar amounts; this is seen from the
chemical reaction. Thus, $[CO_2] = .05 - x$, at equilibrium. These
$x$ moles/liter have been converted to products. Thus, $[H_2O] = [CO] =$
$.05 + x$, at equilibrium. The two products have the same concentration,
since, again, the reaction shows they are formed in equimolar amounts.
As such, you can substitute these values to obtain

$$\frac{(.05+x)(.05+x)}{(.05-x)(.05-x)} = 1.60$$

or

$$\frac{.05 + x}{.05 - x} = \sqrt{1.60} \quad .$$

298

Solving for $x$, you obtain $x = .00585$. Thus, the concentrations become $[H_2] = [CO_2] = .0442$ and $[H_2O] = [CO] = .0558$. If more $H_2$ is injected, the equilibrium is subjected to a stress, one component's concentration has been increased, and according to Le Chatelier's principle, the system will act to relieve the stress by shifting the equilibrium. To do this, more $H_2$ reacts with $CO_2$, thus, decreasing their concentrations, to produce more $H_2O$ and $CO$, thereby, increasing their concentrations.

● **PROBLEM 327**

You are given a box in which $PCl_5(g)$, $PCl_3(g)$, and $Cl_2(g)$ are in equilibrium with each other at $546^{\circ}K$. Assuming that the decomposition of $PCl_5$ to $PCl_3$ and $Cl_2$ is endothermic, what effect would there be on the concentration of $PCl_5$ in the box if each of the following changes were made? (a) Add $Cl_2$ to the box, (b) Reduce the volume of the box, and (c) Raise the temperature of the system.

**Solution:** You are told that the following equilibrium exists in the box $PCl_5 \rightleftarrows PCl_3 + Cl_2$ (all gases) and asked to see what happens to $[PCl_5]$ when certain changes are made. This necessitates the use of Le Chatelier's principle, which states that if a stress is applied to a system at equilibrium, then the system readjusts to reduce the stress. With this in mind, proceed as follows:

(a) Here, you are adding $Cl_2$ to the box. This results in a stress, since one of the components in the equilibrium has its concentration increased. According to Le Chatelier's principle, the system will act to relieve this increased concentration of $Cl_2$ - the stress. It can do so, if the $Cl_2$ combines with $PCl_3$ to produce more $PCl_5$. In this fashion, the stress is reduced, but the concentration of $PCl_5$ is increased.

(b) When the volume of the box is reduced, the concentration of the species is increased, i.e., the molecules are crowded closer together. Thus, a stress is applied. The stress can only be relieved (Le Chatelier's principle) if the molecules could be reduced in number. Notice, in our equilibrium expression you have 2 molecules, 1 each of $PCl_3$ and $Cl_2$, producing 1 molecule of $PCl_5$. In other words, the number of molecules is reduced if the equilibrium shifts to the left, so that more $PCl_5$ is produced. This is exactly what happens. As such, the $[PCl_5]$ increases.

(c) You are told that the decomposition of $PCl_5$ is endothermic (absorbing heat). In other words, it must absorb heat from the surroundings to proceed. If you increase the temperature, more heat is available, and the decomposition proceeds more readily, which means $[PCl_5]$ decreases. This fact can also be seen from the equilibrium constant of the reaction, $K$. This constant measures the ratio of products to reactants, each raised to the power of its coefficients in the chemical reaction. Now, when a reaction is endothermic, $K$ is increased. For $K$ to increase, the reactant's concentration must decrease. Again, therefore, you see that $[PCl_5]$ decreases.

A 1.00-liter reaction vessel containing the equilibrium mixture

$$CO + Cl_2 \rightleftarrows COCl_2$$

was found to contain 0.400 mole of $COCl_2$, 0.100 mole of $CO$, and 0.500 mole of $Cl_2$. If 0.300 mole of $CO$ is added at constant temperature, what will be the new concentrations of each component at equilibrium?

**Solution:** If a stress is placed on a system in equilibrium, whereby the equilibrium is altered, that change will take place which tends to relieve or neutralize the effect of the added stress. Thus, in this reaction when more $CO$ is added after equilibrium has been established more $COCl_2$ will be formed to re-establish the equilibrium. One is given the concentrations of the components at equilibrium, thus the equilibrium constant ($Keq$) can be calculated. The equilibrium constant for this reaction can be stated:

$$Keq = \frac{[COCl_2]}{[CO][Cl_2]} , \quad \text{the ratio of products to}$$

reactants, where [ ] indicate concentration. One can solve for the $Keq$ by using the concentrations given for the first equilibrium,

$$Keq = \frac{[COCl_2]}{[CO][Cl_2]}$$

$$[COCl_2] = 0.400 \text{ mole/liter}$$
$$[CO] = 0.100 \text{ mole/liter}$$
$$[Cl_2] = 0.500 \text{ mole/liter}$$

$$Keq = \frac{(0.400)}{(0.100)(0.500)} = \frac{(0.400)}{(0.05)} = 8.0 .$$

One knows, from the definition of $Keq$, that when more $CO$ is added to the mixture that the concentrations of the components rearrange so that the equation for $Keq$ is equal to 8.

Using Le Chatelier's principle, stated in the first sentence of this explanation, one knows that when more $CO$ is added to this mixture more $COCl_2$ is formed. Let new concentrations of $COCl_2 = [0.400 + x]$. From the equation, one knows that for each mole of $COCl_2$ formed one mole of $Cl_2$ is used, thus the new concentration of $Cl_2 = [0.500 - x]$. One mole of $CO$ is also used for each mole of $COCl_2$ formed, thus the new concentration of $CO$ will be equal to $x$ subtracted from the sum of the number of moles originally present and the number added. The new concentration of $CO = [0.100 + 0.300 - x]$. Because one knows that $Keq = 8$, one can now solve for $x$.

$$Keq = \frac{[COCl_2]}{[CO][Cl_2]} = 8 \quad \text{where}$$

$$[COCl_2] = (0.400 + x) \text{moles}$$
$$[CO] = (0.100 + 0.300 - x) \text{ moles}$$
$$[Cl_2] = (0.500 - x) \text{ moles}$$

Solving,

$$8 = \frac{(0.400 + x)}{(0.100 + 0.300 - x)(0.500 - x)}$$

$$\frac{(0.400 + x)}{(0.2 - 0.9x + x^2)} = 8$$

$$(0.400 + x) = 8(0.2 - 0.9x + x^2)$$
$$(0.400 + x) = (1.6 - 7.2x + 8x^2)$$
$$0 = 1.2 - 8.2x + 8x^2 \ .$$

One uses the quadratic equation to solve for x:

$$x = \frac{-b \pm \sqrt{b^2 - 4ac}}{2a} \ , \ \text{where}$$
$$0 = c + bx + ax^2$$

$$x = \frac{8.2 \pm \sqrt{(-8.2)^2 - 4(8 \times 1.2)}}{2 \times 8}$$

$$x = \frac{8.2 \pm 5.37}{16}$$

$$x = \frac{8.2 + 5.37}{16} = 0.85$$

or

$$x = \frac{8.2 - 5.37}{16} \doteq .177 \ .$$

One cannot use x = .85 in this problem because when it is used, the concentrations of CO and $Cl_2$ will be negative values. Thus x = .177. The new concentrations can now be found.

$$[COCl_2] = (0.400 + x) = .577 \text{ moles}$$
$$[CO] = (0.400 - x) = .223 \text{ moles}$$
$$[Cl_2] = (0.500 - x) = .323 \text{ moles}$$

● **PROBLEM** 329

At a certain temperature, an equilibrium mixture of
$$NO_2 + SO_2 \rightleftarrows NO + SO_3$$
is analyzed and found to contain the following molar concentrations:

$[NO_2] = 0.100$, $[SO_2] = 0.300$, $[NO] = 2.00$, $[SO_3] = 0.600$ .

If 0.500 moles of $SO_2$ are introduced at constant temperature, what will be the new concentrations of reactants and products when equilibrium is re-established?

Solution: Le Chatelier's principle states: If a stress is placed on a system in equilibrium, whereby the equilibrium is altered, that change will take place which tends to relieve or neutralize the effect of the added stress. Thus, in this reaction, if more $SO_2$ is added more NO and $SO_3$ will be formed. If stress is placed on the left side of the equation, the reaction will be forced to the right (and vice versa). One can determine the equilibrium constant (Keq) for this reaction by using the concentrations of the original mixture. The equilibrium constant is defined:

$$Keq = \frac{[NO][SO_3]}{[NO_2][SO_2]}$$

where [ ] indicate concentrations. Solving for Keq with

$$[NO] = 2.00 \qquad\qquad [NO_2] = 0.100$$
$$[SO_3] = 0.600 \qquad\qquad [SO_2] = 0.300 \text{ ,}$$

one obtains

$$Keq = \frac{(2.00)(0.600)}{(0.100)(0.300)} = 40.0 \text{ .}$$

One can solve for the new concentrations by using the Keq. From Le Chatelier's principle, one knows that when $SO_2$ is added to this mixture, the amounts of NO and $SO_3$ will increase. Let x = the number of moles by which NO and $SO_3$ will increase. For each mole of $SO_3$ and NO formed, one mole of $SO_2$ and one mole of $NO_2$ will react, thus the new concentrations of these components will be equal to the original concentrations less x moles. The new concentrations can be stated.

$$[NO] = 2.00 + x$$
$$[SO_3] = 0.600 + x$$
$$[NO_2] = 0.100 - x$$
$$[SO_2] = 0.300 + \text{the amount added} - x$$
$$= 0.300 + 0.500 - x = 0.800 - x \text{ .}$$

Using the formula for the equilibrium constant, one can solve for x.

$$Keq = \frac{[NO][SO_3]}{[NO_2][SO_2]} = 40.0 \text{ .}$$

Substituting,

$$40.0 = \frac{(2.00 + x)(0.600 + x)}{(0.100 - x)(0.800 - x)}$$

$$40.0 = \frac{1.20 + 2.6x + x^2}{.080 - 0.90x + x^2}$$

$$(0.80 - 0.90x + x^2)40 = 1.20 + 2.6x + x^2$$

$$3.20 - 36.0x + 40x^2 = 1.20 + 2.6x + x^2$$

$$2.0 - 38.6x + 38x^2 = 0 \text{ .}$$

One can use the quadratic formula to solve for x.

$$ax^2 + bx + c = 0$$

$$x = \frac{-b \pm \sqrt{b^2 - 4ac}}{2a}$$

$$38x^2 - 38.6x + 2.0 = 0$$

$$x = \frac{38.6 \pm \sqrt{(38.6)^2 - 4 \times 2.0 \times 38}}{2 \times 38}$$

$$x = \frac{38.6 \pm 34.44}{76}$$

$$x = \frac{38.6 + 34.44}{76} = 0.96$$

or

$$x = \frac{38.6 - 34.44}{76} = 0.055$$

One cannot use x = 0.96 because $[NO_2]$ and $[SO_2]$ will be negative.

Concentrations cannot have negative values, which means  x = .055.  One can now find the new concentrations

$$[NO] = 2.00 + x = 2.055 \text{ moles}$$
$$[SO_3] = 0.600 + x = 0.65 \text{ moles}$$
$$[NO_2] = 0.100 - x = 0.045 \text{ moles}$$
$$[SO_2] = 0.800 - x = 0.745 \text{ moles}$$

# CHAPTER 10

# ACID-BASE EQUILIBRIA

> **Basic Attacks and Strategies for Solving Problems in this Chapter. See pages 305 to 366 for step-by-step solutions to problems.**

A simple definition of an acid in an aqueous solution is any compound which produces hydrated hydrogen ions, $H_3O^+$ (also called hydronium ions). Bases are compounds which produce hydroxyl, $OH^-$, ions. There are more sophisticated definitions such as Lewis acids and bases (compounds which can "accept" or "donate," respectively, a share in an electron pair) and Brönsted acids and bases (compounds which act as proton donors and proton acceptors, respectively), but the simple definition works well for most aqueous phase acid-base reactions. While the hydrated hydrogen ion, $H_3O^+$, is always the species that actually exists in aqueous solution, the writing of most reactions is simplified by showing it as the hydrogen ion, $H^+$. Such a simplification is helpful in writing acid-base reactions because the hydrating $H_2O$ molecules are left out.

Water ionizes to form hydrogen (or hydronium) and hydroxyl ions according to the equation

$$H_2O \leftrightarrow H^+ + OH^- \qquad\qquad 10\text{-}1$$

(Note that if the same reaction is written with a hydronium ion, the equation becomes $2H_2O \leftrightarrow H_3O^+ + OH^-$.)

Reaction 10-1, the autoionization of water, has an equilibrium constant of $10^{-14}$. Since the activity of pure water is one by definition, the product of the $H^+$ and $OH^-$ concentrations in an aqueous solution is always $10^{-14}$.

$$(H^+)(OH^-) = 10^{-14} = Kw \qquad\qquad 10\text{-}2$$

Similar reactions in non-aqueous solvents (e.g., $2NH_3 \leftrightarrow NH_4^+ + NH_2^-$) are called autoprotolysis. They are simply reversible ionization reactions resulting from the shift of a proton. Usually they reach equilibrium quickly.

A widely used measure of acidity, pH, is defined as the negative logarithm of the $H^+$ ion concentration.

$$pH = -\log_{10}(H^+) \hspace{3cm} 10\text{-}3$$

Hence, the $H^+$ ion concentration is equal to $10^{(-pH)}$. When both the $H^+$ and $OH^-$ ion concentrations are equal, Equation 10-2 dictates that they both equal $10^{-7}$. The neutral pH, therefore, is 7; acidic solutions have a pH less than 7, basic solutions greater than 7. A similar quantity, pOH, is defined for the $OH^-$ ion concentration, but it is much less common than pH. The sum of pH and pOH will always equal 14 in order to satisfy Equation 10-2.

The ionization constant, dissociation constant, and hydrolysis constant, which are illustrated in Problems 347–363, are all equilibrium constants. The problems can all be solved using the basic principles for solving for equilibrium compositions discussed in Chapter 9. Do not conclude from the different names assigned these equilibria that there are different principles involved. In each case the problem can be solved by the following steps:

1) Write the reaction.
2) Write the equation relating the equilibrium constant to compositions.
3) Write each composition in terms of a single variable from the stoichiometry.
4) Solve the equation for the single variable and calculate the equilibrium compositions.

Acid-base neutralization occurs in aqueous solution when the $H^+$ ions of the acid react with the $OH^-$ ions of the base to form water. This reaction occurs so fast that it is usually considered instantaneous, and the $H^+$ and $OH^-$ concentrations always obey the equilibrium relation of Equation 10-2. The other product of the neutralization of an acid and base — from the cation from the base and the anion from the acid — is called a salt. Normal table salt (NaCl) is the salt formed from the neutralization of hydrochloric acid (HCl) and sodium hydroxide (NaOH). It is, however, only one of numerous salts that form from the neutralization of acids and bases. The key to working the neutralization problems lies in remembering that one mole of $H^+$ ions reacts with one mole of $OH^-$ ions to form a neutral solution. The other ions will form a salt.

Diprotic or triprotic acids possess two or three hydrogens, respectively, that can ionize to form $H^+$ ions. In all such cases, the first $H^+$ ion is more readily formed than the second or third. A very common example is sulfuric acid which can form $H^+$ ions by both of the following reactions.

$$H_2SO_4 \leftrightarrow H^+ + HSO_4^- \tag{10-4}$$

$$HSO_4^- \leftrightarrow H^+ + SO_4^{-2} \tag{10-5}$$

The equilibrium constant for the first $H^+$ ion, Equation 10-4, is very large and all the $H_2SO_4$ reacts. However, the equilibrium constant for the second is .012, so little $H^+$ is formed from this reaction. Let us calculate, for example, the $SO_4^{-2}$ ion concentration in a 1M aqueous solution of sulfuric acid. Follow the procedure outlined above.

1) Write the equation containing $SO_4^{-2}$.

$$HSO_4^- \leftrightarrow H^+ + SO_4^{-2} \tag{10-6}$$

2) Write the equation relating the equilibrium constant to compositions.

$$0.012 = (H^+)(SO_4^{-2})/(HSO_4^-) \tag{10-7}$$

3) Write the composition in terms of a single variable from the stoichiometry.

We can assume, since the equilibrium constant is small, that very little $HSO_4^-$ is consumed and that, since Equation 10-4 goes essentially to completion, $HSO_4^-$ and $H^+$ concentrations are equal and equal to 1M.

$$.012 = (1)(SO_4^{-2})/(1) \tag{10-8}$$

4) Solve the equation for the single variable and calculate the equilibrium compositions.

$$(SO_4^{-2}) = .012M \tag{10-9}$$

The assumption that very little $HSO_4^-$ ion is consumed is reasonable. In this case, .012 moles or 1.2% of the $HSO_4^-$ is consumed.

Aqueous buffered solutions, or aqueous solutions that maintain a very nearly constant pH or constant $H^+$ ion concentration, are formed in two common ways.

1) By dissolving, in water, a weak acid and a soluble ionic salt containing the same anion as the weak acid

2) By dissolving, in water, a weak base and a soluble ionic salt containing the same cation as the weak acid

When the anion of a weak acid is maintained at a constant, relatively large concentration by dissolving a salt, it will control, through equilibrium, the $H^+$ ion concentration and, hence, the pH. For example, acetic acid is a weak acid with a ionization equilibrium constant of $1.76 \times 10^{-5}$. We can write the reaction for the ionization of acetic acid, using the abbreviation $Ac^-$, for the acetate ion.

$$HAc \leftrightarrow H^+ + Ac^- \qquad\qquad 10\text{-}10$$

$$1.76 \times 10^{-5} = (H^+)(Ac^-)/(HAc) \qquad\qquad 10\text{-}11$$

$$H^+ = 1.76 \times 10^{-5} (HAc)/(Ac^-) \qquad\qquad 10\text{-}12$$

The concentration of HAc, unionized acetic acid, is constant because very little ionizes or reacts by Equation 10-10. If the concentration of acetate ion, $Ac^-$, is maintained at a large constant value (large by comparison to that produced by Equation 10-10) by adding a soluble ionic salt of acetic acid, such as sodium acetate, to the solution, then the ratio of HAc and $Ac^-$ will be constant. The addition of small amounts of acids or bases to this solution will not result in significant changes in the $Ac^-$ or HAc concentrations and, from Equation 10-12, the $H^+$ ion concentration or the pH will not change; the solution is said to be buffered against change in the pH.

Indicators are usually large organic molecules that change color by reaction with $H^+$ or $OH^-$ ions. When the concentration of $H^+$, for example, reaches an appropriate level, the equilibrium with the indicator molecule will shift to form a species of a different color. Many acid-base indicators exist; it is possible to select one that will change color at almost any desired pH. The reactions of $H^+$ with indicators reach equilibrium quickly and can be treated like any other equilibrium reaction.

Complex ions are simply ions that form by addition of a molecule (called a complexing agent) to existing ions. They can be treated like any other ion and all of the principles of calculating ionic equilibrium apply. Ammonia, $NH_3$, is a complexing agent with many metallic ions. A typical example is the one given in Problem 379.

$$Cu^{+2} + 4NH_3 \leftrightarrow [Cu(NH_3)_4]^{+2} \qquad\qquad 10\text{-}13$$

Such ions are typically highly colored.

Electrolytes are compounds which form charged ions when dissolved. Most salts are electrolytes, as are most common inorganic acids and bases. Strong electrolytes ionize completely when dissolved; weak electrolytes, on the other hand, have small ionization equilibrium constants. Acetic acid, discussed previously, is a weak electrolyte with an ionization equilibrium constant of $1.76 \times 10^{-5}$. Treatment of the problems involving electrolytes can be solved by following the same principles discussed for other equilibrium reactions.

# Step-by-Step Solutions to Problems in this Chapter, "Acid-Base Equilibria"

## ACIDS AND BASES

● PROBLEM 330

Can $I^+$ (the iodine cation) be called a Lewis base? Explain your answer.

Solution: A Lewis base may be defined as an electron pair donor. Writing out its electronic structure is the best way to answer this question, because it will show the existence of any available electron pairs.

The electronic structure of $I^+$ may be written as

$[: \overset{..}{I} :]^+$. There are three available electron pairs. This might lead one to suspect that it is indeed a Lewis base. But note, $I^+$ does not have a complete octet of electrons, it does not obey the octet rule. According to this rule, atoms react to obtain an octet (8) of electrons. This confers stability.

Therefore, $I^+$ would certainly rather gain two more electrons than lose six. In reality, then, $I^+$ is an electron pair acceptor. Such substances are called Lewis acids.

● PROBLEM 331

Write the equations for the stepwise dissociation of pyrophosphoric acid, $H_4P_2O_7$. Identify all conjugate acid-base pairs.

Solution: Pyrophosphoric acid is an example of a polyprotic acid. Polyprotic acids furnish more than one proton per molecule. From its molecular formula, $H_4P_2O_7$, one can see there exist four hydrogen atoms. This might lead one to suspect that it is tetraprotic, i.e. having 4 protons that can be donated per molecule. This is in fact the case, which means there exist four dissociation reactions. In general, the equation for a dissociation reaction is,

$$HA + H_2O \rightarrow H_3O^+ + A^-.$$

Polyprotic acids follow this pattern. Thus, one can write the following equations for the step-wise dissociation of $H_4P_2O_7$.

(1)  $H_4P_2O_7 + H_2O \rightarrow H_3O^+ + H_3P_2O_7^-$

(2)  $H_3P_2O_7^- + H_2O \rightarrow H_3O^+ + H_2P_2O_7^{-2}$

(3)  $H_2P_2O_7^{-2} + H_2O \rightarrow H_3O^+ + HP_2O_7^{-3}$

(4)  $HP_2O_7^{-3} + H_2O \rightarrow H_3O^+ + P_2O_7^{-4}$

To identify all conjugate acid-base pairs, note the definition of the term. The base that results when an acid donates its proton is called the conjugate base. The acid that results when a base accepts a proton is called the conjugate acid. From these definitions, one sees that in all cases $H_3O^+$ is the conjugate acid of $H_2O$ (the base in these reactions) and $H_3P_2O_7^-$, $H_2P_2O_7^{-2}$, $HP_2O_7^{-3}$ and $P_2O_7^{-4}$ are the conjugate bases of $H_4P_2O_7$, $H_3P_2O_7^-$, $H_2P_2O_7^{-2}$ and $HP_2O_7^{-3}$, respectively.

● **PROBLEM** 332

The dissociation sequence of the polyprotic acid $H_3PO_4$ shows three Bronsted-Lowry acids. Rank them in order of decreasing strengths.

Solution: Polyprotic acids are ones which furnish more than one proton per molecule. From its molecular formula, $H_3PO_4$, it is observed that there are 3 available hydrogen atoms available for release. In general, the equation for a dissociation reaction is

$$HA + H_2O \rightarrow H_3O^+ + A^-,$$

where HA is the acid and water is acting as a weak base. With this in mind, one can write the 3 dissociation reactions as

(1)      $H_3PO_4 + H_2O \rightarrow H_3O^+ + H_2PO_4^-$

(2)      $H_2PO_4^- + H_2O \rightarrow H_3O^+ + HPO_4^{-2}$

(3)      $HPO_4^{-2} + H_2O \rightarrow H_3O^+ + PO_4^{-3}$

From this, one can see that the three acids are

$H_3PO_4$, $H_2PO_4^-$, and $HPO_4^{-2}$. The acids decrease in strength in the order of $H_3PO_4$, $H_2PO_4^-$, $HPO_4^{-2}$.

Ranking can be explained by noting that equivalent H-O bonds are being broken to give off $H^+$. The second and third protons that dissociate leave a progressively more negative ion. This means it is more difficult for the ion to dissociate in order to produce additional $H^+$ ions. This stems from the fact that the increased negativity results in increased attraction for the proton ($H^+$). In summary, as the negative charge of the acid increases, the weaker the acid becomes.

● **PROBLEM** 333

If you place $HClO_4$, $HNO_3$ or $HCl$ in water, you find that they are strong acids. However, they show distinct differences in acidities when dissolved in acetic acid. Such an occurrence is referred to as the leveling effect of the solvent, water: a) Explain the basis for this leveling effect by comparing acid reactions in the water solvent system to the acetic acid solvent system. b) Discuss the leveling effect in terms of basicities instead of acidities.

Solution: (a) To explain the leveling effect, you must consider the relative acidic or basic properties of the species involved. In water, the general reaction for the acid is $HA + H_2O \rightleftharpoons H_3O^+ + A^-$. In acetic acid, however, the reaction is (assuming HA is a stronger acid than acetic acid ($CH_3COOH$), and the three acids given are) $HA + CH_3COOH \rightleftharpoons CH_3COOH_2^+ + A^-$. Let us consider the strengths of the acids in these two solvent systems. Water is less acidic, thus more basic and more strongly proton-attracting, than acetic acid. This means that when strong acids are dissolved the equilibrium will be shifted far to the right in water, but not as far in acetic acid. Thus, more products will be produced in the water solution than in the acetic acid. The acidities of the 3 given acids depend upon how much $H^+$ ion they produce. If the equilibrium is not shifted to the right, less $H^+$ ion is being produced. Thus, the acidicities of the three given acids will be less in acetic acid, since the equilibrium is shifted less to the right, and thus, not much $H^+$ ion is generated.

(b) The leveling effect can also be thought of in terms of basicity. For the two types of bases, you have $B + H_2O \rightleftharpoons BH^+ + OH^-$ in water and $B + CH_3COOH \rightleftharpoons BH^+ + CH_3COO^-$ in acetic acid. Now $OH^-$ is a stronger base than $CH_3COO^-$, and, therefore, the equilibrium is further to the left in water than in acetic acid. Thus, when a base is added to water it will ionize less than in acetic acid.

# THE AUTOIONIZATION OF WATER

A 0.10 M solution of HCl is prepared. What species of ions are present at equilibrium, and what will be their equilibrium concentrations?

Solution: Two processes are occurring simultaneously: the reaction of HCl with $H_2O$ (dissociation of HCl) and the autoionization of $H_2O$.

HCl reacts with $H_2O$ according to the equation

$$HCl + H_2O \rightleftharpoons H_3O^+ + Cl^-.$$

For every mole of HCl that dissociates, one mole of $Cl^-$ and one mole of $H_3O^+$ are produced. The initial concentration of HCl is 0.10 M. Thus, if we assume that HCl dissociates completely,

$$[H_3O^+] = 0.10 \text{ M} \quad \text{and} \quad [Cl^-] = 0.10 \text{ M}.$$

Water autoionizes according to the equation

$$H_2O + H_2O \rightleftharpoons H_3O^+ + OH^-.$$

The water constant for this process is

$$K_w = 10^{-14} \text{ moles}^2/\text{liter}^2 = [H_3O^+][OH^-].$$

Hence, $[OH^-] = \dfrac{10^{-14} \text{ moles}^2/\text{liter}^2}{[H_3O^+]}$

Since $[H_3O^+]$ was determined to be 0.10 M = $10^{-1}$ M = $10^{-1}$ moles/liter,

$$[OH^-] = \frac{10^{-14} \text{ moles}^2/\text{liter}^2}{[H_3O^+]}.$$

$$= \frac{10^{-14} \text{ moles}^2/\text{liter}^2}{10^{-1} \text{ moles/liter}} = 10^{-13} \text{ moles/liter} = 10^{-13} \text{ M}.$$

Hence, at equlibrium, $H_3O^+$, $OH^-$, and $Cl^-$ are present in the concentrations

$$[H_3O^+] = 0.10 \text{ M}, \ [OH^-] = 10^{-13} \text{ M}, \ [Cl^-] = 0.10 \text{ M}.$$

# pH

A) Determine the pH of a solution with a hydrogen ion concentration of $3.5 \times 10^{-4}$.

B) If a solution has a pH of 4.25, what is the hydrogen ion concentration?

Solution:  To determine the acidity or basicity of an aqueous solution, the hydrogen ion concentration must be measured.  The pH of a solution expresses this concentration. pH is defined as the negative logarithm of the hydrogen ion concentration.  In other words, pH = (-log [$H^+$]), where the brackets around $H^+$ signify concentration.  As such, to solve the problem, you must substitute into the equation.  For part "A", you have

$$pH = -\log [3.5 \times 10^{-4}]$$

now $-\log [3.5 \times 10^{-4}] = -\log 3.5 - \log 10^{-4}$

$$= -.54 - (-4)$$

$$= -.54 + 4$$

$$= 3.46$$

It follows, then, that the pH = 3.46 for a hydrogen ion concentration of $3.5 \times 10^{-4}$. Part "B" is similar, but here you are given the pH and asked to find the ion concentration.  Therefore, you have 4.25 = -log [$H^+$] or -4.25 = log [$H^+$].  Now. logarithm numbers give only positive mantissas.  As such, -4.25 must be in the form of -5 + .75. If you take the antilogarithm of each, .75 is 5.6 and -5 is $10^{-5}$, you obtain a hydrogen ion concentration of $5.6 \times 10^{-5}$ mole/liter

Determine the pH of each of the following solutions:
(a) 0.20 M HCl,  (b) 0.10 M NaOH.

Solution:  A pH scale has been devised to express the $H_3O^+$ concentration in solution. By definition,

$$pH = -\log [H_3O^+] \quad \text{or} \quad [H_3O^+] = 10^{-pH}$$

It has been shown that water dissociates to $H_3O^+$ and $OH^-$

A 0.10 M solution of NaOH is prepared. What species of ions are present at equlibrium, and what will be their equilibrium concentrations?

Solution:  Two processes are occurring simultaneously, the dissociation of NaOH, and the autoionization of $H_2O$.

NaOH dissociates according to the equation

$$NaOH \; \rightleftarrows \; Na^+ + OH^-.$$

For every mole of NaOH that dissociates, one mole of $Na^+$ and one mole of $OH^-$ are produced. The initial concentration of NaOH is 0.10 M. Thus, if we assume that NaOH dissociates completely [$Na^+$] = 0.10 M and [$OH^-$] = 0.10 M.

Water autoionizes according to the equation

$$H_2O + H_2O \; \rightleftarrows \; H_3O^+ + OH^-.$$

The water constant for this process is

$$K_w = 10^{-14} \text{ moles}^2/\text{liter}^2 = [H_3O^+][OH^-].$$

Hence, $[H_3O^+] = \dfrac{10^{-14} \text{ moles}^2/\text{liter}^2}{[OH^-]}$

Since [$OH^-$] was determined to be 0.10 M = $10^{-1}$ M = $10^{-1}$ moles/liter, by substitution, we obtain

$$[H_3O^+] = \frac{10^{-14} \text{ moles}^2/\text{liter}^2}{[OH^-]} = \frac{10^{-14} \text{ moles}^2/\text{liter}^2}{10^{-1} \text{ moles/liter}}$$

$$= 10^{-13} \text{ moles/liter} = 10^{-13} \text{ M}.$$

Hence, at equilibrium, $H_3O^+$, $OH^-$, and $Na^+$ are present in the concentrations

$$[H_3O^+] = 10^{-13} \text{ M}, \quad [OH^-] = 0.10 \text{ M}, \quad [Na^+] = 0.10 \text{ M}.$$

## AUTOPROTOLYSIS

Find the equation for the autoprotolysis of water. Indicate which species is the acid, the base, the conjugate acid, and the conjugate base.

Solution:  One can begin by defining autoprotolysis. It may be defined as the donation of a proton from a molecule of one specie to another molecule of the same specie to produce positive and negative ions. Thus, for water, the equation is $H_2O + H_2O \rightarrow H_3O^+ + OH^-$.  An acid is defined as a specie that donates protons. A base is a substance that accepts protons. From the equation, one sees that either water ($H_2O$) molecule can be the base or acid. A conjugate base is a specie obtained by abstracting a proton ($H^+$). If one abstracts a proton from water, one obtains $OH^-$. Thus, $OH^-$ is the conjugate base. The conjugate acid is defined as the base plus a proton. It was stated that either $H_2O$ molecule could be the base. If one adds a proton to one of them, one obtains $H_3O^+$. Thus, $H_3O^+$ is the conjugate acid.

● PROBLEM 337

Indicate the equilibrium equation and constant expression for the autoprotolysis of liquid ammonia.  If $K_{NH_3} = 10^{-22}$, how many molecules of ammonia are ionized in 1 mole of ammonia?  Assume a density of 0.771g/ml for ammonia.

Solution:  Autoprotolysis is that phenomenon whereby an ammonia molecule can donate a proton to another $NH_3$ molecule to form positive and negative charged species.  The equation of the autoprotolysis can be written $NH_3 + NH_3 \rightleftarrows NH_4^+ + NH_2^-$  or

(i)      $2NH_3 \rightleftarrows NH_4^+ + NH_2^-$ .

To find the constant expression, consider the equilibrium constant expression for the reaction:

$$K = \frac{[NH_4^+][NH_2^-]}{[NH_3]^2}$$ .

Note, though, that the concentration of $NH_3$ in pure ammonia is always constant.  By analogy with the autoprotolysis of water (where the $K_w$ expression is written $[OH^-][H_3O^+]$, - without $[H_2O]^2$ in the denominator), the constant expression for the autoprotolysis of $NH_3$ is

(ii)      $K_{NH_3} = [NH_4^+][NH_2^-]$ .

To find the number of molecules of ammonia ionized in 1 mole of ammonia, use the equation

$$K = \frac{[NH_4^+][NH_2^-]}{[NH_3]^2}$$ .

Let  $x$  be the number of moles of ammonia ionized.  Then the $NH_3$ remaining nonionized is  1-x moles.  Since each  2  ammonia molecules must ionize to produce one  $NH_4^+$  and one  $NH_2^-$ , the number of  $NH_4^+$ =

number of  $NH_2^- = \frac{x}{2}$ . Let  V  be the volume of one mole of ammonia. The concentration of  $NH_4^+$  and  $NH_2^-$  can be rewritten as  $\frac{x/2}{V}$ , and the concentration of nonionized  $NH_3$  is  $\frac{1-x}{V}$ .  The equation for  K  can be rewritten as

(iii)      $K = \frac{\left(\frac{x/2}{V}\right)\left(\frac{x/2}{V}\right)}{\left(\frac{1-x}{V}\right)^2} = \frac{x^2/4}{(1-x)^2} \cdot \frac{1/V^2}{1/V^2} = \frac{x^2}{4(1-x^2)}$ .

x, the number of moles of ionized ammonia, can be calculated if  K  is known.  To solve for  K , consider a more general case of the equation

$$K = \frac{[NH_4^+][NH_2^-]}{[NH_3]^2}$$ .

The numerator  $[NH_4^+][NH_2^-]$  is the constant expression for the auto-protolysis of  $NH_3$  and must always equal  $K_{NH_3}$ .  $K_{NH_3}$  is given as $10^{-22}$ .

To find  $[NH_3]^2$, use the fact that the density of ammonia is 0.771g/ml.  The mole weight of ammonia is  17.03g.  Thus, 0.771g is  $\frac{0.771g}{17.03g/mole} = 0.0453$ moles of ammonia; and the density of ammonia is  0.0453 moles/ml. = 45.3 moles/liter.  Thus, $[NH_3] = 45.3M$.

Substitute these results in

$$K = \frac{[NH_4^+][NH_2^-]}{[NH_3]^2}$$

(iv)      $K = \frac{10^{-22}}{45.3^2} = 4.9 \times 10^{-26}$ .

Substitute this value of  K  into (iii)

(v)      $4.9 \times 10^{-26} = \frac{x^2}{4(1-x^2)}$ .

To simplify the problem, note that the dissociation of ammonia is very small and thus  $x \ll 1$.  Thus, approximate  $1-x^2$  as  1. Then, (v) becomes

(vi)      $\frac{x^2}{4} = 4.9 \times 10^{-26}$ .

Solve to obtain  $x = 4.42 \times 10^{-13}$ .  This is the number of moles of  $NH_3$  that ionized.  To find the number of molecules, remember that  1 mole = $6.02 \times 10^{23}$ molecules.  Thus,

No. of molecules ionized = $(4.42 \times 10^{-13}$ moles$)(6.02 \times 10^{23}$ molecules/mole$)$
$= 2.66 \times 10^{11}$ molecules.

$2.66 \times 10^{11}$ molecules of ammonia are ionized.

ions to a small degree.

$$H_2O + H_2O \rightarrow H_3O^+ + OH^-.$$

The equlibrium constant is defined as $K_w$ for this reaction and is expressed as $[H_3O^+][OH^-]$. The $H_2O$ does not appear, since it is presumed to be a constant. From the dissociation equation, it can be seen that the concentration of $H_3O^+$ equals $OH^-$. By experimentation, $K_w$ has been shown to equal $1.0 \times 10^{-14}$. This means that in water, therefore, $H_3O^+$ and $OH^-$ each have a concentration of $1.0 \times 10^{-7}$ M. With this information in mind, one can now solve the problem.

(a) The concentration of HCl is 0.20 M. Since HCl is a strong electrolyte, dissociation is complete. Therefore, the concentration of $H_3O^+$ is also 0.20 M = $2.0 \times 10^{-1}$ M. By definition, then

$$pH = - \log (2.0 \times 10^{-1}) = 1 - 0.3 = 0.7.$$

(b) The $[OH^-]$ equals the concentration of NaOH, since it is also a strong electrolyte. One wants the pH, therefore, employ the expression for $K_w$.

$$[H_3O^+] = \frac{K_w}{[OH^-]} = \frac{1 \times 10^{-14}}{0.10} = 1.0 \times 10^{-13} \text{ M}.$$

Therefore, $pH = - \log (1.0 \times 10^{-13}) = 13.$

A certain solution has pH 3.89 at 0°C. Find pOH and $[OH^-]$.

Solution: pH is a measure of the $[H^+]$ and pOH is a measure of $[OH^-]$. Their product gives $K_w$, the ionization constant of water:

$$[H^+][OH^-] = K_w.$$

pH and pOH are related by the equation,

$$pOH + pH = pK_w.$$

At 0°C, $pK_w = 14.94$. Therefore:

$$pOH = pK_w - pH = 14.94 - 3.89 = 11.05.$$

To find $[OH^-]$ use the equation,

$$pOH = - \log [OH^-]$$

$11.05 = - \log [OH^-]$

$[OH^-] = 10^{-11.05} = 10^{-12+.95} = (10^{-12})(10^{.95})$.

Find the antilog of $10^{.95}$. It is 8.9 which gives

$[OH^-] = 8.9 \times 10^{-12}$.

● **PROBLEM** 341

Assuming complete ionization, calculate (a) the pH of 0.0001 N HCl, (b) the pOH of 0.0001 N KOH.

Solution: (a) pH is defined as the negative log of the hydrogen ion concentration.

$pH = - \log [H^+]$

The normality of an acid is defined as the number of equivalents of $H^+$ per liter of solution. The ionization of HCl can be written

$$HCl \overset{\leftarrow}{\rightarrow} H^+ + Cl^-$$

This means that there is one $H^+$ for every HCl, and that the concentration of $H^+$ equals the concentration of completely ionized HCl.

$[H^+] = [HCl]$

We are told that $[HCl] = 0.0001$ N $= 1 \times 10^{-4}$ N. Therefore, $[H^+] = 1 \times 10^{-4}$ N. We can now solve for pH. Note: In this problem, normality = molarity (concentration), since equivalent weight=M.W.

$pH = - \log [H^+]$

$pH = - \log (1 \times 10^{-4}) = 4.$

The pH of this solution is 4.

(b) The pOH is defined as the negative log of the $OH^-$ ion concentration.

$pOH = - \log [OH^-]$.

The ionization of KOH can be stated

$$KOH \overset{\leftarrow}{\rightarrow} K^+ + OH^-$$

Therefore, one $OH^-$ is formed for every KOH, and when KOH is completely ionized, their concentrations are equal.

$[KOH] = [OH^-]$

We are told that $[KOH] = 0.0001$ N, thus $[OH^-] = 0.0001$ N (again, normality = molarity.)

Solving for pOH:

$$pOH = - \log [OH^-]$$

$$= - \log (0.0001) = - \log (1 \times 10^{-4}) = 4.$$

The pOH of this solution is 4.

● PROBLEM 342

Find the pH of a solution in which $[H^+] = 6.38 \times 10^{-6}$ mole/liter.

Solution: pH indicates the acidity or basicity of a solution. It runs on a scale of 1 to 14. 1 is most acidic, 7 is neutral, and 14 is most basic. pH is a measure of $[H^+]$. The higher the $[H^+]$, the lower the pH, and the stronger the acid. To calculate pH use the equation:

$$pH = - \log [H^+] = - \log [6.38 \times 10^{-6}]$$

$$= - [(- 6) + (\log 6.38)] = 6 + (- .805)$$

$$pH = 5.195.$$

● PROBLEM 343

What is the pH of a neutral solution at 50°C?
$pK_w = 13.26$ at 50°C.

Solution: A neutral solution is defined as $[H^+] = [OH^-]$; an acid solution has $[H^+] > [OH^-]$, and a basic solution has $[H^+] < [OH^-]$. For a solution at 25°, $pK_w = 14$.

$pK_w$ indicates the amount of dissociation of water. To find the neutral pH, one lets pH = pOH = x. Since

$$pH + pOH = 2x = 14 = pK_w,$$

$$x = 7.$$

However, the solution in question is at 50°. At 50° $pK_w = 13.26$. Therefore, to find the neutral pH,

$$pH + pOH = 2x = K_w = 13.26$$

$$x = 6.63 = \text{neutral pH}.$$

Before the advent of pH meters, a urologist collected
1.3 liters of urine from a hospitalized patient over the
course of a day. In order to calculate what the pH was
a laboratory technician determined the number of equivalents of acid present in the sample. A given sample of
urine contains $1.3 \times 10^{-6}$ equivalents of dissociated acid.
What is the pH of this sample?

Solution:  To solve this problem, we have to note that
1 equivalent of $H^+$ is the same as 1 mole of $H^+$. We then
determine $[H^+]$ and from this the pH.

$1.3 \times 10^{-6}$ equivalent of $H^+$ is the same as $1.3 \times 10^{-6}$ mole of $H^+$. The concentration of $H^+$ in the sample is
then

$[H^+]$ = moles of $H^+$/volume

   = $1.3 \times 10^{-6}$ mole/1.3 liter = $10^{-6}$ M.

The pH is defined as pH = $- \log [H^+]$, hence

pH = $- \log [H^+] = - \log (10^{-6}) = - (- 6) = 6$,

which is the pH of normal urine.

A chemist wants to make up a 50 ml solution of pH 1.25
HCl. The HCl is labeled 11.6 M. How should she go
about the preparation?

Solution:  To answer this question, find how many moles of
HCl exist in 50 ml of pH = 1.25. Once this is determined,
one can calculate the volume of concentrated HCl needed.
The total volume of water to be added will be the difference between the total volume of the solution and this
amount of concentrated HCl. Thus, proceed as follows:

One can find the number of moles of HCl present from
the pH. pH = $- \log [H^+]$. As given, pH = 1.25.

Thus, $1.25 = - \log [H^+]$. Solving, $[H^+] = 5.65 \times 10^{-2}$ M.
The concentration of $H^+$ must be equal to the concentration
of $Cl^-$ and HCl, since

HCl $\overset{\rightarrow}{\leftarrow}$ $H^+ + Cl^-$. Thus, $[HCl] = 5.65 \times 10^{-2}$ M.

The solution is to have a volume of 50 ml or 0.05
liters. (Note: 1000 ml = 1 liter.) M = molarity =
moles/liters.

Thus, the number of moles is equal to the volume in liters × molarity or

$$(5.65 \times 10^{-2} \text{ M})(0.05 \text{ }\ell) = 2.82 \times 10^{-3} \text{ moles HCl.}$$

It is given that the molarity of the solution is 11.6 M. One needs $2.82 \times 10^{-3}$ moles. Recalling that molarity = moles/liter, one finds that

$$\frac{2.82 \times 10^{-3} \text{ moles}}{11.6 \text{ M}} = 2.44 \times 10^{-4} \text{ }\ell = 0.24 \text{ ml}$$

of concentrated HCl that is required. If the total volume of the solution is to be 50 ml, then, add (50 - .24 =) 49.76 ml of $H_2O$ to 0.24 ml of the concentrated HCl.

● **PROBLEM** 346

Both HCl and NaOH are strong electrolytes. What is the pH of the solution formed by adding 40 ml of 0.10 M NaOH to 10 ml of 0.45 M HCl?

Solution: We will solve this problem by considering the number of moles of $H_3O^+$ and of $OH^-$ formed by the complete dissociation of HCl and NaOH, respectively.

If we assume that HCl and NaOH dissociate completely, then the concentration of $H_3O^+$ in the HCl solution is equal to the initial concentration of HCl, or $[H_3O^+]$ = 0.45 M, and the concentration of $OH^-$ in the NaOH solution is equal to the initial concentration of NaOH, or $[OH^-]$ = 0.10 M. Since moles = concentration × volume, the number of moles of $H_3O^+$ in 10 ml of the acid solution is

$$\text{moles } H_3O^+ = [H_3O^+] \times \text{volume} = 0.45 \text{ M} \times 10 \text{ ml}$$

$$= 0.45 \text{ M} \times 0.01 \text{ liter}$$

$$= .0045 \text{ mole} = 4.5 \times 10^{-3} \text{ mole.}$$

Similarly, the number of moles of $OH^-$ in 40 ml of the basic solution is

$$\text{moles } OH^- = [OH^-] \times \text{volume} = 0.10 \text{ M} \times 40 \text{ ml}$$

$$= 0.10 \text{ M} \times 0.04 \text{ liter}$$

$$= 0.004 \text{ mole} = 4 \times 10^{-3} \text{ mole}$$

$H_3O^+$ and $OH^-$ neutralize each other according to the reaction

$$H_3O^+ + OH^- \rightarrow 2H_2O.$$

Ignoring the dissociation of water, we can assume that

this reaction is complete. Hence, the $4 \times 10^{-3}$ moles of $OH^-$ will be neutralized by $4 \times 10^{-3}$ moles of $H_3O^+$, leaving $4.5 \times 10^{-3} - 4.0 \times 10^{-3} = 0.5 \times 10^{-3} = 5 \times 10^{-4}$ mole of $H_3O^+$ remaining. Thus, when the two solutions are mixed, no $OH^-$ remains and $5 \times 10^{-4}$ mole of $H_3O^+$ remains. Since the final volume of the solution is 40 ml + 10 ml = 50 ml = 0.05 liter = $5 \times 10^{-2}$ liter, the concentration of $H_3O^+$ is

$$[H_3O^+] = 5 \times 10^{-4} \text{ mole}/5 \times 10^{-2} \text{ liter} = 10^{-2} \text{ mole/liter.}$$

The pH is defined as pH = $- \log [H_3O^+]$, hence

$$pH = - \log [H_3O^+] = - \log (10^{-2}) = - (- 2) = 2.$$

# THE IONIZATION CONSTANT

● PROBLEM 347

The ionization constant for acetic acid is $1.8 \times 10^{-5}$.

a)  Calculate the concentration of $H^+$ ions in a 0.10 molar solution of acetic acid.
b)  Calculate the concentration of $H^+$ ions in a 0.10 molar solution of acetic acid in which the concentration of acetate ions has been increased to 1.0 molar by addition of sodium acetate.

Solution: The ionization constant (Ka) is defined as the concentration of $H^+$ ions times the concentration of the conjugate base ions of a given acid divided by the concentration of unionized acid. For an acid, HA,

$$Ka = \frac{[H^+][A^-]}{[HA]} ,$$

where Ka is the ionization constant, $[H^+]$ is the concentration of $H^+$ ions, $[A^-]$ is the concentration of the conjugate base ions and $[HA]$ is the concentration of unionized acid. The Ka for acetic acid is stated as

$$Ka = \frac{[H^+][\text{acetate ion}]}{[\text{acetic acid}]} = 1.8 \times 10^{-5} .$$

The chemical formula for acetic acid is $HC_2H_3O_2$. When it is ionized, one $H^+$ is formed and one $C_2H_3O^-$ (acetate) is formed, thus the concentration of $H^+$ equals the concentration of $C_2H_3O^-$.

$$[H^+] = [C_2H_3O^-] .$$

The concentration of unionized acid is decreased when ionization occurs. The new concentration is equal to the concentration of $H^+$ subtracted from the concentration of unionized acid.

$$[HC_2H_3O] = 0.10 - [H^+] .$$

Since $[H^+]$ is small relative to 0.10, one may assume that 0.10 -

[$H^+$] is approximately equal to 0.10.

$$0.10 - [H^+] \cong 0.10 .$$

Using this assumption, and the fact that [$H^+$] = [$C_2H_3O^-$] , Ka can be rewritten as

$$Ka = \frac{[H^+][H^+]}{0.10} = 1.8 \times 10^{-5} .$$

Solving for the concentration of $H^+$ :

$$[H^+]^2 = (1.0 \times 10^{-1})(1.8 \times 10^{-5}) = 1.8 \times 10^{-6}$$

$$[H^+] = \sqrt{1.8 \times 10^{-6}} = 1.3 \times 10^{-3} .$$

The concentration of $H^+$ is thus $1.3 \times 10^{-3}$ M.

b)   When the acetate concentration is increased, the concentration of $H^+$ is lowered to maintain the same Ka . The Ka for acetic acid is stated as

$$Ka = \frac{[H^+][C_2H_3O^-]}{[HC_2H_3O]} = 1.8 \times 10^{-5}$$

As previously shown for acetic acid equilibria in a solution of 0.10 molar acid, the concentration of acid after ionization is

$$[HC_2H_3O] = 0.10 - [H^+] .$$

Because [$H^+$] is very small compared to 0.10, $0.10 - [H^+] \cong 0.10$ and $[HC_2H_3O] = 0.10$ .

In this problem, we are told that the concentration of acetate is held constant at 1.0 molar by addition of sodium acetate. Because one now knows the concentrations of the acetate and the acid, the concentration of $H^+$ can be found.

$$\frac{[H^+][C_2H_3O^-]}{[HC_2H_3O]} = 1.8 \times 10^{-5}$$

$$\frac{[H^+][1.0]}{[0.10]} = 1.8 \times 10^{-5}$$

$$[H^+] = 1.8 \times 10^{-6} .$$

● **PROBLEM** 348

Find the hydronium ion concentration of .1M HOAC (acetic acid) solution.   Assume $k_a = 1.75 \times 10^{-5}$ for acetic acid.

Solution: You want to represent the equilibrium constant expression for the reaction, which necessitates a balanced equation. After writing the expression, you want to express the concentrations in terms of the same variables and solve for it. Begin by writing the balanced equation for the reaction of acetic acid in water. The acid will donate a proton ($H^+$) to the only available base, $H_2O$. Thus, HOAC + $H_2O$ → $H_3O^+$ + $OAC^-$. [$H_3O^+$], the hydronium concentration, is the quantity you are looking for. The equilibrium constant expression measures the ratio of the concentrations of the products to the reactants, each raised to the power of their

respective coefficients in the chemical equation. **Thus,**
the constant, Ka, = $\frac{[OAC-]\,[H_3O^+]}{[HOAC]}$. Note: $H_2O$ is omitted,

since it is considered a constant. $Ka = 1.75 \times 10^{-5}$.
Equating, $\frac{[OAC^-]\,[H_3O^+]}{[HOAC]} = 1.75 \times 10^{-5}$. Let x = concen-

tration of $H_3O^+$. According to the reaction, $[H_3O^+]$ =
$[OAC^-]$, thus, x = concentration of $[OAC^-]$, also. If the
initial concentration of HOAC is .1 and X $\frac{moles}{liter}$ of
$[H_3O^+]$ are formed, then you have $(.1-x)$ moles/liter of
HOAC left. Substituting these variables into the
equilibrium constant expression, you have

$$\frac{x^2}{.1-x} = 1.75 \times 10^{-5}.$$

Solving, $x = [H_3O^+] = 0.0013M$

● **PROBLEM** 349

The ionization constant for $NH_4OH$ is $1.8 \times 10^{-5}$.

(a)   Calculate the concentration of $OH^-$ ions in a 1.0
      molar solution of $NH_4OH$.

Solution: The ionization constant ($K_b$) is defined as
the concentration of $OH^-$ ions times the concentration of
the conjugate acid ions of a given base divided by the
concentration of unionized base. For a base, BA,

$$K_b = \frac{[B^-]\,[A^+]}{[BA]},$$

where $K_b$ is the ionization constant, $[B^-]$ is the
concentration of ionized base ions, $[A^+]$ is the
concentration of the conjugate acid, and $[BA]$ is the
concentration of unionized base. The $K_b$ for $NH_4OH$ is
stated as

$$K_b = \frac{[NH_4^+]\,[OH^-]}{NH_4OH} = 1.8 \times 10^{-5}$$

When $NH_4OH$ is ionized, one $NH_4^+$ ion is formed and one
$OH^-$ ion is formed,

$$NH_4OH \rightleftharpoons NH_4^+ + OH^-$$

Thus, the concentrations of each ion are equal.

$$[NH_4^+] = [OH^-]$$

The concentration of unionized base is decreased when
ionization occurs. The new concentration is equal to the
concentration of $OH^-$ subtracted from the concentration of
$NH_4OH$.

$$[NH_4OH] = 1.0 - [OH^-]$$

Since $[OH^-]$ is small relative to 1.0, one may assume that
$1.0 - [OH^-]$ is approximately equal to 1.0

$$[NH_4OH] = 1.0 - [OH^-] \cong 1.0$$

Using this assumption, and the fact that $[OH^-] = [NH_4^+]$, $K_b$ can be rewritten as

$$K_b = \frac{[OH^-][OH^-]}{1.0} = 1.8 \times 10^{-5}$$

Solving for $[OH^-]$:

$$\frac{[OH^-][OH^-]}{1.0} = 1.8 \times 10^{-5}$$

$$[OH^-]^2 = 1.8 \times 10^{-5}$$

$$[OH^-] = \sqrt{1.8 \times 10^{-5}} = 4.2 \times 10^{-3}$$

● **PROBLEM** 350

Find the degree of ionization of 0.05 M $NH_3$ in a solution of pH 11.0. $K_b = 1.76 \times 10^{-5}$.

<u>Solution</u>:  The degree of ionization is the fraction of the total acid or base present ($\alpha$) that ionizes. Namely,

$$\alpha = \frac{x}{c} ,$$

where x is the number of moles of acid or base that dissociate and c is the original number of moles present.

To solve this problem, let it first be stated that $NH_3$ is a base and that $[OH^-]$ can be calculated from the pH value.

The reaction equation is

$$NH_3 + H_2O \underset{\leftarrow}{\rightarrow} NH_4^+ + OH^-.$$

After obtaining $[OH^-]$, $[NH_4^+]$ will be the only unknown in the equilibrium constant equation, which is

$$K_b = \frac{[NH_4^+][OH^-]}{[NH_3]} = 1.76 \times 10^{-5}.$$

To find $[OH^-]$, note that $pOH = -\log [OH^-]$ and that $pH + pOH = 14$. Substituting the given value of pH,

$$pOH = 14 - pH = 14 - 11 = 3.$$
Thus, $[OH^-] = -\log 3 = 1 \times 10^{-3}.$

Next, remember that for each x moles of $NH_3$ that dissociates, x moles of $NH_4^+$ will form. The problem becomes clearer if one observes what is happening from the table below.

$$NH_3 + H_2O \underset{\leftarrow}{\rightarrow} NH_4^+ + OH^-$$

Before Reaction: 0.05      O          O          O

After Reaction: 0.05-x                x      $1 \times 10^{-3}$

Upon substituting these values into the $K_b$ equation,

one obtains:

$$K_b = \frac{[NH_4^+][OH^-]}{[NH_3]} = 1.76 \times 10^{-5} = \frac{x \ (10^{-3})}{0.05 - x} \ .$$

Solving for $[NH_4^+]$,

$$x = [NH_4^+] = 8.65 \times 10^{-4}.$$

$[NH_4^+]$ also represents the unknown value in the degree of ionization equation, since its concentration must be the amount of $NH_3$ that was ionized. Thus,

$$\alpha = \frac{8.65 \times 10^{-4}}{0.05} = 1.73 \times 10^{-2}.$$

● **PROBLEM** 351

Given $K_i$ for acetic acid is $1.8 \times 10^{-5}$, calculate the percentage of ionization of 0.5 M acetic acid. The dissociation reaction is

$$HC_2H_3O_2 \ \rightleftharpoons \ H^+ + C_2H_3O_2^-.$$

Solution: $K_i$ is the ionization constant and indicates to what degree acids and bases will dissociate in solution. Acetic acid is a weak acid because $K_i$ is so small. The larger the value of $K_i$, the greater the % dissociation and the stronger the acid. A useful range of values for $K_i$ is given below.

| Strength | Range |
|---|---|
| Very strong | greater than $1 \times 10^3$ |
| strong | $1 \times 10^3$ to $1 \times 10^{-2}$ |
| weak | $1 \times 10^{-2}$ to $1 \times 10^{-7}$ |
| very weak | less than $1 \times 10^{-7}$ |

$K_i$ is calculated from the ratio of products to reactants, each raised to the power of its coefficient in the reaction equation;

$$K_i = \frac{[\text{products}]^{\text{reaction moles (or coefficient)}}}{[\text{reactants}]^{\text{reaction moles (or coefficient)}}}$$

From the chemical reaction,

$$K_i = \frac{[H^+][C_2H_3O_2^-]}{[HC_2H_3O_2]} = 1.8 \times 10^{-5}$$

To solve this problem, one uses the following method. Before the reaction begins, no dissociation occurs, so that only acetic acid is present, and in its full concentration of 0.5 M. After dissociation occurs, b number of moles/liter has dissociated to the products in equal amounts. Or b moles/liter of each $H^+$ and $C_2H_3O_2^-$ are produced since the coefficients of the reaction indicate that they are formed in equimolar amounts. If b moles/liter dissociate, then there are 0.5 - b moles/liter of acetic acid left.

This can be summarized as follows:

Before:            0.5 moles.              0          0

After:             0.5 - b                 b          b

This gives a $K_i$ of,

$$K_i = \frac{[b][b]}{[0.5 - b]} = 1.8 \times 10^{-5}.$$

Since $K_i$ is so small, very few moles of acetic acid dissociate, which causes b to be insignificantly small. Therefore, 0.5 - b is approximately equal to 0.5, giving

$$K_i = \frac{[b][b]}{[0.5]} = 1.8 \times 10^{-5}.$$

Solving for b, one obtains:

$$b^2 = 9.0 \times 10^{-6}$$

$$b = 3.0 \times 10^{-3} \text{ mole of acetic acid ionized.}$$

Using the following equation to find % ionization:

$$\% \text{ ionization} = \frac{\text{number of moles ionized}}{\text{original number of moles of acetic acid}} \times 100$$

$$= \frac{3.0 \times 10^{-3}}{0.5} \times 100 = 0.60 \%.$$

● **PROBLEM** 352

A solution of 10.0 g of HF in 500 g $H_2O$ freezes at - 1.98°C. Calculate the degree of ionization of HF. (M.W. HF = 20.0   The Freezing point **depression of $H_2O$ is** 1.86°.)

Solution:  When an acid is dissolved in solution, it dissociates into ions. For example, when one mole of HF is dissolved in $H_2O$, one mole of $H^+$ and one mole of $F^-$ are present after complete dissociation. The freezing point of a solution is dependent on the number of particles in the solution, which means that the freezing point of a solution will be lowered more by a compound which ionizes than by the same amount of a compound which does not ionize. The degree of ionization of a compound is a measure of what percent of the compound is ionized when it is placed in a particular solution.

In this problem, one is told that when 10 g of HF is added to 500 g of $H_2O$ the original freezing point of the water (0°) is lowered to - 1.98°C. The freezing point depression is related to the concentration of particles in the solution by the statement

freezing point depression = molality × freezing point
constant

The freezing point constant of water is 1.86°. This means that one mole of a substance (except those substances which ionize) dissolved in 1000 g of water lowers the freezing point 1.86°. Since HF ionizes, we cannot use its molality in this equation. You can use the effective molality, which is the sum of the molalities of $H^+$, $F^-$ and HF. The molality of $H^+$ and $F^-$ will be equal to the degree of ionization of the HF. The concentrations of $H^+$ and $F^-$ will be equal because, when HF ionizes, one $H^+$ and one $F^-$ will be formed. This reaction is given by the equation

$$HF \; \overset{\leftarrow}{\rightarrow} \; H^+ + F^-$$

To solve for the degree of ionization of HF, one must: (a) find the molality of HF as if it were not ionized, (b) define a variable for the molalities of $H^+$ and $F^-$; here, x will be used, (c) find the molality of HF after ionization is taken into account, (d) find the effective molality of the species, (e) use the effective molality in the freezing point depression equation to solve for x, the molality of $H^+$ and $F^-$.

(a) The molality of HF before ionization.

The molality is defined as the number of moles of solute in 1000 g of solvent. The number of moles of HF can be found by dividing the number of grams present by the molecular weight.

$$\text{number of moles} = \frac{\text{number of grams}}{\text{molecular weight}}$$

$$\text{moles of HF} = \frac{10.0 \text{ g}}{20.0 \text{ g/mole}} = 0.5 \text{ moles}$$

The molality can now be found by dividing the number of moles of HF by the number of kg of water present.

$$\text{molality} = \frac{\text{no. of moles of HF}}{\text{no. of kg of } H_2O}$$

$$\text{molality} = \frac{0.5 \text{ moles}}{0.500 \text{ kg}} = 1.0 \text{ moles/kg}$$

(b) $x$ = molality of $H^+$ = molality of $F^-$

(c) After ionization of $x$ molal of $H^+$ and $F^-$, the molality of HF will be $1.0 - x$.

(e) The effective molality of the species will be equal to the sum of the molalities of all of the species present.

The effective molality = molality of $H^+$ + molality of $F^-$ + molality of HF

effective molality = $x + x + (1.0 - x) = 1.0 + x$.

To solve for $x$, the concentration of $H^+$ and $F^-$, the effective molality of the species will be used in the freezing point depression equation.

freezing point depreesion = eff. molality × freezing pt constant

The freezing point depression in this case is $1.98°$ and the freezing point constant of water is $1.86°$. Solving the equation

$$1.98° = (1.0 + x) \times 1.86°$$

$$1.0 + x = \frac{1.98°}{1.86°}$$

$$x = .06$$

The percent of ionization is the molality of $H^+$ and $F^-$ divided by the molality of the unionized HF multiplied by 100.

$$\text{degree of ionization} = \frac{\text{molality of ion}}{\text{molality of unionized species}} \times 100$$

$$\text{degree of ionization} = \frac{0.06}{1.00} \times 100 = 6\%$$

Hence, the HF in this solution is 6 % ionized.

## THE DISSOCIATION CONSTANT

● PROBLEM 353

If 1 mole of HCl and 1 mole of $NaC_2H_3O_2$ are mixed in enough water to make one liter of solution, what will be the concentrations of the species in the final equilibrium? $K_{diss} = 1.8 \times 10^{-5}$ for $NaC_2H_3O_2$.

<u>Solution</u>: To answer this question, you must consider what is happening at equilibrium. This necessitates defining $K_{dissociation}$, which is an equilibrium constant

HCl and $NaC_2H_3O_2$ are strong electrolytes, which means that, in solution, they are completely dissociated. You have, therefore, $H^+$, $Cl^-$, $Na^+$, and $C_2H_3O_2^-$ ions present in the solution. The $Na^+$ and $Cl^-$ do not associate, and need not be considered. Thus, you must only consider the formation of $HC_2H_3O_2$ from $H^+$ and $C_2H_3O_2^-$. The equation for this reaction can be written

$$H^+ + C_2H_3O_2^- \;\rightleftarrows\; HC_2H_3O_2$$

This reaction can proceed in both directions, an equilibrium exists, as the double arrow indicates. The equilibrium constant ($K_{eq}$) for this reaction is equal to

$$\frac{[HC_2H_3O_2]}{[H^+][C_2H_3O_2^-]} \, .$$

$K_{dissociation}$ measures the equilibrium quantitatively. The dissociation reaction for $HC_2H_3O_2$ can be written

$$HC_2H_3O_2 \;\rightleftarrows\; H^+ + C_2H_3O_2^- \, .$$

The dissociation constant,

$$K_{diss} = \frac{[H^+][C_2H_3O_2^-]}{[HC_2H_3O_2]} = 1.8 \times 10^{-5} \, .$$

By examination, you can see that $K_{eq}$ for the association reaction is equal to $1/K_{diss}$. Thus,

$$K_{eq} = \frac{1}{K_{diss}} = \frac{1}{1.8 \times 10^{-5}} = \frac{[HC_2H_3O_2]}{[H^+][C_2H_3O_2^-]}$$

To rewrite into a more convenient form for solving, take the reciprocal of each side.

$$1.8 \times 10^{-5} = \frac{[H^+][C_2H_3O_2^-]}{[HC_2H_3O_2]}$$

The final concentrations of the species, the unknowns, will be those at the equilibrium. Let y be the concentration of $HC_2H_3O_2$ at equilibrium. The concentrations of both $H^+$ and $C_2H_3O_2^-$ can be represented by $1 - y$. Initially, you started with 1 mole/liter of each, therefore, each y mole/liter that associates to form $HC_2H_3O_2$ must be subtracted from the initial concentration. You can now substitute these variables into the expression for $K_{diss}$ to obtain

$$\frac{(1 - y)(1 - y)}{y} = 1.8 \times 10^{-5}$$

Solving for y, using the quadratic formula, you obtain y = .996. Therefore, the concentrations of the species are

$[H^+] = 1 - y = .004$ M

$[C_2H_3O_2^-] = .004$ M

$[HC_2H_3O_2] = .996$ M.

● **PROBLEM** 354

There exists a 0.5 M HF solution for which $K_{diss} = 6.71 \times 10^{-4}$. Determine how much dilution is necessary to double the percent dissociation of HF.

Solution:  Percent dissociation means the ratio of $[H^+]$ to original [HF] concentration times 100. To find the amount of dilution necessary to double the percent dissociation of an acid, first, establish what the percent dissociation is before dilution. This can be determined from $K_{diss}$, the equilibrium dissociation constants, which measures the ratio of products to reactants (i.e. their concentrations), each raised to the power of their coefficients in the balanced chemical equation. The general reaction for the dissociation of an acid, e.g., HA, is

$$HA + H_2O \; \rightleftarrows \; H_3O^+ + A^-.$$

Thus, for HF;     $HF + H_2O \; \rightleftarrows \; H_3O^+ + F^-$

for the dissociation reaction. This means that

$$K_{diss} = 6.71 \times 10^{-4} = \frac{[H_3O^+][F^-]}{[HF]}.$$

It is given that the initial concentration of HF is 0.5. Thus, to find percent dissociation, one needs to know $[H_3O^+]$. To find this, perform the following operations:

Let $x = [H_3O^+]$. Since $H_3O^+$ and $F^-$ are formed in equimolar amounts, as can be seen in the chemical reaction, $[H_3O^+] = [F^-] = x$. If the initial [HF] = 0.5, and x moles/liter dissociate to give $[H_3O^+]$ and $[F^-]$, then, at equilibrium, [HF] = 0.5 - x. Substituting these values:

$$\frac{x \cdot x}{0.5 - x} = 6.71 \times 10^{-4}.$$

Solving for x, using the quadratic formula, x = 0.018 = $[H_3O^+]$. Thus, percentage dissociation becomes

$$\frac{0.018}{0.5} \times 100 = 3.6 \text{ \% HF dissociated.}$$

Then, to determine the final answer, 7.2 % HF dissociation is needed when one dilutes. Recall that $[H^+] = [F^-]$. At 7.2 % dissociation,

$$\frac{[F^-]}{[HF]} = \frac{0.072}{1 - .072} = \frac{0.072}{0.928} .$$

From the $K_{diss}$ expression, one has

$$K_{diss} = 6.71 \times 10^{-4} = \frac{[H^+][F^-]}{[HF]} \quad \text{or}$$

$$\frac{6.71 \times 10^{-4}}{[H^+]} = \frac{[F^-]}{[HF]} .$$

$$\frac{[F^-]}{[HF]} = \frac{0.072}{0.928} , \text{ when percent dissociation is}$$

doubled to 7.2 %. After substitution,

$$\frac{6.71 \times 10^{-4}}{[H^+]} = \frac{0.072}{0.928} .$$

Solving for $[H^+]$, $[H^+] = 8.65 \times 10^{-3} = [F^-]$. Substituting these actual molar concentrations into the equilibrium constant expression, one obtains

$$\frac{6.71 \times 10^{-4}}{8.65 \times 10^{-3}} = \frac{8.65 \times 10^{-3}}{[HF]}$$

Solving:    $[HF] = 0.112$ at equilibrium

$$[H^+] = 8.65 \times 10^{-3}.$$

Before dissociation $[HF]$ = amount at equilibrium plus amount dissociated = $8.65 \times 10^{-3} + 0.112 = 0.121$ M. Thus, when the percent dissociation is doubled, the initial amount equals 0.121 M for $[HF]$. The $[HF]$, when the percent dissociation was unchanged, was 0.5 M (given). Therefore, dilute by factor $0.500/0.121 = 4.13$. Remember that concentration or molarity is a parameter of volume, actually M = moles/liter. One is going from 0.5 M to 0.121 M, which means volume must be increased by a certain factor. When one dilutes a solution, one is adding volume to it and, dividing the solution. This account for using division to obtain the factor of dilution, once initial concentrations are known.

A chemist wants the percent dissociation of $HC_2H_3O_2$ to be 1 %. If the $K_{diss}$ of $HC_2H_3O_2$ is $1.8 \times 10^{-5}$, what concentration of $HC_2H_3O_2$ is required?

**Solution**: To answer this question, consider the dissociation of an acid in $H_2O$. Then write an equilibrium constant expression which relates the equilibrium constant for dissociation, $K_{diss}$, to the ratio of the concentrations of products to reactants, each raised to the power of its coefficients in the reaction. In general, for an acid, there exists the following reaction for dissociation:

$$HA + H_2O \rightarrow H_3O^+ + A^-,$$

where HA is the acid. For this reaction,

$$HC_2H_3O_2 + H_2O \rightarrow H_3O^+ + C_2H_3O_2^- .$$

The equilibrium constant expression is

$$K_{diss} = 1.8 \times 10^{-5} = \frac{[H_3O^+][C_2H_3O_2^-]}{[HC_2H_3O_2]}$$

One wants to find $[HC_2H_3O_2]$ with the knowledge that the percent dissociation is 1 % (note: exclude water from the expression, as it is assumed to be constant). If the percent dissociation is 1 %, then

$$[H_3O^+]/[HC_2H_3O_2] = 1/99.$$

The $H^+$ can only come from $HC_2H_3O_2$. If 1 % of $H^+$ exists, then 100% (initial percentage) - 1% = 99% of $HC_2H_3O_2$ must be left at equilibrium. Thus,

$$1.8 \times 10^{-5} = \frac{1}{99} [C_2H_3O_2^-]. \text{ Solving for } [C_2H_3O_2^-];$$

$$[C_2H_3O_2^-] = 1.78 \times 10^{-3} \text{ M}.$$

This concentration of $C_2H_3O_2^-$ must equal the concentration of $H_3O^+$, since they are formed in equimolar amounts. One can substitute into the equation

$$1.8 \times 10^{-5} = \frac{[H_3O^+][C_2H_3O_2^-]}{[HC_2H_3O_2]}$$

to obtain $\frac{(1.78 \times 10^{-3})(1.78 \times 10^{-3})}{[HC_2H_3O_2]} = 1.8 \times 10^{-5}.$

Solving for $[HC_2H_3O_2]$, $1.76 \times 10^{-1}$ M.

Given a solution of 1.00M $HC_2H_3O_2$ , what is the concentration of all solute species? What is the percentage of acid that dissociated? Assume $K_{diss}$ = 1.8 x $10^{-5}$ .

<u>Solution</u>: Determine the equilibrium equation, write an equilibrium constant expression, and substitute the concentrations of the species into the expression.

By definition, an acid is a substance which donates protons ($H^+$). Since water is the only other species in the solution, it must act as a base; it receives protons.

$$HC_2H_3O_2 + H_2O \rightleftharpoons H_3O^+ + C_2H_3O_2^- .$$

There is a constant called $K_{dissociation}$, which measures the extent of the $HC_2H_3O_2$ donation of protons. For the general reaction,

$$HA + H_2O \rightarrow H_3O^+ + A^- , \quad K_{diss} = \frac{[A^-][H_3O^+]}{[HA]} .$$

For the reaction in this problem, you have

$$\frac{[C_2H_3O_2^-][H_3O^+]}{[HC_2H_3O_2]} = K_{diss} = 1.8 \times 10^{-5} .$$

You are asked to find these concentrations. Let x = the moles per liter of $HC_2H_3O_2$ that dissociated. From the chemical equation, you see that x must also be the concentration of $H_3O^+$ and $C_2H_3O_2^-$, since for each mole of $HC_2H_3O_2$ that dissociates, one mole of $H_3O^+$ and one mole of $C_2H_3O_2^-$ is produced. With this in mind, you can represent $[HC_2H_3O_2]$ as 1-x, since you started with 1M of $HC_2H_3O_2$ and x moles per liter dissociate. You have left 1-x moles per liter. Now substituting, you obtain

$$\frac{[H_3O^+][C_2H_3O_2^-]}{[HC_2H_3O_2]} = 1.8 \times 10^{-5} = \frac{x \cdot x}{1-x} .$$

Solving for x, you obtain x = .0042. Therefore, the concentrations of the species are

$$[HC_2H_3O_2] = 1 - .0042 = .9958M$$

$$[H_3O^+] = x = .0042M$$

$$[C_2H_3O_2^-] = x = .0042M$$

The percentage of acid that dissociated is, thus,

$$\frac{.0042}{1} \times 100 = .42\% .$$

The density of liquid $NH_3$ = 0.68 g/ml. Liquid $NH_3$ dissociates according to the reaction

$$2NH_3(\ell) \rightleftharpoons NH_4^+ + NH_2^- ,$$

for which K = 1.0 × 10⁻³³ at - 33.4°C. Determine the concentrations of $NH_4^+$ and $NH_2^-$ at equilibrium.

<u>Solution</u>:  To solve this problem, write the equilibrium
constant expression for the dissociation of liquid $NH_3$.
This equates the equilibrium constant, K, to the ratio of
the concentrations of products to reactants, each raised
to the power of its coefficient in the chemical reaction.

$$K = 1.0 \times 10^{-33} = \frac{[NH_4^+][NH_2^-]}{[NH_3]^2}$$

One is asked to find $[NH_4^+]$ and $[NH_2^-]$. To do this, one
needs the initial concentration of the liquid ammonia,
which can be found from the density. (density = mass/volume).
It is given that density = 0.68 g/ml. The molecular weight
of ammonia ($NH_3$) is 17 g/mole. For the correct units, convert 0.68 g/ ml  to 680 g/ℓ. Thus concentration
(moles/liter) = (680 g/ℓ)(17 g/mole) = 40 molar $NH_3$. Let x =
$[NH_4^+]$.

From the chemical reaction, it can be seen that $NH_4^+$
and $NH_2^-$ are formed in equimolar amounts, which means
$[NH_4^+] = [NH_2^-]$, so that $[NH_2^-] = x$. If the original
$[NH_3] = 40$ M, and x moles/liter form $[NH_4^+]$ and $[NH_2^-]$
each, then, at equilibrium, $[NH_3]= 40 - x$.

Substituting these values into the equilibrium
constant expression, one obtains

$$1.0 \times 10^{-33} = \frac{x \cdot x}{(40 - x)^2}$$ . Solving for x, using the

quadratic formula:

$$x = 1.26 \times 10^{-15} = [NH_4^+] = [NH_2^-] \text{ at equilibrium.}$$

● **PROBLEM** 358

Given the equilibrium: $ClCH_2COOH + H_2O \rightleftarrows H_3O^+ + ClCH_2COO^-$
exists at 25°C. $K_a = 1.35 \times 10^{-3}$. (a) Determine the $H_3O^+$
concentration for a 0.1 M solution of monochloroacetic
acid ($ClCH_2COOH$) in water ($H_2O$). (b) Can one make the
assumption that the dissociated acid is negligible with
respect to the undissociated acid? (c) Calculate to what
degree this solution is more acidic than 1.0 M acetic
acid ($K_a = 1.8 \times 10^{-5}$ at 25°C).

<u>Solution</u>:  The concentration of $H_3O^+$ can be calculated
through the equilibrium constant expression. This is a
measure of the ratio between the concentrations of products
to reactants, each raised to the power of their respective

coefficients in the chemical equation.

(a) For this reaction, the equilibrium constant,

$$K_a = \frac{[ClCH_2COO^-][H_3O^+]}{[ClCH_2COOH]} \; . \quad \text{Because } K_a = 1.35 \times 10^{-3},$$

$$1.35 \times 10^{-3} = \frac{[ClCH_2COO^-][H_3O^+]}{[ClCH_2COOH]} \; .$$

Letting $x = [H_3O^+]$ at equilibrium, then $ClCH_2COO^- = x$ since the reaction shows that the products are formed in equimolar amounts. The original concentration of the $ClCH_2COOH = 0.1$ M.

If $x$ moles of $H_3O^+$ and $x$ moles of $ClCH_2COO^-$ form, then, at equilibrium, there are $0.1 - x$ moles of $ClCH_2COO^-$ left. Substituting these variables into the equlibrium constant expression:

$$\frac{x \cdot x}{(0.1 - x)} = 1.35 \times 10^{-3}.$$

Solving for $x$, $\quad x = 0.0110$ M.

(b) If the dissociated acid is negligible with respect to the amount left undissociated, then $0.1 - x$ (from previous part) is equal to $0.1$. This means that

$$\frac{x \cdot x}{0.1} = 1.35 \times 10^{-3}.$$

Solving for $x$; $\quad x = 0.0116$ M $= [H_3O^+]$. This means that when you make this assumption, $[H_3O^+] = 0.0116$ M instead of $0.0110$ M, the percent error is over 5 %. The error is too high and, therefore, this assumption cannot be made.

To answer (c), compare the $[H_3O^+]$ in acetic acid solution of 1.0 M with that of the $[H_3O^+]$ calculated in part (a). $[H_3O^+]$ for acetic acid is calculated by using the same procedure as in part (a), except that $K_a = 1.8 \times 10^{-5}$ and the dissociation equation is

$$HOAC + H_2O \; \overset{\rightarrow}{\leftarrow} \quad OAC^- + H_3O^+.$$

After the calculations, $[H_3O^+] = 0.00133$ M. For acetic acid $[H_3O^+] = 0.00133$ M and for $ClCH_2COOH$, $[H_3O^+] = 0.0110$ M. Thus, the $ClCH_2COOH$ has 8.27 times the $H_3O^+$ concentration of $CH_2COOH$ (acetic acid), and is therefore more acidic. This asnwer is found by dividing $[H_3O^+]$ dissociated from $ClCH_2COOH$ by $[H_3O^+]$ dissociated from $CH_3COOH$.

$$\frac{0.0110 \text{ M}}{0.00133 \text{ M}} = 8.27.$$

Sulfuryl chloride decomposes according to the equation,

$$SO_2Cl_2(g) \rightleftarrows SO_2(g) + Cl_2(g) .$$

Given that at 300° K and 1 atm pressure, the degree of dissociation is .121, find the equilibrium constant, $K_p$. What will be the degree of dissociation when pressure = 10 atm?

**Solution:** To find the equilibrium constant, $K_p$, one uses the expression for $K_p$. This is the product of the pressures of the products divided by pressure of the reactant. These pressures are brought to the powers of their coefficients. From this, employ Dalton's law concerning partial pressures to help solve for $K_p$.

At p = 10 atm, assume $K_p$ is pressure-independent because of ideal gas behavior. This, then, allows the determination of the degree of dissociation from the $K_p$ expression.

Thus, one commences by the definition of $K_p$.

$$K_p = \frac{P_{SO_2} \, P_{Cl_2}}{P_{SO_2Cl_2}}$$

To find $K_p$, evaluate these partial pressures. To do this, use Raoult's law, which states partial pressure of gas = mole fraction of gas in mixture times total pressure, i.e., $p = Xp_T$. The $p_T$ is given as 1 atm. Thus, the key is to determine or represent the mole fraction X. Let "a" = the degree of dissociation of $SO_2Cl_2$. From the chemical equation, it is seen that $SO_2$ and $Cl_2$ are formed in equimolar amounts. Thus, at equilibrium there are "a" moles of both $SO_2$ and $Cl_2$. Starting with 1 mole of $SO_2Cl_2$ and "a" moles dissociated, then, at equilibrium, there remain $1 - a$ moles. In the mixture, therefore, total number of moles = $(1-a) + a + a = (1+a)$. Thus, $P_{SO_2Cl_2} = X_{SO_2Cl_2} P_T = \left(\frac{1-a}{1+a}\right) P_T$,

$P_{SO_2} = P_{Cl_2} = \frac{a}{1+a} P_T$. Therefore,

$$K_p = \frac{P_{SO_2} \, P_{Cl_2}}{P_{SO_2} \, Cl_2} = \frac{\left(\frac{a}{1+a}\right) P_T \left(\frac{a}{1+a}\right) P_T}{\left(\frac{1-a}{1+a}\right) P_T}$$

This can be simplified to

$$K_p = \frac{a^2}{1-a^2} P_T .$$

Given the degree of dissociation, a, is equal to .121 and $P_T = 1$ atm; $K_p$ therefore, equal

$$\frac{(.121)^2}{1-(.121)^2} (1) = 1.49 \times 10^{-2} .$$

At 10 atm, then,

$$K_p = \frac{a^2}{1-a^2} (10) = 1.49 \times 10^{-2} .$$

Solving for $a$, $a = .0386$.

# THE HYDROLYSIS CONSTANT

● PROBLEM 360

If the hydrolysis constant of $Al^{3+}$ is $1.4 \times 10^{-5}$, what is the concentration of $H_3O^+$ in 0.1 M $AlCl_3$?

Solution: Hydrolysis refers to the action of the salts of weak acids and bases with water to form acidic or basic solutions. Consequently, to answer this question, write out the reaction, which illustrates this hydrolysis, and write out an equilibrium constant expression. From this, the concentration of $H_3O^+$ can be defined. The net hydrolysis reaction is

$$AlCl_3 \rightarrow Al^{3+} + 3\ Cl^-$$

$$Al^{3+} + 2H_2O \rightleftharpoons AlOH^{2+} + H_3O^+$$

$$K_{hyd} = 1.4 \times 10^{-5} = \frac{[H_3O^+][AlOH^{3+}]}{[Al^{2+}]} .$$

Water is excluded in this expression since it is considered as a constant. Let $x$ = the moles/liter of $[H_3O^+]$. Since $H_3O^+$ and $AlOH^{2+}$ are formed in equal mole amounts, the concentration of $[AlOH^{2+}]$ can also be represented by $x$. If one starts with 0.1 M of $Al^{3+}$, and $x$ moles/liter of it forms $H_3O^+$ (and $AlOH^{2+}$), one is left with $0.1 - x$ at equilibrium. Substituting these representations into the $K_{hyd}$ expression,

$$\frac{x \cdot x}{0.1 - x} = 1.4 \times 10^{-5}.$$

If one solves for $x$, the answer is $x = 1.2 \times 10^{-3}$ M, which equals $[H_3O^+]$.

● PROBLEM 361

Calculate the hydrolysis constants of the ammonium and cyanide ions, assuming $K_w = 1 \times 10^{-14}$ and $K_a = 4.93 \times 10^{-10}$

for HCN and $K_b = 1.77 \times 10^{-5}$ for $NH_3$. For each, determine the percent hydrolysis in a .1M solution.

Solution: To find the hydrolysis constant, you must know what it defines. Hydrolysis is the process whereby an acid or base is regenerated from its salt by the action of water. The hydrolysis constant measures the extent of this process. Quantitatively, it is defined as being equal to $\dfrac{K_w}{K_a \text{ or } K_b}$, where $K_w$ = the equilibrium constant for the autodissociation of water, $K_a$ = dissociation of acid, and $K_b$ = dissociation of base. You are given $K_w$, $K_a$, and $K_b$. Thus, the hydrolysis constants can be easily found by substitution. Let $K_h$ = hydrolysis constant.

For cyanide ion: $K_h = \dfrac{K_w}{K_a} = \dfrac{1 \times 10^{-14}}{4.93 \times 10^{-10}} = 2.02 \times 10^{-5}$

For ammonium ion: $K_h = \dfrac{K_w}{K_b} = \dfrac{1 \times 10^{-14}}{1.77 \times 10^{-5}} = 5.64 \times 10^{-10}$

To find the percent hydrolysis in a .1M solution, write the hydrolysis reaction and express the hydrolysis constant just calculated in those terms. After this, represent the concentrations of the hydrolysis products in terms of variables and solve. For cyanide ion: The hydrolysis reaction is $CN^- + H_2O \rightleftharpoons HCN + OH^-$. Therefore,

$K_h = \dfrac{[HCN][OH^-]}{[CN^-]}$. But you calculated that $K_h = 2.02 \times 10^{-5}$.

Equating, $2.02 \times 10^{-5} = \dfrac{[HCN][OH^-]}{[CN^-]}$. You start with a .1M solution of $CN^-$. Let x = [HCN] formed. Thus, x = $[OH^-]$ also, since they are formed in equimolar amounts. If x moles/liter of substance are formed from $CN^-$, then, at equilibrium you have .1-x moles/liter left. Substituting these values, $2.02 \times 10^{-5} = \dfrac{x \cdot x}{.1-x}$

Solving, $x = 1.4 \times 10^{-3}$M. The percent is just 100 times $\dfrac{x}{.10M}$ since the initial concentration is .10M, so that you have 1.4% hydrolysis in a .1M solution.

For ammonium ion: The hydrolysis reaction is $NH_4^+ + H_2O \rightleftharpoons NH_3 + H_3O^+$. $K_h$ for this reaction = $\dfrac{[NH_3][H_3O^+]}{[NH_4^+]}$
The calculated $K_h = 5.6 \times 10^{-10}$.
Equating, $5.6 \times 10^{-10} = \dfrac{[NH_3][H_2O^+]}{[NH_4^+]}$.

From this point, you follow the same reasoning as was used with the cyanide.

Solving:

Let $x = [NH_3]$

$$5.6 \times 10^{-10} = \frac{(x)(x)}{(.1-x)}$$

$$x^2 = (5.6 \times 10^{-11}) - (5.6 \times 10^{-10})x$$

$$x^2 + (5.6 \times 10^{-10})x - 5.6 \times 10^{-11} = 0$$

Using the quadratic formula one can solve for x, where $ax^2 + bx + c = 0$

$$x = \frac{-b \pm \sqrt{b^2 - 4\,ac}}{2a}$$

$$x = \frac{-5.6 \times 10^{-10} \pm \sqrt{(5.6 \times 10^{-10})^2 - 4(1)(-5.6 \times 10^{-11})}}{2\,(1)}$$

$$x = \frac{-5.6 \times 10^{-10} \pm \sqrt{2.24 \times 10^{-10}}}{2}$$

$$x = \frac{-5.6 \times 10^{-10} \pm 1.50 \times 10^{-5}}{2}$$

$$x = \frac{-1.50 \times 10^{-5}}{2} \text{ or } x = \frac{1.50 \times 10^{-5}}{2}$$

x cannot be negative, because concentration cannot be negative. Thus, $x = 7.5 \times 10^{-6}$

Solving for the percent:

$$\frac{7.5 \times 10^{-6}}{.1} \times 100\% = 7.5\%$$

Thus, you find that the percent hydrolysis is $7.5 \times 10^{-3}\%$.

● **PROBLEM** 362

Find an expression for the hydrolysis constant of the bicarbonate ion $HCO_3^-$, $K_h(HCO_3^-)$, using only the first dissociation constant for $H_2CO_3$, $K_{a_1}(H_2CO_3)$, and the water constant, $K_w$.

Solution: The solution to this problem is a direct application of the definitions of $K_h$, $K_{a_1}$, and $K_w$.

The removal of one proton may be represented by

$$H_2CO_3 \underset{\leftarrow}{\overset{K_{a_1}}{\rightarrow}} H^+ + HCO_3^-$$

Hence, the first dissociation constant is

$$K_{a_1}(H_2CO_3) = \frac{[H^+][HCO_3^-]}{[H_2CO_3]}$$

The water constant is given by

$$K_w = [H^+][OH^-].$$

Hydrolysis of $HCO_3^-$ proceeds according to the reaction

$$HCO_3^- + H_2O \rightleftharpoons H_2CO_3 + OH^-.$$

We are trying to determine an expression for the hydrolysis constant $K_h(HCO_3^-)$ in terms of $K_{a_1}(H_2CO_3)$ and $K_w$. By definition,

$$K_h(HCO_3^-) = \frac{[H_2CO_3][OH^-]}{[HCO_3^-]} \ .$$

Hence,

$$K_h(HCO_3^-) = \frac{[H_2CO_3][OH^-]}{[HCO_3^-]}$$

$$= \frac{[H_2CO_3][OH^-]}{[HCO_3^-]} \times \frac{[H^+]}{[H^+]}$$

$$= \frac{[H_2CO_3]}{[H^+] \ [HCO_3^-]} \times [H^+][OH^-]$$

$$= \frac{1}{K_{a_1}(H_2CO_3)} \times K_w,$$

or,

$$K_h(HCO_3^-) = \frac{K_w}{K_{a_1}(H_2CO_3)} \ .$$

● **PROBLEM** 363

What is the pH of a 1.0 M solution of the strong electrolyte sodium acetate? The dissociation constant of acetic acid is $K_a = 1.8 \times 10^{-5}$ mole/liter.

Solution: The first step is the determination of the hydrolysis constant for sodium acetate. From this we obtain the concentration of hydroxyl contributed by the hydrolysis of sodium acetate. The concentration of hydronium ion, and consequently the pH, is determined by using the water constant.

The dissociation of acetic acid (HAc) into hydronium

ions and acetate ions ($Ac^-$) may be represented by the equation

$$HAc + H_2O \rightleftharpoons H_3O^+ + Ac^-.$$

The dissociation constant for this reaction is

$$K_a = \frac{[H_3O^+][Ac^-]}{[HAc]}$$

This constant and the water constant, $K_w = [H_3O^+][OH^-]$ $= 10^{-14}$ $mole^2/liter^2$ will be used to determine the hydrolysis constant for acetate.

Hydrolysis of acetate proceeds according to the following equation:

$$Ac^- + H_2O \rightleftharpoons HAc + OH^-.$$

The hydrolysis constant is $K_h = [HAc][OH^-]/[Ac^-]$. This may be rewritten in terms of $K_a$ and $K_w$ as follows:

$$K_h = \frac{[HAc][OH^-]}{[Ac^-]} = \frac{[HAc][OH^-]}{[Ac^-]} \times \frac{[H_3O^+]}{[H_3O^+]}$$

$$= \frac{[HAc]}{[Ac^-][H_3O^+]} \times [H_3O^+][OH^-] = \frac{1}{K_a} \times K_w = \frac{K_w}{K_a} . \quad \text{Hence,}$$

$$K_h = \frac{[HAc][OH^-]}{[Ac^-]} = \frac{K_w}{K_a}$$

$$= \frac{10^{-14} \ mole^2/liter^2}{1.8 \times 10^{-5} \ mole/liter} = 5.6 \times 10^{-10} \ mole/liter.$$

Let the equilibrium concentration of HAc formed by the hydrolysis of acetate be x. Since one mole of $OH^-$ is formed per mole of HAc formed, the equilibrium concentration of $OH^-$ is also x. Furthermore, if we assume that sodium acetate dissociates completely, then the initial concentration of $Ac^-$ is equal to the concentration of sodium acetate (1.0 M) and the equilibrium concentration of acetate is 1.0 - x. Note that we have neglected the contribution to $[OH^-]$ from the hydrolysis of water.

Substituting these concentrations into the expression for $K_h$, we obtain

$$K_h = 5.6 \times 10^{-10} \ mole/liter = \frac{[HAc][OH^-]}{[Ac^-]}$$

$$= \frac{x \cdot x}{1.0 - x} = \frac{x^2}{1.0 - x}$$

To avoid use of the quadratic formula, we will assume that x is much smaller than 1.0 so that $1.0 - x \cong 1.0$. (This assumption will be justified later on in the solution). Hence, we obtain

$$5.6 \times 10^{-10} \text{ mole/liter} = \frac{x^2}{1.0 - x} \cong \frac{x^2}{1.0}$$

or  $x = (1.0 \times 5.6 \times 10^{-10})^{\frac{1}{2}} = 2.4 \times 10^{-5}$ mole/liter.

Since $[OH^-] = x$, $[OH^-] = 2.4 \times 10^{-5}$ mole/liter. Hence, x is much smaller than 1.0, justifying our earlier assumption.

We will find $[H_3O^+]$ by use of the water constant, $K_w = [H_3O^+][OH^-]$, or,

$$[H_3O^+] = \frac{K_w}{[OH^-]} = \frac{10^{-14} \text{ mole}^2/\text{liter}^2}{2.4 \times 10^{-5} \text{ mole/liter}}$$

$$= 4.2 \times 10^{-9} \text{ mole/liter}.$$

The pH is then

$$pH = -\log [H_3O^+] = -\log (4.2 \times 10^{-9})$$

$$= -(-9.4) = 9.4.$$


# NEUTRALIZATION

● PROBLEM 364

Assuming complete neutralization, calculate the number of milliliters of 0.025 M $H_3PO_4$ required to neutralize 25 ml of 0.030 M Ca(OH)$_2$.

Solution: This problem can be solved by two methods: mole method or equivalent method.

### Mole Method

This method requires one to write out the balanced equation that illustrates the neutralization reaction. The balanced equation is $3Ca(OH)_2 + 2H_3PO_4 \rightarrow Ca_3(PO_4)_2 + 6H_2O$. From this equation, one can see that 2 moles of $H_3PO_4$ react for every 3 moles of Ca(OH)$_2$. This means that one must first calculate how many moles of Ca(OH)$_2$ are involved. The molarity of the Ca(OH)$_2$ is 0.030. (Molarity = no. of moles/liters.)

As given, the Ca(OH)$_2$ solution is 25 ml or 0.025 liters. Therefore, the number of moles of Ca(OH)$_2$ is

(0.030)(0.025) = 0.00075 moles. As the balanced equation indicates, the number of moles of $H_3PO_4$ is 2/3 the moles of $Ca(OH)_2$ or 2/3 (0.00075) = 0.00050 moles of $H_3PO_4$.

From the definition of molarity for $H_3PO_4$ one has

$$0.025 \text{ M} = \frac{0.00050 \text{ moles}}{\text{liters}} .$$

The molarity, 0.025, is given. Solving for liters, one obtains 0.020 liters or 20 ml. The key to solving this problem with the mole method, is to write a balanced equation, which will indicate the relative amounts of moles required for complete neutralization .

### Equivalent Method

This method requires that one consider normality and the definition of an equivalent. An equivalent is defined as the molecular weight or mass of an acid or base that furnished one mole of protons ($H^+$) or hydroxyl ($OH^-$) ions. For example, the number of equivalents contained in a mole of $H_2SO_4$ is 98/2 or 49. Since each mole of $H_2SO_4$ produces two protons, divide the molecular weight by 2.

The number of equivalents of an acid must equal that of the base in a neutralization reaction. Normality is defined as equivalents of solute per liter. In this problem, it is given that there are 25 ml of 0.03 $Ca(OH)_2$.

To solve the problem, determine how many equivalents are present. The number of moles of $Ca(OH)_2$ is (0.025)(0.03) or 0.00075 from the definition of molarity. The molecular weight of $Ca(OH)_2$ is 74.08. Therefore, there are (0.00075)(74.08) or 0.06 grams of $Ca(OH)_2$.

The number of equivalents per gram is 74.08/2 since two $OH^-$ can be produced. The number of equivalents is

$$\frac{(0.00075)(74.08)\text{g}}{\left( \frac{74.08}{2} \text{ g/equiv} \right)} = 0.0015 \text{ equiv of } Ca(OH)_2 .$$

This indicates that 0.0015 equivalents of $H_3PO_4$ are required. The molarity of $H_3PO_4$ is 0.025 M, which means its normality is 0.075 N, because there are 3 ionizable protons per mole. Recalling the definition of normality, there are

$$0.075 \text{ N} = \textbf{0.0015 equiv./liters}$$

The reason one knows that there is 0.0015 equiv in the $H_3PO_4$ present is because one knows that for the neutralization to occur the number of equivalents of acid must equal the number of equivalents of base. In this problem, one has already calculated that there are

0.0015 equiv of base present. Thus,

volume = 0.20 ℓ or 20 ml.

A 50 ml solution of sulfuric acid was found to contain 0.490 g of $H_2SO_4$. This solution was titrated against a sodium hydroxide solution of unknown concentration. 12.5 ml of the acid solution was required to neutralize 20.0 ml of the base. What is the concentration of the sodium hydroxide solution.

Solution: At the neutralization point, the number of equivalents of acid in the 12.5 ml volume is equal to the number of equivalents of base in the 20.0 ml volume. Since the normality is defined as the number of equivalents per liter of solution, the number of equivalents is equal to the normality times the volume, at the neutralization point we have

$$N_a V_a = N_b V_b$$

where $N_a$ = normality of acid, $V_a$ = volume of acid, $N_b$ = normality of base, $V_b$ = volume of base.

The normality of the 50.0 ml (0.050 ℓ) sulfuric acid solution is

$$N_a = \frac{\text{number of equivalents in 0.050 ℓ}}{0.05 \text{ ℓ}}$$

$$= \frac{\text{mass of acid/gram equivalent weight}}{0.05 \text{ ℓ}}$$

The gram equivalent weight of sulfuric acid is 49.0 g/equivalent, because there are 2 equiv per molecule. The MW of $H_2SO_4$ is 98 g/mole.

$$N_a = \frac{\text{mass of acid/gram equivalent weight}}{0.05 \text{ ℓ}}$$

$$= \frac{0.490 \text{ g}/49.0 \text{ g/equivalent}}{0.05 \text{ ℓ}} = 0.200 \text{ equivalent/ℓ}$$

$$= 0.200 \text{ N}$$

The normality of the base is then found as follows:

$$N_a V_a = N_b V_b$$

$$N_b = \frac{N_a V_a}{V_b} = \frac{0.200 \text{ N} \times 12.5 \text{ ml}}{20.0 \text{ ml}} = 0.125 \text{ N}.$$

Therefore, the sodium hydroxide solution is 0.125 N, which because there is 1 ionizable $OH^-$ in NaOH, is equal to 0.125 M.

● PROBLEM 366

Determine the molarity of an $H_2SO_4$ solution, 15.0 ml of which is completely neutralized by 26.5 ml of 0.100 M NaOH.

Solution:  Acids and bases react with each other to produce salts and water. Such reactions are called neutralizations. To find the molarity of the neutralized acid solution, first write out the balanced reaction.

$$H_2SO_4 + 2NaOH \longrightarrow Na_2SO_4 + 2HOH.$$

When neutralization is complete, there is no longer any acid or base left. It is given that 26.5 ml of 0.100 M NaOH is used. From this, one wants to compute how many moles of NaOH were present.

It is known from the neutralization reaction that the number of moles of acid must be ½ the number of moles of base for a complete neutralization.

From the number of moles of base present, the number of moles of acid needed are determined. Since molarity = moles/liter and one is given the volume, the molarity can be calculated. Proceed as follows:

There exists 26.5 ml of 0.100 M NaOH. There are 1,000 ml in one liter. This means that 26.5 ml is equal to 0.0265 liters. Recalling the definition of molarity,

$$0.1 = \frac{\text{no. of moles of NaOH}}{0.0265 \text{ } \ell} \text{ .} \quad \text{Solving}$$

moles = 0.00265 of NaOH.

This, then, must be twice the number of moles of acid. For the acid, therefore, molarity = (.00265)½/liters. One is told that the volume of the $H_2SO_4$ solution is 15 ml, or 0.015 liters (1000 ml = 1 liter). Molarity of the acid becomes

$$\frac{0.001325}{0.015} = 0.088 \text{ M.}$$

One of the two most common ingredients in medication designed for the relief of excess stomach acidity is aluminum hydroxide (Al(OH)$_3$, formula weight = 78 g/mole). If a patient suffering from a duodenal ulcer displays a hydrochloric acid (HCl, formula weight = 36.5 g/mole), concentration of 80 × 10$^{-3}$ M in his gastric juice and he produces 3 liters of gastric juice per day, how much medication containing 2.6 g Al(OH)$_3$ per 100 ml of solution must he consume per day to neutralize the acid?

<u>Solution</u>: Al(OH)$_3$ neutralizes HCl according to the following reaction:

$$Al(OH)_3 + 3HCl \rightarrow AlCl_3 + 3H_2O.$$

When all the HCl has been neutralized, the number of equivalents of Al(OH)$_3$ is equal to the number of equivalents of HCl. But since the number of equivalents is equal to the product of the normality, N, and the volume, V, this condition may be written as

$$N_{Al(OH)_3} V_{Al(OH)_3} = N_{HCl} V_{HCl}.$$

We must solve for $V_{Al(OH)_3}$

$$V = \frac{N_{HCl} V_{HCl}}{N_{Al(OH)_3}}$$

The molarity of the Al(OH)$_3$ solution is equal to the number of moles of Al(OH)$_3$ divided by the volume in liters. To determine the number of moles corresponding to 2.6 g Al(OH)$_3$, we divide 2.6 g by the formula weight of Al(OH)$_3$. We then divide this by 100 ml = 100 ml × 1 liter/1000 ml = 0.100 liter to obtain the molarity. Hence,

$$\text{concentration of Al(OH)}_3 = \frac{2.6 \text{ g}}{100 \text{ ml}}$$

$$= \frac{2.6 \text{ g}/78 \text{ g/mole}}{0.100 \text{ liter}} = 0.33 \text{ M}.$$

Since HCl has one ionizable proton, its normality is equal to its molarity, or $N_{HCl}$ = 80 × 10$^{-3}$ M = 80 × 10$^{-3}$ N. Al(OH)$_3$ contains three hydroxyl groups, hence, its normality is equal to three times its molarity, or $N_{Al(OH)_3}$ = 3 × 0.33 M = 1.0 N. The required volume of medication is then

$$V_{Al(OH)_3} = \frac{N_{HCl} V_{HCl}}{N_{Al(OH)_3}} = \frac{80 \times 10^{-3} \text{ N} \times 3 \text{ liters}}{1.0 \text{ N}} = 0.240 \text{ liter}$$

$$= 0.240 \text{ liter} \times 1000 \text{ ml/liter} = 240 \text{ ml}.$$

A chemist performs a neutralization reaction. She finds that 1 g of $C_6H_{10}O_4$, an acid, requires 0.768 g of KOH for complete neutralization. Determine the number of neutralizable protons in this molecule.

<u>Solution</u>: To solve this problem, first determine how many equivalents of the acid or base are involved. Once this is known, the number of neutralizable protons will also be known. One equivalent is defined as the mass of the substance, if it is an acid, needed to furnish one mole of $H_3O^+$ or, if it is a base, needed to furnish one mole of $OH^-$.

The number of grams/equiv. for KOH is 56.1, since only 1 mole of $OH^-$ can be furnished. The number of equivalents is

$$\frac{0.768 \text{ g}}{56.1 \text{ g/equiv}} = 0.0137 \text{ equiv of KOH}.$$

The number of equivalents of base must equal that of the acid. The number of equivalents of acid = 0.0137. The number of grams/equiv is

$$\frac{1.00 \text{ g of } C_6H_{10}O_4}{0.0137 \text{ equiv.}} = 73.0 \text{ g/equiv}.$$

One mole of $C_6H_{10}O_4$ weighs 146.1 g, since a mole = weight in grams/molecular weight. Therefore, the number of moles/equiv is

$$\frac{73.0 \text{ g/equiv}}{146.1 \text{ g/mole}} = 0.5 \text{ moles/equiv}.$$

It follows that each mole of acid furnishes two $H_3O^+$. As such, there are two neutralizable protons in $C_6H_{10}O_4$.

A common method for commercially peeling potatoes is to soak them for 1 - 5 minutes in a 10 - 20% solution of NaOH (molecular weight = 40.0 g/mole) at 60 - 88°C, and to spray off the peel once the potatoes are removed from solution. As an economy measure, a manufacturer titrates the NaOH solution with standardized $H_2SO_4$ (molecular weight = 98.0 g/mole) at the end of each day to determine whether the solution is still capable of peeling potatoes. If, at the end of one day, he finds that it takes 64.0 ml of a 0.200 M solution of $H_2SO_4$ to titrate a 10.0 ml sample of NaOH solution to neutrality, what concentration of NaOH did he find?

Solution: At the neutralization point, the number of equivalents of NaOH is equal to the number of equivalents of $H_2SO_4$. But since the number of equivalents is equal to the product of the normality, N, and the volume, V, this condition may be stated as

$$N_{NaOH} \, V_{NaOH} = N_{H_2SO_4} \, V_{H_2SO_4}.$$

To solve this problem, we must solve for $N_{NaOH}$,

$$N_{NaOH} = \frac{N_{H_2SO_4} \, V_{H_2SO_4}}{V_{NaOH}}$$

Since $H_2SO_4$ has two ionizable protons, its normality is equal to twice its molarity, or $N_{H_2SO_4} = 2 \times 0.200$ M = 0.400 N. NaOH has only one hydroxyl group, so that its normality is equal to its molarity. Then

$$N_{NaOH} = \frac{N_{H_2SO_4} \, V_{H_2SO_4}}{V_{NaOH}} = \frac{0.400 \text{ N} \times 64.0 \text{ ml}}{10.0 \text{ ml}} = 2.56 \text{ N}.$$

The concentration of NaOH solution is therefore 2.56 N, or 2.56 M.

● **PROBLEM** 370

A potato peeling solution was found to be 2.56 M in NaOH (formula weight = 40.0 g/mole) at the end of the day. To operate, the solution must be at least 10 % NaOH by weight (100 g NaOH per 1000 g of solution). What weight percent corresponds to 2.56 M? The density of a 2.56 M solution of NaOH is about 1.10 g/ml.

Solution: To solve this problem, the concentration 2.56 M must be converted to a weight-weight basis. 2.56 M = 2.56 moles NaOH/1 liter solution = 2.56 moles NaOH/1000 ml solution. To obtain the mass corresponding to 2.56 moles of NaOH, we multiply by the formula weight of NaOH, or, 2.56 moles × 40.0 g/moles = 102.4 g NaOH.

To obtain the mass of NaOH contained in 1000 ml of solution, we multiply by the density of the solution, 1000 ml × 1.10 g/ml = 1100 g. Hence,

$$2.56 \text{ M} = \frac{2.56 \text{ moles NaOH}}{1000 \text{ ml solution}} = \frac{102.4 \text{ g NaOH}}{1100 \text{ g solution}};$$

$$\frac{102.4 \text{ g NaOH}}{1100 \text{ g solution}} \times 100 \text{ %} = 9.3 \text{ % by weight}.$$

Since this is less than 10 % by weight, the solution is no longer capable of peeling potatoes.

A lab technician prepared a calcium hydroxide solution by
dissolving 1.48 g of $Ca(OH)_2$ in water. How many milliliters
of 0.125 N HCl solution would be required to neutralize
this calcium hydroxide solution?

Solution: For a neutralization reaction to occur, the
number of equivalents of acid must equal the number of
equivalents of base. Therefore, one can solve this problem
by computing how many equivalents are present in 1.48 g of
$Ca(OH)_2$; the equivalents of acid (HCl) must equal this
number.

An equivalent is defined as the weight of a substance
in grams that releases one mole of protons ($H^+$) or hydroxyl
ions. When one mole of $Ca(OH)_2$ dissolves in water, it produces 2 moles of hydroxyl ions ($OH^-$) according to the
equation $Ca(OH)_2 \rightarrow Ca^{+2} + 2OH^-$. The weight of 1 mole of
$Ca(OH)_2$ is equal to its molecular weight. Therefore,
$Ca(OH)_2$ has a weight of 74 grams.

Thus, 74 grams of $Ca(OH)_2$ produces 2 moles of $OH^-$
ions. Recalling the definition of equivalency, 74/2 = 37
grams is the equivalent weight of $Ca(OH)_2$. The number of
equivalents of $Ca(OH)_2$ is:

$$Ca(OH)_2 = \frac{\text{weight in grams}}{\text{equivalent weight}} = \frac{1.48 \text{ g}}{37 \text{ g}} = 0.04 \text{ equiv.}$$

The number of equivalents of HCl must also be .04 for
neutralization to occur. The normality of HCl is given
as 0.125. Normality is defined as the number of equivalents
divided by liters of solution, i.e., N=equivalents/volume.

This means that 0.125 N = 0.04 equivalents/volume for
HCl. The volume in liters required for neutralization can
be obtained by solving this equation.

$$\text{liters} = \frac{0.04}{0.125} = 0.32 \text{ } \ell$$

To convert to milliliters multiply by the conversion factor
1000 ml/$\ell$. One has

$$0.32 \text{ } \ell \times 1000 \text{ ml}/\ell = 320 \text{ ml.}$$

50 ml of rhubarb juice is titrated against 0.25 N NaOH.
20 ml of NaOH solution is required for neutralization.
Assuming the acidity of the juice is due to oxalic acid
($H_2C_2O_4$) determine (a) the weight of oxalic acid per
liter of juice, (b) normality of the juice.

Solution:    (a) For the neutralization to occur, there must be the same number of OH⁻ ions as there are $H^+$ ions present.

$$H^+ + OH^- \rightleftharpoons H_2O$$

The normality of a base is defined as the number of equivalents of base in one liter of solution. An equivalent is the weight of an acid or base that produces one mole of $H^+$ or $OH^-$ ions, respectively. When NaOH ionizes, there is 1 OH⁻ ion formed by each NaOH that ionizes.

$$NaOH \rightleftharpoons OH^- + Na^+$$

Therefore, in a 0.25 N NaOH solution there are 0.25 equivalents of $OH^-$, which, in this case is also the number of moles. 20 ml is 0.02 liters. The number of moles of NaOH present will equal the number of liters of solution times the molarity of the solution. As previously indicated, normality equals molarity. Thus,

no. of moles = no. of liters × normality

$$= 0.02 \text{ liters} \times \frac{.25 \text{ moles}}{\text{liter}} = 0.005 \text{ moles.}$$

Thus, there must be 0.005 moles of $H^+$ ions from the oxalic acid to neutralize the NaOH. When oxalic acid ionizes, there are 2 $H^+$ ions formed for each molecule ionized.

$$H_2C_2O_4 = 2H^+ + C_2O_4$$

Therefore, each mole of oxalic acid ionizes 2 moles of NaOH. Thus, for neutralization to occur, only one half as much oxalic acid is needed as NaOH. There are 0.005 moles of NaOH, present, therefore, 0.0025 moles of oxalic acid is needed to neutralize it. One now knows that there are 0.0025 moles of oxalic acid in the 50 ml of rhubarb juice. One can find the number of moles in 1 liter by multiplying 0.0025 moles/50 ml by the conversion factor 1000 ml/1 liter.

no. of moles/liter of oxalic acid $= \dfrac{0.0025 \text{ moles}}{50 \text{ ml}} \times \dfrac{1000 \text{ ml}}{1 \text{ liter}}$

$$= 0.05 \text{ moles/liter.}$$

Since there are 0.05 moles of $H_2C_2O_4$ in 1 liter of juice, one can find the weight of $H_2C_2O_4$ in this quantity of juice by multiplying 0.05 moles by the MW of $H_2C_2O_4$. (MW = 90).

weight of $H_2C_2O_4$ in 1 liter = 0.05 moles × 90 g/moles

$$= 4.5 \text{ g}$$

(b) For titrations, the following relation is found:

$$N_{acid} V_{acid} = N_{base} V_{base}$$

where $N_{acid}$ is the normality of the acid, $V_{acid}$ is the volume of the acid, $N_{base}$ is the normality of the base, and $V_{base}$ is the volume of the base.

Here, one is given $N_{base}$, $V_{base}$, and $V_{acid}$. One is asked to find $N_{acid}$.

$N_b$ = .25 N

$V_b$ = 20 ml

$N_a$ = ?

$V_a$ = 50 ml

$$N_a V_a = N_b V_b$$

$$N_a \times 50 \text{ ml} = .25 \text{ N} \times 20 \text{ ml}$$

$$N_a = \frac{0.25 \text{ N} \times 20 \text{ ml}}{50 \text{ ml}} = 0.10 \text{ N}.$$

● **PROBLEM** 373

The $^+NH_3CH_2COOH$ ion (glycine$^+$) is a cation of the amino acid glycine ($NH_2CH_2COOH$). Glycine$^+$ undergoes the following two successive dissociations:

$$^+NH_3CH_2COOH \xrightleftharpoons{k_1} H^+ + {}^+NH_3CH_2COO^- \; ; \; pk_1 = 2.35$$

$$^+NH_3CH_2COO^- \xrightleftharpoons{k_2} H^+ + NH_2CH_2COO^- \; ; \; pk_2 = 9.78.$$

What is the isoelectric point of this molecule (the pH at which the number of positive charges on the molecule equals the number of negative charges).

Solution: We will approach this problem by considering the equilibrium constants $k_1$ and $k_2$. For the first reaction, $k_1$ is equal to

$$k_1 = \frac{[H^+][{}^+NH_3CH_2COO^-]}{[{}^+NH_3CH_2COOH]}$$

and for the second reation, $k_2$ is equal to

$$k_2 = \frac{[H^+][NH_2CH_2COO^-]}{[{}^+NH_3CH_2COO^-]}$$

If we take the product of these two constants, the concentration of isoelectric species $^+NH_3CH_2COO^-$ cancels out, giving

$$k_1 k_2 = \frac{[H^+][^+NH_3CH_2COO^-]}{[^+NH_3CH_2COOH]} \times \frac{[H^+][NH_2CH_2COO^-]}{[^+NH_3CH_2COO^-]}$$

$$= \frac{[H^+]^2[NH_2CH_2COO^-]}{[^+NH_3CH_2COOH]}$$

At the isoelectric point, the concentrations of all charged species are equal, that is

$$[NH_2CH_2COO^-] = [^+NH_3CH_2COOH].$$

Hence, $\dfrac{[NH_2CH_2COO^-]}{[^+NH_3CH_2COOH]} = 1$ and

$$k_1 k_2 = \frac{[H^+]^2 \ [NH_2CH_2COO^-]}{[^+NH_3CH_2COOH]} = [H^+]^2.$$

Taking the logarithm of both sides,

$$[H^+]^2 = k_1 k_2$$

$$2 \log [H^+] = \log k_1 k_2 = \log k_1 + \log k_2.$$

Multiplying this equation by negative one and using the definitions $pH = - \log [H^+]$, $pk_1 = - \log k_1$, and $pk_2 = - \log k_2$, we obtain

$$- 2 \log [H^+] = - \log k_1 + (- \log k_2)$$

$$2 \ pH = pk_1 + pk_2, \quad \text{or,} \quad pH = \tfrac{1}{2}(pk_1 + pk_2).$$

Thus, the isoelectric point is

$$pH = \tfrac{1}{2}(pk_1 + pk_2) = \tfrac{1}{2}(2.35 + 9.78) = 6.07.$$

# BUFFERS

● PROBLEM 374

Explain the buffering action of a liter of 0.10 M acetic acid containing 0.1 mole of sodium acetate. In the explanation, use ionic equations.

Solution: By buffering action, one means the ability of a substance to maintain relatively constant conditions of pH in the face of changes that might otherwise affect the acidity or basicity in solution. For example, a weak acid, such as acetic acid, will dissociate according to the following equation:

$$HOAc + H_2O \rightleftharpoons H_3O^+ + OAc^-.$$

The sodium acetate particle will be completely ionized in solution:

$$NaOAc \rightleftharpoons Na^+ + OAc^-.$$

These are the two processes that occur in this buffer. Suppose one increases the acetate concentration. By doing this the equilibrium is shifted to the left, i.e., to acetic acid.

$$k_a = \frac{[H_3O^+][OAc^-]}{[HOAc]} \quad \text{or} \quad [H_3O^+] = \frac{[HOAc]}{[OAc^-]} k_a.$$

The key to the buffering action is this [HOAc] to [OAc$^-$] ratio. If [HOAc] changes, [OAc$^-$] will change accordingly so that the same value of the ratio is obtained. Thus, the ratio is a constant. This means, therefore, that [H$_3$O$^+$] is maintained as a constant. pH = $-$ log [H$^+$], it is also constant.

● **PROBLEM** 375

Design a buffer system that will function in the low, near neutral, and high pH levels, using combinations of the three sodium phosphates (Na$_3$PO$_4$, Na$_2$HPO$_4$, NaH$_2$PO$_4$) and phosphoric acid (H$_3$PO$_4$). The equilibrium constants for this polyprotic acid are $k_1 = 7.5 \times 10^{-3}$, $k_2 = 6.2 \times 10^{-8}$, and $k_3 = 4.8 \times 10^{-13}$. With this information, calculate the pH at both extremes of the buffer, assume 10 : 1 and 1 : 10 ratios, and at the mid-range of the buffer, assume a 1 : 1 ratio. Assume, also, that the acid-salt ratio in making up the buffer is equal to HA/A$^-$ in solution.

Solution: Solutions that contain appreciable amounts of a weak acid, such as phosphoric acid, and its salt or salts, such as sodium phosphates, are called buffers. Their utility is in maintaining a relatively constant pH. This problem requires one to find various pH values for this buffer system. To do this, note that the pH of a buffer is given by

$$pH = pk_a + \log \frac{[A^-]}{[HA]} = pk_a + \log \frac{[salt]}{[acid]}.$$

One knows that to find the pH, one must determine [H$_3$O$^+$], since pH = $-$ log [H$_3$O$^+$]. To do this consider the ionization of the polyprotic acid, phosphoric acid, which can dissociate to produce 3 protons. $k_1$, the first ionization constant, corresponds to H$_3$PO$_4$ + H$_2$O $\rightleftharpoons$ H$_3$O$^+$ + H$_2$PO$_4^-$ so that

$$k_1 = \frac{[H_3O^+][H_2PO_4^-]}{[H_3PO_4]}.$$

$k_2$, the second ionization constant, corresponds to $H_2PO_4^- + H_2O \rightleftharpoons H_3O^+ + HPO_4^{2-}$ so that

$$k_2 = \frac{[H_3O^+][HPO_4^{2-}]}{[H_2PO_4^-]} .$$

$k_3$, the third ionization constant, corresponds to $HPO_4^{2-} + H_2O \rightleftharpoons H_3O^+ + PO_4^{3-}$ and

$$k_3 = \frac{[H_3O^+][PO_4^{3-}]}{[HPO_4^{2-}]} .$$

In general, the $k_a$ of any acid is given by the expression

$$k_a = \frac{[H_3O^+][A^-]}{[HA]} \quad \text{or} \quad [H_3O^+] = k_a \frac{[HA]}{[A^-]} .$$

It is from this equation that the $[H_3O^+]$ must be calculated, and thus yield the pH value wanted. The three systems:

## Buffer system with $k_1$

Consider the ratios of 10 : 1, 1 : 1, and 1 : 10 for the acid : salt ratios, given that the ratio equals $[HA]/[A^-]$ for the low, neutral and high pH values, respectively. This means that $[H_3O^+] = 10\ k_a$, $k_a$ and $0.1\ k_a$ for low, neutral, and high, respectively. Here,

$k_a = k_1$ which is $= 7.5 \times 10^{-3}$. $pk_1 = 2.12$. Since $[H_3O^+] = 10\ k_a$, $k_a$, and $0.1\ k_a$; pH $= -\log[H_3O^+] = (pk_a - 1)$, $pk_a$ and $(pk_a + 1)$ for low, middle, and high pH values, respectively. Thus, with $pk_1 = 2.12$, pH $= 1.12$ (low), 2.12 (middle) and 3.12 (high).

## Buffer system with $k_2$

One still has pH $= (pk_a - 1)$, $pk_a$, and $pk_a + 1$. But now $k_a = k_2 = 6.2 \times 10^{-8}$ $pk_2 = 7.21$. Thus, pH $= 6.21$ (low), 7.21 (middle), and 8.21 (high).

## Buffer system with $k_3$

One still has pH $= (pk_a - 1)$, $pk_a$, and $pk_a + 1$. Now, $k_a = k_3 = 4.8 \times 10^{-13}$ $pk_3 = 12.32$. Thus, pH $= 11.32$ (low), 12.32 (middle), and 13.32 (high).

The following bases, and their conjugate acids (as the chlorides), are available in the lab: ammonia, $NH_3$; pyridine, $C_5H_5N$; ethylamine, $CH_3CH_2NH_2$. A buffer solution of pH 9 is to be prepared, and the total concentration of buffering reagents is to be 0.5 mole/liter. (a) Choose the best acid-base pair. (b) Give the recipe for preparing one liter of the solution. (c) Calculate the pH of the solution after 0.02 mole of NaOH has been added per liter.

Solution:   This problem deals with the preparation of a buffer solution and then to show the effect the addition of a base has on the pH.

If this is a truly buffered system, then, the pH should be only slightly altered  when a small quantity of base is added.

Since we are using bases it will be more convenient to work initially with pOH. pH is converted to pOH by the equation:

14 = pH + pOH or    pOH = 14 - pH.

Substituting the given value of pH, pOH = 14 - 9 = 5.

In part (a) one is asked to choose the strongest base. At a given pOH, the strongest base has pOH about equal to $pK_b$ (the pH at which dissociation occurs). The maximum buffering capacity and minimum sensitivity to pH change occurs when the concentrations of the acid and base are initially equal. From a table of $pK_b$ values, it is found that $pK_{NH_3}$ = 4.75, $pK_{C_5H_5N}$ = 8.81, and $pK_{CH_3CH_2NH_2}$ = 3.3 Therefore, $NH_3$ has a $pK_b$ closest to 5, which means it is the best base.

To prepare a one liter solution, as requested in part (b), find the initial concentrations of the components to be added and their ratio. $NH_3$ is the base and $NH_4Cl$ dissociates to the acid $NH_4^+$. The reaction equation is

$$NH_3 + H_2O \leftrightarrows NH_4^+ + OH^-$$

$$K_b = 1.76 \times 10^{-5}.$$

To find the ratio, set up the equilibrium constant equation:

$$K_b = \frac{[NH_4^+][OH^-]}{[NH_3]} = 1.76 \times 10^{-5}.$$

Rewriting to obtain the ratio, one has:

$$\frac{[\text{NH}_4{}^+]}{[\text{NH}_3]} = \frac{K_b}{[\text{OH}^-]} = \frac{1.76 \times 10^{-5}}{1.0 \times 10^{-5}} = 1.76.$$

( $[\text{OH}^-] = 1 \times 10^{-5}$, since pOH = 5 and pOH = $-\log[\text{OH}^-]$.)

Therefore, $\text{NH}_4\text{Cl}$ and $\text{NH}_3$ should be added in the molar ratio of 1.76 : 1. Since it is given that the sum of the concentrations of the buffering reagents equals 0.5 mole/liter, the following equation can be written which says, in effect, that the sum of the parts is equal to the whole

$$[\text{NH}_3] + [\text{NH}_4{}^+] = 0.5.$$

Let $[\text{NH}_3] = x$ and since the ratio is 1.76 : 1, let $[\text{NH}_4{}^+] = 1.76\ x$.

This results in

$x + 1.76\ x = 0.5$

$x = [\text{NH}_3] = 0.18$ mole/liter

$1.76\ x = [\text{NH}_4{}^+] = 0.32$ mole/liter.

Therefore, to prepare a 1 liter solution, mix 0.18 moles $\text{NH}_3$ and 0.32 moles $\text{NH}_4\text{Cl}$ and then add $\text{H}_2\text{O}$ until a total volume of 1 liter is obtained.

Part (c) tells one to add 0.02 M NaOH. Since it is a strong base, complete dissociation occurs giving 0.02 mole $\text{Na}^+$ and 0.02 mole $\text{OH}^-$. The base $\text{OH}^-$ then reacts fully with the acid $\text{NH}_4{}^+$ and converts 0.02 mole $\text{NH}_4{}^+$ to $\text{NH}_3$.

This is more clearly seen below.

$$\text{NH}_4{}^+ \quad + \quad \text{H}_2\text{O} \quad \overset{\rightarrow}{\leftarrow} \quad \text{NH}_3 \quad + \quad \text{OH}^-$$

Before:    0.32 mole                0.18 mole

After:   0.32-0.02=0.30 mole       0.18+0.02=0.20 mole

Substituting these new values into the equilibrium constant equation, one arrives at $[\text{OH}^-]$ and then pH. One then has

$$K_b = \frac{[\text{NH}_3][\text{OH}^-]}{[\text{NH}_4{}^+]} . \quad \text{Substituting,}$$

$$[\text{OH}^-] = \frac{K_b\ [\text{NH}_4{}^+]}{[\text{NH}_3]} = \frac{1.76 \times 10^{-5}\ (0.20)}{(0.30)}$$

$$= 1.17 \times 10^{-5}\ \text{mole/liter.}$$

Thus,   pOH = - log [OH⁻] = - log [1.17 × 10⁻⁵] = 4.93.

Thus,   pH = 14 - pOH = 14 - 4.93 = 9.07.

# INDICATORS

● **PROBLEM** 377

Using methyl orange as an indicator, a solution of unknown pH was matched to a Bjerrum wedge at a point where the acid wedge was 40 percent of the total thickness of the combined wedges. This point is indicated by the arrow in the diagram. Determine the pH of the solution.

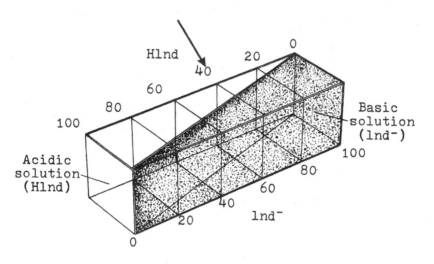

Bjerrum wedge

Solution:  From the Bjerrum wedge, the ratio of the concentration of methyl orange indicator in basic solution ([Ind⁻]) to the concentration in acid solution ([HInd]) will be determined. This ratio will then be used to find the pH of the unknown solution.

A Bjerrum wedge is a glass box divided into two wedge-shaped compartments by a glass plate placed diagonally in the box (see diagram, above). In one compartment is placed an acidic solution of the chosen indicator (HInd) and into the other a basic solution of the indicator (Ind⁻). The concentration of indicator is the same in both wedges. A view from the top will produce a continuum of color varying from that of pure HInd on one side to that of pure Ind⁻ on the other. A solution of indicator of the same concentration as the Bjerrum wedge is prepared using a solution of unknown pH. The resulting solution is then placed in a glass box having a thickness equal to that of the Bjerrum wedge. The boxes are placed side-by-side and the color of the unknown solution is matched to a color on the Bjerrum wedge. At the point where the two colors match, let the acidic wedge be x percent of the total thickness of the box. Then the basic wedge is 100 - x percent of the total thickness. The ratio $[Ind^-]/[HInd]$ is then

$$\frac{[Ind^-]}{[HInd]} = \frac{100 - x}{x} .$$

In this problem, the acidic wedge is 40 percent the total thickness, hence

$$\frac{[Ind^-]}{[HInd]} = \frac{100 - 40}{40} = \frac{60}{40} .$$

To make use of this ratio, we need an expression relating the pH to $[Ind^-]/[HInd]$. Consider the dissociation of acidic indicator:

$$HInd + H_2O \overset{\rightarrow}{\leftarrow} H_3O^+ + Ind^- .$$

The equilibrium constant, $K_a$, for this reaction is

$$K_a = [H_3O^+] \times \frac{[Ind^-]}{[HInd]} .$$

Taking the logarithm of both sides and multiplying by negative one gives

$$- \log K_a = - \log [H_3O^+] - \log \frac{[Ind^-]}{[HInd]} .$$

$$- \log K_a = pK_a \quad \text{and} \quad - \log [H_3O^+] = pH.$$

Hence, $\quad pK_a = pH - \log \frac{[Ind^-]}{[HInd]} ,$

or, $\qquad pH = pK_a + \log \frac{[Ind^-]}{[HInd]} .$

For methyl orange, $pK_a = 3.7$. Using the value of $[Ind^-]/[HInd]$ obtained above, we get

$$pH = 3.7 + \log \left(\frac{60}{40}\right) = 3.9$$

as the pH of our unknown solution.

For the indicator phenolphthalein (In⁻), HIn is colorless and In⁻ is red; $K_{diss}$ is $10^{-9}$. If one has a colorless solution of 50 ml phenolphthalein, how much 0.10 M NaOH solution would one have to add to make it red: $K_w = 10^{-14}$.

Solution: To answer this question, write out the dissociation reaction of phenolphthalein in solution. After this, write out the equilibrium constant expression. This allows one to calculate the $[H^+]$. This informs one of the concentration of $OH^-$ required and allows the calculation of the amount of NaOH to be added.

With this in mind, proceed as follows: An indicator, which changes color, undergoes a dissociation reaction like an acid. For phenolphthalein the reaction is

$$HIn \rightarrow H^+ + In^-.$$

It is given that In⁻ makes the solution red and HIn makes it colorless. The equilibrium constant expression, which measures the ratio of concentrations of products to reactants, each raised to the power of its coefficient in the equation, is

$$K_{diss} = 10^{-9} = \frac{[H^+][In^-]}{[HIn]}$$

The object is to make a **red solution,** from an originally colorless one. In general, a specific color of an indicator will show itself when the concentration of the corresponding species is 10 times as great as the concentration of the other species. Since one wants the solution to be red, and In⁻ = red color, $[In^-]/[HIn] = 10/1$. Substituting this into the equilibrium constant expression,

$$10^{-9} = [H^+] \frac{10}{1}. \text{ Solving for } [H^+], [H^+] = 10^{-10}.$$

From the $K_w$ constant of water, it is known that $[H^+][OH^-] = 10^{-14}$, where $K_w$ is the autodissociation constant of $H_2O$. If $[H^+] = 10^{-10}$, one can calculate $[OH^-]$ and obtain $[OH^-] = 10^{-4}$ M. Thus, to have a red solution, one wants $[OH^-]$ to be $10^{-4}$ M. The NaOH solution being added is 0.1 M. (M = molarity = moles/liter.) The volume of the solution is 50 ml. or 0.05 ℓ. (1000 ml = 1 ℓ.) Thus, one needs $10^{-4}$ M or 0.050 ℓ × $10^{-4}$ M = $5.0 \times 10^{-6}$ moles. Since volume in liters = number of moles/molarity, and molarity = 0.1, liters of NaOH to be added =

$$= \frac{5.0 \times 10^{-6}}{0.1} = 5.0 \times 10^{-5} \text{ liters.}$$

# COMPLEX IONS

> For the complex ion equilibrium for the dissociation of $Cu(NH_3)_4^{2+}$, calculate (1) the ratio of $Cu^{2+}$ ion to $Cu(NH_3)_4^{2+}$ in a solution 1 M in $NH_3$; (2) the equilibrium concentration of $NH_3$ needed to attain 50 % conversion of $Cu^{2+}$ to $Cu(NH_3)_4^{2+}$. $K_a$ for $Cu(NH_3)_4^{2+}$ = 2 × $10^{-13}$.

Solution:  (1) The equilibrium constant for the reaction

$$A \; \underset{\rightarrow}{\leftarrow} \; nB + mC$$

is defined as

$$K = \frac{[B]^n [C]^m}{[A]}$$

where K is the equilibrium constant, [B] is the concentration of B, n is the number of moles of B formed, [C] is the concentration of C, m is the number of moles of C produced and [A] is the concentration of A. The reaction in this problem is

$$Cu(NH_3)_4^{2+} \; \underset{\rightarrow}{\leftarrow} \; Cu^{2+} + 4NH_3$$

This indicates that the equilibrium constant can be stated

$$K_d = \frac{[Cu^{2+}][NH_3]^4}{[Cu(NH_3)_4^{2+}]} = 2 \times 10^{-13}$$

Here, one is trying to find the ratio of $Cu^{2+}$ ion to $Cu(NH_3)_4^{2+}$ in the solution or stated in another way

$\dfrac{[Cu^{2+}]}{[Cu(NH_3)_4^{2+}]}$ .  To obtain this ratio, one can use the equation for $K_d$. In the problem, one is told that the concentration of $NH_3$ is 1 M. Therefore,

$\dfrac{[Cu^{2+}]}{[Cu(NH_3)_4^{2+}]}$  can be found.

$$K_d = \frac{[Cu^{2+}][NH_3]^4}{[Cu(NH_3)_4^{2+}]} = 2 \times 10^{-13}$$

$$\frac{[Cu^{2+}][1.0]^4}{[Cu(NH_3)_4^{2+}]} = 2 \times 10^{-13}$$

$$\frac{[Cu^{2+}]}{[Cu(NH_3)_4^{2+}]} = \frac{2 \times 10^{-13}}{(1.0)^4} = 2 \times 10^{-13}$$

(b) To find the equilibrium concentration of $NH_3$ needed to convert 50 % of the $Cu^{2+}$ ion present to $Cu(NH_3)_4^{2+}$ ion, the equation for the equilibrium constant will again be used.

$$K_d = \frac{[Cu^{2+}][NH_3]^4}{[Cu(NH_3)_4^{2+}]} = 2 \times 10^{-13}$$

When 50 % of $Cu^{2+}$ is converted to $Cu(NH_3)_4^{2+}$, the concentration of $Cu^{2+}$ will equal that of $Cu(NH_3)_4^{2+}$. This is true because there is one $Cu^{2+}$ ion used in the formation of each $Cu(NH_3)_4^{2+}$ ion. If there were 50 $Cu^{2+}$ ions in the solution and 25 of them were converted to $Cu(NH_3)_4^{2+}$, there would be 25 $Cu^{2+}$ ions left and 25 $Cu(NH_3)_4^{2+}$ ions formed. This means that for a 50% conversion

$$\frac{[Cu^{2+}]}{[Cu(NH_3)_4^{2+}]} = 1$$

One can now solve for the $[NH_3]$ by using the equation for $K_d$.

$$K_d = \frac{[Cu^{2+}][NH_3]^4}{[Cu(NH_3)_4^{2+}]} = 2 \times 10^{-13}$$

(1) $[NH_3]^4 = 2 \times 10^{-13}$

$[NH_3] = \sqrt[4]{2 \times 10^{-13}}$

$[NH_3] = 6.7 \times 10^{-3}$ M

A concentration of $6.7 \times 10^{-3}$ M of $NH_3$ will cause a 50 % conversion of $Cu^{2+}$ to $Cu(NH_3)_4^{2+}$.

● PROBLEM 380

You have the four complex ions:

(1) $Cu(NH_3)_4^{2+} \rightleftarrows Cu^{2+} + 4NH_3$      $k = 1.0 \times 10^{-12}$

(2) $Co(NH_3)_6^{2+} \rightleftarrows Co^{2+} + 6NH_3$      $k = 4.0 \times 10^{-5}$

(3) $Co(NH_3)_6^{3+} \rightleftarrows Co^{3+} + 6NH_3$      $k = 6.3 \times 10^{-36}$

(4) $Ag(NH_3)_2^+ \rightleftarrows Ag^+ + 2NH_3$      $k = 6.0 \times 10^{-8}$.

Arrange them in order of increasing $NH_3$ concentration that would be in equilibrium with complex ion, if you start with 1 M of each.

Solution: To solve this problem determine the $[NH_3]$ for each complex ion equilibria. This necessitates writing the equilibrium constant expression, which measures the ratio of concentrations of products to reactants, each raised to the power of their coefficients in the chemical reaction. Proceed as follows:

$$k = 1.0 \times 10^{-12} = \frac{[Cu^{2+}][NH_3]^4}{[Cu(NH_3)_4^{2+}]}$$

Start with 1 M of $Cu(NH_3)_4^{2+}$. Let $x = [NH_3]$ at equilibrium. From the chemical reaction one can see that 1 mole of $Cu^{2+}$ is formed as 4 moles of $NH_3$ are formed. Thus, $[Cu^{2+}] = \frac{1}{4}[NH_3] = 0.25 x$. If x moles/liter of $NH_3$ form, and the only source is $Cu(NH_3)_4^{2+}$, then, at equilibrium, $[Cu(NH_3)_4^{2+}] = 1 - 0.25 x$. It is initial concentration minus $\frac{1}{4}$ of $[NH_3]$, since they have a mole ratio of 1 : 4, as is seen in the chemical reaction. Substituting into the equilibrium equation:

$$\frac{(0.25x)(x)^4}{(1 - 0.25\ x)} = 1.0 \times 10^{-12}.$$

Solving for x,  $x = [NH_3] = 5.3 \times 10^{-3}$.

To obtain this answer the denominator $(1 - 0.025x)$ is approximated as 1, since the value for x is relatively small compared to 1.

To find $[NH_3]$ at equilibrium in (2), (3), and (4), employ the same type of logic and reasoning. The only differences are the values of K and the mole ratios between the substances.

For $Co^{+2}$ : $1/6\ x^7 = 4.0 \times 10^{-5}$

$$[NH_3] = x = 0.3$$

For $Co^{+3}$ : $1/6\ x^7 = 6.3 \times 10^{-36}$

$$[NH_3] = x = 9.4 \times 10^{-6}$$

For $Ag^+$  : $\frac{1}{2}\ x^3\ = 6.0 \times 10^{-8}$

$$[NH_3] = x = 5.0 \times 10^{-3}$$

In increasing order, rank them as

$$Co^{3+},\ Ag^+,\ Cu^{2+},\ Co^{2+}$$

$$\longrightarrow$$

increasing $[NH_3]$

● PROBLEM 381

0.1 moles of $Cu(NO_3)_2$ are added to 1.0 moles of $NH_3$ and then diluted to 1000ml. Find the concentration of $Cu^{2+}$ formed. $K_{eq} = 8.5 \times 10^{-13}$ .

<u>Solution</u>:  The first thing to realize is that the complex $Cu(NO_3)_2$ dissociates to  $Cu^{2+} + 2NO_3^+$  in  $H_2O$

$$Cu(NO_3)_2 \rightleftarrows Cu^{2+} + 2NO_3^+$$

when  $NH$   is added to the solution the following reaction occurs.

$$Cu^{2+} + 4NH_3 \rightleftarrows Cu(NH_3)_4^{2+}$$

$Cu(NH_3)_4^{2+}$  is called a complex ion.  $Cu(NH_3)_4^{2+}$  will be in equilibrium with  $Cu^{2+} + 4NH_3$ .  You are given the  $K_{eq}$  for this reaction which is defined as

$$K = \frac{[Cu^{2+}][NH_3]^4}{[Cu(NH_3)_4^{2+}]} \quad .$$

In general, if you have the equilibrium:  $H(A)_X \rightarrow H^+ + XA^-$ , then
$$K = \frac{[H^+][A^-]^X}{H(A)_X} \quad .$$

That is, the coefficients of the balanced equation serve as the exponents in the  K  expression.  The problem stated  $K = 8.5 \times 10^{-13}$ . Therefore,

$$\frac{[Cu^{2+}][NH_3]^4}{[Cu(NH_3)_4^{2+}]} = 8.5 \times 10^{-13} \quad .$$

You are solving for  $[Cu^{2+}]$ .  Let  $x = [Cu^{2+}]$ , then  $[Cu(NH_3)_4^{2+}] = .1-x$, since it must be the original amount of  $Cu(NO_3)_2$  present minus the  $Cu^{2+}$  formed.  There is originally 1.0 moles  $NH_3$, at equilibrium  4 moles of  $NH_3$  will have reacted with each mole of  $Cu^{2+}$  to form  $Cu(NH_3)_4^{2+}$, thus,  $[NH_3]$  may be represented as  $1.0-4 \times [Cu(NH_3)_4^{2+}]$  $= 1.0-4(.10-x)$.  Since most of the  $Cu^{2+}$  present will be in the form of  $Cu(NH_3)_4^{2+}$  and that there was only  .1M of it in the first place, one can make the assumption that  $x \ll .1$, therefore,  .10-x = .1  and  $1.0 - 4(.10-x) = 1.0 - 4(.1) = 0.6 = [NH_3]$ .  Using this same assumption,  $[Cu(NH_3)_4^{2+}] = .1-x = .1$ .

You now substitute these values into the previous expression. That is,

$$\frac{x(0.6)^4}{.1} = 8.5 \times 10^{-13} , \quad x = 6.5 \times 10^{-13} \quad .$$

Therefore,  $[Cu^{2+}] = 6.5 \times 10^{-13} M$.

An equilibrium solution of the complex ion $Ag(NH_3)_2^+$ contains 0.30 M of $NH_4^+$ and 0.15 M of the actual complex ion. To obtain a concentration of $Ag^+$ equal to $1.0 \times 10^{-6}$ M, what must the pH of this solution be? $K_{diss}$ of $Ag(NH_3)_2^+$ = $6.0 \times 10^{-8}$, $K_b$ of $NH_3$ = $1.8 \times 10^{-5}$ and $K_w = 1 \times 10^{-14}$.

Solution: To find the pH required, calculate the $[H^+]$ (pH = - log $[H^+]$). To find $[H^+]$, the $[OH^-]$ is needed. One is given $K_w$, which is the constant for the autodis-

sociation of $H_2O$. $K_w$ is defined as $[H^+][OH^-]$ for $H_2O \rightleftharpoons H^+ +$

$OH^-$. If $\lfloor OH^- \rfloor$ was known, one could solve for $[H^+]$. To find $[OH^-]$, consider the $K_b$ value given. When bases, such as $NH_3$,

are placed in water, these bases dissociate into positive and negative ions. The $K_b$, the equilibrium constant, measures the ratio of the concentration of products to re-

actants, each raised to the power of its coefficient in the chemical reaction. When $NH_3$ is placed in water, following dissociation exists:

$$NH_3 + H_2O \rightleftharpoons NH_4^+ + OH^- . \qquad \text{Thus,}$$

$$K_b = \frac{[NH_4^+][OH^-]}{[NH_3]} .$$

Note: $H_2O$ is omitted; it is assumed to be a constant. Therefore, if $[NH_4^+]/[NH_3]$ was known, one could find $[OH^-]$. To find $[NH_4^+]/[NH_3]$, calculate $[NH_3]$ since $[NH_4^+]$ is given. To calculate $[NH_3]$, use the equilibrium constant expression for the complex ion dissociation. This constant, $K_{diss}$, is defined in the same way as $K_b$ (i.e., ratio of products to reactants). The problem provides all the information needed to find $[NH_3]$. Proceed as follows:

The complex ion dissociation is similar in nature to other equilibrium dissociations.

$$Ag(NH_3)_2{}^+ \; \rightleftarrows \; 2NH_3 + Ag^+. \qquad \text{As such,}$$

$$K_{diss} = \frac{[Ag^+][NH_3]^2}{[Ag(NH_3)_2{}^+]} \, .$$

As given, $[Ag(NH_3)_2{}^+] = 0.15$ M, one wants $[Ag^+] = 1.0 \times 10^{-6}$ M at equilibrium. Thus, substituting these values, one finds that

$$K_{diss} = 6.0 \times 10^{-8} = \frac{(1.0 \times 10^{-6})[NH_3]^2}{0.15}$$

Solving for $[NH_3]$, one obtains

$$[NH_3] = \sqrt{\frac{(6.0 \times 10^{-8})(0.15)}{1.0 \times 10^{-6}}} = 9.5 \times 10^{-2}.$$

One is told $K_b = 1.8 \times 10^{-5}$. By definition,

$$K_b = \frac{[NH_4{}^+][OH^-]}{[NH_3]} \, \cdot$$

$[NH_3]$ has been found and $[NH_4{}^+]$ is given. Thus, by substitution one can solve for $[OH^-]$.

$$1.8 \times 10^{-5} = \frac{(0.30)[OH^-]}{9.5 \times 10^{-2}} \qquad \text{or}$$

$$[OH^-] = \frac{(1.8 \times 10^{-5})(9.5 \times 10^{-2})}{0.30} = 5.7 \times 10^{-6}.$$

Recalling that $K_w = 1.0 \times 10^{-14} = [OH^-][H^+]$, one has

$$1.0 \times 10^{-14} = [H^+][5.7 \times 10^{-6}].$$

Solving,　　$[H^+] = \dfrac{1.0 \times 10^{-14}}{5.7 \times 10^{-6}} = 1.75 \times 10^{-9}$

Since　　　　$pH = - \log [H^+],$

　　　　　　$pH = - \log [1.75 \times 10^{-9}] = 8.76.$

Thus, bring the pH up to 8.76 to obtain an $[Ag^+] = 1.0 \times 10^{-6}$ M.


# ELECTROLYTES

● **PROBLEM** 383

Describe, by both formal and ionic equations, what takes place when 0.10 mole of sodium sulfite is added to 1 liter of 1 M hydrochloric acid.

<u>Solution</u>:　The formal equation for the reaction is

$$Na_2SO_3 + 2HCl \rightarrow 2NaCl + H_2O + SO_2.$$

The ionic equation gives more detailed information than this. The soluble sodium sulfite salt will react with excess hydrochloric acid according to the ionic reaction

$$(2Na^+, SO_3{}^{2-}) + 2(H_3O^+, Cl^-) \rightarrow 2(Na^+, Cl^-) + H_2SO_3\,(aq) + H_2O.$$

Sulfurous acid, $H_2SO_3$, exists in equilibrium with both sulfur dioxide and hydronium ions:

$$H_2SO_3 \; \rightleftharpoons \; SO_2\,(g) + H_2O,$$

$$SO_3{}^{2-} + 2H_3O^+ \; \rightleftharpoons \; H_2SO_3.$$

$SO_2$ gas will escape from solution, thus pulling the reaction towards completion, to give the products $2(Na^+, Cl^-)$, $SO_2$, and $2H_2O$.


● **PROBLEM** 384

A saturated solution of the strong electrolyte $Ca(OH)_2$ is prepared by adding sufficient water to $5.0 \times 10^{-4}$ mole of $Ca(OH)_2$ to form 100 ml of solution. What is the pH of this solution?

Solution: We need to find $[H_3O^+]$ in order to find the pH. $[H_3O^+]$ is determined by substituting into the expression for the water constant, $K_w = [H_3O^+][OH^-]$ ($K_w = 10^{-14}$ mole²/liter²), or

$$[H_3O^+] = \frac{K_w}{[OH^-]}$$

The hydroxyl concentration, $[OH^-]$, is contributed by the dissociation of $Ca(OH)_2$, which is given by the equation

$$Ca(OH)_2 \rightarrow Ca^{2+} + 2OH^-.$$

$Ca(OH)_2$ is a strong electrolyte, hence we assume that it dissociates completely. Since two hydroxyl ions are formed for every $Ca(OH)_2$ that dissociates, the concentration of $OH^-$ is equal to twice that of $Ca(OH)_2$, or $[OH^-] = 2 \times (0.0005$ mole/100 ml$) = 2 \times (0.0005$ mole/0.10 liter$) = 2 \times 0.005$ mole/liter $= 0.010$ mole/liter $= 10^{-2}$ mole/liter. The hydronium ion concentration is then

$$[H_3O^+] = \frac{K_w}{OH^-} = \frac{10^{-14} \text{ mole}^2/\text{liter}^2}{10^{-2} \text{ mole/liter}} = 10^{-12} \text{ mole/liter}.$$

The pH is defined as $pH = -\log[H_3O^+]$, or

$$pH = -\log[H_3O^+] = -\log(10^{-12}) = -(-12) = 12.$$

● **PROBLEM** 385

Find the pH of a 0.25 M solution of $Na_2CO_3$, a strong electrolyte.

Solution: The crux of this problem is to realize that the solution contains the diprotic base $CO_3^{-2}$. Since $Na_2CO_3$ is a strong electrolyte, it dissociates completely. Diprotic means that 2 hydrogen ions will dissociate. Thus, there are 2 acid equilibrium constants. $H_2CO_3$ is the fully protonated acid. For diprotic acids, the second base ionization constant, $K_{b_2}$, is related to the first acid ionization constant, $K_{a_1}$, and $K_{b_1}$ is related to $K_{a_2}$.

The first $H^+$ ion neutralized behaves as a stronger acid then the second $H^+$ ion, because in each successive stage, ionization is less complete.

The ionization and equilibrium constant equations are:

$$CO_3^{-2} + H_2O \rightleftharpoons HCO_3^- + OH^-$$

$$K_{b_1} = \frac{[HCO_3^-][OH^-]}{[CO_3^{-2}]} \qquad \text{and}$$

$$HCO_3^- + H_2O \rightleftharpoons H_2CO_3 + OH^-$$

$$K_{b_2} = \frac{[H_2CO_3][OH^-]}{[HCO_3^-]}$$

Since it is given that the $[CO_3^{-2}] = 0.25$ M, attention shall be limited to the first ionization.

Each x moles of $CO_3^{-2}$ that reacts with $H_2O$ forms x moles each of $HCO_3^-$ and $OH^-$.

$$CO_3^{-2} + H_2O \rightleftharpoons HCO_3^- + OH^-$$

Before:  0.25  0  0

After:  0.25-x  x  x

If the concentration of $CO_3^{-2}$ is initially 0.25 M, and x moles of it react, then, at equilibrium, there are 0.25 - x.

Before substituting these values into the $K_{b_1}$ equation first find $K_{b_1}$. This is accomplished by referring to a table of $K_a$ ionization constants. $K_{a_2}$ has a value of $5.62 \times 10^{-11}$. One can then use this information in the following equation:

$$K_{a_2} K_{b_1} = K_w$$

which relates the acid and base ionization constants to the constant for the autodissociation of water, $K_w$. Therefore,

$$K_{b_1} = \frac{K_w}{K_{a_2}} = \frac{1.0 \times 10^{-14}}{5.62 \times 10^{-11}} = 1.8 \times 10^{-4}$$

One can find $[OH^-]$ by writing the equilibrium constant expression and substituting the values for the concentrations.

$$K_{b_1} = \frac{[HCO_3^-][OH^-]}{[CO_3^{-2}]}$$

$$\frac{(x)(x)}{0.25} = 1.8 \times 10^{-4}$$

$$x = [OH^-] = 6.7 \times 10^{-3}.$$

The x from $(0.25 - x)$ was eliminated since so few moles react, as $CO_3^{-2}$ is a weak base.

In order to calculate pH, first obtain pOH using

$$pOH = - \log [OH^-]$$

$$= - \log [6.7 \times 10^{-3}] = 3 - \log 6.7 = 2.17.$$

$$pH = pK_w - pOH = 14 - 2.17 = 11.83.$$

It makes sense that $K_a K_b = K_w$ since $K_a$ is a measure of $[H^+]$ and $K_b$ is a measure of $[OH^-]$, and one knows that

$$[H^+][OH^-] = K_w = 10^{-14}.$$

● PROBLEM 386

Explain the following phenomena:

a)  Liquid HCl is a nonelectrolyte, but aqueous HCl is a strong electrolyte b) Liquid HCN is a nonelectrolyte as is aqueous HCN.

<u>Solution</u>:  An electrolyte is a substance which exists as ions.  Both liquid HCN and HCl are not electrolytes because they consist of neutral atoms.  No ions exist. The case with an aqueous solution requires a little more investigation.  In water, HCl is ionized via the reaction $HCl + H_2O \rightleftharpoons H_3O^+ + Cl^-$.  Thus, you have hydronium ions and chloride ions in solution.  Thus, aqueous HCl is an electrolyte.        HCN also dissociate  when placed in water.  But, to what extent? $H_3O^+$ and $CN^-$ ions are only to a very slight extent produced, since HCN's dissociation constant is so small. Thus, while HCN may be called an electrolyte, it is an extremely weak one.

# CHAPTER 11

# SOLUBILITY AND THE ION PRODUCT CONSTANT

Basic Attacks and Strategies for Solving Problems in this Chapter. See pages 367 to 384 for step-by-step solutions to problems.

A solubility product or ion product constant is simply an equilibrium constant for the solubility reaction. The problems in this section can be solved by following the same principles discussed in Chapter 9 and illustrated for acid-base equilibrium in Chapter 10. If ions of several salts are present, the equilibrium solubility of each possible reaction must simultaneously be satisfied. Two examples will illustrate the application of these principles to solubility equilibrium.

First, the solution of a relatively insoluble salt such as silver chloride:

$$AgCl \leftrightarrow Ag^+ + Cl^- \qquad\qquad 11\text{-}1$$

$$K_{sp} = (Ag^+)(Cl^-)/(AgCl) = 1.1 \times 10^{-10} \qquad\qquad 11\text{-}2$$

The activity of solid AgCl is one if the solid is present. Also, from the stoichiometric relationship in Equation 11-1, it is apparent that the concentrations of $Ag^+$ and $Cl^-$ are equal. Let this concentration $= x$.

$$1.1 \times 10^{-10} = (x)(x)/1 = x^2 \qquad\qquad 11\text{-}3$$

$$x = 1.05 \times 10^{-5} = (Ag^+) = (Cl^-) \qquad\qquad 11\text{-}4$$

Second, let us consider the concentration of silver and chloride ions in a solution initially containing 0.1M silver nitrate, a strong electrolyte, and 0.2M sodium chloride, also a strong electrolyte. Several possible ionic equilibrium reactions might be considered. The equilibrium of each must be satisfied.

$$AgNO_3 \leftrightarrow Ag^+ + NO_3^- \qquad\qquad 11\text{-}5$$

$$NaCl \leftrightarrow Na^+ + Cl^- \qquad\qquad 11\text{-}6$$

$$AgCl \leftrightarrow Ag^+ + Cl^- \qquad\qquad 11\text{-}7$$

The equilibrium constant $K_{sp}$, for strong electrolytes, is a large number and so the concentration of $AgNO_3$ and NaCl can be considered to be zero. The 0.2M $Cl^-$ from Equation 11-6 will react with essentially all of the 0.1M $Ag^+$ from Equation 11-7 leaving a $Cl^-$ concentration of approximately 0.1M. To calculate the $(Ag^+)$ concentration in this case, Equation 11-2 can be employed. Note that $(Cl^-)$ is very nearly 0.1 and the AgCl activity is one.

$$1.1 \times 10^{-10} = (Ag^+)(Cl^-)/(AgCl) = (Ag^+)(0.1)/(1) \qquad\qquad 11\text{-}8$$

$$(Ag+) = 1.1 \times 10^{-9} \qquad\qquad 11\text{-}9$$

Step-by-Step Solutions to
Problems in this Chapter,
"Solubility and the Ion
Product Constant"

• PROBLEM 387

Describe, using equations, what takes place when the following solutions are prepared:

A. A solution containing equal amounts of 0.10M lead nitrate and 0.10M potassium chromate;

B. A solution containing 1 mole of sodium chloride and 1 mole of potassium bromide.

Solution: In such problems, two processes can occur: mixing of ions in solution and precipitation.

A. Both lead nitrate $(Pb(NO_3)_2)$ and potassium chromate $(K_2CrO_4)$ are highly soluble, so that four ions are present in solution:

$$Pb^{2+}, \quad K^+, \quad NO_3^-, \quad CrO_4^{2-} \quad .$$

There are four possible combinations of these ions: $Pb(NO_3)_2$, $PbCrO_4$, $KNO_3$, and $K_2Cr_2O_4$. Of these, only lead chromate, $PbCrO_4$, is insoluble, so that the reaction

$$Pb^{2+} + CrO_4^{2-} \rightarrow PbCrO_4$$

is driven to the right, and yields a precipitate, leaving $K^+$ and $NO_3^-$ ions in solution (along with trace amounts of $Pb^{2+}$ and $CrO_4^{2-}$, which exist in equilibrium with solid lead chromate).

B. Both sodium chloride (NaCl) and potassium bromide (KBr) are highly soluble, giving rise to $Na^+$, $K^+$, $Cl^-$, and $Br^-$ ions in solution. The four possible combinations of these ions are NaCl, NaBr, KCl, and KBr, all of which are highly soluble. Therefore no precipitate is formed, but only mixing of the ions according to the equation

$$(Na^+, Cl^-) + (K^+, Br^-) \rightleftarrows (Na^+, Br^-) + (K^+, Cl^-).$$

• PROBLEM 388

Given that $K_{sp}$ for $Mg(OH)_2$ is $1.2 \times 10^{-11}$, calculate the solubility of this compound in grams per 100ml of solution. The reaction equation is

$$Mg(OH)_2 \rightleftharpoons Mg^{+2} + 2OH^-.$$

Solution: $K_{sp}$ is the solubility product constant; it measures the equilibrium established between the ions in the saturated solution and the excess solid phase. Knowing the $K_{sp}$, we can calculate the solubility of the compound. The $K_{sp}$ equation for general compound $A_xB_y$ is

$$K_{sp} = [A]^x[B]^y \ .$$

From the equation, it can be seen that if x moles per liter of $Mg(OH)_2$ dissolves, x moles of $Mg^{+2}$ and 2x moles of OH form per liter,

$$K_{sp} = [Mg^{+2}][OH^-]^2$$

$$1.2 \times 10^{-11} = x(2x)^2 = 4x^3 \ .$$

Solving,

$$x = 1.4 \times 10^{-4} \text{ mole/liter of } Mg(OH)_2 \text{ dissolved.}$$

One is asked, however, for grams per 100ml of solution. $1.4 \times 10^{-4}$ mole/liter can be converted to grams/100ml by the following method: 100ml = .1 liters since there exists 1000ml = 1 liter. If one has $1.4 \times 10^{-4}$ moles in 1 liter, then in .1 liters, there are $(.1)(1.4 \times 10^{-4})$ = $1.4 \times 10^{-5}$ moles. The molecular weight of $Mg(OH)_2$ = 58.312g/mole. Therefore, $1.4 \times 10^{-5}$ moles/100ml translates into

$$(1.4 \times 10^{-5} \tfrac{\text{moles}}{100\text{ml}})(58.312\text{g/mole}) = 8.16 \times 10^{-4} \text{grams/100ml.}$$

● PROBLEM 389

The solubility product constant of magnesium hydroxide is $8.9 \times 10^{-12}$, calculate its solubility in (a) water and (b) .05M NaOH.

Solution: Whenever an ionic solid is placed in water, an equilibrium is established between its ions and the excess solid phase. The solubility constant, $K_{sp}$, measures this equilibrium. For the general reaction $AxBy \rightleftharpoons xA^+ + yB^-$, the $K_{sp}$ is defined as being equal to $[A^+]^x[B^-]^y$. The concentration of solid is always constant, no matter how much is in contact with the ions. This means the solid phase will not appear in the equilibrium constant expression. For this problem, part (a), you have

$$Mg(OH)_2(s) \rightleftharpoons Mg^{2+} + 2OH^- \ .$$

The

$$K_{sp} = [Mg^{2+}][OH^-]^2 = 8.9 \times 10^{-12} \ .$$

You are asked to find these concentrations. From the chemical equation, you find 2 moles of $OH^-$ will be generated per mole of $Mg^{2+}$. Thus, at equilibrium, if the concentration of $Mg^{2+}$ = x (mol/liter), then the concentration of $OH^-$ = 2x (mol/liter). (Note: The dissociation of water contributes some $OH^-$, but this amount is very small, and can be ignored.) Thus, you have

$$[Mg^{2+}][OH^-]^2 = x(2x)^2 = 8.9 \times 10^{-12} = K_{sp} \ .$$

Solving for x, you obtain $x = 1.3 \times 10^{-4}$ mol/liter. Thus, $1.3 \times 10^{-4}$ **mole of** $Mg(OH)_2$ dissolves per liter of water, producing a solution of $1.3 \times 10^{-4}$M $Mg^{2+}$ and $2.6 \times 10^{-4}$M $OH^-$ .

To find the solubility in .05M NaOH, part (b), perform the same process, except you must realize that the NaOH supplies $OH^-$ in addition to the $OH^-$ that dissolves from the salt. This means, as such,
$$[Mg^{2+}][OH^-]^2 = K_{sp} = (x)(2x + .050)^2 = 8.9 \times 10^{-12} \ .$$

If you solve for x, you obtain

$$3.6 \times 10^{-9} \text{ mol/liter.}$$

● **PROBLEM** 390

A chemist dissolves $BaSO_4$ in pure water at $25^{\circ}$C. If its $K_{sp} = 1 \times 10^{-10}$ , what is the solubility of the barium **sulfate** in the water?

Solution: The solubility of a compound is defined as the limiting concentration of the compound in a solution before precipitation occurs. To find the solubility of the barium sulfate, you need to know the concentration of its ions in solution. $BaSO_4$ will dissociate into ions because it is a salt. There will be an equilibrium between these ions and the $BaSO_4$ . The equilibrium can be measured in terms of a constant, $K_{sp}$, called the solubility constant. The $K_{sp}$ is expressed in terms of the concentrations of the ions. As such, to answer the question, you want to represent this $K_{sp}$ . For this reaction, the equation is

$$BaSO_4 \rightleftarrows Ba^{++} + SO_4^{=}$$

$$K_{sp} = [Ba^{2+}][SO_4^{2-}] = 1 \times 10^{-10}.$$

Let $x = [Ba^{2+}]$ . Thus, $x = [SO_4^{2-}]$, also, since both ions will be formed in equimolar amounts. Therefore, $x \cdot x = 1 \times 10^{-10}$ . Solving,

$$x = 1 \times 10^{-5}M = [Ba^{2+}] = [SO_4^{2-}] \ .$$

● **PROBLEM** 391

Determine the approximate solubility of AgCl in 0.10M NaCl solution. $K_{sp}$ for AgCl $= 1.1 \times 10^{-10}$.

Solution: The approximate solubility of AgCl in this solution is equal to the concentration of $Ag^+$ in the solution. This is because the concentration of $Ag^+$ is equal to the amount of AgCl ionized. The ionization of AgCl can be stated

$$AgCl \rightleftarrows Ag^+ + Cl^+ \ .$$

This means that one $Ag^+$ ion is formed each time one molecule of AgCl is ionized. The concentration of $Ag^+$ can be obtained by using the solubility product constant ($K_{sp}$). This constant is equal to the product of the concentration of $Ag^+$ ($[Ag^+]$) times the concentration of $Cl^-$ ($[Cl^-]$).

$$K_{sp} = [Ag^+][Cl^-] = 1.1 \times 10^{-10} .$$

The concentration of the $Cl^-$ ions is also affected by the $Cl^-$ ions from the NaCl solution. Because NaCl is a salt, it is 100% ionized in solution

$$NaCl \rightleftarrows Na^+ + Cl^- .$$

There is one $Cl^-$ ion formed for every NaCl ionized. The original concentration of NaCl is 0.10 molar, and after ionization, the solution is 0.10 molar in $Cl^-$ ions. The $Na^+$ ions can be disregarded because here we are looking at the solubility of AgCl, which does not contain any $Na^+$ ions. One can assume that the total concentration of $Cl^-$ ions in solution will be approximately 0.10, since the amount contributed by the AgCl ionization is small, and can be neglected.

The $[Ag^+]$ can now be found.

$$[Ag^+][Cl^-] = 1.1 \times 10^{-10}$$
$$[Ag^+](0.10) = 1.1 \times 10^{-10}$$
$$[Ag^+] = 1.1 \times 10^{-9} M .$$

● **PROBLEM** 392

Assuming the $K_{sp}$ for radium sulfate is $4 \times 10^{-11}$, what is its solubility in   (a)   pure water, and   (b)   .1M $Na_2SO_4$?

Solution: Whenever an ionic solid is placed in water, an equilibrium is established between its ions and the excess solid phase. A solubility constant, $K_{sp}$, measures this equilibrium. The concentration of solid is always constant, no matter how much it is in contact with the ions. This means the solid phase will not appear in the equilibrium constant expression. The dissociation is

$$RaSO_4(s) \rightleftarrows Ra^{2+} + SO_4^{2-} .$$

The $K_{sp}$ is, thus, $[Ra^{2+}][SO_4^{2-}] = 4 \times 10^{-11}$. From the chemical reaction, it becomes evident that the concentration of $Ra^{2+}$ must equal that of $SO_4^{2-}$, since they are generated in equimolar amounts. Therefore, let $x$ = the solubility of each. Substituting into $[Ra^{2+}][SO_4^{2-}] = 4 \times 10^{-11}$, you have $x \cdot x = 4 \times 10^{-11}$. Solving for $x$, you obtain $x = 6 \times 10^{-6}$ mol/liter. Thus, the solubility of $RaSO_4$ is $6 \times 10^{-6}$ mol/liter, the solution is $6 \times 10^{-6} M$ $Ra^{2+}$ and $6 \times 10^{-6} M$ $SO_4^{2-}$ in water.

b) The equation for this reaction in water is $RaSO_4 \rightleftarrows Ra^{2+} + SO_4^{2-}$. $Na_2SO_4$ is a salt and therefore will ionize completely in water.

The concentration of the $Na_2SO_4$ solution is .1M, which is equivalent to a solution of .2M of $Na^+$ and .1M $SO_4^{=}$.

If one adds more $SO_4^{2-}$ to the system, the equilibrium is forced

to the left. Therefore, there will be even less $Ra^{2+}$ present than in pure water. Thus,
$$[Ra^{2+}] \ll 6.0 \times 10^{-6}M$$
when additional $SO_4^=$ is present. Let $y$ = concentration of $Ra^+$. The solubility of $RaSO_4$ is equal to the product of $[Ra^+]$ and $[SO_4^=]$ in this solution. Solving: $[Ra^+]$ is equal to $y$. Because $RaSO_4$ dissociates into $Ra^{++}$ and $SO_4^=$, the $RaSO_4$ in this contributes $y$ to $[SO_4^=]$. There is also $SO_4^=$ already present in this solution, its concentration is .1M, therefore $[SO_4^=] = (y + .1)$. Writing the equation for the solubility
$$\text{solubility of } RaSO_4 = [Ra^{++}][SO_4^=]$$
$$= y(y + .1).$$
It has been shown that $y$ is much smaller than .1 and the following approximation can be made. $y + .10 \cong .1$ . Solving for $y$:
$$y(y + .1) = 4 \times 10^{-11}$$
$$.1y = 4 \times 10^{-11}$$
$$y = 4 \times 10^{-10}.$$

● **PROBLEM** 393

A saturated solution of $CaF_2$ contains .00168g of $CaF_2$ per 100g of water. Determine the $K_{sp}$.

Solution: When a salt is added to water, it dissociates into ions. Once a saturated solution is obtained, the addition of more salt results in precipitation of the solid. In such a case, an equilibrium is established between the solid phase and the ions in solution. This equilibrium is expressed as the solubility product constant or $K_{sp}$. To determine the $K_{sp}$, you must set up an equation that expresses this equilibrium. You must also determine the molarity of the solution. Molarity refers to the number of moles per liter of solution. The equilibrium reaction is

$$CaF_2(s) \rightleftarrows Ca^{2+}(aq) + 2F^-(aq).$$

The products have their oxidation states expressed because they are ions in solution. The $K_{sp}$ for the reaction is defined as

$$K_{sp} = [Ca^{2+}][F^-]^2.$$

Because this is a heterogeneous equilibrium, there is no need to include $CaF_2(s)$ in the expression for $K_{sp}$. You include only those substances with variable molar concentrations. The fluoride ion concentration is squared because 2 moles of fluoride ion are generated in the reaction. In general, then, the exponent of the concentration will equal the number of moles of that ion that is generated. Now that the $K_{sp}$ is defined, determine the molarity of the solution. Since molarity is defined as moles of solute per liter of solution, you have

$$M = \text{molarity} = \cfrac{\cfrac{.00168g}{78g/\text{mole}} \ (CaF_2)}{\cfrac{100 \ ml}{1000ml/1}}$$

Solving, you obtain $M = .214 \times 10^{-2}M$. You will notice that the 100 grams of water do not appear in this calculation. The density of water is 1g/ml. Since Density = Mass/Volume, 100g $H_2O$ has a volume of 100ml. Now that you know the molarity, go back to

$$CaF_2(s) = Ca^{2+}(aq) + 2F^-(aq).$$

If $CaF_2(s)$ has $M = .214 \times 10^{-2}$, then by looking at the coefficients, you find in solution

$$2.14 \times 10^{-3} M \ Ca^{2+} \quad \text{and} \quad 4.28 \times 10^{-3} M \ F^-.$$

These results are obtained because for every mole/liter of $CaF_2$ in solution, 1 mole/liter of $Ca^{2+}$ and 2 moles/liter of $F^-$ is obtained.

Therefore, recalling

$$K_{sp} = [Ca^{2+}][F^-]^2 ,$$

you have

$$K_{sp} = (.214 \times 10^{-2})(.428 \times 10^{-2})^2$$
$$= 3.92 \times 10^{-8} .$$

● **PROBLEM 394**

Calculate the solubility product constant of pure $PbSO_4$ in water. The solubility of $PbSO_4$ in water at $25^\circ C$ is $1.25 \times 10\text{-}4$ moles/liter.

Solution: The solubility of a substance is the number of moles of that substance which will dissolve in one liter. Since $PbSO_4$ contains one $Pb^{2+}$ ion and one $SO_4^{2-}$ ion per molecule in aqueous solution, the concentration of each of these ions is equal to the solubility. Hence, $[Pb^{2+}] = [SO_4^{2-}] = 1.25 \times 10^{-4}M$.

The solubilization reaction for $PbSO_4$ is

$$PbSO_4 \ (s) \rightleftarrows Pb^{2+} + SO_4^{2-}$$

and the solubility product constant is

$$K_{sp} = [Pb^{2+}][SO_4^{2-}] .$$

Substituting the values for $[Pb^{2+}]$ and $[SO_4^{2-}]$ gives

$$K_{sp} = (1.25 \times 10^{-4}M) \times (1.25 \times 10^{-4}M)$$
$$= 1.6 \times 10^{-8} \ \text{moles}^2/\text{liter}^2 .$$

● **PROBLEM 395**

A chemist dissolves $PbS$ in water. The $K_{sp}$ is found to be $1 \times 10^{-28}$. What is the molarity of each ion in solution?

<u>Solution</u>:    The key to answering this question is the determination of the equilibrium equation.  Since  PbS  is a salt, it dissociates into ions in solution.  Upon inspection of the oxidation states of the periodic table, the equilibrium will be found to be

$$PbS(s) \rightleftarrows Pb^{2+}(Aq) + S^{2-}(Aq) \ .$$

The presence of the equilibrium derives from the fact that a saturated salt solution's ions will exist in equilibrium with any additional salt.  This additional salt is called the solid phase.  $K_{sp}$  measures this equilibrium and can be expressed as

$$K_{sp} = [Pb^{2+}][S^{2-}] \ .$$

This, equals  $1 \times 10^{-28}$ .  Therefore, you have $1 \times 10^{-28} = [Pb^{2+}][S^{2-}]$. The question asks for the concentrations represented on the right side. From the equilibrium equation, you see that whatever the concentration of  $S^{2-}$, it will be equal to  $Pb^{2+}$ .  Therefore, you can represent both by  x .  As such, you now have

$$1 \times 10^{-28} = x \cdot x$$

or

$$1 \times 10^{-28} = x^2 \ .$$

Solving for  x  you obtain  $1 \times 10^{-14}$M.  This means, therefore, that the concentration of each ion in solution is  $1 \times 10^{-14}$M.

<br>

● **PROBLEM** 396

A chemist has a saturated solution of  $CaSO_4$  in  .0100M  $Na_2SO_4$ . The  $K_{sp}$  of  $CaSO_4$  is  $2.4 \times 10^{-5}$, calculate the concentration of. $Ca^{2+}$  ion in this saturated solution.

<u>Solution</u>:  The  $K_{sp}$  is the solubility constant of a substance.  For the general reaction,  $AB_{(s)} \rightarrow A^+ + B^-$ ,  $K_{sp} = [A^+][B^-]$.  The  $K_{sp} =$ the product of the concentrations of the ions.  In this problem, $K_{sp} = [Ca^{++}][SO_4^=]$ .  You are given that  $K_{sp} = 2.4 \times 10^{-5}$ .  Equating, you obtain  $2.4 \times 10^{-5} = [Ca^{++}][SO_4^=]$ .  To solve this problem, you must determine the value of  $[Ca^{++}]$.  Let  x = the moles/liter of $CaSO_4$ .  Since  $CaSO_4 \rightarrow Ca^{++} + SO_4^=$ , this  x  also is the concentration of  $Ca^{++}$ .  The concentration of  $SO_4^=$, however, has two sources. $SO_4^=$  comes from  $CaSO_4$  AND  $Na_2SO_4$ .  The concentration of  $Na_2SO_4$ was given as  .01.  Therefore, the total concentration of  $SO_4^=$  is x + .01, where  x  represents the amount of  $SO_4^=$  contributed by $CaSO_4$ .  Thus,

$$[x][.01 + x] = 2.4 \times 10^{-5} \ .$$

Solving for  x  you obtain,

$$x = 2.0 \times 10^{-3} = [Ca^{++}] \ .$$

How many grams of $Ca(C_2O_4)$ (calcium oxalate) will dissolve in water to form 1.0 liter of saturated solution? The $K_{sp}$ of $Ca(C_2O_4)$ is $2.5 \times 10^{-9}$ mole$^2$ /liter$^2$ .

Solution: This problem is an application of the solubility product constant $(K_{sp})$ expression for dissociation of a solid. In general, if a solid, A, dissociates into ions B,C,D,... according to the equation

$$A \rightarrow bB + cC + dD + ... ,$$

the ion product constant (I.P.) is given by

$$I.P. = [B]^b [C]^c [D]^d ... .$$

At the point where addition of solid will result in precipitation, the ion product constant is equal to the solubility product constant. (When I.P. > $K_{sp}$, precipitation will occur.) Thus, for a solution just on the verge of precipitation,

$$K_{sp} = [B]^b [C]^c [D]^d ... .$$

The dissociation of calcium oxalate is given by the equation

$$Ca(C_2O_4) \rightleftarrows Ca^{2+} + C_2O_4^{2-}$$

and the $K_{sp}$ is

$$K_{sp} = 2.5 \times 10^{-9} \text{ mole}^2 \text{/liter}^2 = [Ca^{2+}][C_2O_4^{2-}] .$$

Let the concentration of the calcium ion be x. Then, since one $C_2O_4^{2-}$ ion is produced for every $Ca^{2+}$ ion produced, $[Ca^{2+}] = [C_2O_4^{2-}] = x$.

Substituting into the expression for $K_{sp}$ gives

$$[Ca^{2+}][C_2O_4^{2-}] = 2.5 \times 10^{-9} \text{ mole}^2/\text{liter}^2$$

$$(x)(x) = 2.5 \times 10^{-9} \text{ mole}^2/\text{liter}^2$$

$$x = (2.5 \times 10^{-9} \text{ mole}^2/\text{liter}^2)^{\frac{1}{2}}$$

$$x = 5.0 \times 10^{-5} \text{ mole/liter} = 5.0 \times 10^{-5} \text{ M} .$$

Thus, the concentration of $Ca(C_2O_4)$ will be $5.0 \times 10^{-5}$ M. To convert to mass, we use the relationship (mass = concentration $\times$ volume $\times$ molecular weight). The molecular weight of $Ca(C_2O_4)$ is 128g/mole. Hence, the required mass of $Ca(C_2O_4)$ is, for one liter of solution,

$$\text{mass} = \text{concentration} \times \text{volume} \times \text{molecular weight}$$

$$= 5.0 \times 10^{-5} \text{ M} \times 1\ell \times 128\text{g/mole}$$

$$= 0.0064\text{g} .$$

● **PROBLEM** 398

Calculate the concentration of calcium ion and fluoride ion in a concentrated solution of calcium fluoride, $CaF_2$ ($K_{sp} = 4.00 \times 10^{-11}$ ).

Solution: Problems of this type are solved by making use of the fact that the $K_{sp}$, which is known, is equal to the product of the concentrations of $Ca^{++}$ and $F^-$ . If we know the concentration of one ion then, from the proportions in which the ions occur in the molecule, we can determine the concentrations of all the ions formed by dissociation of the molecule.

When calcium fluoride dissociates,

$$CaF_2(s) \rightleftarrows Ca^{2+} + 2F^-,$$

one ion of $Ca^{2+}$ and two ions of $F^-$ form per molecule of $CaF_2$. Hence, if we let $x$ denote the concentration of calcium ions ($[Ca^{2+}] = x$), the concentration of fluoride ions is twice as great, ($[F^-] = 2x$). The expression for the solubility product constant is

$$K_{sp} = [Ca^{2+}][F^-]^2 = 4.00 \times 10^{-11}.$$

Substituting $[Ca^{2+}] = x$ and $[F^-] = 2x$ gives

$$(x)(2x)^2 = 4.00 \times 10^{-11},$$
$$(x)(4x^2) = 4.00 \times 10^{-11},$$
$$4x^3 = 4.00 \times 10^{-11},$$

or,

$$x = \left(\frac{4.00 \times 10^{-11}}{4}\right)^{1/3} = 2.15 \times 10^{-4}$$

Hence,

$$[Ca^{2+}] = x = 2.15 \times 10^{-4} \text{ and } [F^-] = 2x = 2 \times 2.15 \times 10^{-4}$$
$$= 4.3 \times 10^{-4}$$

● **PROBLEM** 399

You are given .01 moles of $AgCl$. Assuming the $K_{eq}$ for $Ag(NH_3)_2^+$ is $1.74 \times 10^7$, how much 1M ammonia solution is required to dissolve this $AgCl$? Assume, also, $K_{sp}$ of $AgCl$ is $2.8 \times 10^{-10}$.

<u>Solution</u>: The first thing to do is to write the overall reaction that occurs when ammonia is mixed with $AgCl$. From this, write an equilibrium expression and equate this expression with the equilibrium constant. Next, calculate the concentrations of species in solution. After this point, there is a volume determination. Following this procedure: when $AgCl$ is dissolved in a solution, it dissociates into ions, since it is a salt. The reaction is $AgCl$ (s) $\rightleftarrows Ag^+ + Cl^-$. The $K_{sp} = 2.8 \times 10^{-10}$. The silver ion, $Ag^+$, produced will react with **ammonia** ($NH_3$) to produce the silver ammonia complex, $Ag(NH_3)_2^+$ via the reaction $Ag^+ + 2NH_3 \rightleftarrows Ag(NH_3)_2^+$ with $K = 1.74 \times 10^7$. The overall reaction is the sum of these two. The $K$ for this will be the product of the $K$'s for the two reactions just written. Thus, overall reaction is:

$$AgCl_{(s)} + 2NH_3 \rightleftarrows Ag(NH_3)_2^+ + Cl^-$$

with

$$K = (2.8 \times 10^{-10})(1.74 \times 10^7) = 4.9 \times 10^{-3}.$$

The equilibrium expression indicates the ratio of products to reactants, each raised to the power of their coefficients in the overall reaction. Thus,

$$K = 4.9 \times 10^{-3} = \frac{[Ag(NH_3)_2^+][Cl^-]}{[NH_3]^2}.$$

One must now find these concentrations. Let $x = [Ag(NH_3)_2^+]$, thus $x = [Cl^-]$, since the overall reaction indicates they are formed in equimolar amounts. Since 2 moles of $NH_3$ is consumed per mole of complex, $[NH_3] = 1 - 2x$, where 1 is the original molarity of the ammonia solution. Substituting and solving for $x$ ,

$$4.9 \times 10^{-3} = \frac{x \cdot x}{1 - 2x} \ ;$$

thus,

$$x = .065M.$$

From the initial .01 moles of AgCl, .01 moles of $Cl^-$ must be produced. This concentration of $Cl^-$ is .065M. Since M = molarity = moles/volume, the volume required to dissolve the AgCl is

$$\frac{.01 \text{ moles}}{.065 \text{ moles/liter}} = .154 \text{ liters} = 154 \text{ ml.}$$

● **PROBLEM** 400

Calculate the minimum concentration of $Br^-$ ion necessary to bring about precipitation of AgBr from a solution in which the concentration of $Ag^+$ ion is $1 \times 10^{-5}$ mole per liter.

$$K_{sp} \text{ for AgBr} = 4 \times 10^{-13} .$$

Solution: The product of the concentrations of ions in a saturated solution of a relatively insoluble salt, such as AgBr, at a given temperature is constant. This constant is called the solubility product constant ($K_{sp}$). For AgBr, $K_{sp}$ can be stated as

$$K_{sp} = [Ag^+][Br^-] = 4 \times 10^{-13} .$$

In this problem, we are given the concentration of $Ag^+$ as $1 \times 10^{-5}$M, therefore $[Br^-]$ can be found by using the equation for $K_{sp}$ .

$$[Ag^+][Br^-] = 4 \times 10^{-13}$$
$$[1 \times 10^{-5}][Br^-] = 4 \times 10^{-13}$$
$$[Br^-] = \frac{4 \times 10^{-13}}{1 \times 10^{-5}} = 4 \times 10^{-8}M .$$

If the concentration of $Br^-$ ions is raised to $4 \times 10^{-8}$M, AgBr will precipitate out, for the addition of any more $Br^-$ ions could not be supported by the solubility of the ions in solution, as the $K_{sp}$ indicates.

● **PROBLEM** 401

If you mix 50ml of $5.0 \times 10^{-4}$M $Ca(NO_3)_2$ and 50ml of $2.0 \times 10^{-4}$M NaF to give 100ml of solution, will precipitation occur? The $K_{sp}$ of $CaF_2$ is $1.7 \times 10^{-10}$ .

Solution: Whether or not precipitation will occur, when these two solutions are mixed, depends upon the ion product. If two solutions containing the ions of a salt are mixed, and if the ion product exceeds the $K_{sp}$, then precipitation will occur.

You need to determine the concentrations of $Ca^{2+}$ and $F^-$ and to see if the product exceeds the given $K_{sp}$ of $1.7 \times 10^{-10}$ for $CaF_2$. The equation for this reaction is $CaF_2(s) \rightleftarrows Ca^{2+} + 2F^-$. Thus, the $K_{sp} = [Ca^{2+}][F^-]^2$. You are told that a 50ml solution of $Ca(NO_3)_2$ has a molarity of $5 \times 10^{-4}$M. Each $Ca(NO_3)_2$ dissociates to yield one $Ca^{++}$ ion which means the molarity of $Ca^{++}$ is also $5 \times 10^{-4}$M. However, it is $5 \times 10^{-4}$M in 50ml, and the mixture is 100ml. Therefore, the molarity must be divided by two. The molarity of NaF, $2.0 \times 10^{-4}$, and therefore $Na^+$, because of its dissociation, must also be divided by two for the same reasons. As such, in the mixture,

$$[Ca^{2+}] = 2.5 \times 10^{-4}M$$

and

$$[F^-] = 1.0 \times 10^{-4}M.$$

Recalling that $K_{sp} = [Ca^{2+}][F^-]^2$, you have

$$K_{sp} = (2.5 \times 10^{-4})(1.0 \times 10^{-4})^2 = 2.5 \times 10^{-12}.$$

You were given that the $K_{sp}$ for a saturated solution of $CaF_2$ is $1.7 \times 10^{-10}$. Since $2.5 \times 10^{-12}$ is less than $1.7 \times 10^{-10}$, the mixture is unsaturated, and precipitation should NOT occur.

● **PROBLEM** 402

Will precipitation occur if .01 mole of $Ba^{+2}$ is added to a liter of solution containing .05 mole of $SO_4^{-2}$ ? $K_{sp}$ for $BaSO_4 = 1 \times 10^{-10}$.

Solution: $K_{sp}$ is the solubility product constant. It is defined as the minimum product of the concentration of the ions needed to bring about the formation of a precipitate (i.e., a solid compound). The general form of the $K_{sp}$ equation for compound $A_n B_m$ is:

$$K_{sp} = [A]^n[B]^m$$

Therefore, for $BaSO_4$ the $K_{sp}$ equation is:

$$K_{sp} = [Ba^{+2}][SO_4^{-2}] = 1 \times 10^{-10}.$$

To find out if precipitation occurs, substitute in the given concentration values. If the calculated value is greater than the given value for $K_{sp}$, then precipitation occurs. If it is lower, then no precipitation occurs. Thus,

$$K_{sp} = [Ba^{+2}][SO_4^{-2}]$$

$$= [1 \times 10^{-2}][5 \times 10^{-2}]$$

$$= 5 \times 10^{-4}.$$

Since this value is greater than the given $K_{sp}$ value, precipitation occurs.

A chemist has a solution which contains $Zn^{2+}$ and $Cu^{2+}$, each at .02M. If the solution is made 1M in $H_3O^+$ and $H_2S$ is bubbled in until the solution is saturated, will a precipitate form? $K_{sp}$ of ZnS = $1 \times 10^{-22}$, $K_{sp}$ of CuS = $8 \times 10^{-37}$, and $K_{eq}$ for $H_2S$; when saturated, = $1 \times 10^{-22}$.

Solution: A precipitate will form only when the ion product exceeds the experimentally determined solubility product constant $(K_{sp})$. Therefore, compute the ion product of ZnS and CuS to determine if they exceed the stated $K_{sp}$'s, $1 \times 10^{-22}$ and $8 \times 10^{-37}$, respectively. The equation for the reaction between $H_2S$ and $H_2O$ is

$$H_2S + 2H_2O \rightleftharpoons 2H_3O^+ + S^-.$$

In a saturated solution, the ion product is equal to

$$[H_3O^+]^2[S^{2-}] = 1 \times 10^{-22} = K.$$

You are told that the $H_3O^+$ concentration is 1M. Substituting this into that expression, and solving for $[S^{2-}]$, you obtain

$$[S^{2-}] = 1 \times 10^{-22}M.$$

You are given the $Zn^{2+}$ and $Cu^{2+}$ concentrations, which means that the ion products can be computed. For $Zn^{+2}$,

$$[Zn^{2+}][S^{2-}] = (.02)(1 \times 10^{-22}) = 2 \times 10^{-24}$$

$$= K_{sp} \text{ for ZnS.}$$

$$[Cu^{2+}][S^{2-}] = (.02)(1 \times 10^{-22}) = 2 \times 10^{-24}$$

$$= K_{sp} \text{ for CuS.}$$

Since the ion product of ZnS does not exceed the $K_{sp}$ of ZnS, $1 \times 10^{-22}$, there will be no precipitation. However, the ion product of CuS does exceed the $K_{sp}$ of CuS, $8 \times 10^{-37}$, which means you will see precipitation of CuS.

A chemist mixes equal volumes of .01M $Na_2C_2O_4$ and .001M $BaCl_2$ together. Assuming the $K_{sp}$ of $BaC_2O_4$ is $1.2 \times 10^{-7}$, will a precipitate form?

Solution: A precipitate will only form when the ion concentration product for the particular substances exceed their given $K_{sp}$. In other words, a precipitate will form if $[Ba^{2+}][C_2O_4^{2-}]$ is greater than the $K_{sp} = 1.2 \times 10^{-7}$. Thus, to obtain this ion product, calculate the concentrations of $Ba^{2+}$ and $C_2O_4^{2-}$ in the mixture. One is given that the initial concentrations of $C_2O_4^{2-}$ and $Ba^{2+}$ in the solution are .01M $Na_2C_2O_4$ and .001M $BaCl_2$. When they are mixed, however, the volume doubles. Thus, in the mixture, the concentrations of $BaCl_2$ and $Na_2C_2O_4$ are halved. (Remember, M = moles/liter , so that if the volume increases, M = concentration decreases). Therefore, after mixing, the concentrations are .005M of $Na_2C_2O_4$ and .0005M of $BaCl_2$ . As such, the ion product of

$$[Ba^{2+}][C_2O_4^{2-}] = (.0005)(.005) = 2.5 \times 10^{-6} .$$

$2.5 \times 10^{-6}$ is greater than the $K_{sp}$ of $1.2 \times 10^{-7}$, which indicates that a precipitate will form.

● **PROBLEM** 405

0.01 liter of 0.3M $Na_2SO_4$ is mixed with 0.02 liters of a solution that is initially 0.1M $Ca^{++}$ and 0.1M $Sr^{++}$ . Given that $K_{sp} = 2.4 \times 10^{-5}$ for $CaSO_4$ and $K_{sp} = 7.6 \times 10^{-7}$ for $SrSO_4$ , what is the final concentration of $Ca^{++}$, $Sr^{++}$, and $SO_4^{--}$ in this solution?

Solution: Solve this problem in four steps:
(A)    Find the relationships involving the final concentrations of $Ca^{++}$, $SO_4^{--}$ , and $Sr^{++}$ .

(B)    Find the initial concentrations of the three species.
(C)    Relate the final concentrations to the initial concentrations.

(D)    Solve for $[Ca^{++}]$, $[SO_4^{--}]$, and $[Sr^{++}]$ .

(A) $K_{sp}$, the solubility product constant, equals the product of the concentrations of the ions in a saturated solution. Thus, $K_{sp}$ $CaSO_4 = [Ca^{++}][SO_4^{--}]$ and $Ksp$ $SrSO_4 = [Sr^{++}][SO_4^{--}]$ . Combine this with the given:
    (i)    $[Ca^{++}][SO_4^{--}] = 2.4 \times 10^{-5}$
    (ii)    $[Sr^{++}][SO_4^{--}] = 7.6 \times 10^{-7}$ .

Equation (i) relates $[Ca^{++}]$ and $[SO_4^{--}]$; equation (ii) relates $[Sr^{++}]$ and $[SO_4^{--}]$. To find an equation relating $[Ca^{++}]$ and $[Sr^{++}]$, solve (i) for $[SO_4^{--}]$:

$$[SO_4^{--}] = \frac{2.4 \times 10^{-5}}{[Ca^{++}]} .$$

Substitute this into (ii); and solve for $[Sr^{++}]$.

    (iii)    $[Sr^{++}] = \frac{7.6 \times 10^{-7}}{2.4 \times 10^{-5}} [Ca^{++}] = 0.032 [Ca^{++}]$ .

(B) 0.01 liters of 0.3M $Na_2SO_4$ was initially mixed. The number of moles of $Na_2SO_4$ present is, therefore, (molarity)(volume) = (0.3)(0.01) = 0.003 moles of $Na_2SO_4$ . Since the salt dissociates completely, there are 0.003 moles of $SO_4^{--}$ in the solution. The new volume is (0.02 + 0.01) liters = $0.03\ell$ , after adding the $Ca^{++}$ and $Sr^{++}$ solution. Thus, the initial molarity of $SO_4^{--}$ is:

$$\frac{0.003 \text{ moles}}{0.03 \text{ liter}} = 0.1 \text{ M.}$$

By similar analysis, it follows that the initial concentration of $Sr^{++}$ = initial concentration of $Ca^{++}$ = 0.0667M .

(C)  Let the final concentration of $Ca^{++}$ equal x. Then the final concentration of $Sr^{++}$ , according to equation (iii), must be 0.032 $[Ca^{++}]$ = .032. The change in $Ca^{++}$ concentration is 0.0667 - x; the change in $Sr^{++}$ is 0.0667 - 0.032 x.

To find the final $[SO_4^{--}]$, note that one $SO_4^{--}$ is lost for each $Ca^{++}$ that reacts and one $SO_4^{--}$ is lost for each $Sr^{++}$ that reacts. Thus the change in $[SO_4^{--}]$ is the sum of the changes in $[Ca^{++}]$ and $[Sr^{++}]$. Thus, change in $[SO_4^{--}]$ = (0.0667 - x) + (0.0667 - .032x) = (0.1334 - 1.032x). Since the final $[SO_4^{--}]$ = initial $[SO_4^{--}]$ minus change in $[SO_4^{--}]$, the final $[SO_4^{--}]$ equal 0.10 - (0.1334 - 1.032x) or

(iv)  $[SO_4^{--}]$ = 1.032x - 0.0334

(D)  Substitute the final values for $[Ca^{++}]$ and $[SO_4^{--}]$ in equation (i) and solve for x:

(v)  $[Ca^{++}][SO_4^{--}] = 2.9 \times 10^{-5}$

(x)  $(1.032x - 0.0334) = 2.4 \times 10^{-5}$

(vi)  $1.032x^2 - 0.0334x - 2.4 \times 10^{-5} = 0$ .

Use the quadratic formuls

(vii)  $x = \dfrac{0.0334 \pm \sqrt{(0.0334)^2 - 4(-2.4 \times 10^{-5})(1.032)}}{2(1.032)}$

Thus, x = $3.30 \times 10^{-2}$M = $[Ca^{++}]$. Substitute this value in equation (iii) to obtain

(viii)  $[Sr^{++}]$ = .032 $[Ca^{++}]$ = .032(3.30 $\times 10^{-2}$)
        = $1.045 \times 10^{-3}$M .

Use this value in equation (ii)

(ix)  $[SO_4^{--}] = \dfrac{7.6 \times 10^{-7}}{[Sr^{++}]} = \dfrac{7.6 \times 10^{-7}}{1.045 \times 10^{-3}} = 7.15 \times 10^{-4}$M .

● **PROBLEM** 406

The solubility products of $Fe(OH)_2$ and $Fe(OH)_3$ are $10^{-17}$ and $10^{-38}$, respectively. If the concentrations of $Fe^{2+}$ and $Fe^{3+}$ are each $10^{-5}$ M, at what pH will each hydroxide just begin to precipitate?

**Solution**: To solve this problem, set up the solubility product equations for $Fe(OH)_2$ and $Fe(OH)_3$. Once this is done, substitute the values of $Fe^{2+}$ and $Fe^{3+}$ concentrations and solve for the $OH^-$ ion concentrations. Since $pH + pOH = 14$, then $pH = 14 - pOH$, where $pOH = -\log[OH^-]$. One has, therefore,

$$[Fe^{2+}][OH^-]^2 = 10^{-17} \text{ and}$$

$$[Fe^{3+}][OH^-]^3 = 10^{-33} .$$

For precipitation to occur in these solutions $[Fe^{2+}][OH^-]^2 \geq 10^{-17}$ and $[Fe^{3+}][OH^-]^3 \geq 10^{-33}$. When the product of the concentrations is equal to zero the concentrations are just sufficient for precipitation to occur. Solving for the pH at these concentrations:
For $Fe(OH^-)_2$,

$$[OH^-]^2 = \frac{10^{-17}}{[Fe^{2+}]} = \frac{10^{-17}}{10^{-5}} = 10^{-12} ,$$

so that

$$[OH^-] = 10^{-6}$$

therefore, $pOH = 6$ and thus $pH = 8$.
For $Fe(OH)_3$,

$$[OH^-]^3 = \frac{10^{-38}}{[Fe^{3+}]} = \frac{10^{-38}}{10^{-5}} = 10^{-33} ,$$

so that

$$[OH^-] = 10^{-11} .$$

Consequently,

$$pOH = 11 \text{ and } pH = 3 .$$

● **PROBLEM** 407

Calculate the solubility of carbon dioxide in water at $25^\circ C$, where the partial pressure of $CO_2$ over the solution is 760 torr., using Henry's law constant, $K_{CO_2}$ at $25^\circ C = 1.25 \times 10^6$. Assume that a liter of solution contains 1000 grams of water.

**Solution**: Raoult's law states that the partial pressure, $P_2$, of the component present at lower concentration is directly proportional to its mole fraction, $X_2$, for dilute solutions. Those that don't usually obey Henry's law,

$$P_2 = X_2 K_2 .$$

The subscript 2 indicates that the solute (the component at lower concentration) is being considered. The constant $K_2$ is referred to as Henry's law constant. For nonideal solutions, Henry's law holds for the solute in the same range where Raoult's law holds for the solvent. For ideal solutions $K_2 = P_2^0$ and Henry's law becomes identical with Raoult's law.

To solve this problem one must know three things:
1) the value of $K_{CO_2}$, equal to $1.25 \times 10^6$;

2) the partial pressure $P_2$, equal to 760 torr;

3) the mole fraction of $CO_2$ in water, which has to be determined.

The mole fraction of $CO_2$ in water is

$$X_{CO_2} = \frac{\text{number of moles of } CO_2}{\text{number of moles of } CO_2 + \text{number of moles of } H_2O}$$

The number of moles of $H_2O$ is $\frac{1000g}{18.02g/mole}$ = 55.49 moles $H_2O$.
The number of moles of $CO_2$ has yet to be determined. Thus substituting into Henry's law

$$X_2 = \frac{P_2}{K_2} = \frac{760 \text{ torr}}{1.25 \times 10^6} = \frac{(\text{moles } CO_2)}{(\text{moles } CO_2) + (55.49 \text{ moles } H_2O)} \, .$$

However, since the number of moles of $CO_2$ may be considered negligible in comparison with the number of moles of water, then,

$$\frac{760 \text{ torr}}{1.25 \times 10^6} = \frac{\text{moles } CO_2}{55.49 \text{ moles } H_2O} \, .$$

Rewriting and solving

$$\text{no. of moles of } CO_2 = \frac{(55.49)(760)}{1.25 \times 10^6} = 3.37 \times 10^{-2}$$

There are 1000g of $H_2O$, or 1 liter of solvent, and therefore, the solubility of

$$CO_2 = \frac{3.37 \times 10^{-2} \text{ moles}}{1 \text{ liter}} = 3.37 \times 10^{-2} \text{ M} \, .$$

● **PROBLEM** 408

What is the equilibrium concentration of oxygen gas in fresh water at 25°C when exposed to air at a pressure of 760 torr? The Henry's law constant for oxygen in water at 25°C is 3.30 x 10[7] torrs. Express your answer in milligrams of oxygen per liter.

Solution: Solving this problem requires the use of Henry's law. It is stated

$$P_2 = X_2 K_2 \, ,$$

where $P_2$ is the partial pressure, $X_2$ is the mole fraction, and $K_2$ is the Henry's law constant. The subscript, 2, indicates the solute (or component at lower concentration).
Since $P_2$ = 760 torr and $K_2$ = 3.30 x 10[7] torrs are known, we can solve for $X_2$ or the mole fraction of $O_2$ in water. Thus,

$$X_2 = \frac{P_2}{K_2} = \frac{760 \text{ torr}}{3.3 \times 10^7 \text{ torr}} = \frac{\text{moles of } O_2}{(\text{moles of } O_2) + (\text{moles of } H_2O)}$$

The volume is assume to be 1 liter or 1000 grams of $H_2O$. The number or moles of $H_2O$ is $\frac{1000g}{18.02g/mole}$ = 55.49 moles $H_2O$. In addition, the number of moles of $O_2$ is negligible in comparison with the number of moles of $H_2O$, thus the number of moles of $H_2O$ is

$$\frac{(760 \text{ torr})(55.49 \text{ moles H}_2\text{O})}{3.3 \times 10^7 \text{ torr}} = 1.28 \times 10^{-3} \text{ moles O}_2 \, .$$

The weight in milligrams of $O_2$ is

$$(1.28 \times 10^{-3} \text{ moles})(32\text{g/mole})(10^3\text{mg/g}) = 41 \text{ mg}.$$

Since there is 1 liter of $H_2O$, then the concentration of $O_2$ is $41\text{mg/}\ell$ .

● **PROBLEM** 409

The solubility of iodine in pure water is 0.0279g per 100g of water. In a solution that is originally 0.10M KI, it is possible to dissolve, at the maximum, 1.14g of iodine per 100g of solution. What is K for the reaction $I_2 + I^- \rightleftharpoons I_3^-$ ?

Solution: The concentration of iodine in the given solution is greater than can be dissolved. However, the wording of the problem indicates that there is no precipitate. Thus, the excess iodine must be used up in the production of $I_3^-$ which is soluble. With this in mind, we can solve for K .

The equilibrium constant K can be solved from the following equation,

(i)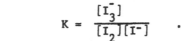
$$K = \frac{[I_3^-]}{[I_2][I^-]} \, .$$

First, solve for the concentrations of $I_3^-$ , $I_2$ , and $I^-$ . The $I^-$ species comes entirely from the dissociation of KI .

Since a salt such as KI totally dissociates in aqueous solution, initial molarity of $I^- $ = molarity of KI = 0.10M .

The initial amount of $I_2$ added is 1.14g per 100g of water or the number of moles added equals

$$\frac{\text{grams of } I_2}{\text{mol.wt. of } I_2} = \frac{1.14\text{g}}{254\text{g/mole}} = 4.49 \times 10^{-3} \text{ moles}$$

per 100g of water. However, the water can only hold 0.0279g of iodine per 100g of water, or

$$\frac{.0279\text{g}}{254\text{g/mole}} = 1.1 \times 10^{-4} \text{ moles per 100g of water.}$$

Since there is to be no precipitate, the excess $I_2$ must be used up in the production of $I_3^-$ . Since one mole $I_2$ yields one mole of $I_3^-$ , we have for the final concentrations:

(1)  concentration of $I_2$ = maximum concentration of $I_2$ that can be dissolved = $1.1 \times 10^{-4}$ moles per 100g of water. Since 1000g of $H_2O = 1\ell$ , $1.1 \times 10^{-4}$ moles/100g $H_2O = \frac{1.1 \times 10^{-3} \text{ moles}}{1000\text{g H}_2\text{O}}$ =

$$\frac{1.1 \times 10^{-3} \text{ moles}}{1 \text{ liter } H_2O}$$ . Thus, $[I_2] = 1.1 \times 10^{-3} M$ .

(2)   concentration of $I_3^-$ = concentration of $I_2$ reacted. Since there were originally $4.49 \times 10^{-3}$ moles of $I_2/100g\ H_2O$ and the final concentration of $I_2$ is $1.1 \times 10^{-4}$ moles/100g $H_2O$, the concentration of $I_2$ that reacted is the difference

$$\frac{4.49 \times 10^{-3} \text{ moles}}{100g\ H_2O} - \frac{1.1 \times 10^{-4} \text{ moles}}{100g\ H_2O} = \frac{4.38 \times 10^{-3} \text{ moles}}{100g\ H_2O}$$

$$= \frac{4.38 \times 10^{-2} \text{ moles}}{1 \text{ liter } H_2O}$$ . Thus, $[I_3^-] = 4.38 \times 10^{-2}\ M$ .

(3)   concentration of $I^-$ = initial concentration $I^-$ minus (concentration of $I^-$ needed to reacted with the $I_2$). From (2), the concentration of reacted $I_2$ was found to be $4.38 \times 10^{-2} M$. Since one mole of $I^-$ is required to react with each mole of $I_2$ , the concentration of $I^-$ needed to reacted with the $I_2$ is also $4.38 \times 10^{-2} M$ .

Thus, $[I^-] = 0.10M - 4.38 \times 10^{-2} M = .0562M$.

To find K, substitute these values in equation (i):

$$K = \frac{(0.0438)}{(0.0011)(0.0562)} = 710 .$$

# CHAPTER 12

# CALCULATIONS USING pH AND DISSOCIATION CONSTANT

Basic Attacks and Strategies for Solving Problems in this Chapter. See pages 385 to 421 for step-by-step solutions to problems.

The principles of equilibrium that have been discussed in the preceding chapters (Chapters 9, 10, and 11) can be applied directly to solve the problems in this chapter. The dissociation constant is simply the equilibrium constant for a dissociation reaction. When the $H^+$ ion concentration is determined from an equilibrium calculation, the pH can be calculated directly from Equation 10-3.

$$pH = -\log_{10}(H^+) \qquad \text{12-1 (also 10-3)}$$

There are also problems in Chapter 12 that ask for values of pK, $pK_a$, $pK_b$, etc. The key to calculating these quantities lies in the definition of pK, which is analogous to the definition of pH.

$$pK_i = -\log_{10}(K_i) \qquad \text{12-2}$$

Dissociation is the ionization or dissolution of an electrolyte and is applied most often to weak electrolytes such as organic acids. Several of the problems in this chapter involve calculation of the concentration of $H^+$ ions resulting from the dissociation of acetic acid. The equilibrium dissociation of acetic acid was outlined in some detail in Equations 10-10 through 10-12 when discussing buffered solutions. Another example of acetic acid dissociation without buffering will illustrate the essential technique for solving such problems. Let us solve for the pH in a 0.1 molar solution of acetic acid.

$$HAc \leftrightarrow H^+ + Ac^- \qquad \text{12-3}$$

$$K_a = 1.8 \times 10^{-5} = (H^+)(Ac^-)/(HAc) \qquad \text{12-4}$$

The key to solving this problem lies in making two observations. First, the stoichiometry of Equation 12-3 dictates that, in the absence of other sources of $H^+$ and $Ac^-$ ions, the concentrations of these two ions will be equal. Second, since the equilibrium constant is a small number, the concentration of undissociated HAc will not be appreciably reduced by the dissociation reaction. We assume that the concentration, (HAc), is equal to the total HAc before dissociation. (We will rework this problem without making this assumption to see if the assumption is valid.)

The concentrations of $H^+$ and $Ac^-$ can be assigned a new variable, $x$. Equation 12-4 becomes

$$1.8 \times 10^{-5} = (x)(x)/(0.1) = 10x^2 \qquad \text{12-5}$$

$$x = (1.8 \times 10^{-6})^{1/2} = 0.00134 = (H^+) = (Ac^-) \qquad \text{12-6}$$

$$pH = -\log_{10}(.00134) = 2.87 \qquad \text{12-7}$$

Note that the pH would be 1 $[-\log(0.1)]$ had dissociation been complete.

Now let us illustrate the same calculation **without** making the assumption that the concentration of undissociated HAc will not be appreciably reduced by the dissociation reaction. In this case, if $x = (H^+) = (Ac^-)$, the concentration, $(HAc) = (0.1 - x)$.

$$1.8 \times 10^{-5} = (x)(x)/(0.1 - x) \qquad \text{12-8}$$

This is a quadratic equation. When arranged in the standard quadratic form, $ax^2 + bx + c = 0$, it appears as follows.

$$x^2 + (1.8 \times 10^{-5}) x - 1.8 \times 10^{-6} = 0 \qquad \text{12-9}$$

$$x = \{-(1.8 \times 10^{-5}) + [(1.8 \times 10^{-5})^2 - 4 (1)(-1.8 \times 10^{-6})]^{1/2}\}/2 \qquad \text{12-10}$$

$$(H^+) = 0.00133 : pH = 2.87 \qquad \text{12-11}$$

The obvious conclusion is that the simplifying assumption is justified.

Step-by-Step Solutions to
Problems in this Chapter,
"Calculations Using pH and the
Dissociation Constant"

Find $K_b$ and $pK_b$ for the acetate ion $CH_3COO^-$. The ionization constant of $CH_3COOH$ is $K_a = 1.75 \times 10^{-5}$; $K_w = 1.00 \times 10^{-14}$.

<u>Solution</u>: This problem is concerned with conjugate acid-base pairs. In the general reaction,

$$\underset{\text{acid}}{HA} + \underset{\text{base}}{H_2O} \rightleftharpoons \underset{\text{acid}}{H_3O^+} + \underset{\text{base}}{A^-},$$

HA is the conjugate acid of $A^-$, or conversely, $A^-$ is the conjugate base of HA. $H_2O$ has acted as a weak base. The acid and base ionization constants for a conjugate pair, as illustrated above, are $K_a$ and $K_b$.

$K_a K_b = K_w$ which means $pK_a + pK_b = pK_w$.

Thus, if given the ionization constant for an acid or base, one can calculate the ionization constant for its conjugate base or acid. Also, remember that the K and pK values are related by the general equation,

pK = - log K.

Therefore, the values of $K_b$ and $pK_b$ are

$$K_b = \frac{K_w}{K_a} = \frac{1.00 \times 10^{-14}}{1.75 \times 10^{-5}} = 5.71 \times 10^{-10}.$$

$$pK_b = - \log K_b = - \log 5.71 \times 10^{-10}$$

$$= - [- 10 + \log 5.71] = 9.244.$$

A chemist mixes .5 moles of acetic acid $(HC_2H_3O_2)$ and .5 moles of HCN with enough water to make a one liter solution. Calculate the final concentrations of $H_3O^+$ and $OH^-$. Assume the following constants:

$$K_{HC_2H_3O_2} = 1.8 \times 10^{-5}, \quad K_{HCN} = 4 \times 10^{-10}, \quad K_w = 1.0 \times 10^{-14}.$$

<u>Solution</u>: To solve this problem, note the simultaneous equilibria. Once this is recognized, you can set up equilibrium constant expressions which measure these equilibria in terms of the concentrations of the species involved.

There are three simultaneous equilibria in the final solution. Two derive from the acids HCN and $HC_2H_3O_2$ donating their protons to form hydronium ions in water. The third stems from the ionization of water to hydronium and hydroxyl ions. You have:

(1) $HC_2H_3O_2 + H_2O \rightleftarrows H_3O^+ + C_2H_3O_2^-$

(2) $HCN + H_2O \rightleftarrows H_3O^+ + CN^-$

(3) $H_2O + H_2O \rightleftarrows H_3O^+ + OH^-$.

Each reaction contributes $H_3O^+$ ions. However, only the acetic acid produces an appreciable concentration of $H_3O^+$. This can be determined by inspection of the dissociation constants. The larger a dissociation constant, the greater the dissociation of the species. The dissociation constant of acetic acid, $1.8 \times 10^{-5}$, is the largest, and, therefore, its $[H_3O^+]$ contribution is the greatest. As such, you can neglect the $H_3O^+$ contribution from HCN and water. Let the $H_3O^+$ contribution from acetic acid = x. If you started with a concentration of .5M for acetic acid, then, at equilibrium, it becomes .5 - x. From the dissociation of acetic acid in water, it also becomes evident that the concentration of $C_2H_3O_2^-$ can be represented by x, since the equation states $H_3O^+$ and $C_2H_3O_2^-$ will be produced in equimolar quantities. You have

$$HC_2H_3O_2 + H_2O \rightleftarrows H_3O^+ + C_2H_3O_2^-,$$

thus,

$$K_{HOAc} = \frac{[H_3O^+][C_2H_3O_2^-]}{[HC_2H_3O_2]} = \frac{x \cdot x}{.5-x} = 1.8 \times 10^{-5}.$$

Solving for x, using the quadratic equation, you obtain
$$x = 3.0 \times 10^{-3}M = [H_3O^+].$$
To find $[OH^-]$, use the fact that $K_w = [OH^-][H_3O^+] = 1 \times 10^{-14}$. Since you know $[H_3O^+]$, you can substitute to find $[OH^-]$. Thus, $[OH^-] = 3.3 \times 10^{-12}M$.

Pentobarbital, sometimes prescribed for the treatment of chronic insomnia under the trade name Nembutal, has the following structure:

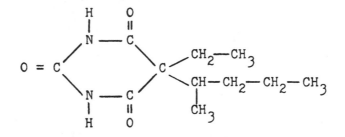

pentobarbital

It is toxic in doses of 0.5 g and lethal in doses of 1.5 g. Barbituric acid, denoted by H Bar, and which may be used in the synthesis of pentobarbital, has the following structure:

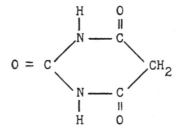

barbituric acid

H Bar dissociates according to the following equation:

H Bar $\rightleftharpoons$ H$^+$ + Bar$^-$.

It is a weak acid, having a $pK_a$ of only 4.01. Approximately what percent of H Bar molecules are dissociated in a 0.10 N solution?

Solution: We will solve this problem by determining the concentrations of H$^+$ and Bar$^-$ formed from the dissociation of H Bar.

If we let x denote the equilibrium concentration of H$^+$, then, since one mole of Bar$^-$ is formed per mole of H$^+$ formed, the equilibrium concentration of Bar$^-$ is also x. Furthermore, the equilibrium concentration of H Bar is 0.10 - x. Then

$$K_a = \frac{[H^+][Bar^-]}{[H\ Bar]} = \frac{x^2}{0.10 - x}$$

Note that we have neglected the contribution to $[H^+]$ from the dissociation of water. Assuming that x is much smaller than 0.10 (a justifiable assumption, since H bar is so weak an acid), the $0.10 - x \overset{\sim}{=} 0.10$ and we have

$$K_a = \frac{x^2}{0.10 - x} \overset{\sim}{=} \frac{x^2}{0.10}$$

or, $\quad x^2 = 0.10 \times K_a$

$$x = (0.10 \times K_a)^{\frac{1}{2}}.$$

The value of $K_a$ is determined from the $pK_a$. By definition, $pK_a = - \log K_a$. Hence $K_a = 10^{-pK_a} = 10^{-4.01}$, and

$$x = (0.10 \times K_a)^{\frac{1}{2}} = (0.10 \times 10^{-4.01})^{\frac{1}{2}} = (10^{-5.01})^{\frac{1}{2}}$$

$$= 10^{-2.50}$$

$$\overset{\sim}{=} 3.2 \times 10^{-3}.$$

The equilibrium concentrations of $H^+$, $Bar^-$, and H Bar are then $[H^+] = x = 3.2 \times 10^{-3}$ M, $[Bar^-] = x = 3.2 \times 10^{-3}$ M, and $[H\ Bar] = 0.10 - x = 0.10 - 3.2 \times 10^{-3} \overset{\sim}{=} 0.10$ M.

The percent dissociation of H Bar is given by

$$\%\ dissociation = \frac{[H^+]}{[H\ Bar]} \times 100\ \% = \frac{[Bar^-]}{[H\ Bar]} \times 100\ \%$$

$$= \frac{3.2 \times 10^{-3}\ M}{0.10\ M} \times 100\ \%$$

$$\overset{\sim}{=} 3\ \%.$$

● **PROBLEM** 413

Assuming $K_w = 1.0 \times 10^{-14}$ and $K_a = 4.0 \times 10^{-10}$, find the pH of a 1.0 M NaCN solution.

<u>Solution</u>: NaCN is a salt that exists in water as $Na^+$ and $CN^-$. As such, this problem requires considering the hydrolysis of a salt of a weak acid and strong base. Since water is a weak acid or weak base (which dissociates to $H^+$ and $OH^-$ ions), it reacts with ions from either weak acids or bases.

$$CN^- + H_2O \rightleftarrows HCN + OH^-.$$

Inherent in this equation is the ionization of water, that is, $H_2O \rightleftharpoons HO^- + H^+$ so that it is actually,

$CN^- + OH^- + H^+ \rightleftharpoons HCN + OH^-$.

The hydrolysis of the NaCN salt can be expressed by $K_{hyd}$, a hydrolysis constant. $K_{hyd} = K_w/K_a$, where $K_a$ is the ionization of the acid formed and $K_w$ is the ionization constant of water to $OH^- + H^+$, as previously mentioned.

$$K_{hyd} = \frac{K_w}{K_a} = \frac{[HCN][OH^-]}{[CN^-]} = \frac{1.0 \times 10^{-14}}{4.0 \times 10^{-10}}.$$

It is given that $K_a = 4.0 \times 10^{-10}$. $K_w$ is always $1.0 \times 10^{-14}$, since a pure water solution must be neutral. As such, the molar concentrations of $H^+$ and $OH^-$ are both $1.0 \times 10^{-7}$ moles/liter. This stems from the fact that $[H^+] = [OH^-]$ with $[H^+][OH^-] = K_w = 1.0 \times 10^{-14}$. Therefore,

$$[H^+]^2 = 1.0 \times 10^{-14} \quad \text{or} \quad [H^+] = 1.0 \times 10^{-7}.$$

One now has:

$$\frac{1.0 \times 10^{-14}}{4.0 \times 10^{-10}} = \frac{[HCN][OH^-]}{[CN^-]}$$

The object is to calculate $[OH^-]$. From $[OH^-]$, one can determine pOH, since pOH = $-$ log $[OH^-]$. Once one knows pOH, one can find pH, since 14 = pH + pOH. To find $[OH^-]$, let the concentration of $OH^-$ be x. If this is the case, then $[HCN]$ can also be represented by x. Recall, that $CN^- + H_2O \rightleftharpoons HCN + OH^-$. For every mole of $OH^-$ generated, a mole of HCN must also be produced. The concentration of $[CN^-]$ can be represented as $1 - x$. It is given that the original concentration of NaCN or $CN^-$, is 1.0 M. When placed in water, some of $CN^-$ becomes HCN. If HCN is x, then the original concentration of $CN^-$ must be reduced by x. Therefore, $[CN^-]$ has a concentration of $1 - x$. It follows that

$$\frac{1.0 \times 10^{-14}}{4.0 \times 10^{-10}} = \frac{x \cdot x}{1 - x} \text{ , if one substitutes the values}$$

in the equation $\dfrac{1.0 \times 10^{-14}}{4.0 \times 10^{-10}} = \dfrac{[HCN][OH^-]}{[CN^-]}$.

The procedure now is to solve for x.

$x = 5.0 \times 10^{-3}$ M = $[OH^-]$.

As previously stated, pOH = $-$ log $[OH^-]$.

pOH = 2.30    and pH = 14 $-$ pOH = 11.70.

A laboratory technician has a solution of 0.1 M $NaC_2H_3O_2$. Find its pH and the percent hydrolysis. Assume $K_w$ = $1.0 \times 10^{-4}$ and $K_{diss}$ = $1.8 \times 10^{-5}$.

Solution: $NaC_2H_3O_2$ is a salt that exists as ions $Na^+$ and $C_2H_3O_2^-$ in water. Since water is a weak acid and weak base, which dissociates to $H^+$ and $OH^-$ ions, it can react with ions from weak acids and bases. Such a process, termed hydrolysis, occurs in this problem. Because the problem asks for pH, one must determine the concentrations of the $H_3O^+$ or $OH^-$ in this problem. This means that the hydrolysis reaction must be written and an equilibrium constant set up. The net hydrolysis reaction may be written as

$$C_2H_3O_2^- + H_2O \ \rightleftarrows \ HC_2H_3O_2 + OH^-.$$

The $K_{hyd}$ for this reaction is

$$\frac{[OH^-][HC_2H_3O_2]}{[C_2H_3O_2^-]}$$

However, $K_{hyd}$ is defined as $K_w/K_{diss}$, where $K_w$ is the dissociation constant for water and $K_{diss}$ is the dissociation constant for the acid formed. One can equate the two to obtain

$$\frac{[OH^-][HC_2H_3O_2]}{[C_2H_3O_2^-]} = \frac{K_w}{K_{diss}} \ .$$

But the value of $K_w$ and $K_{diss}$ is known. Therefore,

$$\frac{[OH^-][HC_2H_3O_2]}{[C_2H_3O_2^-]} = \frac{1.0 \times 10^{-14}}{1.8 \times 10^{-5}} = 5.6 \times 10^{-10}.$$

If one can find $[OH^-]$, one can determine $[H_3O^+]$ from $[H_3O^+][OH^-] = K_w = 1.0 \times 10^{-14}$. Therefore, let x = moles/liter of $OH^-$. From the hydrolysis reaction, one can see that for each mole of $OH^-$ generated, a mole of $HC_2H_3O_2$ is also generated. Therefore, let x = $[HC_2H_3O_2]$ also. If one starts with 0.1 M of $C_2H_3O_2^-$, and x moles/liter reacts to form $OH^-$ (and $C_2H_3O_2^-$), then, at equilibrium, one has 0.1 - x moles/liter of $C_2H_3O_2^-$ left. Substituting these values into the previous expression:

$$\frac{x \cdot x}{0.1 - x} = 5.6 \times 10^{-6}. \text{ Solving for x,}$$

$$x = 7.5 \times 10^{-4} \text{ M.}$$

(assuming in this calculation that x is small, therefore approximating 0.1 - x as 0.1.) To find $[H^+]$, solve

$$[H_3O^+] = \frac{K_w}{[OH^-]} = \frac{1.0 \times 10^{-14}}{7.5 \times 10^{-4}} = 1.33 \times 10^{-11}.$$

Since $pH = -\log [H_3O^+]$

$$pH = -\log [1.33 \times 10^{-11}] = 10.88.$$

The percent hydrolysis of this reaction is the number of moles of hydrolyzed acetate ion, divided by the number of moles available for hydrolysis, times 100.

$$\% \text{ Hydrolysis} = \frac{\text{Moles } C_2H_3O_2^- \text{ hydrolyzed}}{\text{Moles } C_2H_3O_2^- \text{ available}} \times 100$$

$$= \frac{7.5 \times 10^{-4}}{0.1} \times 100 = 0.75\%.$$

● **PROBLEM** 415

What is the pH of a 0.001 M aqueous solution of the strong electrolyte KOH? The dissociation constant for water is $K_w = 10^{-14}$ mole$^2$/liter$^2$.

Solution: We need to find $[H_3O^+]$ in order to find the pH. $[H_3O^+]$ can be determined by substituting into the expression $K_w = [H_3O^+][OH^-]$, or,

$$[H_3O^+] = \frac{K_w}{[OH^-]}$$

The hydroxyl concentration, $[OH^-]$, is contributed by the dissociation of KOH. This equation for this reaction is

$$KOH \rightarrow K^+ + OH^-.$$

Since KOH is a strong electrolyte, we assume that it dissociates completely. Hence, the concentrations of $OH^-$ and of $K^+$ are equal to the initial concentration of KOH, or $[OH^-] = 0.001$ M $= 10^{-3}$ M. The hydronium ion concentration is then

$$[H_3O^+] = \frac{K_w}{[OH^-]} = \frac{10^{-14} \text{ mole}^2/\text{liter}^2}{10^{-3} \text{ M}}$$

$$= \frac{10^{-14} \text{ mole}^2/\text{liter}^2}{10^{-3} \text{ mole}/\text{liter}} = 10^{-11} \text{ mole}/\text{liter} = 10^{-11} \text{ M}.$$

The pH is defined as $pH = -\log [H_3O^+]$. Hence

$$pH = -\log [H_3O^+] = -\log (10^{-11}) = -(-11) = 11.$$

Find the pH of a 0.2 M solution of formic acid. $K_a = 1.76 \times 10^{-4}$.

*Solution*:  To solve the problem, first note that the acid dissociates according to the general equation

$$HA + H_2O \rightleftarrows H^+ + A^-,$$

where HA is the acid and $H^+$ and $A^-$ are the dissociation products. To find the pH, calculate $[H^+]$.

One can see from the equation that equal concentrations of $H^+$ and $A^-$ form when HA dissociates. At equilibrium, x moles of HA dissociate into x moles of $H^+$ and x moles of $A^-$.

$$HA + H_2O \rightleftarrows H^+ + A^-$$

Before:      0.2                    0        0

After:     0.2-x                    x        x

Now, set up the equilibrium constant equation, which is

$$K_a = \frac{[products]}{[reactants]} = \frac{[H^+][A^-]}{[HA]} = 1.76 \times 10^{-4}.$$

Before substituting in the values, simplify the calculation by assuming that $0.2 - x \cong 0.2$. This is a valid assumption since the acid is very weak and only a very small percentage dissociates. Thus, one has

$$K_a = \frac{(x)(x)}{(0.2)} = 1.76 \times 10^{-4}. \quad \text{Solving:}$$

$$x = [H^+] = 5.93 \times 10^{-3}.$$

$H^+$ concentration is then converted to pH according to the equation,

$$pH = - \log [H^+] = - \log 5.93 \times 10^{-3}$$

$$= - [- 3 + \log 5.93] = 3 - 0.77 = 2.23.$$

Lemon juice is very acidic, having a pH of 2.1. If we assume that the only acid in lemon juice is citric acid (represented by HCit), that HCit is monoprotic and that no citrate salts are present, what is the concentration of citric acid in lemon juice? The dissociation constant for citric acid

$$HCit + H_2O \rightleftarrows H_3O^+ + Cit^-$$

is $K_a = 8.4 \times 10^{-4}$ mole/liter.

Solution: The solution to this problem involves determination of [HCit] from the expression for $K_a$ of the
reaction

$$HCit + H_2O \rightleftarrows H_3O^+ + Cit^-.$$

By definition,

$$K_a = \frac{[H_3O^+][Cit^-]}{[HCit]} \text{, or, } [HCit] = \frac{[H_3O^+][Cit^-]}{K_a}$$

Since one mole of $Cit^-$ is produced per mole of $H_3O^+$
produced, $[Cit^-] = [H_3O^+]$. $[H_3O^+]$ can be determined from
the pH. By definition, pH $= -\log [H_3O^+]$, hence

$[H_3O^+] = 10^{-pH} = 10^{-2.1}$ mole/liter. Therefore,

$[H_3O^+] = [Cit^-] = 10^{-2.1}$ mole/liter and we have

$$[HCit] = \frac{[H_3O^+][Cit^-]}{K_a}$$

$$= \frac{10^{-2.1} \text{ mole/liter} \times 10^{-2.1} \text{ mole/liter}}{8.4 \times 10^{-4} \text{ mole/liter}}$$

$$= \frac{(10^{-2.1})^2}{8.4 \times 10^{-4}} \text{ mole/liter} = \frac{10^{-4.2}}{8.4 \times 10^{-4}} \text{ mole/liter}$$

$$\cong \frac{6.3 \times 10^{-5}}{8.4 \times 10^{-4}} \text{ mole/liter}$$

$$= 7.5 \times 10^{-2} \text{ mole/liter} = 7.5 \times 10^{-2} \text{ M.}$$

● **PROBLEM** 418

Find the pH of a solution of 0.3 M acetic acid. Assume a
$K_a$ value of $1.8 \times 10^{-5}$.

Solution: The dissociation of acetic acid into positive
and negative ions proceeds as follows:

$$HC_2H_3O_2 \rightleftarrows H^+ + C_2H_3O_2^-$$

For this particular dissociation the equilibrium

constant, $K_a$, is written as

$$K_a = \frac{[H^+][C_2H_3O_2^-]}{[HC_2H_3O_2]} = 1.8 \times 10^{-5},$$

where the brackets represent the concentrations of the substances within them.

The pH of a solution is defined as the negative log of the hydrogen ion concentration. Mathematically, it is written as

$$pH = -\log [H^+].$$

Thus, to answer this problem, the $[H^+]$ in this solution must be calculated from the equilibrium constant expression and then substituted into the pH equation.

From the stoichiometry of the dissociation reaction, for every x molar concentration of $H^+$ formed there are x molar concentration of $C_2H_3O_2^-$ also formed. Therefore, let

$$x = [H^+] = [C_2H_3O_2^-].$$

The x molar concentration of these ions comes from the dissociation of the 0.3 M acetic acid. Thus, the exact amount of acetic acid in solution at equilibrium is 0.3- x However, x is a very small amount as compared with 0.3 M of the acetic acid and is, therefore, approximated as 0.3 M. Substituting these values into the equilibrium expression

$$1.8 \times 10^{-5} = \frac{x \cdot x}{0.3 - x} \cong \frac{x^2}{0.3}$$

Solving for x; $\quad x = [H^+] = 2.3 \times 10^{-3}$

Substituting this value of $[H^+]$ into the pH equation:

$$pH = -\log (2.3 \times 10^{-3}) = 2.63.$$

● **PROBLEM** 419

For HF, $K_{diss} = 6.7 \times 10^{-4}$, what is the $H_3O^+$ and $OH^-$ concentrations of a 0.10M HF solution? Assume $K_w = 1 \times 10^{-14}$.

Solution: To answer this question, you must set up the equations for the equilibria involved. HF is an acid, which means it donates protons ($H^+$). The only base present is $H_2O$. They react as shown in the following equation:

$$HF + H_2O \rightleftharpoons H_3O^+ + F^-.$$

This is not the only source of $H_3O^+$. Recall that water also can dissociate into ions.

$$H_2O + H_2O \rightleftarrows H_3O^+ + OH^- \ .$$

This means that in the consideration of the $H_3O^+$ concentration you must look at both sources. Because you are asked the concentrations, you must employ equilibrium constants, $K_{diss}$ and $K_w$. Let $x =$ the number of moles/liter of $H_3O^+$ from the reaction of HF with $H_2O$, then $x$ is also equal to the number of moles/liter of $F^-$ produced. They have the same mole ratio, as indicated by the equilibria. Since HF is the stronger acid, some of its protons ionize and then join with $H_2O$ to form $H_3O^+$. Thus, the concentration of HF is $(0.1 - x)$. If you let $y =$ moles/liter of $OH^-$, then $y =$ moles/liter of $H_3O^+$ formed by the dissociation of water. These variables that represent the concentrations at equilibrium must satisfy the two equilibrium conditions:

$$K_{diss} = \frac{[H_3O^+][F^-]}{[HF]} = 6.7 \times 10^{-4} = \frac{(x+y)(x)}{0.1-x}$$

$$K_w = [H_3O^+][OH^-] = 1 \times 10^{-14} = (x+y)(x) \ .$$

For both conditions, the $H_3O^+$ concentration is represented by $(x+y)$ since, at equilibrium, the concentration must be the sum from both sources of $H_3O^+$ production. Solve these equations simultaneously for $x$ and $y$, you can avoid cumbersome calculations by noting which contribution is the dominant one. In this way, the other can be neglected to give a fairly good approximation. A general rule for determining which is dominant is to compare the dissociation constants. The larger the dissociation constant, the more it dominates the final equilibrium state. From inspection of these constants, you find that the contribution of $H_3O^+$ from HF dominates over that from water. You need consider only $HF + H_2O \rightleftarrows H_3O^+ + F^-$, therefore,

$$K_{diss} = \frac{[H_3O^+][F^-]}{[HF]} = 6.7 \times 10^{-4} = \frac{(x+y)(x)}{0.1-x} \ .$$

Solving, you obtain $x = 7.7 \times 10^{-3}M = [H_3O^+]$. To find $y$, go back to $K_w = [H_3O^+][OH^-] = 1 \times 10^{-14} = (x+y)(y)$ and substitute in your value for $x$. Assuming that $y$ is negligible in comparison to $x$, the answer becomes

$$\frac{1.0 \times 10^{-14}}{7.7 \times 10^{-3}} = 1.3 \times 10^{-12}M = [OH^-] \ .$$

● **PROBLEM** 420

There exists the equilibrium $CO_2 + 2H_2O \rightleftarrows H_3O^+ + HCO_3^-$. The $K_a = 4.3 \times 10^{-7}$ and pH = 7.40. Find the ratio of concentrations of bicarbonate ($HCO_3^-$) to carbon dioxide ($CO_2$).

<u>Solution</u>: The equlibrium constant expression, which indicates the ratio of the products to reactants, each

raised to the power of their respective coefficients, can be written as follows:

$$K_a = \frac{[H_3O^+][HCO_3^-]}{[CO_2]} \ .$$

Note, $[H_2O]$ is omitted, since it is considered to be constant. Since $K_a = 4.3 \times 10^{-7}$, then

$$4.3 \times 10^{-7} = \frac{[H_3O^+][HCO_3^-]}{[CO_2]}$$

One is solving for $[HCO_3^-]/[CO_2]$. If one knows $[H_3O^+]$, then one can calculate this ratio.

$[H_3O^+]$ can be found, since, as given, the pH = 7.40.

$$pH = -\log [H_3O^+] \quad \text{or}$$

$$[H_3O^+] = \text{antilog} (-pH) = \text{antilog} (-7.40)$$

$$= 4.0 \times 10^{-8} \ M.$$

Thus, $\quad 4.3 \times 10^{-7} = \dfrac{4.0 \times 10^{-8} [HCO_3^-]}{[CO_2]}$ , and

$$\frac{[HCO_3^-]}{[CO_2]} = \frac{4.3 \times 10^{-7}}{4.0 \times 10^{-8}} = 10.75.$$

● **PROBLEM** 421

100 ml of 0.1 M NaOH is mixed with 100 ml of 0.15 M HOAC. $K_a = 1.75 \times 10^{-5}$ for acetic acid, what is the pH of this mixture?

Solution: To find the pH of the mixture, first determine the $H_3O^+$ concentration, since pH = $-\log [H_3O^+]$. To find $H_3O^+$, consider what happens when acetic acid is added to NaOH, a base. When an acid is added to a base, a neuralization reaction occurs such that water and a salt are produced. Thus,

$$\underset{\text{base}}{\underbrace{NaOH}} + \underset{\text{acid}}{\underbrace{HOAC}} \rightarrow \underset{\text{salt}}{\underbrace{Na^+OAC^-}} + \underset{\text{water}}{\underbrace{HOH}}$$

This reaction does not show the presence of any $H_3O^+$ ions, which may lead one to incorrectly conclude that its concentration was zero. There are 100 ml of 0.15 M HOAC. M = Molarity = moles/liter. Thus, 0.15 = moles/0.10 liters (1 liter = 1000 ml). Therefore, (0.15)(0.10)= 0.015 moles of HOAC. For NaOH, M = 0.10 = moles/0.10 liters. Thus,

the number of moles of NaOH = (0.10)(0.10) = 0.010. From
the previously written neutralization equation, one mole
of acid reacts for every mole of base. Only 0.01 moles of
HOAC could react with NaOH. Thus, 0.015 - 0.010 = 0.005
moles of HOAC was NOT neutralized by the base, NaOH. This
excess 0.005 moles of HOAC is, in the presence of its salt
NaOAC. Thus, it is a buffer solution. Solutions that contain a weak acid, such as HOAC, and its salt are called
buffers. Therefore, consider the dissociation of acetic
acid; it is in excess and is the source of the hydronium
ion concentration.

$$HOAC + HOH \; \underset{\leftarrow}{\overset{\rightarrow}{}} \; H_3O^+ + OAC^-.$$

The equilibrium constant for this reaction measures the
ratio of the concentration of the products to reactants,
each raised to a power based on its coefficient in the
chemical equation. Thus,

$$K_a, \text{ the equilibrium constant, } = \frac{[H_3O^+][OAC^-]}{[HOAC]} \; .$$

Water is not included since it is a constant. One is solving
for $[H_3O^+]$ to find pH, let $x = [H_3O^+]$. From the dissociation
equation $x = [OAC^-]$ also, since they are formed in equimolar
amounts. Recall, however, that the neutralization reaction
produced $\overline{OAC^-}$ also. There were 0.010 moles each of NaOH and
HOAC which reacted. From the equation, 0.010 moles of $OAC^-$
should have been produced. The volume was 200 ml. Thus,

$$M = \text{concentration} = \frac{0.01 \text{ moles}}{\frac{200 \text{ ml}}{1000 \text{ ml}/\ell}} = 0.05 \text{ M}.$$

The total concentration of $OAC^-$, from the acid
dissociation and neutralization, is $x + .05$. The acetic
acid concentration is the initial amount minus the amount
that dissociated. Recall, there were 0.005 moles of acetic
acid in 200 ml. to give

$$\frac{0.005 \text{ moles}}{\frac{200 \text{ ml}}{1000 \text{ ml}/\ell}} = 0.025 \text{ M}.$$

x moles/liter of it dissociated. Thus, the final concentration
is (0.025 - x).

$$K_a = \frac{[H_3O^+][OAC^-]}{[HOAC]} \; .$$

$$1.75 \times 10^{-5} = \frac{[H_3O^+][OAC^-]}{[HOAC]} \; .$$

Substituting final concentrations,

$$1.75 \times 10^{-5} = \frac{(x)(x + 0.05)}{(0.025 - x)} \; .$$

The excess $OAC^-$, from the dissociation of HOAC, x, is small. Thus,

$$\frac{(x)(x + 0.05)}{(0.025 - x)} \approx \frac{(0.050\ x)}{(0.025)} = 1.75 \times 10^{-5}.$$

Solving for x, one finds

$$[H_3O^+] = x = 8.8 \times 10^{-6}\ M.$$

Thus,     $pH = -\log [H_3O^+] = -\log [8.8 \times 10^{-6}] = 5.06.$

● **PROBLEM** 422

Allow exactly 100 ml of 1.5 N NaOH solution to be mixed with 100 ml 3.0 N $H_3PO_4$ solution and allow them to reach equilibrium. (1) Determine what species will be present at equilibrium. (2) Find the pH of the solution. $k_2 = 6.2 \times 10^{-8}$.

<u>Solution</u>:  To solve this problem, first consider the type of reaction, if any, that occurs and to what extent it has proceeded. Once this is done, the answers to (1) and (2) follow directly. One is adding a base of a given amount to an acid of a given amount. This is a neutralization; i.e., the reaction of an acid and base to produce a salt and water. One is told there are 100 ml of 1.5 N NaOH and 100 ml of 3.0 N $H_3PO_4$ solution. N = normality, which is defined as equivalents/liter. An equivalent = the number of hydronium ions that an acid can produce or hydroxide ions that a base can produce per mole of that substance. In a neutral solution the number of equivalents of base is equal to the number of equivalents of acid. Thus, to see to what extent the reaction has proceeded, calculate the equivalents of base and acid. If they are equal, the reaction went to completion. Recall, N = equivalents/liter. Thus, equivalents NaOH = (1.5 equiv/$\ell$)100 m$\ell$/1000 m/$\ell$ = 0.15 equiv. Equivalents $H_3PO_4$ = (3.0 equiv/$\ell$)100 m$\ell$/1000$\ell$ = 0.30 equiv.

The equivalents of acid do NOT equal those of the base. The $H_3PO_4$ has 0.30 - 0.15 = 0.15 equivalents that do not react. The rest has been neutralized by the base to form a salt and water. $H_3PO_4$ is a polyprotic acid, which means it can donate 3 protons when a mole of it exists. Since only one-half of the equivalents are neutralized, only one-half of the protons were used. This means that the reaction

$$H_3PO_4 + OH^- \overset{\rightarrow}{\leftarrow} H_2PO_4^- + H_2O$$

goes to completion, but

$$H_2PO_4^- + OH^- \overset{\rightarrow}{\leftarrow} HPO_4^{2-} + H_2O$$

will only be half complete. With this information, (1) can be answered. One started with 1.5 N of 100 ml NaOH solution, which indicates that 0.15 moles of $Na^+$ is present. NaOH can react with only 1 mole of hydronium ion. Thus, its equivalents correspond to its moles. 0.30 equivalents of $H_3PO_4$ correspond to 0.10 moles, since each mole generates 3 hydronium ions. In

$$(0.1)H_3PO_4 + (0.15)OH^- \overset{\rightarrow}{\leftarrow} (0.1)H_2PO_4^- + (0.1)H_2O + (.05)OH^-$$

there is a total 0.1 moles of $H_2PO_4^-$ generated, since $H_3PO_4$ is the limiting reagent with only 0.1 moles compared with $OH^-$'s 0.15 moles. This means that for the reaction;

$$(0.1)H_2PO_4^- + (0.05)OH^- \overset{\rightarrow}{\leftarrow} (0.05)HPO_4^{2-} + (0.05)H_2O + (0.05)H_2PO_4^-,$$

0.05 moles of $OH^-$ is left to react with the 0.1 moles of

$H_2PO_4^-$ generated previously. Thus, only 0.05 moles of $H_2PO_4^-$ is consumed in producing 0.05 moles of $HPO_4^{2-}$. Thus, an unreacted amount of 0.05 moles of $H_2PO_4^-$ is left. In summary, the species in solution are 0.05 moles each of $H_2PO_4^-$ and $HPO_4^{2-}$ with small amounts of $H_3O^+$, $H_3PO_4$, $PO_4^{3-}$, $OH^-$ and 0.15 moles $Na^+$. One can now calculate the pH of this. There exists some $H_2PO_4^-$ in solution. This is the source of hydronium ions, whose pH is measured by pH = $-\log [H_3O^+]$. The reaction is

$$H_2PO_4^- + H_2O \overset{\rightarrow}{\leftarrow} H_3O^+ + HPO_4^{2-}.$$

The equilibrium constant,

$$k_2 = \frac{[H_3O^+][HPO_4^{2-}]}{[H_2PO_4^-]} = 6.2 \times 10^{-8}.$$

One can calculate $[HPO_4^{2-}]$ and $[H_2PO_4^-]$, since one calculated the mole amounts. The volume of the solution is 0.2 liters, and the concentration in the brackets, are in moles/liter. Substituting these values,

$$6.2 \times 10^{-8} = \frac{[H_3O^+](0.05/.2)}{(0.05/.2)}.$$

Solving for $[H_3O^+]$, one obtains

$[H_3O^+] = 6.2 \times 10^{-8}$ M. pH = $-\log [H_3O^+]$. Thus,

$$pH = -\log [6.2 \times 10^{-8}] = 7.21.$$

● **PROBLEM** 423

0.001 mol of NaOH is added to 100 ml. of a solution that is 0.5 M $HC_2H_3O_2$ and 0.5 M $NaC_2H_3O_2$. Determine the pH of this solution; the $K_{diss}$ of acetic acid is $1.8 \times 10^{-5}$.

<u>Solution</u>:  This problem requires an understanding of the concept of buffer solution. In some processes, a solution of constant pH is desired. This constancy is maintained by the buffering action of an acid-base equilibrium. A buffer contains both an acid and a base and responds to the addition of either $H_3O^+$ or $OH^-$ to maintain the pH. In this problem, the 0.5 M solutions of $HC_2H_3O_2$ and $NaC_2H_3O_2$ act as a buffer that must respond to the addition of base, NaOH. One has equal numbers of moles of $HC_2H_3O_2$ and $NaC_2H_3O_2$ in the buffer, which means that the

$\dfrac{[HC_2H_3O_2]}{[C_2H_3O_2^-]}$ ratio is one. In solution, one has

$$HC_2H_3O_2 + H_2O \overset{\rightarrow}{\leftarrow} H_3O^+ + C_2H_3O_2^-, \quad \text{which}$$

indicates that each mole/liter of $HC_2H_3O_2$ must produce the same number of mole/liter of $C_2H_3O_2^-$. For this reaction,

$$k = \dfrac{[H_3O^+][C_2H_3O_2^-]}{[HC_2H_3O_2]} \quad \text{or} \quad [H_3O^+] = \dfrac{k\ [HC_2H_3O_2]}{[C_2H_3O_2^-]},$$

which is an expression for the concentration of hydronium ions in the buffer. Recalling that pH = - log $[H_3O^+]$ and that pK = - log K, by taking the negative log of both sides of the above equation, one can write

$$- \log [H_3O^+] = - \log K - \log \dfrac{[HC_2H_3O_2]}{[C_2H_3O_2^-]}.$$

$$pH = pK - \log \dfrac{[HC_2H_3O_2]}{[C_2H_3O_2^-]}$$

For this solution, it is given that K = $1.8 \times 10^{-5}$ and that $[HC_2H_3O_2]/[C_2H_3O_2^-]$ is equal to one. Thus,

$$pH = pK - \log \dfrac{[HC_2H_3O_2]}{[C_2H_3O_2^-]}$$

$$pH = (- \log 1.8 \times 10^{-5}) - \log (1)$$

$$pH = 4.74 - 0 = 4.74$$

One adds 0.001 moles of NaOH. This base will convert an equal number of moles of the acid, $HC_2H_3O_2$ to $C_2H_3O_2^-$, by a neutralization reaction. If one started with 0.5 M $HC_2H_3O_2$ in 100 ml, one had 0.05 mole of it. Upon addition of NaOH, however, 0.001 moles of it is converted to $C_2H_3O_2^-$. This means one has 0.05 - 0.001 = 0.049 moles of $HC_2H_3O_2$ left. There were 0.05 moles of $NaC_2H_3O_2$ to start. $NaC_2H_3O_2$ exists as the ions $Na^+$ and $C_2H_3O_2^-$. There were 0.001 moles of $C_2H_3O_2^-$ produced upon addition of NaOH, which means one has a total of 0.001 + 0.05 = 0.051 moles of $C_2H_3O_2^-$. The volume of the solution remained at 100 ml. Therefore, these mole amounts are 0.49 M concentration of $HC_2H_3O_2$ and 0.51 M concentration of $C_2H_3O_2^-$. To find the new pH, one need only substitute these values into

$$pH = pK - \log \frac{[HC_2H_3O_2]}{[C_2H_3O_2^-]} \quad .$$

$$pH = - \log (1.8 \times 10^{-5}) - \log \frac{.49}{.51} = 4.74 + 0.017 = 4.76.$$

● **PROBLEM** 424

Find the pH of a 0.1 M solution of ammonia, $NH_3$. $pK_b$ = $1.76 \times 10^{-5}$.

Solution:  This problem involves the dissociation of a weak base, which proceeds by the following general equation:

$$B + H_2O \rightleftarrows BH^+ + OH^-,$$

where B is the base and $BH^+$ and $OH^-$ are the dissociation products. x moles of B dissociate into x moles of $BH^+$ and x moles of $OH^-$. This is more clearly seen below:

$$NH_3 + H_2O \rightleftarrows NH_4^+ + OH^-$$

Before:  0.1 M            O      O

After: 0.1 - x            x      x

If x moles of $NH_3$ dissociate, then, at equilibrium, one has 0.1 - x left, since 0.1 is the initial concentration. $NH_3$ is a weak base and very little dissociation occurs, which means x is extremely small. This allows one to approximate 0.1 - x as 0.1. By now, it should be evident that $[H^+]$ does not appear in the equilibrium e-quation. The strategy will be to first find $[OH^-]$, then calculating the pOH, and then using the equation pK = pOH + pH = 14 to find pH.

In order to calculate the value of $[OH^-]$, set up the equilibrium constant equation for which $K_b = 1.76 \times 10^{-5}$.

Thus, $K_b = \dfrac{[NH_4^+][OH^-]}{[NH_3]}$ . Substituting,

$$K_b = \frac{(x)(x)}{0.1} = 1.76 \times 10^{-5}, \text{ so that}$$

$$x = [OH^-] = 1.33 \times 10^{-3}.$$

Then, one can calculate pOH by:

$$pOH = - \log [OH^-] = - \log [1.33 \times 10^{-3}]$$

$$= 3 - \log 1.33 = 2.88.$$

Therefore, pH = 14 - pOH

$$= 14 - 2.88 = 11.12.$$

Find the pH of 0.15 M $H_2SO_4$ solution, assuming $K_2 = 1.26 \times 10^{-2}$.

Solution: Begin this problem by noting that pH = - log
$[H^+]$. Thus, to find pH, calculate the $[H^+]$ in 0.15 M $H_2SO_4$.
$H_2SO_4$ is an acid and dissociates into $H^+$ (= $H_3O^+$) ions.
The general reaction for acid dissociation can be written
as

$$HA + H_2O \rightleftarrows H_3O^+ + A^-.$$

$H_2SO_4$ undergoes this reaction, as

$$H_2SO_4 + H_2O \rightleftarrows H_3O^+ + HSO_4^-.$$

But note, $HSO_4^-$ is also an acid, it still possesses a
hydrogen that can dissociate. Thus, one also has

$$HSO_4^- + H_2O \rightleftarrows H_3O^+ + SO_4^=.$$

In other words, there exist two dissociation reactions. As
given, the dissociation constant for the second reaction,
$K_2$, measures the ratio of the concentrations of products
to reactants. It is from this expression that $[H^+]$ can
be determined. $H_2SO_4$ is a very strong acid in its first
dissociation reaction. This means it is completely ionized;
i.e. 100% of the $H^+$ comes off (the first $H^+$ only). Thus,
if there are 0.15 M of $H_2SO_4$, then, there are also 0.15 =
$[H^+]$ for the first dissociation.

$[H^+] = [HSO_4^-]$, as can be seen from the first
dissociation reaction and are formed in equimolar amounts.
Also, $[HSO_4^-] = 0.15$ (initially). For the second ionization,
one is given $K_2 = 1.26 \times 10^{-2}$, indicating dissociation is
not complete. From the prior explanation of what $K_2$
indicates, one can write

$$K_2 = 1.26 \times 10^{-2} = \frac{[H_3O^+][SO_4^=]}{[HSO_4^-]}.$$

Note: Water is not included since it is assumed to be a
constant. From the first dissociation, it is known that
$[H_3O^+] = 0.15$. If x moles/liter of $H^+$ is produced in the
second dissociation, the total $[H_3O^+] = 0.15 + x$. $[SO_4^=] =$
$[H_3O^+]$ from the second dissociation, since here they are
formed in equimolar amounts. Thus, $[SO_4^=] = x$. If one
started with $[HSO_4^-] = 0.15M$ (from the first dissociation)
and x moles/liter dissociate in the second, one has
$[HSO_4^-] = 0.15 - x$ left.

Substituting these values,

$$1.26 \times 10^{-2} = \frac{[H_3O^+][SO_4^=]}{[HSO_4^-]} = \frac{(0.15 + x)(x)}{(0.15 - x)} .$$

Solving for x (using the quadratic formula), x = .011. Thus, $[H^+]$ = 0.15 + 0.011 = 0.161. pH = $-$ log $[H^+]$, so that pH = $-$ log [.161] = 0.79.

● **PROBLEM** 426

Find the pH of 0.10 M HOAc solution that has 0.20 M NaOAc dissolved in it. The dissociation constant of HOAc is $1.75 \times 10^{-5}$.

Solution: To find the pH of this solution, calculate the $H_3O^+$ concentration; pH is defined as $-$ log $[H_3O^+]$. This necessitates considering the dissociation of acetic acid, HOAc, since the only source of $H_3O^+$ is from this dissociation. The dissociation reaction may be written

$$HOAc + H_2O \ \underset{\leftarrow}{\rightarrow} \ H_3O^+ + OAc^-.$$

To find the $H_3O^+$ concentration, the equilibrium constant expression must be written. This expression measures the ratio of the concentrations of products to reactants, each raised to the power of their respective coefficients in the chemical equation. Thus,

$$K_a = \frac{[OAc^-][H_3O^+]}{[HOAc]} .$$

(Note: water is excluded since it is considered a constant.)

Let x = $[H_3O^+]$ at equilibrium. The $[OAc^-]$ at equilibrium will be from two sources: the dissociation of HOAc and the presence of NaOAc, which exists as ions $Na^+$ and $OAc^-$. From the dissociation, the concentration is x, since the reaction indicates that $H_3O^+$ and $OAc^-$ are formed in equimolar amounts. Since one starts with 0.2 M NaOAc, one has a concentration of 0.2 M for $OAc^-$. The total is x + 0.2. If one starts with 0.1 M HOAc and x M of $H_3O^+$ or $OAc^-$ form, then, at equilibrium, one has 0.1 $-$ x left of HOAc, Substituting these values into the equilibrium constant expression, and noting that $K_a$ = 1.75 $\times$ $10^{-5}$ (given),

$$K_a = \frac{x (0.2 + x)}{(0.1 - x)} \approx \frac{0.20 \, x}{0.1} = 1.7 \times 10^{-5}.$$

The acetic acid is weak; its dissociation to $H_3O^+$ is low, thus, one can make the above approximations. Solving for x;

$$x = 8.5 \times 10^{-6} M = [H_3O^+].$$

$$pH = - \log [H_3O^+] = - \log [8.5 \times 10^{-6}] = 5.07.$$

Assuming pD = - log $[D_3O^+]$ in analogy to pH, what is pD of pure $D_2O$? K = 2 × $10^{-15}$.

Solution:  This question can be solved with the knowledge that $D_2O$ undergoes dissociation in a similar manner to $H_2O$. The only difference is that $D_2O$ is composed of deuterium instead of hydrogen. Therefore, one can write:

$$D_2O + D_2O \rightleftharpoons D_3O^+ + OD^-$$

$$K = [D_3O^+][OD^-] = 2 \times 10^{-15}.$$

$[D_2O]$ is a constant and does not appear in the above equilibrium expression. From the reaction, it becomes apparent that the concentrations of $D_3O^+$ and $OD^-$ are equal. Therefore, both concentrations can be represented by x, x · x = 2 × $10^{-15}$. Solving for x,

$$x = 4.47 \times 10^{-8}.$$

This, then, represents the concentration of $D_3O^+$ (and $OD^-$).

$$pD = - \log [D_3O^+] = - \log 4.47 \times 10^{-8} = 7.35.$$

Calculate the pH of a 0.2 M $NH_3$ solution for which $K_b$ = 1.8 × $10^{-5}$ at 25°C. The equation for the reaction is

$$NH_3 + H_2O \rightleftharpoons NH_4^+ + OH^- .$$

Solution:  pH is defined in terms of $[H^+]$, (pH = - log $[H^+]$), but this reaction shows the production of $OH^-$. However, $[H^+]$ and $[OH^-]$ are related by the definition, $[H^+] = 10^{-14}/[OH^-]$. Therefore, one can find $[H^+]$ and pH after solving for $[OH^-]$. One solves for $[OH^-]$ by using the fact that $K_b$ is equal to $\dfrac{[NH_4^+][OH^-]}{[NH_3]}$

$[H_2O]$ is excluded from this expression because it is assumed to be constant.

In solving for $[OH^-]$, one can assume that 1 ℓ of the 0.2 M $NH_3$ solution is present. This means that there were originally 0.2 moles of $NH_3$ present. For each $NH_3$ that dissociates, 1 mole of $OH^-$ and 1 mole of $NH_4^+$ is formed. Let x = $[OH^-]$ = $[NH_4^+]$ and 0.2 - x = $[NH_3]$

$$K_b = \frac{[NH_4^+][OH^-]}{[NH_3]} = 1.8 \times 10^{-5} = \frac{(x)(x)}{(0.2 - x)}$$

One can assume that x is negligible compared to 0.2. Therefore, 0.2 - x = 0.2. Using this assumption, one can now solve for x.

$$1.8 \times 10^{-5} = \frac{x^2}{0.2}$$

$$x^2 = 3.6 \times 10^{-6}; \qquad x = 1.9 \times 10^{-3}.$$

$[NH_4^+] = 1.9 \times 10^{-3}$ M, $[OH^-] = 1.9 \times 10^{-3}$ M and

$[NH_3] = 0.2 - 1.9 \times 10^{-3} \sim 0.2$.

One can now solve for $[H^+]$ using $[OH^-]$.

$$[H^+] = \frac{10^{-14}}{[OH^-]} = \frac{10^{-14}}{1.9 \times 10^{-3}} = 5.26 \times 10^{-12} \text{ M}$$

Using this value one can solve for pH.

$$pH = - \log [H^+] = - \log (5.25 \times 10^{-12})$$

$$= - (.72 - 12) = 11.28.$$

● **PROBLEM** 429

What is the pH of 0.500 M $NaHSO_3$? The $K_{eq}$ for $NaHSO_3$ is $1.26 \times 10^{-2}$.

Solution: The pH is defined as $- \log [H^+]$, where $[H^+]$ is the concentration of $H^+$ ions. $NaHSO_3$ ionizes as shown in the equation:

$$NaHSO_3 \; \overset{\rightarrow}{\leftarrow} \; Na^+ \; + \; HSO_3^-$$

Thus, for each $NaHSO_3$ that ionizes, one $H^+$ and one $HSO_3^-$ are formed. Therefore, $[H^+] = [HSO_3^-]$. The equilibrium constant ($K_{eq}$) for $NaHSO_3$ is $1.26 \times 10^{-2}$. The equation for $K_{eq}$ is

$$K_{eq} = \frac{[H^+][SO_3^=]}{[NaHSO_3]} = 1.26 \times 10^{-2}$$

One knows that $[NaHSO_3] = 0.500$ and that for each $H^+$ and $NaSO_3^-$ formed, one $NaHSO_3$ is ionized. Therefore, at equilibrium, the concentration of $NaHSO_3$ is $.500 - [H^+]$. One can now solve for $[H^+]$, using the equation for the equilibrium constant.

$$K_{eq} = \frac{[H^+][HSO_3^-]}{[NaHSO_3]} = 1.26 \times 10^{-2}$$

and because $[H^+] = [SO_3^=]$, you have

$$K_{eq} = \frac{[H^+][H^+]}{(.500 - [H^+])} = 1.26 \times 10^{-2}$$

$$[H^+]^2 = (.500 - [H^+]) \times 1.26 \times 10^{-2}$$

$$[H^+]^2 = (6.3 \times 10^{-3}) - 1.26 \times 10^{-2}[H^+]$$

Rewriting, $[H^+]^2 + 1.26 \times 10^{-2}[H^+] - 6.3 \times 10^{-3} = 0$.

Using the quadratic formula:

$$ax^2 + bx + c = 0 \qquad\qquad a = 1$$

$$b = 1.26 \times 10^{-2}$$

$$c = -6.3 \times 10^{-3}$$

$$x = \frac{-b \pm \sqrt{b^2 - 4ac}}{2a}$$

$$x = \frac{-1.26 \times 10^{-2} \pm \sqrt{(1.26 \times 10^{-2})^2 - [4 \times 1 \times (-6.3 \times 10^{-3})]}}{2 \times 1}$$

$$x = \frac{-1.26 \times 10^{-2} \pm \sqrt{1.59 \times 10^{-4} + (2.52 \times 10^{-2})}}{2}$$

$$x = \frac{-1.26 \times 10^{-2}}{2} \pm \frac{\sqrt{2.54 \times 10^{-2}}}{2}$$

$$x = -6.3 \times 10^{-3} \pm \frac{1.59 \times 10^{-1}}{2}$$

$$x = -6.3 \times 10^{-3} \pm 7.97 \times 10^{-2}$$

$$x = -6.3 \times 10^{-3} + 7.97 \times 10^{-2} = 7.34 \times 10^{-2} \qquad \text{or}$$

$$x = -6.3 \times 10^{-3} - 7.97 \times 10^{-2} = -8.60 \times 10^{-2}$$

x must equal $7.34 \times 10^{-2}$ because $[H^+]$ cannot be negative. One can now find pH.

$$pH = -\log [H^+] \qquad\qquad [H^+] = 0.734$$

$$pH = -\log 0.734$$

$$pH = 1.13.$$

● **PROBLEM** 430

At normal body temperature, 37°C (98.6°F), the ionization constant of water, $K_w$, is $2.42 \times 10^{-14}$ moles$^2$/liter$^2$.

A physician injects a neutral saline solution into a patient. What will be the pH of this solution when it has come into thermal equilibrium with the patient's body?

Solution: To solve this problem we must employ the definition of $K_w$, i.e. $K_w = [H_3O^+][OH^-]$. At 37°C, we are given

$$K_w = 2.42 \times 10^{-14} \text{ mole}^2/\text{liter}^2 = [H_3O^+][OH^-].$$

Since, for neutral solution, $[H_3O^+] = [OH^-]$, we can set $x = [H_3O^+] = [OH^-]$. Then,

$$x^2 = [H_3O^+][OH^-] = 2.42 \times 10^{-14} \text{ mole}^2/\text{liter}^2 \quad \text{or}$$

$$x = (2.42 \times 10^{-14} \text{ mole}^2/\text{liter}^2)^{\frac{1}{2}} = 1.56 \times 10^{-7} \text{ mole/liter}.$$

Hence, $[H_3O^+] = x = 1.56 \times 10^{-7}$ mole/liter and the pH of the solution is

$$pH = -\log [H_3O^+] = -\log (1.56 \times 10^{-7})$$

$$= -(-6.807) = 6.807 \overset{\sim}{=} 6.8.$$

● **PROBLEM** 431

$K_{diss}$ of water for $2H_2O \overset{\rightarrow}{\leftarrow} H_3O^+ + OH^-$ changes from $1.0 \times 10^{-14}$ at 25°C to $9.62 \times 10^{-14}$ at 60°C. Does the pH of water or its neutrality change when the temperature is increased from 25°C to 60°C?

Solution: To answer this question, it is necessary to find $[H^+]$, since $pH = -\log [H^+]$. To do this, write out the equilibrium dissociation expression for water, which equates the dissociation constant, $K_{diss}$, with the ratio of concentrations of products to reactants, each raised to the power of its coefficient in the chemical reaction. Thus, $K_{diss} = [H_3O^+][OH^-]$. The concentration of water is omitted; it is assumed to be constant. For the chemical reaction one can see that $OH^-$ and $H_3O^+$ are formed in equimolar amounts, which means $[OH^-] = [H_3O^+]$. Thus, when $K_{diss} = 1.0 \times 10^{-14}$, one has

$$1.0 \times 10^{-14} = [H_3O^+][OH^-] = [H_3O^+]^2 = [OH^-]^2. \text{ Solving}$$

for either, $[H_3O^+] = [OH^-] = 1.0 \times 10^{-7}$. Thus, $pH = -\log [H_3O^+] = 7$. When $K_{diss} = 9.62 \times 10^{-14}$ at 60°C,

$$9.62 \times 10^{-14} = [H_3O^+][OH^-] = [H_3O^+]^2 = [OH^-]^2. \text{Solving}$$

for either, $[H_3O^+] = [OH^-] = 3.1 \times 10^{-7}$. Thus, $pH = -\log [H_3O^+] = 6.51$. Thus, the pH does change. What about neutrality? Neutrality occurs when $[H^+] = [OH^-]$. In both cases, such a condition exists, so that the neutrality was not altered when the temperature was changed.

A solution of hydrogen peroxide is 30% by weight $H_2O_2$. Assuming a density of 1.11 g/cm$^3$ and a dissociation constant of $1.0 \times 10^{-12}$ for $H_2O_2$, what is the pH of the solution?

**Solution**: To obtain the pH of any solution, calculate $[H^+]$, since pH = $- \log [H^+]$. To find $[H^+]$, write a balanced chemical equation that expresses the reaction, and from this, set up a dissociation constant expression. Begin by writing the reaction.

$$H_2O_2 \; \overset{\rightarrow}{\leftarrow} \; H^+ + HO_2^-.$$

The dissociation constant for this reaction, K, equals $1.0 \times 10^{-12}$. K also = $[H^+][HO_2^-]/[H_2O_2]$. Equating, one obtains

$$\frac{[H^+][HO_2^-]}{[H_2O_2]} = 1.0 \times 10^{-12}.$$

Solving this expression for $[H^+]$, one obtains the pH. To do that it is necessary to know $[H_2O_2]$. Given the density, one can assume a volume of 1 liter solution. This means that the mass of the solution is 1.11 g/cm$^3$ × 1000 ml (or 1 liter) = 1110 g/$\ell$. However, the percent by weight of $H_2O_2$ is 30%. Therefore, there is only 0.30 1110 g/$\ell$ of $H_2O_2$ or 333 g/$\ell$. The molecular weight of $H_2O_2$ is 34 g/mole. The molarity of the $H_2O_2$ solution is, then,

$$\frac{333 \text{ g/liter}}{34 \text{ g/mole}} = 9.79 \text{ M } H_2O_2,$$

since molarity equals moles per liter. If one lets x = $[H^+]$, then $[H_2O_2]$ = 9.79 - x, since, at equilibrium, the original amount of $H_2O_2$ must be decreased by the amount of $H^+$ (or $O_2H^-$) formed. The equation,

$$H_2O_2 \; \overset{\rightarrow}{\leftarrow} \; H^+ + HO_2^-$$

indicates that $H^+$ will be formed in equal mole amounts with $HO_2^-$. Thus, $[HO_2^-]$ can equal x. Substituting these values into the expression:

$$\frac{[H^+][HO_2^-]}{[H_2O_2]} = 1.0 \times 10^{-12},$$

$$\frac{x \cdot x}{9.79 - x} = 1.0 \times 10^{-12}. \text{ Solving for x,}$$

$$x = 3.13 \times 10^{-6} = [H^+].$$

Since pH = $- \log [H^+]$, pH = 5.5.

Calculate the pH at 25°C of a solution containing 0.10 M sodium acetate and 0.03 M acetic acid. The apparent pK for acetic acid at this ionic strength is 4.57.

Solution: This problem involves the calculation of $[H^+]$ $(pH = - \log [H^+])$ in a mixture of a monoprotic acid (acetic acid) and its completely dissociated salt (sodium acetate). Because the addition of a salt represses the ionization of the acid, the concentration of undissociated acid is approximately equal to the molar concentration of added weak acid ($C_a$). The addition of the acid represses the hydrolysis of the salt so that the concentration of anions of the weak acid is approximately equal to the molar concentration of the salt $C_s$.

Using several approximations, the pK for a weak acid is the pH of a solution containing equimolar quantities of salt and acid. However, if the solution does not contain equimolar quantities of salt and acid, then the following approximation holds:

$$pH = pK + \log \frac{C_s}{C_a} \quad ,$$

where $C_s$ and $C_a$ are the concentrations of the salt and the acid, respectively.

Thus, in this problem,

$$C_s = 0.10 \text{ M} \qquad\qquad C_a = 0.03 \text{ M}$$

$$pK = 4.57$$

and $\quad pH = 4.57 + \log \frac{0.10}{0.03} = 4.57 + 0.52 = 5.09.$

10 ml of 0.200 M $HC_2H_3O_2$ is added gradually to 25 ml of 0.200 M NaOH. Calculate the pH of the initial 0.200 M NaOH solution and after each successive addition of 5 ml of acid. $K_w = 10^{-14}$.

Solution: This is a titration problem. One is asked to find the pH of the 0.200 M NaOH at different concentration levels. First one wants to calculate the pH before any acid is added. To do this, consider the fact that the equilibrium constant for the autodissociation of water, $K_w$, is defined as

$$K_w = [H_3O^+][OH^-].$$

Since, pH = $-$ log $[H_3O^+]$, one needs to calculate $[H_3O^+]$, which can be determined from $K_w$, once $[OH^-]$ is known. One knows $[OH^-]$ from the fact that one has a 0.2 M NaOH solution. M = molarity = moles/liter. In solution, NaOH dissociates according to

$$NaOH \overset{\rightarrow}{\leftarrow} Na^+ + OH^-.$$

Thus, $[OH^-] = [NaOH]$. Since $[NaOH] = 0.2$, $[OH^-] = 0.2$, which allows for substitution into the equilibrium expression,

$$K_w = 1.0 \times 10^{-14} = [H_3O^+](0.2), \quad \text{or}$$

$$[H_3O^+] = \frac{1.0 \times 10^{-14}}{0.2} = 5.0 \times 10^{-14}, \text{ which means}$$

$$pH = - \log [5.0 \times 10^{-14}] = 13.3.$$

What is the pH when 5 ml of acid ($HC_2H_3O_2$) is added? To answer this, note that a neutralizing reaction occurs, (i.e. a reaction between an acid and base to produce a salt and water) when the acid is added to the NaOH solution.

$$HC_2H_3O_2 + NaOH \longrightarrow HOH + NaC_2H_3O_2.$$

It is necessary to find how much base is consumed, since this will reflect the remaining $[OH^-]$. From this, return to the expression for $K_w$ and substitute this new value for $[OH^-]$ to determine $[H_3O^+]$ from which the pH can be calculated. To find the amount of NaOH that was consumed, use Molarity = moles/liter.

If the acid is 0.200 M and one adds in 5 ml or 0.005 $\ell$ (1000 ml = 1 $\ell$), one has $(0.2)(0.005) = 0.001$ moles of acid. Amount of base present in moles equals:

molarity $\times$ volume = $(0.2)(0.025) = (0.005)$ moles.

According to the reaction, they will react in equimolar amounts. Thus, since only 0.001 moles of acid is present, only 0.001 moles of base react. This leaves 0.004 moles of base (NaOH), and as such 0.004 moles of $OH^-$. Since

volume = base ($\ell$) + acid ($\ell$) = 0.03,

$$M = \frac{0.004}{0.03} = 0.13 \text{ M}.$$

Recalling, $K_w = 1.0 \times 10^{-14} = [H_3O^+][OH^-]$ and that $[OH^-] = 0.13$, such that

$$[H_3O^+] = \frac{1.0 \times 10^{-14}}{0.13} = 7.5 \times 10^{-14}.$$

Thus,    pH = - log [7.5 × $10^{-14}$] = 13.13.

If one adds another 5 ml of acid, the total acid added is 10 ml. The same procedure is followed, except that volume of acid = 10 ml or 0.01 ℓ. Molarity of acid = 0.2, so that (0.2)(0.01) = 0.002 moles of acid are present.

From the previous calculation, it can be seen that 0.005 - 0.002 = 0.003 moles of $OH^-$ are now left in a volume of 0.025 ℓ from base + 0.010 ℓ from acid or 0.035 liters. Thus, its molarity is

$$\frac{0.003}{0.035} = 0.086 \text{ M.}$$

$K_w$ = 1.0 × $10^{-14}$ = [$H_3O^+$][$OH^-$] = [$H_3O^+$][0.086]. Solving,

[$H_3O^+$] = 1.17 × $10^{-13}$.        This means,

pH = - log [$H_3O^+$] = - log [1.17 × $10^{-13}$] = 12.9.

● **PROBLEM** 435

Calculate the pH of (a) a 0.5 M solution with respect to $CH_3COOH$ and $CH_3COONa$; (b) the same solution after 0.1 mole HCl per liter has been added to it. Assume that the volume is unchanged. $K_a$ = 1.75 × $10^{-5}$.

Solution:    Part (a) involves the determination of the concentration of acid that has dissociated. Since one is adding a strong acid, part (b) is a buffer type problem.

For part (a), $CH_3COOH$ is the acid and for each x moles of it that dissociates, x moles of $H_3O^+$ and x moles of $CH_3COO^-$ are formed. This is seen more clearly below.

$$CH_3COOH + H_2O \rightleftarrows H_3O^+ + CH_3COO^-$$

After reaction  0.5-x              x        x

There is also $CH_3COONa$ present in this solution. Its effect on the equilibrium must be also taken into account. Since it is a salt it will completely dissociate, according to the following equation, in this dilute solution.

$$CH_3COONa \rightleftarrows CH_3COO^- + Na^+$$

It is 0.05 M, thus, in the solution there will be 0.05 M $CH_3COO^-$ and $Na^+$ contributions from it. The concentrations of the various species found in this solution are shown in the table below.

| | |
|---|---|
| $CH_3COOH$ | $(0.05 - x)M$ |
| $H_3O^+$ | $x$ M |
| $CH_3COO^-$ | $(0.05 + x)M$ |

(contributions are from both the $CH_3COOH$ and $CH_3COONa$ equilibria.)

| | |
|---|---|
| $Na^+$ | 0.05 M |

pH is defined as being equal to $- \log [H_3O^+]$. Thus, to solve for pH, one must first determine $[H_3O^+]$. This can be done by setting up the equation for the equilibrium of the reaction

$$CH_3COOH + H_2O \rightleftharpoons CH_3COO^- + H_3O^+$$

$$K_a = \frac{[H_3O^+][CH_3COO^-]}{[CH_3COOH]} = 1.75 \times 10^{-5}.$$

$$1.75 = \frac{(x)(0.05 + x)}{(0.05 - x)}$$

Since $CH_3COOH$ is a weak acid, there will be little dissociation and x will be very small. Thus one can approximate $0.05 + x \cong 0.05$ and $0.05 - x \cong 0.05$.

• Solving:

$$\frac{x (0.5)}{(0.5)} = 1.75 \times 10^{-5}$$

$$x = [H_3O^+] = 1.75 \times 10^{-5}$$

$$pH = - \log [H_3O^+] = - \log (1.75 \times 10^{-5}) = 4.76.$$

Part (b):

A buffer solution contains an acid and its conjugate base, both are moderately weak. What this type of solution does is to prevent a large change in pH when more acid or base is added. In a nonbuffered solution, the pH will change drastically. If a strong acid is added to a buffered solution, it reacts with the base $A^-$ to produce a weak acid.

$$H_3O^+ + A^- \rightleftharpoons HA + H_2O.$$

When a strong base is added the same effect occurs. It reacts with the acid ro produce a weak base.

$$OH^- + HA \rightleftharpoons A^- + H_2O.$$

The acid-base ratio is altered somewhat but not to any great extent, thus allowing the pH to stay relatively stable. However, if the number of moles of acid or base added is greater than the number of moles of base or acid present, then the pH will change, thereby destroying the buffer.

It is given that 0.1 M HCl is added. Since HCl is a strong acid, and dissociates completely, the $H_3O^+$ and $Cl^-$ formed each have an initial concentration of 0.1 M. However, the $H_3O^+$ will react with the $CH_3COO^-$ and drive the equilibrium to the left by Le Châtelier's principle and produce a weak acid. Therefore, the buffer has converted the acid ions so that they cannot change the pH.

Le Châtelier's principle states that when a stress is brought to bear on a system at equilibrium, the system tends to change so as to relieve the stress. The stress is relieved by driving the reaction to the other side of the equilibrium sign to prevent excesses of ions from building up. The net effect can be seen below.

$$CH_3COOH + H_2O \rightleftharpoons H_3O^+ + CH_3COO^-$$

Before:      0.5-x              x       0.5 + x

After add of    0.5+0.1-x       x       0.5 - 0.1 + x
0.1 M $H_3O^+$    (0.6-x)                (0.4 + x)

At the new equilibrium, there will be x moles of $H_3O^+$ still in solution, which were formed from the redissociation of $CH_3COOH$. x moles of $CH_3COO^-$ are also formed resulting in the new concentrations of

$$CH_3COOH + H_2O \rightleftharpoons H_3O^+ + CH_3COO^-$$

0.6-x                        x       0.4 + x

Upon setting up the equilibrium constant equation, one obtains $[H_3O^+]$.

$$K_a = \frac{[H_3O^+][CH_3COO^-]}{[CH_3COOH]} = 1.75 \times 10^{-5}$$

$$1.75 \times 10^{-5} = \frac{(x)(0.4 + x)}{(0.6 - x)}$$

Using the assumption outlined above, $0.6 - x \approx 0.6$, and $0.4 + x \approx 0.4$.

Solving:

$$1.75 \times 10^{-5} = \frac{x (0.4)}{(0.6)}$$

$$x = 2.63 \times 10^{-5} = [H_3O^+]$$

Since, pH $= - \log [H_3O^+]$, one finds by substitution,

$$pH = - \log 2.63 \times 10^{-5} = 4.58.$$

Calculate the number of grams of $NH_4Cl$ that must be added to 2 ℓ of 0.1 M $NH_3$ to prepare a solution of pH = 11.3 at 25°C. The $K_b$ for $NH_3$ = 1.8 × $10^{-5}$.The equilibrium equation

is:     $NH_3 + H_2O \overset{\leftarrow}{\rightarrow} NH_4^+ + OH^-$.

**Solution**: To determine how many grams of $NH_4Cl$ must be added to this solution to maintain a pH of 11.3, one must first calculate the $[NH_4^+]$ needed, because $[NH_4^+]$ = $[NH_4Cl]$ in the solution. This is done by using the expression that describes the dissociation,

$$NH_3 + H_2O \overset{\leftarrow}{\rightarrow} NH_4^+ + OH^-.$$

The dissociation constant ($K_b$) for this reaction is 1.8 × $10^{-5}$. $K_b$ is equal to the product of the concentrations of the substances formed divided by the concentration of the reactants. These concentrations are raised to the power of their coefficients in the equation for the reaction. $[H_2O]$ is not included because it is assumed to remain constant.

$$K_b = \frac{[NH_4^+][OH^-]}{[NH_3]} = 1.8 \times 10^{-5}.$$

To solve for $[NH_4^+]$, one must know $[OH^-]$ and $[NH_3]$ first. One is given that $NH_3$ solution is .1 M, thus $[NH_3]$ = 1. One can find $[OH^-]$ by using the definition of pOH. pOH = 14.0 - pH = - log $[OH^-]$. The pH of this solution is 11.3, therefore pOH= 14.0 - 113. = 2.7. One solves for $OH^-$ by using the relation pOH = - log $[OH^-]$. Solving for $[OH^-]$

$$2.7 = - \log [OH^-]$$

$$10^{-2.7} = [OH^-]$$

$$2.0 \times 10^{-3} = [OH^-]$$

One can now use the expression for $K_b$ to solve for $[NH_4^+]$

$$1.8 \times 10^{-5} = \frac{[NH_4^+][OH^-]}{[NH_3]} = \frac{[NH_4^+][2.0 \times 10^{-3}]}{[.1]}$$

$$[NH_4^+] = \frac{(.1)(1.8 \times 10^{-5})}{(2.0 \times 10^{-3})} = 9.0 \times 10^{-4} M$$

One is given that 2 ℓ of this solution has been made up. Because, Molarity = number of moles/volume in liters;

number of moles = molarity × volume

$$= \left(9.0 \times 10^{-4} \; \frac{moles}{liter}\right)(2 \; liters)$$

$$= 18 \times 10^{-4} \; moles.$$

$18 \times 10^{-4}$ moles of $NH_4Cl$ must be added to the solution. Because moles = weight in grams/molecular weight

$$weight \; in \; grams = moles \times M.W. \; (MW \; of \; NH_4Cl = 53.5)$$

$$= 18 \times 10^{-4} \; moles \; (53.5 \; g/mole)$$

$$= 0.096 \; g \; of \; NH_4Cl \; to \; be \; added.$$

● **PROBLEM** 437

Calculate the pH of a 0.25 M solution of the salt ethylamine hydrochloride, $C_2H_5NH_3Cl$. The dissociation constant for the base, ethylamine ($C_2H_5NH_2$) is $K_b = 5.6 \times 10^{-4}$.

Solution:  The pH will be determined by three processes occurring simultaneously: dissociation of ethylamine, ionization of ethylamine hydrochloride, and hydrolysis of the ethylammonium cation, $C_2H_5NH_3^+$. All three species are present in an aqueous solution of the salt.

Ethylamine hydrochloride ionizes according to the equation

$$C_2H_5NH_3Cl \; \overset{\rightarrow}{\leftarrow} \; C_2H_5NH_3^+ + Cl^-,$$

giving rise to the ethylammonium cation. The ethylammonium cation hydrolysis according to the equation

$$C_2H_5NH_3^+ + H_2O \; \overset{\rightarrow}{\leftarrow} \; H_3O^+ + C_2H_5NH_2,$$

forming the ethylamine base. Ethylamine base dissociates according to

$$C_2H_5NH_2 + H_2O \; \overset{\rightarrow}{\leftarrow} \; C_2H_5NH_3^+ + OH^-.$$

Let $K_w$ denote the water constant, $K_w = [H_3O^+][OH^-] = 10^{-14}$, and $K_b$ the dissociation constant of ethylamine,

$K_b = [C_2H_5NH_3^+][OH^-]/[C_2H_5NH_2] = 5.6 \times 10^{-4}$.  Then,

$$\frac{K_w}{K_b} = \frac{[H_3O^+][OH^-]}{[C_2H_5NH_3^+][OH^-]/[C_2H_5NH_2]} = \frac{[H_3O^+][C_2H_5NH_2]}{[C_2H_5NH_3^+]}$$

$$\frac{10^{-14}}{5.6 \times 10^{-4}} = 1.8 \times 10^{-11} = \frac{[H_3O^+][C_2H_5NH_2]}{[C_2H_5NH_3^+]}$$

Note that this is the equlibrium constant for hydrolysis of the ethylammonium cation.

Let $x = [C_2H_5NH_2]$ be the number of moles per liter of ethylamine formed by hydrolysis of the ethylammonium cation. Then $[H_3O^+] = x$, since the balanced equation for hydrolysis states that for every mole of ethylamine formed, one mole of $H_3O^+$ is formed and one mole of ethylammonium cation is consumed. If we assume that ethylamine hydrochloride dissociates completely, then the number of moles per liter of ethylammonium chloride not hydrolyzed is 0.25 M (from the salt) minus the number of moles per liter which is hydrolyzed (x), so that $[C_2H_5NH_3^+] = 0.25 - x$. Hence,

$$1.8 \times 10^{-11} = \frac{[H_3O^+][C_2H_5NH_2]}{[C_2H_5NH_3^+]} = \frac{x \cdot x}{0.25 - x}.$$

Assuming that the number of moles hydrolyzed is much less than 0.25 M, then $0.25 - x \cong 0.25$ and the algebra simplifies to

$$1.8 \times 10^{-11} = \frac{x \cdot x}{0.25} = \frac{x^2}{0.25}, \quad \text{or}$$

$$x = (0.25 \times 1.8 \times 10^{-11})^{\frac{1}{2}} = 2.1 \times 10^{-6}$$

(which is much smaller than 0.25, justifying our assumption). But $x = [H_3O^+] = 2.1 \times 10^{-6}$, hence

$$pH = - \log [H_3O^+] = - \log 2.1 \times 10^{-6} = 5.7.$$

The pH of 0.25 M ethylamine hydrochloride is thus 5.7.

● **PROBLEM** 438

Calculate the pH of a 0.10 M solution of sodium acetate, NaOAc. The equilibrium constant for the dissociation of acetic acid is $1.8 \times 10^{-5}$.

Solution: The pH will be governed by two processes: the ionization of NaOAc, and the hydrolysis of acetate ion, $OAc^-$.

NaOAc dissociates according to the equation

$$NaOAc \; \rightleftarrows \; Na^+ + OAc^-.$$

If we assume that dissociation is complete (a reasonable assumption for dilute salt solutions), the concentrations of $Na^+$ and $OAc^-$ are equal to the initial concentration of NaOAc: $[Na^+] = [OAc^-] = 0.10$ M.

Hydrolysis of **acetate** ion proceeds according to the equation

$$OAc^- + H_2O \overset{\rightarrow}{\leftarrow} HOAc + OH^-.$$

Let x denote the concentration of undissociated HOAc formed by hydrolysis. Then, since one mole of $OH^-$ is formed and one mole of $OAc^-$ is consumed per mole of HOAc formed, the concentration of hydroxide ion formed is x, and the concentration of acetate ion remaining is 0.10 - x. That is,

$$[OH^-] = x, \qquad [OAc^-] = 0.10 - x.$$

These concentrations can be substituted into the expression for the hydrolysis constant,

$$K_h = \frac{[HOAc][OH^-]}{[OAc^-]}$$

to find $x = [OH^-]$. $K_h$ can be found by using the water constant, $K_w = 1.0 \times 10^{-14} = [H_3O^+][OH^-]$ and the equilibrium constant for dissociation of acetic acid (HOAc),

$$K_a = 1.8 \times 10^{-5} = \frac{[H_3O^+][OAc^-]}{[HOAc]} . \quad \text{Then,}$$

$$K_h = \frac{[HOAc][OH^-]}{[OAc^-]} = [H_3O^+][OH^-] \times \frac{[HOAc]}{[H_3O^+][OAc^-]} = \frac{K_w}{K_a}$$

$$= \frac{1.0 \times 10^{-14}}{1.8 \times 10^{-5}} = 5.6 \times 10^{-10}.$$

Substituting the unknown concentrations into the expression gives

$$K_h = 5.6 \times 10^{-10} = \frac{[HOAc][OH^-]}{[OAc^-]}$$

$$= \frac{(x)(x)}{0.10-x} \cong \frac{x^2}{0.10} = 10 \ x^2,$$

where we have assumed that x is much smaller than 0.10 (hence $0.10 - x \cong 0.10$). Then

$$10 \ x^2 = 5.6 \times 10^{-10}$$

$$x = (5.6 \times 10^{-11})^{\frac{1}{2}} = 7.5 \times 10^{-6}.$$

Therefore, $[OH^-] = x = 7.5 \times 10^{-6}$. Using the water constant,

$$[H_3O^+][OH^-] = 1.0 \times 10^{-14}$$

$$[H_3O^+] = \frac{1.0 \times 10^{-14}}{[OH^-]} = \frac{1.0 \times 10^{-14}}{7.5 \times 10^{-6}} = 1.3 \times 10^{-9}$$

and the pH of the solution is

$$pH = - \log [H_3O^+] = - \log (1.3 \times 10^{-9}) = 8.9.$$

● PROBLEM 439

A chemist mixes together 100 ml of 0.5 M sodium acetate, 100 ml of 0.25 M hydrochloric acid (HCl), and 100 ml of a 1.0 M salt solution. She dilutes this to 1000 ml. Determine the concentrations of all the ions present, undissociated acetic acid, and the final pH of the solution, using $K_a$ of acetic acid = $1.7 \times 10^{-5}$ and $K_w = 1.0 \times 10^{-14}$.

Solution: To start, determine if any reaction took place in this mixture. If no reaction took place, then the molar concentrations will be numerically the same in 1 liter as the computed moles from the initial concentrations. However, a reaction does take place. Namely,

$$HOAc \text{ (acetic acid)} + H_2O \;\overset{\rightarrow}{\leftarrow}\; H_3O^+ + OAc^-.$$

The $H_3O^+$ (from HCl) reacts with the acetate ion ($OAc^-$) from the NaOAc (sodium acetate) to produce acetic acid and water. Start the analysis of the reaction by computing the number of moles involved. M = molarity = moles/liter. Thus, in 100 ml of 0.5 M NaOAc, there are 0.5 (100/1000) = 0.05 moles of NaOAc. Similar calculations show that 0.025 moles of HCl and 0.1 moles of NaCl (the salt solution) are present. The problem asks one to find the pH, the concentration of the ions present, and the concentration of undissociated acetic acid. This necessitates writing the equilibrium constant expression for this reaction, from which these concentrations can be calculated. Thus, write that $K_a$, the equilibrium constant, equals

$$\frac{[H_3O^+][OAc^-]}{[HOAc]}.$$

Suppose the reaction went to completion. If it did, 0.25 moles each of HOAc and excess $OAc^-$ would have been produced. One starts with 0.5 moles and only 1 mole of $OAc^-$ can react per mole of $H_3O^+$. If there exist 0.25 moles $H_3O^+$, then, only 0.25 moles of the $OAc^-$ can react, leaving 0.25 moles in excess. If the reaction, however, stops x moles short of completion, x moles of $H_3O^+$ would remain and the excess of $OAc^-$ would increase from 0.25 to 0.25 + x. The HOAc would be reduced to (0.25 - x) moles. Since they are in 1000 ml or 1 liter, these become concentrations that can be substituted into

$$K_a = \frac{[H_3O^+][OAc^-]}{[HOAc]}, \text{ to give } K_a = \frac{x\,(0.025 + x)}{(0.025 - x)}.$$

It is given that $K_a = 1.75 \times 10^{-5}$. Equating,

$$\frac{x (0.025 + x)}{(0.025 - x)} = 1.75 \times 10^{-5}.$$

When the $K_a$ for an acid is small, one knows that dissociation is minimal and therefore the concentration of $H_3O^+$ is very small compared to the concentration of $OAc^-$ and $HOAc$. This means that one can assume that their concentrations will not be appreciably changed by the dissociation. Thus,

$$0.025 - x \overset{\sim}{=} 0.025 \quad \text{and} \quad 0.025 + x \overset{\sim}{=} 0.025.$$

The equation for the $K_a$ can be rewritten as

$$K_a = \frac{x (0.025)}{0.025} = 1.75 \times 10^{-5}$$

$$\frac{x (0.025)}{(0.025)} = 1.75 \times 10^{-5}$$

$$x = 1.75 \times 10^{-5}.$$

Therefore, $\quad [H_3O^+] = 1.75 \times 10^{-5}$ M.

Since, $\quad\quad\quad\quad pH = - \log [H_3O^+]$,

$$pH = - \log [1.75 \times 10^{-5}] = 4.76.$$

$$[OAc^-] = 0.025 + x = 0.025 + 1.75 \times 10^{-5}$$

$$\cong 0.025.$$

$$[HOAc] = 0.025 - x = 0.025 - 1.75 \times 10^{-5}$$

$$\cong 0.025.$$

The other ions: $[Na^+] = 0.150$ M from NaOAc and NaCl. $[Cl^-] = 0.125$ M from HCl and NaCl. Since

$$[H_3O^+][OH^-] = K_w = 1.0 \times 10^{-14}$$

and $\quad [H_3O^+] = 1.7 \times 10^{-5}$,

$$[OH^-] = \frac{1.0 \times 10^{-14}}{1.7 \times 10^{-5}} = 5.9 \times 10^{-10} \text{ M}.$$

● **PROBLEM** 440

(a) A chemist prepares a 0.01 M solution of $NaC_2H_3O_2$. Find its pH. $K_A = 1.8 \times 10^{-5}$. (b) Determine the pH of 0.1 M solution of $NH_4Cl$. $K_b = 1.8 \times 10^{-5}$.

Solution: Both parts of this problem involve the

hydrolysis of a salt of a weak acid or base to form an acidic or basic solution. This occurs because water can dissociate to $H^+$ and $OH^-$ ions, and these can react with ions from weak acids and bases. This necessitates a consideration of $K_{hyd}$, the hydrolysis constant. Because both $NaC_2H_3O_2$ and $NH_4Cl$ are salts, they exist as ions in aqueous solution. The situation in part (a) is

$$C_2H_3O_2^- + H_2O \underset{\leftarrow}{\rightarrow} HC_2H_3O_2 + OH^-$$

The $K_{hyd}$ for this is

$$\frac{[HC_2H_3O_2][OH^-]}{[C_2H_3O_2^-]} .$$

$K_{hyd}$ is also defined as $K_w/K_a$. Therefore,

$$K_{hyd} = \frac{K_w}{K_a} = \frac{[HC_2H_3O_2][OH^-]}{[C_2H_3O_2^-]} .$$

From the equation that depicts the reaction one can see that the $[OH^-] = [HC_2H_3O_2]$, since for every molecule of $OH^-$, a molecule of $HC_2H_3O_2$ must be generated. One can say

$$\frac{K_w}{K_a} = \frac{[OH^-][OH^-]}{[C_2H_3O_2^-]} \quad \text{or} \quad [OH^-]^2 = \frac{K_w}{K_a} [C_2H_3O_2^-].$$

$K_w/K_a$ is small enough so that $[C_2H_3O_2^-]$ at equilibrium will have a value as if hydrolysis had not occurred.

Letting B represent $[C_2H_3O_2^-]$ and taking the square root;

$$[OH^-] = \sqrt{\frac{K_w}{K_a} B} .$$

Next take the negative log of the equation. (Remember that $pOH = -\log [OH^-]$, and that $K_w$ is the ionization constant of water, which is $1.0 \times 10^{-14}$).

$$pOH = \tfrac{1}{2}pK_w - \tfrac{1}{2} pK_a + \tfrac{1}{2} pB$$

However, $\tfrac{1}{2} pK_w = 7$ and $pOH = 14 - pH$. As stated,

$$K_w = 1.0 \times 10^{-14} = [H^+][OH^-] \text{ and } [H^+] = [OH^-].$$

Therefore, $[OH^-]^2$ (or $[H^+]^2$) $= 1.0 \times 10^{-14}$,

$$[OH^-] = 1.0 \times 10^{-7}, \text{ and } [H^+] = 1.0 \times 10^{-7}$$

$$pOH = -\log [1 \times 10^{-7}] = 7 \qquad pOH + pH = 14.$$

$$pH = -\log [1 \times 10^{-7}] = 7. \text{ Therefore, } pOH = 14 - pH.$$

One can now substitute to obtain

$$pH = 14 - (7 - \tfrac{1}{2}pK_a + \tfrac{1}{2}pB) \qquad \text{or}$$

$$pH = 7 + \tfrac{1}{2}pK_a - \tfrac{1}{2}pB.$$

This is the equation wanted. It is known that $B = [C_2H_3O_2^-]$. Only when this union is at equilibrium $\underset{\sim}{=} B$ can this procedure be used (i.e. the $K_a$ of the acid must be $>> 10^{-10}$). The $K_a$ is given as $1.8 \times 10^{-5}$ and meets these requirements. The $pK_a = -\log[1.8 \times 10^{-5}] = 4.74$. $B = 0.01$ M (given) and pB = 2.00.

Now go back to $pH = 7 + \tfrac{1}{2}pK_a - \tfrac{1}{2}pB$ and substitute these values to obtain:

$$pH = 7 + \tfrac{1}{2}(4.74) - \tfrac{1}{2}(2.00)$$

$$pH = 8.37$$

Part (b) is worked out in exactly the same way. One ends up having

$$pH = 7 - \tfrac{1}{2}pK_b + \tfrac{1}{2}pB$$

$$= 7 - \tfrac{1}{2}(4.74) + \tfrac{1}{2}(1.00)$$

$$pH = 5.13$$

where $\tfrac{1}{2}pB = -\tfrac{1}{2}\log[0.1] = \tfrac{1}{2}(1.00)$.

# CHAPTER 13

# CHEMICAL KINETICS

> **Basic Attacks and Strategies for Solving Problems in this Chapter. See pages 423 to 457 for step-by-step solutions to problems.**

The rate of a homogeneous chemical reaction typically depends on the temperature and the concentrations of the species which enter the reaction. For a typical general homogeneous reaction,

$$A + B \rightarrow \text{Products} \qquad\qquad 13\text{-}1$$

experimental results have shown that the rate of disappearance of the reactant A, defined as $(-d[A]/dt)$, can be written as

$$\text{rate} = -k[A]^n [B]^m \qquad\qquad 13\text{-}2$$

In this expression, $k$, the specific reaction rate constant, is a function of the temperature, but not of the concentrations, and $n$ and $m$ are called the order of the reaction with respect to species $A$ and $B$ respectively. It is important to understand that, while the order and the stoichiometric coefficients for a reaction may be the same, the order is **not** necessarily equal to the stoichiometric coefficient. However, the order is equal to the stoichiometric coefficient for each species in an elementary reaction — a reaction that represents the actual molecular path (or mechanism) the reaction follows. This is discussed briefly under chain reactions.

The Arrhenius equation is the most commonly used equation to express how the specific reaction rate constant, $k$, varies with temperature.

$$k = k_o \exp(-E/RT)\{(t^{-1})(\text{conc})^{(1-\text{order})}\} \qquad\qquad 13\text{-}3$$

In this equation, $k_o$ (the pre-exponential constant) has the same number of units as $k$ and $E$ (the activation energy) has units of energy (e.g., joules). $R$ is the universal gas constant and $T$ is the absolute temperature measured in kelvins. A common error is the failure to use an absolute temperature for $T$; this mistake can lead to very serious errors in reaction rate calculations.

To calculate the order of a reaction, it is important to examine the data. If rate data vs. concentrations are available (presumably at a constant temperature), simply try different orders until Equation 13-2 is satisfied. For example, if the rate at which $A$ is consumed, $-r_A$, is tabulated as a function of the concentration of $A$, the different values of n can be tried until the correct one is determined.

$$-r_A = k\,[A]^n \tag{13-4}$$

However, it is more common to measure the concentration as a function of time. It is then necessary to integrate Equation 13-2 to obtain an expression, from the rate equation, giving concentration as a function of time (again, presumably at a constant value of temperature).

$$-d[A]/dt = k\,[A]^n \tag{13-5}$$

In order to integrate Equation 13-5, the value of $n$ must be known. The procedure, therefore, is to guess a value of $n$, say $n = 1$, integrate the equation, and compare the result to the data. If $n = 1$,

$$[A] = [A_0]\,e^{(-kt)} \tag{13-6}$$

where $[A_0]$ is the concentration of $A$ when time, $t$, equals zero. Equation 13-6 can also be written

$$-\ln\{[A/[A_0]\} = kt \tag{13-7}$$

$$\text{or } k = (-\ln\{[A/[A_0]\})/t \tag{13-8}$$

If $k$ is the same for each data point, then the guess that $n = 1$ was correct. If not, other values of $n$ must be tried. If $n \neq 1$, the integrated form of Equation 13-5 is

$$[A]^{(1-n)} = [A_0]^{(1-n)} + kt \tag{13-9}$$

A plot of $[A]^{(1-n)}$ vs. $t$ is a straight line of slope $k$.

If the half-life is measured, these data can also be used to determine the order and specific rate constant of a reaction. The half-life is defined as the time for one half of a reactant to disappear. For first-order reactions, it is independent of the initial concentration; hence the half-life is a direct measure of the reaction rate constant.

$$k = (\ln 2)/t_{1/2} \text{ (for first-order reactions)} \qquad 13\text{-}10$$

Equation 13-10 is easily derived from Equation 13-8 where $[A] = [A_o]/2$ at $t = t_{1/2}$. For reactions of any other order (i.e., $n \neq 1$), the half-life depends on the initial concentration and the rate constant and is given by the equation.

$$k = \{2^{(n-1)}-1\}/\{(n-1)[A_o]^{(n-1)}t_{1/2}\} \qquad 13\text{-}11$$

The activation energy of a reaction, which can be loosely interpreted as the energy the reactants must possess to react, can be determined by measuring the rate constant at several different temperatures. If the Arrhenius equation (Equation 13-3) is written differently,

$$\ln k = \ln k_o - E/RT \qquad 13\text{-}12$$

it becomes apparent that a plot of $\ln k$ vs. $1/T$ will be a straight line with a slope of $-E/R$. If the specific rate constant is known at two temperatures, $T_1$ and $T_2$, $E$ can be calculated from the equation

$$E = R \ln(k_1/k_2)/(1/T_2 - 1/T_1) \qquad 13\text{-}13$$

A common "rule of thumb" states that reaction rates double with a 10°C increase in temperature. For this rule to apply for reactions at or near room temperature, let us calculate what the activation energy must be. For the rate to double between 25°C and 35°C, the specific reaction rate constant must double.

$$k_{308K} = 2\, k_{298K} \qquad 13\text{-}14$$

$$k_o e^{-[E/308R]} = 2\, k_o e^{-[E/298R]} \qquad 13\text{-}15$$

$$-E/(308R) = (\ln 2) - E/(298R) \qquad 13\text{-}16$$

$$E = R\, (\ln 2)/\{1/298 - 1/308) = 52.87 \text{ KJ} \qquad 13\text{-}17$$

Problem 469 asks you to repeat the calculation above for different temperatures; its solution follows the procedure above with different values of temperature. Problem 468 asks what activation energy will result if a tripling of the rate occurs for a 10°C increase in temperature. The solution is identical to that just completed except that 3 replaces 2 in Equation 13-14 and is carried through to the subsequent equations.

Chain reactions are reactions that take place in sequence. Two types are discussed in the problems of Chapter 13. The first, illustrated in Problem 474, is two consecutive first-order reactions. The quantitative solutions for the concentrations of $A$, $B$, and $C$ are given here, which expands on the qualitative answers requested in the problem.

$$A \xrightarrow{k_1} B \xrightarrow{k_2} C \qquad\qquad \text{13-18}$$

$$A = [A_o]\, e^{(-k_1 t)} \qquad\qquad \text{13-19}$$

$$B = \{k_1/(k_2 - k_1)\}\, [A_o]\{e^{(-k_1 t)} - e^{(-k_2 t)}\} \qquad\qquad \text{13-20}$$

$$C = \{A_o/(k_2 - k_1)\}\{k_2[1 - e^{(-k_1 t)}] - k_1[1 - e^{(-k_2 t)}]\} \qquad\qquad \text{13-21}$$

Note also that

$$C = A_o - A - B. \qquad\qquad \text{13-22}$$

The second type of chain reaction discussed is a reaction mechanism. Mechanisms are series of elementary reactions that represent the actual path followed on a molecular scale in an overall reaction. The key to solving the problems is to assume, for short-lived intermediates, that the concentration does not change with time (i.e., $dC/dt = 0$ for short-lived intermediates). This is called the steady state assumption and permits the rate equations written for any step in the mechanism containing a short-lived species to be greatly simplified.

Step-by-Step Solutions to
Problems in this Chapter,
"Chemical Kinetics"

## THE RATE LAW

A group of mountain climbers set up camp at a 3 km altitude and experience a barometric pressure of 0.69 atm. They discover that pure water boils at 90°C and that it takes 300 minutes of cooking to make a "three-minute" egg. What is the ratio of the rate constant $k_{100°C}$ and $k_{90°C}$?

Solution:   Since we do not know the rate expression for cooking an egg, we will assume one of the form

$$rate = k \, [A]^m \, [B]^n \, ...$$

where k is the rate constant, A, B, ... are the reactants, and the overall order of the reaction is m + n + ..., We will write the rate equations at the normal boiling point of water (100°C) and at 90°C as

$$rate_{100°C} = k_{100°C} \, [A]^m [B]^n \, ... \qquad and$$

$$rate_{90°C} = k_{90°C} \, [A]^m [B]^n \, ...$$

Dividing the first of these by the second gives

$$\frac{rate_{100°C}}{rate_{90°C}} = \frac{k_{100°C} \, [A]^m \, [B]^n ...}{k_{90°C} \, [A]^m [B]^n ...} = \frac{k_{100°C}}{k_{90°C}}$$

Since the egg cooks 100 times faster at 100°C than at 90°C (300 min/3 min = 100), $rate_{100°C}/rate_{90°C} = 100$.

Hence,

$$\frac{rate_{100°C}}{rate_{90°C}} = 100 = \frac{k_{100°C}}{k_{90°C}} , \quad or, \quad \frac{k_{100°C}}{k_{90°C}} = 100.$$

● **PROBLEM** 442

The reaction

$$A + B \rightarrow C$$

was studied kinetically and the following data was obtained.

| Experiment | A | B | Rate (mole/liter - min) |
|---|---|---|---|
| 1 | 1.0 M | 1.0 M | 0.15 |
| 2 | 2.0 M | 1.0 M | 0.30 |
| 3 | 1.0 M | 2.0 M | 0.15 |

Determine the rate expression.

Solution: The rate of reaction is equal to some rate constant, k, multiplied by the concentrations of A and B raised to the appropriate powers. That is,

$$rate = k [A]^m [B]^n$$

where the exponents m and n are to be determined.

Comparing experiments 2 and 1, we see that holding [B] constant while doubling [A] doubles the rate of reaction (from 0.15 to 0.30). Hence, the rate is directly proportional to [A] and m = 1. Thus, if we hold [B] constant and triple [A], the rate triples.

Comparing experiments 3 and 1, we see that holding [A] constant and changing [B] (from 1.0M to 2.0M) has no effect on the rate. Hence, the rate is independent of [B] and n = 0 (so that $[B]^0 = 1$ and [B] does not appear in the rate expression).

Substituting m = 1 and n = 0 into the rate expression gives

$$rate = k [A]^1 [B]^0, \quad or, \quad rate = k [A].$$

Two different molecules, A and B, were reacted to give products according to the equation

$$A + B \rightarrow \text{products.}$$

The rate of reaction was measured for fixed concentrations of A and B and the following data was obtained:

| Experiment | [A] (moles/liter) | [B] (moles/liter) | Rate (moles$^2$/liter$^2$ -sec) |
|---|---|---|---|
| 1 | 1.0 | 1.0 | 0.05 |
| 2 | 1.0 | 2.0 | 0.10 |
| 3 | 3.0 | 1.0 | 0.45 |

Write an expression for the rate of reaction.

Solution: The rate of reaction is equal to some rate constant, k, multiplied by the concentrations of A and B raised to appropriate powers. That is,

$$\text{rate} = k \, [A]^m \, [B]^n$$

where the exponents m and n are to be determined.

Comparing experiment 2 with experiment 1, we see that when the concentration of B is doubled from 1.0 M to 2.0 M, the concentration of A being held constant at 1.0 M, the rate of reaction doubles from 0.05 to 0.10 moles$^2$/liter$^2$-sec. Hence, the rate of reaction is directly proportional to the concentration of B, and the exponent of [B] in the rate expression is n = 1. If we were to triple [B], holding [A] constant, the rate would triple, and so on.

Comparing experiment 3 with experiment 1, we see that when the concentration of A is tripled from 1.0 M to 3.0 M, the concentration of B being held constant at 1.0 M, the rate of reaction is multiplied by 9 (9 × 0.05 = 0.45 moles$^2$/liter$^2$ - sec). Since the factor by which the rate is multiplied (9) is the square of the factor by which the concentration of A is multiplied (3), the exponent of [A] in the rate expression must be m = 2. Thus if [A] is doubled, [B] being held constant, the rate is multiplied by 4.

Substituting the exponents of [A] and [B] determined above into the rate expression gives

$$\text{rate} = k \, [A]^2 \, [B].$$

For the reaction

$$2NO(g) + H_2(g) \rightarrow N_2O(g) + H_2O(g)$$

at 1100°K, data, as shown in the following table, were obtained. Find the rate law and the numerical value of the specific rate constant.

| Initial Pressure of NO, atm | Initial Pressure of $H_2$, atm | Initial rate of pressure decrease, atm/min |
|---|---|---|
| 0.150 | 0.400 | 0.020 |
| 0.075 | 0.400 | 0.005 |
| 0.150 | 0.200 | 0.010 |

<u>Solution</u>:    The rate of a homogeneous reaction, i.e. one which occurs in only one phase, depends on the concentration of reactants in that phase. This reaction is a homogeneous reaction in the gaseous phase. It can be investigated kinetically by following the change in pressure of the gaseous mixture as the reaction proceeds. The pressure drops because 3 moles of gaseous reactants are converted to two moles of gaseous products. Since the reactants are being used up during the course of the reaction, their concentrations and their rate if reaction are constantly changing. The concentrations and rates listed are those at the very beginning of the reaction, when little change has occurred. From the data given, one can see that when the initial pressure of NO is halved with the pressure of $H_2$ remaining constant, the initial rate is quartered.  When the pressure of $H_2$ is halved and the NO remains constant, the initial rate is halved.  When the pressure of $H_2$ is doubled and NO remains constant, the rate doubles.  We conclude that the rate is proportional to $[H_2]$. When the $H_2$ pressure is kept constant and the NO concentration is doubled, the rate is quadrupled, this means that the rate is proportional to $[NO]^2$. The equation for the rate can thus be written

$$\text{Rate} = k \, [H_2][NO]^2$$

where k is the rate constant.

Using the values in the first trial of the table one can solve for k.

$$0.020 \text{ atm/min} = k \, (0.400 \text{ atm})(0.150 \text{ atm})^2$$

$$\frac{0.020 \text{ atm/min}}{(0.400 \text{ atm})(0.150 \text{ atm})^2} = k$$

$$k = 2.22 \text{ atm}^{-2} \text{ min}^{-1}.$$

Assume that an A molecule reacts with two B molecules in a one-step process to give $AB_2$. (a) Write a rate law for this reaction. (b) If the initial rate of formation of $AB_2$ is $2.0 \times 10^{-5}$ M/sec and the initial concentrations of A and B are 0.30 M, what is the value of the specific rate constant?

Solution:    (a) The overall equation for this reaction is

$$A + 2B \rightarrow AB_2$$

Since no other information is provided about the reaction, the rate law for the reaction is assumed to be written

$$\text{Rate} = k \, [A][B]^2,$$

where k is the rate constant and [] indicates concentration.

(b) One can solve for k using the rate law when the rate, [A] and [B] are given as they are in this problem.

$$\text{Rate} = k \, [A][B]^2$$

$$2.0 \times 10^{-5} \text{ M/sec} = k(0.30 \text{ M})(0.30 \text{ M})^2$$

$$\frac{2.0 \times 10^{-5} \text{ M/sec}}{(0.30 \text{ M})(0.30 \text{ M})^2} = k$$

$$k = 7.41 \times 10^{-4} \text{ M}^{-2} \text{ sec}^{-1}.$$

A reacts with B in a one-step reaction to give C. The rate constant for the reaction is $2.0 \times 10^{-3}$ $M^{-1}$ $sec^{-1}$. If 0.50 mole of A and 0.30 mole of B are placed in a 0.50 liter-box, what is the initial rate of the reaction?

Solution:    The equation for this reaction can be written:

$$A + B \rightarrow C$$

From this, one can write the rate law assuming that the reaction is first order in both A and B. When a reaction is first order in a particular reactant, it means that the rate is proportional to the concentration of the reactant. Thus, the rate law is written

$$\text{Rate} = k[A][B],$$

where k is the rate constant and [] indicates concentration. In this problem, one is given k, [A] and [B]. Thus,

Rate = $2.0\times10^{-3}$ $M^{-1}$ $sec^{-1}$

    × (1 M)(.60M)

    = $1.20\times10^{-3}$ M/sec

$k = 2.0\times10^{-3}$ $M^{-1}$ $sec^{-1}$

[A]= 0.50 mole/0.5 liter

   = 1 M

[B]= 0.30 mole/0.5 liter

   = .60 M.

● **PROBLEM** 447

In studying the kinetics of the reaction

$$X(g) + Y(g) \rightarrow Z(G)$$

at 800°K, the data in the following table were observed. (a) Write the rate law for this reaction. (b) What is the numerical value of the specific rate constant? (c) What would be the initial rate of Z formation starting with 0.15 M X and 0.15 M Y? (d) How would the rate in (c) be changed if, after the reaction had just begun, the volume of the container were abruptly doubled?

| Initial concentration of X, M | Initial concentration of Y, M | Initial rate of formation of Z, M/min |
|---|---|---|
| 0.10 | 0.10 | 0.030 |
| 0.20 | 0.20 | 0.240 |
| 0.20 | 0.10 | 0.120 |

Solution:  (a) All of the components in this system are in the gaseous phase, thus the system is homogeneous. The rate of a reaction in a gaseous homogeneous system is dependent upon the concentration of the reactants. In a gaseous system, concentration is a function of the pressure of the system. From the data given in the table, one can see that when the concentration of Y is held constant and the concentration of X is doubled, the rate is quadrupled. Therefore, the rate is proportional to $[X]^2$. When the concentration of X is held constant and the concentration of Y is doubled, the rate is doubled and is thus proportional to [Y]. The rate law for this reaction can thus be written  Rate = k $[X]^2[Y]$  where k is the rate constant.

(b) The numerical value of k is found by substituting in values for [X], [Y] and the rate into the rate law. Concentration values for X and Y  may be obtained from any one line of the table. The values obtained from the first line of the table are

$$0.030 \text{ M/min} = k \ (0.10 \text{ M})^2 (0.10 \text{ M})$$

$$\frac{0.030 \text{ M/min}}{(0.10 \text{ M})^2 (0.10 \text{ M})} = k$$

$$30.0 \text{ M}^{-2} \text{ min}^{-1} = k$$

(c) One can solve for the initial rate of the formation of Z by using the rate law and k when [X] and [Y] are given. Thus,

$$\text{Rate} = k[X]^2[Y]$$

$$\text{Rate} = (30.0 \text{ M}^{-2} \text{ min}^{-1})(0.15 \text{ M})^2 (0.15 \text{ M})$$

$$= .101 \text{ M/min}.$$

(d) According to Boyle's Law the pressure is inversely proportional to the volume. Therefore, if the volume is doubled the pressure is halved. Because the reactants in this system are gases, when the pressure is halved, the concentration is halved. One can find the rate of the reaction described in (c) by using the rate law with the concentrations of the reactants halved

$$\text{Rate} = 30.0 \text{ M}^{-2} \text{ min}^{-1} \ (\tfrac{1}{2}(0.15 \text{ M}))^2 (\tfrac{1}{2}(0.15 \text{ M})$$

$$\doteq 30.0 \text{ M}^{-2} \text{ min}^{-1} \ (.075 \text{ M})^2 (.075 \text{ M})$$

$$= 1.27 \times 10^{-2} \text{ M/min}$$

The percent of the first rate that the second rate is equal to is found by dividing the second rate by the first and then multiplying the quotient by 100.

$$\frac{\text{Rate}_2}{\text{Rate}_1} \times 100 = \frac{1.27 \times 10^{-2}}{1.01 \times 10^{-1}} \times 100 = 12.5\% \text{ or } 1/8.$$

The second rate is, therefore, 1/8 of the first.

● **PROBLEM** 448

Photochemical smogs are thought to be partially initiated by the photolytic decomposition of $NO_2$ to form NO and O. When the subsequent reaction

$$2NO + O_2 \rightarrow 2NO_2$$

is studied kinetically, it is found that when the initial concentration of $O_2$ is doubled and that of NO held constant, the initial reaction rate doubles; and when the initial concentration of NO is doubled and that of $O_2$ held constant, the initial reaction rate quadruples. Write the rate expression for this reaction.

Solution:   The rate of reaction is equal to some rate constant k multiplied by the concentrations of NO and $O_2$ raised to the appropriate powers. That is,

$$\text{rate} = k\,[NO]^m\,[O_2]^n$$

where the exponents m and n are to be determined.

When [NO] is held constant and $[O_2]$ is doubled, the rate is doubled. Hence, the rate is directly proportional to $[O_2]$ and n = 1. Thus, if we were to hold [NO] constant and triple $[O_2]$, the rate is tripled.

When $[O_2]$ is held constant and [NO] is doubled, the reaction rate is quadrupled. Hence, the rate is second order in [NO] and m = 2. Thus, if we were to hold $[O_2]$ constant and triple [NO], the rate is multiplied by $3^2 = 9$.

Substituting m = 2 and n = 1 into the rate expression above gives

$$\text{rate} = k\,[NO]^2\,[O_2]^1, \quad \text{or, rate} = k\,[NO]^2\,[O_2].$$

● **PROBLEM** 449

The reaction

$$A + 2B \rightarrow \text{products}$$

was studied kinetically and the following data was obtained.

| Experiment | [A] (moles/liter) | [B] (moles/liter) | Rate (mole/liter - sec) |
|---|---|---|---|
| 1 | $2.0 \times 10^{-3}$ | $4.0 \times 10^{-3}$ | $5.7 \times 10^{-7}$ |
| 2 | $2.0 \times 10^{-3}$ | $8.0 \times 10^{-3}$ | $11.4 \times 10^{-7}$ |
| 3 | $4.0 \times 10^{-3}$ | $4.0 \times 10^{-3}$ | $22.8 \times 10^{-7}$ |

Determine the rate law for this reaction.

Solution:   The rate of reaction is equal to some rate constant, k, multiplied by the concentrations of A and B raised to the appropriate powers. That is,

$$\text{rate} = k\,[A]^m\,[B]^n$$

where the exponents m and n are to be determined. The presence of the coefficient 2 in the reaction equation

$$A + 2B \rightarrow \text{products}$$

in no way affects the form of the rate equation.

Comparing experiments 2 and 1 we see that doubling [B] while holding [A] constant doubles the rate of reaction ($2 \times 5.7 \times 10^{-7} = 11.4 \times 10^{-7}$). Hence, the rate is directly proportional to [B] and n = 1.

Comparing experiments 3 and 1 we see that doubling [A] while holding [B] constant quadruples the rate ($4 \times 5.7 \times 10^{-7} = 22.8 \times 10^{-7}$). Thus, the rate has a square dependence on [A] ($2^2 = 4$), that is, it is second order in [A], and m = 2.

Substituting m = 2 and n = 1 into the rate expression gives

$$\text{rate} = k\,[A]^2\,[B]^1, \quad \text{or} \quad \text{rate} = k\,[A]^2\,[B].$$

● **PROBLEM** 450

You have the general reaction: A + 2B + C = AB + BC. You collect the following data:

Concentration(moles/liter)

| | [A] | [B] | [C] | -dA/dt (moles/liter-sec) |
|---|---|---|---|---|
| (1) | 1.00 | 1.00 | 2.00 | 1.00 |
| (2) | 2.00 | 1.00 | 2.00 | 2.00 |
| (3) | 2.00 | 2.00 | 2.00 | 8.00 |
| (4) | 2.00 | 2.00 | 4.00 | 8.00 |

(a) Determine the experimental rate law expression. (b) Find the specific reaction rate constant. (c) Calculate the rate of reaction, if the [A], [B], and [C] are, respectively, 1.0, 2.0, and 3.0: (d) Speculate on the rate-controlling step, i.e., the slow step.

Solution: First write a general rate law for the reaction, and then write a specific law based on the given data. The specific rate constant can be found from a given set of data. The rate with given concentrations is obtainable from the specific rate law. From this, a rate-controlling step can be guessed at. Proceed as follows:

(a) To write the rate law, note that, in general, the rate = r = $k[A]^x[B]^y[C]^z$, where the brackets indicate concentrations of A, B, and C, k = specific rate constant and the exponents (x, y, and z) are the orders of the reaction with respect to each reactant. The order of a chemical reaction is the number of particles needed to form the activated complex or transition state. To find the actual law, you must determine these exponents. To do this, consult the data. You are told the actual rate of the reaction, since you are given - dA/dt, which is an indication of rate. Thus, to find the order, see what happens when one of the concentrations is changed. In going from (1) to (2), the concentration of A is doubled from 1.00 to 2.00. Notice: - dA/dt also doubles, from 1.00 to 2.00, which means the reaction is first order, x = 1, with respect to A. In going from (2) to (3), [B] goes from 1.00 to 2.00 (doubles). But notice, - dA/dt quadruples going from 2.00 to 8.00. Thus, the reaction is second order, y = 2, with respect to B. In going from (3) to (4), [C] is doubled going from 2.00 to 4.00, while - dA/dt remains constant at 8.00. Thus, the reaction is zero order, z = 0, with respect to C, i.e., rate doesn't depend on [C]. You then have, r = $k[A][B]^2[C]^0$ or $r = k[A][B]^2$ for the experimental rate law expression.

(b) To find k, substitute any of the four sets of data in the experimental rate law expression.

$$\text{rate} = \frac{- dA}{dt} = k[A][B]^2$$

For example, let us take the data from line (1). You have 1.00 mole/liter-sec = k[1.00 mole/liter][1.00 mole/liter]$^2$, or k = 1.00. Thus, the value of the specific rate constant k is 1.00 liter$^2$/mole$^2$ - sec.

(c) To find the rate of the reaction at the concentrations [A] = 1, [B] = 2, and [C] = 3, substitute these values into the rate law, r = k [A][B]$^2$. Thus, r = $(1)[1][2]^2$ = 4.0 moles/liter-sec.

(d) You can use previous information gained in this problem to speculate about what the rate-controlling step must be. From part (a) you know the rate does not depend on [C]. Thus, C cannot be involved in the slow step. The rate of the reaction is influenced by [A] and [B], however. The rate-controlling step is probably the combination of $B_2$ and A, i.e., $B_2 + A \xrightarrow{\text{slow}} AB_2$.

# THE ORDER OF REACTIONS

Under certain conditions, the rate equation for the formation of water

$$2H_2 + O_2 \rightarrow 2H_2O$$

is given by

$$\text{rate} = k \ [H_2]^2 \ [H_2O]$$

where k is the rate constant. What is the overall order of this rate equation?

Solution:    The overall order of a rate equation is equal to the sum of the exponents to which the concentrations are raised. In the equation

$$\text{rate} = k \ [H_2]^2 \ [H_2O],$$

$[H_2]$ is raised to the second power and $[H_2O]$ is raised to the first power. Hence, the rate is second order in $[H_2]$, first order in $[H_2O]$, and 2 + 1 = 3, or third order overall.

● **PROBLEM** 452

You have the general reaction: A + B → products. Consider the following reaction rate data and then determine the rate law for each case: (a) When [A] is doubled, the initial rates doubles; when [B] is doubled, the initial rate doubles, (b) When [A] doubles, initial rate doubles; when [B] doubles, the initial rate is halved, (c) Doubling [A], doubles initial rate; but doubling [B] leaves the rate unchanged.

Solution:    The given reaction is one with two different molecules being converted to products. The rate of any such reaction is:

$$r = \text{rate} = k \ [A]^x \ [B]^y, \qquad \text{where } k = \text{a rate}$$

constant and [A] = concentration of A and [B] = concentration of B. x and y are the number of each A and B molecules, respectively, that are present in the activated complex or transition state.    The determination of these exponents is the key to writing the rate law for each case. You proceed as follows:

(a) You are told that when A or B is doubled, the rate is doubled. Thus, when A → 2A, r → 2r, which means $2r = k \ [2A]^x \ [B]^y$. Now, if x = 1, then $[2A]^x = [2A]$ and dividing by 2 you come back to $r = k[A][B]^y$. In other words, r → 2r if A → 2A, implies r varies directly as A; that is, x = 1. The same type of situation exists with B, so that y = 1. Thus, the rate law becomes r = k[A][B].

(b) Here, when A → 2A, r → 2A, which means, again, A varies directly as r, and so x = 1. When B → 2B, r → ½r, however, in this case, $½r = [A][2B]^y$. This can only come about if y = - 1. If y = - 1, then $[2B]^{-1} = 1/[2B]$, so that ½r = [A] 1/[2B]. The halves can be cancelled to return

433

a direction proportion. Thus, the rate law becomes r = k[A]/[B].

(c) Here, when [A] → [2A], r → 2r but when [B] → [2B], r → r. For A: you have, again, a direction proportion between A and r, which requires x = 1. For B: when B's concentration is doubled, nothing happens to r; it remains constant. This means r does not depend upon the [B]. For r not to depend upon [B]$^y$, y must equal zero, thus, r = k $\overline{[A]}$ [B]$^0$ = k [A](1). As such, the rate law becomes r = k [A].

● **PROBLEM** 453

From collision theory, it is found that when the rate-determining step involves collision of two A molecules, the rate will be proportional to the square of the A concentration. Explain why.

Solution:   Collision theory makes the assumption that for a chemical reaction to take place, particles must collide. The rate will depend upon the number of collisions per unit time between the particles involved and the fraction of these collisions that are effective. If you increase [A], there are more A molecules which can be hit. Also, the average A molecule gets hit more often. As such, the number of A with A collisions varies as the square of concentration due to this double effect.

● **PROBLEM** 454

For the general reaction A + B → C + D, what is the effect on the number of collisions between A and B of (a) tripling the concentration of each, (b) quadrupling the concentration of each?

Solution:   For any given chemical reaction, the number of collisions is proportional to the rate. The Law of Mass Action states that the speed of a chemical reaction is proportional to the product of the concentrations of the reacting molecules. The number of collisions in the above reaction depends upon the concentration of A and B. If the concentration of A is tripled the number of collisions is tripled, then if the concentration of B is tripled the number of collisions will triple again. Thus the number of collisions is increased 3 × 3 times or 9 times. When the concentrations of A and B are quadrupled a similar method is used. The number of collisions is increased 4 × 4 or 16 times.

The following data was collected for the reaction

$OH^- + CH_3COOCH_2CH_3 \rightarrow CH_3COO^- + CH_3CH_2OH$ :

| Time, min | $[CH_3COOCH_2CH_3]$ | $[OH^-]$ |
|---|---|---|
| 0 | .02000 | .02000 |
| 5 | .01280 | .01280 |
| 15 | .00766 | .00766 |
| 25 | .00540 | .00540 |
| 35 | .00426 | .00426 |
| 55 | .00289 | .00289 |
| 120 | .00137 | .00137 |

(a) Determine whether the reaction is first or second order. (b) Write the rate law expression.

Solution:    The order of a reaction is the number of particles needed to form the transition state. One can tell whether the reaction is first or second order by an investigation into the constancy of the half-life of this reaction over time. It is a characteristic feature of first order reactions that the half-life, the time necessary for half the particular reactant present initially to decompose, is constant over time. Thus, by determining whether the half-life is constant or not, the order of this reaction can be found.

From the data, half the material is decomposed between 5 and 15 minutes. In the period, 5 to 15 minutes, the concentration of both $[CH_3COOCH_2CH_3]$ and $[OH^-]$ goes from .0128 to .00766, which means half has decomposed. One can approximate that the half-life is about 10 minutes. To see if this time of half-life remains constant, notice the time needed for the concentration to go from .0054 to .00270, which is another ½ being decomposed. This occurs between t = 25 minutes to approximately t = 55 minutes, a difference of 30 minutes. Thus, the half-life is not constant. It changed from 10 to 30 minutes. Thus, the reaction is not first order. It must be second order.

(b) To solve this part, use the information from (a). If the reaction is second order, the rate depends upon both the concentration of $CH_3COOCH_2CH_3$ and $OH^-$. The rate law expression is, thus,

$$\text{rate} = \frac{- d[CH_3COOCH_2CH_3]}{dt} = \frac{- d[OH^-]}{dt} = k[OH^-][CH_3COOCH_2CH_3],$$

where k = rate constant. This equation is, by definition, the rate law expression for a second order reaction with 2 different molecules.

The reaction $C_2H_4O \rightarrow CH_4 + CO$ is a first-order reaction with a specific rate constant of $.0123$ $min^{-1}$ at $415°C$. Calculate the percent of original $C_2H_4O$ present that will be decomposed after heating at $415°C$ for 1 hour.

Solution:    First find the percentage of starting material left (as compared to the initial amount $C_0$), $C/C_0 = (100)$; then the percent decomposition equals $1 - C/C_0$ $(100)$. To find $C/C_0$ $(100)$, employ the fact that $2.303 \log C_0/C = kt$, where $C_0 =$ initial concentration, $C =$ existing concentration, $k =$ specific rate constant, and $t =$ time in minutes for a first order reaction. The formula can be rewritten as $C/C_0 = $ antilog $(- kt/2.303)$.

Substitute the values for k and t to obtain

$$\frac{C}{C_0} = \text{antilog} \left[ \frac{(- .0123)(60)}{2.303} \right] = \text{antilog} (- .320) = .479.$$

Thus, $C/C_0(100) = 100 \times .479 = 47.9\%$. That is, $.479$ or $47.9\%$ of the original concentration remains after 60 minutes, so that $1 - 47.9\%$ or $52.1 \%$ has decomposed.

The following observations were made for the reaction $NH_4CNO \rightleftharpoons NH_4^+ + CNO^- \rightarrow (NH_2)_2CO$: 1) Addition on an equimolar concentration of KCNO with respect to $NH_4CNO$ doubled the initial rate of the reaction. 2) Addition of an equimolar concentration of $NH_4Cl$ doubled the initial rate of reaction. From these observations, (a) Determine the order of the reaction. (b) Write the kinetic rate law expression for the reaction. (c) Discuss the influence, if any, of the following equilibrium upon the reaction kinetics: $NH_4 + CNO \rightleftharpoons NH_3 + HCNO$ with

$$K_{equil} = \frac{[NH_3][HCNO]}{[CNO^-][NH_4^+]} = 10^{-4}.$$

Solution:    (a) The order of a chemical reaction is the number of distinct particles that must come together to form the activated complex. This factor can also be called the molecularity of the reaction. To see which particles directly determine the reaction rate, note the effect on the rate when more of a certain type of particle is added. You are told that when KCNO is added, that is, the KCNO concentration is doubled, the rate doubles. This means the reaction must be first order with respect to $CNO^-$ (since, in solution, KCNO dissociates to $CNO^-$). When $[NH_4Cl]$ is doubled, the reaction rate is again doubled. Thus, the reaction is first order with respect to $NH_4^+$, since, in solution, $NH_4Cl$ dissociates into $NH_4^+$. Since doubling the amount of either $NH_4^+$ or $CNO^-$ doubles the reaction rate, the reaction must be second-order overall.

(b) By definition, a second order reaction with two different molecules has rate = $- d[A]/dt = - d[B]/dt = k[A][B]$, where k = rate constant and A and B are the two different molecules in A + B → products. Thus, for this reaction, the rate law becomes

$$\text{rate} = \frac{- d\,[NH_4CNO]}{dt} = \frac{d\,[(NH_2)_2CO]}{dt} = k\,[NH_4^+][CNO^-]$$

(c) To answer this part, notice the following: The equilibrium constant for the formation of $NH_3$ and HCNO is extremely small, which means that, in solution, the concentration of these species in pure $NH_4CNO$ solution should also be small, i.e.,

$$\frac{[NH_3]}{[NH_4^+]} = \frac{[HCNO]}{[CNO^-]} = (10^{-4})^{\frac{1}{2}} = .01.$$

If either $NH_3$ or HCNO is added to the solution, the reaction is driven to the left. But, even if all of the other reactants available were consumed, the concentration of $NH_4^+$ and $CNO^-$ would increase by only 1%, which means that the reaction rate would increase by 2%, which is not easily detected.

● **PROBLEM** 458

When mixed at 700°K, $H_2$(g) and $I_2$(g) react to produce HI(g). The reaction is first order in $H_2$ and first order in $I_2$. Suppose at time t = 0, one mole of $H_2$ and one mole of $I_2$ are simultaneously injected into a 1-liter box. One second later, before the reaction is complete, the contents of the box are examined for the number of moles of HI. What would be the probable effect on this number if each of the following changes were made in the initial conditions? (a) Use two moles of $H_2$ instead of one. (b) Use two moles of $I_2$ instead of one. (c) Use a 2-liter box. (d) Raise the temperature to 750°K. (e) Add a platinum catalyst. (f) Add enough neon gas to double the initial pressure.

<u>Solution</u>: When a reaction is said to be first order in a particular reactant, it means that the rate of the reaction is proportional to the concentration of that reactant. Thus, for this reaction the rate law can be written

Rate = k $[H_2][I_2]$,

where k is the rate constant and [] indicates concentration. When 1 mole of each $H_2$ and $I_2$ are used, one can solve for the rate by substituting these values into the rate law.

Rate = k (1)(1) = 1k

Solving for the rate when $[H_2]$ is 2 moles

Rate = k[2][1] = 2k

Thus, when the concentration of $H_2$ is doubled the rate is doubled.

(b) One can solve for the rate when $[I_2]$ = 2 moles by substituting into the rate law.

Rate = k(1)(2) = 2 k

Thus when $[I_2]$ doubled the rate is also doubled.

(c) Boyle's law states that the pressure is inversely proportional to the volume. Thus, when this reaction which originally takes place in a 1-liter box is moved to a 2-liter box, the pressure of the reactants is halved. Solving for the rate:

Rate = k ($\frac{1}{2}$)($\frac{1}{2}$) = $\frac{1}{4}$ k      or      .25 k.

Thus, for this reaction, when the pressure of the reactants is halved, the rate is $\frac{1}{4}$ of the original rate.

(d) Product is formed when an atom of H collides with an atom of I, thus the rate of the reaction will increase with a rise in temperature because the atoms will move more quickly creating a greater chance for collisions.

(e) By definition, a catalyst will increase the rate of the reaction by lowering the amount of energy needed for the reaction to proceed.

(f) When neon gas is added to increase the pressure in the 1-liter box, the rate of the reaction does not change because the concentrations of $H_2$ and $I_2$ remain unchanged. When the pressure in the system is doubled by decreasing the volume by $\frac{1}{2}$ the rate is increased because the concentrations of $H_2$ and $I_2$ are increased. If one starts out with one mole per liter of $H_2$ at 1 atm and then increases the pressure to 2 atm by adding neon, the concentration of $H_2$ is still 1 mole per liter. When the pressure in a container containing 1 mole per liter of $H_2$ is increased from 1 to 2 atm by halving the volume of the container, the concentration of $H_2$ is now 1 mole per $\frac{1}{2}$ liter or 2 moles per liter. Thus when neon is added to the system in this problem, the rate is unchanged.

Interpret the following: The intensity of the deep purple color of an acid $KMnO_4$ solution can be used as an indicator of the extent to which oxidation has taken place. Often there is a considerable time lapse before any visual evidence of reaction is noted; but once begun, the process proceeds vigorously. If, on the other hand, a small amount of the essentially colorless $MnCl_2$ salt is dissolved in this solution, the process is immediately vigorous, no latency period being noted.

Solution: The presence of the $Mn^{2+}$ in the solution is the key to the reaction. When $MnCl_2$ is dissolved in the solution, the reaction is vigorous since the salt immediately dissociates to produce $Mn^{2+}$ ions.

$KMnO_4$ also immediately dissociates into $K^+$ and $MnO_4^-$. With $KMnO_4$, however, a latency period is noticed since it takes time for $Mn^{2+}$ to be produced from the reduction of the $MnO_4^-$ ion.

The fact that once $Mn^{2+}$ does exist, whether from $MnO_4^-$ or $MnCl_2$, the reaction proceeds vigorously suggests that $Mn^{2+}$ is a catalyst. Thus, $Mn^{2+}$ increases the rate of the reaction without being consumed. Because $Mn^{2+}$ is also a reaction product, the reaction is autocatalytic.

# HALF-LIFE

What is the half-life of an unstable substance if 75% of any given amount of the substance decomposes in one hour?

Solution: The half-life is defined as the time it takes for one half of the amount of a substance to decompose. When given the time elapsed and the percent of decomposition, one can find the half-life. One knows that ½ of the substance decomposes in the time equal to the half-life. This leaves ½ of the substance, ½ of this decomposes in the next span of time elapsed equal to the half-time. This leaves ¼ of the substance. 75% of it has decomposed after two half-lifes have elapsed. Thus 2 half-lifes equal one hour or the half-life of the substance is ½ hour.

For the first order decomposition of azomethane at 600°K, it takes 30 minutes for the original concentration to decrease to half its value. After 60.0 minutes have elapsed, what percentage of the azomethane originally present remains?

<u>Solution</u>:  The decomposition is first order which means rate = k [A], where k is a constant and [A] is the concentration of azomethane. The half-life is independent of time, i.e., it is a constant and is defined as the time necessary for half the particular reactant present to decompose. Thus, if you knew the half-life of azomethane, you could calculate the percentage of azomethane that remains after a certain period of time. The half-life of azomethane is 30 minutes. Thus, after 30 minutes, you have half of the original material left. After 60 minutes half of <u>this</u> original material is left or ¼ of the original material remains.

In the reaction $N_2O_5 \rightarrow N_2O_4 + \frac{1}{2}O_2$, the $N_2O_5$ decomposes by a first-order mechanism. At 298°K, the half-life is 340 minutes. Find the value of the reaction rate constant. Calculate the number of minutes required for the reaction to proceed 70 percent towards completion.

<u>Solution</u>:  Half-life may be defined as the time necessary for half the particular reactant present initially, in this case $N_2O_5$, to decompose. For a first order reaction, the rate constant can be expressed in terms of half-life, $t_{\frac{1}{2}}$.

Namely, $t_{\frac{1}{2}} = \frac{.693}{k}$, where k is the rate constant. (Caution: This expression is only true for a first order reaction.) Solving this expression for k to obtain

$$k = \frac{.693}{t_{\frac{1}{2}}} = \frac{.693}{340} = 2.04 \times 10^{-3} \text{ min}^{-1}.$$

To find the amount of time required for the reaction to proceed to 70% completion, use the fact that

$$t = \frac{2.303}{k} \log \frac{C_0}{C}$$ in a first order reaction, where

t = time, k = rate constant, $C_0$ = initial concentration and C = existing concentration.

Suppose the initial concentration is X, then at 70% completion, the existing concentration is .30 X. Having calculated k, substitute to find t = minutes.

$$t = \frac{-2.303}{k} \log \frac{C_0}{C} = \frac{-2.303}{2.04 \times 10^{-3}} \log \left( \frac{.30X}{X} \right)$$

$$= (1129)(.523) = 590 \text{ min.}$$

The half-life for the first order reaction:

$$SO_2Cl_2 \rightarrow SO_2 + Cl_2$$

is 8.0 minutes. In what period of time would the concentration of $SO_2Cl_2$ be reduced to 1.0% of the original?

<u>Solution</u>:     If a reaction is first order, it means that the rate of the reaction is proportional to the concentration of a single reactant. For example, if substance A decomposes into the products B and C,

A → B + C,

the rate is proportional to the concentration of A which is present at any time.

Rate = k [A],

where k is the rate constant for the reaction and [A] is the concentration of A.

The rate at which the reaction proceeds is equal to the decrease in concentration of A with change of time. This can be written in the integrated form

$$2.303 \log \frac{[A]}{[A_0]} = - kt,$$

where [A] is the concentration of A after time t has elapsed, $[A_0]$ is the original concentration of A and k is the rate constant for the reaction. In this problem, the final ratio of [A] to $[A_0]$ is given as 1.0% or 0.010. Therefore, this equation can be used to find the time it takes for this process to occur, once one has calculated the rate constant k. The rate constant can be found if one remembers the relationship between the half-life of a reaction and k. Namely,

$$t_{\frac{1}{2}} = \frac{0.693}{k} \,,$$

where $t_{\frac{1}{2}}$ is the half life and k is the rate constant. The $t_{\frac{1}{2}}$ for this reaction is given as 8.0 minutes, therefore this relation can be used to obtain k.

$$k = \frac{0.693}{t_{\frac{1}{2}}}$$

$$k = \frac{0.693}{8.0 \text{ minutes}} = 0.087/\text{minute}$$

Once k is known, it can be used in the original rate equation to determine the time it takes for this reaction to proceed until the concentration of A is 1.0% that of the original concentration.

$$2.303 \log \frac{[A]}{[A_0]} = -kt. \qquad \text{Substituting,}$$

$$2.303 \log 0.010 = -0.087/\text{minute} \times t$$

$$t = \frac{2.303 \log 0.010}{-0.087/\text{min}} = \frac{(2.303)(-2)}{-0.087/\text{min}}$$

$$= 53 \text{ minutes.}$$

● PROBLEM 464

At a certain temperature, the half-life for the decomposition of $SO_2Cl_2$ is 4.00 hours. (a) In what period of time would the concentration of $SO_2Cl_2$ be reduced to 10% of the original? (b) Starting with 100 mg of $SO_2Cl_2$, what weight would be left at the end of 6.50 hours? The reaction is first order, therefore the equation $\ln[A]/[A_0] = -kt$ can be used. [A] is the concentration of A at time t, $[A_0]$ is the original concentration of A and k is rate constant.

Solution: (a) When the concentration of A is 10% of the original, the ratio of $[A]/[A_0]$ equals .10. Therefore, one can use the equation given to solve for t after first determining k for the reaction. k is determined from the half-life in a first order reaction. The relation between k and $t_{\frac{1}{2}}$ (half-life) is

$$t_{\frac{1}{2}} = \frac{0.693}{k}$$

One is given the $t_{\frac{1}{2}}$. Solving for k:

$$k = \frac{0.693}{t_{\frac{1}{2}}} = \frac{0.693}{4.0 \text{ hrs}} = .173 \text{ hr}^{-1}$$

One can now solve for t:

$$\frac{\ln[A]}{[A_0]} = -kt$$

$$\ln .10 = -.173 \text{ hr}^{-1} \, t$$

$$t = \frac{\ell n \ \ .10}{- \ .173 \ hr^{-1}} = \frac{- \ 2.303}{- \ .173 \ hr^{-1}} = 13.3 \ hrs.$$

(b) One can solve for the [A] after 6.5 hours by use of the equation:

$$\frac{\ell n[A]}{[A_0]} = - \ kt$$

One is given that there is originally 100 mg of A. 1 g = 1000 mg, thus 100 mg = 0.1 g. $[A_0]$ must be in moles/liter. One finds the number of moles of $SO_2Cl_2$ in 0.1 g by dividing 0.1 g by the molecular weight of $SO_2Cl_2$ (MW = 135).

$$[A_0] = \frac{no. \ of \ moles}{liter} = \frac{0.1 \ g}{\dfrac{135 \ g/mole}{liter}} = 7.41 \times 10^{-4} \ \frac{moles}{liter}$$

Solving for [A] using the value of k found in part a;

$$\ell n \ \frac{[A]}{7.41 \times 10^{-4} \ moles/liter} = - \ (.173 \ hr^{-1})(6.5 \ hr)$$

$$\ell n \ [A] - \ell n[7.41 \times 10^{-4} \ moles] = - \ (.173 \ hr^{-1})(6.5 \ hr)$$

$$\ell n \ [A] = - \ (1.12) + \ell n \ (7.41 \times 10^{-4})$$

$$\ell n \ [A] = - \ 1.12 - 7.21$$

$$\ell n \ [A] = - \ 8.33$$

$$[A] = 2.41 \times 10^{-4} \ moles$$

One can find the number of grams by multiplying the number of moles by the molecular weight.

$$weight = 2.41 \times 10^{-4} \ moles \times 135 \ g/mole$$

$$= .0326 \ g = 32.6 \ mg$$

● **PROBLEM** 465

The ketone acid $(CH_2CO_2H)_2CO$ undergoes a first-order decomposition in aqueous solution to yield acetone and carbon dioxide:

$$(CH_2CO_2H)_2CO \rightarrow (CH_3)_2CO + 2CO_2$$

(a) Write the expression for the reaction rate. (b) The rate constant k has been determined experimentally as $5.48 \times 10^{-2}$/sec at 60°C. Calculate $t_{\frac{1}{2}}$ at 60°C. (c) The rate constant at 0°C has been determined as $2.46 \times 10^{-5}$/sec. Calculate $t_{\frac{1}{2}}$ at 0°C. (d) Are the calculated half-lives in accord with the stated influence of temperature on reaction rate?

Solution:    (a) For a chemical decomposition, the rate of the reaction is equal to the product of the rate constant (k) and the concentration of the compound decomposing. Thus,

$$\text{Rate} = k\ [(CH_2CO_2H)_2CO]$$

(b) Because the rate is only proportional to $[(CH_2CO_2H)_2CO]$, the reaction is first-order. For a first-order reaction, the half-life ($t_{\frac{1}{2}}$) is related to k by the following equation:

$$t_{\frac{1}{2}} = \frac{0.693}{k}$$

Solving for $t_{\frac{1}{2}}$:

$$t_{\frac{1}{2}} = \frac{0.693}{5.48 \times 10^{-2}/sec} = 12.65\ \text{sec}.$$

(c) One can solve for $t_{\frac{1}{2}}$ at $0°C$ using the same equation.

$$t_{\frac{1}{2}} = \frac{0.693}{2.46 \times 10^{-5}/sec} = 2.82 \times 10^4\ \text{sec}.$$

(d) In general, the speed of a chemical change is approximately doubled for each ten degrees rise in temperature. The temperature rises $60°$, from $0°C$ to $60°C$, therefore, the rate should double six times or the ratio of the $t_{\frac{1}{2}}$ at $0°C$ to the $t_{\frac{1}{2}}$ at $60°C$ is $2^6$. $2^6 = 64$.

$$\frac{t_{\frac{1}{2}}\ 0°}{t_{\frac{1}{2}}\ 60°} = \frac{2.82 \times 10^4\ \text{sec}}{12.65\ \text{sec}} = 2.23 \times 10^3$$

This is much greater than the expected ratio of 64.

● **PROBLEM** 466

When heated to $600°C$, acetone ($CH_3COCH_3$) decomposes to give CO and various hydrocarbons. The reaction is found to be first order in acetone concentration with a half-life of 81 sec. Given at $600°C$ a 1-liter container into which acetone is injected at 0.48 atm, approximate how long would it take for the acetone pressure to drop to 0.45 atm?

Solution:    When a reaction is said to be first order in a particular reactant it means that the rate of the reaction is proportional to the concentration of that reactant. Thus the rate law for this reaction can be written:

$$\text{Rate} = k\ [CH_3COCH_3]$$

where k is the rate constant and [] indicate concentration. The rate constant is related to the half-life ($t_{\frac{1}{2}}$) by the following equation:

$$t_{\frac{1}{2}} = \frac{0.693}{k}$$

Solving for k:    $t_{\frac{1}{2}}$ = 81 sec

$$81 \text{ sec} = \frac{0.693}{k}$$

$$k = 8.56 \times 10^{-3} \text{ sec}^{-1}$$

To solve for the time it takes for a reaction to proceed to a certain degree one uses the equation:

$$d\ln[A] = - k \, dt$$

where $d\ln[A]$ is the natural logarithm of the original concentration subtracted from the natural log of the final concentration of acetone, k is the rate constant and dt is the time elapsed. Because acetone is a gas in this experiment the pressure is proportional to the concentration. Therefore one can let $[A]_{original}$ = 0.48 and $[A]_{final}$ = 0.45. Solving for dt:

$$\ln 0.48 - \ln 0.45 = (- 8.56 \times 10^{-3} \text{ sec}^{-1})(dt)$$
$$(-.734) - (-.799) = (- 8.56 \times 10^{-3} \text{ sec}^{-1})(dt)$$
$$\frac{(-.734) - (-.799)}{- 8.56 \times 10^{-3} \text{ sec}^{-1}} = dt$$

$$- 7.5 \text{ sec} = dt$$

Therefore 7.5 sec have elapsed since the beginning of the reaction.


# THE ARRHENIUS EQUATION; RELATING TEMPERATURE AND REACTION RATE

● **PROBLEM** 467

Using a specific case, show that the effect of a 10°K rise in temperature will have a greater effect on the rate constant, k, at low temperatures than it does at high temperatures.

Solution:    The best way to demonstrate this fact is to use the natural logarithm form of Arrhenius' equation, which gives an indication of the effect of temperature on reaction rate. The expression may be written $\ln k = \ln A - E_a/2.303\ RT$, where k is the rate constant, A = the Arrhenius constant, $E_a$ = activation energy, R = universal gas constant ($8.314\ J\ mol^{-1}\ deg^{-1}$) and T = temperature in Kelvin (Celsius plus 273). For a low temperature, consider room temperature (T = 300°K). For a first-order reaction, A might be $1.0 \times 10^{14}\ sec^{-1}$ and E might be 80 k J/mole. Thus, at T = 300°K, you have log k = log ($1 \times 10^{14}$) − $80,000/(2.303)(8.314)(300) = .07$. Thus, $k = 1.2\ sec^{-1}$. If you increase the temperature 10°K to 310°K, you find by exactly the same method of calculation, that k = $3.3\ sec^{-1}$. Thus, at a temperature of 10°K higher (in a low temperature range), k was increased nearly three times.

Now consider a higher temperature range, say T = 900°K. For this, log k = ($1 \times 10^{-14}$) − $80,000/(2.303)$ $(8.314)(900) = 9.358$. Solving log k = 9.358, k = 2.28 × $10^9$. Again, let us increase the temperature by 10°K to 910°K. Using the same type of calculations, you find that k becomes 2.56 × $10^9$. The percent change =

$$\frac{2.56 \times 10^9 - 2,28 \times 10^9}{2.28 \times 10^9} \times 100 = 12.28\%.$$

Thus, at the higher temperatures, the rate constant was increased only 12% as compared to 300% at the low temperatures.

● **PROBLEM** 468

What activation energy should a reaction have so that raising the temperature by 10°C at 0°C would triple the reaction rate?

Solution:    The activation energy is related to the temperature by the Arrhenius equation which is stated

$$k = Ae^{-\ E/RT}$$

where A is a constant characteristic of the reaction; e is the base of natural loragithms, E is the activation energy, R is the gas constant ($8.314\ J\ mol^{-1}\ deg^{-1}$) and T is the absolute temperature. Taking the natural log of each side:

$$\ln k = \ln A - E/RT$$

For a reaction that is 3 times as fast, the Arrhenius equation becomes

$$3\ k = Ae^{-\ E/R(T\ +\ 10°)}$$

Taking the natural log:

$$\ln 3 + \ln k = \ln A - E/R(T + 10°)$$

Subtracting the equation for the final state from the equation for the initial state:

$$\ln k = \ln A - E/RT$$
$$- (\ln 3 + \ln k = \ln A - E/R(T + 10))$$
$$\overline{\quad\quad - \ln 3 = - E/RT + E/R(T + 10)\quad\quad}$$

Solving for E:

$$-\ln 3 = - E/RT + E/R(T + 10) \qquad R = 8.314 \text{ J/mole } °K$$
$$T = 0 + 273 = 273$$

$$- \ln 3 = - E/(8.314 \text{ J/mole-K})(273K) + E/(8.314 \text{ J/mole-K})(283K)$$

$$- 1.10 = - E/2269.72 \text{ J/mole} + E/2352.86 \text{ J/mole}$$

$$(2269.72 \text{ J/mole})(2352.86 \text{ J/mole}) \times - 1.10 =$$

$$(- E/2269.72 \text{ J/mole} + E/2352.86 \text{ J/mole})(2269.72 \text{ J/mole})(2352.86 \text{ J/mole})$$

$$- 5.874 \times 10^6 \text{ J}^2/\text{mole}^2 = (- E)(2352.86 \text{ J/mole}) + E (2269.72 \text{ J/mole})$$

$$- 5.874 \times 10^6 \text{ J}^2/\text{mole}^2 = - 8.314 \times 10^1 \text{ J/mole} \times E$$

$$7.06 \times 10^4 \text{ J/mole} = E$$

● **PROBLEM** 469

It has been suggested that a 10° rise in temperature results in a twofold response in the rate of a chemical reaction. What is the implied activation energy at 27°C?; at 1000°C?

Solution:    The activation energy, $E_A$, is the difference between the heat concentration of the active molecules and the inert molecules. The activation energy can be related to the rate constants of the reaction at two different temperatures by one form of the Arrhenius equation:

$$(i) \quad \log \frac{k_2}{k_1} = - \frac{E_a}{2.303 \text{ R}} \left( \frac{1}{T_2} - \frac{1}{T_1} \right)$$

(where R = universal gas constant; $k_1$ is the rate constant at temperature $T_1$; $k_2$ is the rate constant at $T_2$).

This equation can be rewritten as

(ii)   $E_a = \dfrac{2.303 \; R}{\dfrac{1}{T_1} - \dfrac{1}{T_2}} \; \log \dfrac{k_2}{k_1}$

Thus to solve for $E_a$, you find $T_1$, $T_2$, and $k_2/k_1$ and substitute. Let $T_1$ = initial temperature. If there is a $10°$ rise in temperature, then $T_2 = T_1 + 10°K$. You are told that when the temperature is increased, the original rate is doubled. Thus, if $k_1$ = original rate, $k_2 = 2k_1$. Thus, $\log k_2/k_1 = \log 2 = .301$. For the case, $T_1 = 27°C = 300°K$, $T_2 = 310°K$ you have

$$E_a = \frac{-\;2.303 \; R}{\left(\dfrac{1}{T_2} - \dfrac{1}{T_1}\right)} \; \log \frac{k_2}{k_1} = \frac{-\;2.303 \; (1.987) \left(\dfrac{1 \; k \; cal}{1000 \; cal}\right)}{\left(\dfrac{1}{310} - \dfrac{1}{300}\right)} \; (.301)$$

= 12.8 k cal/mole

For the case with $T_1 = 1000°C = 1273°K$, $T_2 = 1283°K$. You have,

$$E_a = \frac{-\;2.303 \; (1.987) \; (\; 1 \; k \; cal/1000 \; cal)}{\left(\dfrac{1}{1283} - \dfrac{1}{1273}\right)} \; (.301)$$

=   226 k cal/mole

(In both cases, 1 k cal/1000 cal is a conversion factor.)

● **PROBLEM** 470

(1) A reaction proceeds five times as fast at 60°C as it does at 30°C. Estimate its energy of activation. (2) For a gas phase reaction with $E_A$ = 40,000 cal/mole, estimate the change in rate constant due to a temperature change from 1000°C to 2000°C.

Solution:   The actuation energy $E_A$ can be related to the rate constants $k_1$ (at temperature $T_1$) and $k_2$ (at temperature $T_2$) by the Arrhenius equation:

$$\log \frac{k_2}{k_1} = -\;\frac{E_a}{2.303 \; R} \left(\frac{1}{T_2} - \frac{1}{T_1}\right),$$

where R = universal gas constant.

(1) You are told a reaction proceeds five times as fast at 60°C as it does at 30°C. Therefore, if $k_1$ = rate constant at 30°C = 303°K with $T_1 = 303°K$, then $k_2 = 5k_1$ at 60°C = 333°K with $T_2 = 333°K$. You are given R. Substitute these values into the Arrhenius equation, and

solve for $E_A$. Rewriting and substituting,

$$E_a = \frac{-2.303 \ R}{\frac{1}{T_2} - \frac{1}{T_1}} \ \log \frac{k_2}{k_1} = \frac{(-2.303)(1.987)\left(\frac{1 \ k \ cal}{1000 \ cal}\right)}{\left(\frac{1}{333} - \frac{1}{303}\right)} \ \log 5$$

$$= (15.4 \ k \ cal/mole)(.699) = 10.8 \ k \ cal/mole.$$

[Note: 1 k cal/1000 cal is a conversion factor to obtain the correct units.] To answer (2) find $k_2/k_1$ from the Arrhenius equation. Rewriting and substituting,

$$\frac{k_2}{k_1} = antilog \ \left[\frac{E_a}{2.303 \ R} \left(\frac{1}{T_2} - \frac{1}{T_1}\right)\right]$$

$$= antilog\left[\frac{-40{,}000}{(2.303)(1.987)}\left(\frac{1}{2{,}273} - \frac{1}{1{,}273}\right)\right]$$

$$= antilog \ 3.02 = 1.05 \times 10^3.$$

That is, the rate should be about 1050 times as great at 2000°C as at 1000°C.

● **PROBLEM** 471

For the gas-phase decomposition of acetaldehyde, $CH_3CHO \rightarrow CH_4 + CO$, the second-order rate constant changes from 0.105 $m^{-1}$ $sec^{-1}$ at 759°K to 0.343 $m^{-1}$ $sec^{-1}$ at 791°K. Calculate the activation energy that this corresponds to. What rate constant would you predict for 836°K?

Solution: The rate constant is related to the activation energy by the Arrhenius equation, which is stated

$$k = Ae^{-E/RT}$$

where k is the rate constant, A is a constant characteristic to the reaction, e is the base of natural logarithms, E is the activation energy, R is the gas constant (8.314 J/mole °K) natural log:

$$\ln k = \ln A - E/RT$$

One is given k, R and T in the problem for two trials.

For 759°K:     k = 0.105 $m^{-1}$ $sec^{-1}$

a) $\ln(0.105) = \ln A - E/(8.314 \ J/mole \ °K)(759°K)$

For 791°K:     k = 0.343 $m^{-1}$ $sec^{-1}$

b) $\ln(0.343) = \ln A - E/(8.314 \ J/mole \ °K)(791°K)$

Subtract equation b from equation **a** to solve for E.

$\ln(0.105) = \ln A - E/(8.314 \text{ J/mole}°K)(759°K)$

$- [\ln(0.343) = \ln A - E/(8.314 \text{ J/mole}°K)(791°K)]$

c) $\ln(0.105) - \ln(0.343) = -E/(8.314 \text{ J/mole}°K)(759°K) +$
$E/(8.314 \text{ J/mole}°K)(791°K)$

$- 2.25 - (- 1.07) = - E/6310.33 \text{ J/mole} + E/6576.37 \text{ J/mole}$

$(6310.33 \text{ J/mole})(6576.37 \text{ J/mole})(- 1.18) = E/6310.33 \text{ J/mole}$
$+ E/6576.37 \text{ J/mole}(6310.33 \text{ J/mole})(6576.37 \text{ J/mole})$

$- 4.8969 \times 10^7 \text{ J}^2/\text{mole}^2 = - 6576.37 \text{ E J/mole}$
$+ 6310.33 \text{ E J/mole}$

$- 4.8969 \times 10^7 \text{ J}^2/\text{mole}^2 = - 2.6604 \times 10^2 \text{ JE/mole}$

$E = \dfrac{- 4.8969 \times 10^7 \text{ J}^2/\text{mole}^2}{- 2.6604 \times 10^2 \text{ J/mole}}$

$E = 1.8407 \times 10^5 \text{ J/mole}$

One can solve for k at 836°K by replacing the values of this third trial for the values found at 791°K.

$\ln(0.105) - \ln k = - \dfrac{1.8407 \times 10^5 \text{ J/mole}}{(8.314 \text{ J/mole}°K)(759°K)}$

$+ \dfrac{1.8407 \times 10^5 \text{ J/mole}}{(8.314 \text{ J/mole}°K)(836°K)}$

$- 2.25 - \ln k = - 2.917 \times 10^1 + 2.648 \times 10^1$

$- \ln k = - 4.4 \times 10^{-1}$

$k = 1.55 \text{ m}^{-1} \text{ sec}^{-1}.$

● **PROBLEM** 472

A chemist found the value of the specific rate constant for the decomposition of nitrous oxide, $2NO + N_2 + O_2$, at two separate temperatures: $k_1$ = .14 liter/mole-sec at 970°K, $k_2$ = 3.7 liter/mole-sec at 1085°K. (a) Calculate the activation energy, $E_a$, for the reaction, (b) Calculate the A factor in the Arrhenius equation, (c) Calculate the specific rate constant at 800°K.

Solution: This problem deals with the effect of temperature on the kinetics of a reaction.

(a) $E_a$ is the activation energy, the difference between the heat content of the active molecules and that of the inert molecules. To find the value of $E_a$ for the decomposition of nitrous oxide, use the Arrhenius equation:

$$E_a = \frac{-2.303\ R}{\left[\frac{1}{T_2} - \frac{1}{T_1}\right]}\ \log\left(\frac{k_2}{k_1}\right)$$

where T = temperature in degrees Kelvin (Celsius plus 273°), k = specific rate constant and R = universal gas constant. Substitute known values to obtain ($T_1 = 970°$, $T_2 = 1085°$)

$$E_a = \frac{(-2.303)(1.987)(1/1000)}{(1/1085 - 1/970)}\ \log\left(\frac{3.7}{.14}\right)$$

$$= 59.2\ k\ cal/mole$$

[Note: 1/1000 is a conversion factor of 1 k cal/1000 cal.]

(b) The Arrhenius equation can also be written in the form

$$\ell_n k = \ell_n A - \frac{E_A}{RT}\ ,\qquad \text{where A is called the frequency}$$

factor. To find A, first solve the equation in terms of A.

$$\ell_n A = \ell_n k + \frac{Ea}{RT}\quad \text{or}\quad \log A = \log k + \frac{Ea}{2.303\ RT}\ .$$

Substitute for k and T and calculate

$$\log A = \log 3.7 + \frac{(59.2)\left(\frac{1000\ cal}{1\ k\ cal}\right)}{(2.303)(1.987)(1085)} = 0.57 + 11.9 = 12.5.$$

Since antilog 12.5 = A, $A = 3 \times 10^{12}$ liters/mole-sec. (Notice: $k_2$ and $T_2$ values were used in the calculations but $k_1$ and $T_1$ could have been used instead.)

(c) To find k at 800 solve the Arrhenius equation for k

$$\log k = \log A - \frac{E_a}{2.303\ RT}\ .$$

Substitute in known values for A, $E_a$, R, and T to obtain:

$$\log (3 \times 10^{12}) - \frac{(59.2)\left(\frac{1000\ cal}{1\ k\ cal}\right)}{2.303\ (1.987)(800)} = 12.5 - 16.2 = -3.7$$

k.= antilog $(-3.7) = 2 \times 10^{-4}$ liters/mole-sec.

# CHAIN REACTIONS

Define the term "chain reaction". Using a specific example, distinguish between a chain-starting, a chain-propagating, and a chain-terminating step. Discuss the energy absorbed or released in the first and last of these.

Solution:   In chemical kinetics, a chain reaction occurs when an intermediate species that is consumed in one step is regenerated in a later step. As a result, there is a sequence of steps which endlessly repeat themselves, like links in a chain, until the chain is ended or the starting material exhausted. A specific example of a chain reaction, is the formation of methylchloride from methane ($CH_4$) and chlorine gas ($Cl_2$). Overall, the reaction is:

$$CH_4 + Cl_2 \rightarrow CH_3CL + HCl$$

  methyl
  chloride

The mechanism of this chain reaction is as follows:

(1)     $Cl_2 \rightarrow 2Cl\cdot$

(2) $CH_4 + Cl\cdot \rightarrow CH_3\cdot + HCl$

(3) $CH_3\cdot + Cl_2 \rightarrow CH_3Cl + Cl\cdot$

A chain-starting step encompasses one or more stable molecules and produces highly reactive species, such as single atoms. In this example, (1) fits this description. A stable $Cl_2$ molecule forms single chlorine radicals ($Cl\cdot$) that immediately react in step (2). $Cl_2$ does not go to $Cl\cdot$ spontaneously. Energy must be supplied in the form of heat or light. Chain-starting steps generally require energy. A chain-propagating step involves the reaction between a molecule and one of the highly reactive species, producing at least one highly reactive species. Reactions (2) and (3) fit this description. In (2), the reactive species $CH_3\cdot$, methyl radical, is produced via the reaction between the molecule $CH_4$ and reactive species $Cl\cdot$ . In (3), this species, i.e., $CH_3\cdot$ reacts with $Cl_2$ to produce another reactive species, chlorine radical ($Cl\cdot$). Note: this chlorine radical can go back to react with $CH_4$ to repeat the entire process, i.e., (2) and (3) over and over again, until you have the chain-terminating step. Chain termination occurs when one of the highly reactive species reacts with another highly reactive species, or with the wall and no energetic species result. In our example, such a step would be the reaction of two chlorine radicals to produce chlorine gas, $Cl\cdot + Cl\cdot \rightarrow Cl_2$. Since $Cl_2$ is not a reactive species, the chain is terminated. Notice that this is the reverse of $Cl_2 \rightarrow 2Cl\cdot$, which required energy. Thus, a chain-terminating step will generally release energy.

Given the consecutive reaction A $\xrightarrow{k_1}$ B $\xrightarrow{k_2}$ C with $k_1 = k_2$. Draw a graph for the time variation of the concentrations of A, B, and C.

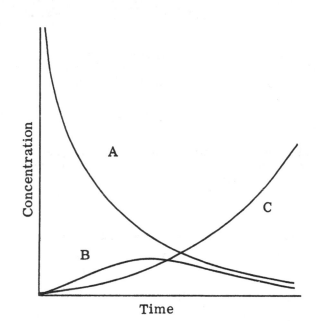

**Solution:** To draw this graph, you need to determine the concentration with respect to time of each species in this reaction. This will necessitate writing rate law expressions for each of the species. The rate law states that rate is equal to a proportionality constant k times the concentration of one species or more than one species, each raised to a power. The power is equal to the number of moles reacting. You are told that

$$A \xrightarrow{k_1} B \xrightarrow{k_2} C \qquad \text{with } k_1 = k_2$$

The rate at which A disappears can be written as $\dfrac{-d[A]}{dt} = k_1[A]$. A will disappear at a certain rate as given by the rate constant. How long it takes to disappear will reflect how much was initially present: the concentration. The rate at which C appears may be written as $\dfrac{d[C]}{dt} = k_2[B]$. C forms from B at a certain rate as indicated by $k_2$. Thus, the amount of B initially present determines the time it take for C to appear. What about the concentration of B? It increases with time as A disappears, yet decreases as C appears. The net increase in B will be the differences in these processes. Thus, $\dfrac{d[B]}{dt} = k_1[A] - k_2[B]$. If you start with pure A at t = 0,

$\dfrac{d\,[B]}{dt}$ is positive, since $k_1[A] \gg k_2[B]$. After a time, how-ever, A is used up, so that $\dfrac{d[B]}{dt}$ becomes negative. This means the concentration of B will peak sometime along in the reaction. At that instant,

$\dfrac{d\,[B]}{dt} = 0$, so that $k_1[A] - k_2[B] = 0$, which means $\dfrac{[A]}{[B]} = \dfrac{k_2}{k_1}$.

Thus, the behavior of the system depends on $k_2/k_1$. As $k_1$ grows small, B becomes less notable. But, you are told that $k_2 = k_1$, which indicates that $k_2/k_1 = 1$.

[C] will rise as [B] falls. Thus, you have the following graph.

If all reactions proceed to completion C will be the only species present at the conclusion of the process.

● **PROBLEM** 475

$N_2O_5$ decomposes according to the following equation: $N_2O_5 \rightarrow 2NO_2 + \frac{1}{2}O_2$. The rate expression =

$-\dfrac{d[N_2O_5]}{dt} = k\,[N_2O_5]$. The following mechanism has been proposed: $N_2O_5 \underset{\leftarrow}{\overset{k}{\rightleftharpoons}} NO_2 + NO_3$

$\qquad NO_2 + NO_3 \overset{k_1}{\rightarrow} NO_2 + O_2 + NO$

$\qquad NO + NO_3 \overset{k_2}{\rightarrow} 2NO_2$

Show the rate of $O_2$ formation is directly proportional to $[N_2O_5]$.

Solution:   To relate $O_2$ formation with the concentration of $N_2O_5$ investigate the mechanism of this reaction. In the mechanism, oxygen ($O_2$) is formed only from the second re-action, the rate being

$\qquad \dfrac{d[O_2]}{dt} = k_1[NO_2][NO_3]$. To relate $\dfrac{d[O_2]}{dt}$ to $[N_2O_5]$,

express $[NO_2][NO_3]$ in terms of $[N_2O_5]$.

This can be done from the first reaction, which is at equilibrium. The equilibrium constant of this reaction, k, measures the ratio of products to reactants, each raised to the power of its coefficient in the chemical equation. Thus,

$\qquad k_{eq} = \dfrac{[NO_2][NO_3]}{[N_2O_5]}$    or $[NO_2][NO_3] = k_{eq}\,[N_2O_5]$

Recall, $\frac{d[O_2]}{dt} = k_1 [NO_2][NO_3]$. Thus $[NO_2][NO_3]$ can be replaced by $k_{eq} [N_2O_5]$, so that

$\frac{d[O_2]}{dt} = k_1 k_{eq} [N_2O_5]$, which means $O_2$ formation is directly proportional to $[N_2O_5]$.

● **PROBLEM** 476

You have the reaction $HCrO_4^- + 3Fe^{2+} + 7H^+ \rightarrow Cr^{3+} + 3Fe^{3+} + 4H_2O$ with the rate law expression: rate = $k [HCrO_4^-] [Fe^{2+}][H^+]$. Explain why the rate of reaction is not proportional to the number of each species reacting according to the stoichiometry of the balanced equation.

Solution: Stoichiometry predicts only the amount of products expected; it cannot predict when they will be produced. The rate of production is dependent only on the number of molecules in the activated complex and their concentrations in the system. A reaction can occur only when the species come in contact with each other. Thus, the concentration reflects the probability of a collision among the species. Stoichiometry cannot provide any information in these areas, and, as such, the rate of the reaction is not necessarily reflective of the stoichiometry of the balanced equation.

● **PROBLEM** 477

The rate law for the reaction $H_2(g) + Br_2(g) \rightarrow 2HBr(g)$ is $\frac{d[HBr]}{dt} = \frac{k[H_2][Br_2]^{\frac{1}{2}}}{1 + k' [HBr]/[H_2]}$ . Derive this law from the reaction mechanism, which consists of chain-starting, chain-propagating, and chain-terminating steps. Use steady-state approximation.

Solution: A chain-reaction comes about when an intermediate species that is consumed in one step is regenerated in a later step. As a consequence of this, a sequence of steps is set up such that they endlessly repeat themselves until the chain is terminated or the starting material is used up. By various experiments and observations, the mechanism has been found to be the following:

Chain-starting step     : (1) $Br_2 \xrightarrow{k_1} 2Br\cdot$

Chain-propagating step :
$\begin{cases} (2) \ Br\cdot + H_2 \xrightarrow{k_2} HBr + H\cdot \\ (3) \ H\cdot + Br_2 \xrightarrow{k_3} HBr + Br\cdot \\ (4) \ H\cdot + HBr \xrightarrow{k_4} H_2 + Br\cdot \end{cases}$

Chain-terminating step : (5) $2Br\cdot \xrightarrow{k_5} Br_2$

   To derive the rate law for this reaction, the formation
of HBr, use these individual reactions:

   Notice, first, that only those reactions that dealt
with HBr in some way were included. In other words,(1) +
 (5) did not have this molecule, so that they cannot be
included. The whole idea of writing the rate in this
fashion comes from the fact that when, say A + B → products,

$$\text{rate} = \frac{-d[A]}{dt} = \frac{-d[B]}{dt} = k\,[A][B].$$   This is called the
expression for the rate law of a second order reaction. By
order,you mean the number of atoms or molecules whose concentrations directly determine the reaction rate. In
other words, for (2), $d[HBr]/dt = k_2[Br][H_2]$. You can write
similar expressions for (3) and (4). The overall rate of
HBr formation is the sum of the rates of formation of the
HBr in each of these reactions. In (2) and (3), you are
making HBr, so that their rate law expressions are added.
In (4), however, HBr is being eliminated, so that you
subtract. To repeat, you now have

$$\frac{d[HBr]}{dt} = k_2[Br][H_2] + k_3[H][Br_2] - k_4[H][HBR]$$

   To obtain the desired formula, you need to find [Br]
and [H]. You can do this by writing overall rate expressions
for the formation of [H] and [Br]. To do this, use the same
reasoning employed for [HBr]. You find that

$$\frac{d[H]}{dt} = k_2[Br][H_2] - k_3[H][Br_2] - k_4[H][HBr]$$

   Note: Only (2), (3), and (4) dealt with H, which
explains why only they are used. You subtract both (3) and
(4) from (2), since they consume H instead of producing it
as (2) does. For Br, you find

$$\frac{d[Br]}{dt} = 2k_1[Br_2] - k_2[Br][H_2] + k_3[H][Br_2] +$$

$$k_4[H][HBr] - 2k_5[Br]^2.$$        Again, use the

same type of reasoning and technique to obtain this expression. You are told steady-state conditions exist. This
means that the concentrations of the reactive intermediates
that are regenerated do not change with time. This means,
therefore that $d[Br]/dt = 0$ and $d[H]/dt = 0$, since each
is being regenerated. Thus, you can equate the above expressions with zero. You obtain

456

$$\frac{d[Br]}{dt} = 0 = 2k_1[Br_2] - k_2[Br][H_2] + k_3[H][Br] +$$
$$k_4[H][HBr] - 2k_5[Br]^2 .$$

$$\frac{d[H]}{dt} = 0 = k_2[Br][H_2] - k_3[H][Br_2] - k_4[H][HBr].$$

These two equations can be solved simultaneously to find [H] and [Br]. If you perform this operation, you find $[Br] = [k_1/k_5[Br]]^{\frac{1}{2}}$ and

$$[H] = k_2 \frac{(k_1/k_5)^{\frac{1}{2}} [H_2][Br_2]^{\frac{1}{2}}}{k_3[Br_2] + k_4[HBR]} .$$ Recalling, now, that

$$\frac{d[HBr]}{dt} = k_2[Br][H_2] + k_3[H][Br_2] - k_4[H][HBr],$$ you can

substitute these values for [H] and [Br] to obtain

$$\frac{d[HBr]}{dt} = \frac{2\,k_3\,k_2\,k_4^{-1}\,k_1^{\frac{1}{2}}\,k_5^{-\frac{1}{2}}\,[H_2][Br]^{\frac{1}{2}}}{k_3\,k_4^{-1} + [HBr][Br_2]^{-1}} .$$ Now, if you

let $k = 2k_2\,k_1^{\frac{1}{2}}\,k_5^{-\frac{1}{2}}$ and $k' = \dfrac{1}{k_3\,k_4^{-1}}$ , you obtain

$$\frac{d[HBr]}{dt} = \frac{k[H_2][Br_2]^{\frac{1}{2}}}{1 + k'\,[HBr]/[H_2]} .$$

# CHAPTER 14

# THERMODYNAMICS I

> **Basic Attacks and Strategies for Solving Problems in this Chapter. See pages 459 to 502 for step-by-step solutions to problems.**

The key to all of the problems in this chapter is an understanding of the First Law of Thermodynamics. The First Law says, simply, that energy can neither be created nor destroyed. Therefore, in any reaction or process, the energy must balance. This is true of bond energies in chemical reactions, of energy changes associated with chemical reactions, and, with allowances, for changes between mass and energy (i.e., $E = mc^2$) in nuclear reactions.

In chemical reactions, there are two energies that are very common and enter into almost all energy calculations associated with reactions. The first is the internal energy, $E$. This is the bond energy associated with chemical bonds plus the kinetic energy of molecular species. The second is the $PV$ energy. Since so many reactions take place at conditions that cause the system to either expand or contract against the atmospheric surroundings, the effect of $PV$ work on the atmosphere becomes important. The enthalpy is the thermodynamic quantity defined to include both the internal energy and the $PV$ work against the atmosphere.

$$H = E + PV \qquad\qquad 14\text{-}1$$

where $H$ is the enthalpy, $E$ the internal energy, and $PV$ the pressure and volume terms. For condensed phases (i.e., liquids and solids) the volume change is often negligible and $H \approx E$, but for gases the difference is significant. Only changes in enthalpy and internal energy have physical significance; there is no absolute zero point for these quantities. Therefore, tabulation of enthalpy is the difference between the state for which it is tabulated and some reference state either implied or specified.

Important definitions used in the calculations in this chapter follow.

$\Delta H°_r$ = enthalpy change in a reaction (typically joules/mole)

$\Delta E°_r$ = internal energy change in a reaction (typically joules/mole)

$\Delta H°_f$ = enthalpy change for formation of a species from the elements (joules/mole)

$C_p$ = heat capacity at constant pressure (typically joules/mole/°C)

$C_v$ = heat capacity at constant volume (typically joules/mole/°C)

The superscript ° indicates that the value is at some defined standard state. Typically for gases, this is one atmosphere pressure at the specified temperature; for liquids and solids it is the pure species at the specified temperature. If temperature is not specified, it is usually 25°C or 298K.

Since for any balanced reaction, the same number of moles of each element exist in the products and reactants:

$$\Delta H°_r = \Sigma (\Delta H°_f)_{products} - \Sigma (\Delta H°_f)_{reactants} \qquad 14\text{-}2$$

where $\Delta H_f$ is the enthalpy associated with forming a species from the elements. It is simply the value of $\Delta H_r$ where all the reactants are elements. Extensive tables of the values of $\Delta H_f$ are found in handbooks. By convention, $\Delta H$ is positive for endothermic changes (changes which consume energy) and negative for exothermic changes (changes which produce energy).

Similarly, the First Law of Thermodynamics makes it possible to calculate the energy change associated with a reaction from bond energies. The energy, usually the enthalpy, is associated with a reaction.

$$\Delta H°_r = \Sigma (\text{bond energies})_{products} - \Sigma (\text{bond energies})_{reactants} \qquad 14\text{-}3$$

Heat capacities are defined as the energy change required to change the temperature. Again, for liquids and solids, there is little difference in the heat capacity at constant pressure or constant volume and often it is not specified. However, gases expand appreciably when heated and the energy to push back the atmosphere ($P\Delta V$) is significant. For gases, therefore, it is important to specify whether the heat capacity applies for a constant pressure or constant volume process. For ideal gases there is a simple relationship between $C_p$ and $C_v$.

$$C_p - C_v = nR \text{ (the universal gas constant)} \qquad 14\text{-}4$$

Heats of fusion and vaporization are heat changes associated with the solid-liquid phase change and the liquid-vapor phase change, respectively. The solid to liquid phase change and the liquid to vapor phase change are always endothermic — i.e., energy must be supplied to cause the phase change to occur. Energy is required to break physical bonds between molecules. If the phase change proceeds in the opposite direction, as in condensation or freezing, the numerical value of the associated energy change is the same, but the sign is opposite.

● **PROBLEM** 478

For the system described by the following diagram, is the forward reaction exothermic or endothermic?

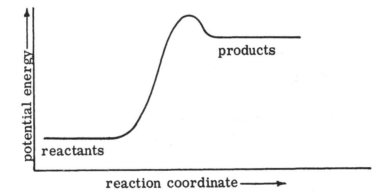

*Solution:* By observing the diagram, we see that the products are at a higher potential energy than the reactants, thus energy must have been absorbed during the course of the reaction. The only mechanism by which this can occur is absorption of heat and subsequent conversion of heat into potential energy. Hence, the reaction absorbs heat and is therefore endothermic.

A steam engine is operated reversibly between two heat reservoirs, a primary reservoir and a secondary reservoir at 35°C. Determine the efficiency of the engine.

Solution: By definition, the primary reservoir is the reservoir from which the engine absorbs heat and the secondary reservoir is the reservoir to which the engine releases heat. Since the engine is steam operated, the primary reservoir must be a source of steam (such as a boiler) operating at a temperature $T_p$ = 100°C ($T_p$ = 100 + 273 = 373°K). The secondary reservoir operates at a temperature $T_s$ = 35 + 273 = 308°K. The efficiency of such an engine is defined as

$$\text{efficiency} = \frac{T_p - T_s}{T_p}$$

where $T_p$ and $T_s$ are in degrees Kelvin. Thus, the efficiency of the engine is

$$\text{efficiency} = \frac{T_p - T_s}{T_p} = \frac{373°K - 308°K}{373°K} = 0.174 \text{ or } 17.4\%.$$

# BOND ENERGIES

Given for hydrogen peroxide, $H_2O_2$:

$$HO + OH(g) \rightarrow 2OH(g) \qquad \Delta H^o_{diss} = 51 \text{ Kcal/mole}$$

From this value and the following data calculate:
(a) $\Delta H^o_f$ of OH(g). (b) C-O bond energy; $\Delta H^o_{diss}$ in $CH_3OH(g)$. $\Delta H^o_f$ of $H_2O_2(g)$ = $-$ 32.58 Kcal/mole. $\Delta H^o_f$ of $CH_3(g)$ = 34.0 Kcal/mole. $\Delta H^o_f$ of $CH_3OH(g)$ = $-$47.96 Kcal/mole.

Solution: (a) $\Delta H^o_{diss}$ for this reaction is equal to the $\Delta H^o_f$ of the reactants subtracted from the $\Delta H^o_f$ of the products, where $\Delta H^o$ equals enthalpy or heat content.

$$\Delta H^o_{diss} = 2 \times \Delta H^o_f \text{ OH} - \Delta H^o_f \text{ HO-OH} = 51 \text{ Kcal/mole}$$

$$2 \times \Delta H_f^o \text{ OH} = 51 \text{ Kcal/mole} + \Delta H_f^o \text{ HO-OH}$$

$$\Delta H_f^o \text{ OH} = \frac{51 \text{ Kcal/mole} - 32.58 \text{ Kcal/mole}}{2}$$

$$\Delta H_f^o \text{ OH} = \frac{18.42 \text{ Kcal/mole}}{2} = 9.21 \text{ Kcal/mole.}$$

(b) $CH_3OH$ dissociates by breaking the C-O bond:

$$CH_3OH(g) \rightarrow CH_3^+(g) + OH^-(g)$$

Therefore, the bond energy of the C-O bond equals $\Delta H_{diss}^o$. $\Delta H_{diss}^o$ equals the sum of the $\Delta H$'s of the products minus the $\Delta H$'s of the reactants. Thus,

bond energy of C-O = $\Delta H_f^o$ of $CH_3^+(g)$ + $\Delta H_f^o$ of $OH^-(g)$ -

$$\Delta H_f^o \text{ of } CH_3OH(g).$$

bond energy of C-O = 34.0 Kcal/mole + 9.21 Kcal/mole

+ 47.96 Kcal/mole

= 91.17 Kcal/mole.

● **PROBLEM** 481

Using the following table of bond energies, calculate the energy change in the following:

(a) $2H_2(g) + O_2(g) \rightarrow 2H_2O(g)$
(b) $CH_4(g) + 2O_2(g) \rightarrow CO_2(g) + 2H_2O(g)$
(c) $CH_4(g) + Cl_2(g) \rightarrow CH_3Cl(g) + HCl(g)$
(d) $C_2H_6(g) + Cl_2(g) \rightarrow C_2H_5Cl(g) + HCl(g).$

**Bond Energies (in kcal per mole)**

| | | | |
|---|---|---|---|
| H–H | 104 | C–O | 83 |
| H–F | 135 | C=O | 178 |
| H–Cl | 103 | C–Cl | 79 |
| H–Br | 88 | C–F | 105 |
| H–I | 71 | Si–O | 106 |
| Li–H | 58 | Si–F | 136 |
| Cl–Cl | 58 | C–C | 83 |
| C–H | 87 | C=C | 146 |
| O–H | 111 | C≡C | 199 |
| O=O | 118 | N≡N | 225 |
| P–Cl | 78 | | |

Solution: When using the bond energies to calculate the net energy change in a reaction, the net energy change is equal to the total bond energy of the bonds formed subtracted from the bond energy of bonds broken. This method is illustrated in the following examples.

a) $2H_2(g) + O_2(g) \rightarrow 2H_2O(g)$

There are two $H_2$ bonds and 1 $O_2$ bond broken and 4 O-H bonds formed here.

$2H-H(g) + O = O(g) \rightarrow 2 H-O-H(g)$

net bond energy = (2× bond energy of H-H) + (1 × bond energy of O=O) - (4 × bond energy of O-H)

The following bond energies can be found in the table.

| Bond | Bond energy (Kcal/mole) |
|------|-------------------------|
| H — H | 104 |
| O = O | 118 |
| O — H | 111 |

The net energy or heat evolved in the reaction can be found using these values.

net bond energy = [(2×104)+(1×118)]-(4×111)

$$\Delta H = (208 + 118) - 444 = -118 \text{ Kcal.}$$

Thus when this reaction occurs 118 Kcal are released. The following examples will be solved in a similar manner.

b) $CH_4(g) + 2O_2 \rightarrow CO_2(g) + 2H_2O(g)$

This reaction can be rewritten

$$H - \overset{\overset{\displaystyle H}{|}}{\underset{\underset{\displaystyle H}{|}}{C}} - H \ (g) + 2 \ O=O \ (g) \rightarrow O=C=O(g) + 2H-o-H(g)$$

This means that there are 4 C-H bonds and 2 O=O bonds broken and 2 O=C bonds and 4 O-H bonds formed.

net energy = [(4 × bond energy of C-H) + (2 × bond energy of O=O)] - [(2 × bond energy of C=O) + (4 × bond energy of O-H)]

The following bond energies are used.

| Bond | Bond energy (Kcal/mole) |
|------|-------------------------|
| O — H | 111 |
| O = C | 178 |
| O = O | 118 |
| C — H | 87 |

One can now determine the heat evolved in this reaction.

net energy = [(4 × 87) + (2 × 118)] - [(2 × 178) + 4 × 111]

$\Delta H = -216$ Kcal.

c) $CH_4(g) + Cl_2(g) \rightarrow CH_3Cl(g) + HCl(g)$

This equation can be rewritten

$H_3C-H$ (g) + $Cl-Cl$ (g) $\rightarrow H_3C-Cl$ (g) + $H-Cl$ (g)

This means that there is 1 C-H and 1 Cl-Cl bond broken and 1 C-Cl and 1 H-Cl bond formed.

net energy = (1 × bond energy of C-H + 1 × bond energy of Cl-Cl) - (1 × bond energy of C-Cl + 1 × bond energy of H-Cl)

The following bond energies are used.

| Bond | Bond energy (Kcal/mole) |
|------|-------------------------|
| C - H | 87 |
| Cl - Cl | 58 |
| C - Cl | 79 |
| H - Cl | 103 |

The heat evolved can now be determined.

net energy = $\Delta H = [(1 \times 87) + (1 \times 58)] - [(1 \times 79) + (1 \times 103)]$

$= (87 + 58) - (79 + 103)$

$= -37$ Kcal.

d) $C_2H_6(g) + Cl_2(g) \rightarrow C_2H_5Cl(g) + HCl(g)$

This can be rewritten as

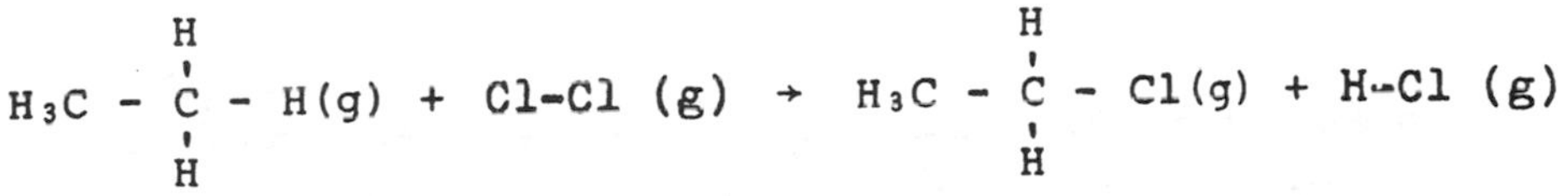

$$H_3C - \overset{\overset{\displaystyle H}{|}}{\underset{\underset{\displaystyle H}{|}}{C}} - H(g) + Cl-Cl \text{ (g)} \rightarrow H_3C - \overset{\overset{\displaystyle H}{|}}{\underset{\underset{\displaystyle H}{|}}{C}} - Cl(g) + H\text{-}Cl \text{ (g)}$$

Thus, there is 1 C-H and 1 Cl-Cl bond broken and 1 C-Cl and 1 H-Cl bond formed.

The net energy for the reaction can be determined by using the following values.

| Bond | Bond energy (Kcal/mole) |
|------|-------------------------|
| C - H | 87 |
| Cl - Cl | 58 |
| C - Cl | 79 |
| H - Cl | 103 |

The net energy can now be found.

net energy = $\Delta H = [(1 \times 87) + (1 \times 58)] - [(1 \times 79) + (1 \times 103)]$

$$= (87 + 58) - (79 + 103)$$

$$= - 37 \text{ Kcal.}$$

# HEAT CAPACITY

● **PROBLEM** 482

At constant volume, the heat capacity of gas differs from the heat capacity at constant pressure. Why?

Solution:  Heat capacity is defined as the quantity of heat that will result in a temperature rise of $1°C$ per mole of material. When heat is added at constant volume, all the heat goes into increasing the kinetic energy of the molecules. However, with constant pressure, the volume can expand. This means the added heat must do two things. It must be able to increase the kinetic energy of the molecules AND do work against the pressure to expand the volume. The constant pressure case would require more heat. In fact, the difference between $C_p$ and $C_v$, where $C_p =$ heat capacity at constant pressure, and $C_v =$ heat capacity at constant volume, is the universal gas constant R. In summary, then, the difference stems from the fact that $C_p$ must increase kinetic energy and do work to expand the volume, while $C_v$ only has to increase the kinetic energy.

● **PROBLEM** 483

What is the heat capacity of uranium? Its atomic heat is 26 J/mole and its atomic weight is 238 g/mole.

Solution:  The atomic heat of a solid element at room temperature is defined as the amount of heat required to raise the temperature of one mole of an element by one degree Celsius. The heat capacity, or specific heat, is the amount of heat required to raise the temperature of one gram of a substance by one degree Celsius. These two quantities can be related by the equation,

atomic heat = atomic weight × heat capacity.

Thus, to find the heat capacity of uranium, substitute the given values of atomic heat and weight.

heat capacity of uranium = 26 J/mole C°/238 g/mole

$$= 0.11 \text{ J/g°C.}$$

How many calories of heat must be added to 3.0 liters of water to raise the temperature of the water from 20°C to 80°C?

Solution:   A calorie is defined as the amount of heat necessary to raise the temperature of one gram of water one degree Centigrade. One can find the mass of 3.0 liters of water by using the density. The density of water is 1 g/ml. 1 liter = 1000 ml. Therefore, 3.0 liters = 3000 ml.

Mass = density × volume

Mass = 1 g/ml × 3000 ml = 3000 g

The temperature change is found by subtracting 20 from 80, which is 60°C.

One can find the calories necessary to raise 3000 g of water 60°C by multiplying 1 cal/g°C by 3000 g by 60°C.

No. of calories = 1 cal/g°C × 3000 g × 60°C

= 180000 cal. = 180 Kcal.

Calculate the quantity of heat required to raise the temperature of one gallon of water (3.78 liters) from 10° to 80°C.

Solution:  The unit of heat is called the calorie. It is defined as the quantity of heat necessary to raise the temperature of one gram of water one degree centigrade. In this problem, one is told that 3.78 liters of water are heated from 10°C to 80°C. If one subtracts the final temperature from the original temperature, one can find the number of degrees the temperature of the water is raised.

number of °C the temperature is raised =

final temperature - original temperature

no. of °C the temperature is raised = 80°C - 10°C = 70°C.

One must now determine the number of grams in 3.78 liters of water. Because 1 milliliter of water weighs one gram, one must determine the number of milliliters in 3.78 liters of water. There are 1000 milliliters in 1 liter; therefore, liters can be converted to milliliters by multiplying the number of liters present by the factor 1000 ml/1 liter.

    3.78 liters × 1000 ml/l liter = 3780 ml

    Because number of ml = number of grams for water, there are 3780 g of water present in 3.78 liters. One now knows that 3780 g of water have been raised 70°C in temperature. Remembering the definition for calorie, one can find the number of calories absorbed in this process by multiplying the number of grams of water by the number of degrees the temperature was raised by the factor 1 calorie/1g - 1°C, the specific heat of water.

number of calories absorbed = 3780 g × 70°C ×

$$1 \text{ calorie}/1g - 1°C$$

$$= 265,000 \text{ calories}$$

    There are 1000 calories in 1 kilocalorie, so that calories can be converted to kilocalories by multiplying the number of calories by the factor 1 Kcal/1000 cal.

    265,000 cal × 1 Kcal/1000 cal = 265 Kcal.

265 Kcal are absorbed in this process.

● **PROBLEM** 486

A piece of iron weighing 20.0 g at a temperature of 95.0°C was placed in 100.0 g of water at 25.0°C. Assuming that no heat is lost to the surroundings, what is the resulting temperature of the iron and water? Specific heats: iron = .108 cal/g-C°; water = 1.0 cal/g-C°.

Solution:   The heat lost by the iron must be equal to the heat gained by the water. One solves for the heat lost by the iron by multiplying the number of grams of Fe by the number of degrees the temperature dropped by the specific heat of iron. The specific heat of a substance is defined as the amount of heat energy required to raise the temperature of 1 g of a substance by 1°C. The specific heat for iron is .108 cal/g°C. Let t = the final temperature of the system.

amount of heat lost by the iron = .108 cal/g-°C × 20.0 g

$$× (95.0°C - t)$$

The amount of heat gained by the water is the specific heat of water multiplied by the weight of the water multiplied by the rise in the temperature. The specific heat of water is 1.0 cal/g-C°. Let t = final temperature of the system.

amount of heat gained by water = 1.0 cal/g-°C × 100 g

$$× (t - 25°C)$$

Solving for t:

amount of heat lost by the iron = amount of heat gained by the water. Therefore,

$$(.108 \text{ cal/g}°C)(20.0 \text{ g})(95°C-t) = (1 \text{ cal/g}°C)(100 \text{ g})(t-25°C)$$

$$205.2 \text{ cal} - (2.16 \text{ cal/}°C)t = (100 \text{ cal/}°C)t - 2500 \text{ cal}$$

$$2705.2 \text{ cal} = (102.16 \text{ cal/}°C)t$$

$$\frac{2705.2 \text{ cal}}{102.16 \text{ cal/}°C} = t$$

$$26.48°C = t.$$

● **PROBLEM** 487

A container has the dimensions 2.5 m × 50 cm × 60 mm. If it is filled with water at 32°F, how many kilocalories will be necessary to heat the water to the boiling point? (212°F)

Solution: A kilocalorie is defined as the amount of heat necessary to raise the temperature of 1 kg of water 1°C. Thus, to solve for the number of kilocalories necessary, one must first solve for the number of grams present and the number of °C that the temperature is raised. One cubic cm of water weighs 1 gram. Thus, if one solves for the volume in terms of cubic centimeters, the weight is then quickly determined. One can solve for the volume in cubic centimeters after converting the lengths of dimensions to centimeters. 1 m = 100 cm, therefore 2.5 m = 250 cm.

If 1 cm = 10 mm, then 60 mm = 6 cm.

Solving for the volume:

volume = 250 cm × 50 cm × 6 cm = $7.50 \times 10^4$ cc.

Therefore, the weight of the water is $7.5 \times 10^4$ g. 1 kg = 1000 g, thus $7.5 \times 10^4$ g = 75 kg.

To convert from °F to °C one uses the following formula:

$$°C = 5/9 \ (°F - 32)$$

Solving for °C when t = 32°F:

$$°C = 5/9 \ (32 - 32) = 0°C.$$

When t = 212°F.

$$°C = 5/9 \ (212-32) = 5/9 \ (180) = 100°C.$$

Therefore, the change in temperature is 100° - 0° or 100°C.

Solving for the number of kilocalories needed:

$$\text{no. of Kcal} = \frac{1 \text{ Kcal}}{°C \text{ kg}} \times 100°C \times 75 \text{ kg} = 7.5 \times 10^3 \text{ Kcal.}$$

● **PROBLEM** 488

The following reaction using hydrogen and oxygen is carried out in a bomb calorimeter: $2H_2(g) + O_2(g) \rightarrow 2H_2O(\ell)$. The following data are recorded: Weight of water in calorimeter = 2.650 kg., Initial temperature of water = 24.442°C., Final temperature of water after reaction = 25.635°C., Specific heat of reaction vessel is 0.200 Kcal/°C-kg, the weight of the calorimeter is 1.060 kg, and the specific heat of water is 1.00 Kcal/°C-kg, calculate the heat of reaction. Assuming the 0.050 mole of water was formed in this experiment, calculate the heat of reaction per mole of liquid water formed. Neglect the specific heat of the thermometer and stirrer.

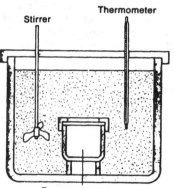

Stirrer  Thermometer

Reaction chamber

Solution:   The heat emitted during the reaction must be equal to the heat absorbed by the system, including the water and the reaction vessel. Therefore, the heat of reaction can be found by calculating the heat absorbed by the water and the heat absorbed by the reaction vessel.

1) To find the heat absorbed by the water one must be concerned with the specific heat of water, the amount of water present and the number of degrees the temperature is raised. The specific heat is defined as the number of kilocalories absorbed when a kilogram of mass is raised one degree. The specific heat of water is 1 Kcal/kg-•C. To find the total heat absorbed by the water, the weight of the water and the number of degrees that the temperature is raised must be multiplied by the specific heat.

Number of Kcal absorbed by water = 2.650 kg × 1 kg/°C - kg
                                    × (25.635° - 24.442°C)
                                  = 3.161  Kcal.

    2) To determine the number of Kcal absorbed by the reaction vessel a similar procedure is used. The specific heat of the vessel is multiplied by the weight of the vessel and the number of degrees the temperature was raised.

Number of Kcal absorbed by the reaction vessel =

= 1.060 kg × 0.200 Kcal/kg-° × (25.635° - 24.442°)

= 0.253 Kcal.

    3) The heat of reaction is the sum of the heat absorbed by the water and the reaction vessel.

heat of reaction = (3.161 + 0.253)Kcal = 3.414 Kcal.

A ratio can be set up to determine the heat of reaction per mole of $H_2O(\ell)$ if 3.414 Kcal is the heat of reaction of 0.05 mole of water. The heat of reaction of 1 mole of water will be related to one as 3.415 Kcal is related to 0.05 moles. Let x = the heat of reaction of 1 mole of water.

$$\frac{3.415 \text{ Kcal}}{0.05 \text{ mole}} = \frac{x}{1 \text{ mole}}$$

$$x = 68.3 \text{ Kcal.}$$

    Therefore the heat formed when one mole of water is formed is 68.3 Kcal.

# ENTHALPY

● PROBLEM 489

Determine $\Delta H°$ for the following reaction of burning ethyl alcohol in oxygen:

$$C_2H_5OH(\ell) + 3O_2(g) \rightarrow 2CO_2(g) + 3H_2O(\ell)$$

$\Delta H_f°$ of $C_2H_2OH(\ell)$ = - 65.9 Kcal/mole
$\Delta H_f°$ of     $CO_2(g)$ = - 94.1 Kcal/mole
$\Delta H_f°$ of     $H_2O(\ell)$ = - 68.3 Kcal/mole

Solution:   The heat reaction ($\Delta H°$) may be found from the heats of formation ($\Delta H_f°$ ) by subtracting the sum of the heats of formation of all reactants from the sum of the heats of formation of all products. The heat of formation of all pure elements is zero. When more than one mole of a compound is either reacted or formed, the heat of formation is multiplied by the stoichiometric coefficient for the specific compound in the equation.

$$\text{heat of reaction} = \left[ 2 \times \Delta H_f^o \text{ of } CO_2 + 3 \times \Delta H_f^o \text{ of } H_2O \right] -$$

$$\left[ \Delta H_f^o \text{ of } C_2H_2OH + 3 \text{ moles} \times \Delta H_f^o \text{ of } O_2 \right]$$

heat of reaction = [2 moles × (- 94.1 Kcal/mole) + 3 moles ×

(- 68.3 Kcal/mole) - 1 mole ×

(- 65.9 Kcal/mole + 3 moles × 0]

= - 327.2 Kcal.

● **PROBLEM** 490

Using the information in the following table, determine $\Delta H^o$ for the reactions:
a) $3Fe_2O_3(s) + CO(g) \rightarrow 2Fe_2O_4(s) + CO_2(g)$
b) $Fe_3O_4(s) + CO(g) \rightarrow 3FeO(s) + CO_2(g)$
c) $FeO(s) + CO(g) \rightarrow Fe(s) + CO_2(g)$

Heats of Formation

| Compound | $\Delta H^o$ (Kcal/mole) |
|---|---|
| CO (g) | - 26.4 |
| $CO_2$ (g) | - 94.1 |
| $Fe_2O_3$ (s) | - 197 |
| $Fe_3O_4$ (s) | - 267 |
| FeO (s) | - 63.7 |

Solution:  The energy change involved in the formation  of one mole of a compound from its elements in their normal state is called the heat of formation. Thus, $\Delta H^o$ for a reaction can be found by subtracting the sum of the $\Delta H^o$'s for the reactants from the sum of the $\Delta H^o$'s for the products.

$\Delta H^o = \Delta H^o$ of the products - $\Delta H^o$ of the reactants

a) 3 $Fe_2O_3(s) + CO(g) \rightarrow$ 2 $Fe_3O_4(s) + CO_2(g)$

When more than one mole of a compound is either reacted or formed in a reaction, the $\Delta H^o$ for that compound is multiplied by the number of moles present in solving for the $\Delta H^o$ of the reaction.

For this reaction:

$\Delta H^o = (2 \times \Delta H^o \text{of } Fe_3O_4 + \Delta H^o \text{ of } CO_2) - (3 \times \Delta H^o \text{of } Fe_2O_3 + \Delta H^o \text{ of } CO)$

$$\Delta H° = [2 \times (-267) + (-94.1)] - [3 \times (-197) + -26.4)]$$

$$= -10.7 \text{ Kcal.}$$

b) $Fe_3O_4(s) + CO(g) \rightarrow 3FeO(s) + CO_2(g)$

$$\Delta H° = (3 \times \Delta H° \text{ of FeO} + \Delta H° \text{ of } CO_2) - (\Delta H° \text{ of } Fe_3O_4 + \Delta H° \text{ of } CO)$$

$$\Delta H° = (3 \times (-63.7) + (-94.1)) - ((-267) + (-26.4))$$

$$= 8.2 \text{ Kcal.}$$

c) $FeO(s) + CO(g) \rightarrow Fe(s) + CO_2(g)$

$$\Delta H° = (\Delta H° \text{ of Fe} + \Delta H° \text{ of } CO_2) - (\Delta H° \text{ of FeO} + \Delta H° \text{ of } CO)$$

The $\Delta H°$ of any element is 0. Thus, the $\Delta H°$ of Fe is 0.

$$\Delta H° = (0 + (-94.1)) - ((-63.7) + (-26.4))$$

$$= -4.0 \text{ Kcal.}$$

● **PROBLEM** 491

Given the following reactions:
$$S(s) + O_2(g) \rightarrow SO_2(g) \qquad \Delta H = -71.0 \text{ Kcal}$$
$$SO_2(g) + \tfrac{1}{2}O_2(g) \rightarrow SO_3(g) \qquad \Delta H = -23.5 \text{ Kcal}$$
calculate $\Delta H$ for the reaction:
$$S(s) + 1\tfrac{1}{2}O_2(g) \rightarrow SO_3(g)$$

Solution: The heat of a given chemical change is the same whether the reaction proceeds in one or several steps; in other words, the energy change is independent of the path taken by the reaction. The heat for a given reaction is the algebraic sum of the heats of any sequence of reactions which will yield the reaction in question. For this problem, one can add the 2 reactions together to obtain the third. The $\Delta H$ is found for the third reaction by adding the $\Delta H$'s of the other two.

$$\Delta H = -71.0 \text{ Kcal}$$
$$+ \Delta H = -23.5 \text{ Kcal}$$
$$\overline{\phantom{+ \Delta H =} -94.5 \text{ Kcal}}$$

$$S(s) + O_2(g) \rightarrow SO_2(g)$$
$$+ \quad SO_2(g) + \tfrac{1}{2}O_2(g) \rightarrow SO_3(g)$$
$$\overline{S(s) + O_2(g) + SO_2(g) + \tfrac{1}{2}O_2(g) \rightarrow SO_2(g) + SO_3(g)}$$

This equation can be simplified by subtracting $SO_2(g)$ from each side and adding $O_2$ and $\tfrac{1}{2}O_2$ together. You have, therefore

$$S(s) + [O_2(g) + \tfrac{1}{2}O_2(g)] + SO_2(g) - SO_2(g) \rightarrow SO_3(g) + SO_2(g) - SO_2(g)$$

or $\quad S(s) + 1\tfrac{1}{2}O_2(g) \rightarrow SO_3(g)$

The $\Delta H$ for this reaction is $-94.5$ Kcal, as shown above.

Given that $\Delta H^\circ_{CO_2(g)} = -94.0$, $\Delta H^\circ_{CO(g)} = -26.4$, $\Delta H^\circ_{H_2O(\ell)} = -68.4$ and $\Delta H^\circ_{H_2O(g)} = -57.8$ in Kcal/mole, determine the heats of reaction of (1) $CO(g) + \frac{1}{2}O_2(g) \rightarrow CO_2(g)$, (2) $H_2(g) + \frac{1}{2}O_2(g) \rightarrow H_2O(\ell)$ and (3) $H_2O(\ell) \rightarrow H_2O(g)$.

Solution: To solve this problem, you need to know what heat of reaction means and how to determine it quantitatively. The heat of reaction may be defined as the heat released or absorbed as a reaction proceeds to completion. It is measured quantitatively from the heats of formation of products minus heats of formation of reactants. Namely, heat of reaction, $\Delta H^\circ = \Delta H^\circ_{products} - \Delta H^\circ_{reactants}$. (Note: By convention, the $\Delta H^\circ$ of any element is defined as being equal to zero.) You proceed as follows:

(1) $\Delta H^\circ = \Delta H^\circ_{CO_2(g)} - \Delta H^\circ_{CO(g)} - \frac{1}{2}\Delta H^\circ_{O_2(g)}$

$= -94 - (-26.4) - \frac{1}{2}(0) = -67.6$ Kcal/mole.

(2) $\Delta H^\circ = \Delta H^\circ_{H_2O(\ell)} - \Delta H^\circ_{H_2(g)} - \frac{1}{2}\Delta H^\circ_{O_2(g)}$

$= -68.4 - 0 - \frac{1}{2}(0) = -68.4$ Kcal/mole.

(3) $\Delta H^\circ = \Delta H^\circ_{H_2O(g)} - \Delta H^\circ_{H_2O(\ell)}$

$= -57.8 - (-68.4) = +10.6$ Kcal/mole.

You are given the following reactions at $25°C$:
$2NaHCO_3(s) \rightarrow Na_2CO_3(s) + CO_2(g) + H_2O(\ell)$, $\Delta H = 30.92$ Kcal/mole. $\Delta H^\circ_{Na_2CO_3(s)} = -270.3$, $\Delta H^\circ_{CO_2(g)} = -94.0$ and $\Delta H^\circ_{H_2O(\ell)} = -68.4$ Kcal/mole, what is the standard enthalpy of formation for $NaHCO_3(s)$?

Solution: The $\Delta H^\circ$ for a reaction is an indication of the amount of heat released (or absorbed) when the reactants are converted to products. In general, $\Delta H^\circ$ for a reaction will be the sum of the heats of formation of products minus the sum of the heats of formation of reactants, each of which is multiplied by its coefficient in the equation. Because you are given this sum and the $\Delta H^\circ$ (heats of formation) of all products, the standard enthalpy of formation of $NaHCO_3$ can be found. Proceed as follows:

$$\Delta H° = \Delta H°_{Na_2CO_3(s)} + \Delta H°_{CO_2(g)} + \Delta H°_{H_2O(\ell)} - 2\Delta H°_{NaHCO_3(s)}$$

Substituting known values,

$$30.92 = (-270.3) + (-94.0) + (-44.8) - 2\Delta H°_{NaHCO_3(s)}.$$

Solving for $\Delta H°_{NaHCO_3(s)}$, you obtain

$$\Delta H°_{NaHCO_3(s)} = (-270.3 - 94.0 - 44.8 - 30.9)/2$$

$$= -440.0/2 = -220.0 \text{ Kcal/mole},$$

which is its standard enthalpy of formation.

● **PROBLEM** 494

Calculate the standard enthalpy change, $\Delta H°$, for the combustion of ammonia, $NH_3(g)$, to give nitric oxide, $NO(g)$, and water $H_2O(\ell)$. The enthalpies of formation, $\Delta H°_f$, are -68.32 Kcal/mole for $H_2O(\ell)$, -11.02 Kcal/mole for $NH_3(g)$, and 21.57 Kcal/mole for $NO(g)$.

Solution: The enthalpy change, $\Delta H$, refers to the change in heat content between the products and the reactants of a chemical reaction. The symbol $\Delta H°$ is the enthalpy change for a reaction in which each reactant and product is in its standard state at a specified reference temperature (the common reference temperature is 25°C).

The standard enthalpy of formation, $\Delta H°_f$, of a substance is defined as the change in enthalpy for the reaction in which one mole of the compound is formed from its elements at standard conditions.

For any chemical reaction, the change in enthalpy, $\Delta H$, may be expressed as

$$\Delta H_{reaction} = \sum H_{products} - \sum H_{reactants}$$

(where the symbol $\sum H$ means the summation of the enthalpies of each substance).

To solve this problem, one must first write and balance the equation for the combustion of $NH_3(g)$:

$$4 \text{ NH}_3(g) + 5O_2(g) \rightarrow 4NO(g) + 6 \text{ H}_2O(\ell)$$

From the balanced equation, one knows that 4 moles of $NH_3(g)$ reacts with 5 moles of $O_2(g)$ to form 4 moles of $NO(g)$ and 6 moles of $H_2O(\ell)$. For the combustion of 4 moles of $NH_3(g)$:

$$4 \cdot \Delta H°_{f, NH_3(g)} = 4\left(-11.02 \frac{Kcal}{mole}\right) = -44.08 \text{ Kcal}$$

$$5 \cdot \Delta H^{\circ}_{f, \ O_2(g)} \quad = 5 \ (0) \qquad\qquad = 0$$

$$4 \cdot \Delta H^{\circ}_{f, \ NO(g)} \quad = 4 \left(21.57 \ \frac{Kcal}{mole}\right) \quad = 86.28 \ Kcal$$

$$6 \cdot \Delta H^{\circ}_{f, \ H_2O(\ell)} \ = 6 \left(- \ 68.32 \ \frac{Kcal}{mole}\right) = - \ 409.92 \ Kcal$$

Note: The $\Delta H^{\circ}_{f}$ of all elements is zero. As such $\Delta H^{\circ}_{f, \ O_2(g)} = 0.$

Thus,

$$\Delta H^{\circ} = \left(4 \cdot \Delta H^{\circ}_{f, NO(g)} + 6 \cdot \Delta H^{\circ}_{f, H_2O(\ell)}\right) - \left(4 \cdot \Delta H^{\circ}_{f, NH_3(g)} + 5 \cdot \Delta H^{\circ}_{f, O_2(g)}\right)$$

$$= \ [(86.28 \ Kcal) + (-409.92 Kcal)] - [(-44.08 Kcal) + 0]$$

$$= \ - \ 279.56 \ Kcal.$$

However, since this result is the standard enthalpy change for the combustion of 4 moles of $NH_3(g)$, one must divide - 279.56 Kcal/mole by 4 to give the combustion of 1 mole of $NH_3$,

$$\Delta H^{\circ} = \frac{- \ 279.56 \ Kcal}{4 \ moles} = - \ 69.89 \ Kcal/mole.$$

● **PROBLEM** 495

Find the $\Delta H^{\circ}$, heat of reaction, for the hydrolysis of urea to $CO_2$ and $NH_3$. Although the reaction does not proceed at a detectable rate in the absence of the enzyme, urease, $\Delta H$ may still be computed from the following thermodynamic data:

| Compound | $\Delta H^{\circ}_{f}$ (Kcal/mole) |
|---|---|
| Urea (aq) | - 76.30 |
| $CO_2$ (aq) | - 98.69 |
| $H_2O$ ($\ell$) | - 68.32 |
| $NH_3$ (aq) | - 19.32 |

The hydrolysis reaction is

$$H_2O \ (\ell) + H_2N - CO - NH_2 \rightarrow CO_2(aq) + 2NH_3(aq)$$
$$\text{(urea)}$$

Solution: The heat of reaction, $\Delta H^{\circ}$, may be found from the heats of formation, $\Delta H^{\circ}_{f}$, of the reactants and products.

Heat of formation may be defined as the heat absorbed or evolved in the synthesis of one mole of a compound from its elements, all components being in their standard states.

The heat of reaction is the sum of the heats of formation of the products minus the sum of the heats of formation of the reactants. In other words,

$$\Delta H_{reaction} = \sum \Delta H_{formation\ of\ products}$$

$$- \sum \Delta H_{formation\ of\ reactants}$$

Therefore, for the hydrolysis of urea to $CO_2$ and $NH_3$

$$\Delta H_{reaction} = \left[ \left( \Delta H^\circ_{f,\ CO_2} + 2\Delta H^\circ_{f,\ NH_3} \right) - \left( \Delta H^\circ_{f,\ H_2N-CO-NH_2} + \Delta H^\circ_{f\ H_2O} \right) \right]$$

Given these values, one can substitute to obtain:

$$\Delta H_{reaction} = [(- 98.69 - 2 \times 19.32) - (- 76.30 - 68.32)]$$

From this equation one obtains

$$\Delta H_{reaction} = 7.29\ Kcal/mole.$$

● PROBLEM 496

Using the data from the accompanying figure, calculate the heat of reaction for the following process at 25°C:

a)  $CaCl_2 (s) + Na_2CO_3 (s) \rightarrow 2NaCl (s) + CaCO_3 (s)$

b)  $H_2SO_4 (l) + 2NaCl (s) \rightarrow Na_2SO_4 (s) + 2HCl (g)$

Standard Heats of Formation, $\Delta H^\circ$, in kcal/mole at 25°C.

| Substance | $\Delta H^\circ$ |
|---|---|
| $CaCl_2$ (s) | -190. 0 |
| $Na_2CO_3$ (s) | -270. 3 |
| NaCl (s) | - 98. 2 |
| $CaCO_3$ (s) | -288. 4 |
| $H_2SO_4$ (l) | -193. 9 |
| HCl (g) | - 22. 1 |
| $Na_2SO_4$ (s) | -330. 9 |

Calculate the amount of heat released (or absorbed) from the heats of formation. In general, $\Delta H$ for a reaction will be the sum of the heats of formation of products minus the heats of formation of reactants, each of which is multiplied by its coefficient in the equation. Once $\Delta H$ is determined, $\Delta E$ can be found from $\Delta E = \Delta H - (\Delta n)RT$, where $\Delta n$ = change of moles, R = universal gas constant and T = temperature in kelvin (Celsius plus 273°)

Proceed as follows:

(a) $CaCl_2 (s) + Na_2CO_3 (s) \rightarrow 2NaCl(s) + CaCO_3(s)$

Thus,

$$\Delta H = 2\Delta H^{O}_{NaCl(s)} + \Delta H^{O}_{CaCO_3(s)} - \Delta H^{O}_{CaCl_2(s)} - \Delta H^{O}_{Na_2CO_3(s)}$$

$$= 2(-98.2) + (-288.4) - (-190.) - (270.3)$$

$$= -24.5 \text{ Kcal/mole.}$$

$\Delta E = \Delta H - \Delta nRT$. But, in this reaction, no gases appear. This means $\Delta nRT$ becomes zero. As such, $\Delta E = \Delta H = -24.5$ Kcal/mole.

(b) $H_2SO_4 (1) + 2NaCl(s) \rightarrow Na_2SO_4 (s) + 2HCl(g)$

$$\Delta H = \Delta H^{O}_{Na_2SO_4(s)} + 2\Delta H^{O}_{HCl(g)} - \Delta H^{O}_{H_2SO_4 (1)} - 2\Delta H^{O}_{NaCl(s)}$$

$$= (-330.9) + 2(-22.1) - (-193.9) - 2(-98.2) = +15.2 \text{ kcal/mole.}$$

$\Delta E$ = heat of reaction = $\Delta H - (\Delta n)RT$.

A gas, HCl, is involved in this reaction. $\Delta n$ = moles products - moles reactants = 2 - 0 = 2

T = 25 + 273 = 298°K; R is in terms of cal. and $\Delta H$ in terms of Kcal, so that you must use the conversion factor of 1Kcal/1000 cal.

Thus, $\Delta E = -15.2 - (2)(1.987)(298)(1/1000)$

$$= - 15.2 - 1.2$$

$$= - 16.4 \text{ Kcal/mole.}$$

● **PROBLEM** 497

Using the data from the accompanying table find the enthalpy change for the combustion of a mole of $C_2H_4$(g) to form $CO_2$(g) and $H_2O$(g) at 298°K and 1 atm standard state conditions.

<u>Solution</u>:  When a chemical reaction occurs heat is usually absorbed or given off to the environment.  This means that the enthalpy or heat content of the system will also change.  If one knows the enthalpy change of the combustion of each reactant and product, one can find the change in enthalpy of a reaction.  The reaction in this problem can be written as

$$C_2H_4(g) + 3O_2(g) \rightarrow 2CO_2(g) + 2H_2O(g)$$

For the reaction,

$$\Delta H_{combustion} = \Sigma \Delta H_{products} - \Sigma \Delta H_{reactants}.$$

Remembering that the enthalpy change for the formation of of an element (i.e. $O_2$) in its standard state is zero, one then has,

$$\Delta H_{combustion} = 2\Delta H_{CO_2} + 2\Delta H_{H_2O} - \Delta H_{C_2H_4} - 3\Delta H_{O_2}$$

$$= 2(-394) + 2(-242) - (52) - 3(0)$$

$$= -1324 \text{ KJ}$$

Enthalpies of formation in kilojoules per mole from the elements for various compounds at 298°k and 1 atm pressure

| Compound | $\Delta H$ |
|----------|------|
| $C_2H_4$ (g) | + 52 |
| $CO_2$ (g) | -394 |
| $H_2O$ (g) | -242 |
| $O_2$ (g) | 0 |

● **PROBLEM** 498

Calculate the quantity of heat required to (a) convert a liter of $H_2O$ at 30°C to a liter of water at 60°C, and (b) heat a 1 kg block of aluminum from 30° to 60°C. Assume the specific heat of water and aluminum is, respectively, 1 cal/g°C and .215 cal/g°C.

<u>Solution</u>:  When heat is added to a mass of unspecified substance, the temperature will rise. The quantitative

relationship between the quantity of heat, expressed in kilocalories, and the rise in temperature - provided there is no phase change - is:

  quantity of heat = mass × specific heat × ΔT,

where ΔT is the change in temperature and the specific heat equals the amount of heat required to raise the temperature of 1 g of any substance by 1°C.

(a)   To find the mass of $H_2O$, remember that density = mass/volume, and the density of $H_2O$ is one. Thus, mass = (1 liter $H_2O$)(1000 ml/liter)(1.00 g $H_2O$)/ml. $H_2O$). Therefore,

$$\text{quantity of heat} = (\text{mass})(\text{specific heat})(\triangle T)$$

$$\text{mass} = (1\ell\ H_2O)(1000\ ml/\ell)(1\ g/ml)$$

$$= 1000\ g$$

$$\text{specific heat} = 1\ cal/g°C$$

$$\triangle T = (60 - 30)°C$$

$$= 30°C$$

$$\text{quantity of heat} = (1000\ g)(1\ cal/g°C)(30°C)$$

$$= 30,000\ cal = 30\ Kcal$$

(b)   A similar calculation can be used to determine the amount necessary to raise the temperature of the aluminum from 30 to 60°C.

$$\text{mass} = (1\ kg\ Al)(1000\ g/kg) = 1000\ g$$

$$\text{specific heat} = .215\ cal/g°C$$

$$\triangle T = (60 - 30)°C$$

$$= 30°C$$

$$\text{quantity of heat} = (1000\ g)(.215\ cal/g°C)(30°C)$$

$$= 6450\ cal = 6.45\ Kcal.$$

● PROBLEM 499

In a pound-for-pound comparison, which would you prefer as a rocket fuel, hydrogen ($H_2$) or dimethylhydrazine $(CH_3)_2NNH_2$? $H_2(g) + \frac{1}{2}O_2(g) \rightarrow H_2O(\ell)$, $\Delta H = -68.3 Kcal/mole$ $(CH_3)_2NNH_2 + 4O_2(g) \rightarrow N_2(g) + 4H_2O(g) + 2CO_2(g)$, $\Delta H = -404.9\ Kcal/mole.$

Solution: The material that is the better fuel will release more energy per unit weight in its reaction with oxygen. Calculate the energy released per unit weight for $H_2$ and $(CH_3)_2NNH_2$. You can do this by using the heats of formation and enthalpy changes given. Calculate the total mass or weight of materials from moles that react and divide the $\Delta H$ by it to obtain the amount of energy released per unit weight.

For hydrogen:

1 mole of $H_2$ reacts with $\frac{1}{2}$ mole of oxygen (1 mole $H_2$)(1.008 g $H_2$/mole $H_2$) + ($\frac{1}{2}$ mole $O_2$)(16.00 g $O_2$/mole $O_2$)= 1.008 g + 8.00 g = 9.01 g. $\Delta H = -68.3$ Kcal. Thus,

$$\frac{-68.3}{9.01} = -7.58 \text{ Kcal/g.}$$

For dimethylhydrazine:

[1 mole $(CH_3)_2NNH_2$][60.10 g $(CH_3)_2NNH_2$/mole $(CH_3)_2NNH_2$] + (4 moles $O_2$)(16.00 g. $O_2$/mole $O_2$) = 60.10 g + 64.00 g = 124.10 g. $\Delta H = -404.9$. The amount of heat released per unit weight with $(CH_3)_2NNH_2 + O_2 =$

$$\frac{-404.9 \text{ Kcal}}{124.10 \text{ g}} = -3.26 \text{ Kcal/g.}$$

Hydrogen yields over twice as much energy per gram of fuel mass, and is thus preferable in this respect.

● **PROBLEM** 500

Calculate $\Delta H^o_r$ for the combustion of methane, $CH_4$. The balanced reaction is

$$CH_4(g) + 2O_2(g) \rightarrow CO_2(g) + 2H_2O(\ell)$$
$\Delta H^o_f$ in KCal/mole=-17.89, 0 , -94.05 , 2(-68.32)

Solution: $\Delta H^o_r$ is the Standard Enthalpy Change of Reaction. Standard conditions are defined as 25°C and 1 atm.

Enthalpy is the heat content of a system. If the overall change in enthalpy is negative, then heat is given off to the surroundings and the reaction is called exothermic. When the change is positive, heat is absorbed and the reaction is endothermic. Endothermic compounds are often unstable and can sometimes explode. An endothermic compound, however, is a more efficient fuel because, upon combustion, it yields more heat energy.

$\Delta H^o_r$ is calculated using the enthalpies of formation, $\Delta H^o_f$. The sum of enthalpies of formation of products minus the sum of the enthalpies of formation of reactants, where each product and reactant is multiplied by its molar amount in the reaction as indicated by the coefficients, gives the value of $\Delta H_r^o$. In other words,

$$\Delta H_r^o = \sum \Delta H_f^o \text{ products} - \sum \Delta H_f^o \text{ reactants}$$

$\Delta H_f^o$ for elements is always zero. In this reaction, therefore, $\Delta H_f^o$ for $O_2$ is zero.

$$\sum \Delta H_f^o \text{ products} = -94.05 - 136.64$$

$$\sum \Delta H_f^o \text{ reactants} = -17.89 \text{ Kcal/mole}$$

$$\Delta H_r^o = -94.05 - 136.64 - (-17.89)$$

$$= -212.8 \text{ Kcal/mole } CH_4 \text{ burned}$$

This reaction is exothermic, because the $\Delta H_r^o$ = a negative value.

● **PROBLEM** 501

Using the information supplied in the accompanying table, determine the comparative heats of combustion of water gas ($CO + H_2$) and of methane ($CH_4$) per unit weight and per unit volume. The products are $CO_2$(g) and $H_2O$(g) in both reactions.

Enthalpies of formation in kilojoules per mole from the elements for various compounds at 298°K and 1 atm pressure.

| Compound | $\Delta H$ |
|----------|------------|
| CO   (g) | - 110 |
| $H_2O$ (g) | - 242 |
| $CO_2$ (g) | - 394 |
| $CH_4$ (g) | - 75 |

Solution: Write out the chemical reactions and calculate the enthalpy change. Enthalpy change is a measure of the amount  of heat absorbed or released in the reaction. For the water gas, the reaction is $CO + H_2 + O_2 \rightarrow CO_2$(g) $+ H_2O$(g). The overall enthalpy is the sum of the enthalpies of formation of the products minus the sum of the enthalpies of formation of the reactants. Thus,

$$\Delta H^o = -394 \text{ KJ} + (-242 \text{ KJ}) - (-110 \text{ KJ}) = -526 \text{ KJ}.$$

Notice that the $\Delta H$'s of $H_2$ and $O_2$ are not included. Elements are considered to have $\Delta H$'s of zero.

The question asks for $\Delta H$ by unit weight and volume. Therefore, $\Delta H$ by unit weight =

$$\frac{\Delta H°}{mol.wt.CO + mol.wt.H_2} = \frac{-526\ KJ}{30\ g} = -17.5\ KJ/g.$$

The sum of the molecular weights of $CO + H_2$ is 30. To find $\Delta H$ by molar volume, note that two molar volumes are involved, one from $H_2$ and one from $CO$, which means

$$\Delta H\ by\ volume = \frac{\Delta H}{no.\ of\ molar\ volumes}$$

$$= \frac{-526\ KJ}{2\ molar\ volumes} = -263\ KJ/molar\ volume.$$

An analogous method is used for methane. The reaction is $CH_4 + O_2 \rightarrow CO_2(g) + 2H_2O(g)$.

$$\Delta H° = -394 + 2(-242) - (-75) = -803\ KJ$$

The molecular weight of $CH_4 = 16$. Therefore,

$$\Delta H\ per\ unit\ weight = \frac{-803\ KJ}{16\ g} = -50.2\ KJ/g.$$

The only molar volume is that of methane. Then

$$\frac{\Delta H}{molar\ volume} = \frac{-803\ KJ}{1}$$

The comparative heats can thus be related as

$$\frac{263}{803} = \frac{.33}{1}\ by\ volume\ and\ \frac{17.5}{50} = \frac{.35}{1}\ by\ weight.$$

● **PROBLEM** 502

A common rocket propellant is a mixture of one part liquid hydrazine ($N_2H_4$) as fuel and two parts liquid hydrogen peroxide ($H_2O_2$) as oxidizer. These reactants are hypergolic, that is, they ignite on contact. The reaction proceeds according to the equation
$N_2H_4(\ell) + 2H_2O_2(\ell) \rightarrow N_2(g) + 4H_2O(g)$.
If reactants are mixed at 25°C, what is the heat of reaction?

Solution: We will make use of the following expression for the heat of reaction at 25°C (298°K), $\Delta H°_{298}$:

$\Delta H°_{298}$ = sum of heats of formation of products
— sum of heats of formation of reactants.

Denoting the heat of formation of species s by $\Delta H°_{f,s}$, this becomes

$$\Delta H^o_{298} = \Delta H^o_{f, N_2(g)} + 4\Delta H^o_{f,H_2O(g)} - \left[\Delta H^o_{f,N_2H_4(\ell)} + 2\Delta H^o_{f, H_2O_2(\ell)}\right]$$

From a table of the standard heats of formation at 298°K, we find

$\Delta H^o_{f,N_2(g)} = 0$ Kcal/mole, $\Delta H^o_{f,H_2O(g)} = -57.8$ Kcal/mole,

$\Delta H^o_{f, N_2H_4(\ell)} = 12$ Kcal/mole, and $\Delta H^o_{f,H_2O_2(\ell)} = -46$ Kcal/mole.

Then

$$\Delta H^o_{298} = \Delta H^o_{f,N_2(g)} + 4\Delta H^o_{f,H_2O(g)} - \left[\Delta H^o_{f,N_2H_4(\ell)} + 2\Delta H^o_{f,H_2O_2(\ell)}\right]$$

$$= 0 + 4(-57.8) - (12 + 2(-46))$$

$$= -151 \text{ Kcal/mole.}$$

● **PROBLEM** 503

The hydrogen-filled dirigible "Hindenberg" had a gas volume of $1.4 \times 10^8$ liters. Assuming that it was filled to a pressure of 1 atm at 0°C, how much energy was released when it burned to gaseous products at a reaction temperature of 25°C? The standard heat of formation of water vapor at 25°C is $-57.8$ Kcal/mole.

<u>Solution:</u>   The solution to this problem involves calculation of the heat of reaction for the combustion of hydrogen and the number of moles of hydrogen in the dirigible.

The combustion reaction is written

$$H_2(g) + \tfrac{1}{2} O_2(g) \rightarrow H_2O(g).$$

The heat of reaction at 25°C (298°K) is written

$\Delta H^o_{298}$ = sum of **heats** of formation of products - sum of **heats** of formation of **reactants**

$$= \Delta H^o_{f, H_2O} - \left[\Delta H^o_{f, H_2} + \Delta H^o_{f, O_2}\right]$$

Since the heats of formation of $O_2$ and $H_2$ are zero by definition, $\Delta H^o_{f,O_2} = \Delta H^o_{f, H_2} = 0$ Kcal/mole, the heat of reaction is equal to the heat of formation of water vapor:

$$\Delta H^o_{298} = \Delta H^o_{f, H_2O} = -57.8 \text{ Kcal/mole.}$$

Since $\Delta H^o_{298}$ is negative, the reaction is exothermic and - 57.8 Kcal are released per mole of water produced. But since one mole of water is formed per mole of hydrogen burned, - 57.8 Kcal are released per mole of hydrogen burned.

To calculate the number of moles of hydrogen present in the dirigible, we make use of the fact that at 0°C and 1 atm pressure, one mole of gas occupies 22.4 ℓ (molar volume of gas). Then

$$\text{number of moles of } H_2 = \frac{\text{volume of dirigible}}{\text{molar volume of } H_2}$$

$$= \frac{1.4 \times 10^8 \ \ell}{22.4 \ \ell/\text{mole}} = \frac{1.4 \times 10^8}{22.4} \ \text{moles}$$

The heat released is then the absolute value of the heat of reaction (per mole) multiplied by the number of moles of gas.

$$\text{heat released} = |\Delta H^o_{298}| \times \text{number of moles of } H_2$$

$$= 57.8 \ \text{Kcal/mole} \times \frac{1.4 \times 10^8}{22.4} \ \text{moles}$$

$$= 360 \times 10^6 \ \text{Kcal}.$$

(We use the absolute value of $\Delta H^o_{298}$ because the word "released" already takes into account the fact that the reaction is exothermic and therefore the sign of $\Delta H^o_{298}$ is negative.) Note that we have assumed that all of the hydrogen is burned.

● PROBLEM 504

It is known that 1 ton of TNT can produce $4.6 \times 10^9$ J of energy. A chemist wanted to obtain this much energy from acetylene ($C_2H_2$) and oxygen ($O_2$). How much $C_2H_2$ and $O_2$ would be required? The $\Delta H^o$ = - 2510 KJ for the reaction between $C_2H_2$ and $O_2$.

Solution: The equation for this reaction is

$2 C_2H_2 + 5O_2 \rightarrow 4CO_2 + 2H_2O$, $\Delta H^o$ = - 2510 KJ. In a chemical reaction, energy is usually liberated or absorbed. The $\Delta H^o$ enthalpy, expresses this energy liberation or absorption quantitatively. The negative sign denotes the release of energy.

Two moles of acetylene and 5 moles of oxygen react to form 2510 KJ or $2510 \times 10^3$ J of energy. Similarly, 2x moles of acetylene and 5x moles of oxygen produce $2.510 \times 10^6$ x J of energy. In this case, the energy re-

leased must equal $4.6 \times 10^9$ J. Therefore,
$$2.510 \times 10^6 \, x = 4.6 \times 10^9, \quad \text{or} \quad x = 1.83 \times 10^3.$$

From the equation for the reaction one knows that 2 moles of $C_2H_2$ and 5 moles of $O_2$ react to produce 2510 KJ. The calculations already done show that to produce as much energy as 1 ton of TNT there must be $1.83 \times 10^3 \times 2$ moles of $C_2H_2$ and $1.83 \times 10^3 \times 5$ moles of $O_2$ to form $4.6 \times 10^9$ J. To solve for the amount of each needed, multiply the number of moles by the molecular weight of the substance. (MW of $C_2H_2 = 26$).

$$\text{wt of } C_2H_2 = (1.83 \times 10^3 \times 2) \text{moles} \times 26 \text{ g/moles}$$

$$= 9.52 \times 10^4 \text{ g} = 9.52 \times 10^4 \text{ g} \times 1.10 \times 10^{-6} \text{tons/g}$$

$$= .1044 \text{ tons.}$$

(MW of $O_2 = 32$)

$$\text{wt of } O_2 = (1.83 \times 10^3 \times 5) \text{ moles} \times 32 \text{ g/mole}$$

$$= 5.86 \times 10^4 \text{ g} = 5.86 \times 10^4 \text{ g} \times 1.10 \times 10^{-6} \text{tons/g}$$

$$= 0.646 \text{ tons.}$$

● **PROBLEM** 505

In a hydrogen bomb, deuterium nuclei fuse to produce a helium nucleus. In the process, 5MeV of energy is re-leased per deuterium. If, 1/7000 of natural H is D, how many liters of gasoline ($C_8H_{18}$) are needed to be burned to produce the same amount of energy as derived from 1 liter of water. Assume the following: the heat of combustion of gasoline is 5100 kJ/mole; the density of **gasoline** is .703 g/cm$^3$; leV = $1.60 \times 10^{-19}$ J.

Solution: The answer to this question involves several calculations. To determine the amount of gasoline which will yield an equal amount of energy upon combustion as one liter of water, first compute how much energy can be obtained from each. For gasoline, this entails calcula-tion of the number of moles per liter of the substance. Multiplication by the heat of combustion yields the total energy. For water, first determine, how many deuterium ions are present per liter and then, using conversion factors, multiply this by the energy re-leased per deuterium ion fused.

To obtain the energy per liter of water, first find the number of deuterium ions. To obtain this, you need to first find the number of atoms of H per liter of water, since the density of water is defined as one and density = mass/volume, 1 liter $H_2O$ =1000 g $H_2O$. The molecular weight of one mole of water is 18. Therefore, the number of moles of water in one liter is

$$\frac{1000}{18} = 55.5 \ \frac{\text{moles water}}{\text{liter}} \ . \quad \text{There are two moles}$$

of H atoms, per mole of water, since its molecular formula is $H_2O$. Therefore, the number of moles of H atoms

$$= \frac{55.5 \ \text{moles water}}{\text{liter}} \times \frac{2 \ \text{mole H atoms}}{\text{moles } H_2O} = \frac{111 \ \text{mole H atoms}}{\text{liter}}.$$

However, in a mole of anything, there are $6.02 \times 10^{23}$ particles, (Avogadro's number). Therefore, the total number of H atoms per liter of water is

$$\frac{111 \ \text{mole H atoms}}{\text{liter}} \times 6.02 \times 10^{23} \ \frac{\text{atoms}}{\text{mole}} = 6.68 \times 10^{25} \ \frac{\text{H atoms}}{\text{liter}}$$

since you have 1 deuterium atom per 7000 H atoms. Therefore, the number of deuterium atoms per liter of water is

$$6.68 \times 10^{25} \ \frac{\text{H atoms}}{\text{liter}} \times \frac{1 \ \text{atom deuterium}}{7000 \ \text{H atoms}} = 9.55 \times 10^{21} \frac{\text{D atoms}}{\text{liter}}$$

Since each deuterium atom yields 5MeV or $5 \times 10^6$ ev of energy, when it fuses to produce helium, the amount of energy per liter of water is

$$9.55 \times 10^{21} \ \frac{\text{D atoms}}{\text{liter}} \times 5 \times 10^6 \ \frac{\text{ev}}{\text{D atom}} = 47.75 \times 10^{27} \ \frac{\text{ev}}{\text{liter}} \ .$$

Since, $1.60 \times 10^{-19}$ J = one ev = $1.60 \times 10^{-22}$ kJ,

$$47.75 \times 10^{27} \frac{\text{ev}}{\text{liter}} = 47.75 \times 10^{27} \ \frac{\text{ev}}{\text{liter}} \times 1.60 \times 10^{-22} \ \text{kJ/ev}$$

$$= 7.64 \times 10^6 \ \text{kJ/liter}$$

of energy per liter of water.

To compute the energy in one liter of gasoline, remember that the density of gasoline equals mass of gasoline divided by its volume. Therefore, the mass of one liter of gasoline is (density × volume) = .703 g/cm$^3$ × 1000 cm$^3$ = 703 g. Thus, there are 703 g/liter of gasoline. The molecular weight of gasoline is 142 grams per mole. Therefore, the number of moles of gasoline per liter is

$$\frac{703 \ \text{g/}\ell}{142 \ \text{g/mole}} = 6.16 \ \text{moles/liter for gasoline. The heat}$$

of combustion of gasoline is 5100 kJ/mole. Therefore, the energy per liter of gasoline is

6.16 moles/liter × 5100 kJ/mole = 31,400 kJ/liter

$$= 3.14 \times 10^4 \ \text{kJ/liter}.$$

To find out how may liters of gasoline is equivalent to water in terms of energy, divide the energy of 1 liter $H_2O$ by the energy of 1 liter gasoline.

For water, it is $7.64 \times 10^6$ kJ/$\ell$.

For gasoline, it is $3.14 \times 10^4$ kJ/$\ell$.

$$\frac{7.64 \times 10^6 \text{ kJ/}\ell \text{ water}}{3.14 \times 10^4 \text{ kJ/}\ell \text{ gasoline}} = 243 \frac{\text{liters gasoline}}{\text{liter water}}$$

# ENTHALPY CALCULATIONS USING THE FIRST LAW OF THERMODYNAMICS

● **PROBLEM** 506

You have 1 liter of an ideal gas at 0°C and 10 atm pressure. You allow the gas to expand against a constant external pressure of 1 atm, while the temperature remains constant. Assuming, 24.217 cal/liter-atm, find q, w, $\Delta E$ and $\Delta H$ in calories, (a) in these values, if the expansion took place in a vacuum and (b) if the gas were expanded to 1 atm pressure.

Solution: The solutions to the parts of this problem require a combination of thermodynamics and ideal gas law theory. The gas expands because its pressure is greater than the external pressure. When the pressure of the gas falls to 1 atm, which is the pressure being applied externally, gas expansion will terminate. This allows for determination of the volume change using Boyle's Law. You want $\Delta V$ (volume change), since w = work = $P\Delta V$. Boyle's law states that PV = constant for ideal gas. Thus, $P_1 V_1 = P_2 V_2$. Originally, $P_1$ = 10 atm with $V_1$ = 1 liter. You end up with $P_2$ = 1 atm. Thus,

$$V_2 = \frac{P_1 V_1}{P_2} = \frac{(10 \text{ atm})(1 \text{ liter})}{(1 \text{ atm})} = 10 \text{ liters}.$$

$\Delta V$ = 10 liters (final) - 1 liter (original) = 9 liters. Therefore, w = $P\Delta V$ = (1 atm)(9 liters) = 9 liters-atm. But, there are 24.217 cal per liter-atm, so that, in calories,

w = 9 liter-atm × 24.217 cal/liter-atm = 218 cal.

q = the heat absorbed by the system. To find its value, you employ the first law of thermodynamics, which says $\Delta E$ = q - w, where $\Delta E$ = change in energy of system and w = work performed. Since the gas is ideal, E = energy is only a function of temperature. As such, $\Delta E$ = 0, since the temperature is constant. You have, therefore 0 = q - w or q = w. But, you just found w, which means q = 218 cal.

To find $\Delta H$, the enthalpy change, remember that $\Delta H = \Delta E + \Delta(PV)$. You know that $\Delta E$ = 0. To find $\Delta(PV)$, recall that $\Delta(PV) = P_2 V_2 - P_1 V_1$, as derived from Boyle's law. However, $P_1 V_1 = P_2 V_2$ thus; $P_2 V_2 - P_1 V_1 = 0$. This means $\Delta H$=0.

In a vacuum, there is no external pressure. This means that no work can be done, so that w = 0. This is derived from the fact that work is defined as performing an action against a surrounding environment. If there is no environment, there cannot be any work. Since $\Delta E$ remains 0 and q = $\Delta E$+ w, then q = 0.

Since there is zero pressure, PV = 0. Since $\Delta$(PV) = 0, $\Delta E$ = 0 and $\Delta H$ = $\Delta E$ + $\Delta$(PV), then $\Delta H$ = 0.

You now find the work (w), required in going from zero pressure to 1 atm pressure; use the equation for the reversible expansion of a gas in terms of work.

w = 2.303 n RT log $V_2/V_1$,   where  $V_1$ = initial

volume, $V_2$ = final volume, R = universal gas constant, T = temperature in Kelvin (Celsius plus 273°) and n = number of moles. $V_2$ was calculated from Boyle's law. The rest is given. Thus, you substitute to obtain

w = 2.303 (1 mole)(1.987)(273)log (10/1) = 1249 cal.

$\Delta E^\circ$ and $\Delta H^\circ$ are still zero, but since q = $\Delta E$ + w, q = 0 + 1249 = 1249 cal.

● **PROBLEM** 507

Calculate $\Delta E^\circ$ and $\Delta H^\circ$ for the following reactions at 25°C:
(a)  $4NH_3(g) + 5O_2(g) \rightarrow 4NO(g) + 6H_2O(\ell)$
(b)  $H_2S(g) + 1\tfrac{1}{2}O_2(g) \rightarrow SO_2(g) + H_2O(\ell)$
Use the following values for $\Delta H^\circ$ of the compounds

| Compound | H° (Kcal/mole) |
|---|---|
| $H_2O(\ell)$ | - 68.3 |
| $SO_2(g)$ | - 71.0 |
| $H_2S(g)$ | - 5.3 |
| NO(g) | 21.6 |
| $NH_3(g)$ | - 11.0 |

Solution:   $\Delta H^\circ$ may be defined as the heat released or absorbed as a reaction proceeds. $\Delta H^\circ$ for a particular reaction is found by subtracting the sum of the $\Delta H^\circ$'s of the compounds reacting from the sum of the $\Delta H^\circ$'s of the products being formed, i.e., $\Delta H^\circ$ = ($\Delta H^\circ$ of products) - ($\Delta H^\circ$ of reactants).

(1) Determination of the $\Delta H^\circ$'s for the reactions:

(a)  $4NH_3(g) + 5O_2(g) \rightarrow 4NO(g) + 6H_2O(\ell)$. When more than one mole of a compound is reacted or formed, the $\Delta H^\circ$ of the compound is multiplied by the number of moles present.

$$\Delta H^\circ = (4 \times \Delta H^\circ \text{ of NO} + 6 \times \Delta H^\circ \text{ of } H_2O) - (4 \times \Delta H^\circ \text{ of } NH_3 +$$

$$5 \times \Delta H^\circ \text{ of } O_2)$$

The $\Delta H^\circ$ of any element is always 0. Thus, the $\Delta H^\circ$ of $O_2$ is zero. Substituting,

$$\Delta H^\circ = \Big[(4 \times 21.6) + 6 \times (-68.3)\Big] - \Big[(4 \times (-11.0)) + (5 \times 0)\Big]$$

$$= -279.40 \text{ Kcal.}$$

(b) $H_2S(g) + 1\tfrac{1}{2}O_2(g) \rightarrow SO_2(g) + H_2O(\ell)$

$$\Delta H^\circ = (\Delta H^\circ \text{ of } SO_2 + \Delta H^\circ \text{ of } H_2O) - (\Delta H^\circ \text{ of } H_2S + 1\tfrac{1}{2} \Delta H^\circ \text{ of } O_2)$$

$$\Delta H^\circ = \Big[(-71.0) + (-68.3)\Big] - \Big[(-5.3) + (1\tfrac{1}{2} \times 0)\Big]$$

$$= -134.0 \text{ Kcal}$$

(2) Determination of $\Delta E$: $\Delta E^\circ$ is the change in energy of a given reaction. It is defined as the difference of the $\Delta H^\circ$'s of the reaction and the pressure times the change in volume occurring during the reaction, i.e.,

$$\Delta E^\circ = \Delta H^\circ - P\Delta V,$$

where P is the pressure and $\Delta V$ is the change in volume. One is not given the values for pressure or volume here, thus one uses the Ideal Gas Law to substitute $\Delta nRT$ for $P\Delta V$ where $\Delta n$ is the change in the number of moles present, R is the gas constant (1.99 cal/mole-°K) and T is the absolute temperature. The Ideal Gas Law states that

$$P\Delta V = \Delta nRT$$

Here, one is told that the reactions both occur at 25°C. This temperature can be converted to the absolute scale by adding 273 to 25°C.

$$T = 25 + 273 = 298°K.$$

$\Delta n$ is found by subtracting the number of moles of compounds reacting from the number of moles formed. Caution: Only moles of gases are taken into account. $\Delta n$ = number of moles formed - number of moles reacted.

One can solve for $\Delta E^\circ$ in these two reactions.

(a) $4NH_3(g) + 5O_2(g) \rightarrow 4NO(g) + 6H_2O (\ell)$

$\Delta H^\circ$ for this reaction was found to be $-279.40$ Kcal. $\Delta H^\circ$ must be converted to calories when R is used. Kcal are converted to cal by multiplying the number of Kcal by 1000 cal/1 Kcal.

$\Delta n$ = (4 moles NO) - (4 moles $NH_3$ + 5 moles $O_2$)

$$= (4) - (4 + 5) = -5 \text{ mole}$$

$$\Delta E^\circ = \Delta H^\circ - \Delta nRT$$

$$\Delta H^\circ = -279.40 \text{ Kcal}$$
$$\Delta n\cdot = 5 \text{ moles}$$
$$R = 1.99 \text{ cal/mole-}^\circ K$$
$$T = 298^\circ K$$

$$\Delta E^\circ = -279.4 \text{ Kcal} \times \left(\frac{1000 \text{ cal}}{1 \text{ Kcal}}\right) - (-5 \text{ mole}$$
$$\times 1.99 \text{ cal/mole } ^\circ K \times 298^\circ K\Big)$$

$$= -279,400 \text{ cal} + 2965 \text{ cal}$$

$$= -276,435 \text{ cal} = -276.4 \text{ Kcal}.$$

(b) $H_2S(g) + 1\frac{1}{2}O_2(g) \rightarrow SO_2(g) + H_2O(\ell)$

$$\Delta H^\circ = -134.0 \text{ Kcal}$$

$\Delta H^\circ$ should be expressed in calories

$$\Delta H^\circ \text{ in cal} = -134.0 \text{ Kcal} \times 1000 \text{ cal/1 Kcal}$$

$$= -134,000 \text{ cal}$$

$\Delta n = $ (no. of moles of $SO_2$)- (no. of moles of $H_2S$ + no. of moles of $O_2$)

$\Delta n = $ (1 mole of $SO_2$) - (1 mole of $H_2S$ + 1 $\frac{1}{2}$ moles of $O_2$)

$$= (1) - (1 + 1\tfrac{1}{2}) = -1.5 \text{ moles}$$

$R = 1.99 \text{ cal/mole } ^\circ K$

$T = 298^\circ K$

$\Delta E^\circ = \Delta H^\circ - \Delta nRT$

$$= -134,000 \text{ cal} - (-1.5 \text{ moles} \times 1.99 \text{ cal/mole}^\circ K \times 298^\circ K)$$

$$= -134,000 \text{ cal} + 890 \text{ cal}$$

$$= -133,110 \text{ cal} = -133.1 \text{ Kcal}.$$

● **PROBLEM** 508

Calculate $\Delta H$ for the reaction at 25°C.

$$CO(g) + \tfrac{1}{2}O_2(g) \rightarrow CO_2(g) \qquad \Delta E = -67.4 \text{Kcal}.$$

Solution:   The change in enthalpy ($\Delta H$) is defined by the first law of thermodynamics as the sum of the change of the internal energy ($\Delta E$) and the work done by a particular system. Work is defined as the product of the pressure (P) and the change in volume ($\Delta V$) of a system, where only pressure-volume work is done. Much of the experimentation done by chemists is pressure-volume work because the work

is usually done in an open container in the laboratory.
The pressure, in this case, is constant and the volume of
the materials is allowed to change.

$$\Delta H = \Delta E + P\Delta V$$

In this problem, one is given $\Delta E$ but not P or $\Delta V$,
so that one must calculate $P\Delta V$ in some other way. The
Ideal gas Law is defined as:

$$P\Delta V = \Delta nRT,$$

where P is the pressure, $\Delta V$ is the change in volume, $\Delta n$
is the change in the number of moles reacting, R is the gas
constant (1.99 cal/mole-°K), and T is the absolute temperature. Here, one knows R and can calculate T and $\Delta n$.
Therefore, $\Delta nRT$ can be substituted for $P\Delta V$ in the equation
to find $\Delta H$. $\Delta n$ is found by subtracting the number of moles
reacting from the number of moles formed as products. Here,
1 mole of CO and 0.5 mole of $O_2$ form 1 mole of $CO_2$.

$\Delta n$ = 1 mole $CO_2$ - (1 mole CO + 0.5 mole $O_2$)

= - 0.5 mole.

The absolute temperature is found by adding 273 to
the temperature in °C.

T = 25 + 273 = 298°K.

Therefore, $\Delta nRT$ = (- 0.5 mole) × 1.99 cal/mole-°K × 298°K

= - 0.3 Kcal.

One is now ready to find $\Delta H$, via substitution.

$$\Delta H = \Delta E + \Delta nRT$$

$$\Delta E = - 67.4 \text{ Kcal}$$

$$\Delta nRT = - 0.3 \text{ Kcal}$$

$$\Delta H = - 67.4 \text{ Kcal} + (- 0.3 \text{ Kcal}) = - 67.7 \text{ Kcal}.$$

● **PROBLEM** 509

The equation for the burning of naphthalene is
$C_{10}H_8$ (s) + $12O_2$ (g) → $10CO_2$ + $4H_2O$ (ℓ). For every mole of
$C_{10}H_8$ burned, - 1226.7 Kcal is evolved at 25° in a fixed-
volume combustion chamber. $\Delta H°$ for $H_2O$(ℓ) =
- 64.4 Kcal/mole and $H_2O$ (g) = - 57.8 Kcal/mole. Calculate
(a) the heat of reaction at constant temperature and
(b) $\Delta H$ for the case where all of the $H_2O$ is gaseous.

Solution:  This problem deals with the heat evolved

when a compound is heated with oxygen to form carbon dioxide and water (the heat of combustion). The heat of reaction, or combustion in this case, $\Delta H$, is given by the formula $\Delta H = \Delta E + \Delta n R T$, where $\Delta E$ = amount of heat released per mole, $\Delta n$ = change in moles, R = universal gas constant and T = temperature in Kelvin (Celsius plus 273°). Thus, to answer (a) substitute these values and solve for $\Delta E$. $\Delta n$ is moles of gas produced - moles of gas reacted, which is 10 moles- 12 moles= -2 moles based on the coefficients in the equation. T = 25° + 273 = 298°K. Use R in terms of kilocalories. As such

$$\Delta H = - 1226.7 \text{ kcal} + (- 2 \text{ moles})(1.987 \text{ cal/mole °K})$$
$$(298°K)(1 \text{ Kcal/1000 cal})$$

$$= - 1226.7 - 1.2 = - 1227.9 \text{ Kcal/mole.}$$

To find (b), note that the reaction is the same, except that $H_2O$ is gaseous not liquid. You have, therefore, $4H_2O(\text{liq}) \rightarrow 4H_2O(\text{g})$. The $\Delta H$, change in enthalpy, for this conversion is $4\Delta H°(\text{liq}) -4H°\Delta(\text{g})$ of $H_2O$ = $4(- 57.8) - 4(- 68.4) = 42.4$ Kcal/mole. It follows, then, that the resulting

$$\Delta H = - 1227.9 + 42.4 = - 1185.5 \text{ Kcal/mole.}$$

● **PROBLEM** 510

In the reaction, $CaCO_3(s) \rightarrow CaCO(s) + CO_2(g)$ at 950°C and $CO_2$ pressure of 1 atm, the $\Delta H$ is found to be 176 kJ/mole. Assuming that the volume of the solid phase changes little by comparison with the volume of gas generated, calculate the $\Delta E$ for this reaction.

Solution: To solve this problem, you must know that $\Delta E$, the change in energy, can be related quantitatively to the change in enthalpy, $\Delta H$, the pressure (P) and the change in volume ($\Delta V$), by the formula $\Delta E = \Delta H - P\Delta V$. This equation means that if heat is added to a system at constant pressure, part of it goes into increasing the internal energy of the system ($\Delta E$) and the rest is used to do work on its surroundings, $P\Delta V$. Therefore, the total of both $\Delta E$ and $P\Delta V$ is the heat content of the system, $\Delta H$. To solve the problem, therefore, you need only substitute in the values for P, $\Delta V$ and $\Delta H$.

$$\Delta V = V_{products} - V_{reactants}$$

There is little change in the volume of the solid by comparison with the gas. This means, therefore, that $\Delta V$ is approximately equal to $V_{gas}$. $V_{gas}$ can be found from the equation of state; $PV = nRt$ or $V = nRt/P$, where n = number of moles, P = pressure, R = universal gas constant and t = temperature in Kelvin (Celsius temp + 273 ). From the reaction, 1 mole of $CO_2$ gas is produced, thus n = 1. R is defined as being equal to 8.31 J mol$^{-1}$ deg$^{-1}$. You are given P and t. Therefore

$$V = \frac{1(8.31)(1223)}{1} = 10.2 \text{ kJ}$$

Recalling, $\Delta E = \Delta H - P\Delta V$, you have

$$\Delta E = 176 - 1(10.2) = 166 \text{ kJ.}$$

● **PROBLEM** 511

You have the reaction $H_2O(\ell) \rightarrow H_2O(g)$. For both states, 1 mole of water is at 100°C and 1 atm pressure. The volume of 1 mole of water = 18 ml, $\Delta H$ = 9710 cal/mole and there are 24.2 cal/liter-atm. Calculate the work done in this conversion and the value of $\Delta E$.

Solution: In the conversion from liquid water to steam, the pressure and temperature remain constant, work (w) = P (pressure) times $\Delta V$(change in volume). You are given the pressure.

To find the volume, note $\Delta V$ = volume of gas - volume of liquid. Thus, to find the work done in the conversion, find the volume that the steam occupies. This can be done by using the equation of state, which indicates PV = n RT, where P = pressure, V = volume, n = moles, R = universal gas constant and T = temperature in Kelvin (Celsius plus 273°). You know all values, except V. Substituting and re-writing,

$$V = \frac{n\ RT}{P} = \frac{(1)(.0821)(373)}{1} = 30.62 \text{ liters.}$$

This is the volume of the steam. The volume of the liquid = 18 ml or .018 liters. Thus, $\Delta V$ = 30.62 - .018 = 30.602 $\ell$ and w = work = (1 atm)(30.602 liter) = 30.602 liter-atm. 24.2 cal. exist per liter-atm. Thus, w = work = 24.2 cal/liter-atm × 30.602 liter-atm = 740.56 cal. To find $\Delta E$, remember that $\Delta E = \Delta H - P\Delta V$, where $\Delta E$ = change of energy, $\Delta H$ = change of enthalpy and P$\Delta V$ = work, which was calculated. You are given $\Delta H$, therefore

$$\Delta E = 9710 \text{ cal} - 740.56 \text{ cal} = 8969.44 \text{ cal.}$$

● **PROBLEM** 512

A chemist expands an ideal gas against a constant external pressure of 700 mmHg, and finds its volume changes from 50 to 150 liters. He finds that 1.55 Kcal of heat have been absorbed in the process. Determine the internal energy change that took place. 24.217 cal = 1 liter-atm.

Solution: The internal energy change of a system at constant pressure is given by the formula $\Delta E = \Delta H - P\Delta V$, where $\Delta E$ = change in internal energy, $\Delta H$ = enthalpy or heat absorbed or released, P = pressure and $\Delta V$ = change in volume.

In this problem, $\Delta V = 150 - 50 = 100$ liters. The pressure, in atms, = (700 mm)(1 atm)/(760 mm) = .921 atm. To convert, liter-atm to kilocalories, multiply by 24.217 cal./liter-atm and then multiply by Kcal/1000 cal. Therefore,

$\Delta E = 1.55$ Kcal $- (.921$ atm$)(100$ ℓ$)(24.217$ cal/liter-atm$)$

(Kcal/1000 cal)

$= 1.55 - 2.23 = -.68$ Kcal.

● **PROBLEM** 513

Exactly one mole of gaseous methane is oxidized at fixed volume and at 25°C according to the reaction

$CH_4(g) + 2O_2(g) \rightarrow CO_2(g) + 2H_2O(ℓ)$.

If 212 Kcal is liberated, what is the change in enthalpy, $\Delta H$?

Solution: We must first find the change in internal energy, $\Delta E$, before applying the equation,

$\Delta H = \Delta E + \Delta nRT$,

to find the enthalpy change ($\Delta n$ = number of moles of gaseous products - number of moles of gaseous reactants, R = 1.987 cal/deg-mole, T = absolute temperature).

Since heat was evolved (exothermic reaction), $\Delta E$ must be negative. Furthermore, since no useful work was done ($\Delta V \neq 0$ if work is done), $\Delta E$ is equal in magnitude to the quantity of heat evolved. Thus $\Delta E = -212$ Kcal. Since there is one mole of gaseous product (1 $CO_2(g)$) and three moles of gaseous reactants ($2O_2(g) + 1\ CH_4(g)$), $\Delta n = 1 - 3 = -2$ mole. The absolute temperature is T = 25 + 273 = 298°K. Hence

$\Delta H = \Delta E + \Delta nRT$

$= -212$ Kcal $+ (-2$ mole$)(1.987$ cal/deg-mole$)(298°K)$

$= -212$ Kcal $- 1180$ cal

$= -212$ Kcal $- 1.180$ Kcal

$\cong -213$ Kcal.

For the following exothermic reaction at 25°C,

$C_6H_6(\ell) + 7\frac{1}{2} O_2(g) \rightarrow 6CO_2(g) + 3H_2O(\ell)$,

the change in energy, $\Delta E$, is - 780 Kcal/mole. Find the change in enthalpy, $\Delta H$.

Solution: The solution is based on the following equation for the enthalpy change

$$\Delta H = \Delta E + \Delta nRT,$$

where $\Delta n$ is the number of moles of gaseous products minus the number of moles of gaseous reactants, R is the gas constant (1.987 cal/deg-mole), and T is the absolute temperature. Since there are 6 moles of gaseous products ($6CO_2(g)$) and $7\frac{1}{2}$ moles of gaseous reactants ($7\frac{1}{2}O_2(g)$), $\Delta n = 6 - 7\frac{1}{2} = - 1.5$. The absolute temperature is T = 25° + 273 = 298°K.

Hence, $\Delta H = \Delta E + \Delta nRT$

$= - 780$ Kcal $+ (- 1.5$ mole$)(1.987$ cal/deg-mole$)\times$
$\qquad\qquad\qquad\qquad (298°K)$

$= - 780$ Kcal $- 888$ cal

$= - 780$ Kcal $- 0.888$ Kcal

$\cong - 781$ Kcal.

# HEATS OF FUSION AND VAPORIZATION

It is known that the heat of vaporization of water is 5 times as great as the heat of fusion. Explain this fact.

Solution: The heat of vaporization is the quantity of heat necessary to vaporize 1 g. of a liquid substance at its boiling point at constant temperature. The heat of fusion is the quantity of heat necessary to liquefy 1 g. of a solid substance at constant temperature at its melting point. With this in mind, let us consider what goes on in each of these phase changes. The volume-increase of the gas in going from a liquid state to a vapor state is much greater than the volume-increase accompanying a solid to liquid transformation. When you increase volume, work is required to overcome the existing external pressure that hinders the volume expansion. Work is defined as the product of the pressure and the change in volume. Since, the volume increase is larger in converting from liquid to vapor, more energy is necessary because more work is done.

More importantly, however, is that the molecules of a substance in the gaseous state are so much further apart than those in the liquid state. To bring about this separation of particles requires tremendous energy. This is the main reason that the heat of vaporization exceeds the heat of fusion.

● **PROBLEM** 516

What weight of ice could be melted at 0°C by the heat liberated by condensing 100 g of steam at 100°C to liquid. Heat of vaporization = 540 cal/g, heat of fusion = 80 cal/g.

Solution: The quantity of heat necessary to convert 1 g of a liquid into a vapor is termed the heat of vaporization. For water, 540 calories are necessary to change liquid water at 100°C into vapor at 100°C. In this problem vapor is condensed, thus 540 cal of heat are evolved for each gram of liquid condensed.

$$\text{no. of cal evolved in condensation} = 540 \text{ cal/g} \times \text{weight of vapor}$$

Here 100 g of vapor is condensed. Thus,

$$\text{no. of cal evolved} = 540 \text{ cal/g} \times 100 \text{ g} = 54000 \text{ cal}$$

When ice melts, heat is absorbed. About 80 calories of heat are required to melt 1 g of ice, the heat of fusion. Here, 54000 cal are evolved in the condensation. Therefore, this is the amount of heat available to melt the ice. Because 80 cal are needed to melt 1 g of ice, one can find the number of grams of ice that can be melted by 54000 cal, by dividing 54000 cal by 80 cal/g.

$$\text{no. of grams of ice melted} = \frac{54000 \text{ cal}}{80 \text{ cal/g}} = 675 \text{ g.}$$

● **PROBLEM** 517

The following are physical properties of methyl alcohol, $CH_3OH$: freezing point - 98°C; boiling point 65°C; specific heat of liquid 0.570 cal/g - degree; heat of fusion 22.0 cal/g; and heat of vaporization 263 cal/g. Calculate the number of kilocalories required to convert one mole of methyl alcohol solid at - 98°C to vapor at 65°C.

<u>Solution</u>:   There are three steps in this process:
(1) melting the solid, (2) heating the liquid from - 98°C
to 65°C, (3) vaporizing the liquid at 65°C.

Heat will be absorbed in each of these processes,
thus one must calculate the heat absorbed in each process
and then take the total.

(1) Heat absorbed in melting the solid.

To find the amount of heat absorbed when a compound
is melted, one uses the heat of fusion. The heat of fusion
for methyl alcohol is 22.0 cal/g. This means that for each
gram of solid melted 22.0 cal of heat are absorbed. Here,
one is melting one mole of $CH_3OH$. The molecular weight of
$CH_3OH$ is 32, thus one will calculate the heat absorbed
when 32 g of $CH_3OH$ is melted.

no. of cal absorbed = heat of fusion × no. of grams

no. of cal absorbed = 22.0 cal/g × 32 g = 704 cal.

(2) Heat absorbed in heating the liquid.

In determining the amount of heat absorbed in heating
the liquid, one uses the specific heat of the liquid. For
$CH_3OH$ the specific heat is 0.570 cal/g-degree. This means
that for each gram of $CH_3OH$ raised 1 degree, 0.570 calories
is absorbed. Here, the liquid is raised 65° - (- 98°) or
163°C. There are 32 g of $CH_3OH$ heated.

no. of cal absorbed = no. of degrees × weight × specific
heat

no. of cal absorbed = 163°× 32 g × 0.570 cal/g-°

= 2973 cal.

(3) Heat absorbed when the liquid is vaporized.

To calculate the amount of heat absorbed when the
liquid is vaporized, one uses the heat of vaporization. The
heat of vaporization for $CH_3OH$ is 263 cal/g. This means that
for each gram of $CH_3OH$ vaporized, 263 calories are absorbed.

no. of cal absorbed = heat of vaporization × weight

no. of cal absorbed = 263 cal/g × 32 g = 8416 cal.

(4) To find the heat absorbed by the whole process,
the heat absorbed in these three steps must be added together.

total heat absorbed = heat absorbed in melting + heat
    absorbed in heating the liquid + heat absorbed in
    vaporization

total heat absorbed = 704 cal + 2973 cal + 8416 cal

= 12093 cal

Calories can be converted to kilocalories by multiplying the number of calories by 1 Kcal/1000 cal.

no. of Kcal = no. of cal × 1 Kcal/1000 cal

no. of Kcal = 12093 cal × 1 Kcal/1000 cal = 12.09 Kcal.

● **PROBLEM** 518

40 g of ice at 0°C is mixed with 100 g of water at 60°C. What is the final temperature after equilibrium has been established? Heat of fusion of $H_2O$ = 80 cal/g, specific heat = 1 cal/g-°C.

Solution: In a determination of this kind, the heat lost by the water in cooling must be balanced by the heat gained by the ice in melting and in warming the resulting water (from the melted ice) to the final temperature. This means that the heat absorbed by the ice must equal the heat lost by the water at 60°C.

The heat absorbed by the ice is the combination of the heat absorbed by the ice in melting and the heat absorbed by the cold water to the new temperature. The heat absorbed when the ice melts can be calculated by taking into account the heat of fusion. The heat of fusion is defined as the number of calories absorbed when 1 g of solid melts. For water, 80 calories are absorbed for each gram of ice melted. Here 40 g of ice is melted. Therefore, the amount of heat absorbed by the ice is 40 times the heat of fusion.

40 g × 80 cal/g = 3200 cal.

As such, 3200 calories are absorbed when the ice melts. To find the amount of heat absorbed, when this water (from the melted ice) is heated to a new temperature, you must consider the specific heat of water. The specific heat is defined as the number of calories needed to raise 1 g of liquid 1 degree. The specific heat of water is 1, which means that for every gram of water raised 1 degree, 1 calorie is absorbed. In calculating the amount of heat absorbed by the water to the new temperature, let t = new temperature. To find the amount of heat absorbed, multply the specific heat by 40 g, because 40 g of water is present, and by the new temperature, t, because the water will be raised t degrees from zero.

Amount of heat absorbed by liquid = 40 g × 1cal/g-degree)×t

= 40 t cal.

Therefore, 40 t calories will be absorbed by the melted ice. Thus, the amount of heat absorbed by the ice when it melts, and by the water when it is heated, is equal to the total amount of heat absorbed by the system.

Total amount of heat absorbed = 3200 cal + 40 t cal

$$= (3200 + 40 t) \text{ cal.}$$

To find the amount of heat lost by the 100 g of water at 60°C, when it is cooled to a new temperature, t, the specific heat of water is also used. Here, it means that for every gram of water that is lowered one degree, 1 calorie of heat is released. Therefore, to find the amount of heat released by this water, the specific heat will be multiplied by 100 g, the amount of water present, and by 60-t°, which is the number of degrees that the temperature will drop.

(60-t)° × (1 cal/g-degree) × (100 g) = (6000 - 100t)cal.

To calculate t, one must set the amount of heat absorbed by the ice equal to the amount of heat lost by the water at 60°C.

amount of heat absorbed = amount of heat lost

amount of heat absorbed = (3200 + 40 t) cal

amount of heat lost = (6000 - 100 t) cal

(3200 + 40 t) cal = (6000 - 100 t) cal

140 t = 2800          t = 20°

Therefore, the final temperature is 20°C.

● **PROBLEM** 519

Determine the quantity of heat required to convert 10 g of ice at 0°C to vapor at 100°C. For water, heat of fusion = 80 cal/g, heat of vaporization = 540 cal/g, and specific heat = 1 cal/g -°C.

Solution:    Heat is absorbed when ice is transformed from the solid to the liquid state. The heat absorbed when 1 g of solid melts is called the heat of fusion. The heat of fusion for water is 80 cal/g. In calculating the amount of heat needed to raise the temperature of the liquid water, the specific heat must be used. The specific heat is defined as the number of calories of heat necessary to raise the temperature of a gram of a substance 1°C. The specific heat of water is 1°, thus 1 calorie is needed to raise the temperature of 1 gram of water 1°C. The quantity of heat necessary to convert 1 g of a liquid into a vapor is termed the heat of vaporization. For water, 540 calories are necessary to change 1 g of liquid water at 100°C into vapor (steam) at 100°C.

In this problem three steps are involved.

1) The ice must be melted.

2) The water must be heated from 0°C to 100°C.

3) The water must be vaporized to steam.

Heat is absorbed in each of these steps, so that each must be accounted for.

Melting the ice:

Because the heat of fusion is 80 cal/g for water, one knows that for every gram of ice melted, 80 cal of heat must be absorbed. Here, there are 10 grams of ice being melted, so the specific heat must be multiplied by 10 to find the amount of heat absorbed.

10 g × 80 cal/g = 800 cal.

Therefore, 800 cal of heat are absorbed when the ice is melted.

Heating the liquid:

To calculate the amount of heat absorbed when the water is heated, the specific heat should be taken into account. The specific heat of water is 1 cal/g°C. This means that for every degree of temperature each gram of water is raised, 1 cal of heat is absorbed. In this case, 10 grams of water are raised 100°. Therefore, the specific heat is multiplied by 10 to take into account the weight of the water and then multiplied (again) by 100 to account for the fact that the water is being raised 100°.

1 cal/g-deg × 10 g × 100° = 1000 cal.

Therefore, 1000 calories are absorbed when the liquid is heated to 100°C.

Vaporization:

To find the amount of heat absorbed when the water is vaporized, the heat of vaporization must be used. For water, the heat of vaporization is 540 cal/g, which means that for every gram of water vaporized, 540 calories are absorbed. Here, 10 g of water is vaporized so that the heat of vaporization must be multiplied by 10 g to find the amount of heat absorbed.

10 g × 540 cal/g = 5400 cal.

To calculate the total amount of heat absorbed, the heat absorbed in all three steps must be added together.

```
melting        800 cal
heating       1000 cal
vaporization  5400 cal
```
7200 cal or 7.20 Kcal.

Therefore, 7200 calories are needed to melt 10 g of ice, heat it to 100°C and vaporize it to steam.

● **PROBLEM** 520

500 g of ice at 0°C is added to 500 g of water at 64°C. When the temperature of the mixture is 0°C, what weight of ice is still present? Heat of fusion of $H_2O$ = 80cal/g.

Solution:  The amount of heat used to melt the ice is equal to the amount of heat lost by the water. The amount of heat lost by the water is determined by using the specific heat of water. The specific heat of water is 1 cal/g-degree. This means that for each gram of water cooled 1degree, 1 calorie of heat will be evolved. Here, 500 g of $H_2O$ is lowered 64° - 0° or 64°. The heat evolved by the water can now be found.

no. of calories evolved = 1 cal/g-degree × 500 g × 64°

= 32000 cal.

The heat of fusion is the quantity of heat necessary to liquefy 1 g of a solid substance at constant temperature at its melting point. Therefore, if the heat of fusion of water is 80 cal/g, then it takes 80 cal to melt one gram of ice. You found that 32000 calories are absorbed by the ice. The weight of ice melted by 32000 cal is found by dividing 32000 cal by 80 cal/g.

no. of grams of ice = $\frac{32000 \text{ cal}}{80 \text{ cal/g}}$ = 400 g

Thus 400 g of the ice is melted. Originally, there was 500 g of ice, therefore 100 g of ice is left.

● **PROBLEM** 521

Determine the heat needed to raise 60 g of Pb from 20°C to 360°C, given its specific heats are 0.0306 (solid) and 0.0375 (liquid). Its heat of fusion is 5.86 cal/g; its melting point is 327°C.

Solution:  There are three heats involved in finding the amount of heat absorbed in raising the temperature of Pb from 20°C to 360°C. First, the amount of heat absorbed in raising the temperature of the solid from 20°C to 327°C (its melting point); then, the amount of heat absorbed in melting the compound; last, the amount of heat absorbed in raising the temperature of the liquid from 327°C to 360°C. The heat absorbed in these three processes are added to-gether to find the amount of heat needed to bring Pb from

20°C to 360°C.

1) Raising the temperature of the solid from 20°C to 327°C.

The specific heat of solid Pb is used to calculate the amount of heat absorbed in this process. The specific heat is defined by the number of calories absorbed when one gram of mass is raised one degree. The specific heat of Pb as a solid is 0.0306. Here, 60 g of Pb is heated 307°C (327 - 20°C). Thus, the specific heat is multiplied by 60 and 307.

heat absorbed by solid = 60 × 307 × .0306 = 564 cal.

2) Heat needed to melt the solid.

To calculate the amount of heat needed to melt the solid, one needs to use the heat of fusion. The heat of fusion is the number of calories necessary to melt one gram of solid. The heat of fusion of Pb is 5.86 cal/g. There are 60 g of Pb, here. Thus, the heat of fusion must be multiplied by 60 g to find the amount of heat needed to melt the Pb.

heat needed to melt Pb = 60 g × 5.86 cal/g = 352 cal.

3) Heat absorbed in raising the temperature of the melted liquid from 327°C to 360°C.

Here the specific heat is used again, but because the Pb is now liquid, the specific heat for liquid Pb must be used. This is the amount of heat necessary to raise one gram of liquid one degree. 60 g Pb is heated 33°C (360°-327°); therefore, the specific heat must be multiplied by 60 and 33 to find the heat absorbed. The specific heat for liquid Pb is 0.0375.

heat absorbed by liquid Pb = 60 × 33 × 0.0375 = 74 cal.

The heat absorbed by the total process is the combination of the heats absorbed in these three processes.

| | |
|---|---|
| Heat absorbed by solid | 564 cal |
| Heat absorbed to melt solid | 352 cal |
| Heat absorbed by liquid | 74 cal |
| Total amount of heat needed | 990 cal. |

● **PROBLEM** 522

Using data from the accompanying table   find the increase of enthalpy for 100 g of $H_2O$ going from ice at - 10°C to liquid water at + 15°C. Assume that the molar heat of fusion of ice is 6.02 kJ.

| | |
|---|---|
| $H_2O(s)$ *at* $239°K$ | *33.30* |
| $H_2O(s)$ *at* $271°K$ | *37.78* |
| $H_2O(l)$ *at* $273°K$ | *75.86* |
| $H_2O(l)$ *at* $298°K$ | *75.23* |
| $H_2O(l)$ *at* $373°K$ | *75.90* |
| $H_2O(g)$ *at* $383°K$ | *36.28* |

<u>Solution</u>: Enthalpy is an indication of the changes in heat content of a system due to changes in the system. To solve this problem, one must be aware that in going from ice at - 10°C to water at 15°C, three things happen: (1) the ice is heated, (2) the ice is melted, (3) the liquid is heated. For the heating processes, (1) + (3), the change in enthalpy is quantitatively related to the heat capacity, $C_p$, and the change in temperature, $\Delta T$ by the relation is $\Delta H = C_p \Delta T$. $C_p$, the heat capacity, is defined as the amount of heat required to raise one mole of material by one degree. The melting process, (2), is expressed in terms of the molar heat of fusion, which is the amount of heat necessary to melt one mole of solid. The summation of the heat involved in all three processes will give the change in enthalpy. Because each step is defined in terms of moles, one must first calculate the number of moles involved. The molecular weight of $H_2O$ is 18. Therefore, there are 100/18 or 5.55 moles of $H_2O$ in 100 g. Thus,

   (1) To heat ice, one needs,

(5.55 moles) (37.78 J mol$^{-1}$ deg$^{-1}$) (10 deg) = 2,100 J

   (2) To melt ice, one requires

(5.55 moles) (6.02 kJ/mole) (1000 J/kJ) = 33,400 J

   (3) To heat liquid, one needs

(5.55 moles) (75.86 J mol$^{-1}$ deg$^{-1}$) (15 deg) = 6,300 J.

   The total increase in enthalpy is

$\Delta H$ = 2,100 + 33,400 + 6,300 = 41,800 J.

# CHAPTER 15

# THERMODYNAMICS II

> **Basic Attacks and Strategies for Solving Problems in this Chapter. See pages 503 to 538 for step-by-step solutions to problems.**

The Second Law of Thermodynamics is not nearly as easy to state as the First Law, but is nonetheless very important in understanding how spontaneous changes occur. For example, the First Law is satisfied when water falls over a waterfall and the temperature increases to account for the change in potential energy from the top to the bottom of the waterfall. But the First Law would also be satisfied if the water went up the waterfall and was simultaneously cooled. Experience, of course, indicates that this does not happen; the Second Law of Thermodynamics provides a formal explanation.

The Second Law of Thermodynamics states that the change in entropy, defined as

$$\Delta S = \Delta q/T \tag{15-1}$$

is positive when contributing to spontaneous change. In Equation 15-1, $\Delta S$ is the entropy change, $\Delta q$ is the change in heat (positive, by convention, for heat entering the system and negative for heat leaving), and $T$ is the absolute temperature. Unlike enthalpy, there is an absolute zero for entropy — the elements at 0K. Therefore, absolute values of entropy are tabulated for various compounds — usually at 298K. But entropy change for a reaction is calculated in the same way as the enthalpy changes.

$$\Delta S^\circ_r = \Sigma\, S^\circ_{products} - \Sigma\, S^\circ_{reactants} \tag{15-2}$$

A very important thermodynamic quantity involving both entropy and enthalpy is called the Gibbs Free Energy ($G$) or, sometimes, simply the free energy. It is defined as

$$G = H - TS \qquad\qquad 15\text{-}3$$

Since it involves enthalpy, only changes in the free energy have physical significance. Changes in the free energy associated with a reaction are calculated in a manner analogous to changes in enthalpy.

$$\Delta G^\circ_r = \Sigma\,(\Delta G^\circ_f)_{products} - \Sigma\,(\Delta G^\circ_f)_{reactants} \qquad\qquad 15\text{-}4$$

$\Delta G^\circ_f$ is defined, in a manner analogous to $\Delta H^\circ_f$, as the free energy associated with the formation of a species from the elements. Extensive tables of the values of $\Delta G^\circ_f$ are found in a variety of handbooks.

The importance of the free energy of reaction, $\Delta G^\circ_r$, lies in its relationship to the equilibrium constant for a reaction. For any change, whether it be a chemical reaction or a physical change such as dissociation, the relationship between $\Delta G^\circ_r$ and the equilibrium constant (K) is given by the following equivalent relationships.

$$K = e^{(-\Delta G^\circ r /RT)} \qquad\qquad 15\text{-}5$$

$$-RT \ln K = \Delta G^\circ_r \qquad\qquad 15\text{-}6$$

Since values of $\Delta G^\circ_f$ are tabulated and values of $\Delta G^\circ_r$ can be determined for any reaction from Equation 15-4, values of the equilibrium constant are readily available for most reactions without the necessity of making additional experimental measurements.

Another important relationship that follows from Equation 15-5 and relates $K$ to $\Delta H^\circ_r$ is the van't Hoff Equation.

$$[d\,(\ln K)/dT]_p = \Delta H^\circ_r /RT^2 \qquad\qquad 15\text{-}7$$

For modest temperature changes, the value of $\Delta H^\circ_r$ can usually be considered constant and the integrated form of Equation 15-7 permits the calculation of $K$ as a function of temperature.

$$\ln\,(K_2/K_1) = (\Delta H^\circ_r /R)\,(1/T_1 - 1/T_2) \qquad\qquad 15\text{-}8$$

Once the value of the equilibrium constant has been determined, the computation of equilibrium compositions follows the same procedures outlined in Chapters 9–12. Some examples are shown in Chapters 9–12. A second, slightly more complicated example follows. Let us compute the equilibrium composition of a stoichiometric mixture of nitrogen and hydrogen used to produce ammonia at 500K and one atm pressure.

$$N_2 + 3H_2 \leftrightarrow 2NH_3 \qquad\qquad 15\text{-}9$$

The tabulated value of $\Delta G°_f$ for ammonia at 500K is 1114 cal/mole. The reaction in Equation 15-9 is, in fact, the reaction that forms ammonia from the elements, $\Delta G°_f = \Delta G°_r$. But observe that 2 moles of ammonia are formed in Equation 15-9. Therefore, for the reaction as written $\Delta G°_f = 2(1114) = 2228$, $K$ from Equation 15-5 equals 0.106. The basic steps are:

1) Write the reaction (Equation 15-9).

2) Write the equation relating the equilibrium constant to compositions.

$$.106 = p_{NH_3}^2/(p_{N_2}p_{H_2}^3) \qquad\qquad 15\text{-}10$$

$$.106 = (n_{NH_3})^2/[(n_{N_2})(n_{H_2})^3][P/n_T]^{-2} \qquad\qquad 15\text{-}11$$

3) Write the composition in terms of a single variable from the stoichiometry. Note that a basis, e.g., 1 mole of $N_2$ initially present, must be selected. Let $x$ = moles (or the fraction) of $N_2$ consumed at equilibrium.

At equilibrium:

$$
\begin{aligned}
n_{N_2} &= 1 - x \\
n_{H_2} &= 3(1 - x) \\
n_{NH_3} &= 2x \\
n_T &= 4 - 2x
\end{aligned}
$$

4) Solve the equation for the single variable and calculate the equilibrium compositions.

$$.106 = (2x)^2/[3^3(1-x)^4][P/(4-2x)^2] \qquad\qquad 15\text{-}12$$

The solution to this equation, since it is nonlinear, is typically obtained by trial and error.

$$x = .1035$$

The resulting mole fractions ($N_i = n_i/n_T$) are:

$$
\begin{aligned}
N_{N_2} &= .236 \\
N_{H_2} &= .709 \\
N_{NH_3} &= \underline{.055} \\
\Sigma &= 1.000
\end{aligned}
$$

Step-by-Step Solutions to
Problems in this Chapter,
"Thermodynamics II"

## ENTROPY

For the following reaction at 25°C

$$CuO(s) + H_2(g) \rightarrow Cu(s) + H_2O(g)$$

Values of $S°$, the absolute entropies for the substances are:

| | | |
|---|---|---|
| $CuO(s)$ | = | 10.4 cal/mole |
| $H_2(g)$ | = | 31.2 cal/mole |
| $Cu(s)$ | = | 8.0 cal/g-atm |
| $H_2O(g)$ | = | 45.1 cal/mole |

Assuming standard conditions find out if the reaction will proceed spontaneously.

Solution: Entropy change is often used to predict the spontaneity of a reaction. A process will occur spontaneously if there is an increase in entropy i.e. $\Delta S°$ is positive.

$$\Delta S° = S°_{(products)} - S°_{(reactants)}$$

For the above reaction

$$\Delta S° = S°_{Cu(s)} + S°_{H_2O(g)} - S°_{CuO(s)} - S°_{H_2(g)}$$

$$= (8 + 45.1 - 10.4 - 31.2) \text{ cal/deg-mole}$$

$$= +11.5 \text{ cal/deg-mole}$$

$\Delta S°$ is positive thus this is a spontaneous reaction.

When mercury is vaporized at its boiling point at standard pressure, the entropy change is 20.7 cal/mole-°K. Determine the boiling point of Hg if the heat of vaporization is 65 cal/g.

Solution: When a process occurs at constant temperature, the change in entropy, $\Delta S$, is equal to the heat absorbed divided by the absolute temperature at which the change occurs, i.e.,

$$\Delta S = \frac{\Delta H}{T} \ ,$$

where $\Delta H$ is the heat of vaporization (in this case) and T is the absolute temperature. T will be equal to the boiling point of mercury in this problem.

Using this equation, one can solve for T after either converting $\Delta H$ to cal/moles or converting $\Delta S$ to cal/g. Here, one will convert $\Delta H$ from cal/g to cal/mole by multiplying $\Delta H$ by the molecular weight of Hg. (MW of Hg = 200.6.)

$\Delta H$ in cal/moles = 65 cal/g × 200.6 g/mole

= 13039 cal/mole

One can now solve for T.

$$T = \frac{\Delta H}{\Delta S}$$

$\Delta H = 13039$ cal/mole

$\Delta S = 20.7$ cal/mole-°K

$$T = \frac{13039 \text{ cal/mole}}{20.7 \text{ cal/mole-}^{\circ}K} = 630^{\circ}K$$

T in °C = 630 - 273 = 357°C.

Calculate $\Delta S$ for the conversion of one mole of liquid water to vapor at 100°C. Heat of vaporization = 540 cal/g.

Solution: When a process occurs at constant temperature, the change in entropy ($\Delta S$) is equal to the heat absorbed divided by the absolute temperature at which the change occurs.

$$\Delta S = \frac{\Delta H}{T}$$

In this problem, one is given $\Delta H$ and one can find T.

The absolute temperature is calculated by adding 273 to the temperature in °C.

$$T = 273 + 100 = 373°K$$

Because one mole of water is reacting here, and the heat of vaporization is given in cal/g, one must multiply the $\Delta H$ given by the molecular weight of water to find the $\Delta H$ in cal/mole.

The molecular weight of water is 18.

$$\Delta H = 540 \text{ cal/g} \times 18 \text{ g/mole} = 9720 \text{ cal/mole}$$

One can now calculate $\Delta S$.

$$\Delta S = \frac{\Delta H}{T} \qquad\qquad \Delta H = 9720 \text{ cal/mole}$$
$$T = 373°K$$

$$\Delta S = \frac{9720 \text{ cal/mole}}{373°K} = 26.1 \text{ cal/mole-}°K$$

The change in entropy, when one mole of water is vaporized at 100°C, is 26.1 cal/mole-°K.

● **PROBLEM** 526

A chemist knows that the $\Delta H° = 485$ kJ for the reaction $2H_2(g) + O_2(g) \rightarrow 2H_2O(g)$ and that $\Delta H° = -537$ kJ for $H_2(g) + F_2(g) \rightarrow 2HF(g)$. With this information, he calculated the $\Delta H°$ for $2H_2O(g) + 2F_2(g) \rightarrow 4HF(g) + O_2(g)$ and predicted whether $\Delta S°$ was positive or negative. How?

Solution: Hess' Law states that the net heat change resulting from a particular chemical reaction is the same, independent of the steps involved in the transformation. Thus, $\Delta H°$ of the sum of two reactions equals the sum of the $\Delta H°$'s of each reaction. The chemist knows that

(i) $2H_2(g) + O_2(g) \rightarrow 2H_2O(g)$ $\qquad \Delta H° = 485$ kJ

(ii) $H_2(g) + F_2(g) \rightarrow 2HF(g)$ $\qquad \Delta H° = -537$ kJ

By doubling (ii) and adding to (i), the chemist obtains the desired equation

$$2H_2O \rightarrow 2H_2 + O_2 \quad \Delta H° = +485 \text{ kJ}$$

$$\underline{2(H_2 + F_2 \rightarrow 2HF) \quad \Delta H° = -2(537)\text{kJ}}$$

$$2H_2O + 2F_2 \rightarrow 4HF + O_2$$

The $\Delta H°$ for this reaction, according to Hess' Law, equals

$\Delta H^{\circ}_{(i)} + \Delta H^{\circ}_{(ii)} \times 2 = 485 + 2(-537) = -589$ kJ/moles.

Because the $\Delta H^{\circ}$ for the overall reaction is negative, the reaction is spontaneous. The $\Delta S^{\circ}$ is positive because as the reaction proceeds from reactants to products there is an increase in the number of moles present in the system. There are 4 moles of reactants and 5 moles of products.

● PROBLEM 527

Determine the entropy change that takes place when 1 mole of ammonia (a) Passes from the liquid state to the gaseous state at its boiling point, - 33°C; $\Delta H_{VAP}$ = 5570 cal/mole

(b) as a gas at - 33°C comes to room temperature, 25°C. Assume heat capacity is constant at 8.9 cal/deg-mole for this range.

Solution: Entropy, S, may be defined as an indication of the randomness of a system. The change in entropy for any change in the state of a system may be written as

$\Delta S = S_{final} - S_{initial}$. For a reversible process,

$\Delta S = \dfrac{q_{rev}}{T}$ , where $q_{rev}$ = amount of heat involved in a

reversible process and T = the temperature in °K at which the amount of heat is reversibly absorbed or evolved. $\Delta S$ can also be defined in terms of heat capacity ($C_p$) and

temperature ranges, say $T_1 \rightarrow T_2$, by the following equation: $\Delta S = 2.303 \ C_p \log T_2/T_1$. With this in mind, proceed as follows:

   (a) 1 mole of $NH_3$ (ammonia) goes from liquid to gas at - 33°C. The $\Delta H_{VAP}$ is the heat necessary to vaporize 1 mole of liquid material at its boiling point at constant temperature. A gas can also be liquefied, thus, this change of state is reversible. Recalling that $\Delta S = q_{rev}/T$, you can say, therefore, that $\Delta S = \Delta H_{VAP}/T$. Since $\Delta H_{VAP}$ is the only heat involved, $q_{rev} = \Delta H_{VAP}$. By substitution into $\Delta S = \Delta H_{VAP}/T$, you obtain

$\Delta S = \dfrac{5570 \text{ cal/mole}}{240°K}$    or  $\Delta S = 23.2$ cal/deg-mole.

   (b) This question uses heat capacity, which demands use of the expression $\Delta S = 2.303 \ C_p \log T_2/T_1$. Since all the values are known, you can substitute them into the above equation:

506

$$\Delta S = 2.303 \ C_p \ \log \frac{T_2}{T_1} = (2.303)(8.9) \ \log \frac{298°K}{240°K}$$

$$= 20.5 \ \log 1.24 = (20.5)(.0934) = 1.91 \ \text{cal/deg-mole.}$$

Given, for acetic acid that $\Delta H_{fus}$ = 2592 cal/mole at its melting point, 16.6°C and $\Delta H_{VAP}$ = 5808 cal/mole at its boiling point, 118.3°C, calculate the change in entropy that takes place when 1 mole of the vapor is condensed at its boiling point and changed to a solid at its melting point, all under constant pressure, taken as 1 atm. Assume that the molar heat capacity of acetic acid is 27.6 cal/deg-mole.

Solution: Entropy measures the randomness of a system. In this problem, there are two phase changes. As such, there exist three changes in entropy to calculate. The entropy change from vapor to liquid at the boiling point, the cooling entropy change from the liquid at the boiling point to liquid at the freezing point, and the entropy change from the liquid to solid. Measure these by employing the following equations:

$$\Delta S = \frac{q_{rev}}{T} \ ,$$

which states that for a reversible reaction, the change in entropy ($\Delta S$) equals the heat absorbed or released by a system ($q_{rev}$) divided by the temperature in Kelvin (Celsius plus 273°). This equation can be used for the condensation and freezing.

To measure cooling, use the fact that

$$\Delta S = 2.303 \ C_p \ \log \frac{T_2}{T_1} \ .$$

Proceed as follows:

$$\Delta S_{condensation} = - \ \Delta S_{VAP} = - \ \Delta H_{VAP}/T.$$

Condensation is the opposite of vaporization. Thus, condensation = - vaporization. Since $\Delta H$ is the only heat in this reaction, its value becomes $q_{rev}$. Substitute in values to obtain

$$\Delta S_{condensation} = - \ \frac{(5808 \ \text{cal/mole})}{391.5°K}$$

$$= - \ 14.84 \ \text{cal/deg-mole.}$$

For $\Delta S_{cooling}$,

$$\Delta S = 2.303\ C_p\ \log\frac{T_2}{T_1} = (2.303)(27.6\ cal/deg\text{-}mole)$$

$$\log\left(\frac{289.8°}{391.5°}\right)$$

$$= -8.30\ cal/deg\text{-}mole.$$

[Note: $T_1$ = b.p. in Kelvin, $T_2$ = m.p. in Kelvin.]

For freezing, $\Delta S_{freezing} = -\Delta S_{fus} = -\Delta H_{fus}/T$.

Fusion is the opposite of freezing. Thus, freezing = - fusion. $\Delta H$ is the only heat involved, so its value becomes $q_{rev}$. Therefore,

$$\Delta S = -(2592\ cal/mole)/298.8°K = -8.67\ cal/deg\text{-}mole.$$

The total entropy change is

$$\Delta S_{condensation} + \Delta S_{cooling} + \Delta S_{freezing} =$$

$$-14.84 - 8.30 - 8.67 = -31.81\ cal/deg\text{-}mole.$$

# FREE ENERGY

● PROBLEM 529

Hydrogen peroxide, $H_2O_2$, can be synthesized in two ways. The first method involves reduction of oxygen by hydrogen,

$$H_2(g) + O_2(g) \rightarrow H_2O_2(\ell).$$

The second method involves oxidation of water:

$$2H_2O(\ell) + O_2(g) \rightarrow 2H_2O_2(\ell).$$

Find the free energy of formation, $\Delta G°$, for both processes and predict which process is more efficient for the commercial preparation of hydrogen peroxide.

Solution: The standard free energy change for a reaction is equal to the difference between the standard free energy of formation of the products and the standard free energy of formation of the reactants,

$$\Delta G° = \Delta G°_{products} - \Delta G°_{reactants}.$$

The reaction with the more negative free energy change will be the more efficient, since it proceeds spontaneously as written. We will require the following standard free energies of formation:

$$\Delta G^o_{H_2O_2(\ell)} = -27.2 \text{ Kcal/mole}$$

$$\Delta G^o_{H_2O(\ell)} = -56.7 \text{ Kcal/mole}$$

$$\Delta G^o_{H_2(g)} = \Delta G^o_{O_2(g)} = 0 \text{ Kcal/mole}$$

For the first process

$$H_2(g) + O_2(g) \rightarrow H_2O_2(\ell),$$

the standard free energy change is

$$\Delta G^o = \Delta G^o_{products} - \Delta G^o_{reactants}$$

$$= \Delta G^o_{H_2O_2(\ell)} - \left[ \Delta G^o_{H_2(g)} + \Delta G^o_{O_2(g)} \right]$$

$$= -27.2 \text{ Kcal/mole} - (0 \text{ Kcal/mole} + 0 \text{Kcal/mole})$$

$$= -27.2 \text{ Kcal/mole}$$

For the second process,

$$2H_2O(\ell) + O_2(g) \rightarrow 2H_2O_2(\ell),$$

the standard free energy change is

$$\Delta G^o = \Delta G^o_{products} - \Delta G^o_{reactants}$$

$$= 2\Delta G^o_{H_2O_2(\ell)} - \left[ 2\Delta G^o_{H_2O(\ell)} + \Delta G^o_{O_2(g)} \right]$$

$$= 2(-27.2 \text{ kcal/mole}) - (2 \times (-56.7)$$
$$\text{kcal/mole} + 0 \text{ kcal/mole})$$

$$= 59.0 \text{ Kcal/2 moles } H_2O_2 \text{ produced.}$$

$$= 29.5 \text{ Kcal/mole.}$$

Since the first process proceeds spontaneously as written (negative $\Delta G^o$) and the second process requires energy to proceed as written (positive $\Delta G^o$), the first method for preparing $H_2O_2$ is more efficient than the second.

Determine $\Delta G^\circ$ for the reaction

$$4\,NH_3(g) + 5O_2(g) \rightarrow 4NO(g) + 6H_2O(\ell)$$

$\Delta G^\circ_f$ of $NH_3(g)$ = $-4.0$ Kcal/mole

$\Delta G^\circ_f$ of $NO(g)$ = $20.7$ Kcal/mole

$\Delta G^\circ_f$ of $H_2O(\ell)$ = $-56.7$ Kcal/mole

<u>Solution</u>: The change in free energy ($\Delta G^\circ$) may be found by subtracting the sum of free energies ($\Delta G^\circ_f$) of the reactants from the free energies of the products. The free energy of formation of pure elements is always 0. When more than 1 mole of a compound is either reacted or formed, the $\Delta G^\circ_f$ of that compound must be multiplied by the stoichiometric coefficient for the specific compound.

$\Delta G^\circ$ = (4 moles $\times$ $\Delta G^\circ_f$ of $NO(g)$ + 6 moles $\times$ $\Delta G^\circ_f$ of $H_2O(\ell)$)

$\qquad$ - (4 moles $\times$ $\Delta G^\circ_f$ of $NH_3(g)$ + 5 moles $\times$ $\Delta G^\circ_f$ of $O_2$)

$G^\circ$ = $\Big($4 moles x 20.7 kcal/mole + 6 moles x ($-56.7$ kcal/mole)$\Big)$
$\qquad$ $-$(4 moles x ($-4.0$ kcal/mole) + 5 moles x 0 kcal/mole)

$\quad$ = $\;$ $-241.4$ kcal.

For sublimation of iodine crystals,

$$I_2(s) \overset{\rightarrow}{\leftarrow} I_2(g),$$

at 25°C and atmospheric pressure, it is found that the change in enthalpy, $\Delta H$ = 9.41 Kcal/mole and the change in entropy, $\Delta S$ = 20.6 cal/deg-mole. At what temperature will solid iodine be in equilibrium with gaseous iodine?

<u>Solution</u>: Use the fact that the system is in a state of equilibrium. The change in Gibb's free energy is related to $\Delta H$ and $\Delta S$ by the equation

$$\Delta G = \Delta H - T\Delta S,$$

where T is the absolute temperature of the system. At equilibrium, $\Delta G = 0$ and T is the equilibrium temperature, $T_{equil}$. Hence

$$\Delta G = \Delta H - T\Delta S,$$

$$0 = \Delta H - T_{equil} \; \Delta S,$$

or, $\quad T_{equil} = \dfrac{\Delta H}{\Delta S}$ .

Therefore,

$$T_{equil} = \frac{\Delta H}{\Delta S} = \frac{9.41 \; Kcal/mole}{20.6 \; cal/deg\text{-}mole} = \frac{9410 \; cal/mole}{20.6 \; cal/deg\text{-}mole}$$

$$= 457 \; K,$$

or, $\quad T_{equil} = 457 - 273 = 184°C.$

● **PROBLEM** 532

At the melting point of a solid (or the freezing point of a liquid), the free energies of the solid state and the liquid state are equal, $\Delta G = 0$. Likewise, at the boiling point of a liquid, where there is an equilibrium between the liquid and vapor phases, the free energy is equal in the two states. Calculate the change in entropy for the following process at 0°C if the heat of fusion of $H_2O$ = 80 cal/g. $H_2O(s) \rightarrow H_2O(\ell)$.

Solution:  The free energy, $\Delta G$, can be expressed as $\Delta G = \Delta H - T\Delta S$, where $\Delta H$, in this case, is the heat of fusion, T is the absolute temperature and $\Delta S$ is the change in the entropy. One is given that $\Delta G = 0$ for this process and thus, $\Delta H - T\Delta S = 0$. One is told that the process takes place at 0°C. To use the free energy equation, the temperature must be in °K. To convert from °C to °K, add 273 to the temperature in °C.

$$T = 0° + 273 = 273°K$$

$\Delta H$ is given as 80 cal/g. When solving for the entropy, convert $\Delta H$ to cal/mole. There are 18 g of $H_2O$ in one mole, thus the $\Delta H$ can be converted by multiplying it by 18 g/mole

$$\Delta H = 80 \; cal/g \times 18 \; g/mole = 1440 \; cal/mole$$

It is now possible to solve for $\Delta S$.

$$0 = \Delta H - T\Delta S \qquad\qquad \Delta H = 1440 \; cal/mole$$

$$0 = 1440 \; cal/mole - 273°K \times \Delta S$$

$$- 1440 \; cal/mole = - 273°K \times \Delta S$$

$$\frac{- 1440 \; cal/mole}{- 273°K} = \Delta S = 5.27 \; cal/mole\text{-}°K$$

2 moles of hydrogen chloride are to be made from 1 mole each of hydrogen ($H_2$) and chlorine ($Cl_2$) at 25° and 1 atm.

Calculate the $\Delta G$ for this chemical reaction.

STANDARD ENTHALPIES OF FORMATION, $\Delta H_f^\circ$, AT 25°C

| Compound | $\Delta H_f^\circ$ |
|----------|--------------------|
| NO($g$) | 21.600 kcal/mole |
| NO$_2$($g$) | 8.091 |
| HF($g$) | −64.2 |
| HCl($g$) | −22.063 |
| HBr($g$) | −8.66 |
| HI($g$) | 6.20 |

ENTROPIES OF CERTAIN SUBSTANCES AT 25°C AND 1 ATM

| Substance | $S$(cal deg$^{-1}$ mole$^{-1}$) |
|-----------|----------------------------------|
| H$_2$ | 31.21 e.u. |
| O$_2$ | 49.00 |
| Cl$_2$ | 53.31 |
| N$_2$ | 45.77 |
| HCl | 44.61 |

<u>Solution</u>: Free energy ($\Delta G$) may be expressed quantitatively as $\Delta G = \Delta H - T\Delta S$, where $\Delta H$ is the change in enthalpy, $\Delta S$ is the change in entropy, and T is temperature in degrees kelvin. Entropy measures the randomness of the system, while enthalpy is the heat content.

The reaction may be written $H_2(g) + Cl_2(g) \rightarrow 2HCl(g)$

Because free energy is defined in terms of $\Delta H$, $\Delta S$, and T, you can see these values to solve for $\Delta G$. The table of standard enthalpies of formation, states that $\Delta H^\circ_{f,HCl} =$ − 22.063 kcal/mole. Two moles of HCl are formed, $\Delta H$ for the reaction is 2 **moles** x (−22,063) cal/$_{mole}$. The $\Delta S$ is the final minus the inital entropy;

$$S_f - S_i = S_{2HCl} - S_{H_2} - S_{Cl_2}$$
$$= 2(44.61) - 31.21 - 53.31$$
$$= 4.70 \text{ cal deg}^{-1}$$

Substituting into $\Delta G = \Delta H - T\Delta S$;

$$G = -44,123 \text{ cal} - (298)(4.70 \text{ eu}) = -45,523 \text{ cal.}$$

Determine the entropy difference between solid and liquid states for a substance melting at 100°C and having a heat of fusion of 10,000 J/mol.

<u>Solution</u>: At the melting point, the liquid and solid are in equilibrium, which means the change in free energy

($\Delta G$), i.e., the energy available for useful work, is zero. This is due to the fact that $\Delta G$ is defined as the difference of $\Delta G$ of the liquid and the $\Delta G$ of the solid. Because the system is in equilibrium,

$$\Delta G_{liq} = \Delta G_{solid}.$$

Therefore, $\Delta G$ for the whole system is

$$\Delta G_{solid} - \Delta G_{liq} = 0.$$

With this in mind, you can compute the entropy change ($\Delta S$), i.e., the change in randomness of the system from the equation

$$\Delta G = \Delta H - T\Delta S,$$

where $\Delta H$ = enthalpy or heat content and T = temperature in degrees Kelvin (Celsius plus 273°).

For this problem you are given $\Delta H$ in the form of heat of fusion, which is the amount of heat necessary to melt one mole of solid. You are also told T. $\Delta G$ is equal to zero. Substituting:

$$\Delta G = 0 = \Delta H - T\Delta S \qquad \text{or} \qquad \Delta S = \frac{\Delta H}{T}$$

$$\Delta S = \frac{10,000 \text{ J/mole}}{373°K} = 26.8 \frac{J}{mole \ °K}$$

● **PROBLEM** 535

If it takes 30.3 KJ/mole of heat to melt NaCl, calculate the melting point of NaCl, assuming the entropy increase is 28.2 J $mol^{-1}$ $deg^{-1}$.

Solution: To solve this problem, relate free energy, enthalpy, and entropy. The formula, $\Delta G = \Delta H - T\Delta S$, does this. $\Delta G$ = free energy, $\Delta H$ = enthalpy, T = temperature in Kelvin and $\Delta S$ = entropy. Let T = the melting point of NaCl. Entropy is a measure of randomness. Reactions are favored that increase the randomness of the system. Reactions will also be favored that proceed to a lower energy state. Above the melting point, randomness increases, but so does energy. However, randomness predominates, and so that solid melts, when the temperature is greater than the melting point. The situation below the melting point is exactly the opposite; here the energy consideration predominates. Only at the melting point, or when the solid and liquid are in equilibrium, will $\Delta H$ and $T\Delta S$ be equal. As such, at the melting point $\Delta G = 0$. You have, therefore, $\Delta G = 0 = \Delta H - T_{Mp}\Delta S$ or

$\Delta H = T_{Mp}\Delta S$ or $T_{MP} = \Delta H/\Delta S$. Because it takes 30.3 KJ/mole to melt NaCl, this must be the heat content or enthalpy. The change of entropy was given as 28.2 J mol$^{-1}$ deg$^{-1}$. You have therefore,

$$T_{Mp} = \frac{\Delta H}{\Delta S} = \frac{30,300 \text{ J/mole}}{28.2 \text{ J mole}^{-1} \text{ deg}^{-1}} = 1070°K.$$

● **PROBLEM** 536

The density of ice at 0°C is .917 g/cm$^3$ and has an entropy of 37.95 J mol$^{-1}$ deg$^{-1}$. The density of liquid water at this temperature is .9998 g/cm$^3$ and has an entropy of 59.94 J mol$^{-1}$ deg$^{-1}$. Given these data, calculate the change of entropy $\Delta S$, change of enthalpy $\Delta H$, and the change of energy $\Delta E$ for the conversion of one mole of ice to liquid water at the normal melting point.

Solution: The solution of this problem involves the ability to relate energy, free energy, entropy, and enthalpy.

One must first calculate the change in entropy. You are told the entropies of both ice and water. The change in entropy is their difference or

$\Delta S = S_{liq} - S_{solid} = 59.94 - 37.95 = 21.99$ J mol$^{-1}$ deg$^{-1}$.

To calculate the change in enthalpy at the melting point use the fact that $\Delta G = 0$ at the melting point of a substance. $\Delta G = \Delta H - T_{Mp}\Delta S$, where $\Delta H$ = change in enthalpy, $T_{Mp}$ = temperature of melting point in Kelvin and $\Delta S$ = change in entropy. Since $\Delta G = 0$ at the melting point, $\Delta H = T_{Mp}\Delta S$.

From the previous calculation, you know $\Delta S$. $T_{Mp}$ is given. Thus,

$\Delta H = T\Delta S$

$\Delta H = (273.15K)(21.99$ J mol$^{-1}$ deg$^{-1}) = 6010$ J/mole.

To calculate the $\Delta E$ for the conversion of one mole of ice to liquid water at the normal melting point, employ the equation $\Delta E = \Delta H - P\Delta V$, where P = pressure, V = volume. $\Delta H = 6010$ J/mole. To determine P$\Delta V$, note that $\Delta V$ is $V_{liq} - V_{solid}$. You are given the densities of ice and water. Recalling that density = mass/volume, and that the molecular weight (mass) of ice or water is 18.015 grams/mole, you have for $V_{liq}$: 18.015/.9998. For $V_{solid}$, you have 18.015/.917. Therefore

$$\Delta V = V_{liq} - V_{solid} = \frac{18.015 \text{ g/mole}}{.9998 \text{ g/cm}^3} - \frac{18.015 \text{ g/mole}}{.917 \text{ g/cm}^3}$$

$$= -1.63 \text{ cm}^3/\text{mole} = -.00163 \text{ liter/mole}.$$

Since one is asked to make these calculations for 1 mole of $H_2O$, the change in volume, $\Delta V$, equals $-.00163$ liter.

Because you are asked to calculate at the normal melting point, the pressure must be 1 atm (by definition, the "normal" is considered to be under a pressure of one atm.)

Therefore,

$$P\Delta V = (1.00 \text{ atm})(-0.00163 \text{ liter}) = -0.00163 \text{ liter-atm.}$$

Since 1 liter-atm equals 101.3 J, $P\Delta V$ is equal to

$$(-0.00163 \text{ liter-atm})(101.3 \text{ J liter}^{-1} \text{ atm}^{-1}) = -0.1651 \text{ J.}$$

Therefore,

$$\Delta E = \Delta H - P\Delta V = 6010 + .1651 = 6010.1651 \text{ J.}$$

● **PROBLEM** 537

Determine $\Delta S^\circ$, $\Delta H^\circ$, and $\Delta G^\circ$ for the following reaction at 25°C.

$$CO(g) + Cl_2(g) \rightarrow COCl_2(g)$$

Use the following table.

| Substance | $S^\circ$ | $\Delta H_f^\circ$ |
|-----------|-----------|--------------------|
| CO | 47.3 | -26.4 |
| $Cl_2$ | 53.3 | 0 |
| $COCl_2$ | 69.1 | -53.3 |

$S^\circ$ is expressed in cal/mole °K and $\Delta H_f^\circ$ is in kcal/mole.

Solution: The change in entropy ($\Delta S^\circ$) or randomness of a reaction is obtained by subtracting the entropy ($S^\circ$) of the reactants from the $S^\circ$ of the products. If more than one mole of a compound reacts or is formed in a reaction, the S° should be multiplied by the appropriate stoichiometric coefficient. Here, one mole of CO reacts with one mole of $Cl_2$ to form one mole of $COCl_2$.

$$\Delta S° = S° \text{ of } COCl_2 - (S° \text{ of } CO + S° \text{ of } Cl_2)$$

$$\Delta S° = 69.1 - (47.3 + 53.3) = -31.5 \text{cal/mole-}°K$$

The change in enthalpy ($\Delta H°$) of a reaction is determined by subtracting the enthalpy of formation ($\Delta H_f^o$) of the reactants from the $\Delta H_f^o$ of the products. $\Delta H_f^o$ of any pure element is always $0$. When more than one mole of a compound either reacts or is formed in a reaction, the $\Delta H_f^o$ must be multiplied by the corresponding stoichiometric coefficient when calculating $\Delta H°$.

$$\Delta H° = \Delta H_f^o \text{ of } COCl_2 - \left| \Delta H_f^o \text{ of } CO + H_f^o \text{ of } Cl_2 \right|$$

$$\Delta H° = -53.3 \text{ kcal} - (-26.4 \text{ kcal} + 0) = -26.9 \text{ kcal.}$$

The change in free energy ($\Delta G°$) can be obtained if one remembers how $\Delta G°$ is related to $\Delta S°$, T and $\Delta H°$. Namely, $\Delta G° = \Delta H° - T\Delta S°$, where $\Delta G°$ is the change in free energy, $\Delta H°$ is the change in enthalpy, $\Delta S°$ is the change in entropy and T is the absolute temperature. In this problem, one has already obtained $\Delta H°$ and $\Delta S°$, and one can quickly obtain T. The absolute temperature can be found by adding 273 to the temperature in °C.

$$T = 25 + 273 = 298°K$$

In the previous part of this problem, one calculated $\Delta S°$ in cal/mole-°K and $\Delta H°$ in kcal/mole. To use $\Delta S°$ and $\Delta H°$ in the formula to find $\Delta G°$, one must have them both in the same units. Thus, one should change the $\Delta S°$ term from cal to kcal by multiplying it by the conversion factor 1. kcal/1000 cal.

One can now obtain $\Delta G°$.

$$\Delta G° = \Delta H° - T\Delta S°$$
$$\Delta G° = -26.9 \text{ kcal} - (298K)(-31.5 \text{ cal/}°K)(1 \text{ kcal/1000 cal})$$

$$= -26.9 + 9.4 = -17.5 \text{ kcal.}$$

● **PROBLEM** 538

Determine the free energy change, $\Delta G$, for transforming liquid water at 100°C and 1 atm. to vapor at the same conditions. $\Delta H = 9720$ cal (molar enthalpy of vaporization).

Solution: This problem deals with the energies of reactions: enthalpy, entropy and free energy. The entropy of a system is a measure of the state of randomness. As more thermal energy is added to a system, the more random it becomes. Entropy can be measured quantitatively. Entropy or $\Delta S = q_{rev}/T$, where $q_{rev}$ is the amount of heat added to the system that can be recovered if the reaction were reversed. For example, the energy needed to melt ice can be recovered by freezing the liquid back to ice.

Enthalpy may be defined as the heat content of a system.

The concept of energy and entropy may be combined to define the driving force of a reaction. This driving force is referred to as Gibbs free energy (G) where $\Delta G = \Delta H - T\Delta S$, where $\Delta H$ is the enthalpy. This energy must be equal to the heat content of the system ($\Delta H$) minus the energy necessary to increase the randomness, or entropy, of the same system.

Therefore, to solve this problem, you need to substitute into this equation the following values:

$$T = 100^\circ C = 373^\circ K$$

$$q = \Delta H = 9720 \text{ cal}$$

Therefore,

$$\Delta S = \frac{q}{T} = \frac{9720 \text{ cal}}{373 \, ^\circ K}$$

Now,

$$\Delta G = 9720 \text{ cal} - 373 \, ^\circ K \left( \frac{9720 \text{ cal}}{373^\circ K} \right) = 0.$$

The answer is zero.

The system can derive no usable energy to perform work from this change. For any system when $\Delta G = 0$, the system is at equilibrium.

# EQUILIBRIUM CALCULATIONS

• PROBLEM 539

Calculate the equilibrium constant for the following reaction at 25°C or 298°K

$$C_{(graphite)} + 2H_2(g) \rightarrow CH_4(g)$$

$\Delta H^\circ$ for this reaction is - 17,889 cal.

Solution: At equilibrium the equilibrium constant of any

517

reaction is independent of the amount of pure solid (or liquid) phase.

To solve this problem (1) determine $\Delta S°$, the entropy change or randomness of the system. (2) Determine $\Delta G°$, the free energy or the energy available to do work, by using the formula: $\Delta G° = \Delta H° - T\Delta S°$, where $\Delta H°$ = the change in enthalpy or heat content, and T = absolute temperature (Celsius plus 273°). (3) Determine $K_p$, the equilibrium constant, by using $\Delta G° = - RT \ln K_p$.

The entropy change (from any entropy table) for this reaction is

$$\Delta S° \text{ of reaction} = \Delta S°_{products} - \Delta S°_{reactants}$$

$$= \Delta S°_{CH_4} - 2\Delta S°_{H_2} - \Delta S°_C$$

$$= 44.50 - 2(31.211) - 1.3609$$

$$= - 19.28 \text{ cal/k}$$

$$\Delta G° = \Delta H° - T\Delta S°$$

$$= - 17,889 - (298)(- 19.28)$$

$$= - 12,143 \text{ cal.}$$

The equilibrium constant is thus

$$\Delta G° = - RT \ln K_p = - (1.987 \text{ cal}°K^{-1} \text{ mole}^{-1})(298°K)\times$$
$$(2.303)\log K_p$$

$$- 12,143 \text{ cal} = - 1364 \log K_p$$

$$\log K_p = 8.90$$

$$K_p = 7.94 \times 10^8.$$

● **PROBLEM** 540

Assuming $\Delta H°$ remains constant, calculate the equilibrium constant, K, at 373°K, if it equals $1.6 \times 10^{12}$ at 298°K for the reaction $2NO(g) + O_2(g) \rightleftarrows 2NO_2(g)$. The standard enthalpy change for this reaction is - 113 kJ/mole.

Solution: This problem can be solved by employing the van't Hoff equation. This equation allows a determination of an equilibrium constant, if the $\Delta H°$ is known (and constant) and the value of the equilibrium constant at another temperature is known. It states

$$\log \frac{K_2}{K_1} = \frac{\Delta H^\circ}{19.15} \left[ \frac{1}{T_1} - \frac{1}{T_2} \right] \ ,$$

where $K_2$ and $K_1$ are equilibrium constants, $\Delta H^\circ$ = the standard enthalpy formation, and $T_1$ and $T_2$ are temperatures in Kelvin. Let $K_1 = 1.6 \times 10^{12}$ at $T_1 = 298^\circ K$; $T_2 = 373^\circ$; and $\Delta H = -113$ J/mole. Substitute these values into the equation and solve for $K_2$, which will be the solution to this problem. Thus

$$\log \frac{K_2}{1.6 \times 10^{12}} = \frac{-113,000}{19.15} \left( \frac{1}{298} - \frac{1}{373} \right)$$

Solving for $K_2$, one obtains $K_2 = 1.7 \times 10^8$.

● **PROBLEM** 541

Given that k = 8.85 at 298°K and k = .0792 at 373°K, calculate the $\Delta H^\circ$ for the reaction of the dimerization of $NO_2$ to $N_2O_4$. Namely, $2NO_2$ (g) $\rightleftarrows$ $N_2O_4$ (g).

Solution: $\Delta H^\circ$ is the standard enthalpy change, a measure of the heat content of the system. It is quantitatively related to the equilibrium constants of a system at different temperatures by the van't Hoff equation:

$$\log \frac{k_2}{k_1} = \frac{\Delta H^\circ}{19.15} \left[ \frac{1}{T_1} - \frac{1}{T_2} \right] \ ,$$

where $k_2$ and $k_1$ are equilibrium constants, and $T_1$ and $T_2$ are temperatures in Kelvin. Given that $k_1 = 8.85$ at $T_1 = 298^\circ K$ and $k_2 = 0.0792$ at $373^\circ K$, one can find $\Delta H^\circ$ by the substitution of these values in the equation. Thus,

$$\log \frac{.0792}{8.85} = \frac{\Delta H^\circ}{19.15} \left( \frac{1}{298} - \frac{1}{373} \right) \ .$$

Solving for $\Delta H^\circ$, one obtains

$\Delta H^\circ = -58,200$ J/mole    for the reaction.

● **PROBLEM** 542

Ammonia is synthesized commercially according to the equation

$$\tfrac{1}{2} N_2 (g) + 1\tfrac{1}{2} H_2 (g) \rightarrow NH_3 (g),$$

under a constant total pressure of 50 atm. At 350°C the equilibrium constant has been determined as 0.0278. The heat of reaction for this process is $\Delta H =$

- 11.04 Kcal/mole. Determine the value of equilibrium constant at 450°C.

Solution: For a reaction with equilibrium constant $k_1$ at absolute temperature $T_1$ and equilibrium constant $k_2$ at temperature $T_2$,

$$\log \frac{k_2}{k_1} = \frac{-\Delta H°}{2.303\ R} \left[ \frac{1}{T_2} - \frac{1}{T_1} \right],$$

where R is the gas constant.

Let $T_1 = 350°C = 623°K$, $k_1 = 0.0278$, and $T_2 = 450°C = 723°K$. $\Delta H° = -11.04$ Kcal/mole $= -11,040$ cal/mole. Then

$$\log \frac{k_2}{k_1} = \frac{-\Delta H°}{2.303\ R} \left[ \frac{1}{T_2} - \frac{1}{T_1} \right]$$

$$\log \left( \frac{k_2}{0.0278} \right) = \frac{11,040\ \text{cal/mole}}{2.303 \times 1.987\ \text{cal/mole-deg}} \left( \frac{1}{723°K} - \frac{1}{623°K} \right)$$

$$= -0.535.$$

Solving for $k_2$,

$$\log \left( \frac{k_2}{0.0278} \right) = -0.535$$

$$\frac{k_2}{0.0278} = 10^{-0.535} = 0.292$$

$$k_2 = 0.0278 \times 0.292 = 0.00812.$$

Thus, the value of the equilibrium constant at 450°C. is 0.00812.

● PROBLEM 543

Calculate the enthalpy change, $\Delta H°$, for the reaction

$$N_2(g) + O_2(g) = 2NO\ (g),$$

given the equilibrium constants $4.08 \times 10^{-4}$ for a temperature of 2000°K and $3.60 \times 10^{-3}$ for a temperature of 2500°K.

Solution: The effect of temperature on chemical equilibrium is determined by $\Delta H°$ (enthalpy or heat content); over moderate ranges in temperature, $\Delta H°$ is relatively independent of temperature.

If, as in the case of this problem, $\Delta H°$ is independent of temperature then,

(1)  $\Delta G^\circ = -RT \ln k$,  where

$\Delta G^\circ$ = standard free energy change,

R = gas constant,

T = absolute temperature

and k the equilibrium constant.


(2)  $\Delta G^\circ = \Delta H^\circ - T\Delta S^\circ$


(3)  $\Delta H^\circ - T\Delta S^\circ = -RT \ln K$         or

$$-\left(\frac{\Delta H^\circ - T\Delta S^\circ}{RT}\right) = \ln k,$$ where $\Delta S^\circ$ = the change in

entropy or randomness of system.


(4)  $\dfrac{-\Delta H^\circ}{RT} + \dfrac{\Delta S^\circ}{R} = \ln k$

For two different temperatures, $T_1$ and $T_2$, equation
(4) becomes

$$\ln k_2 - \ln k_1 = \left[\frac{-\Delta H^\circ}{RT_2} + \frac{\Delta S^\circ}{R}\right] - \left[\frac{-\Delta H^\circ}{RT_1} + \frac{\Delta S^\circ}{R}\right] \quad \text{or}$$

$$\ln \frac{k_2}{k_1} = \frac{-\Delta H^\circ}{RT_2} + \frac{\Delta H^\circ}{RT_1}$$

$$\ln \frac{k_2}{k_1} = \frac{\Delta H^\circ (T_2 - T_1)}{RT_2 T_1} \quad \text{or, finally,}$$

$$\log \frac{k_2}{k_1} = \frac{\Delta H^\circ (T_2 - T_1)}{(2.303)\ R\ T_2\ T_1}$$

Thus,    $\log \dfrac{k_{2500^\circ K}}{k_{2000^\circ K}} = \log \dfrac{3.60 \times 10^{-3}}{4.08 \times 10^{-4}}$

$$= \frac{\Delta H^\circ\ (2500 - 2000)^\circ K}{(2.303)(1.987\ cal^\circ K^{-1} mole^{-1})(2500)^\circ K (2000)^\circ K}$$

$$\log 8.82 = .945 = \frac{\Delta H^\circ\ (500^\circ K)}{(2.303)(1.987)(5 \times 10^6) cal^\circ K^{-1} mole^{-1}\ {}^\circ K^2}$$

$$\frac{(0.945)(2.303)(1.987)(5 \times 10^6)}{(500)\ mole\ cal^{-1}} = \Delta H^\circ$$


$\Delta H^\circ = 43,240$ cal/mole.

For the reaction $PbSO_4(s) \rightarrow Pb^{2+} + SO_4^{2-}$, $\Delta H = +2990$ cal/mole. Will the solubility of $PbSO_4$ increase or decrease with increasing temperature? $K_{sp} = 1.8 \times 10^{-8}$ at 25°C. Find its $k_{sp}$ at 55°C.

Solution: Whether the solubility of a salt increases or decreases with an increase in temperature can be determined by an investigation of its $\Delta H$, the enthalpy change. The formula for the temperature dependence on k is

$$\log k = \frac{\Delta H°}{2.303\ RT} + \frac{\Delta S°}{2.303\ R} .$$

A positive $\Delta H°$ suggests that solubility increases with increasing temperature. When $\Delta H°$ is positive, the first term remains a negative value. When T increases, the first term becomes a smaller negative number. Thus, k increases. Since k measures solubility, the solubility increases. To find the $k_{sp}$ of $PbSO_4$ at 55°C, use the fact that

$$\log \frac{k_2}{k_1} = \frac{-\ \Delta H°}{2.303\ R} \left( \frac{1}{T_2} - \frac{1}{T_1} \right),$$

where $k_2$ = solubility constant at the temperature in Kelvin (Celsius plus 273°) of $T_2$; $k_1$ = solubility constant at a temperature of $T_1$. Thus, if the $k_{sp}$ at one temperature is known, the $k_{sp}$ at another temperature can be found given $\Delta H°$. Let $T_2 = 55°C$ or 328°K and $T_1 = 298°K$. (25°C). Thus

$$\log \frac{k_{328}}{k_{298}} = \frac{-\ \Delta H°}{2.303\ R} \left( \frac{1}{328} - \frac{1}{298} \right)$$

$$\log \frac{k_{328}}{1.8 \times 10^{-8}} = \frac{-\ (2990\ \text{cal/mole})(-\ 3.07 \times 10^{-4}/K)}{(2.303)1.987\ \text{cal/mole K})}$$

$$= 0.200$$

$$k_{328} = (1.8 \times 10^{-8})(10^{0.200})$$

$$= (1.8 \times 10^{-8})(1.58)$$

$$= 2.8 \times 10^{-8}.$$

The equilibrium constant and standard Gibbs free energy change for ammonia synthesis at 400°C or 673°K are $1.64 \times 10^{-4}$ and 11,657 cal/mole, respectively. The equation for this reaction is

(1) $\qquad N_2(g) + 3H_2(g) \rightarrow 2NH_3(g)$

Calculate the equilibrium constants and standard free energy changes for

(2) $\qquad \frac{1}{2}N_2(g) + 3/2\ H_2(g) = NH_3(g) \qquad$ and

(3) $\qquad 2NH_3(g) \rightarrow N_2(g) + 3H_2(g)$.

Solution: This problem illustrates the necessity of knowing the balanced chemical equation in order to determine the numerical value of the equilibrium constant.

For ideal gases, the equilibrium constant, $K_p$, is the product of the partial pressure of the product gases, each raised to the power that corresponds with its stoichiometric coefficient in the balanced equation, divided by the product of the partial pressure of the reacting gases, each raised to the power that corresponds with its stoichiometric coefficient in the balanced equation.

The quantity $\Delta G°$ is the standard change in Gibbs free energy. It is the difference between the Gibbs free energy of the products and the Gibbs free energy of the reactants. The Gibbs Free Energy and the equilibrium constant are related in the equation

$$\Delta G = - RT \ln K_p$$

where R is the gas constant, T the absolute temperature and $K_p$ the equilibrium constant calculated using partial pressures.

The equilibrium constant for reaction 1 is

$$K_1 = \frac{P^2_{NH_3}}{P_{N_2} P^3_{H_2}} = 1.64 \times 10^{-4}$$

Thus, for (2) $\frac{1}{2}N_2(g) + (3/2)H_2(g) = NH_3(g)$

$$K_2 = \frac{P_{NH_3}}{P^{\frac{1}{2}}_{N_2}\ P^{3/2}_{H_2}} = K_1^{\frac{1}{2}} = (1.64 \times 10^{-4})^{\frac{1}{2}} = 1.28 \times 10^{-2}$$

By definition, at equilibrium,

$$\Delta G° = - RT \ln K_2$$

$$= - (1.987 \; cal^\circ K^{-1} \; mole^{-1})(673^\circ K)(2.303) \times$$

$$\log(1.28 \times 10^{-2})$$

$$= 5829 \; cal/mole.$$

And for (3)    $2NH_3(g) = N_2(g) + 3H_2(g)$

$$K_3 = \frac{P_{N_2} \; P_{H_2}^3}{P_{NH_3}^2} = \frac{1}{K_1} = \frac{1}{1.64 \times 10^{-4}} = 0.6098 \times 10^4$$

$$\Delta G_3^\circ = - RT \; \ln K_3 = - (1.987 \; cal^\circ K^{-1} \; mole^{-1})(673^\circ K)(2.303) \times$$

$$\log (0.6098 \times 10^4)$$

$$= - 11,657 \; cal/mole.$$

If a reaction is reversed, the equilibrium constant becomes the reciprocal of that for the first reaction; the standard Gibbs free energy change has the same magnitude but the opposite sign.

● **PROBLEM** 546

Calculate $\Delta G^\circ$ for the reaction

$$N_2(g) + 3H_2(g) = 2NH_3(g)$$

at $400^\circ C = 673^\circ K$ using $K_c$. The value of $K_p$ for this reaction is $1.64 \times 10^{-4}$. Interpret the different values of $\Delta G^\circ$ calculated from $K_p$ and $K_c$.

Solution:  Given the equilibrium reaction:

(i)      $aA_{(g)} + bB_{(g)} + \ldots = cC_{(g)} + dD_{(g)} + \ldots$

there are two constants that measure the equilibrium: $K_c$ relates the concentrations.

(ii)      $K_c = \dfrac{[C]^c \; [D]^d \; \ldots}{[A]^a \; [B]^b \; \ldots}$

$K_p$ relates the pressures

(iii)      $K_p = \dfrac{P_C^c \; P_D^d \; \ldots}{P_A^a \; P_B^b \; \ldots}$

To derive $K_c$ from $K_p$ substitute for each pressure $P = cRT$, where R is the gas constant; T = absolute temperature; and c is the number of moles of a gas per liter.

For the reaction,

$$N_2(g) + 3H_2(g) = 2NH_3(g),$$

$$K_p = \frac{P^2_{NH_3}}{P_{N_2} P^3_{H_2}} = \frac{C^2_{NH_3} (RT)^2}{C_{N_2} C^3_{H_2} (RT)^4} = \frac{K_c}{(RT)^2},$$

where $K_c = \dfrac{C^2_{NH_3}}{C_{N_2} C^3_{H_2}} = (RT)^2 K_p$. Thus,

$$K_c = \left[\left(0.082 \frac{liter\text{-}atm}{{}^\circ K\text{-}mole}\right)(673\,{}^\circ K)\right]^2 (1.64 \times 10^{-4}) = 0.500$$

and $\Delta G^\circ_c = - RT \ln K_p$

$$= - (1.987 \text{ cal}\,{}^\circ K^{-1} \text{ mole}^{-1})(673\,{}^\circ K)(2.303) \log(0.500)$$

$$= 927 \text{ cal/mole.}$$

$$\Delta G^\circ_p = - RT \ln K_p$$

$$= - (1.987 \text{ cal}\,{}^\circ K^{-1} \text{ mole}^{-1})(673\,{}^\circ K)(2.303)$$

$$\log (1.64 \times 10^{-4})$$

$$= 11,657 \text{ cal/mole.}$$

$\Delta G^\circ_c$ is the change in Gibbs free energy when 1 mole of $N_2$ in the ideal gas state at the concentration of 1 mole per liter and 3 moles of $H_2$ in the ideal gas state at 1 mole per liter react to form 2 moles of $NH_3$ in the ideal gas state at 1 mole per liter.

$\Delta G^\circ_p$ is the change in Gibbs free energy when 1 mole of $N_2$ in the ideal gas state at 1 atm, and 3 moles of $H_2$ in the ideal gas state at 1 atm react to form 2 moles of $NH_3$ in the ideal gas state at 1 atm.

● **PROBLEM** 547

Calculate $\Delta G^\circ$ and $K_p$ at 25°C for

$$CO(g) + H_2O(g) \rightarrow CO_2(g) + H_2(g),$$

$\Delta G^\circ_f$ for $H_2(g)$, $CO_2(g)$, $H_2O(g)$, and $CO(g)$ are 0, $- 94.2598$, $- 54.6357$, and $- 32.8079$ Kcal/mole, respectively.

Solution: The Gibbs free energy of formation ($\Delta G_f^o$) of a substance is the Gibbs free energy change, i.e., the amount of free energy available for work, for the reaction in which the substance in its standard state is formed from its elements in their standard states.

By use of the following equation,

$$\Delta G^o = \Delta G_{f,\ products}^o - \Delta G_{f,\ reactants}^o$$

the Gibbs free-energy change for a reaction may be calculated from the Gibbs free energies of formation. The equilibrium constant for the reaction may then be calculated using the standard Gibbs free energy.

$$\text{Thus,} \quad \Delta G^o = \left( \Delta G_{f,CO_2}^o + \Delta G_{f,H_2}^o \right) - \left( \Delta G_{f,CO}^o + \Delta G_{f,H_2O}^o \right)$$

$$= (-94.2598 + 0) - (-32.8079 - 54.6357)$$

$$= -6.8162 \text{ Kcal/mole} = -6816.2 \text{ cal/mole.}$$

$$\Delta G^o = - RT \ln K_p,$$

where R = gas constant, T = absolute temperature (Celsius plus 273°) and $K_p$ = equilibrium constant, which is to be calculated. As such, one can, via substitution, write

$$\Delta G^o = -6816.2 \text{ cal/mole} = - RT \ln K_p$$

$$= -(1.987 \text{ cal}^o\text{K}^{-1} \text{ mole}^{-1})(298^o\text{K})(2.303)\log K_p$$

$$\log K_p = 4.997$$

$$K_p = \frac{P_{H_2} P_{CO_2}}{P_{CO} P_{H_2O}} = 9.931 \times 10^4.$$

● **PROBLEM** 548

If the standard free energy of formation of HI from $H_2$ and $I_2$ at 490°C is - 12.1 kJ/mole of HI, what is the equilibrium constant for this reaction? Assume R = 8.31 J mol$^{-1}$ deg$^{-1}$.

Solution: You are given the chemical reaction $H_2 + I_2 \rightleftarrows 2HI$ and the standard free energy formation $\Delta G^o$ = - 12.1 kJ/mole of HI and are asked to find the equilibrium constant K. The equilibrium constant and the standard free energy are related by the formula:

(i) $\Delta G_{490}^o = - RT \ln K$

where R = universal gas constant and T = temperature in Kelvin. Solving for K, you obtain:

(ii)    $\ln K = \dfrac{\Delta G^\circ_{490}}{-RT}$

In this problem, R = 8.31 J mol$^{-1}$ deg$^{-1}$, and T = 490°C = (490° + 273°)K = 763°K. $\Delta G^\circ_{490}$ is the amount of free energy needed for the reaction H$_2$ + I$_2$ $\overset{\rightarrow}{\leftarrow}$ 2 HI, i.e., the amount of free energy needed for the production of 2 moles of HI at 490°C. Given that - 12.1 kJ will produce I mole of HI at 490°C, it follows that

$\Delta G^\circ_{490}$ = 2( - 12.1 kJ/mole)= -24.2 kJ/mole

= -24,200 J/mole.

Substituting these values in equation (ii), you obtain

(iii)  $\ln K = -\dfrac{\Delta G^\circ}{RT} = \dfrac{-(-24,200 \text{ J/mole})}{(8.31 \text{ J mol}^{-1} \text{ deg}^{-1})(763 \text{ deg})}$

(iv)   $\ln K = 3.82$

(v)    $K = 46.$

● **PROBLEM** 549

Phosgene is formed at 25°C according to the equation

$$CO(g) + Cl_2(g) \rightarrow COCl_2(g).$$

Calculate the free energy change $\Delta G^\circ$ and the equilibrium constant k for this reaction.

Solution: The free energy change for the reaction is the difference between the sum of the free energies of formation of the products and the sum of the free energies of formation of the reactants.

The free energies of formation of CO(g), Cl$_2$(g) and COCl$_2$(g) are

$\Delta G^\circ_{CO}$     = - 32.8 Kcal/mole

$\Delta G^\circ_{Cl_2}$   = 0 Kcal/mole

$\Delta G^\circ_{COCl_2}$ = - 50.3 KCal/mole.

Hence,    $\Delta G^\circ = \Delta G^\circ_{COCl_2} - \left( \Delta G^\circ_{CO} + \Delta G^\circ_{Cl_2} \right)$

$$= - 50.3 \text{ Kcal/mole} - (- 32.8 \text{ Kcal/mole} +$$
$$0 \text{ Kcal/mole})$$

$$= - 17.5 \text{ Kcal/mole} = - 17,500 \text{ cal/mole.}$$

The equilibrium constant is calculated from

$$\Delta G^\circ = - 2.303 \text{ RT log k,}$$

where R is the gas constant and T is the absolute temperature (T = 25°C + 273 = 298°K). Then,

$$\Delta G^\circ = - 2.303 \text{ RT log k}$$

$$- 17,500 \text{ cal/mole} = - 2.303 \times 1,987 \text{ cal/mole-deg} \times 298°K$$

$$\times \text{ log k.}$$

Solving for k,

$$\log k = \frac{- 17,500 \text{ cal/mole}}{- 2.303 \times 1,987 \text{ cal/mole-deg} \times 298°K} = 12.8,$$

$$k = 10^{12 \cdot 8} = 6 \times 10^{12}.$$

● **PROBLEM** 550

Sulfur exists as $S_2$ vapor at temperatures between 700°C and 1500°C. At 1473 k it combines with hydrogen according to the equation

$$H_2(g) + \tfrac{1}{2}S_2(g) \longrightarrow H_2S(g).$$

At 750°C the equilibrium constant is $1.07 \times 10^2$ and at 1200°C it is 4.39. Determine the heat of reaction in the temperature range 750°C to 1200°C, and the change in free energy at each of these temperatures.

Solution: The change in free energy $\Delta G$ is determined from the equation

$$\Delta G^\circ = - 2.303 \text{ RT log k,}$$

where R is the gas constant, T the absolute temperature, and k the equilibrium constant. From this equation and the expression

$$\Delta G^\circ = \Delta H^\circ - T\Delta S^\circ,$$

where $\Delta H^\circ$ is the change in the heat of reaction and $\Delta S^\circ$ the change in entropy, a relation between $\Delta H$ and T will be found.

At  T = 750°C = 1023°K,  k = 1.07 × 10² and

528

$$\Delta G^\circ = -2.303 \ RT \ \log k$$

$$= -2.303 \times 1.987 \ \frac{cal}{mole\text{-}deg} \times 1023^\circ K \times$$

$$\log (1.07 \times 10^2)$$

$$= -9500 \ cal/mole$$

$$= -9.5 \ Kcal/mole.$$

At $T = 1200^\circ C = 1473^\circ K$, $k = 4.39$ and

$$\Delta G^\circ = -2.303 \ RT \ \log k$$

$$= -2.303 \times 1.987 \ \frac{cal}{mole\text{-}deg} \times 1473^\circ K \times \log (4.93)$$

$$= -4330 \ cal/mole$$

$$= -4.33 \ Kcal/mole.$$

To find an expression relating $\Delta H$ to $T$, consider the following two equations:

$$\Delta G^\circ = -2.303 \ RT \ \log k, \quad \text{and} \quad \Delta G^\circ = \Delta H^\circ - T\Delta S^\circ.$$

Setting these equal gives

$$-2.303 \ RT \ \log k = \Delta H^\circ - T\Delta S^\circ,$$

or, 
$$\log k = \frac{-\Delta H^\circ}{2.303 \ RT} + \frac{\Delta S^\circ}{2.303 \ R} \ .$$

For two different temperatures, $T_1$ and $T_2$ with equilibrium constants $k_1$ and $k_2$, respectively,

$$\log k_1 = \frac{-\Delta H^\circ}{2.303 \ RT_1} + \frac{\Delta S^\circ}{2.303 \ R} \quad \text{and}$$

$$\log k_2 = \frac{-\Delta H^\circ}{2.303 \ RT_2} + \frac{\Delta S^\circ}{2.303 \ R} \ .$$

Subtracting the first equation from the second,

$$\log k_2 - \log k_1 = \left( \frac{-\Delta H^\circ}{2.303 \ RT_2} + \frac{\Delta S^\circ}{2.303 \ R} \right)$$

$$- \left( \frac{-\Delta H^\circ}{2.303 \ RT_1} + \frac{\Delta S^\circ}{2.303 \ R} \right),$$

or, 
$$\log \frac{k_2}{k_1} = \frac{-\Delta H^\circ}{2.303 \ R} \left( \frac{1}{T_2} - \frac{1}{T_1} \right).$$

Let $T_1 = 1023^\circ K$, $k_1 = 1.07 \times 10^2$, $T_2 = 1473^\circ K$, and $k_2 = 4.39$. Then

$$\log \frac{k_2}{k_1} = \frac{-\Delta H^\circ}{2.303 \ R} \left( \frac{1}{T_2} - \frac{1}{T_1} \right)$$

$$\log \frac{4.39}{1.07 \times 10^2} = \frac{-\Delta H^\circ}{2.303 \times 1.987 \ cal/mole\text{-}deg} \left( \frac{1}{1473^\circ K} - \frac{1}{1023^\circ K} \right)$$

$$\log \frac{4.39}{1.07 \times 10^2} = \frac{-\Delta H^\circ}{2.303 \times 1.987 \ cal/mole\text{-}deg} (0.000679^\circ k^{-1}$$
$$- 0.000978^\circ k^{-1})$$

or,

$$\Delta H^\circ = \frac{\log (4.39/107)}{2.303 \times 1.987 \ cal/mole\text{-}deg} (0.000978^\circ k^{-1} - 0.000679^\circ k^{-1})$$

$$= -21,300 \ cal/mole$$

$$= -21.3 \ Kcal/mole.$$

Thus, the heat of reaction in the temperature range 1023°K to 1473°K (750°C to 1200°C) is - 21.3 Kcal/mole.

● **PROBLEM** 551

In the human body, the enzyme phosphoglucomutase catalyzes the conversion of glucose-1-phosphate into glucose-6-phosphate:

glucose-1-phosphate $\overset{\rightarrow}{\leftarrow}$ glucose-6-phosphate.

At 38°C, the equilibrium constant, k, for this reaction is approximately 20. Calculate the free energy change, $\Delta G^\circ$, for the equilibrium conversion. Calculate the free energy change $\Delta G$ for the nonequilibrium situation in which [glucose-1-phosphate] = 0.001 M and [glucose-6-phosphate] = 0.050 M.

Solution: The nonequilibrium free energy change $\Delta G$ is related to the standard free energy change $\Delta G^\circ$ by

$$\Delta G = \Delta G^\circ + 2.303 \ RT \log k,$$

where R is the gas constant and T the absolute temperature.

At equilibrium, $\Delta G = 0$, hence

$$\Delta G^\circ = - 2.303 \ RT \log k.$$

Thus, for the equilibrium conversion with k = 20 and T = 38°C = 311°K.

$$\Delta G^\circ = - 2.303 \ RT \log k$$

$$= -\ 2.303 \times 1.987 \text{ cal/mole-deg} \times 311^\circ K \times \log 20$$

$$= -\ 1850 \text{ cal/mole.}$$

The equilibrium constant for the conversion

glucose-1-phosphate $\rightleftharpoons$ glucose-6-phosphate

is    $k = \dfrac{[\text{glucose-6-phosphate}]}{[\text{glucose-1-phosphate}]}$

In the case where [glucose-6-phosphate] = 0.050 M and [glucose-1-phosphate] = 0.001 M

$$k_n = \dfrac{[\text{glucose-6-phosphate}]}{[\text{glucose-1-phosphate}]} = \dfrac{0.050 \text{ M}}{0.001 \text{ M}} = 50,$$

where the subscript "n" has been added to distinguish this ratio from the equilibrium constant. Then

$$\Delta G = \Delta G^\circ + 2.303 \text{ RT} \log k_n$$

$$= -\ 1850 \text{ cal/mole} + (2.303)(1.987 \text{ cal/mole-deg})$$

$$(311^\circ K)(\log 50)$$

$$= -\ 1850 \text{ cal/mole} + 2420 \text{ cal/mole}$$

$$= +\ 570 \text{ cal/mole.}$$

● **PROBLEM** 552

---

Calculate the equilibrium constant at 25°C for the reaction:

$$S + 3/2\ O_2 \rightleftharpoons SO_3$$

The heat formation of $SO_3$ at 25°C is - 94.45 Kcal/mole and the standard molar entropy changes for S, $O_2$, and $SO_3$ at 25°C are 7.62, 49.0 and 61.24 cal/mole°K, respectively.

---

Solution:  One is given the values for $\Delta H$ and S which can be related by the equation

$$\Delta G^\circ = \Delta H^\circ - T\Delta S,$$

where $\Delta G$ is the change in free energy, $\Delta H$ is the change in the heats of formation, T is the absolute temperature, and $\Delta S$ is the change in entropy.

After calculating $\Delta G^\circ$, one can solve for the equilibrium constant (K) by using the equation

$$\Delta G^\circ = - RT \ln (K)$$

where R = 1.987 cal/mole °K.

One can find $\Delta S$ by remembering that it is equal to the sum of the $\Delta S$ values of the reactants subtracted from the $\Delta S$ values of the products. In solving for $\Delta S$ one multiplies the $\Delta S$ values by its coefficient in the equation for the reaction. Solving for $\Delta S$:

$$\Delta S = \Delta S^\circ_{products} - \Delta S^\circ_{reactants}$$

$$\Delta S = \Delta S^\circ_{SO_3} - \left[ \Delta S^\circ_S + 3/2 \Delta S^\circ_{O_2} \right]$$

$$\Delta S = 61.24 \text{ cal/mole}^\circ K$$
$$-(7.62 \text{ cal/mole}^\circ K + 3/2(49 \text{ cal/mole}^\circ K))$$

$$= 61.24 - 7.62 - 73.5 = -19.88 \text{ cal/mole}^\circ K$$

Solving for $\Delta G^\circ$: T = 25°C + 273 = 298°K

$$\Delta G^\circ = \Delta H^\circ - T\Delta S^\circ$$

$$\Delta G^\circ = - 94,450 \text{ cal/mole} - (298°K)(- 19.88 \text{ cal/mole} - °K)$$

$$= - 94,450 \text{ cal/mole} + 5924 \text{ cal/mole}$$

$$= - 88,525 \text{ cal/mole}$$

Solving for K:

$$\Delta G^\circ = - RT \ln K$$

$$- 88,525 \text{ cal/mole} = (- 1.987 \text{ cal/mole °K})(298°K)(\ln K)$$

$$\ln K = (-88,525 \text{ cal/mole})/(-592 \text{ cal/mole})$$

$$\ln K = 150$$

$$K = 1.39 \times 10^{65}.$$

● PROBLEM 553

Calculate, $E^\circ$, $\Delta G^\circ$, and K for the following reaction at 25°C.

$$\tfrac{1}{2} Cu(s) + \tfrac{1}{2}Cl_2(g) = \tfrac{1}{2}Cu^{2+} + Cl^-.$$

Solution: The magnitude of the standard electrode potential, ($E^\circ$) is a measure of the tendency of the half-reaction to occur in the direction of reduction. The standard EMF (Electromotive Force) of a cell is equal to the standard electrode potential of the right-hand electrode minus the standard electrode potential of the left-hand electrode (by convention),

$$E^\circ = E^\circ_{right} - E^\circ_{left}$$

The reaction taking place at the left electrode is written as an oxidation reaction and the reaction taking place at the right electrode is written as a reduction reaction. The cell reaction is the sum of these two reactions. Thus,

Oxidation: $\frac{1}{2} Cu(s) = \frac{1}{2}Cu^{2+} + e$

Reduction: $\frac{1}{2} Cl_2(g) + e = Cl^-$

Since the oxidation reaction is to occur at the left electrode and the reduction at the right electrode, the cell is written as

$$Cu \mid Cu^{2+} \mid\mid Cl^- \mid Cl_2(g) \mid Pt$$

$$E^\circ = E^\circ_{Cl^-\mid Cl_2\mid Pt} - E^\circ_{Cu^{2+}\mid Cu}$$

$E^\circ$ can be found on any Standard Electrode Potential Table. $E^\circ_{Cl^-\mid Cl_2\mid Pt} = 1.360$ $E^\circ_{Cu^{2+}\mid Cu} = 0.337$; Thus, $E^\circ = 1.360 - 0.337$

$$E^\circ = 1.023 V$$

Now that $E^\circ$ is known, $\Delta G^\circ$, standard Gibbs free energy, i.e., the energy available to do useful work, can be calculated from $\Delta G^\circ = - n F E^\circ$, where n = number of electrons transferred and F = the value of a Faraday of electricity, (23,060 cal/V-mole). From the redox half-reactions written, one sees that 1 electron is being transferred. Consequently, n = 1.

Substituting these values one obtains

$$\Delta G^\circ = - (1) \left(23,060 \frac{cal}{V\text{-}mole}\right)(1.023 V)$$

$$= - 23,590 \text{ cal.}$$

The equilibrium constant, K, can be calculated using the Nernst equation for unit activity. In other words, $E^\circ = .0591/N \log K$. Substituting the previously calculated values for $E^\circ$ and n,

$$1.023 V = .0591 \log K$$

Solving, $\log K = 17.31$    or $K = 2 \times 10^{17}$.

Determine the equilibrium constant for the following reaction at 25°C:

$$Mg(s) + Sn^{2+} \rightarrow Mg^{2+} + Sn(s)$$

Solution: The equilibrium constant, k, is related to the cell potential, $\Delta E°$, by means of the following two formulas:

$$\Delta G° = - n F \Delta E°, \quad \text{and} \quad \Delta G° = - 2.303 \ RT \ \log k,$$

where n is the number of electrons transferred, F the Faraday constant, R the gas constant, $\Delta G°$ the free energy, and T the absolute temperature. $\Delta E°$ will be determined from the half-cell potentials, E°.

First, the cell potential is determined from the half-cell reactions:

$$Mg(s) \rightarrow Mg^{2+} + 2e^- \qquad E° = + 2.37 \ v$$

$$Sn^{2+} + 2e^- \rightarrow Sn(s) \qquad E° = - 0.14 \ v.$$

Adding the two half-cell reactions and their respective values for E°,

$$Mg(s) + Sn^{2+} + 2e^- \rightarrow Mg^{2+} \ 2e^- + Sn(s)$$

$$\Delta E° = (2.37 \ v) + (- 0.14 \ v)$$

or, $Mg(s) + Sn^{2+} \rightarrow Mg^{2+} + Sn(s) \quad \Delta E° = 2.23 \ v.$

The cell potential for the reaction is therefore 2.23 v.

Equating the two expressions

$$\Delta G° = - n F \Delta E°$$

and $\Delta G° = - 2.303 \ RT \ \log k$

gives $- 2.303 \ RT \ \log k = - n F \Delta E°,$

or, $\log k = \dfrac{n \ F \ \Delta E°}{2.303 \ RT}$ .

Since two electrons are transferred from Mg(s) to $Sn^{2+}$, n = 2. The absolute temperature is T = 25°C + 273 = 298°K. Hence

$$\log k = \frac{n \ F \ \Delta E°}{2.303 \ RT}$$

$$= \frac{(2)(2.30609 \times 10^4 \ cal/mole\text{-}v)(2.23 \ v)}{(2.303)(1.987 \ cal/mole\text{-}deg)(298°K)} = 75.6,$$

or,   $k = 10^{75.6} = 4 \times 10^{75}$.

The equilibrium constant for this reaction is therefore $4 \times 10^{75}$.

The following reaction is performed at 298°K.
$2NO(g) + O_2(g) \rightleftarrows 2NO_2(g)$. The standard free energy of formation of $NO(g)$ is 86.6 kJ/mole at 298°K; find the standard free energy of formation of $NO_2(g)$ at 298°K.
$k_p = 1.6 \times 10^{12}$.

Solution: Free energy, $\Delta G°$, is a measure of the useful work that a system can perform. There is a quantitative relation between standard free energy and the equilibrium constant at a constant temperature:

$$\Delta G° = - RT \ln k_p,$$

where R = the universal gas constant, which equals 8.31 J/mole deg, T = temperature in Kelvin (Celsius plus 273°) and $k_p$ = the equilibrium constant for the reaction. You have all the information to find the $\Delta G°$ for the entire reaction.  Therefore, to find $\Delta G°_{NO_2}$, you (1) find $\Delta G°$ for the reaction. (2) Use the fact that

$$\Delta G°_{react} = 2\Delta G°_{NO_2} - 2\Delta G°_{NO} - \Delta G°_{O_2} \text{ to find } \Delta G°_{NO_2}.$$

$$\Delta G° = - RT \ln k_p$$

$$= - (8.31 \text{ J mol}^{-1} \text{ deg}^{-1})(298) \ln 1.6 \times 10^{12}$$

$$= - (8.31 \text{ J mol}^{-1} \text{ deg}^{-1})(298)(2.303) \log (1.6 \times 10^{12})$$

$$= - 69.6 \text{ kJ/mole.}$$

$$\Delta G°_{react} = 2\Delta G°_{NO_2} - 2\Delta G°_{NO} - \Delta G°_{O_2}$$

The $\Delta G°$ for the entire reaction must just be the $\Delta G°$ of the product minus the $\Delta G°$'s of the reactants.

$$\Delta G°_{react} = 2\Delta G°_{NO_2} - 2\Delta G°_{NO} - \Delta G°_{O_2}$$

You have calculated $\Delta G°$ overall and are given $\Delta G°_{NO}$. $\Delta G°_{O_2}$ is zero, since the $\Delta G°$ of any element in its standard state is taken as zero. To find $\Delta G°_{NO_2}$, therefore, substitute in these values and solve for it. Thus,

$$\Delta G^{\circ}_{NO_2} = \tfrac{1}{2}\left[\Delta G^{\circ} + 2\Delta G^{\circ}_{NO}\right]$$

$$= \tfrac{1}{2}\,(-\,69.6 + 2 \times 86.6)$$

$$= 51.8 \text{ kJ/mole.}$$

● **PROBLEM** 556

Calculate the change in Gibbs free energy ($\Delta G$) for the production of 2 $NO_2$(g) at 1 atm from $N_2O_4$(g) at 10 atm at 25°C. The standard Gibbs free energy for the reaction

$$N_2O_4\,(g) = 2NO_2\,(g)$$

is + 1161 cal at 25°C.

Solution: If $\Delta G^{\circ}$ has a positive value, the reactants in their standard states will not react spontaneously to give products in their standard states. By increasing the pressures or concentrations of the reactants or decreasing the pressures or concentrations of the products, however, it may be possible to make $\Delta G$ negative so that the reaction will proceed spontaneously.

To calculate $\Delta G$ under nonstandard conditions, the following formula is used:

$$\Delta G = \Delta G^{\circ} + RT \ln \underset{i}{\pi}\ p_i^{\nu i}, \qquad \text{where R = gas constant,}$$

T = absolute temperature (Celsius + 273), and $\underset{i}{\pi}\ p_i^{\nu i}$ represents the product of the partial pressures of the component gases, each raised to the power that corresponds with its stoichiometric coefficient in the balanced equation. Thus,

$$\Delta G = \Delta G^{\circ} + RT \ln \frac{p_{NO_2}^2}{p_{N_2O_4}}$$

$$= +\ 1161 + (1.987 \text{ cal}^{\circ}K^{-1}\text{ mole}^{-1})(298^{\circ}K)$$

$$(2.303)\ \log\ (1 \text{ atm})^2/(10 \text{ atm})$$

$$= 1161 - 1364 = -\ 203 \text{ cal.}$$

The production of $NO_2$(g) at 1 atm from $N_2O_4$(g) at 10 atm is a spontaneous process due to the fact that a negative $\Delta G$ is obtained. Thus, a continuous process would be thermodynamically feasible if $N_2O_4$ was maintained at 10 atm and $NO_2$ was withdrawn by some method so that its partial pressure was maintained at 1 atm.

Write out complete equations for the molar free energy of
formation of nitric oxide, ammonia, and water vapor.
$4NH_3(g) + 5O_2(g) \rightarrow 4NO(g) + 6H_2O(g)$. Using the data from the
accompanying table, determine the $\Delta G°$ for this reaction.
If $\Delta S° = 42.94cal$, what is $\Delta H°$ for the reaction?  Calculate the value of its equilibrium constant at 298°k;  at
798°k. Calculate $\Delta G°$ at 798°k.  Refer to the following table

Standard Heats of Formation, $\Delta H°$,
and Standard Free Energies of
Formation, $\Delta G°$, in Kcal/mole at 25°C

| Substance | $\Delta H°$ | $\Delta G°$ |
|---|---|---|
| $NH_3$ (g) | -11. 0 | - 4. 0 |
| NO (g) | +21. 6 | +20. 7 |
| $H_2O$ (g) | -57. 8 | -54. 6 |

Solution:  ΔG may be defined as net energy (free energy)
available for the production of useful work.  It can
predict spontaneity, maximum work obtainable, and maximum yields in an equilibrium reaction.

There exist a number of equations that allow measurement
of ΔG quantitatively.  You will employ these in answering
some parts of this question.  You commence as follows:

To write the equations for molar free energy of formation
of the given substances, consider their components and, with
the appropriate coefficients, write a balanced equation
so that only 1 mole is formed.  You have, then,
Ammonia: $1/2 N_2(g) + 3/2 H_2(g) \rightarrow NH_3(g)$

Nitric oxide: $1/2 N_2(g) + 1/2 O_2(g) \rightarrow NO(g)$

Water vapor:  $H_2(g) + 1/2 O_2(g) \rightarrow H_2O(g)$

The ΔG°'s for these are, respectively, -4.0, + 20.7, and
-54.6 kcal/mole.

The ΔG° for the overall reaction = $\Delta G°_{products}$ -
ΔG°reactants, each multipled by its coefficients in the
chemical equation.  Therefore, $\Delta G°_{overall} = 4\Delta G°_{NO(g)} +$

$6\Delta G^\circ_{H_2O(g)} - 4\Delta G^\circ_{NH_3(g)} - 5\Delta G^\circ_{O_2(g)} = 4(20.7) + 6(-54.6) -$

$4(-4.0) - 5(0) = -228.8$ kcal/mole. | Note: by definition the $\Delta G^\circ$'s of all elements are zero, which explains why

$\Delta G^\circ_{O_2(g)} = 0.$ | To find $\Delta H^\circ$, remember that $\Delta H^\circ =$

$\Delta G^\circ + T\Delta S^\circ$, where $\Delta H^\circ$ = change in enthalpy, T = temperature in kelvin (Celsius plus 273°) and $\Delta S^\circ$ = change in entropy; T = 25°C or 298°K since all $\Delta G^\circ$'s are measured at 25°C. By substituting, you find $\Delta H^\circ = -228.8 +$ (298)×(42.94)×(1kcal/1000cal) = -228.8 + 12.8 = ─216.0 kcal/mole.(The 1 kcal/1000 cal is a conversion factor.)

To find the equilibrium constant at a given temperature, remember that $\Delta G^\circ = -2.303RT \log k$, where R = universal gas constant, T = temperature in kelvin ( Celsius plus 273°) and k = the equilibrium constant. Thus, at 298°k, you have $-228.8 = -2.303(1.987) \times (298) \log k$. Solving,

$$k = \text{antilog}\left(\frac{-\Delta G^\circ}{2.303 \ RT}\right) = \text{antilog}\left[\frac{(228.8)(1000)}{(2.303)(1.987)(298)}\right] =$$

antilog 167.8 = $6 \times 10^{167}$ .

To find k at 798°k, use the equation $\log \frac{k_2}{k_1} = \frac{-\Delta H^\circ}{2.303R}\left(\frac{1}{T_2} - \frac{1}{T_1}\right)$

Substituting, $\log \frac{k_{798}}{k_{298}} = \frac{-\Delta H^\circ}{2.303R}\left(\frac{1}{798} - \frac{1}{298}\right) =$

$\frac{-(-216.0)(1000 \text{ cal/kcal})}{(2.303)(1.987)}$. Rewriting and solving,

$k_{798} = (k_{298})$ antilog $(-99.2) = (6 \times 10^{167})(6 \times 10^{-100})$

$\qquad = 4 \times 10^{68}$

To find $\Delta G^\circ$ at 798°k, use the equation $\Delta G^\circ = -2.303RT \log k$:
$\Delta G^\circ = -2.303RT \log k_{789}$.

$\Delta G^\circ = (-2.303)(1.987)(798) \log 4 \times 10^{68} = -2.5 \times 10^5$ cal/mole

$\qquad = -250$ kcal/mole.

# CHAPTER 16

# ELECTROCHEMISTRY

> **Basic Attacks and Strategies for Solving Problems in this Chapter. See pages 539 to 579 for step-by-step solutions to problems.**

Electrochemistry is that branch of chemistry which deals with reactions that involve the production and consumption of electrons. Hence electric currents are consumed or produced in electrochemical reactions. Normally the required electric current is introduced into the reactor, called a cell, through electrodes.

The conduction of electric currents in solutions of electrolytes is effected by the motion of ions within the solution. Positively charged ions migrate toward the anode (the negative electrode) while negatively charged ions migrate toward the cathode (the positive electrode). Another way to identify the difference between the cathode and anode is to remember that reduction reactions always occur at the cathode and oxidation reactions at the anode.

In electrochemical reactions, the charge must be balanced as though it were an elemental species in the reaction, and overall, the electrons produced in one part of the reaction must be consumed in another part. If this were not so, tremendous electric charges would develop for even small amounts of reaction. A typical example is the industrially important electrolysis of molten sodium chloride to produce chlorine gas and sodium metal.

The electrolyte contains sodium ($Na^+$) and chloride ($Cl^-$) ions. The chloride ions react at the anode and sodium ions at the cathode according to the equations

$$2Cl^- \rightarrow Cl_2 + 2e^- \text{ (oxidation)} \qquad 16\text{-}1$$

$$Na^+ + e^- \rightarrow Na \text{ (reduction)} \qquad 16\text{-}2$$

Since the same number of electrons enter the electrolyte at the cathode and leave at the anode, the overall reaction is obtained by multiplying the second half-reaction by two and adding the two reactions to obtain

$$2Na^+ + 2Cl^- \rightarrow 2Na + Cl_2 \qquad\qquad 16\text{-}3$$

Note that no electrons appear in the overall reaction. It is significant that electrons cannot be stored in sufficient numbers to permit the appreciable independent conduction of electrochemical half-reactions.

The equivalent weight is defined for electrochemical reactions in a manner analogous to that outlined in Chapter 8 for equivalent weights of solution. The equivalent weight of a species is the molecular (or atomic) weight divided by the charge or number of electrons which react. Michael Faraday, in 1832, observed that the amount of a substance undergoing oxidation or reduction at an electrode in an electrochemical cell was directly proportional to the amount of electric current that passed through the cell. This observation is known today as Faraday's Law and the unit of the "faraday" is the charge equal to that in one mole $(6.022 \times 10^{23})$ of electrons. It is a huge charge, equal to 96,487 coulombs (c), or 96,487 ampere seconds (A-sec).

Electrochemical reactions are often analyzed as oxidation-reduction couples, abbreviated as redox reactions. The key to the balancing of redox reactions is to balance the charge as well as the elements on each side of the reaction. An example of balancing such a reaction is shown later for the displacement of $Cu^{+2}$ ions from solution by iron metal (Equation 16-4).

An important concept in electrochemistry is that of the half-cell standard potential or emf. Table 1 lists a sample of important half-cell reactions and their standard potential. This is the potential produced by the half-cell reaction when balanced against a standard hydrogen electrode with all species present at standard (1 molar) concentrations.

## Table 1: Sample of Standard Aqueous Electrode Potentials at 298K

| Element | Electrode Half Reaction | E° (volts) |
|---------|------------------------|------------|
| Li | Li $\leftrightarrow$ Li$^+$ + e$^-$ | 3.045 |
| Na | Na $\leftrightarrow$ Na$^+$ + e$^-$ | 2.714 |
| Zn | Zn $\leftrightarrow$ Zn$^{++}$ + 2e$^-$ | .763 |
| Fe | Fe $\leftrightarrow$ Fe$^{++}$ + 2e$^-$ | .41 |
| H$_2$ | H$_2$ $\leftrightarrow$ 2H$^+$ + 2e$^-$ (reference) | 0.000 |
| Cu | Cu $\leftrightarrow$ Cu$^{++}$+ 2e$^-$ | $-.337$ |
| Au | Au $\leftrightarrow$ Au$^{+3}$ + 3e$^-$ | $-1.50$ |

This table, also called an emf (for electromotive force) table, is shown for illustration. More extensive tables exist in numerous texts and handbooks. The higher a species is in the emf table, the more likely the reaction is to proceed. If a species is higher in the table, it will displace the ion of a lower species from solution. For example, since iron is higher than copper in the table, iron metal will displace copper ions from solution according to the equation

$$Fe + Cu^{++} \rightarrow Cu + Fe^{++} \qquad \text{16-4}$$

This reaction is the basis for the commercial recovery of copper from dilute solutions. Note that very reactive metals, like sodium, are near the top of the table while unreactive metals, like gold, are near the bottom.

The Nernst equation relates the electrode potential at nonstandard conditions to that measured (and widely tabulated) for standard conditions

$$E = E° - (RT/nF) \ln Q \qquad \text{16-5}$$

where $n$ is the number of electrons in the reaction, $F$ is Faraday's constant (96,487 cl/mol), $R$ is the universal gas constant, and $T$ the absolute temperature. $Q$ is the ratio of product concentrations to reactant concentrations each raised to the power of their stoichiometric coefficient in the reaction. It is much like the equilibrium constant $K$ except that the concentrations may not be the equilibrium concentrations. If common (base 10) logarithms are used, the factor 2.303 must be included in the second term of the Nernst equation. Hence it is frequently written as

$$E = E° - 2.303RT/nF \log Q \qquad \text{16-6}$$

Let us use the Nernst equation to calculate the zinc half-cell potential if

$Zn^{+2}$ ions are present at a concentration of 0.01M. $E°$, from Table 1, is 0.763 volts.

$$E = .763 - .0256/n \ln (0.01/1) = 0.822 \text{ volts} \qquad 16\text{-}7$$

In using the Nernst equation, the molar concentrations are used for dissolved species, the partial pressure in atmospheres for gases, and the activity equals one for pure condensed phases.

Step-by-Step Solutions to
Problems in this Chapter,
"Electrochemistry"

## CONDUCTION

● **PROBLEM** 558

Design an experiment which demonstrates that both positive and negative ions move in electrolytic conduction.

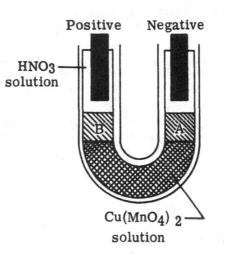

Positive    Negative

HNO3 solution

$Cu(MnO_4)_2$ solution

<u>Solution</u>:  The accompanying figure illustrates the set up of an experiment which will demonstrate the migration of both positive and negative ions in electrolytic conductivity.  A  U-tube is initially half filled with a deep purple solution of  $Cu(MnO_4)_2$  in water.  The color of the blue hydrated  $Cu^{2+}$  ions is masked by the purple of the  $MnO_4^-$ ions.  A colorless aqueous solution of  $HNO_3$  floats on top of the $Cu(MnO_4)_2$  solution in each arm of the  U-tube.  If an electric field is established for a period of time across the solution by 2 electrodes, the blue color, characteristic of hydrated  $Cu^{2+}$  ions, moves into the region marked  A,  suggesting a migration of positive ions toward the negative electrode.  At the same time, the purple color, characteristic of  $MnO_4^-$ , moves into the  B  region, indicating a migration of negative ions toward the positive electrode.  Thus, there is empirical evidence of the migration of both positive and negative ions.

Would you expect the mobility of an electrolytic ion to increase or decrease for each of the following? (a) increasing the size of the ion; (b) increasing charge on the ion; (c) increasing temperature and (d) placing it in a solvent of lower viscosity.

Solution: To see how the mobility of the electrolytic ion changes when you alter these parameters, you must know what happens to these ions when an electric field is applied. The ions are free to move randomly around in solution before an electric field is applied. However, when the field is present, the positive ions experience a force in one direction, while the negative ions experience a force in the opposite direction. With this in mind, you proceed as follows:

(a) when you increase the size of the ion, you would anticipate a decrease in mobility. Remember, the ion must move when the electric current is applied; additional mass and volume, therefore, inhibits this movement, since there is greater resistance from the solvent molecules.

(b) If you increase the charge on the ion, you expect mobility to increase due to the fact that the force which results in movement is directly proportional to the attraction of the ions to the poles. Thus, by increasing the charge, you increase this force, which, in turn, increases movement.

(c) When you increase the temperature, you increase the mobility of the ions. By increasing the temperature, the average kinetic energy of the ion increases. Kinetic energy is a measure of movement.

(d) A solution of lower viscosity is expected to have an increase in mobility. This stems from the fact that viscosity measures internal resistance to flow. Thus, if the viscosity decreases, there is less resistance to flow, which means that the ions can move more freely.

# EQUIVALENT WEIGHT

Calculate the theoretical quantity of chlorine obtainable by the electrolysis of 2.0 kg of a 20% sodium chloride solution. What other products would be obtained and what would be the weight of each?

Solution: The equation for this reaction is as follows:

$$2NaCl + 2H_2O \xrightarrow{\text{electric current}} 2NaOH + H_2 + Cl_2 \ .$$

One mole of $Cl_2$ will be formed from every 2 moles of $NaCl$ reacted. Therefore, to find the quantity of $Cl_2$ formed one must first know the amount of $NaCl$ reacted. This can be determined by solving for the weight of $NaCl$ in the solution and dividing this number by the mole-

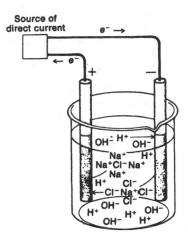

Source of
direct current

cular weight to find the number of moles present. Since the solution
is 20% NaCl, it means that 20% of its weight is made up by NaCl.

weight of NaCl = 0.20 × 2.0 kg = 0.4 kg .

Therefore, there are 0.4 kg or 400 g of NaCl present.

One can now solve for the number of moles present by dividing 400g
by the molecular weight (MW of NaCl = 58.5)

$$\text{no. of moles} = \frac{400\text{g}}{58.5\text{g/mole}} = 6.84 \text{ moles} .$$

From the equation one sees that $\frac{1}{2}$ of this amount is equal to the number
of moles of $Cl_2$ formed.

no. of moles of $Cl_2 = \frac{1}{2} \times 6.84$ moles = 3.42 moles .

The weight of this number of moles of $Cl_2$ will be equal to the molecular weight of $Cl_2$ times the number of moles.

(MW of $Cl_2$ = 71.0)

weight = 3.42 moles × 71.0g/mole = 242.8g
of $Cl_2$

From the reaction one can also see that if 2 moles of NaCl react,
2 moles of NaOH are formed. Here, 6.84 moles of NaCl is reacted,
therefore, 6.84 moles of NaOH are formed. The weight of this quantity
is equal to the number of moles × the molecular weight. (MW of NaOH = 40).

weight of NaOH = 40g/mole × 6.84 moles = 273.6g .

It is also seen from the reaction, that if 2 moles of NaCl are reacted,
1 mole of $H_2$ is formed. Thus, if 6.84 moles of NaCl are reacted, 3.42
moles of $H_2$ are formed. The weight of $H_2$ can then be found. (MW of $H_2$ =
2).

weight of $H_2$ = 2g/mole × 3.42 moles = 6.84g .

● **PROBLEM** 561

.0324 Faradays (F) liberated .651g of Calcium. What is the atomic
weight of Calcium?

Solution: A faraday may be defined as the number of coulombs or charges of electricity that liberate one gram-equivalent weight of an element in solution. Note, the equivalent weight of a substance is the amount of substance which 1 mole of electrons can oxidize or reduce into a neutral species.

Calcium has an equivalent weight of $\frac{1}{2}$ its atomic weight, since its oxidation state is +2. In other words, one mole of electrons will neutralize only half of the calcium ions present, since each ion needs two electrons. Thus, if you can find the gram-equivalent weight of calcium, you can find its atomic weight by multiplying it by two. However, you can determine the number of gram-equivalents given the number of faradays of electricity used. You are told that .0324F deposits .651g. By definition, 1 F deposits the gram-equivalent weight. You have, then, the following proportion:

$$\frac{.651g\ Ca}{.0324\ F} = \frac{g\text{-eq. wt. of Ca}}{1\ F} = 20.1 \ .$$

Hence, 20.1g is the gram-equivalent of calcium. Thus, the atomic weight is found to be $2 \times 20.1 = 40.2$.

● **PROBLEM** 562

Two electrolytic cells were placed in series. One was composed of $AgNO_3$ and the other of $CuSO_4$ . Electricity was passed through the cells until 1.273g of Ag had been deposited. How much copper was deposited at the same time?

Solution: To find out how much copper was deposited, you need to know how many faradays were contained in the electricity that passed through the solution. One faraday, be definition, deposits one gram-equivalent of material. The equivalent weight of a substance is that amount of the substance which 1 mole of electrons (1 faraday) can oxidize or reduce to a neutral species.
You are told that 1.273g of Ag are deposited. From this, you can compute the number of faradays. The oxidation state of Ag is +1. Thus, the reaction that deposited Ag had to be $Ag^+ + e^- \rightarrow Ag$ . Thus, 1 mole of electrons yields 1 gram-equivalent weight of Ag . In this case, the gram-equivalent weight equals the atomic weight, since 1 mole of Ag is produced with 1 mole of electrons. Recalling the definition of a faraday, you can say that 1 faraday can deposit 107.87g of Ag (the atomic weight). You are told, however, that only 1.273g has been deposited. Thus, the following proportion can be used to find the number of faradays used:

$$\frac{1.273g\ Ag}{X\ \text{Faradays}} = \frac{107.87g\ Ag}{1\ \text{Faraday}}$$

Solving, X = .01180 Faradays. Thus, the amount of faradays passed is .01180F, which means the amount of copper deposited depends on this amount of electricity.
To deposit copper, you must perform the reaction $Cu^{2+} + 2e^- \rightarrow Cu$ you see that it takes 2 moles of electrons to deposit one mole of copper. Thus, one faraday can only deposit one half of a mole copper. Thus, 1 faraday reduces $(\frac{1}{2})(63.54) = 31.77$ grams. You have, however, only .01180 faradays. Thus, the amount of Cu deposited can be found using

the proportion:

$$\frac{1 \text{ Faraday}}{31.77g} = \frac{.01180 \text{ Faraday}}{X} \quad .$$

Solving for X, X = .3749g. Therefore, 0.3749g of Cu will be deposited at the same time.

● **PROBLEM** 563

The same quantity of electricity was passed through two separate electrolytic cells. The first of these contained a solution of copper sulfate ($CuSO_4$) and exhibited the following cathode reaction (reduction):

$$Cu^{2+} + 2e^- \rightarrow Cu(s) \quad .$$

The second of these contained a solution of silver nitrate ($AgNO_3$) and exhibited the following cathode reaction:

$$Ag^+ + e^- \rightarrow Ag(s) \quad .$$

If 3.18g of Cu were deposited in the first cell, how much Ag was deposited in the second cell?

Solution: The solution to this problem involves calculating how many Faradays were passed through the first cell (and similarly, the second cell) and, then, using this value to calculate the amount of Ag that will be deposited in the second cell.

By definition, one Faraday ( 1F ) will deposit the equivalent weight of any element. Also, 1 F is equal to the charge of one mole of electrons. From the reaction

$$Cu^{2+} + 2e^- \rightarrow Cu(s)$$

we see that it takes two moles of electrons (2 F) to deposit one mole of copper. Hence, 1 F will deposit one-half mole of Cu and the equivalent weight of Cu is, therefore, ½ the atomic weight, or ½ x 63.6g/mole = 31.8 g/mole. From the reaction

$$Ag^+ + e^- \rightarrow Ag(s)$$

we see that one mole of electrons (1 F) will deposit one mole of silver, so that the equivalent weight of Ag is equal to its atomic weight, or 108g/mole.

Since 3.18g of Cu were deposited and this is equal to one-tenth the atomic weight (0.1 x 31.8 = 3.18), one-tenth of a Faraday (0.1 F) was passed through the $CuSO_4$ cell. Since the same quantity of electricity was passed through both cells, 0.1 F was also passed through the $AgNO_3$ cell.

The amount of Ag deposited is then

amount Ag = number of moles of electrons x equivalent weight of Ag
= 0.1 F x 108g/mole electrons
= 0.1 mole electrons x 108g/mole electrons
= 10.8g.

Two electrochemical cells were fitted with inert platinum electrodes and connected in series. The first cell contained a solution of a metal nitrate; the second cell, metal sulfate. After operation of the cell for a prescribed period of time, it was observed that 675 mg of the metal had been deposited in the first cell and 76.3 ml of hydrogen gas ($25^\circ$ and 1 atm) evolved in the second. Determine the equivalent weight of the metal.

Solution: Equivalent weight may be defined as that amount of substance which one mole of electrons will reduce or oxidize.

Since the electrodes are connected in series, the same current passes through both cells. Thus, equal numbers of equivalents of the metal (m) and hydrogen must be liberated. The cathode reactions for the two cells are: first cell: $M^{n+} + ne^- \rightarrow M(s)$ ; second cell: $2H^+ + 2e^- \rightarrow H_2(g)$, where n is the number of electrons needed to form the M(s) from the metal ion. One can solve for the number of moles of $H_2$ formed by use of the Ideal Gas Law.

$$n = \frac{PV}{RT}$$

where n is the number of moles, R is the gas constant, 0.082 . liter-atm/mole$^\circ$K, T is the absolute temperature, V is the volume and P is the pressure. One is given the temperature in $^\circ$C, to convert to $^\circ$K add 273 to the temperature in $^\circ$C .

$$T = 25^\circ + 273 = 298^\circ K .$$

Now, solving for n:

$$n = \frac{(1 \text{ atm})(.0763 \text{ liters})}{(0.082 \text{ liter-atm/mole}^\circ K)(298^\circ K)} = 3.12 \times 10^{-3} \text{ moles} .$$

From the half equations, one can see that there are two equivalents for each mole of $H_2$ formed. Thus, there must be $2 \times (3.12 \times 10^{-3})$ moles of equivalents of the metal present.

Hence, equivalents of $M = 2 \times 3.12 \times 10^{-3} = 6.24 \times 10^{-3}$ equiv. One is given that 675mg or .675 g of metal are formed, thus, in .675 g of the metal there are $6.24 \times 10^{-3}$ equiv . The weight of one equivalent can be found by dividing .675 g by $6.24 \times 10^{-3}$ equiv .

$$\text{equiv. wt.} = \frac{.675g}{6.24 \times 10^{-3} \text{equiv.}} = 108.2 \text{g/equiv.}$$

In an electrolytic cell, a liter of a 1M aqueous solution of $MnO_4^-$ is reduced at the cathode. Determine the number of faradays required for each of the following to be made;
a)    a solution that is .01 M $MnO_4^{2-}$ ;
b)    1 gram of $MnO_2$ ;
c)    1 gram-equivalent of Mn metal.

Solution: For each part, you need to write the reaction that occurs. The masses of substances produced during the process are proportional to their equivalent weights. The unit electrical equivalent = 1 Faraday or 1F. One Faraday = 96,500 coul and is capable of reducing one equivalent positive charge, an equivalent positive charge contains Avogadro's number of individual unit charges. You proceed as follows:

a)  You want to produce $.01$ M $MnO_4^{2-}$ from a $MnO_4^-$ solution. The reaction is $MnO_4^- + e^- \rightarrow MnO_4^{2-}$. One electron is being transferred. You want $[MnO_4^{2-}] = .01$ M. Since the solution has a volume of 1 liter, $.01$ M $= .01$ moles, since M = molarity $= \frac{moles}{liter}$. Due to the fact that only 1 electron is transferred, only one Faraday is needed to reduce a mole of $MnO_4^-$, since it would require only a mole of electrons. You need only, however, $.01$ moles of $MnO_4^{2-}$. Since a mole of $MnO_4^{2-}$ is formed from every mole of $MnO_4^-$ oxidized you need to reduce $.01$ moles of $MnO_4^{2-}$. To do this, requires, therefore, $.01$ F.

b)  The reaction to produce $MnO_2$ is

$$MnO_4^- + 4H^+ + 3e^- \rightarrow MnO_2 + 2H_2O .$$

Here, 3 electrons are required to reduce $MnO_4^-$ to $MnO_2$. Thus, one Faraday can only reduce $1/3$ of a mole of $MnO_4^-$. Recall, one Faraday is Avogadro's number of electrons and this reaction requires three moles, or $3F$ per mole of $MnO_4^-$. You want 1g of $MnO_2$ (atomic wt. = 86.94 g/mole), which is

$$\frac{1g}{86.94g/mole} \quad \text{or} \quad .0115 \text{ moles of } MnO_2 .$$

Since one mole of $MnO_2$ will be produced for every mole of $MnO_4^-$, you must reduce $.0115$ moles of $MnO_4^-$. One faraday will reduce $1/3$ of a mole of $MnO_4^-$. To reduce all of it, you want 3 Faraday's/mole. Thus, number of Faraday's required = $(3F/mole)(.0115 \text{ mole}) = .0345F$.

c)  To obtain Mn metal, you need the reaction

$$MnO_4^- + 8H^+ + 7e^- \rightarrow Mn + 4H_2O \quad \text{to occur.}$$

7 electrons are to be transferred. Thus, 1 Faraday can only reduce $1/7$ mole of material. However, you want 1 gram-equivalent of material, which is the amount deposited by 1F. Hence, 7 equiv. of Mn = 1 mole Mn. Thus, to deposit this amount, you use

$$(1.00 \text{ equiv. Mn})(F/\text{equiv. Mn}) = 1.00F .$$

# REDOX REACTIONS

● PROBLEM 566

Balance the following reaction in basic aqueous solution:

$$SO_3^{2-} + CrO_4^{2-} \rightarrow SO_4^{2-} + Cr(OH)_3$$

<u>Solution</u>: Three rules can be used to balance oxidation-reduction reactions: (1) Balance charge by adding $H^+$ (in acid) or $OH^-$ (in base). (2) Balance oxygen by adding water. (3) Balance atoms (of hydrogen) by adding hydrogen to the appropriate side. These 3 rules will balance the redox equation. You proceed as follows:

Reduction:   $CrO_4^{2-} \rightarrow Cr(OH)_3$.

Add $2OH^-$, to the right side so that charge is balanced. You obtain

$$CrO_4^{2-} \rightarrow Cr(OH)_3 + 2OH^-.$$

Balance oxygens by adding one water molecule to left side. Thus,

$$H_2O + CrO_4^{2-} \rightarrow Cr(OH)_3 + 2OH^-.$$

Balance H's by adding three H's to  left  side. You have

$$\frac{3}{2} H_2 + H_2O + CrO_4^{2-} \rightarrow Cr(OH)_3 + 2OH^-$$

Oxidation:   $SO_3^{2-} \rightarrow SO_4^{2-}$.

Charges are already balanced. To balance oxygen, add water to left side. As such,

$$H_2O + SO_3^{2-} \rightarrow SO_4^{2-}.$$

Now balance hydrogens to obtain

$$H_2O + SO_3^{2-} \rightarrow SO_4^{2-} + H_2.$$

In summary, the balanced half-reactions are

oxid:     $H_2O + SO_3^{2-} \rightarrow SO_4^{2-} + H_2$

red:    $\frac{3}{2} H_2 + H_2O + CrO_4^{2-} \rightarrow Cr(OH)_3 + 2OH^-$

So that no free H's appear in overall reaction, multiply the oxidation reaction by 3 and red by 2. You obtain

oxid:          $3H_2O + 3SO_3^{2-} \longrightarrow 3SO_4^{2-} + 3H_2$

red:    $\underline{3H_2 + 2H_2O + 2CrO_4^{2-} \longrightarrow 2Cr(OH)_3 + 4OH^-}$

overall (oxid + red):

$$5H_2O + 3SO_3^{2-} + 2CrO_4^{2-} \longrightarrow 3SO_4^{2-} + 2Cr(OH)_3 + 4OH^-.$$

Notice:  The $H_2$'s dropped out.

Balance the following reaction in acidic aqueous solution:

$$ClO_3^- + Fe^{2+} \rightarrow Cl^- + Fe^{3+}$$

Solution: Reactions in which electrons are transferred from one atom to another are known as oxidation-reduction reactions or as redox reactions. To balance this type of reaction, you want to conserve charge and matter, i.e., one side of the equation must not have an excess of charge or matter. To perform this balancing, you need to (1) Balance charge by adding electrons. (2) Balance oxygen by adding water. (3) Balance atoms (of hydrogen) by adding $H^+$ (in acid) or $OH^-$ (in base). These three rules will balance the redox equation. These rules apply to balancing only the half-reactions. The overall reaction, the sum of these, will be balanced by their addition. Proceed as follows; $Fe^{2+}$ goes to $Fe^{3+}$. It lost an electron thus, it's the oxidation half reaction. To balance charge, add $e^-$. Thus,

$$Fe^{2+} \rightarrow Fe^{3+} + e^-.$$

The reduction must be $ClO_3^- \rightarrow Cl^-$. Chlorine changes oxidation state from +5 to -1 so $6e^-$ must be added. To balance the 3 oxygen atoms on left side, add 3 water molecules on right side. You obtain

$$ClO_3^- + 6e^- \rightarrow Cl^- + 3H_2O$$

Since we know the reaction occurs in acidic media, add $6H^+$ to the reactants as the source of hydrogen in the water produced.

$$ClO_3^- + 6e^- + 6H^+ \rightarrow Cl^- + 3H_2O$$

In summary, you have

oxid: $\quad Fe^{2+} \rightarrow Fe^{3+} + e^-$

red: $\quad ClO_3^- + 6e^- + 6H^+ \rightarrow Cl^- + 3H_2O$

To balance the number of electrons appearing in the equations, multiply the oxidation reaction by six. You obtain

$$6Fe^{2+} \rightarrow 6Fe^{3+} + 6e^-$$

Thus, oxid: $\quad 6Fe^{2+} \rightarrow 6Fe^{3+} + 6e^-$

red:     $ClO_3^- + 6e^- + 6H^+ \rightarrow Cl^- + 3H_2O$

overall:     $6Fe^{2+} + ClO_3^- + 6H^+ \rightarrow 6Fe^{3+} + Cl^- + 3H_2O$

Notice; The electrons dropped out.

● **PROBLEM** 568

Balance the equation for the following reaction taking place in aqueous acid solution:

$$Cr_2O_7^{2-} + I_2 \rightarrow Cr^{3+} + IO_3^-$$

<u>Solution</u>:   The equation in this problem involves both an oxidation and a reduction reaction. It can be balanced by using the following rules: (1) Separate the net reaction into its two major components, the oxidation process (the loss of electrons) and the reduction process (the gain of electrons). For each of these reactions, balance the charges by adding $H^+$, if the reaction is occurring in an acidic medium, or $OH^-$ in a basic medium. (2) Balance the oxygens by addition of $H_2O$. (3) Balance hydrogen atoms by addition of H. (4) Combine the two half reactions, so that all charges from electron transfer cancel out. These rules are applied in the following example.

The net reaction is

$$Cr_2O_7^{2-} + I_2 \rightarrow Cr^{3+} + IO_3^-$$

The oxidation reaction is

$$I_2^0 \rightarrow 2IO_3^- + 10e^-$$

The I atom went from an oxidation number of 0 in $I_2$ to + 5 in $IO_3^-$, because O always has a - 2 charge. You begin with $I_2$, therefore, 2 moles of $IO_3^-$ must be produced and 10 electrons are lost, 5 from each I atom. Recall, the next step is to balance the charges. The right side has a total of 12 negative charges. Add 12 $H^+$'s to obtain

$$I_2 \rightarrow 2IO_3^- + 10e^- + 12H^+$$

To balance the oxygen atoms, add $6H_2O$ to the left side, since there are 6 O's on the right, thus,

$$I_2 + 6H_2O \rightarrow 2IO_3^- + 10e^- + 12H^+.$$

Hydrogens are already balanced. There are 12 on each side. Proceed to the reduction reaction:

$$Cr_2O_7^{2-} + 6e^- \rightarrow 2Cr^{3+}$$

Cr began with an oxidation state of + 6 and went to + 3. Since $2Cr^{3+}$ are produced, and you began with $Cr_2O_7^{2-}$, a total of 6 electrons are added to the left. Balancing charges: the left side has 8 negative charges and right side has 6 positive charges. If you add 14 $H^+$ to left, they balance. Both sides now have a net + 3 charge. The equation can now be written.

$$Cr_2O_7^{2-} + 6e^- + 14H^+ \rightarrow 2Cr^{3+}.$$

To balance oxygen atoms, add $7H_2O$'s to right. You obtain

$$Cr_2O_7^{2-} + 6e^- + 14H^+ \rightarrow 2Cr^{3+} + 7H_2O.$$

The hydrogens are also balanced, 14 on each side. The oxidation reaction becomes

$$I_2 + 6H_2O \rightarrow 2IO_3^- + 10e^- + 12H^+$$

The reduction reaction is

$$Cr_2O_7^{2-} + 6e^- + 14H^+ \rightarrow 2Cr^{3+} + 7H_2O.$$

Combine these two in such a manner that the number of electrons used in the oxidation reaction is equal to the number used in the reduction. To do this, note that the oxidation reaction has $10e^-$ and the reduction $6e^-$. Both are a multiple of 30. Multiply the oxidation reaction by 3, and the reduction reaction by 5, obtaining

oxidation: $\qquad 3I_2 + 18H_2O \rightarrow 6IO_3^- + 30e^- + 36H^+$

reduction: $\quad 5Cr_2O_7^{2-} + 30e^- + 70H^+ \rightarrow 10Cr^{3+} + 35H_2O$

Add these two half-reactions together.

$$3I_2 + 18H_2O \rightarrow 6IO_3^- + 30e^- + 36H^+$$
$$+ \quad 5Cr_2O_7^{2-} + 30e^- + 70H^+ \rightarrow 10Cr^{3+} + 35H_2O$$
$$\overline{3I_2 + 18H_2O + 5Cr_2O_7^{2-} + 30e^- + 70H^+ \rightarrow 10Cr^{3+} + 35H_2O + 30e^- + 36H^+}$$

Simplifying, you obtain:

$$3I_2 + 5Cr_2O_7^{2-} + 34H^+ \rightarrow 6IO_3^- + 10Cr^{3+} + 17H_2O$$

This is the balanced equation.

● **PROBLEM** 569

Balance the equation for the following reaction taking place in aqueous basic solution:

$$MnO_4^- + H_2O_2 \rightarrow MnO_2 + O_2$$

Solution: The equation in this problem involves both oxidation and reduction. When balancing it, you can use

549

the following rules. Separate the net reaction into its
two major components, the oxidation reaction (the loss
of electrons) and the reduction reaction (the gain of
electrons). For each half-reaction, balance the charges
with $H^+$, if the medium is acidic, or $OH^-$, if the medium
is basic. Next, balance the oxygens by the addition of
$H_2O$. Balance total atoms by the addition of H atoms.
Finally, combine the two half reactions, so that all
charges from electron transfer cancel out.

You employ these rules as follows:

The net reaction is

$$MnO_4^- + H_2O_2 \rightarrow MnO_2 + O_2$$

The oxidation process is

$$H_2O_2 \rightarrow O_2 + 2e^-$$

The oxygen atoms in $H_2O_2$ go from $-1$ to zero in $O_2$. Thus,
you have a loss of two electrons. To balance charges, add
$2\ OH^-$ to the left side, since there exist 2 negative charges
on the right side. You obtain

$$H_2O_2 + 2OH^- \rightarrow O_2 + 2e^-$$

You now have 4 oxygens on the left, but only 2 on the
right. Thus, add 2 water molecules ro the right, obtaining

$$H_2O_2 + 2OH^- \rightarrow O_2 + 2e^- + 2H_2O$$

There are the same number of H's on each side. Proceed,
now, to reduction half-reaction. Here,

$$MnO_4^- + 3e^- \rightarrow MnO_2$$

Mn begins with a $+7$ oxidation number and ends up with $+4$
in $MnO_2$. Therefore, 3 electrons must be added to the left
side of the equation. To balance the charges, add $4\ OH^-$

ions to right, since you have a total of 4 negative charges
on left. Rewriting the equation

$$MnO_4^- + 3e^- \rightarrow MnO_2 + 4OH^-.$$

Add 2 water molecules to the left, so that oxygen atoms
can be balanced, obtaining

$$2H_2O + MnO_4^- + 3e^- \rightarrow MnO_2 + 4OH^-.$$

The hydrogens are balanced. Thus,

oxidation:     $H_2O_2 + 2OH^- \rightarrow O_2 + 2e^- + 2H_2O$

reduction:     $2H_2O + MnO_4^- + 3e^- \rightarrow MnO_2 + 4OH^-$

To combine these two so that electrons cancel out. Select
a multiple of 3 and 2, since these are the number of

electrons involved in the half-reactions. This multiple is six. Multiply the oxidation by 3 and the reduction by 2, obtaining

oxidation: $3H_2O_2 + 6OH^- \rightarrow 3O_2 + 6e^- + 6H_2O$

reduction: $4H_2O + 2MnO_4^- + 6e^- \rightarrow 2MnO_2 + 8OH^-$

The net reaction is the total. Thus, adding you obtain:

$3H_2O_2 + 4H_2O + 6OH^- + 2MnO_4^- + 6e^- \rightarrow 3O_2 + 6e^- + 6H_2O + 8OH^- + 2MnO_2$

Cancel the electrons, subtract $OH^-$ ions and $H_2O$'s to obtain:

$2MnO_4^- + 3H_2O_2 \rightarrow 2MnO_2 + 3O_2 + 2H_2O + 2OH^-$

which is the balanced equation.

● **PROBLEM** 570

Determine the volume in milliliters of .20 M $KMnO_4$ required to oxidize 25.0 ml of .40 M $FeSO_4$ in acidic solution. Assume the reaction which occurs is the oxidation of $Fe^{2+}$ by $MnO_4^-$ to give $Fe^{+3}$ and $Mn^{2+}$.

Solution: This problem can be solved by two methods: the mole and the equivalent methods. The mole method requires consideration of the balanced equation that illustrates the reaction. From the data provided, this equation becomes

$$5Fe^{2+} + MnO_4^- + 8H_3O^+ \rightarrow 5Fe^{3+} + Mn^{2+} + 12 H_2O.$$

Now 25.0 ml of .40 M $FeSO_4$ furnishes (.025 liters)(.40 mol/liter) = .010 moles of $Fe^{2+}$, since the definition of molarity is

$$M = \frac{\text{number of moles of solute}}{\text{number of liters}}.$$

The balanced equation indicates that the number of moles of $MnO_4^-$ will be 1/5 that of $Fe^{2+}$. As such, the number of moles of $MnO_4^-$ = (.010)(1/5) = .002 moles. Since the $KMnO_4$ solution has a concentration of .2 M, then the number of liters required is

$$\frac{.002 \text{ mol of } MnO_4^-}{.2 \text{ mol/liter}} = .01 \text{ liters, which equals 10 ml.}$$

The equivalent method functions differently. An equivalent is defined as that mass of oxidizing or reducing agent that picks up or releases the Avogadro number of electrons. Normality is defined as the number of equivalents per liter. Since, in going from $Fe^{2+}$ to $Fe^{3+}$, you lose 1 electron, .40 M $FeSO_4$ is equal to .40 N $FeSO_4$. Recalling the definition of normality, you have

$$(.025 \ \text{liter})(.40 \ \text{equiv/liter}) = .01 \ \text{equiv. of Fe}^{2+},$$

the reducing agent. In an oxidation-reduction reaction, the number of equivalents of oxidizing agent must equal that of the reducing agent. This means you must have .01 **equiv.** of $MnO_4^-$. You know that for $KMnO_4$, there exists 1 equiv/liter. Therefore, the number of liters equals

$$\frac{.01 \ \text{equiv.}}{1.0 \ \text{equiv./liter}} = .01 \ \text{liters or 10 ml.}$$

# FARADAY'S LAW OF ELECTROLYSIS

● **PROBLEM** 571

You want to plate out 50g of copper from an aqueous solution of $CuSO_4$. Assuming 100% efficiency, how many coulombs are required?

Solution: In this question, you are dealing with the phenomenon of electrolysis. When an electric current is applied to a solution containing ions, the ions will either be reduced or oxidized to their electronically neutral state.

To answer this question, you must realize that $Cu^{2+}$ ions exist in solution. To plate out copper, 2 electrons must be added to obtain the copper atom, $Cu^0$. Since the $Cu^{2+}$ must gain electrons, it must be reduced. The amount of electricity that produces a specific amount of reduction (or oxidation) is related by $q = nF$ (Faraday's Law). Where $q$ = the quantity of electricity in coulombs, $n$ = number of equivalents oxidized or reduced and $F$ = Faradays. The number of equivalents equals the weight of material oxidized or reduced (m) divided by the gram-equivalent weight of the material ($M_{eq}$) i.e.,

$$N = \frac{m}{M_{eq}} .$$

A faraday = 96,490 coulombs or one mole of electrons.

Since copper ion requires two electrons for reduction, the gram-equivalent weight is one half of the atomic weight or 31.77g-equiv. You have, therefore,

$$q = \frac{(50)}{(31.77)}(96,490) = 1.52 \times 10^5 \ \text{coul.}$$

required.

● **PROBLEM** 572

How much electricity will be required to decompose 454g of water by electrolysis? The overall reaction is

$$2H_2O \rightarrow 2H_2 + O_2 .$$

Solution: Whenever a problem deals with weights and electricity, the solution involves an application of Faraday's Law: The passage of 1 faraday of electricity (96,500 coulombs) causes 1 equivalent weight of matter to be oxidized (the loss of 1 electrons) at one electrode and the reduction (the gain of 1 electrons) of 1 equivalent weight at the other electrode. Equivalent weight may be defined as molecular weight divided by number of moles of hydrogen transferred. To solve this problem, therefore, calculate the number of equivalents present in 454g of water. Water has a molecular weight of 18g/mole, but since 2 H's are transferred, water has an equivalent weight of 9g. Therefore, the number of equivalents is

$$\frac{\text{total weight}}{\text{equivalent weight}} = \frac{454}{9} = 50.4 \text{ equiv.}$$

Recalling that 1 faraday of electricity is used per equivalent, 50.4 equivalents times 1 Faraday/equivalent = 50.4 faradays of electricity required to decompose 454g of water by electrolysis.

● **PROBLEM** 573

The flashlight battery is a zinc and manganese dioxide ($MnO_2$) dry cell. The anode consists of a zinc can and the cathode of a carbon rod surrounded by a moist mixture of $MnO_2$, carbon, ammonium chloride ($NH_4Cl_2$), and zinc chloride ($ZnCl_2$). The cathode reaction is written

$$2MnO_2(s) + Zn^{2+} + 2e^- \rightarrow ZnMn_2O_4(s) .$$

If the cathode in a typical zinc and manganese dioxide dry cell contains 4.35g of $MnO_2$, how long can it deliver a steady current of 2.0 milliamperes (mA) before all its chemicals are exhausted?

Solution: The problem is solved by calculating the amount of charge required to exhaust the supply of $MnO_2$ and, from this, determining the lifetime of the battery using the relationship charge (coulombs) = current (A) × time (sec),
or,

$$\text{time (sec)} = \frac{\text{charge (coulombs)}}{\text{current (A)}} .$$

The cathode reaction indicates that two moles of $MnO_2$, 2F $MnO_2$, are consumed for every two moles of electrons present (2F $e^-$). The number of moles of $MnO_2$ present is

$$\frac{\text{mass, } MnO_2}{\text{molecular weight, } MnO_2} = \frac{4.35g}{87g/\text{mole}} = 0.05 \text{ mole } MnO_2 .$$

Hence, it requires 0.05 mole of electrons (or 0.05F) to consume the 0.05 mole of $MnO_2$ in the cathode. Converting Faraday's to coulombs (there are 96,500 coulombs in 1F), 0.05 F is equivalent to

$$0.05F \times 96,500 \text{ coulombs/F} = 4.8 \times 10^3 \text{ coulombs.}$$

The battery is supposed to deliver $2.0 \times 10^{-3}$ amp. Therefore, the lifetime of the battery is

$$\text{time} = \frac{\text{charge}}{\text{current}} = \frac{4.8 \times 10^3 \text{ coulombs}}{2.0 \times 10^{-3} \text{ amp}} = 2.4 \times 10^6 \text{ sec.}$$

Therefore the battery lasts $2.4 \times 10^6$ sec (about 30 days).

You pass a 1.0 amp current through an electrolytic cell for 1.0 hr. There are 96,500 coul in a Faraday (F). Calculate the number of grams of each of the following that would be deposited at the cathode: (1) Ag from an $Ag^+$ solution, (2) Cu from a $Cu^{+2}$ solution and (3) Al from an $Al^{3+}$ solution.

_Solution_: To answer this problem, you can use Faraday's Laws of Electrolysis. Electrolysis is the phenomenon that occurs when electricity is passed through a solution, such that ions are generated and move toward an anode or cathode. The laws are as follows: Masses of substances involved are proportional to the quantity of electricity that flows through the electrolytic cell. Masses of different substances produced during the process are proportional to their equivalent weight. The electrical equivalent is defined as a Faraday (F). It is capable of reducing one equivalent of positive charge, i.e., Avogadro's number of individual unit electric charges. With this information, you can calculate how many Faraday's of electricity were passed through the solution. This tells you the equivalents of the substance that are reduced (recall, one Faraday reduces one equivalent of positive charge). From this number, the weight deposited can be determined.

For all 3 parts, the Faradays generated $= (1.00 \text{ amp})\left(\dfrac{1 \text{ coul/sec.}}{\text{amp}}\right)(1 \text{ hr})$ times $\left(\dfrac{F}{96,500 \text{ coul}}\right) = .0373$ F.

(1) Ag from an $Ag^+$ solution. 1 electron is transferred in $Ag^+ + e^- \rightarrow Ag(s)$. Since one electron is transferred per Ag atom, 1 mole of Ag atoms requires one mole of electrons or 1F. 1 mole of Ag atoms weighs 107.87g (atomic weight - see Periodic Table). You have, though, only .0373 F. Thus, (.0373 mole)(107.87g/mole) = 4.02g of Ag is deposited.

(2) Here, 2 electrons are transferred in $Cu^{2+} + 2e^- \rightarrow Cu(s)$. Thus, 1F is required for ½ of a mole of Cu(s) to be deposited. 1 mole weighs 63.55, ½ of a mole = 63.55/2 = 31.775g. Thus, with .0373 F, you can deposit .0373(31.775) = 1.19g.

(3) Here, 3 electrons are transferred. Thus, 1F can only deposit 1/3 of a mole.

weight deposited $= \left(\dfrac{mw}{3}\right) \times (.0373) = \left(\dfrac{26.98}{3}\right)(.0373) = .335$g.

When a current of 5.36 A is passed for 30 min. through an electrolytic cell containing a solution of a zinc salt, it is found that 3.27g of zinc metal is deposited at the cathode. What is the equivalent weight of zinc in the salt?

_Solution_: To solve this problem one must recall that 1 coulomb = 1A-sec. This means that the number of coulombs passed through the cell is equal to the product of the current ( in amperes ) and the time (in seconds). In our case, since 5.36 A was passed for 30 min. (30 min. × 60 sec/min = 1800 sec), the number of coulombs is 5.36A × 1800 sec = 9650 coulombs. One Faraday (1 F) is equivalent to 96,500 coulombs. Hence, converting 96500 coulombs to Faradays gives

$$\frac{9650 \text{ coulombs}}{96,500 \text{ coulombs}/\digamma} = 0.10 \, \digamma.$$

The equivalent weight of zinc is the weight deposited by $1\digamma$ of charge. Since $0.10 \, \digamma$ deposits 3.27g of zinc, the equivalent weight of zinc is $10 \times 3.27g = 32.7g$ (since $10 \times 0.10 \, \digamma = 1.0 \, \digamma$).

(a)  You desire to plate 5g of Ag from a solution of $Ag^+$. Determine the number of amperes required over a period of 1000 seconds.
(b)  With a current of 0.5 ampere, how much time is required to plate out 10g of gold from an $Au^{3+}$ solution?

Solution:  Both of these problems can be solved using the equation

$$q = I(t_2 - t_1) = n\digamma,$$

where  $q$  is the charge in coulombs,  $I$  the current in amperes, $(t_2 - t_1)$ the difference in time, $n$  the number of equivalents oxidized or reduced, and  $F$  Faraday's constant, 96,490 coulombs.

For part (a), the time span , $t_2 - t_1$ , and the weight of material to be reduced is known.  The number of equivalents to be reduced (or oxidized) equals the weight of material divided by the gram-equivalent weight.
The atomic weight of Ag = 107.88 g/mole and the value of

$$n = \frac{5g \text{ of } Ag^+}{107.88 \text{ g/equiv}} \, .$$

Substituting the known values into the equation

$$I = \frac{n\digamma}{t_2 - t_1} = \frac{(5/107.88)(96,490)}{1000}$$
$$= 4.45 \text{ amps.}$$

For part (b), the time span, $t_2 - t_1$, is unknown.  The atomic weight of Au = 197 g/mole.  Using the same equation

$$n = \frac{10g \, Au^{+++}}{65.7g/equiv.} \, ;$$

$Au^{3+}$ requires 3 electrons to be reduced and the atomic weight must be divided by 3.  Therefore,

$$t_2 - t_1 = \frac{n\digamma}{I} = \frac{\left(\frac{10g}{65.7}\right)(96,490)}{0.5 \text{ coulombs/sec}} = 29,400 \text{ seconds.}$$

A chemist wants to produce chlorine from molten NaCl.  How many grams could be produced if he uses a current of one amp for 5.00 minutes?

Solution: When electrical energy is used to produce a chemical change, it is called electrolysis. Here, electricity is being used to produce chlorine from NaCl. Michael Faraday discovered the laws that illustrate this process quantitatively. You must make use of them to answer this question.

You know that chlorine exists as $Cl_2$ gas. In NaCl, however, it is in the form of the $Cl^-$ ion. Therefore, to obtain $Cl_2$ from NaCl, you must cause the following reaction to occur: $2Cl^- \rightarrow Cl_2 + 2e^-$ A loss of 2 electrons is involved.

Faraday's law states that the amount of gas produced is proportional to the amount of electricity. One mole of electrons is defined as the faraday (F) and equals 96,500 coulombs of electricity. From the equation, $2Cl^- \rightarrow Cl_2 + 2e^-$, you see that 1 mole of $Cl_2$ is produced for every 2 faradays of electricity.

We can determine the amount of $Cl_2$, by relating it to the number of Faradays present. First determine the number of coulombs. You used one ampere, which is just equal to 1 coulomb/sec, for 5 minutes. Therefore, number of coulombs =

(1.00 coulomb/sec)(5.00 minutes)(60 sec/min) = 300 coulombs.

Since there are 96,500 coulombs per faraday, then the number of faradays = $\dfrac{300 \text{ coul}}{96,500 \text{ coul/F}}$ = 0.00311F.

Recalling that 1 mole of $Cl_2$ is obtained for every two faradays, you have

$$\frac{0.00311}{2} = .00155$$

moles of chlorine formed, which translates into 0.111g of $Cl_2$. Atomic weight of $Cl_2$ = 70.9g/mole .

wt. of $Cl_2$ = (mole)(g/mole)

weight of $Cl_2$ = .00155 moles $\times$ 70.9g/mole = 0.111g .

● PROBLEM 578

In the electroplating of nickel from a solution containing $Ni^{+2}$ ion, what will be the weight of the metal deposited on the cathode by a current of 4.02 amperes flowing for 1000 minutes?

Solution: When $Ni^{2+}$ plates out of solution, the reaction can be stated as

$$Ni^{2+} + 2e^- \rightarrow Ni .$$

This means that for every $Ni^{2+}$ ion that deposits on the cathode, 2 electrons must be used. These electrons will be obtained from the current. One mole of electrons is commonly called one Faraday. One can see from the reaction that 2 Faradays are needed to convert one mole of $Ni^{2+}$ to Ni . (1 Faraday = 96,500 coulombs). The MW. of Ni is 58.7, therefore, 2 Faradays will plate out 58.7g of Ni. Since the number of coulombs determines the amount of Ni that plates onto the cathode, the number of coulombs that flow in this current during 1000 minutes needs to be calculated. It is stated in the problem that

a current of 4.02 amperes flows for 1000 minutes. One ampere is defined as one coulomb per second

$$ampere = coulomb/second.$$

To find the total number of coulombs that were transmitted to the $Ni^{2+}$
ions, the amperage of the current must be multiplied by the time the
current was flowing. The time must be converted from minutes to
seconds by use of the conversion factor, 60 secs/1 minute.

$$no. \ of \ coulombs = 4.02 \ coulombs/sec. \times 1000 \ min.$$

$$\times \ 60 \ sec/minute = 24,120 \ coulombs.$$

Since one knows that 2 Faradays or 193,000 coulombs (2 × 96,500 coul)
plate out 1 mole, or 58.7g, of Ni, one can set up the following ratio
to determine the number of grams that 24,120 coulombs plate out. Let
$x$ = the number of grams of $Ni^{2+}$ that 24,120 coulombs will convert
to Ni .

$$\frac{58.7 \ grams}{193,000 \ coulombs} = \frac{x}{24,120 \ coulombs}$$

$$x = \frac{24,120 \times 58.7 \ g}{193,000} = 7.34g. .$$

7.34g of Ni will be deposited.

● **PROBLEM** 579

A meter reads that a battery is putting out .450 amp in the external
circuit of a cell. The cell is involved in the electrolysis of a copper
sulfate solution. During the 30.0 min that the current was allowed to
flow, a total of .30g of copper metal was deposited at the cell's cathode.
Was the meter an accurate measurement of current? (F = 96,500 coul.)

Solution: The key to solving this problem is the determination of the
actual current given the amount of copper deposited in 30.0 minutes.
This can be done by employing Faraday's Laws of Electrolysis. Electrolysis is the phenomenon that occurs when electricity is passed through
a solution, such that ions are generated and moved toward an anode or
cathode. The laws that govern this are as follows: Masses of substances
involved are proportional to the quantity of electricity that flows
through the cell. Masses of different substances produced during the
process are proportional to their equivalent weight. The electrical
equivalent is defined as a Faraday (F). It is capable of reducing one
equivalent of positive charge (Avogadro's number of individual unit
electric charges). The problem can now be solved.

You know how much material is deposited in a given time. Thus, a
certain amount of electricity based on current had to be used. The
reaction that occurred at the cathode was

$$Cu^{2+} + 2e^- \rightarrow Cu.$$

Thus, each mole of copper ion requires 2 moles of electrons or, in other
words, 2 faradays (1 Faraday = 96,500 coul). Thus, a total of 193,000
coul. is required per mole of copper. Current = charge/time. If you
had one mole of copper, the charge would be 193,000 coul. You have only

$$\frac{0.30g(wt. \ of \ Cu)}{63.55(m.w. \ of \ Cu)} = 4.72 \times 10^{-3} \ moles.$$

Thus, charge = $4.72 \times 10^{-3}$ (193,000) = 911.075 coul. To determine the current, time must be considered. The time must be in seconds, since

$$current \ (amps) = \frac{charge \ (couls)}{time \ (secs)} \ .$$

You are given the time as 30.0 min, which equals 1800 secs. Thus,

$$current = \frac{911.075}{1800} = .506 \ amp.$$

The meter said .450, thus, it was inaccurate by

$$\left(\frac{.506 - .450}{.506}\right) \times 100 = 11.1\% \ .$$

● **PROBLEM** 580

Discuss the separate half-reactions and net cell reaction that occur in a flashlight battery. If you have a current of .01 amp, how long can the current flow if the battery contains 25.0g of a paste of $MnO_2$, $NH_4Cl$, and $ZnCl_2$ of which 20% is $MnO_2$ ?

Solution: A standard flashlight dry cell, or Leclanché cell, is composed of a graphite rod in a moist $MnO_2$ - $ZnCl_2$ - $NH_4Cl$ paste, all in a zinc wrapper. The zinc wrapper serves as the electrode for the oxidation half-cell (the anode), while the graphite rod is the electrode for the reduction half-cell (the cathode). Thus, the probable half-reactions are: at the anode:

$$Zn(s) \rightarrow Zn^{2+} + 2e^{-}$$

and at the cathode:

$$2MnO_2(s) + 2H_3O^{+} + 2e^{-} \rightarrow 2MnO(OH) + 2H_2O \ .$$

The net reaction for the overall process is $Zn(s) + 2MnO_2 \rightarrow ZnO \cdot Mn_2O_3$ .

To determine how long the current can flow, calculate the number of moles of $MnO_2$ present and the number of Faraday's required to reduce it per mole. You are told that 20% of 25 grams of paste is $MnO_2$ . Thus, you have (.20)(25g) = 5.0g of $MnO_2$ . The molecular weight of $MnO_2$ = 86.94g . Thus, you have

$$5g/86.94g/mole = .0575 \ moles \ of \ MnO_2 \ .$$

From the net equation, you see that 2 electrons are transferred. A Faraday = 96,500 coul and is capable of reducing one equivalent positive charge, which means Avogardro's number of individual unit charges (one mole of electrons). Since 2 electrons were transferred per 2 moles, it takes 2 Faraday's to reduce 2 moles of $MnO_2$ . According to the net equation 2 moles of $MnO_2$ are reduced for every mole of $ZnO \cdot Mn_2O_3$ obtained. Thus, you can write that the current will last for

$$(.0575 \ moles \ MnO_2)\left(\frac{2F}{2 \ moles \ MnO_2}\right)\left(\frac{96,500 \ coul}{F}\right)\left(\frac{amp}{coul/sec}\right)\left(\frac{1}{.01 \ amp}\right)$$

$$= 154 \ hours.$$

558

For 1000 seconds, a current of 0.0965 amp is passed through a 50 ml. of 0.1 M NaCl. You have only the reduction of $H_2O$ to $H_2$ at the cathode and oxidation of $Cl^-$ to $Cl_2$ at the anode. Determine the average concentration of $OH^-$ in the final solution.

**Solution:** To find out how much $OH^-$ is in the final solution, you must determine the number of faradays that were produced. Then calculate the number of moles of $OH^-$ generated per faraday. This problem involves electrolysis, since electrical energy was used to produce a chemical change.

Let us compute how many faradays were produced by this current, for according to Faraday's law, this will indicate the number of moles of $OH^-$ that will be produced. There are 96,500 coulombs per faraday. If the current is 0.0965 amp, then you have 0.0965 coulombs/sec. You used the current for 1000 secs, which means the total number of coulombs is 0.0965 × 1000. Therefore, you have

$$\frac{(0.0965 \text{ coulomb/sec})(1000 \text{ sec})}{96,500 \text{ coulomb/Faraday}} = .001 \text{ faraday.}$$

Now, to generate $OH^-$ from $H_2O$, the reaction to be followed is $2e^- + 2H_2O \rightarrow H_2(g) + 2OH^-$. A faraday is defined as Avogard's number of electrons. This means, therefore, that 2 faradays liberate two moles of $OH^-$, since two moles of electrons are involved. It follows, therefore, that 0.001 faraday would liberate 0.001 mole of $OH^-$. Assuming the volume of the solution remains 50.00 ml., the final concentration is

$$\frac{0.001 \text{ mol}}{0.05 \text{ liter}} = 0.02 \text{ M}.$$

# ELECTRODE POTENTIAL

Using the tables of standard electrode potentials, arrange the following substances in decreasing order of ability as reducing agents: Al, Co, Ni, Ag, $H_2$, Na.

**Solution:** The tables of standard electrode potentials list substances according to their ability as oxidizing agents. The greater the standard electrode potential, $E^o$, of a substance, the more effective it is as an oxidizing agent and the less effective it is as a reducing agent. From the table of standard electrode potentials,

$$Al^{3+} + 3e^- \rightleftarrows Al(s) \qquad E^o = -1.66v$$

$$Co^{2+} + 2e^- \rightleftarrows Co(s) \qquad E^o = -0.28v$$

$$Ni^{2+} + 2e^- \rightleftarrows Ni(s) \qquad E^o = -0.25v$$

$$Ag^+ + e^- \rightleftarrows Ag(s) \qquad E^o = +0.80v$$

$$2H^+ + 2e^- \rightleftarrows H_2(g) \qquad E^o = 0 \ v$$

$$Na^+ + e^- \rightleftarrows Na(s) \qquad E^o = -2.71v$$

Thus, in increasing ability as oxidizing agents,

$$Na^+ < Al^{3+} < Co^{2+} < Ni^{2+} < H^+ < Ag^+ \ .$$

But if $Na^+$ has a greater tendency to oxidize (gain electrons) than $Al^{3+}$, then, from looking at the reverse reactions, the "conjugate oxidant" Na must have a greater tendency to reduce (lose electrons) than the "conjugate oxidant" Al. Thus, Na, is a better reducing agent than Al, and so on. The substances, in order of decreasing ability as reducing agents, are therefore

$$Na > Al > Co > Ni > H_2 > Ag \ .$$

● **PROBLEM** 583

Using the tables, of standard electrode potentials, list the following ions in order of decreasing ability as oxidizing agents: $Fe^{3+}$, $F_2$, $Pb^{2+}$, $I_2$, $Sn^{4+}$, $O_2$ .

| Half-reaction | $E^o$, V |
|---|---|
| $Li^+ + e^- \rightleftharpoons Li$ | −3.05 |
| $K^+ + e^- \rightleftharpoons K$ | −2.93 |
| $Na^+ + e^- \rightleftharpoons Na$ | −2.71 |
| $Mg^{2+} + 2e^- \rightleftharpoons Mg$ | −2.37 |
| $Al^{3+} + 3e^- \rightleftharpoons Al$ | −1.66 |
| $Mn^{2+} + 2e^- \rightleftharpoons Mn$ | −1.18 |
| $Zn^{2+} + 2e^- \rightleftharpoons Zn$ | −0.76 |
| $Cr^{3+} + 3e^- \rightleftharpoons Cr$ | −0.74 |
| $Fe^{2+} + 2e^- \rightleftharpoons Fe$ | −0.44 |
| $Cd^{2+} + 2e^- \rightleftharpoons Cd$ | −0.40 |
| $Co^{2+} + 2e^- \rightleftharpoons Co$ | −0.28 |
| $Ni^{2+} + 2e^- \rightleftharpoons Ni$ | −0.250 |
| $Sn^{2+} + 2e^- \rightleftharpoons Sn$ | −0.14 |
| $Pb^{2+} + 2e^- \rightleftharpoons Pb$ | −0.13 |
| $Fe^{3+} + 3e^- \rightleftharpoons Fe$ | −0.04 |
| $2 H^+ + 2e^- \rightleftharpoons H_2$ | 0 (definition) |
| $Sn^{4+} + 2e^- \rightleftharpoons Sn^{2+}$ | 0.15 |
| $Cu^{2+} + 2e^- \rightleftharpoons Cu$ | 0.34 |
| $Fe(CN)_6^{3-} + e^- \rightleftharpoons Fe(CN)_6^{4-}$ | 0.46 |
| $I_2 + 2e^- \rightleftharpoons 2 I^-$ | 0.54 |
| $O_2 + 2 H^+ + 2e^- \rightleftharpoons H_2O_2$ | 0.68 |
| $Fe^{3+} + e^- \rightleftharpoons Fe^{2+}$ | 0.77 |
| $Hg_2^{2+} + 2e^- \rightleftharpoons 2 Hg$ | 0.79 |
| $Ag^+ + e^- \rightleftharpoons Ag$ | 0.80 |
| $2 Hg^{2+} + 2e^- \rightleftharpoons Hg_2^{2+}$ | 0.92 |
| $Br_2 + 2e^- \rightleftharpoons 2 Br^-$ | 1.09 |
| $O_2(g) + 4 H^+ + 4e^- \rightleftharpoons 2 H_2O$ | 1.23 |

| Reaction | $E^o$ |
|---|---|
| $Cr_2O_7{}^{2-} + 14\,H^+ + 6e^- \rightleftharpoons 2\,Cr^{3+} + 7\,H_2O$ | 1.33 |
| $Cl_2 + 2e^- \rightleftharpoons 2\,Cl^-$ | 1.36 |
| $MnO_4{}^- + 8\,H^+ + 5e^- \rightleftharpoons Mn^{2+} + 4\,H_2O$ | 1.51 |
| $Ce^{4+} + e^- \rightleftharpoons Ce^{3+}$ | 1.61 |
| $MnO_4{}^- + 4\,H^+ + 3e^- \rightleftharpoons MnO_2(s) + 2\,H_2O$ | 1.68 |
| $H_2O_2 + 2\,H^+ + 2e^- \rightleftharpoons 2\,H_2O$ | 1.77 |
| $O_3 + 2\,H^+ + 2e^- \rightleftharpoons O_2 + H_2O$ | 2.07 |
| $F_2 + 2e^- \rightleftharpoons 2\,F^-$ | 2.87 |

<u>Solution</u>:    The best oxidizing agent will be the one with the greatest
ability to gain electrons (be reduced) and will therefore have the most
positive standard electrode potential, $E^o$ .  From the tables,

$$Fe^{3+} + e^- \rightleftharpoons Fe^{2+} \qquad\qquad E^o = +0.77v$$

$$F_2(g) + 2e^- \rightleftharpoons 2F^- \qquad\qquad E^o = +2.87v$$

$$Pb^{2+} + 2e^- \rightleftharpoons Pb(s) \qquad\qquad E^o = -0.13v$$

$$I_2(s) + 2e^- \rightleftharpoons 2I^- \qquad\qquad E^o = +0.54v$$

$$Sn^{4+} + 2e^- \rightleftharpoons Sn^{2+} \qquad\qquad E^o = +0.15v$$

$$O_2(g) + 2H^+ + 2e^- \rightleftharpoons H_2O_2(\ell) \qquad E^o = +0.68v$$

Thus the substances, in order of decreasing ability as oxidizing agents,
are

$$F_2 > Fe^{3+} > O_2 > I_2 > Sn^{4+} > Pb^{2+} \; .$$

● **PROBLEM** 584

Calculate $\Delta E^o$ for the following cells:  (1)  Cadmium and Hydrogen,
(2)  Silver and Hydrogen and  (3)  Cadmium and silver, using the following data:

| Reaction | $E^o$ volts |
|---|---|
| $Cd \rightarrow Cd^{+2} + 2e^-$ | + .403 |
| $H_2 \rightarrow 2H^+ + 2e^-$ | 0.00 |
| $Ag \rightarrow Ag^+ + e^-$ | - .799 |

<u>Solution</u>:    You are asked to calculate the standard cell potential
$(\Delta E^o)$ for each of the given pairs.  To do this, you must realize that
in such cells, you have 2 half-reactions.  Namely, an oxidation reaction (loss of electrons) and a reduction reaction   ( gain of electrons).  The sum of these half-reactions yields the overall reaction
and the $\Delta E^o$ of the whole cell.  Thus, to find the $\Delta E^o$ for each of
these pairs, you need to know the $E^o$ of the half-reactions.  There
is one other important fact to be kept in mind.

In cells, if the reaction is to proceed spontaneously, $\Delta E^o$ must
have a positive value.  This means, therefore, that you must choose
the half-reactions such that their sum always gives a positive $\Delta E^o$ .
You are told the $E^o$ of the oxidation half-reaction for each element
in each pair.  The reduction half-reaction is the reverse of the
oxidation reaction for each element, with a change in sign of $E^o$ .
For example, if you have  $A \rightarrow A^{+1} + e^-$ with an $E^o = B$ for oxida-

tion, the reduction is $A^{+1} + e^- \rightarrow A$ with a $E^o = -B$ .

With this in mind, the procedure is as follows:
(1) Cadmium (Cd) and hydrogen ($H_2$).

The reaction for this cell must be the sum of the oxidation and reduction such that the $\Delta E^o$ is positive. This can only occur if the anode (oxidation) has the higher oxidation potential. Thus, you calculate $\Delta E^o$ as $+ .403 - (0.000) = +.403v$. Similarly, for (2) and (3), a positive $\Delta E^o$ can only be obtained with the anode having the higher oxidation potential. Thus, for (2), $E^o = (0.000) - (-.799) = +.799v$. For (3), $E^o = (.403) - (-.799) = +1.202v$.

● **PROBLEM** 585

If $I_2$ and $Br_2$ are added to a solution containing $I^-$ and $Br^-$, what reaction will occur if the concentration of each species is 1 m?

*Some half-reactions and their standard reduction potentials.*

| | Half-reaction | Standard reduction potential, V |
|---|---|---|
| | $2e^- + F_2(g) \rightarrow 2F^-$ | +2.87 |
| | $2e^- + Cl_2(g) \rightarrow 2Cl^-$ | +1.36 |
| | $4e^- + 4H_3O^+ + O_2(g) \rightarrow 6H_2O$ | +1.23 |
| | $2e^- + Br_2 \rightarrow 2Br^-$ | +1.09 |
| | $e^- + Ag^+ \rightarrow Ag(s)$ | +0.80 |
| | $2e^- + I_2 \rightarrow 2I^-$ | +0.54 |
| | $2e^- + Cu^{2+} \rightarrow Cu(s)$ | +0.34 |
| | $2e^- + 2H_3O^+ \rightarrow H_2(g) + 2H_2O$ | zero |
| | $2e^- + Fe^{2+} \rightarrow Fe(s)$ | −0.44 |
| | $2e^- + Zn^{2+} \rightarrow Zn(s)$ | −0.76 |
| | $3e^- + Al^{3+} \rightarrow Al(s)$ | −1.66 |
| | $2e^- + Mg^{2+} \rightarrow Mg(s)$ | −2.37 |
| | $e^- + Na^+ \rightarrow Na(s)$ | −2.71 |
| | $e^- + Li^+ \rightarrow Li(s)$ | −3.05 |

*Increasing strength as oxidizing agents* / *Increasing strength as reducing agents*

**Solution**: When $I_2$ and $Br_2$ are added to the solution, a dynamic equilibrium is attained between the non-charged element and its ion form. These reactions, redox reactions, are written as:

$$2e^- + I_2 \rightarrow 2I^- \qquad + 0.54V$$

$$2e^- + Br_2 \rightarrow 2Br^- \qquad + 1.09V$$

The voltages listed indicate the electric potential between two electrodes. In other words, it is a measurement of the work done in moving a unit charge (an electron) from one electrode to the other. For any reaction to occur, the half reactions must be written such that the overall voltage is positive. It is written positively to indicate that the reaction tends to go spontaneously. Any redox reaction for which the overall potential is positive has the tendency to take place as written. Therefore one must pick the larger voltage in the half reaction to be positive and the other negative. Thus,

$$2e^- + Br_2 \rightarrow 2Br^- \qquad\qquad + 1.09V$$

$$\underline{2I^- \rightarrow I_2 + 2e^- \qquad\qquad\qquad - 0.54V}$$

$$2I^- + Br_2 \rightarrow I_2 + 2Br^-$$

The half-reactions are written so that when added together, the electrons cancel out. Because the reaction,

$$2I^- + Br_2 \rightarrow I_2 + 2Br^-,$$

has a positive voltage, it will proceed as written.

● **PROBLEM** 586

For the following oxidation-reduction reaction, (a) write out the two half-reactions and balance the equation, (b) calculate $\Delta E^\circ$, and (c) determine whether the reaction will proceed spontaneously as written;

$$Fe^{2+} + MnO_4^- + H^+ \rightarrow Mn^{2+} + Fe^{3+} + H_2O.$$

(1) $\qquad\qquad Fe^{3+} + e^- \rightleftharpoons Fe^{2+}, \quad E^\circ = 0.77 \text{ eV}$

(2) $\quad MnO_4^- + 8H^+ + 6e^- \rightleftharpoons Mn^{2+} + 4H_2O, \quad E^\circ = 1.51 \text{ eV}.$

<u>Solution</u>: (a) The two half-reactions of an oxidation-reduction reaction are the equations for the oxidation process (loss of electrons) and the reduction process (gain of electrons). In the overall reaction, you begin with $Fe^{2+}$ and end up with $Fe^{3+}$. It had to lose an electron to accomplish this. Thus you have oxidation:

$$Fe^{2+} \rightarrow Fe^{3+} + e^-.$$

Notice: this is the reverse of the reaction given with $E^\circ = .77$. As such, the oxidation reaction in this problem has $E^\circ = - .77$ eV. The reduction must be

$$MnO_4^- + 8H^+ + 5e^- \rightarrow Mn^{2+} + 4H_2O,$$

since in the overall reaction, you see $MnO_4^- + H^+$ go to $Mn^{2+}$, which suggests a gain of electrons. This is the same reaction as the one given in the problem, $E^\circ = 1.51$ eV. To balance the overall reaction, add the oxidation reaction to the reduction reaction, such that all electron charges disappear. If you multiply the oxidation reaction by 5, you obtain:

$$5Fe^{2+} \longrightarrow 5Fe^{3+} + 5e^-$$

$$MnO_4^- + 8H^+ + 5e^- \longrightarrow Mn^{2+} + 4H_2O$$

$$\overline{5Fe^{2+} + MnO_4^- + 8H^+ \longrightarrow 5Fe^{3+} + Mn^{2+} + 4H_2O}$$

Notice: Since both equations contained $5e^-$ on different sides, they cancelled out. This explains why the oxidation reaction is multiplied by five. Thus you have written the balanced equation.

(b) The $\Delta E^\circ$ for the overall reaction is the sum of the $E^\circ$ for the half-reactions, i.e., $\Delta E^\circ = E_{red} + E_{oxid}$. You know that $E_{red}$ and $E_{oxid}$; $\Delta E^\circ = 1.51 - .77 = 0.74$ eV.

(c) A reaction will only proceed spontaneously when $\Delta E^\circ$ = a positive value. You calculated a positive $\Delta E^\circ$, which means the reaction proceeds spontaneously.


# ELECTROCHEMICAL CELL REACTIONS

● **PROBLEM** 587

For the following voltaic cell, write the half-reactions, designating which is oxidation and which is reduction. Write the cell reaction and calculate the voltage of the cell made from standard electrodes. The cell is  Co; $Co^{+2}$ ‖ $Ni^{+2}$ ; Ni .

<u>Solution</u>:   The cell reaction is the algebraic sum of the reactions that take place at the electrodes. Every cell has 2 electrodes an anode and a cathode. Oxidation, which is the loss of electrons, occurs at the anode. Reduction, which is the gain of electrons, takes place at the cathode.
The cell is always written as solid; ion in solution ‖ ion in solution; solid (anode)                    (cathode)

Oxidation and reduction are the half reactions that take place in the cell. For this cell, they are

$$Co \rightarrow Co^{+2} + 2e^- \text{ (oxidation at anode)}$$

$$Ni^{+2} + 2e^- \rightarrow Ni \text{ (reduction at cathode)}$$

Sum: $Ni^{+2} + Co \rightarrow Ni + Co^{+2}$ (Cell reaction) .

Since  Co  is losing electrons, it provides the oxidation reaction and  $Ni^{+2}$ , gaining these electrons, takes part in the reduction reaction.
The voltage of a cell is the sum of the oxidation and reduction potentials in units of volts and is designated by  $E^\circ$  (under standard conditions) .

$$E^\circ_{cell} = E^\circ_{oxidation} + E^\circ_{reduction} .$$

The voltages of half cell reactions are usually given as the reduction potentials. The oxidation potential is opposite in sign to the reduction potential. The potentials can be obtained from a table of standard reduction potentials.

For
$$Ni^{+2} + 2e^- \rightarrow Ni \text{ , the potential is } -.25 \text{ v, } E^o_{red} = -.25v.$$
$$Co^{+2} + 2e^- \rightarrow Co, \text{ the potential is } -.277 \text{ v.}$$

Since $Co \rightarrow Co^{+2} + 2e^-$ is the oxidation reaction, $E^o_{ox}$ equals the negative of $-.277$ v, or $E^o_{ox} = .277$ v.

Substituting these values into the equation $E^o = E^o_{oxid} + E^o_{red}$, one obtains
$$E^o = +.277 + (-.25) = .027 \text{ v .}$$
Since $E^o$ is positive, the reaction proceeds spontaneously and can be used to supply current.

● **PROBLEM** 588

For the following voltaic cell, write the half reactions, designating which is oxidation and which reduction. Write the cell reaction and calculate the voltage ($E^o$) of the cell from the given electrodes. The cell is
$$Cu; Cu^{+2} \parallel Ag^{+2} \text{ ; Ag .}$$

Solution: In a voltaic cell, the flow of electrons creates a current. Their flow is regulated by 2 types of reactions occur concurrently, oxidation and reduction. Oxidation is a process where electrons are lost and reduction where electrons are gained. The equation for these are the half-reactions. From the cell diagram, the direction of the reaction is always left to right.

$$Cu \rightarrow Cu^{+2} + 2e^- \qquad \qquad \text{oxidation}$$
$$Ag^{+2} + 2e^- \rightarrow Ag \qquad \qquad \text{reduction .}$$

Therefore, the combined cell reaction is
$$Cu + Ag^{+2} \rightarrow Cu^{+2} + Ag .$$

To calculate the total $E^o$, look up the value for the $E^o$ of both half-reactions as reductions. To obtain $E^o$ for oxidation, reverse the sign of the reduction $E_o$. Then, substitute into $E^o_{cell} = E^o_{red} + E^o_{ox}$. If you do this, you find
$$E^o_{cell} = -(E^o_{red} Cu) + E^o_{red} Ag^{+2}$$
$$= -.34 + .80$$
$$= .46 \text{ volt .}$$

● **PROBLEM** 589

Calculate $E^o$ for the cell in which the following reaction occurs.
$$2Al + 3NiCl_2 \rightarrow 2AlCl_3 + 3Ni.$$
First, indicate the direction of the electron flow.

Solution: This problem is solved by finding the oxidation and reduction reactions which occur. Knowing this, we can solve both questions.

First, we must determine the metals involved. They are Al and

Ni, for which the half-reactions are

$$Al \rightarrow Al^{+3} + 3r^- \qquad E^o_{oxid} = 1.66$$
$$Ni^{+2} + 2e^- \rightarrow Ni \qquad E^o_{red} = -.25$$

Since Al is losing electrons (oxidation), and Ni is gaining electrons (reduction), the electron flow is from Al to Ni

$E^o$ , the standard electrode potential is the sum of the $E^o$'s for the half-reactions listed above

$$E^o = E^o_{ox} + E^o_{red}$$
$$= 1.66 + (-.25)$$
$$= 1.41 \text{ volts} .$$

● **PROBLEM** 590

Given the following standard electrode potentials at $25^o$ C: $Sn^{4+} + 2e^- \rightarrow Sn^{2+}$, $E^o = 0.15$ ev and $Fe^{3+} + e^- \rightarrow Fe^{2+}$, $E^o = +0.77$ ev will the reaction $Sn^{2+} + 2Fe^{3+} \rightleftarrows Sn^{4+} + 2Fe^{2+}$ proceed spontaneously?

Solution; A reaction will proceed spontaneously only if it has a positive $\Delta E^o$ as written. Thus, to answer this question, you want to calculate the $\Delta E^o$ and see whether it is positive. This can be done by considering its half-reactions.— The sum of the electrode potentials for these reactions will be the $\Delta E^o$.

The overall reaction is
$$Sn^{2+} + 2Fe^{3+} \rightleftarrows Sn^{4+} + 2Fe^{2+} .$$

The half-reactions are, oxidation (loss of electrons) and reduction (gain of electrons) reactions. You have then,

$$\text{oxidation: } Sn^{2+} \rightarrow Sn^{4+} + 2e^- .$$

This, has the reverse of the standard potential of the reaction given. Thus, its $E^o = -(0.15) = -.15$ ev.

$$\text{Reduction: } Fe^{3+} + e^- \rightarrow Fe^{2+} .$$

This has the same standard potential as in the given reaction. Thus, its $E^o = 0.77$ ev. Adding these two equations together, one obtains the desired net reaction.

$$S_n{}^{2+} \rightarrow Sn^{4+} + 2e^-$$
$$\underline{2Fe^{3+} + 2e^- \rightarrow 2Fe^{2+}}$$
$$2Fe^{3+} + Sn^{2+} \rightarrow Sn^{4+} + 2Fe^{2+}$$

The $\Delta E^o = E^o_{oxid} + E^o_{red} = -.15 + .77 = +.62$ volt. The fact that it is positive means the reaction proceeds spontaneously.

● **PROBLEM** 591

Construct a galvanic cell based on the reaction
$$3Fe(s) + 2Au^{+3} \rightarrow 3Fe^{+2} + 2Au(s)$$
with $\Delta H$ negative.

566

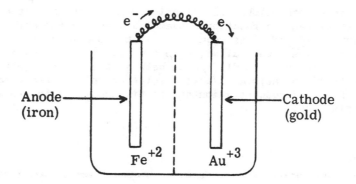

Anode
(iron)

Cathode
(gold)

$Fe^{+2}$

$Au^{+3}$

Solution: A galvanic cell may be defined as one in which spontaneous chemical reactions occur at each of the electrodes of the cell to produce an electric current. Thus, to construct this cell, you need to know those reactions that occur at the electrodes.

The two electrodes are the anode and cathode. At the anode oxidation occurs, which means you have the loss of electrons. At the cathode, reduction occurs, which means you have the gain of electrons. The oxidation and reduction reactions are each called half-reactions. These are the reactions that occur at the electrodes. From the given reaction, it becomes apparent that oxidation is given by $Fe(s) \rightarrow Fe^{+2} + 2e^-$ and reduction by $Au^{+3} + 3e^- \rightarrow Au(s)$. Thus, the anode half-cell would consist of an iron electrode in a solution containing $Fe^{+2}$ ions, while the cathode half-cell would consist of a gold electrode in a solution containing $Au^{+3}$ ions.

A compound such as $FeSO_4$ could serve as a source of $Fe^{+2}$ ions, and $AuCl_3$ as a source of $Au^{+3}$ ions.

The cell might appear as in the figure shown .

# NERNST EQUATION

● PROBLEM 592

Show how Le Chatelier's principle for oxidation-reduction reactions corresponds to the Nernst equation.

Solution: Le Chatelier's principle predicts that increasing the concentration of a reactant favors its tendency to react and decreasing the concentration of a reactant diminishes its tendency to react. Similarly, decreasing the concentration of a product favors the tendency towards formation of that product. The Nernst equation relates E, the potential for a reaction or half-reaction at non-standard conditions, to $E^\circ$, the standard potential for that reaction or half-reaction at unit activities. Namely,

$$E = E^\circ - \frac{.0591}{n} \log Q,$$

where n = the number of electrons transferred in the reaction and Q is the equilibrium expression. Q gives the ratio of products to reactants, each raised to the power of their coefficients in the chemical equation.

Notice, that both Le Chatelier's principle and the Nernst equation relate what happens to reaction tendencies when concentrations are changed. For example, in the Nernst equation, if the concentrations are such that yield a positive potential, E, then the reaction is favorable in the direction written. The difference between the Nernst equation and Le Chatelier's principle is that the latter relates reaction tendencies with a change of concentration qualitatively, while the former discusses it quantitatively.

● **PROBLEM** 593

A cell possesses two electrodes. Both half-cells are .01M $MnO_4^-$ ion. One cell is .01M $H_3O^+$ ion, while the other has a $H_3O^+$ concentration of .10M. The electrode reaction for the reduction half-cell may be written:

$$MnO_4^- + 4H^+ + 3e^- \rightarrow MnO_2 + 2H_2O .$$

The oxidation half-cell is the reverse of this reaction. 1) Write the net equation for the spontaneous cell process taking place; 2) Find $\Delta E$ for the reaction; 3) Find the value of the equilibrium constant.

Solution: 1) The net reaction, in such a situation, is the sum of the balanced half-reactions, i.e., oxidation reaction plus reduction reaction. Since, you are given both reactions, add the equations together to find the overall reaction. You have, with concentrations included,

ox: $MnO_2 + 2H_2O \rightarrow MnO_4^- (.01M) + 4H^+(.01M) + 3e^-$

red: $MnO_4^- (.01M) + 4H^+ (.10M) + 3e^- \rightarrow MnO_2 + 2H_2O$

Net Reaction: $4H^+(.10M) \rightarrow 4H^+(.01M)$ .

Notice: all species cancelled out, except $H^+$ (actually $H_3O^+$). This is the net equation for the spontaneous reaction taking place.

2) To find $\Delta E$ for the reaction, use the Nernst equation, which states

$$\Delta E = \Delta E^o - \frac{.059}{n} \log K$$

for a temperature at $25^oC$, where $\Delta E$ = potential for cells under other than standard conditions, $\Delta E^o$ = standard cell potential, n = number of electrons transferred and K = equilibrium concentration expression. n = 3, since from either the oxidation or reduction reaction, 3 electrons are being transferred. $\Delta E^o$ = 0, since $\Delta E^o = E^o_{prod} - E^o_{reactants}$ and both the product and reactant are the same species. K is defined as the ratio of products to reactants, each raised to the power of their coefficients in the net equation. Substituting these values into the Nernst equation, you have

$$\Delta E = 0 - \frac{.059}{3} \log \frac{(.01)^4}{(.10)^4} = 0 - .0197(\log 10^{-4})$$

$$= (-.0197)(-4) = 0.079 \text{ volt.}$$

3) To find the value of the equilibrium constant, note that there exists a relationship between $\Delta E^o$ and the constant at $25^o$c. Namely, $\Delta E^o = \frac{.059}{n} \log K$. From part 2, you found $\Delta E^o = 0$ . Thus, K = unity (one), since this is the only value that permits log K = 0, which, then, allows $\Delta E^o = 0$, as it must.

You have the following cell process:

$$Fe(s) + Co^{2+}(.5M) \rightarrow Fe^{2+}(1.0M) + Co(s) \ .$$

$Fe^{2+} + 2e^- \rightleftarrows Fe(s)$ with $E^\circ = -.44e$ and $Co^{2+} + 2e^- \rightleftarrows Co(s)$ with $E^\circ = -.28$, find the standard cell potential $\Delta E^\circ$, the cell potential $\Delta E$ and the concentration ratio at which the potential generated by the cell is exactly zero.

Solution: Assume that the reaction proceeds spontaneously. This means, therefore, that the reaction must have a positive value for $\Delta E^\circ$. With this in mind, you proceed as follows: You are given two half-reactions:

$$Fe^{2+} + 2e^- \rightleftarrows Fe(s) \qquad E^\circ = -.44 \text{ ev}$$
$$Co^{2+} + 2e^- \rightleftarrows Co(s) \qquad E^\circ = -.28 \text{ ev}$$

Both reactions represent reduction (gain of electrons). But, the overall reaction in a cell is a combination of both a reduction and an oxidation reaction. Thus, you must reverse one, keeping in mind that the $\Delta E^\circ$ must be a positive value. Recall, also, that $\Delta E^\circ = E^\circ_{oxid} + E^\circ_{red}$. You can write

$$Fe(s) \rightleftarrows 2e^- + Fe^{2+} \ ; \qquad E^\circ = -(-.44) = .44 \text{ ev}$$
$$\underline{Co^{2+} + 2e^- \rightleftarrows Co(s) \ : \qquad E^\circ = -.28 \text{ ev}}$$
$$Fe(s) + Co^{2+} \rightleftarrows Fe^{2+} + Co(s) \text{ (overall reaction)}$$

with $\Delta E^\circ = .44 - .28 = .16$ ev. Notice: By reversing the Fe reaction and combining it with the other, you obtained the overall reaction with a $\Delta E^\circ = .16$ ev, a positive value, which indicates that the reaction proceeds spontaneously. To find $\Delta E$, use the Nernst equation, which states

$$\Delta E = \Delta E^\circ - \frac{.059}{n} \log K,$$

where $n =$ number of electrons transferred and $K =$ equilibrium constant of reaction. In this problem, the number of electrons transferred is 2, so that $n = 2$.

$$K = \frac{[Fe^{2+}]}{[Co^{2+}]} \ , \text{ i.e., the ratio of the concentrations}$$

of products to reactants, each raised to the power of its respective coefficient in the chemical equation. Note, Co(s) and Fe(s) are omitted, because they are solids and, thus, considered constants themselves. You are given $[Fe^{2+}]$ and $[Co^{2+}]$ and you have calculated $\Delta E^\circ$. Therefore,

$$\Delta E = \Delta E^\circ - \frac{.059}{n} \log \frac{[Fe^{2+}]}{[Co^{2+}]} = .16 - \frac{.059}{2} \log \frac{(1.0)}{(.5)} = .16 - .0295 \log 2$$

$$= .16 - .01 = .15 \text{ ev.}$$

To find the concentration ratio, K, when $\Delta E = 0$, use the Nernst equation;

$$\Delta E = \Delta E^\circ - \frac{.059}{n} \log K.$$

$\Delta E = 0$, $\Delta E^\circ = .16$, $n = 2$ and $K = \frac{[Fe^{2+}]}{[Co^{2+}]}$ , so that

$$0 = .16 - \frac{.059}{2} \log \frac{[Fe^{2+}]}{[Co^{2+}]} \quad \text{or}$$

$$0 = .16 - .0295 \log \frac{[Fe^{2+}]}{[Co^{2+}]} .$$

$$\frac{[Fe^{2+}]}{[Co^{2+}]} = \text{antilog} \left(\frac{.16}{.0295}\right) = \text{antilog } 5.4 = 2.5 \times 10^5 \quad \text{thus,}$$

$$\frac{[Fe^{2+}]}{[Co^{2+}]} = 2.5 \times 10^5, \text{ when } \Delta E = 0 .$$

● **PROBLEM** 595

You are given the following Daniell cell:

$$Zn, \, Zn^{+2}(.50m) \| Cu^{2+}(0.20m), \, Cu .$$

The concentrations of the ions are given in parentheses. Find the E for this cell at $25°C$. Assume the following standard oxidation potentials:

$$Zn \rightarrow Zn^{+2} + 2e^- \qquad E^o = .763$$
$$Cu \rightarrow Cu^{+2} + 2e^- \qquad E^o = -.337 .$$

<u>Solution</u>: First of all, it is important to note that a Daniell cell is a typical galvanic cell, which means that spontaneous chemical reactions occur at each electrode of the cell to produce an electric current. To calculate E, the potential at other than standard conditions ($25°C$ and 1 molar concentrations are the standard conditions), you can employ the Nernst equation. It is stated

$$E = E^o - \frac{.0592}{n} \log K,$$

where $E^o$ = standard electrode potential, n = number of electrons gained or lost as shown in the overall reaction and K is the ratio of concentration of ions on right side of equation to those on left side. The overall equation for this cell is given by

$$Zn + Cu^{+2} \rightarrow Zn^{+2} + Cu .$$

Thus, for this problem,

$$K = \frac{[Zn^{+2}]}{[Cu^{+2}]} .$$

First, let us try to find the values of these parameters of the Nernst equation, so that E can be found. $E^o = E^o_{\text{oxidation}} + E^o_{\text{reduction}}$. For the reaction to be spontaneous, $E^o$ must equal a positive value. Thus, you want to choose the oxidation half-reaction (loss of electrons) and reduction half-reaction (gain of electrons) carefully, so that their sum, $E^o$, is a positive value.

You are given the standard oxidation potentials, $E^o_{ox}$, of all species in question. The reduction will be the reverse of this reaction, with a reduction potential, $E^o_{red}$, **equals** to the negative of the $E^o_{ox}$ given. The electrode potential of the overall reaction is the sum of these two half-reactions. For the $E^o$ (overall) to be positive, it follows that the oxidation is

$$Zn \rightarrow Zn^{+2} + 2e^- \qquad E^o = .763$$

and the reduction is
$$Cu^{2+} + 2e^- \rightarrow Cu \qquad E^o = -(-.337)$$

overall reaction
$$Zn + Cu^{2+} \rightarrow Zn^{+2} + Cu \quad \text{with}$$
$$E^o = .763 + [-(-.337)] = 1.1v \ .$$

From the overall reaction, recall
$$K = \frac{[Zn^{+2}]}{[Cu^{+2}]} \ .$$

(The concentrations of solids are not included in a dissociation constant). You are told $Zn^{+2} = .5m$ and $Cu^{+2} = .2m$, so that $K = \frac{.50}{.20}$ . In the overall reaction, 2 electrons are lost and gained, which means $n = 2$. If you substitute these values in the Nernst equation, you obtain
$$E = 1.1 - \frac{.-592}{2} \log \frac{.50}{.20} \ .$$

Solving, $E = 1.088v$ for this Daniell cell.

● **PROBLEM** 596

Calculate the voltage of the cell $Fe; Fe^{+2} \| H^+ ; H_2$ if the iron half cell is at standard conditions but the $H^+$ ion concentrations is .001M .

Solution: The voltage (E) of a cell is found using the Nernst Equation because it involves the use of concentration factors. It is stated
$$E = E^o - \frac{RT}{NF} \ln Q,$$
where R is 8.314 joules per degree, F is 96,500 coulombs, N is the number of moles of electrons transferred and Q is the concentration term. $T = 25^o$ C, by definition of standard conditions, in this equation (or $298^o$ K).

But to solve the problem we must first obtain $E^o$ . This is done by writing down the appropriate half-reactions. Oxidation is the loss of electrons and reduction is the gain of electrons.

| Reaction | Type | $E^o_{red}$ |
|---|---|---|
| $Fe \rightarrow Fe^{+2} + 2e^-$ | Oxidation | + .44 |
| $2H^+ + 2e^- \rightarrow H_2$ | Reduction | 0 |

Next, take the algebraic sum of the $E^o_{red}$ and $E^o_{oxid}$ , which gives $E^o$ .
$$E^o = + .44 + 0 = + .44 \ .$$

Now set up the concentration term.
$$\ln Q = \ln \frac{[H^+]^2}{[Fe^{+2}]} = \ln \frac{[.001]^2}{[1]}$$

Standard conditions always means a concentration of 1 M. Substituting all these terms into the Nernst equation, one calculates E to be
$$E = +.44 - \frac{.059}{2} \log \frac{10^{-6}}{1} = 0.617v \quad \text{or} \quad E = +.44 - \frac{0.0257}{2} \ln \frac{10^{-6}}{1} = 0.617v$$

Calculate the voltage (E) of a cell with $E^o = 1.1$ volts, if the copper half-cell is at standard conditions but the zinc ion concentration is only .001 molar. Temperature is $25°$ C . The overall reaction is

$$Zn + Cu^{+2} \rightarrow Cu + Zn^{+2} .$$

**Solution**: To solve this problem requires a relation between the cell potential (voltage), E, and the standard cell potential, $E^o$ . The equation used is the Nernst equation, which states

$$E = E^o - \frac{RT}{nF} \ln(Q),$$

where n is the number of mole of electrons transferred, R is the constant 8.314 joules per degree, F is 96,500 coulombs, and Q is the equilibrium constant expression. In other words,

$$Q = \frac{[product]}{[reactant]}$$

You are told the concentration of $Zn^{+2}$ is .001.

$$= \frac{[Zn^{+2}]}{[Cu^{+2}]} = \frac{.001}{1}$$

Standard conditions indicate that the concentration of the substance is 1M and at a temperature of $25°$ C . Thus $[Cu^{+2}] = 1$ and therefore, from the reaction equation, it can be seen that 2 electrons are transferred. Thus, n = 2.

Substituting these values into the Nernst equation, one obtains

$$E = 1.1 - \frac{8.314(298)}{2(96,500)} \log \frac{.001}{1}$$

$$= 1.1 - .0295(-3)$$
$$= 1.19 \text{ volts.}$$

Given $Zn \rightarrow Zn^{+2} + 2e^-$ with $E^o = + .763$, calculate E for a Zn electrode in which $Zn^{+2} = .025M$.

**Solution**: This problem calls for the use of the Nernst equation, which relates the effect of concentration of ions in a cell on the voltage. It is stated

$$E = E^o - \frac{.0592}{n} \log K , \text{ at } 25° C$$

where E = potential under conditions other than standard, $E^o$ = standard electrode potential, n = number of electrons gained or lost in the reaction, and K = ratio of concentration of products to reactants. You are given $E^o$ and n = 2, since Zn loses 2 electrons to become $Zn^{+2}$ . For this problem,

$$K = \frac{[Zn^{+2}]}{[Zn]} .$$

You are given that $[Zn^{+2}] = .025M$.

To find [Zn] , you have to know the meaning of standard electrode potential. Standard conditions are defined as $25^\circ C$ and 1 molar in concentration. Thus, if you assume that Zn is in the standard state, then [Zn] = 1M . To find E, substitute these values into the Nernst equation.

$$E = .763 - \frac{.0592}{2} \log \frac{.025}{1}$$

$$= .763 - [(.0296)(-1.602)]$$

$$= .810 \text{ V.}$$

• **PROBLEM** 599

The standard $E^o$ for $\frac{1}{2}F_2(g) + e^- \rightarrow F^-(aq)$, which is +2.87V, applies when the flouride ion concentration is 1M . What would the corresponding E be in 1M $H_3O^+$, i.e., for the electrode reaction $\frac{1}{2}F_2(g) + e^- + H_3O^+ \rightarrow HF(aq) + H_2O$? (The $K_{diss}$ of HF is $6.7 \times 10^{-4}$).

Solution: One can solve for E for the electrode reaction by using the Nernst equation. This is stated

$$E = E^o - \frac{0.0591}{n} \log Q \quad,$$

where E is the actual electrode potential for the half-reaction, $E^o$ is the standard potential for the half-reaction, n is the number of electrons transferred in the reaction, Q is the dissociation constant. Thus, one must find the concentration of $F^-$, in order to evaluate Q, before solving for $E^o$ . The dissociation constant is equal to

$$\frac{[H_3O^+][F^-]}{[HF]} \quad.$$

One is given that the dissociation constant ($K_{diss}$) for 1M HF is $6.7 \times 10^{-4}$. Thus,

$$K_{diss} = \frac{[H_3O^+][F^-]}{[HF]} = 6.7 \times 10^{-4} = Q \quad.$$

Solving for the E of the electrode reaction:

For the half-reaction $\frac{1}{2}F_2(g) + e^- \rightarrow F^-(aq)$, one electron is transferred, thus n = 1. One is given that $E^o = 2.87V$ and one has determined Q. Thus,

$$E = 2.87V - \frac{0.0591}{1} \log(6.7 \times 10^{-4})$$

$$E = 2.87V - 0.0591 \times (-3.17)$$

$$E = 2.87V + .188 = +3.06V \quad.$$

• **PROBLEM** 600

Knowing that the $K_{sp}$ for AgCl is $1.8 \times 10^{-10}$ , calculate E, the electrode potential, for a silver-silver chloride electrode immersed in 1M KCl. The standard oxidation potential for the (Ag,Ag$^+$)

half reaction is  -0.799 volts.

**Solution**: The silver-silver chloride electrode process is a special case of the  $(Ag,Ag^+)$  couple, except that silver ions collect as solid $AgCl$  on the electrode itself.  However, solid $AgCl$  has some $Ag^+$ in equilibrium with it in solution.  This  $[Ag^+]$  can be calculated from the  $K_{sp}$  equation:

$$[Ag^+][Cl^-] = K_{sp} \quad \text{or} \quad [Ag^+] = \frac{K_{sp}}{[Cl^-]}$$

$$[Ag^+] = \frac{1.8 \times 10^{-10}}{1} = 1.8 \times 10^{-10} \quad .$$

This value for  $[Ag^+]$  can be inserted into the Nernst equation for the $(Ag,Ag^+)$  half-reaction

$$Ag \rightarrow Ag^+ + e^- \quad E^° = -0.799$$

The Nernst equation is

$$E = E^° - \frac{0.05916 \log Q}{n} \quad , \text{ where } E^° = \text{standard}$$

electrode potential ($25°C$ and $1M$), n = number of electrons transferred, and  Q = the mass-action expression or equilibrium constant.  For this reaction, one sees that one electron is transferred.  Hence, n = 1. Also, $Q = \frac{[Ag^+]}{[Ag]}$ , but, since  Ag  is solid, its concentration is assumed to be constant; it can be removed from the expression, so that  $Q = [Ag^+]$. Thus,

$$E = E^° - \frac{.05916 \log[Ag^+]}{1} = -0.799 - 0.05916 \log(1.8 \times 10^{-10})$$

$$= -0.799 + 0.576 = -0.223 \text{ volts} \quad .$$

● **PROBLEM** 601

What will be the relative concentrations of  $Zn^{+2}$  and  $Cu^{+2}$  when the cell  $Zn; Zn^{+2} \| Cu^{+2}; Cu$  has completely run down?  ($E^° = 1.10$ volts).

**Solution**: The key phrase to understand before solving this problem is "has completely run down".  That means there is no apparent current, leading us to assume a value of zero for the electrode potential, E.

Next, one should realize that whenever concentrations are involved one should apply the Nernst Equation, which states

$$E = E^° - \frac{RT}{nF} \ln Q \quad ,$$

where  Q  is the equilibrium constant expression, R is the gas constant (8.314 joules per degree), F is 96,500 coulombs, n is the number of moles of electrons transferred and  T  is $25°c$  (or  $298°K$).

Since the overall cell reaction is

$$Zn + Cu^{+2} \rightarrow Zn^{+2} + Cu \quad ,$$

2 moles of electrons are transferred and  $\log Q$  equals $\log \frac{[Zn^{+2}]}{[Cu^{+2}]}$  .

Remember that solids are omitted from all concentration equations.
This yields (via substitution):

$$E = E^° - \frac{RT}{nF} \ln Q$$

and

$$0 = 1.1 \text{ volts} - \frac{.059}{2} \log \frac{[Zn^{+2}]}{[Cu^{+2}]} \ ,$$

which means $\log \dfrac{[Zn^{+2}]}{[Cu^{+2}]} = 37.3$ . Solving,

$$\frac{Zn^{+2}}{Cu^{+2}} = 2 \times 10^{37} \ .$$

This means that the concentration of $Zn^{+2}$ is far greater than $Cu^{+2}$.

● **PROBLEM 602**

For a cell reaction of the type $T + R^{+2} \rightleftarrows R + T^{+2}$, calculate the value that $E^{o}$ must have so that the equilibrium concentration of $T^{+2}$ is a thousand times the $R^{+2}$ concentration.

Solution:    In this question, we are asked to obtain the standard cell potential ($E^{o}$), not the actual cell potential (E). Since there is no net electron flow at equilibrium, E = 0.

There exists an expression which relates $E^{o}$, E and concentrations. This expression, called the Nernst equation, states $E = E^{o} - .059/n$ log Q, at a temperature of $25^{o}c$  where  n = number of electrons transferred and  Q = mass action expression.

From the chemical reaction, you see  T  goes from an oxidation state of  0  to  +2 . It lost 2 electrons. Thus, $R^{+2}$ gained 2 electrons. Therefore, the number of electrons transferred, n, equals 2.  From the same chemical reaction, Q  becomes

$$\frac{[T^{+2}]}{[R^{+2}]} \ .$$

You are asked to determine $[T^{+2}]$ to be  1000 times  $[R^{+2}]$, which means Q = 1000.  You already determined that  E = 0.  To find  $E^{o}$, substitute these values in the Nernst equation.  You obtain

$$0 = E^{o} - \frac{0.059}{2} \ln 1000.$$

Solving, $E^{o}$ = .088 volt.

● **PROBLEM  603**

You are given the reaction:

$$H_2(g) + 2AgCl(s) + 2H_2O \rightarrow 2Ag(s) + 2H_3O^{+}(Ag) + 2Cl^{-}(Ag) \ .$$

It has been determined that at  $25^{o}C$  the standard free energy of formation of  AgCl(s)  is  -109.7 kJ/mole, that of  $H_2O(\ell)$  is  -237.2 kJ/mole, and that of  $(H_3O^{+} + Cl^{-})(ag)$  is  -368.4 KJ/mole.  Now assuming the reaction is run at  $25^{o}c$  and  1 atm of pressure in a cell in which the  $H_2(g)$  activity is unity and  $H_3O^{+}(ag)$  and  $Cl^{-}(ag)$  activities are .01, what will be the cell voltage?

<u>Solution</u>:   The answer to this question requires the use of the Nernst equation and the relationship between overall free energy change ($\Delta G$) and the cell voltage  E.

The Nernst equation implies that the voltage of a cell depends on the concentrations or activity of the species involved.  It states

$$E = E^\circ - \frac{RT}{nF} \ln Q \ ,$$

where  E  is the potential for a reaction at nonstandard conditions, $E^\circ$ = the standard potential, n = number of faradays of electricity, F = the value of a faraday, (for conversion to kilojoules per mole is 96.49 kJ $V^{-1}$ equiv$^{-1}$), T = temperature in kelvin (celsius plus 273$^\circ$), R = universal gas constant, (8.31 J mol$^{-1}$ deg$^{-1}$), and  Q  is the mass-action expression.  The relationship between the overall free energy and cell voltage is  $\Delta G = -nFE$ , where  $\Delta G$  is the free energy.  If all the species are in their standard states, then you can say  $\Delta G^\circ = -nFE^\circ$. The sum of  $\Delta G^\circ$  must be the free energy contribution from each component. Therefore,

$$\Delta G^\circ = 2\Delta G^\circ_{H_3O^+Cl^-} - 2\Delta G^\circ_{AgCl} - 2\Delta G^\circ_{H_2O}$$

$$= 2(-368.4) - 2(-109.7) - 2(-237.2)kJ$$

$$= -43.0 \text{ kJ per mole of reaction.}$$

$$E^\circ = \frac{-\Delta G^\circ}{nF} = -\left[\frac{-43.0 \text{ kJ}}{2(96.49 \text{ kJ/v})}\right] = .223v \ .$$

Since the activities of solid  Ag, of solid  AgCl  and of  $H_2O$  are unity, they can be eliminated from the mass-action expression:

$$\frac{[H_3O^+]^2[Cl^-]^2}{[H_2]} \ .$$

Thus,

$$E = E^\circ - \frac{RT}{nF} \ln \frac{[H_3O^+]^2[Cl^-]^2}{[H_2]} \ .$$

Therefore,

$$E = .223 - \frac{(8.31)(298)(2.303)}{2(96490)} \log \frac{(.0100)^2(.0100)^2}{1.00}$$

$$E = .223 + .236 = .459v.$$

● **PROBLEM** 604

You have two half-cells coupled together to form a concentration cell: $H_2$ (1 atm) $\rightarrow 2H^+$(.1M) + 2e$^-$  and $H_2$ (1 atm) $\rightarrow 2H^+$(1.0M) + 2e$^-$. Calculate the free energy and electrical work obtainable. F = 96,500 coulombs.

<u>Solution</u>:  To find the free energy and electrical work obtainable, calculate the maximum cell voltage in this cell, since $\Delta G$ = free energy = $- n$ F$\Delta E$, where n = number of electrons transferred, F = Faraday constant, and $\Delta E$ = cell voltage.

To find $\Delta E$, use the Nernst equation, which states $\Delta E = \Delta E^\circ - .059/N \log K$, where $\Delta E^\circ$ = voltage at standard conditions, and K = equilibrium constant of reaction.

To find $\Delta E°$, note that the half-reactions are given as oxidation processes (loss of electrons). In a cell, however, the overall reaction is the sum of the oxidation and reduction process (gain of electrons). Therefore, you reverse one of these half-reactions to obtain a reduction. If you reverse

$$H_2 (1 \text{ atm}) \rightarrow 2H^+ (1.0M) + 2e^-, \qquad \text{you obtain}$$

$$2H^+ (1.0M) + 2e^- \rightarrow H_2 (1 \text{ atm}): \qquad \text{reduction}$$

oxid: $\underline{H_2 (1 \text{ atm}) \rightarrow 2H^+ (.1M) + 2e^-}$

Total (overall) $H^+ (1.0M) \rightarrow H^+ (.1M)$

From this, you see that only one electron is being transferred per $H^+$, which means N = 1. $\Delta E° =$ standard potential = $E°_{products} - E°_{reactants}$. But the product and reactant are the same species, which means the difference is zero. Thus, $\Delta E° = 0$.

To find K, equilibrium constant, remember that the ratio of concentrations of products to reactants, each raised to the power of its coefficient in the equation. Thus

$$K = \frac{[H^+]_{product}}{[H^+]_{reactant}} \text{ . for this reaction. Thus,}$$

$$K = \frac{(.1M)}{(1.0M)}$$

Recalling, $\Delta E = \Delta E° = -.059/N \log K$, you can substitute to obtain the following:

$$\Delta E = 0 - \frac{.059}{1} \log \frac{(.1)}{(1.0)} = 0 - (.059)(-1) = +0.59 \text{ V.}$$

Recalling $\Delta G = -n F\Delta E$, you have all the values to find $\Delta G$. Substituting and solving,

$$\Delta G = -(1)(96500)(.059) = 5693.5 \text{ coul-volt/mole}$$

$$\text{or } \Delta G = 5693.5 \left( \frac{1 \text{ cal}}{4.184 \text{ coul-volt}} \right) \left( \frac{Kcal}{1000 \text{ cal}} \right) = -1.36 \text{ Kcal/mole}$$

Thus, the maximum work obtainable, the free energy of the system, equals $-1.36$ Kcal/mole.

● **PROBLEM** 605

A voltaic cell is made up of a silver electrode in a 1M silver nitrate solution and an aluminum electrode in a 1M aluminum nitrate solution. The half reactions are

(1) $Ag^+ + e^- \rightarrow Ag(s)$  $E° = .80$ volt, and (2) $Al^{3+} +$ $3e^- \rightarrow Al(s)$  $E° = -1.66$ volt, calculate the standard free energy change, $\Delta G°$, in Kcal/mole. 1 cal = 4.184 joule 1 F = 96,500 coul, 1 Kcal = 1000 cal.

Solution: The standard free energy change, $\Delta G°$, can be related to standard cell potential change, $\Delta E°$, via $\Delta G° = n F\Delta E°$, where n = number of electrons transferred per mole. To find $\Delta G°$, calculate (1) $\Delta E°$, (2) n.

(1) You are told a voltaic cell exists. This means that electricity is generated. You are given the $E°$ values for the half-reactions. For electricity to be generated, however, the $\Delta E°$ for the overall reaction must be positive. Both half-reactions represent reduction processes (gain of electrons). But $\Delta E° = E°_{oxid} + E°_{red}$. Thus, to find $\Delta E°$, you must reverse one of these reactions to oxidation (loss of electrons) so that $\Delta E°$ for the overall reaction, is positive. This condition can only be met if (2) is reversed to $Al(s) \rightarrow Al^{3+} + 3e^-$, $E° = -(-1.66) = 1.66$ and (1) remains as $Ag^+ + e^- \rightarrow Ag(s)$. Before adding these two to obtain the overall reaction, multiply (1) by 3, so that all electrons in the overall equation can be cancelled. Thus

$$Al(s) \rightarrow Al^{3+} + 3e^-$$

$$\underline{3Ag^+ + 3e^- \rightarrow 3Ag(s)}$$

$$Al(s) + 3Ag^+ \rightarrow Al^{3+} + Ag(s) \text{ with}$$

$\Delta E° = E°_{oxid} + E°_{red} = 1.66 + .80 = 2.46$ volts.

(2) The number of electrons being transferred = 3, so that n = 3.

$\Delta G° = n F\Delta E°$.

Substitute in values to obtain

$\Delta G° = (3)(96500)(2.46)(cal/4.184 \text{ joules}) \times (Kcal/1000 \text{ cal})$

$= + 170$ Kcal/mole.

● **PROBLEM** 606

Calculate the standard $E°$, electrode potential, of the hydrogen-oxygen fuel cell. The standard Gibbs free energy of formation $(\Delta G^0_f)$ of $H_2O(\ell) = -56,690$ cal/mole.

Solution: The standard Gibbs free energy of formation $(\Delta G^0_f)$ is a measure of the energy available to do work when compounds are formed from their elements in the standard state. The electrode potential,

578

i.e., the voltage across the two electrodes in a galvanic cell, such as the hydrogen-oxygen cell, can be computed from $\Delta G_f^0$, by using the equation $\Delta G^0 = -nF E^0$, where $n$ = number of electrons transferred and $F$ = a faraday of electricity, which is 23,060 cal $v^{-1}$ mole$^{-1}$. In a hydrogen-oxygen fuel cell, water is produced from the transfer of 2 electrons. Thus, $n = 2$. Solving for $E^0$:

$$-56,690 \frac{cal}{mole} = -2(23,060 \text{ cal } v^{-1} \text{ mole}^{-1}) E^0 .$$

$$E^0 = 1.229 \text{ v} \quad \text{for this fuel cell.}$$

# CHAPTER 17

# ATOMIC THEORY

Basic Attacks and Strategies for Solving Problems in this Chapter. See pages 581 to 611 for step-by-step solutions to problems.

The atomic weight is defined as the weight of an atom compared to $^{12}C$ being exactly 12 atomic mass units (amu). This translates into a mole of $^{12}C$ having a mass of exactly 12 grams. Tabulated values of atomic weights are the weighted averages of the naturally occurring isotopes. (Isotopes have the same number of protons but different numbers of neutrons.) For example, the atomic weight of chlorine is tabulated as 35.45, natural chlorine is 75.5% $^{35}Cl$ (atomic weight 34.97), and 24.5% $^{37}Cl$ (atomic weight 36.97).

$$\text{Avg. At. Wt.} = .755(34.97) + .245(36.97) = 35.45 \qquad 17\text{-}1$$

Many periodic tables show the electronic configuration for each atomic species. This will be discussed in more detail in the following paragraphs and in Chapter 18, but since chemistry is the study, principally, of the interaction of electrons in the outer or valence orbitals, it is often simplified by considering only the numbers of valence or reactive electrons. The key is to note that in the first orbital there are two valence electrons. The neutral atom with one electron is hydrogen; the one with two is helium. No more electrons can enter the first orbital for reasons to be outlined in Chapter 18. For all other orbitals, up to eight electrons can exist. The tendency or driving force for reaction is to fill the outermost orbital by sharing or acquiring electrons from other atomic species. Therefore, atoms react in a way to get eight electrons in the outermost or valence orbital (except for hydrogen which reacts to get two).

If the valence orbital is filled by sharing electrons the bond is called covalent; if it is filled by acquiring an electron from another atom, the bond is called ionic. For large differences in electronegativity, ionic bonds are formed; for moderate differences, covalent bonds are formed. The relative electronegativity of the elements is tabulated. An arbitrary scale, called the Pauling scale, is used, and the value of electronegativity for each of the elements is shown in a table in Problem 642. If the difference in electronegativity between two elements is greater than 1.7, the bond between them will be ionic; if less, the bond will be covalent. However, if the difference in electronegativity is near 1.7, bonds may have both covalent and ionic character. These bonds are polar in that the electrons are preferentially attracted to the more electronegative atom. The greater the difference in electronegativity between bonded atoms, the more polar the bond will be until eventually, when the electron is completely attracted to one atom, it becomes ionic.

There are several examples in the problem list showing electron dot diagrams and putting atoms together to obtain the desired number of electrons in the valence orbit. When atoms in Groups I and II of the Periodic Table donate their one or two valence electrons, the electronic configuration that results is that of the filled stable orbit under the valence electrons.

Electrons are numbered for each element by the following convention which is explained by quantum theory and discussed in Chapter 18.

a) the first number is the principal orbit (e.g., 1, 2, 3, etc.).

b) the second digit is designated by a letter (e.g., $s$, $p$, $d$, $f$).

c) the superscript following the letter designates the number of electrons of that type.

The maximum number of each type of electrons in any orbital is 2 for $s$ electrons, 6 for $p$ electrons, 10 for $d$ electrons, and 14 for $f$ electrons. The $s$ and $p$ electrons are the principal reactive electrons for small atoms. The $d$ electrons are somewhat less reactive and account for the multiple valence states in many transition metals.

In the first row of the Periodic Table, only $s$ electrons exist. In the second and third rows, $s$ plus $p$ electrons exist, and in subsequent rows the $d$ and $f$ electrons may also exist. Table 1 shows the electron configuration of a sample of elements from the first three rows of the Periodic Table.

## Table 1: Sample of the Electronic Configurations of Elements

| Element | At. No. | 1s | 2s | 2p | 3s | 3p | Electronic Configuration* |
|---------|---------|-----|-----|-----|-----|-----|---------------------------|
| H | 1 | 1 | | | | | $1s^1$ |
| He | 2 | 2 | | | | | $1s^2$ |
| C | 6 | 2 | 2 | 2 | | | $1s^2, 2s^2, 2p^2$ |
| Na | 11 | 2 | 2 | 6 | 1 | | $1s^2, 2s^2, 2p^6, 3s^1$ |
| Cl | 17 | 2 | 2 | 6 | 2 | 5 | $1s^2, 2s^2, 2p^6, 3s^2, 3p^5$ |
| Ar | 18 | 2 | 2 | 6 | 2 | 6 | $1s^2, 2s^2, 2p^6, 3s^2, 3p^6$ |

* The superscripts of the electronic configuration add up to the atomic number.

To calculate bond distances and angles, the key is to rely on the geometry of the bond and calculate the relationship between the distances and angles from standard trigonometric principles. For example, for a molecule with a tetrahedral conformation, such as methane, $CH_4$, or carbon tetrachloride, $CCl_4$, the H-C-H angle or Cl-C-Cl angle is the tetrahedral angle of 109.5°. The bond distance is, to a first approximation, the sum of the atomic radii of the two atoms between which the bond is formed.

Step-by-Step Solutions to
Problems in this Chapter,
"Atomic Theory"

## ATOMIC WEIGHT

● PROBLEM 607

If the atomic weight of carbon 12 is exactly 12 amu, find the mass of a single carbon-12 atom.

Solution:  To solve this problem, one must first define the mole concept. A mole is defined as the weight in grams divided by the atomic weight (or molecular weight) of the atom or compound.

$$mole = \frac{weight\ in\ grams}{atomic\ or\ molecular\ weight}$$

If one has a mole of carbon 12, then there are 12 grams of it present. One mole of any substance contains $6.022 \times 10^{23}$ particles.

The mass of a single carbon atom is found by dividing 12 g/mole by $6.022 \times 10^{23}$ atoms/mole.

$$mass\ of\ 1\ C\ atom = \frac{12\ g/mole}{6.022 \times 10^{23}\ atoms/mole}$$

$$= 2.0 \times 10^{-23}\ g/atom.$$

● PROBLEM 608

The atomic weight of iron is 55.847 amu. If one has 6.02 g of iron, how many atoms are present?

Solution:  A mole is defined as the weight in grams of a

substance divided by its atomic weight:

$$\text{mole} = \frac{\text{amount in grams}}{\text{atomic weight}}$$

If one calculates the number of moles of iron, then the number of atoms present can be calculated. There are $6.02 \times 10^{23}$ atoms per mole.

$$\text{no. of moles of Fe} = \frac{6.02 \text{ g}}{55.847 \text{ g/mole}} = 1.08 \times 10^{-1} \text{ moles}$$

no. of Fe atoms present = $(1.08 \times 10^{-1}$ moles) x

$$(6.02 \times 10^{23} \text{ atoms/mole})$$

$$= 6.49 \times 10^{22} \text{ atoms.}$$

● **PROBLEM** 609

Nitrogen reacts with hydrogen to form ammonia ($NH_3$). The weight-percent of nitrogen in ammonia is 82.25. The atomic weight of hydrogen is 1.008. Calculate the atomic weight of nitrogen.

<u>Solution</u>: When elements combine to form a given compound, they do so in a fixed and invariable ratio by weight.

In a given amount of $NH_3$, 82.25% of its weight is contributed by nitrogen. The weight-percent of hydrogen in ammonia is, then

100 - 82.25 = 17.75%.

One mole of nitrogen and 3 moles of hydrogen combine to form one mole of ammonia. Therefore, one mole of nitrogen constitutes 82.25% and 3 moles of hydrogen constitute 17.75% by weight of one mole of ammonia. The atomic weight of an element is equal to the weight of one mole of that element. The following ratio can be set up to solve for the atomic weight (AW) of nitrogen:

$$\frac{AW_N}{3 \times AW_H} = \frac{\text{weight-percent N}}{\text{weight-percent H}}$$

Solving for $AW_N$

$$AW_N = \frac{(3 \times AW_H)(\text{weight-percent N})}{(\text{weight-percent H})}$$

$$= \frac{3(1.008)(82.25)}{17.75} = 14.01.$$

John Dalton found water to be 12.5% hydrogen by weight. Calculate the atomic weight of oxygen if Dalton assumed water contained two hydrogen atoms for every three oxygen atoms. Assume 1 H atom weighs 1 amu.

Solution: This problem can be solved if one sets up a ratio. If it is assumed water is 12.5% hydrogen by weight, then water is 100-12.5 or 87.5% oxygen by weight. Dalton assumed that there are two hydrogen atoms for every three oxygen atoms. Therefore,

$$\frac{12.5\%}{87.5\%} = \frac{1}{7} = \frac{2 \text{ H's}}{3 \text{ O's}}.$$

It is given that 2 H atoms weight 2 amu. Thus,

$$\frac{1}{7} = \frac{2 \text{ amu}}{3 \times \text{atomic weight of O}}.$$

Solving for the atomic weight of O,

$$\text{atomic weight of O} = \frac{2 \text{ amu} \times 7}{3} = 4.67 \text{ amu}.$$

## VALENCE AND ELECTRON DOT DIAGRAMS

On the basis of valence, predict the formulas of the compounds formed between the following pairs of elements: (a) Sn and F. (b) P and H. (c) Si and O.

Solution: Valence may be defined as a number which represents the combining capacity of an atom or radical, based on hydrogen as a standard.

For molecules containing two kinds of atoms, the product of the number of times one kind of atom appears in a molecule and the valence of that kind of atom, must be equal to the product of the number of times the other kind of atom appears multiplied by the valence of this second kind of atom.

(a) The valence of Sn is 4 and that of F is 1. Hence, the compound is $SnF_4$ (1 atom Sn × 4 = 4 atoms F × 1).

(b) The valence of P is 3 and that of H is 1. Hence, the compound is $PH_3$ (1 atom P × 3 = 3 atoms H × 1).

(c) The valence of Si is 4 and that of O is 2. Hence, the compound is $SiO_2$ (1 atom Si × 4 = 2 atoms O × 2).

● **PROBLEM** 612

The atomic weight of element X is 58.7. 7.34 g of X displaces 0.25 g of hydrogen from hydrochloric acid (HCl). What is the valence of element X?

Solution: Chlorine has a valence of -1, thus to determine the valence of X, one must calculate the number of moles of $Cl^-$ that will bind to each mole of X. The valence will be equal to this number. There are 2 moles of $Cl^-$ present for every mole of $H_2$ formed. To find the number of moles of $H_2$ formed, one must divide 0.25 g by the molecular weight of $H_2$ (MW of $H_2$ = 2).

$$\text{moles of } H_2 = \frac{0.25 \text{ g}}{2 \text{ g/mole}} = .125 \text{ moles of } H_2.$$

no. of moles of $Cl^-$ = 2 × .125 = .250 moles.

One should now determine the number of moles of X present. This is done by dividing the number of grams by the molecular weight. (MW of X = 58.7).

$$\text{no. of moles} = \frac{7.34 \text{ g}}{58.7 \text{ g/mole}} = .125 \text{ moles.}$$

From this, one sees that .125 moles of X combines with .250 moles of $Cl^-$. The number of moles of $Cl^-$ that bind to each mole of X is equal to the number of moles of $Cl^-$ present divided by the number of moles of X.

$$\begin{array}{c} \text{no. of } Cl^- \text{ that combine} \\ \text{with each X} \end{array} = \frac{.250 \text{ moles of } Cl^-}{.125 \text{ moles of X}}$$

$$= 2 \text{ moles/}Cl^-\text{/mole X}$$

The formula for the resulting compound is $XCL_2$. Because $Cl^-$ has a valence of -1 and 2 $Cl^-$ combine with each X, X must have a valence of +2 for a neutral molecule to be formed.

● **PROBLEM** 613

The faint light sometimes seen over Marshland at night, the "will-o'-the-wisp", is believed to come about as a result of the burning of a compound of phosphorus (P) and hydrogen (H). What is the formula of this compound?

H
x•
H $\overset{x}{\underset{\bullet\bullet}{\bullet}}$ P $\overset{x}{\underset{\bullet\bullet}{\bullet}}$ H

Solution: To find the formula of this compound, it is necessary to determine the valence of the elements from which it is composed. The valence of an element is the number of electrons that are involved in chemical bonding.

To find the valence of phosphorus and hydrogen, consider their atomic number and electronic configuration.

Hydrogen: Atomic number = 1.

Electronic configuration = $1s^1$

Phosphorus: Atomic number = 15.

Electronic configuration = $1s^2 2s^2 2p^6 3s^2 3p^3$.

The outer electrons are $1s^1$ for hydrogen, and $3p^3$ for phosphorus. It takes one additional electron to fill hydrogen's s orbital; its valence is one. It takes 3 more electrons to fill phosphorus' p orbital; its valence is three. Elements react with the purpose of filling all their orbitals with the maximum number of electrons by either a transfer of electrons or by sharing electrons.

It would take three hydrogen atoms to complete phosphorus' outer orbital. In turn, each electron of phosphorus would serve to complete the outer orbital of each hydrogen atom. This can be pictured in an electron-dot formula as shown above.

In this figure, the X's represent the outer electrons of hydrogen and **dots** represent the electrons in the outer shell of phosphorus. The formula of this compound is, thus, $PH_3$.

● **PROBLEM** 614

$H_2O_3$, hydrogen trioxide, a close relative of hydrogen peroxide, has recently been synthesized. It is extremely unstable and can be isolated only in very small quantities. Write a Lewis electron dot structure for $H_2O_3$.

Solution: A Lewis dot structure of a compound shows the arrangement of valence electrons. Valence electrons are defined as an element's outer electrons which participate in chemical bonding. Thus, to write an electron dot structure for $H_2O_3$, calculate the total number of valence electrons.

This can be done by considering the electronic con-figurations of H (hydrogen) and O (oxygen).

Hydrogen possesses one valence electron while oxygen has 6 valence electrons. In $H_2O_3$, a total of (2)1 + (3)6 = 20 valence electrons are involved.

With this in mind, the Lewis electron dot structure becomes

$$H : \overset{\cdot\cdot}{\underset{\cdot\cdot}{O}} : \overset{\cdot\cdot}{\underset{\cdot\cdot}{O}} : \overset{\cdot\cdot}{\underset{\cdot\cdot}{O}} : H$$

Notice, you have represented the required 20 valence electrons.

● **PROBLEM** 615

Using electron-dot notation, show for each of the following the **outer** shell electrons for the uncombined atoms and for the molecules or ions that result:

(a)  H  +  H   →   hydrogen molecule
(b)  Br  +  Br   →   bromine molecule
(c)  Br  +  Cl   →   bromine chloride
(d)  Si  +  F   →   silicon fluoride
(e)  Se  +  H   →   hydrogen selenide
(f)  Ca  +  O   →   calcium oxide

<u>Solution</u>:  When electrons are transferred from one atom to another, ions are formed, which gives rise to ionic bonding. Two atoms, both of which tend to gain electrons, may combine with each other by sharing one or more pairs of electrons. These two atoms form a covalent bond.

To solve this problem, one must know the number of valence electrons, in each of the atoms in the equations. The valence number reflects the combining capacity of an atom. Next, one must know which atoms combine to form ionic bonds and which form covalent bonds. The only ionic bond formed in these equations is for Ca + O; the other bonds are covalent, and electrons are shared to form an isoelectronic electron cloud such as a noble gas.

Thus,

(a)   H·  +   H·      →      H : H

(b)  : $\overset{\cdot\cdot}{\underset{\cdot}{Br}}$ : + : $\overset{\cdot\cdot}{\underset{\cdot}{Br}}$ :    →     : $\overset{\cdot\cdot}{\underset{\cdot\cdot}{Br}}$ : $\overset{\cdot\cdot}{\underset{\cdot\cdot}{Br}}$ :

(c)  :B̈r: + :C̈l:   →   :B̈r:C̈l:
     ·         ·         ··  ··

(d)  ·S̈i· + :F̈:   →        :F̈:
     ·        ·         ·· ··  ··
                        :F̈:S:F̈:
                        ·· ··  ··
                           :F̈:
                            ··

(e)  ·S̈e: + H·   →   H:S̈e:
     ·                      H

(f)  Ca· + ·Ö:   →   Ca⁺⁺ :Ö:̈
          ·                ··

The sulfate ion consists of a central sulfur atom with four equivalent oxygen atoms in a tetrahedral arrangement. Keeping in mind the octet rule, draw the electronic structure for the ion. What should the internal O-S-O bond angle be?

**Figure a**          **Figure b**

**Figure c**          **Figure d**

Solution:  The formula for the sulfate ion is $SO_4^=$ . To write the electronic structure of the ion in this problem, one must consider the definition of valence and the octet rule. Valence electrons refer to those outer electrons that participate in chemical bonding. The octet rule states that for stability, there can be no more than eight electrons in the outer orbit of an atom, either as a re-

sult of transfer or sharing. Thus, after determining the
number of valence electrons present, arrange them so that
they obey the octet rule. The electronic configuration of
sulfur is $1s^2 2s^2 2p^6 3s^2 3p^4$. It has 6 valence electrons; they
are found in the 3s and 3p orbitals. The electronic configuration of oxygen is $1s^2 2s^2 2p^4$, which means it also has
six valence electrons; the first orbital contains only 2
electrons. The electronic structure of each of these atoms
can be represented as shown in figure a and figure b, where
x's and dots indicate valence electrons. The electronic
structure of the ion can be pictured as figure c. Here,
one can see that the sulfur and two of the oxygen atoms
are surrounded by eight electrons. Because two of the
oxygen atoms are only surrounded by seven electrons, each
of these atoms possesses a negative charge. This structure
may also be drawn as shown in figure d.

For greatest stability the oxygen atoms must take
positions as far apart as possible. It has been shown that
when four atoms surround a fifth the most stable arrangement
is tetrahedral. In this arrangement, the O-S-O bonds form
a 109.5° angle.

# IONIC AND COVALENT BONDING

● **PROBLEM** 617

Distinguish a metallic bond from an ionic bond and from a
covalent bond.

Solution: The best way to distinguish between these bonds
is to define each and provide an illustrative example of
each.

When an actual transfer of electrons results in the
formation of a bond, it can be said that an ionic bond is
present. For example,

$$2K° \quad + \quad \overset{..}{S} : \quad \to \quad 2K^+ \quad + \quad : \overset{..}{\underset{..}{S}} :^{2-} \quad \to \quad K_2S$$

| potassium atoms | sulfur atom | potassium ions | sulfur ion | ionic bond due to the attraction of unlike ions |
|---|---|---|---|---|
| | | (unlike ions due to trans- fer of electrons from potassium to sulfur) | | |

When a chemical bond is the result of the sharing of
electrons, a covalent bond is present. For example:

$$: \overset{..}{\underset{..}{Br}} \cdot \; + \; \cdot \overset{..}{\underset{..}{F}} : \quad \to \quad : \overset{..}{\underset{..}{Br}} : \overset{..}{\underset{..}{F}} :$$    These electrons are shared by
both atoms

A pure crystal of elemental metal consists of millions of atoms held together by metallic bonds. Metals possess electrons that can easily ionize, i.e., they can be easily freed from the individual metal atoms. This free state of electrons in metals binds all the atoms together in a crystal. The free electrons extend over all the atoms in the crystal and the bonds formed between the electrons and positive nucleus are electrostatic in nature. The electrons can be pictured as a "cloud" that surrounds and engulfs the metal atoms.

● **PROBLEM** 618

An ionic bond is established between positive ion A and negative ion B. How would one expect the strength of the bond to be affected by each of the following changes: (a) Doubling the charge on A, (b) Simultaneously doubling the charge on A and B, (c) Doubling the radius of B, and (d) Simultaneously doubling the radius of A and B?

Solution: When two ions come together to form an ionic bond, energy is released. The more energy released, the stronger the ionic bond. The amount of energy released is determined by the equation

$$E = 1.44 \ \frac{q_1 q_2}{r^2}$$

where E = energy in electron volts, $q_1$ = charge on the positive ion, $q_2$ = charge on the negative ion, and r = the distance between the nuclear centers. If one changes one parameter, another must adjust to maintain this e-quality.

Solving:

(a) If the charge on A, the positive ion, is doubled, the energy released is doubled. That is,

$$2E = 1.44 \ \frac{2q_1 q_2}{r^2} \quad \text{is equivalent to} \quad E = \frac{1.44 \ q_1 q_2}{r^2}.$$

If E is doubled, the strength of the chemical bond is doubled.

(b) Using a similar line of reasoning, it is found that if one doubles both the positive and negative charges, four times E is released. The strength of the bond is, thus, 4 times as great as the original bond.

(c) If one increases the radius of the ions then, the distance between their nuclei must also increase. To maintain the equality, E must decrease because the square of r is inversely proportional to E. Therefore, the bond strength decreases.

(d) Both radii are doubled. E is diminished by a factor of 4 and the strength of the bond is decreased by a factor of 4.

● **PROBLEM** 619

Compare the ionic bond strength of NaCl to KF; assuming that the radii of $Na^+$, $K^+$, $F^-$, and $Cl^-$ are, respectively, 0.097, 0.133, 0.133, and 0.181 NM.

Solution:  The ionic bond strength is directly proportional to the energy released when positive and negative ions form chemical bonds. To solve this problem, therefore, compute how much energy is released when $Na^+$ and $Cl^-$ ions form NaCl and when $K^+$ and $F^-$ ions form KF. Employ the expression that relates energy to charge and radii:

$$E = 1.44 \frac{q_1 q_2}{r^2}$$

where E = energy in electron volts, r = distance between the nuclei of the ions in the molecule, and $q_1$ and $q_2$ are the charges on the ions. Note that the charge on each ion is the same. The distance between the nuclei, r, is the sum of the radii of the ions. With these facts in mind, one can determine the amount of energy released in the formation of NaCl.

$$E_{NaCl} = 1.44 \frac{q_1 q_2}{0.0773} . \text{ For KF, } E_{KF} = 1.44 \frac{q_1 q_2}{0.0708}$$

Comparing the amount of energy released in each case

$$\frac{E_{NaCl}}{E_{KF}} = \frac{0.0708}{0.0773}$$

$$\frac{0.266}{0.278} \times 100 = 91.6\%.$$

Thus, the bond strength of NaCl is 91.6% as strong as the bond strength of KF.

● **PROBLEM** 620

A chemist possesses KCl, $PH_3$, $GeCl_4$, $H_2S$, and CsF. Which of these compounds do not contain a covalent bond?

Solution:  A covalent bond is defined as one in which electrons are shared. The stability of covalent bonds in molecules depends on the difference in electronegativity values of the two atoms which make up the molecule. Electronegativity refers to the tendency of an atom to attract shared electrons in a chemical bond. If the electronegativity difference of two elements is greater than 1.7, an ionic bond is formed; if it is less than 1.7, a covalent bond is formed.

To solve this problem, consult a table of electronegativity values, and compute the electronegativity difference of the atoms in each of the given compounds. Proceed as follows:

| | Electronegativity Values | | Difference |
|---|---|---|---|
| KCl | K = 0.8 | Cl = 3.0 | 2.2 |
| $PH_3$ | P = 2.1 | H = 2.1 | 0 |
| $GeCl_4$ | Ge = 1.8 | Cl = 3.0 | 1.2 |
| $H_2S$ | H = 2.1 | S = 2.5 | 0.4 |
| CsF | Cs = 0.7 | F = 4.0 | 3.3 |

Thus, only KCl and CsF exceed 1.7. They possess ionic bonds. The remainder of the molecules possess covalent bonds.

● **PROBLEM** 621

Consider a covalent bond between hydrogen and arsenic. It is known that the radii of hydrogen and arsenic atoms are respectively: 0.37 and 1.21 Angstroms. What is the approximate length of the hydrogen-arsenic bond?

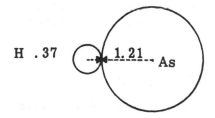

Solution: One may assume that the two atoms are spheres. Thus, as in the figure shown, the bond length defined as the distance between the two nuclei is the sum of the lengths of the two radii.

bond length = length of H radii + length of As radii.

bond length = 0.37 Å + 1.21 Å = 1.58 Å.

# ELECTRONEGATIVITY

● **PROBLEM** 622

What is the meaning and significance of the Pauling electronegativity scale?

Solution: In any chemical bond, electrons are shared between the bonding atoms. In covalent bonds, the valence electrons are shared almost equally - the shared electrons spend about the same amount of time with each atom. In ionic bonds, the valence electrons are monopolized by one atom. The degree to which a bond will be ionic or covalent is dependent upon the relative electron-attracting ability of the bonding atoms.

The Pauling electronegativity scale provides a measure for the relative electron-attracting abilities or electronegativity of each element. The most electronegative element, fluorine, is assigned the highest number, 4.0. The least electron attracting elements (and consequently, those most willing to lose electrons), cesium and francium, are assigned the lowest number, 0.7. Numbers are assigned to the remaining elements so that no element has a higher number than an element more electronegative than it. Furthermore, the numbers are assigned so that the differences between the electronegativities of two elements is indicative of the ionic quality of the bond that forms between them. In particular, if the difference is 1.7 or greater, then the shared electrons are monopolized by the more electronegative atom to such a great extent that the bond is said to be ionic. If the difference is less than 1.7, the bond is covalent.

Thus, when given a compound, the Pauling electronegativity scale can be used to determine the species that acts as the electron donor, the species that acts as the electron acceptor, and the degree of electron polarization of the bond.

● **PROBLEM** 623

Classify the bonds in the following as ionic, ionic-covalent, covalent-ionic and covalent: $I_2$, $LiCl$, $MgTe$, $Cl_2O$, $H_2S$.

Solution: The bonds in these molecules can be classified by considering the electronegativities of the atoms involved.

The electronegativity of any element is the tendency of that element to attract electrons. The greater the electronegativity, the stronger the attraction. When 2 atoms join, the type of bond formed can be determined by calculating the difference in electronegativity of the two atoms involved. When the difference is greater than 1.7, the bond is classified as ionic, if less than 1.7, then the bond is covalent. An ionic bond involves the complete transfer of electrons from 1 atom to another, while in a covalent bond, the electrons are shared between like or similar atoms. Covalent-ionic indicates a mixed character which is more covalent in nature. Ionic-covalent bonds have more of an ionic character.

Consider the cases in point. For $I_2$, there is no difference in electronegativity because the 2 atoms involved in the bond are the same, **and** hence contains a covalent bond with equal sharing of the electrons.

In LiCl, the electronagetivities are Li = 1.0 and Cl = 3.0. (The electronegativity values are obtained from an electronegativity table). The electronegative difference is 2.0. Therefore, the bond is ionic.

One proceeds in a similar manner for the remainder of the compounds.

| Compounds and Electronegativities of the Elements | Difference in Electronegativity | Type of Bond |
|---|---|---|
| Mg – Te<br>1.2    2.1 | 2.1 - 1.2 = 0.9 | Covalent-ionic |
| $Cl_2$ – O<br>3.0    3.5 | 3.5 - 3.0 = 0.5 | Covalent-ionic |
| $H_2$ – S<br>2.1    2.5 | 2.5 - 2.1 = 0.4 | Covalent-ionic |

● **PROBLEM** 624

In a given chemical compound, there is a bond between "A" and "B". The electronegativities of the involved elements are, respectively, 1.2 and 2.1 What is the probable nature of the bond between "A" and "B"?

Solution: ELectronegativity is a measure of the tendency of an atom to attract shared electrons in a molecule. The greater the difference in electronegativity between two elements, the greater the ionic character of the bond. When the difference in electronegativities is greater than 2.0, the bond is considered ionic.

The electronegativity difference in this problem is 2.1 - 1.2 or 0.9. Thus, while the bond is not ionic, the difference suggests a degree of polarity in this covalent bond.

What is the difference between electronegativity and electron affinity? Use specific examples.

<u>Solution</u>: Electron affinity represents the energy released when an electron is added to an isolated neutral gaseous atom. Electronegativity is the tendency of an element to attract electrons, not in the gaseous state, but from a nucleus joined to it by a covalent bond. The electron affinity is a precise quantitative term while the electronegativity is a more qualitative concept.

Assuming the ionization potential of sodium is 5.1 eV and the electron affinity of chlorine is 3.6 eV, calculate the amount of energy required to transfer one electron from an isolated sodium (Na) atom to an isolated chlorine (Cl) atom.

<u>Solution</u>: Ionization potential is the amount of energy required to pull an electron off an isolated atom. Electron affinity is the amount of energy released when an electron is added to an isolated neutral atom. In this problem, one must add energy to remove an electron from Na and energy will be released upon the addition of an electron to Cl. 5.1 eV are needed to expel an electron from Na and 3.5 eV are released when Cl accepts an electron. Thus, the amount of energy required for the overall process to occur is the difference between the ionization potential of Na and the electron affinity of Cl.

The energy necessary to be added to the system for this reaction to occur is 5.1 eV - 3.6 eV or 1.5 eV.

It is a fact that the second ionization potential of alkali atoms falls off more rapidly with increasing atomic number than does the first ionization potential. Why?

<u>Solution</u>: The first ionization potential is the energy necessary to remove one electron from a neutral atom in the gasesous state. After this first electron is removed, the atom becomes an ion with a +1 ionization state. Second ionization potential refers to the energy necessary to remove an electron from the +1 ion in the gaseous state.

As the atomic numbers of elements increase the atomic radii of the atoms increase. The atomic radius determines the distance from the nucleus to the outermost electron of the atom.

The first and second ionization potentials should fall off as the atomic radius increases. As the radius increases, the outer electron, which is the one that will be removed, is at a greater distance from the nucleus. Thus, the "pull" on it by the nucleus, is less than that of an electron closer to the nucleus. Hence, less energy should be required to remove an electron that is far away from the nucleus than one that is close.

In considering the second ionization potential, one has an added factor involved; the element is charged. The concentration of charge will be decreased as the atomic radius increases. This increased diffusion of charge with the increased atomic number is the reason for the added fall off of the second ionization potential as compared to first ionization potential.

● **PROBLEM** 628

40.0 kJ of energy is added to 1.00 gram of magnesium atoms in the vapor state. What is the composition of the final mixture? The first ionization potential of Mg is 735 kJ/mole and the second ionization potential is 1447 kJ/mole

Solution:  Ionization potential may be defined as the energy required to pull an electron away from an isolated atom. The second ionization potential is the amount of energy required to pull off a second electron after the first has been removed.

The composition of the final mixture is determined by calculating the number of electrons that will be removed from the magnesium ions. To do this one must determine the number of moles of Mg present in 1 g. From this one can determine the number of electrons that will be liberated by using the values for the first and second ionization potentials of Mg.

The atomic weight of Mg is 24.3. Since moles = grams/atomic weight, there are in 1 gram of Mg, 1/24.30 or $4.11 \times 10^{-2}$ moles present. The first ionization potential of Mg is 735 kJ/mole.  Therefore, $4.11 \times 10^{-2}$ moles of Mg requires $4.11 \times 10^{-2}$ moles x 735 kJ/mole or 30.2 kJ to ionize all of the atoms once. 40 kJ was added to the system leaving 40 kJ - 30.2 kJ or 9.8 kJ to remove the second electron. If one has 9.8 kJ and 1447 kJ/mole is required to remove the second electrons, then

$$\frac{9.8 \text{ kJ}}{1447 \text{ kJ/mole}} = 6.77 \times 10^{-3} \text{ moles of atoms can}$$

have their second electron removed. $4.11 \times 10^{-2}$ moles of Mg are present.

$$\frac{6.77 \times 10^{-3}}{4.11 \times 10^{-2}} \times 100 = 16.5\%.$$

This means that 16.5% of the atoms can have a second electron removed. Therefore, the composition of the mixture is: $Mg^{++}$ 16.5%, $Mg^{+}$ 100 - 16.5 or 83.5%.

● **PROBLEM** 629

Consider the formation of an ionic molecule, AB, by the donation of an electron from atom A to atom B. If the ionization potential of A is 5.5 eV and the electron affinity of B is 1.5 eV, and the ion $B^-$ has a radius of 0.20 nm, what is the maximum size of $A^+$ that would lend itself to the formation of an energetically stable bond?

Solution: When two atoms form an ionic bond, the energy difference between the ionization potential of one and the electron affinity of the other goes into the electrostatic attraction between the ions of the two atoms. This energy difference is related to the internuclear separation between ionized atoms by the following relationship:

$$E = 1.44 \times \frac{q_1 q_2}{r^2}$$

where E is the energy difference in electron-volts (eV), $q_1$ and $q_2$ are the charges on the ions in electrostatic units (esu), and r is the internuclear separation in nanometers (nm). In this problem we can determine E and use this to find r.

When A ionizes to $A^+$, it absorbs 5.5 eV of energy. When B accepts an electron to form $B^-$, it gives up 1.5 eV of energy. Hence, a total energy of E = 1.5 eV - 5.5 eV = - 4.0 eV goes into the formation of the AB ionic bond.

The charge on $A^+$ is $q_1$ = + 1 esu and that on $B^-$ is $q_2$ = -1 esu. One can now determine the internuclear separation of this bond by solving the equation

$$E = 1.44 \frac{q_1 q_2}{r^2} \text{ for r to obtain:}$$

$$r = \sqrt{\frac{1.44 \ q_1 q_2}{E}} = \sqrt{\frac{(1.44)(1 \ esu)(-1 \ esu)}{-4.0 eV}} = 0.6 nm$$

Hence, the nucleus of A is separated from the nucleus of B by 0.6 nm. Since the radius of $B^-$ is 0.20 nm, the maximum radius of $A^+$ (at the point where it just "touches" $B^-$) is 0.6 nm - 0.2 nm = 0.4 nm.

A chemist has one mole of X atoms. He finds that when half of the X atoms transfer one electron to the other half, 409 kJ must be added. If all of the resulting X⁻ ions are subsequently converted to X⁺ ions, an additional 733 kJ must be added. Find the ionization potential, and the electron affinity of X.

Solution: Two facts must be known before this problem can even be attempted:

(1) The ionization potential is the energy required to pull off an electron from an isolated atom, and (2) electron affinity is the energy released when an electron adds to an isolated neutral atom. In this problem, there are two different reactions. The first is one half of a mole of neutral atoms losing electrons, and one half of a mole of neutral atoms gaining electrons. The amount of energy needed for this process to occur is the amount of energy expended to remove the electrons minus the amount of energy released by the atoms accepting the electrons. Energy required = ionization potential of X atoms − electron affinity of X atoms. Let x = ionization potential of entire mole, and let y = electron affinity of entire mole. It is given that the energy required for this process is 409 kJ and that only half the mole of atoms is involved. Thus, one can write $\frac{1}{2}x - \frac{1}{2}y = 409$ kJ.

In the second process, one half of a mole of X⁻ ions becomes X⁺ ions. To do this, one must remove an electron from each X⁻ ion to form an X atom. This amount of energy needed for this process to occur is equal the opposite of the electron affinity. An electron from the X atom is then removed to form the X⁺ ion. The amount of energy necessary for this reaction to occur is equal to the ionization potential. The sum of the reverse of the electron affinity and the ionization potential is equal to the energy for the overall process required. One can therefore write: $\frac{1}{2}x + \frac{1}{2}y = 733$.

The two processes in this problem are represented by two equations using two variables. These equations are solved simultaneously by adding them together.

$$\frac{1}{2}x - \frac{1}{2}y = 409$$

$$\underline{\frac{1}{2}x + \frac{1}{2}y = 733}$$

$$x = 1142 \text{ kJ/mole}$$

Substituting this value back into one of the equations, one can solve for y. y = 324 kJ/mole. 96.48 kJ/mole = 1 eV/particle. Therefore,

$x = 11.84$ eV/atom,         $y = 3.36$ eV/atom.

When the following reaction

$$F_2 + Cl_2 \rightarrow 2ClF$$

occurs, 26 Kcal/mole are released. The bond energies of $F_2$ and $Cl_2$ are 37 and 58 Kcal/mole, respectively. Calculate the dissociation energy of the ClF bond.

Solution: The dissociation energy is the amount of energy needed to break the chemical bonds of a molecule. The reaction for the dissociation of 2ClF bonds is

$$2ClF \rightarrow F_2 + Cl_2$$

This is the opposite of the reaction described in the problem, therefore, 26 Kcal/mole are consumed. The amount of energy consumed is equal to twice the dissociation energy of ClF (because 2 moles of ClF are present) subtracted from the sum of the dissociation energies of $F_2$ and $Cl_2$

$$\Delta H = \left( \Delta H_{F_2} + \Delta H_{Cl_2} \right) - 2\Delta H_{Cl-F} = -26 \text{ Kcal/mole}$$

$$26 \text{ Kcal/mole} = 2\Delta H_{Cl-F} - 37 \text{ Kcal/mole} - 58 \text{ Kcal/mole}$$

$$2\Delta H_{Cl-F} = 26 \text{ Kcal/mole} + 37 \text{ Kcal/mole} + 58 \text{ Kcal/mole}$$

$$= \frac{121}{2} \text{ Kcal/mole} = 60.5 \text{ Kcal/mole}.$$

Thus, 60.5 Kcal are needed to break one mole of Cl-F bonds.

Based upon the following thermochemical data, show that ozone, $O_3$, is considerably more stable than a cyclic structure would suggest. The enthalpy for the O-O bond is approximately 33 Kcal/mole.

$$1\tfrac{1}{2} O_2 \rightarrow O_3 \qquad \Delta H_{formation} = +34.5 \text{ Kcal}$$

$$O_2 \rightarrow 2O \qquad \Delta H_{dissociation} = +119 \text{ Kcal}.$$

Solution: The bond energy is the amount of energy needed to separate atoms joined in a chemical bond. The greater the bond energy the more stable the bond. The stability of a molecule is proportional to the sum of the bond energies of the bonds in the molecule.

If $O_3$ assumes a cyclic conformation, there are 3 O-O bonds present. Thus, the stability of this structure is proportional to 3 times the bond energy of an O-O bond, 33 Kcal/mole.

stability of the cyclic structure = 3 × 33Kcal/mole or 99 Kcal/mole

One can calculate the actual stability of $O_3$ by using the data supplied in the problem. The overall reaction for the dissociation of $O_3$ is

(i)      $O_3 \rightarrow 3O$.

One can derive this equation from the two given in the problem. The reverse of the equation and the Δ dissociation for the formation of $O_3$ is

(ii)     $O_3 \rightarrow 1\frac{1}{2} O_2$      $\Delta H_{diss}$ = - 34.5 Kcal/mole

If one multiplies the second equation by 3/2 it and its ΔH becomes:

(iii)    $3/2\ O_2 \rightarrow 3O$      $\Delta H_{diss}$ = 178.5 Kcal/mole

One obtains the overall equation (i) by adding together equations (ii) and (iii). One finds the overall ΔH by adding together the ΔH's of these reactions.

(ii)          $O_3 \rightarrow 3/2O_2$    ΔH = - 34 Kcal/mole

(iii) +    $3/2\ O_2 \rightarrow 3O$    + ΔH = 178.5 Kcal/mole

$O_3$+ $3/2O_2 \rightarrow (3/2)O_2$+3O          144.5 Kcal/mole

or        $O_3 \rightarrow 3O$

The actual stability of $O_3$ is, therefore, proportional to 144.5 Kcal/mole. This is greater than that which would be expected if $O_3$ had a cyclic **structure**.

● **PROBLEM** 633

What is the explanation for the following trends in lattice energies?

| | | | |
|---|---|---|---|
| NaF | - 260 Kcal/mole | NaCl | - 186 Kcal/mole |
| NaCl | - 186 Kcal/mole | KCl | - 169 Kcal/mole |
| NaBr | - 177 Kcal/mole | CsCl | - 156 Kcal/mole |

Solution: Lattice energy is an indication of the stability of the species.

In the first column, since the sodium (Na) ion is common to all species, one should concentrate on $F^-$, $Cl^-$, and Br⁻ ions. Is there any change in any property when these ions are compared? Notice that in going from F⁻ to Cl⁻ to Br⁻, there is an increase in ionic size. The same situation exists in the second column in going from $Na^+$ to $K^+$ to $Cs^+$, with Cl⁻ being common to all. How does this relate to lattice energy? The energy derived from two ions in a compound stems from the mutual attraction of the oppositely charged ions. The closer the ions, the greater the attraction and the larger the lattice energy. In each column in this problem, as the size of one ion increases, the ions in the lattice become further apart. As such, there is a decrease in attraction between adjacent ions and, thus, a lowering of the magnitude of the lattice energy.

● **PROBLEM** 634

Write the Born-Haber cycle for the formation of crystalline sodium fluoride ($Na^+F^-$), starting with solid Na and gaseous F. Then, using the thermochemical data supplied below, determine its heat of formation:

(1)  $Na(s)$  →  $Na(g)$  $\Delta H = +26.0$ Kcal:sublimation

(2)  $F_2(g)$  →  $2F(g)$  $\Delta H = +36.6$ Kcal: dissociation

(3)  $Na(g)$  →  $Na^+(g) + e^-$  $\Delta H = +120.0$ Kcal: ionization

(4)  $F(g) + e^- \rightarrow F^-(g)$  $\Delta H = -83.5$ Kcal: electron addition

(5)  $Na^+(g) + F^-(g) \rightarrow Na^+,F^-(s)$  $\Delta H = -216.7$ Kcal: lattice formation.

Solution:  The Born-Haber cycle shows that the heat of the reaction is the sum of the $\Delta H$'s for the sequence of the reactions used in the formation of crystalline sodium fluoride.

The Born-Haber cycle relates reactions which produce the gaseous atoms from solids, those that produce the gaseous ions, and those that indicate the union of these ions to form the crystalline product. Thus, one can write the Born-Haber cycle for this reaction as

$$Na(s) \quad + \quad \tfrac{1}{2}F_2(g) \xrightarrow{(1)+(2)} Na(g) \quad + \quad F(g)$$

$$\downarrow \qquad\qquad (3)\downarrow \qquad\qquad (4)\downarrow$$

$$Na+, F^-(s) \xleftarrow{(5)} Na^+(g) \quad + F^-(g)$$

(The numbers above the arrows correspond to the numbered reactions in the problem.)

One is given the energies, i.e., the **thermodynamic** data, for these processes. If one numbers the reactions in

600

this problem (1) to (5), the reaction becomes (1)+½(2) + (3) + (4) + (5). The ½ value for (2) is derived from the fact that one mole of NaF is required, but, in (2), 2 moles of F(g) atoms are generated. Thus, take ½ this number.

Therefore, heat of reaction is:

$$\Delta H = \Delta H(1) + \tfrac{1}{2}\Delta H(2) + \Delta H(3) + \Delta H(4) + \Delta H(5)$$

$$= 26 + 18.3 + 120 - 83.5 - 216.7 = -135.9 \text{ Kcal/mole.}$$

# BOND LENGTH AND ANGLES

● **PROBLEM** 635

Calculate the interatomic nonbonded distance between 2 bromine nuclei in a molecule of carbontetrabromide (CBr₄). The C-Br distance is 1.94 Å.

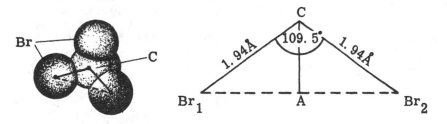

Diagram A              Diagram B

Solution: From diagram A, one can see that CBr₄ is in a tetrahedral conformation. Thus, one knows that the Br-C-Br angle is equal to 109.5°. As such, diagram B is then true.

One is solving for the length of $\overline{Br_1Br_2}$.

Angle $Br_1$-C-A is 109.5°/2 or 54.75°, since one can let line $\overline{CA}$ be an angle bisector. To find $\overline{Br_1A}$, note that $\sin 54.75 = B_1A/1.94$ Å.  Therefore.

$$\overline{Br_1A} = 1.94 \text{ Å} \times \sin 54.75$$

$$\overline{Br_1A} = 1.94 \text{ Å} \times .8166 = 1.58 \text{ Å}$$

$$\overline{Br_1B_2} = 2 \times \overline{Br_1A} = 2 \times 1.58 \ \overset{o}{A} = 3.17 \ \overset{o}{A},$$

since $\overline{Br_1A} = \overline{Br_2A}$ for $\overline{CA}$ is also a median.

● **PROBLEM** 636

Given 0.207 nm for the S-S distance and 105° for the S-S-S bond angle, calculate the S-to-S distance across the $S_8$ ring.

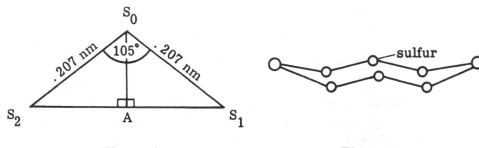

Figure A                    Figure B

Solution:  One is asked to find the distance from $S_1$ to $S_2$ as seen in figure A. This can be done by using trigonometry. Draw a line perpendicular to $S_1S_2$ to A from $S_0$. One can solve for the length of $S_2A$ with the knowledge that the angle $S_2$-$S_0$-A  is equal to ½ of the S-S-S angle or 52.5°. This is true because the triangle $S_0S_1S_2$ is isosceles.

$$\sin 52.5° = \frac{S_2A}{S_2S_0} = 0.793$$

$$S_2A = 0.793 \times S_2S_0 = (0.793) \times (0.207 \ nm) = 0.164 \ nm.$$

$$S_2S_1 = 2 \times S_2A = 2 \times 0.164 \ nm = 0.328 \ nm.$$

● **PROBLEM** 637

Given that the H-to-H distance in $NH_3$ is 0.1624 nm and N-H distance is 0.101 nm, calculate the bond angle H-N-H.

Solution:  Because the NH distances are equal, an isosceles triangle can be drawn as shown. The altitude NX is drawn to form two right triangles. The magnitude of the angle HNX can be found trigonometrically from its sine. The sine of an angle is equal to the length of the opposite side divided by the hypotenuse.

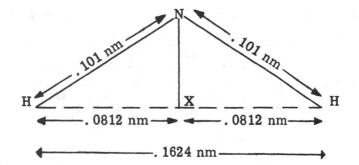

$$\sin HNX = \frac{0.0812 \text{ nm}}{0.101 \text{ nm}} = 0.804.$$

Using a table of trigonometric functions, one can find the angle whose sin is 0.804.

$$\text{sine}^{-1}\ 0.804 = 53.50°.$$

The angle H-N-H is twice HNX. As such, the measure of angle H-N-H is 2 × 53.50° or 107°. It is known that H-N-H is twice HNX because the altitude of an isosceles triangle bisects the vertex angle.

● **PROBLEM** 638

In the trans form of nitrous acid, all the atoms are in the same plane, but the HO is directed away from the other O. The O-H distance is 0.098 nm; the distance from the central N to the hydroxyl O is 0.146 nm; and the distance to the other O is 0.120 nm. If the H-O-N bond angle is 105° and the O-N-O bond angle is 118°, how far is the H from the other O?

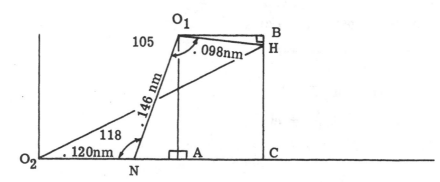

bond distance - molecular structure

<u>Solution</u>:  The formula for nitrous acid is $HNO_2$. The

structure of the trans form as described in the problem is:

One is asked to find the distance from $O_2$ to H. This can be done by solving for the length of $O_2C$ and CH. The sum of the squares of these two lengths is equal to the square of the length of $O_2H$. This can be done through the use of trigonometry. To find $O_2C$, one must find NA and AC. $O_2C$ is equal to the sum of $O_2N$, NA, and AC. The measure of angle ONA is equal to $180° - 118°$ or $62°$, since they are supplementary.

$$\cos 62° = 0.469 = \frac{NA}{NO_1} = \frac{NA}{0.146 \text{ nm}} \quad \left(\frac{\text{adjacent}}{\text{hypotenuse}}\right)$$

$$0.469 = \frac{NA}{0.146 \text{ nm}}$$

$$NA = 0.069 \text{ nm}$$

$$O_2A = NA + O_2N = 0.069 \text{ nm} + 0.120 \text{ nm} = 0.189 \text{ nm}$$

$\overline{AC} = \overline{O_1B}$, since they are the opposite sides of a rectangle.

$$\angle BO_1H = 90° - \angle HO_1A \qquad\qquad (\angle = \text{angle})$$

$$\angle NO_1A = 90° - \angle O_1NA = 90° - 62° = 28°$$

$$\angle HO_1A = \angle HO_1N - \angle NO_1A = 105° - 28° = 77°$$

$$\angle BO_1H = 90° - 77° = 13° \text{ from above.}$$

$$\cos 13° = \frac{\overline{OB}}{\overline{OH}} = 0.974$$

$$\frac{\overline{O_1B}}{\overline{O_1H}} = \frac{\overline{O_1B}}{0.098 \text{ nm}} = 0.974$$

$$\overline{O_1B} = (0.098 \text{ nm}) \times (0.974) = 0.095 \text{ nm}$$

Thus, $\overline{AC} = 0.095$ nm.

Solving for $\overline{O_2C}$:

$$\overline{O_2C} = \overline{O_2N} + \overline{NA} + \overline{AC}$$

$$= 0.120 \text{ nm} + 0.069 \text{ nm} + 0.095 \text{ nm} = 0.284 \text{ nm}$$

$$\overline{HC} = \overline{OA} - \overline{BH}$$

$$\sin \angle ONA = \frac{\overline{O_1A}}{\overline{NO_1}} \qquad\qquad \angle ONA = 62°$$

$$\sin 62° = \frac{\overline{O_1A}}{\overline{NO_1}} = 0.883$$

$$\frac{\overline{O_1A}}{\overline{NO_1}} = 0.883$$

$$\overline{OA} = (0.883) \times \overline{NO_1} = (0.883) \times (0.146 \text{ nm})$$

$$= 0.129 \text{ nm}$$

$$\sin \angle BOH = \frac{\overline{BH}}{\overline{O_1H}} \qquad \angle BOH = 13°$$

$$\sin 13° = \frac{\overline{BH}}{0.098 \text{ nm}} = 0.225$$

$$\overline{BH} = (0.098 \text{ nm}) \times (0.225) = 0.022 \text{ nm}$$

$$\overline{HC} = 0.129 \text{ nm} - 0.022 \text{ nm} = 0.107 \text{ nm}$$

$$(\overline{O_2H})^2 = (\overline{O_2C})^2 + (\overline{HC})^2$$

$$(\overline{O_2H})^2 = (0.284 \text{ nm})^2 + (0.107 \text{ nm})^2$$

$$\overline{O_2H}^2 = 0.081 \text{ nm}^2 + 0.011 \text{ nm}^2$$

$$= 0.092 \text{ nm}^2$$

$$\overline{O_2H} = 0.303 \text{ nm}.$$

● **PROBLEM** 639

What effect do bond angles have on bond strain? What is the influence of bond strain on bond energy?

Solution: All molecules have an optimum configuration of their atoms in space. Bond angles are a function of this configuration. For example, carbon atoms that have four bonds assume a tetrahedral arrangement with bond angles of 109.5°. When the arrangement of the atoms is forced to differ from this configuration by external forces, the bond angles must, therefore, adjust and are said to be strained.

The bond energy is the amount of energy necessary to break a bond between two atoms. When the bonds of a molecule are strained, less energy is needed to break the bonds. Hence, the greater the bond strain the lower the bond energy.

# POLARITY OF BONDS

Determine which of the atoms in each pair possess a
partial positive charge and which a partial negative.
(a) the O-F bond, (b) the O-N bond, (c) the O-S bond.
Electronegativity values for these elements can be
found from a table of electronegativities.

<u>Solution</u>: Electronegativity is the tendency of an
atom to attract shared electrons in a chemical covalent
bond. Since electrons are negatively charged, to find
the partial charge on each atom of the pair consult
a table for the electronegativity values of the atoms
in the molecules. The atom with the higher value will
have a greater tendency to attract electrons, and will,
thus, have a partial negative charge. Because the overall bond is neutral, the other atom must have a partial
positive charge.

For part (a), F has a higher electronegativity
than O, which means that F will have a partial negative
charge ($\delta^-$) and O will have a partial positive charge
($^+\delta$). To show this the molecule can be written
$\left( \overset{\delta^+}{O} - \overset{\delta^-}{F} \right)$. Similar logic is used in working out parts b
and c.

(b) O-N bond:

The electronegativity of O is 3.5 and of N is 3.0,
thus O is more negative than N. The molecule is written
$\left( \overset{\delta^-}{O} - \overset{\delta^+}{N} \right)$.

(c) O-S bond:

The electronegativity of O is 3.5 and of S is 2.5;
therefore, O is the more negative of this pair,
$\left( \overset{\delta^-}{O} - \overset{\delta^+}{S} \right)$.

Which molecule of each of the following pairs would
exhibit a higher degree of polarity. HCl and HBr, $H_2O$
and $H_2S$; BrCl and IF?

<u>Solution</u>: Polarity indicates that there is an uneven
sharing of electrons between 2 atoms. This creates a

charge distribution in the molecule where one atom is partially positive and the other is partially negative. The degree of polarity is measured by finding the difference in the abilities of the two atoms to attract electrons. This tendency to accept electrons is called the electronegativity. The greater the electronegativity difference, the greater the degree of polarity.

From the table of electronegativity values, the following electronegativities can be obtained:

| | Compounds | Electronegativity Difference |
|---|---|---|
| (1) | HCl | $3.0 - 2.1 = .9$ |
| | HBr | $2.8 - 2.1 = .7$ |
| (2) | $H_2O$ | $3.5 - 2.1 = 1.4$ |
| | $H_2S$ | $2.5 - 2.1 = .4$ |
| (3) | BrCl | $3.0 - 2.8 = .2$ |
| | IF | $4.0 - 2.5 = 1.5$ |

In pair (1) (HCl and HBr), HCl has the larger electronegativity. Hence, HCl has a greater degree of polarity than Hbr. For the same reasons, $H_2O$ in pair (2) and IF in pair (3) have the greater degrees of polarity.

● **PROBLEM** 642

Of the following pairs, which member should exhibit the largest dipole moment. Use the data from the accompanying table. (a) H-O and H-N; (b) H-F and H-Br; (c) C-O and C-S.

*Pauling electronegativities*
*(H = 2.1)*

| Li | Be | B | | | | | | | | | | | C | N | O | F |
|---|---|---|---|---|---|---|---|---|---|---|---|---|---|---|---|---|
| 1.0 | 1.5 | 2.0 | | | | | | | | | | | 2.5 | 3.0 | 3.5 | 4.0 |
| Na | Mg | Al | | | | | | | | | | | Si | P | S | Cl |
| 0.9 | 1.2 | 1.5 | | | | | | | | | | | 1.8 | 2.1 | 2.5 | 3.0 |
| K | Ca | Sc | Ti | V | Cr | Mn | Fe | Co | Ni | Cu | Zn | Ga | Ge | As | Se | Br |
| 0.8 | 1.0 | 1.3 | 1.5 | 1.6 | 1.6 | 1.5 | 1.8 | 1.8 | 1.8 | 1.9 | 1.6 | 1.6 | 1.8 | 2.0 | 2.4 | 2.8 |
| Rb | Sr | Y | Zr | Nb | Mo | Tc | Ru | Rh | Pd | Ag | Cd | In | Sn | Sb | Te | I |
| 0.8 | 1.0 | 1.2 | 1.4 | 1.6 | 1.8 | 1.9 | 2.2 | 2.2 | 2.2 | 1.9 | 1.7 | 1.7 | 1.8 | 1.9 | 2.1 | 2.5 |
| Cs | Ba | La–Lu | Hf | Ta | W | Re | Os | Ir | Pt | Au | Hg | Tl | Pb | Bi | Po | At |
| 0.7 | 0.9 | 1.1–1.2 | 1.3 | 1.5 | 1.7 | 1.9 | 2.2 | 2.2 | 2.2 | 2.4 | 1.9 | 1.8 | 1.8 | 1.9 | 2.0 | 2.2 |
| Fr | Ra | Ac | Th | Pa | U | Np | | | | | | | | | | |
| 0.7 | 0.9 | 1.1 | 1.3 | 1.5 | 1.7 | 1.3 | | | | | | | | | | |

The values given in the table refer to the common oxidation states of the elements. For some elements variation of the electronegativity with oxidation number is observed, for example, Fe(II) 1.8, Fe(III) 1.9; Cu(I) 1.9, Cu(II) 2.0; Sn(II) 1.8, Sn(IV) 1.9.

Solution: A dipole consists of a positive and a negative charge separated by some distance. Quantitatively, a dipole is described by giving its dipole moment, which is equal to the charge times the distance between the positive and negative centers. The polarity of a bond is measured as the magnitude of the moment of the dipole. Thus, to find which member has the higher dipole moment, determine which bond has the greatest polarity. The polarity of the bond is indicated by the difference in electronegativities of the atoms (i.e., the difference in their tendency to attract shared electrons in a chemical covalent bond). The greater the difference in electronegativities, the greater the bond polarity, giving it the greater dipole moment. From the table, note that in part

(a) the difference in electronegativities of H-O is 3.5 - 2.1 or 1.4. For H-N, the difference is 3.0 - 2.1 or 0.9. Therefore, H-O has the largest dipole moment. Using similar calculations, one can determine the bonds with the larger dipole moments in parts b and c.

(b) H-F and H-Br

For H-F:   4.0 - 2.1 = 1.9

For H-Br: 2.8 - 2.1 = 0.7.

Thus, H-F has the larger dipole moment.

(c) C-O and C-S

For C-O: 3.5 - 2.5 = 1.0

For C-S; 2.5 - 2.5 = 0.

Thus, C-O has the larger dipole moment here.

● PROBLEM 643

Find the net dipole moment in Debyes for each of the following situations: (a) One + 1 and one - 1 ion separated by $2 \times 10^{-8}$ cm, and (b) one +2 and one - 2 ion separated by $2 \times 10^{-8}$ cm. One charge = $4.80 \times 10^{-10}$ esu and 1 Debye = $1 \times 10^{-18}$ esu-cm.

Solution: A dipole consists of a positive and a negative charge separated by a distance. Quantitatively, a dipole is described by its dipole moment, which is equal to the charge times the distance between the positive and negative centers. For both situations, substitute the values into this quantitative expression:

net dipole moment = charge × distance

(a) Net dipole moment

$$= (4.80 \times 10^{-10} \text{ esu})(2 \times 10^{-8} \text{ cm})$$

$$= 9.6 \times 10^{-18} \text{ esu-cm.}$$

1 Debye = $1 \times 10^{-18}$ esu-cm.

$$\frac{9.6 \times 10^{-18} \text{ esu-cm}}{1 \times 10^{-18} \text{ esu-cm/Debye}} = 9.6 \text{ Debyes}$$

(b) Net dipole moment

$$= (2)(4.80 \times 10^{-10})(2 \times 10^{-8})$$

$$= 19.2 \times 10^{-18} \text{ esu-cm}$$

Converting to Debyes

$$\frac{19.2 \times 10^{-18} \text{ esu-cm}}{1 \times 10^{-18} \text{ esu-cm/Debye}} = 19.2 \text{ Debyes.}$$

● **PROBLEM** 644

The net dipole moment of water is 1.84 debyes and the bond angle is 104.45°. What moment can be assigned to each O-H bond?

(a)

(b)

(c)

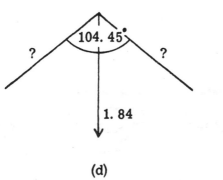

(d)

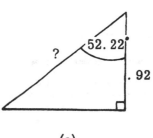

(e)

609

<u>Solution</u>: To answer this problem, consider the definition of a dipole, how it is expressed quantitatively, and the geometry of a water molecule.

A dipole consists of a positive and negative charge separated by some distance. Quantitatively, it is expressed as dipole moment, which is the charge times the distance between the positive and negative centers. The units are debyes.

The geometry of a water molecule is not linear. It is represented by FIG. a.

The oxygen atom shares its electrons with the hydrogen atoms; covalent bonding. Because the oxygen atom is electronegative, which means it has a tendency to attract electrons, the shared electrons will spend more time on the oxygen atom than on the hydrogen atoms. Thus, a polar covalent bond exists. The oxygen tends to develop a negative charge because the electrons spend more time on the oxygen atom. Since a water molecule is neutral (overall), the charge on the hydrogen atoms become positively charged (Fig. b). Therefore, a dipole exists. The net dipole is represented by Fig. c. As given, the dipole moment is 1.84 D. One is asked to compute the moment of each OH bond. One can represent the net moment and the OH moment as in Fig. d. The net dipole moment is the longer line and bisects the bond angle of 104.45°. The shorter lines represent the components of the net dipole, which comes from each bond.

This can be simplified and drawn as Fig. e. This is the representation for one OH bond. The net dipole must be divided by two (1.84/2 = 0.92), since the total number O-H bonds is 2. The bisected angle of 104.45° yields 52.22°. To find X, the unknown OH dipole, use trigonometry.

The $\cos 52.22°$ is $= \dfrac{\text{adjacent leg}}{\text{hypotenuse}} = \dfrac{.92}{X}$ , since

a right triangle is present. $\cos 52.22° = 0.6127$. Thus, equating $0.6127 = 0.92/X$. Therefore,

$$X = \frac{.92}{.6127} = 1.50 \text{ Debyes.}$$

● **PROBLEM** 645

Given that the electronegativities of F, Cl, Br, and I are, respectively, 4.0, 3.0, 2.8, and 2.5, account for the fact that the dipole moment decreases in the sequence HF, HCl, HBr, and HI, even though bond length and the number of electrons increase.

<u>Solution</u>:  A dipole is defined as a molecule which is
electrically unsymmetrical - that is, the centers of
positive and negative charges are not located at the same
point within the molecule.

Quantitatively, a dipole is described by its dipole
moment, which is equal to the charge times the distance
between the positive and negative centers. The decrease
in dipole moment will reflect a decrease in either the
value of the charges or the distance between them. It is
given that the bond length increases in this sequence,
which means that distance increases. This suggests an increase in dipole moment within the sequence. But the dipole
moments decrease, this means that the charge must be decreasing. This can be explained by noting that the electronegativity values are decreasing within the sequence.
Electronegativity measures the tendency of an atom to
attract shared electrons in a molecule. By attracting
these electrons, it develops a negative charge. If the
electronegativity decreases, so does the charge, which
would then account for the decrease in dipole moment.

# CHAPTER 18

# QUANTUM CHEMISTRY

Basic Attacks and Strategies for Solving Problems
in this Chapter. See pages 613 to 669 for step-by-
step solutions to problems.

The name "quantum" comes from the concept that an electron (or any other particle) can only have discrete energies. Most values of the energy are forbidden so that if an electron changes energy it must do so in a quantum jump.

All electrons in the atoms can be described by four quantum numbers and no two electrons can have the same set of quantum numbers. The quantum numbers are, for historical reasons, called respectively $n$, $l$, $M_l$, and $M_s$. $n$ is the principal quantum number, $l$ is the orbital quantum number, $M_l$ is the magnetic quantum number, and $M_s$ is the spin. The allowed quantum numbers are as follows:

| Quantum No. | Permitted Values |
|:---:|:---:|
| $n$ | 1, 2, 3,... etc. |
| $l$ | 0, 1, 2, ... $n-1$ |
| $M_l$ | $+l$ to $-l$ |
| $M_s$ | $+1/2$, $-1/2$ only |

In Chapter 17, the electronic configuration designations, for example $1s^2$, $2s^2$, $2p^2$ for carbon, are, in fact, a tabulation of the quantum numbers of each of the electrons in the atom. Tables 1 and 2 show the relationship between the designation of Chapter 17 and the quantum numbers associated with each electron. In Table 1 the following electronic configuration for neon is tabulated by the rules discussed in Chapter 17.

| Table 1: Electronic Configuration of Neon | | | | | | | |
|---|---|---|---|---|---|---|---|
| Element | At. No. | 1s | 2s | 2p | 3s | 3p | Electronic Configuration |
| Ne | 10 | 2 | 2 | 6 | | | $1s^2, 2s^2, 2p^6$ |

| Table 2: Quantum Numbers for Electrons in Neon Atom | | | | |
|---|---|---|---|---|
| $n$ | $l$ | $M_l$ | $M_s$ | Designation of Chapter 17 |
| 1 | 0 | 0 | +1/2 | 1s |
| 1 | 0 | 0 | −1/2 | 1s (two total 1s electrons) |
| 2 | 0 | 0 | +1/2 | 2s |
| 2 | 0 | 0 | −1/2 | 2s (two total 2s electrons) |
| 2 | 1 | +1 | +1/2 | 2p |
| 2 | 1 | +1 | −1/2 | 2p |
| 2 | 1 | 0 | +1/2 | 2p |
| 2 | 1 | 0 | −1/2 | 2p |
| 2 | 1 | −1 | +1/2 | 2p |
| 2 | 1 | −1 | −1/2 | 2p (six total 2p electrons) |

The "exclusion" of two electrons having the same set of quantum numbers is called the Pauli exclusion principle. The order in which the quantum numbers occur in the ground, or lowest, energy state is governed by Hund's rule. This rule is outlined in detail in Problem 650.

Two principal types of bonds, sigma ($\sigma$) and pi ($\pi$) bonds, can be formed in a molecular orbital. The sigma bond is formed when the molecular orbital (of shared electrons) is symmetrical around the nucleus while the pi bond is formed from parallel $p$ orbitals. A succinct description of $\sigma$ and $\pi$ bonds is given in Problem 657, and sketches of these molecular orbitals are shown in the problems which follow.

An early attempt to describe the quantum nature of electrons was undertaken by Niels Bohr, and his model is called the Bohr atom. It is simplistic by modern understanding but, for his time, represented a giant step forward in understanding atomic structure. Bohr postulated that the hydrogen atom consisted of a nucleus and a single electron in orbit around the nucleus much as a planet of the solar system orbits the sun. The Coulombic attractive forces are balanced by the centrifugal force of the electron in orbit.

$$e^2/r^2 = mv^2/r \qquad\qquad\qquad\qquad \text{18-1}$$

Where $e$ = the charge on one electron, $r$ = distance between the nucleus and the electron, and $mv^2/r$ = centrifugal force. The quantum nature of the electronic energy was introduced by postulating that the angular momentum could equal only quantized values.

$$mvr = nh/2\pi \qquad\qquad\qquad\qquad \text{18-2}$$

where $h$ is Planck's constant ($6.626 \times 10^{-27}$ erg sec) and $n$ is the principal quantum number. From Bohr's theory it is possible to calculate the orbital radius from each quantum number $n$.

The quantum theory indicates matter has both wave and particle characteristics. The deBroglie equation relates the momentum of a particle to its wavelength.

$$\lambda = h/p \qquad\qquad\qquad\qquad \text{18-3}$$

where $\lambda$ is the wavelength and $p$ the momentum ($m \bullet v$) of a particle. An electron, because of its wave character, is not a particle as hypothesized by Bohr. Its position and energy are uncertain as specified by the Heisenberg uncertainty principle. The Heisenberg uncertainty principle states that it is impossible to simultaneously determine the exact position and momentum of a particle and is given by:

$$(\Delta x)(m\Delta V_x) \geq h/4\pi \qquad\qquad\qquad\qquad \text{18-4}$$

where $m\Delta V_x$ is equal to the momentum in the $x$ direction and $x$ is equal to the position of the particle.

The energy of a quantum of light or other electromagnetic radiation is given by the equation

$$E = h\nu = \frac{hc}{\lambda} \qquad\qquad\qquad\qquad \text{18-5}$$

where $\nu$ is the frequency, $c$ is the speed of light, and $\lambda$ is the wavelength. Because of the quantum nature of energies which an electron can possess, when it moves from one energy state to another, it emits or absorbs energy of a wavelength given by Equation 18-5. This principle forms the basis of the science of atomic spectroscopy, which is the study of atomic structure from light emissions. A diagram of the energy levels showing relative energies for an atom and one electron is shown in Problem 692.

Step-by-Step Solutions to
Problems in this Chapter,
"Quantum Chemistry"

PAULI EXCLUSION PRINCIPLE, HUND'S RULE AND
ELECTRONIC CONFIGURATION

● **PROBLEM** 646

Explain the following: Pauli exclusion principle and
Hund's rule.

<u>Solution</u>: An atom is described by four quantum numbers:
the principal quantum number N, the angular momentum
quantum number ℓ, the magnetic quantum number m, and
the spin quantum number s. According to the Pauli exclusion principle, no two electrons in an atom can have
the same four quantum numbers. If two electrons did, they
would have identical fingerprints, a situation forbidden
by nature.

Each orbital can accommodate a maximum of two
electrons. With this in mind, Hund's rule states that
once an electron is in an orbital, a second electron will
not enter into that same orbital if there exist other
orbitals in that subshell that contain zero electrons.
In other words, all orbitals in a subshell must contain
one electron, before a second one can enter. Hund's rule
also states that single electrons in their separate
orbitals of a given subshell will have the same spin
quantum number.

● **PROBLEM** 647

Discuss the following statement: The Pauli exclusion
principle is the main reason why atoms do not collapse to
a point.

<u>Solution</u>: The Pauli exclusion principle states that no
two electrons in the same atom can be completely identical:

that is, have the same values for all four quantum numbers. The quantum numbers measure the energy level of the electron, the order of increasing distance of the average electron distribution from the nucleus, the angular shape of electron distribution, (the electronic magnetism), and the two possible orientations of electron spin. An electron in an atom is completely described by its four quantum numbers. Inherent in this description is the point of location of an electron at a specific time. If two electrons have the same four quantum numbers, they would have the same point of location at a specific time. This means, therefore, that all the electrons would collapse to a point at a specific time. The Pauli exclusion principle states that no two electrons can have the same four quantum numbers and therefore the atom cannot collapse to a point.

● **PROBLEM** 648

Given the ground-state oxygen atom, tabulate each of the electrons by its quantum number.

<u>Solution</u>: Begin by writing the electronic configuration of the ground-state oxygen atom. An electron is in its ground state when it is in its lowest energy level, this is the normal state of the electrons. The atomic number of oxygen is 8. Thus, the oxygen atom has 8 electrons. The electronic configuration of the oxygen atom is $1s^2 2s^2 2p^4$. The quantum numbers are N, $\ell$, M, and s. N is the principal quantum number which, in the electronic configuration, is the integer in front of the subshells (s and p) and corresponds to the energy level occupied by the electron. The other quantum numbers can be found once N is determined. $\ell$ is equal to 0, 1, 2, ... N-1, depending upon N. M is equal to + $\ell$ to -$\ell$ and S is equal to +$\frac{1}{2}$ or - $\frac{1}{2}$. $\ell$ is the orbital quantum number which denotes the subshell and angular shape of electron distribution. M is the magnetic quantum number and s denotes the spin of electron. All four are needed to describe each electron. To tabulate the electrons according to their quantum numbers proceed as follows: the electron configuration is $1s^2 2s^2 2p^4$. Take the first shell, or N = 1, $1s^2$. The superscript indicates that 2 electrons are present in this shell. Thus, 2 sets of quantum numbers are needed. For both electrons, N = 1. Because $\ell = 0 \ldots$ n - 1, $\ell = 0$ for both electrons in the s subshell. M + $\ell$ to - $\ell$, thus, M = 0 for both also. The four quantum numbers for each electron must be different. Therefore, one electron has a spin of + $\frac{1}{2}$ and one has a spin of - $\frac{1}{2}$. The sets of quantum numbers for these two electrons are 1, 0, 0, + $\frac{1}{2}$ and 1, 0, 0, - $\frac{1}{2}$. For the second shell, N = 2. There are two subshells, s and p. Consider the s subshell first. In the $2s^2$ orbital, N = 2 and there are two electrons to be described. For both electrons $\ell = 0$, because both are in the s subshell. M = 0, because M = + $\ell$ to - $\ell$. For one electron s = +$\frac{1}{2}$, for the other s = - $\frac{1}{2}$. The sets of quantum numbers for these two electrons

are 2, 0, 0, + ½ and 2, 0, 0, - ½. In the 2p⁴ orbital
there are 4 electrons to be described. For all, N = 2,
and ℓ = 1 (ℓ is always equal to 1 for a p subshell).
M = + ℓ to - ℓ. Thus, one electron has M = + 1, the
second electron has M = 0, and the third electron has M =
- 1. The fourth electron will have an M = - 1, 0, or + 1.
Whatever it is, it will differ in spin from the electron
with the same M. The spins of the others can be + ½ or
- ½. The sets of the quantum numbers for the first three
electrons in a 2p orbital can be written (2, 1, - 1, +½),
(2, 1, 0. +½), and (2, 1, 1, + ½). The fourth might be
written (2, 1, - 1, - ½), (2, 1, 0, - ½), or
(2, 1, 1, - ½).

● **PROBLEM** 649

Write possible sets of quantum numbers for electrons in
the second main energy level.

**Solution**: In wave mechanical theory, four quantum numbers
are needed to describe the electrons of an atom. The first
or principal quantum number, n, designates the main energy
level of the electron and has integral values of 1, 2, 3,
... . The second quantum number, ℓ, designates the energy
sublevel within the main energy level. The values of ℓ
depend upon the value of n and range from zero to n - 1.
The third quantum number, $m_\ell$, designates the particular
orbital within the energy sublevel. The number of orbitals
of a given kind per energy sublevel is equal to the number
of $m_\ell$ values (2ℓ + 1). The quantum number $m_\ell$ can have any
integral value from + ℓ to - ℓ including zero. The fourth
quantum number, s, describes the two ways in which an e-
lectron may be aligned with a magnetic field (+ ½ or - ½).

The states of the electrons within atoms are described
by four quantum numbers, n, ℓ, $m_\ell$, s. Another important
factor is the Pauli exclusion principle which states that
no two electrons within the same atom may have the same
four quantum numbers.

To solve this problem one must use the principles of
assigning electrons to their orbitals.

If n = 2; ℓ can then have the values 0, 1; $m_\ell$ can
have the values of + 1, 0 or - 1, and s is always + ½ or
- ½.

Thus, the answer is:

$$n = 2$$

$$\ell = 0, 1$$

$$m_\ell = +1, 0, -1$$

$$s = +\tfrac{1}{2}, -\tfrac{1}{2}$$

| | $\underline{n}$ | $\underline{\ell}$ | $\underline{m_\ell}$ | $\underline{m_s}$ |
|---|---|---|---|---|
| 2s | 2 | 0 | 0 | $+\tfrac{1}{2}$ |
| | 2 | 0 | 0 | $-\tfrac{1}{2}$ |
| | 2 | 1 | +1 | $+\tfrac{1}{2}$ |
| | 2 | 1 | +1 | $-\tfrac{1}{2}$ |
| 2p | 2 | 1 | 0 | $+\tfrac{1}{2}$ |
| | 2 | 1 | 0 | $-\tfrac{1}{2}$ |
| | 2 | 1 | -1 | $+\tfrac{1}{2}$ |
| | 2 | 1 | -1 | $-\tfrac{1}{2}$ |

● **PROBLEM** 650

Apply Hund's rules to obtain the electron configuration for Si, P, S, Cl, and Ar.

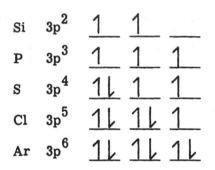

Si $3p^2$

P $3p^3$

S $3p^4$

Cl $3p^5$

Ar $3p^6$

Solution: The ground state of an atom is that in which the electrons are in the lowest possible energy level. Each level may contain two electrons of opposite spin. When there are several equivalent orbitals of the same energy, Hund's rules are used to decide how the electrons are to be distributed between the orbitals:

1) If the number of electrons is equal to or less than the number of equivalent orbitals, then the electrons are assigned to different orbitals.

2) If two electrons occupy two different orbitals, their spins will be parallel in the ground state. Hund's rules states that the electrons attain positions as far apart as possible which minimizes the repulsion obtained from interelectronic forces.

To solve this problem one must:

(1) Find the total number of electrons within the atom

(2) Determine the number of valence electrons

(3) Find the number of electrons in the highest equivalent energy orbital.

The total number of electrons in an atom is equal to that atom's atomic number.

Thus,      Si  has  14 electrons

           P   has  15 electrons

           S   has  16 electrons

           Cl  has  17 electrons

           Ar  has  18 electrons.

Next, from the orbital configuration:

$_{14}$Si     $1s^2 2s^2 2p^6 3s^2 \mid 3p^2$

$_{15}$P      $1s^2 2s^2 2p^6 3s^2 \mid 3p^3$

$_{16}$S      $1s^2 2s^2 2p^6 3s^2 \mid 3p^4$

$_{17}$Cl     $1s^2 2s^2 2p^6 3s^2 \mid 3p^5$

$_{18}$Ar     $1s^2 2s^2 2p^6 3s^2 \mid 3p^6$

One knows that the highest equivalent energy orbital is the 3p orbital. The number of electrons in this orbital increases by 1 starting with 2 for Si, then, 3 for P, 4 for S, 5 for Cl, and 6 Ar. Thus, Ar closes this orbital. Using Hund's rules, the electron configurations can be written as shown in the accompanying figure.

● **PROBLEM** 651

Under ordinary conditions of temperature and pressure, nitrogen gas exists as diatomic molecules ($N_2$). As a result of its electronic configuration, $N_2$ is very inert and requires extreme conditions before it will react with any other species. What is the electronic ground state configuration of a nitrogen atom (atomic number = 7)?

**1s**

**2s**

**2p**

Solution: The solution to this problem involves the application of the Pauli Exclusion Principle and of Hund's Rule.

Since the atomic number of nitrogen is 7 and it is a neutral atom, there are 7 electrons to place in atomic orbitals. The two orbitals of lowest energy are the 1s and the 2s. These are filled according to the Pauli Exclusion Principle - 2 electrons of unpaired spin in each orbital. The orbitals with the lowest energy of those remaining are the three 2p orbitals, $2p_x$, $2p_y$, $2p_z$. Hund's rule dictates that the remaining three e-lectrons enter these orbitals singly and with parallel spin.

The ground state electronic distribution for nitrogen can therefore be pictured as shown in the accompanying Figure.

● **PROBLEM** 652

Predict the total spin for each of the following electronic configurations: (a) $1s^2 2s^1$; (b) $1s^2 2s^2 2p^3$; and (c) $1s^2 2s^2 2p^6 3s^2 3p^6 3d^5 4s^2$.

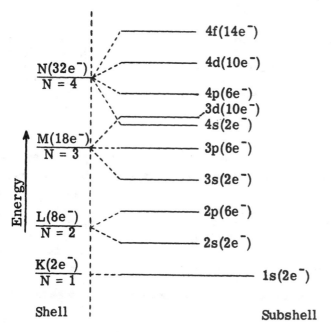

Figure A

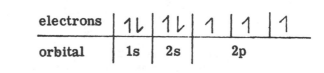

Figure B

Figure C

Solution: One determines the total spin of an atom by calculating the sum of all the s quantum numbers of the electrons. There are four quantum numbers needed to describe each electron. They are the principal quantum number n, the azimuthal quantum number $\ell$, the magnetic quantum number m and the spin quantum number s.

In each orbital of the atom two electrons can be accommodated. These electrons will have opposing spins of $+\frac{1}{2}$ and $-\frac{1}{2}$. The total spin is calculated by determining the total number of unpaired electrons and multiplying this number by $\frac{1}{2}$. It is assumed that all unpaired electrons will have the same spin. Solving for the total spins: Figure A shows the total number of electrons that can occupy each orbital.

(a) $1s^2 2s^2$. Filled s orbitals each contain 2 electrons. One writes this electronic configuration **diagrammatically as shown in figure B.**

Because there are no unpaired electrons the total spin for this atom is 0.

(b) $1s^2 2s^2 2p^3$. Here the two s orbitals are filled but the p orbital is only half filled. There are three subshells in the p orbital and there will be one electron occupying each of these. The configuration of lowest e-nergy is assumed when each of these 3 electrons has the same spin. This configuration is written diagrammatically as shown in figure C.

This configuration contains 3 unpaired electrons. The total spin is, thus $3 \times \frac{1}{2}$ or 3/2.

(c) $1s^2 2s^2 2p^6 3s^2 3p^6 3d^5 4s^2$. From figure A one sees that the only unfilled orbital in this configuration is the 3d. Therefore, unpaired electrons can only occur in this shell. From figure A, one sees that the 3d orbital can contain 10 electrons. Because each subshell contains 2 electrons, there are 5 subshells in this orbital. In this configuration there are 5 electrons occupying the 3d orbital, therefore, each subshell contains 1 electron and there are 5 unpaired electrons in the configuration. The total spin of this configuration is, thus, $5 \times \frac{1}{2}$ or 5/2.

● **PROBLEM** 653

Assuming all other orbitals to have zero net spin, what would be the net spin and the multiplicity of an atom having each of the following outer shell configurations; $2p^1$, $2p^3$, $2p^5$, $3d^1$, $3d^3$, $3d^5$? Rank these net spins in order of increasing paramagnetism.

Solution: We proceed by first defining "total spin" and

| Configuration | Orbital Filling | $n_\uparrow$ | $n_\downarrow$ | $S = \left\lvert \dfrac{n_\uparrow}{2} - \dfrac{n_\downarrow}{2} \right\rvert$ | $m = 2S + 1$ |
|---|---|---|---|---|---|
| $p^1$ | $P_x$ (↑), $P_y$ ( ), $P_z$ ( ) | 1 | 0 | 1/2 | 2 (doublet) |
| $p^3$ | $P_x$ (↑), $P_y$ (↑), $P_z$ (↑) | 3 | 0 | 3/2 | 4 (quartet) |
| $p^5$ | $P_x$ (↑↓), $P_y$ (↑↓), $P_z$ (↑) | 1 | 0 | 1/2 | 2 (doublet) |
| $d^1$ | $d_{xy}$ (↑), $d_{yz}$ ( ), $d_{zx}$ ( ), $d_{x^2-y^2}$ ( ), $d_{z^2}$ ( ) | 1 | 0 | 1/2 | 2 (doublet) |
| $d^3$ | $d_{xy}$ (↑), $d_{yz}$ (↑), $d_{zx}$ (↑), $d_{x^2-y^2}$ ( ), $d_{z^2}$ ( ) | 3 | 0 | 3/2 | 4 (quartet) |
| $d^5$ | $d_{xy}$ (↑), $d_{yz}$ (↑), $d_{zx}$ (↑), $d_{x^2-y^2}$ (↑), $d_{z^2}$ (↑) | 5 | 0 | 5/2 | 6 (sextet) |

"multiplicity". Let the number of electrons having "spin up" be denoted by $n_\uparrow$ and the total number of electrons having "spin down" be denoted by $n_\downarrow$. Assigning a value of $+ \frac{1}{2}$ to spin up and $- \frac{1}{2}$ to spin down, the total spin s is given by the absolute value of $n_\uparrow (+ \frac{1}{2}) + n_\downarrow (- \frac{1}{2}) = n_\uparrow/2 + n_\downarrow/2$. The multiplicity is then defined as $m = 2s + 1$.

The configuration with the highest spin (highest multiplicity) will have the greatest paramagnetism. (Substances that are weakly attracted to magnets are paramagnetic.) The configuration with the lowest spin (lowest multiplicity), will be the least paramagnetic.

The p orbital can contain six electrons and the d orbital can contain 10 electrons. When these orbitals are filled, electrons are sequentially placed in each sub-orbital (e.g. the $p_x$ or the $d_{xy}$) until there is one e-lectron in each suborbital, all electrons having the same spin. Only when there is one electron in each suborbital does another electron enter a suborbital already containing an electron, this time with a spin opposite to that of the electron already in the suborbital (the spins are then "paired") in accordance with the Pauli Exclusion Principle.

In the accompanying table we show the orbital filling corresponding to each configuration. Counting $n_\uparrow$ and $n_\downarrow$, we then calculate s and m.

The configurations $p^1$, $p^5$, and $d^1$ have the same electron paramagnetism (all have spin $\frac{1}{2}$). The next most paramagnetic configurations are $p^3$ and $d^3$ (spin 3/2) and the most paramagnetic configuration is $d^5$ (spin 5/2).

● **PROBLEM** 654

Given the following electron configurations: (a) (Z = 11) $1s^2 2s^2 2p^6$, (b) (Z = 25) $1s^2 2s^2 2p^6 3s^2 3p^6 3d^5$, and (c) (Z = 29) $1s^2 2s^2 2p^6 3s^2 3p^6 3d^{10}$; provide the proper ionic symbol.

Solution: The ionic symbol of an atom is equal to the charge on the atom. This charge is determined by comparing the atomic number Z and the number of electrons as shown in the electronic configuration. The atomic number corresponds to the net positive charge on the nucleus and the number of electrons indicates magnitude of the negative charge of the electron cloud.

One writes the electronic configuration of hydrogen as $1s^1$, 1s indicates the atomic orbital and the superscript 1 indicates that there is one electron in the orbital. Thus, one can determine the number of electrons present by taking the sum of the superscripts. The net charge on an atom is found by adding the net negative charge (the sum of the electrons) and the net positive charge (the atomic number, which is equal to the number of protons) For hydrogen, (Z = 1) $1s^1$, the net negative charge is - 1 and the net positive charge is + 1, thus, the atom is neutral and no ionic symbol is used. One uses this method to determine the ionic symbols for the atoms described in the problem.

(a) (Z = 11) $1s^2 2s^2 2p^6$. The net negative charge is equal to (- 2) + (- 2) + (- 6) or - 10. The net positive charge is 11. The net charge on the atom is 11 - 10 = + 1. From the periodic table one sees that this atom is $Na^+$, because the atomic number of sodium is 11.

(b) (Z = 25) $1s^2 2s^2 2p^6 3s^2 3p^6 3d^5$. Total number of electrons in configuration is 2 + 2 + 6 + 2 + 6 + 5 = 23. The net charge on the atom is, then 25 - 23 or + 2. From the periodic table one can determine that this atom is $Mn^{+2}$.

(c) (Z = 29) $1s^2 2s^2 2p^6 3s^2 3p^6 3d^{10}$. Total number of electrons in configuration = 2 + 2 + 6 + 2 + 6 + 10 = 28. Z = 29 = number of electrons in a neutral atom. Difference: 29 - 28 = + 1. The atom is $Cu^{+1}$.

● **PROBLEM** 655

Assume that 90% of the electron density is representative of the volume of the atom. The following atomic radii have been obtained for 7 elements across the second period of the periodic table and for six elements down through the first family:

| | | | | | | |
|---|---|---|---|---|---|---|
| H | | | | | | |
| 0.37 | | | | | | |
| Li | Be | B | C | N | O | F |
| 1.23 | .89 | .80 | .77 | .74 | .74 | .72 |
| Na | | | | | | |
| 1.57 | | | | | | |
| K | | | | | | |
| 2.03 | | | | | | |
| Rb | | | | | | |
| 2.16 | | | | | | |
| Cs | | | | | | |
| 2.35 | | | | | | |

Explain the observed trends of these atomic radii.

Solution: Atomic radii increase going down a family, but decrease going across a period. To explain this, one must consider the change in the elements in going across a period or down a family and see how it could affect atomic radii. In moving across a period, the atomic number increases, which means the nuclear positive charge increases. But the principal quantum number of the outside electrons remains constant. This means that the outer electrons, assuming all other factors, are constant, will remain at the same distance from the nucleus.

Not all factors are constant, though. With increasing atomic number, the nuclear positive charge increases, and, thus, the force pulling the electrons towards the nucleus increases. Therefore, the atomic radii decreases as one moves across a period. Moving down a family presents a similar case of increasing atomic number. But, here, the principal quantum number is also increasing, which means the outermost electrons are farther and farther away from the nucleus. The principal quantum number has a stronger effect than atomic number in determining atomic radii. Thus, the radii increase going down a family.

● **PROBLEM** 656

You are given H, N, O, Ne, Ca, Al, and Zn. Determine which of these atoms (in their ground state) are likely to be paramagnetic. Arrange these elements in the order of increasing paramagnetism.

<u>Solution</u>: Paramagnetic substances possess permanent magnetic moments. An electric current flowing through a wire produces a magnetic field around the wire. Magnetic fields are thus produced by the motion of charged particles. Then, a single spinning electron, in motion around the nucleus, should behave like a current flowing in a closed circuit of zero resistance and therefore should act as if it were a small bar magnet with a characteristic permanent magnetic moment. The magnetism of an isolated atom results from two kinds of motion: the orbital motion of the electron around the nucleus, and the spin of the electron around its axis. Two spin orientations are permitted for electrons, $+\frac{1}{2}$ and $-\frac{1}{2}$. Two electrons occupy each filled orbital and their opposing spins cancel out the magnetic moments, thus, for an atom to be paramagnetic it must contain unpaired electrons.

H (Z = 1): $1s^1$ → The subshell, s, has only 1 electron as indicated by the superscript number. In the s subshell, you have only 1 orbital. Each orbital can hold 2 electrons. Therefore, this electron is unpaired and H is paramagnetic.

N (Z = 7): $1s^2 2s^2 2p^3$ → The p subshell has 3 orbitals that contain a total of 3 electrons. Because electrons have the same charge, they try to avoid each other, if possible. Thus, each electron is in a different orbital. Therefore, they are unpaired and N is paramagnetic.

O (Z = 8): $1s^2 2s^2 2p^4$ → The p subshell has 3 orbitals with 4 electrons. Recalling the information given above, this means that 1 orbital has two electrons. The other 2 orbitals possess one electron each. Thus, there are two unpaired electrons. O is paramagnetic.

Ne (Z = 10): $1s^2 2s^2 2p^6$   Here, all orbitals contain two electrons each. No electron is unpaired. Thus, Ne is not paramagnetic.

Ca (Z = 20): $1s^2 2s^2 2p^6 3s^2 3p^6 4s^2$.   Again, all orbitals have two electrons. Therefore, calcium is not paramagnetic.

Al (Z = 13): $1s^2 2s^2 2p^6 3s^2 3p^1$ The 3p subshell has only 1 electron for three orbitals. It must be unpaired, as such. It is paramagnetic.

Zn (Z = 30): $1s^2 2s^2 2p^6 3s^2 3p^6 3d^{10} 4s^2$.Each orbital has two electrons. Thus, no paramagnetism exists.

The order of paramagnetism (increasing) is proportional to the number of unpaired electrons. Thus, H, Al, O and N is the order of increasing paramagnetism.


# MOLECULAR ORBITAL THEORY

● **PROBLEM** 657

What is a sigma bond? What is a pi bond? What are their basic differences?

Solution:  A molecular orbital that is symmetrical around the line passing through two nuclei is called a sigma ($\sigma$) orbital. When the electron density in this orbital is con-centrated in the bonding region between the two nuclei, the bond is called a sigma bond. The covalent bonds in $H_2$ and HF are sigma bonds.

In the formation of the bonding orbital between two fluorine atoms, the 2p orbitals overlap in a head-to-head fashion to form a sigma bond. However, there is a second way in which half-filled p orbitals of two different atoms may overlap to form a bonding orbital.

If the two p orbitals are situated perpendicular to the line passing through the two nuclei, then the lobes of p orbitals will overlap intensively sideways to form an electron cloud that lies above and below the two nuclei. The bond resulting from this sideways or lateral overlap is called a pi ($\pi$) bond; the bonding orbital is called a pi orbital. It differs from a sigma orbital in that it is not symmetrical about a line joining the two nuclei. Pi bonds are present in molecules having two atoms connected by a double or triple bond. The sigma bond has greater orbital overlap and is usually the stronger bond; a pi bond, with less overlap, is generally weaker.

Compare the bond order of $He_2$ and $He_2^+$.

Solution: The bond order, or number of bonds in a molecule, is equal to the difference in the sum of the number of bonding electrons and the number of antibonding electrons divided by two.

Bond order = $\dfrac{\text{no. of bonding electrons} - \text{no. of antibonding electrons}}{2}$

This means that the number of bonding and antibonding electrons must be determined. There are 2 electrons in He, thus in $He_2$ there are 4. These electrons are all in the 1s level. For each level, there exists bonding and antibonding orbitals, each of which holds 2 electrons. Thus, in $He_2$, 2 electrons are bonding and 2 are antibonding. From this, the

Bond order = $\dfrac{2 - 2}{2} = 0$.

Thus, there are no bonds in $He_2$; and two He atoms will not bond together to form a molecule of $He_2$.

In $He_2^+$, one electron is removed from $He_2$, which means that there are now three electrons present. They are all in the 1s level. This is the lowest energy level that an electron can assume. The three electrons are distributed so that two are in bonding orbitals and one is in an antibonding orbital. Thus,

no. of bonds = $\dfrac{2 - 1}{2} = 0.5$ = bond order.

Because the bond order is not zero, this molecule can form.

Consider an octahedral complex having a univalent negative ion at each vertex of the octagon. Explain why the $d_{x^2 - y^2}$ orbital of the central atom is less stable relative to the $d_{xy}$ orbital.

Solution: Both the $d_{x^2-y^2}$ and the $d_{xy}$ orbitals are centered about the x-y plane; hence, the difference in the stability of the two orbitals must be related to their different orientations along the x- and y-axes.

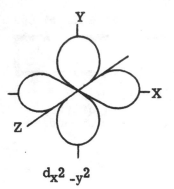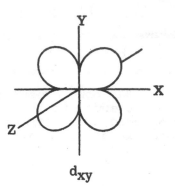

$d_{x^2-y^2}$                                   $d_{xy}$

Figure A

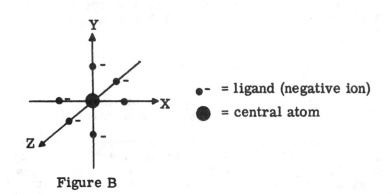

Figure B

In the $d_{x^2-y^2}$ orbital, the lobes are directed along the axes; whereas in the $d_{xy}$ orbital, the lobes are directed along the bisectors of the angles between the axes (see Figure A). In the octahedral configuration the ligands (the negatively charged ions electrostatically bound to the central atom) lie on the axes (see Figure B).

Hence, the $d_{x^2-y^2}$ orbital brings the negatively charged electrons in the lobes closer to the negatively charged ligands than does the $d_{xy}$ orbital. Since the negative charge of the electrons repels that of the ligands, and this electrostatic repulsion falls off as the reciprocal of the square of the distance between the charges, the greater degree of repulsion between the $d_{x^2-y^2}$ orbital and the ligands, relative to the repulsion between the $d_{xy}$ orbital and the ligands, decreases the stability of the $d_{x^2-y^2}$ orbital.

How would one expect the bond strength of NO to compare
with that of $O_2$?

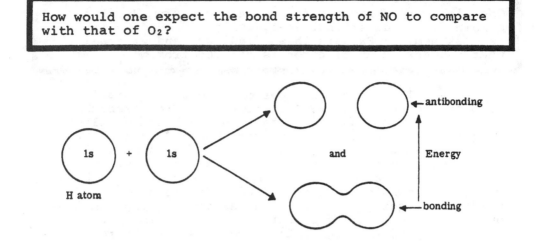

Solution:  The molecular orbital is the sum of the atomic
orbitals. For example, when two atomic orbitals are
combined, two molecular orbitals are formed. One orbital
is called a bonding orbital and the other is an antibonding orbital. The bonding orbital is at a lower energy
level than the antibonding orbital. If possible, electrons
seek out the bonding orbital rather than the antibonding
orbital. The orbitals in the production of $H_2$ from H
atoms can be visualized as shown in the accompanying
figure.

The greater the number of antibonding orbitals in a
molecule, the weaker the bond. Bond order is a quantity
that indicates the strength of bonding orbitals. It is
defined as half the number of bonding electrons minus half
the number of antibonding electrons. Thus, the higher the
bond order, the stronger the chemical bond.

To compare the bond strength of NO with $O_2$, compare
their bond orders. To do this, consider the total number
of valence electrons in each element. Valence electrons
are the outer electrons, which participate in bonding.
For NO, the total number of valence electrons is 3 + 4 = 7.
There exist 3 bonding p orbitals, which accommodate 6 e-
lectrons. The 1 unpaired electron must be in an antibonding orbital. This means that the bond order of NO is
$\frac{1}{2}(6) - \frac{1}{2}(1) = 3 - 0.5 = 2.5$.

In $O_2$ there is a total of 8 valence electrons. This
means 2 electrons must be in antibonding orbitals, since
the 3 bonding orbitals can accommodate only 6 electrons.
The bond order of $O_2 = \frac{1}{2}(6) - \frac{1}{2}(2) = 2$. The bond order
of NO is higher, which means, its chemical bond is stronger
than that of $O_2$.

One electron is removed from $O_2$ and one from $N_2$. The bonding in $O_2$ is strengthened, while the bonding in $N_2$ is weakened. 1) Explain these findings, and 2) predict what happens if an electron is removed from NO.

Solution: There are two types of molecular orbitals: bonding and antibonding. A chemical bond is strengthened by electrons in bonding orbitals and weakened by electrons in antibonding orbitals. For every bonding orbital, there is a corresponding antibonding orbital. Each orbital can hold 2 electrons. Bond order measures bond strength by giving an indication of the number of electrons in bonding versus antibonding orbitals. Bond order is defined as one half the number of electrons in the bonding orbital less one half the number of electrons in the antibonding orbital. Thus, the higher the bond order, the stronger the bond.

To find the original bond order, consider the valence electrons, the outermost electrons, since they are the only ones that participate in bonding. For N, Z = 7. Its electron configuration is $1s^2 2s^2 2p^3$. The outermost electrons are in $2p^3$. This means that in $N_2$ there are a total of six valence electrons. There exist 3 bonding p orbitals. They can accommodate the six valence electrons. Since no electrons need be in antibonding orbitals, bond order = $\frac{1}{2}(6) - \frac{1}{2}(0) = 3$. When an electron is removed, the bonding orbitals have only 5 electrons. Thus, the bond order becomes $\frac{1}{2}(5) - \frac{1}{2}(0) = 2.5$. Since the bond order went from 3 to 2.5, the bond is weakened by removing the electron from $N_2$.

For O, Z = 8. Since its electron configuration is $1s^2 2s^2 2p^4$, each O atom has 4 valence electrons. $O_2$ has a total of 8 valence electrons. The three bonding p orbitals can hold only 6 of these electrons. Thus, two electrons are in antibonding orbitals. The bond order is $\frac{1}{2}(6) - \frac{1}{2}(2) = 2$. When one removes an electron, it is removed from the antibonding orbitals, if electrons exist in such orbitals. Therefore, the $O_2$, after removal of the electron, has only 1 electron in an antibonding electron. This means the bond order becomes $\frac{1}{2}(6) - \frac{1}{2}(1) = 2.5$. The bond increased from 2 to 2.5, which means that bond strength increases when one removes an electron from $O_2$.

To predict what happens to NO bond strength, consider the bond order before and after the electron removal. Recall, an O atom has 4 valence electrons and N has 3 valence electrons. The total in NO is seven. The 3 p bonding orbitals can hold six of these electrons. This means 1 electron is in on antibonding orbital. Bond order = $\frac{1}{2}(6) - \frac{1}{2}(1) = 2.5$. If one removes one electron, the antibonding orbitals contain zero electrons. Thus, the bond order becomes $\frac{1}{2}(6) - \frac{1}{2}(0) = 3$. The bond increased from 2.5 to 3, which

means the bond strength increases. Therefore, one can predict the chemical bond strength of NO increases when an
electron is removed.

● PROBLEM 662

Describe the bonding in linear, covalent $BeCl_2$ and planar,
covalent $BCl_3$. What is the difference in the hybrid
orbitals used?

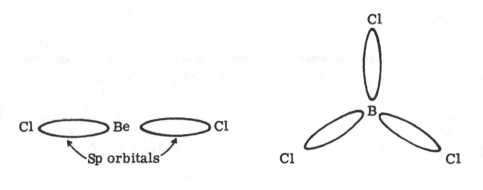

Figure A                                 Figure B

Solution:  The solution to this problem involves the
hybridization of orbitals. Once this is clear, the bonding
in $BeCl_2$ and $BCl_3$ will follow.

Quantum theory deals with independent orbitals, such
as 2s and 2p. This can be applied to a species, like
hydrogen, with only 1 electron. However, with atoms that
contain more than one electron, different methods must be
used. For example, the presence of a 2s electron perturbs
a 2p electron, and vice versa, such that a 2s electron
makes a 2p electron take on some s-like characteristics.
The result is that the hydrogenlike 2s and 2p orbitals
are replaced by new orbitals, that contain the combined
characteristics of the original orbitals.  These new
orbitals are called hybrid orbitals.  The number of
hybrid orbitals resulting from hybridization equals the
number of orbitals being mixed together. For example, if
one mixes an s and p orbital, one obtains two sp hybrid
orbitals. One s and two p = three $sp^2$ orbitals. One s
with three p = four $sp^3$ orbitals. sp orbitals are linear.
$sp^2$ orbitals assume a planar shape and $sp^3$ orbitals
assume a tetrahedral shape.

Solving:  It is given that $BeCl_2$ is linear and covalent. Since sp orbitals are linear, Be undergoes sp
hybridization. If something is linear, the bond angle
is 180°. By understanding hybridization, one also knows

the geometry of the molecule. A diagram of the bonding resembles Fig. A.

Given that $BCl_3$ is planar, and since $sp^2$ hybridization yields a planar structure, B has $sp^2$ hybridized bonding with angles of 120°(Fig. B).

# EARLY QUANTUM CHEMISTRY

● **PROBLEM** 663

The accompanying figures show a cathode-ray tube (figure A) and the deflection of an electron in a region of magnetic field (figure B). If deflection in the tube is being produced by the magnetic field alone, predict the effect on the observed deflection by increasing (a) mass of particles, (b) velocity of the particles, (c) magnetic field, and (d) charge on the particles.

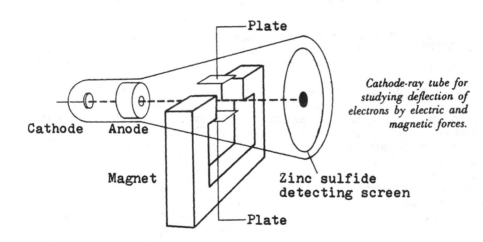

Cathode-ray tube for studying deflection of electrons by electric and magnetic forces.

Figure A

Solution: The magnetic field, charge, velocity, mass and the deflection of a particle are related by the following equation,

$$Hev = \frac{mv^2}{r}$$

where H is the strength of the magnetic field, e the

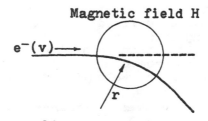

Magnetic field H

e⁻(v)⟶

r

*Schematic representation
showing how an electron
with velocity v is deflected
in a region of magnetic
field H.*

### Figure B

charge on the particles, r the radius of deflection, v
the velocity of the particles, and m the mass of the
particles. The derivation of this equation need not be
considered in solving this problem.

Rewriting the given equation and solving for e/m,
the charge to mass ratio,

$$\frac{e}{m} = \frac{v}{Hr} \ .$$

Since e/m is a constant if (a) the mass is increased,
then the radius of deflection must also increase. This
is seen more clearly by solving the given equation for r,

$$r = \frac{vm}{He} \ .$$

Here, if m increases (keeping v, H, and e constant),
then r increases.

Following this type of reasoning, one can predict:
(b) the radius of deflection increases with increasing
velocity; (c) the radius of deflection decreases with increasing magnetic field strength; and (d) the radius of
deflection increases with increasing charge on the
particles. Note that the radius of curvature is inversely
proportional to the distance the particle is actually
deflected from its original path.

● **PROBLEM** 664

Consider the experimental apparatus, illustrated in
figure A which is used to measure the deflection of an
electron. The source of electrons is a heated filament
and the detector consists of a screen painted with a
material which phosphoresces when struck by energetic
particles. The magnetic field is confined to the spherical

region. During a particular run of this experiment, an electron enters the magnetic field at a distance of 30 cm away from the detecting screen and travels 25 cm after leaving the field. If the strength of the magnetic field is 0.18 g/coulomb-sec and the beam deflection on the screen is 4.0 mm, what is the velocity of the electrons in the beam?

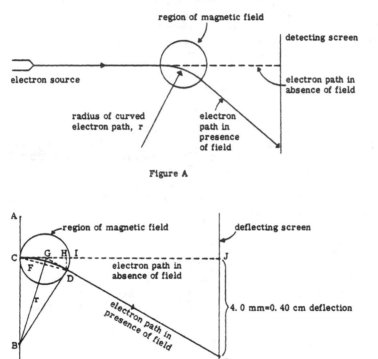

Figure A

Figure B

Solution: We are looking for a relationship between the magnetic field strength H, the deflection of the electron and the electron velocity v. Such a relationship is provided by the equation of motion of an electron in a magnetic field, $v = Her/m$, where e is the electronic charge (e = $1.602 \times 10^{-19}$ coulomb), r is the radius of curvature, H is the strength of the magnetic field, and m is the electron mass (m = $9.11 \times 10^{-28}$ g). In order to apply this equation to solve for v, we must first determine r.

In the following discussion to determine r, we will make reference to figure B (which is not drawn to scale for reasons of clarity). In the problem, we are given that the distance from the point at which the electron beam leaves the field to the deflecting screen is 25 cm. The point at which the electron leaves the field is D. H and D lie on the same vertical line, hence the distance

632

from D to the screen is the same as the distance from H to the screen, or HJ = 25 cm. Since the electron deflection is small, we can assume that the path of the electron is not far from the path the electron would have in the absence of a field. Hence, we can assume that HI is much smaller than CI and HJ, and that FG is much smaller than FB and BG. If we make this assumption that HI is negligible, then IJ = HJ - HI $\cong$ HJ = 25 cm. The distance from the point at which the electron enters the field (point C) to the screen is 30 cm, or CJ = 30 cm. Then CI = CJ - IJ = 30 cm - 25 cm = 5 cm. Denoting the center of the circle containing the magnetic field by G, we have CG = GI = ½CI = 5 cm/2 = 2.5 cm. Using our assumption on the smallness of HI, we obtain GH = GI - HI $\cong$ GI = 2.5 cm. In summary, so far we have obtained the following distances:

$$CG = GI \cong GH = 2.5 \text{ cm}, \qquad IJ \cong HJ = 25 \text{ cm}.$$

The strategy from this point on will be as follows. We will use the similarity of triangle GJE and triangle GHD to determine HD. Then, using the similarity of triangle CHD and triangle CFB, we will obtain the ratio BF/CF, and from this, the radius of curvature, r.

There is a theorem from elementary plane geometry which states that when a triangle is cut by a line parallel to one of its sides, the two resulting triangles are similar. Since HD and JE are both drawn vertically, and are therefore parallel, triangle GJE is similar to triangle GHD. Hence,

$$\frac{GH}{GJ} = \frac{HD}{JE} \qquad \text{or}$$

$$HD = JE \times \frac{GH}{GJ} = 4 \text{ mm} \times \frac{2.5 \text{ cm}}{27.5 \text{ cm}} = 0.4 \text{ cm} \times \frac{2.5 \text{ cm}}{27.5 \text{ cm}}$$

$$= 0.036 \text{ cm}.$$

Next we note that, using the theorems of plane geometry, it can be shown that triangle CHD is similar to CGB. Then

$$\frac{HD}{CG} = \frac{CD}{BG}$$

Because r = BG, we solve this ratio for BG. We have already determined HD and CG and can find CD by using the Pyhthagorean Theorem, thus, we can find BG. Solving for CD:

$$CD^2 = CH^2 + HD^2 = (5.0 \text{ cm})^2 + (0.036 \text{ cm})^2 = 25.001 \text{ cm}^2;$$

$$CD = 5.00 \text{ cm}.$$

Solving for BG;

$$BG = \frac{CD \cdot CG}{HD} = \frac{(5.0 \text{ cm})(2.5 \text{ cm})}{(0.036 \text{ cm})} = 347.2 \text{ cm.}$$

We can now substitute r into the equation of motion of an electron in a magnetic field. We thus obtain the electron velocity,

$$v = \frac{Her}{m} = \frac{0.18 \text{ g/coulomb-sec} \times 1.602 \times 10^{-19} \text{ coulomb} \times 347.2 \text{ cm}}{9.11 \times 10^{-28} \text{ g}}$$

$$= 1.10 \times 10^{10} \text{ cm/sec.}$$

● **PROBLEM** 665

If the gold foil in a Rutherford type experiment is 1/10,000 in. thick, what is a probable minimum number of gold atoms an alpha particle passed through before hitting the fluorescent screen? 1 in = 2.54 cm, 1 cm = $10^8 \overset{\circ}{A}$. The radius of a gold atom is 1.5 $\overset{\circ}{A}$.

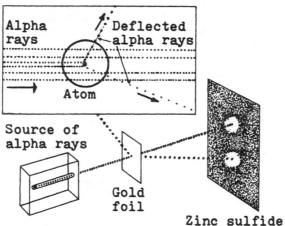

The Rutherford
experiment

Solution: In Rutherford's experiment, a beam of alpha particles is aimed at a gold foil of given thickness. Most of the alpha particles pass through undeflected. The deflections that occur suggest that the alpha particles collided with a solid body. This body is the gold nucleus. This experiment elucidated some important properties of the nucleus. One of these properties is the positive charge of the nucleus.

If one knows the thickness of the foil and the atomic radius of the particles that compose the foil, one can calculate the number of particles which make up the width of the foil.If one assumes cubic packing in the foil, the sum of the lengths of the diameters of the gold atoms will be equal to the thickness of the foil. Proceed as follows:

The radius, and therefore the diameter, of the gold atom is given in terms of angstroms. The thickness of the foil must be converted to the same units.

The thickness = $(10^{-4}$ in$)(2.54$ cm/in$)(1 \times 10^{8} \overset{o}{A}$/cm$)$

$$= 2.54 \times 10^{4} \overset{o}{A}.$$

Diameter of gold nucleus = 2 × Radius = $2 \times 1.5 \overset{o}{A} = 3 \overset{o}{A}$.

The minimum number of gold atoms passed by alpha particles is equal to

$$\frac{2.54 \times 10^{4} \overset{o}{A}}{3 \overset{o}{A}} = 8.466 \times 10^{3}$$

● **PROBLEM** 666

A chemist has a piece of foil that is approximately $5 \times 10^{4}$ atoms thick. If an alpha particle must come within $10^{-12}$ cm of a nucleus for deflection to occur, what is the probability that an alpha particle will be deflected, assuming the nuclei are not directly behind one another? Assume that the area of one atom is $4 \times 10^{-16}$ cm$^{2}$.

Solution: To calculate the chance of deflection occurring in any one layer of the foil, assume that each layer is one atom thick. The foil is, therefore, $5 \times 10^{4}$ layers thick. If the number of deflections in any one layer is known, it can be multiplied by $5 \times 10^{4}$ to find the total number of possible deflections as an alpha particle moves through the foil. To find the chance of a deflection occurring in one layer, assume that the atoms are spherical. At $10^{-12}$ cm from the center of the circle, a deflection can occur. The radius of the circle for which a deflection will occur is $10^{-12}$ cm. The area of a circle is equal to $\pi r^{2}$ where $\pi = 3.14$ and r = radius.

The area of one atom in which deflection can occur is equal to $3.14 \times (10^{-12})^{2}$ or $3.14 \times 10^{-24}$ cm$^{2}$. The area of each atom is $4 \times 10^{-16}$ cm$^{2}$. The thickness of each layer is one atom. Therefore, the chance of getting a deflection in any given layer of atoms is

$$\frac{3.14 \times 10^{-24} \text{ cm}^{2}}{4 \times 10^{-16} \text{ cm}^{2}} = 7.85 \times 10^{-9}.$$

There are $5 \times 10^4$ layers. Thus, a chance of deflection in one of the $5 \times 10^4$ layers is

$$5 \times 10^4 \times 7.85 \times 10^{-9} = 4 \times 10^{-4} \text{ or 1 in 2500.}$$

● **PROBLEM** 667

During run of the Millikan oil-drop experiment, the following charges were observed on five different oil droplets: $1.44 \times 10^{-18}$, $2.56 \times 10^{-18}$, $4.80 \times 10^{-19}$, and $9.60 \times 10^{-19}$ coulomb. Does this support the charge of an electron as being $1.60 \times 10^{-19}$ coulomb?

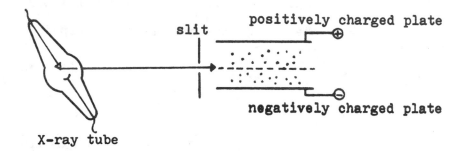

slit    positively charged plate

negatively charged plate

X-ray tube

Solution: This experiment, illustrated schematically in the above diagram, was first performed in 1909 by Robert A. Millikan. Oil droplets are sprayed between a top plate having a positive charge and a bottom plate having a negative charge. A beam of X-rays is simultaneously directed at this spray. The X-rays will knock electrons out of molecules in the air and these electrons will be picked up by the oil droplets, which will acquire a charge that is an integral multiple of the electronic charge. This occurs because each oil droplet will pick up an integral number of electrons. By adjusting the charge on the plates, the upward pull on the electrons due to attraction by the positive plate on the top and repulsion by the negative plate on the bottom can be balanced against the downward pull of gravity. The forces are balanced when the oil drop is stationary, as viewed through a telescope. By equating the gravitational and the electrostatic forces one can determine the charge on an oil droplet.

It is given that the following charges were observed on oil droplets: $1.44 \times 10^{-18}$, $2.56 \times 10^{-18}$, $4.80 \times 10^{-19}$, and $9.60 \times 10^{-19}$ coulomb. If the elementary electronic charge is $1.60 \times 10^{-19}$ coulomb, each of these observed charges must be an integral multiple of the elementary charge. Hence, each of these charges must be exactly divisible by $1.60 \times 10^{-19}$ coulombs.

$$\frac{1.44 \times 10^{-18} \text{ coulomb}}{1.60 \times 10^{-19} \text{ coulomb}} = 9.00,$$

$$\frac{2.56 \times 10^{-18} \text{ coulomb}}{1.60 \times 10^{-19} \text{ coulomb}} = 16.00,$$

$$\frac{4.80 \times 10^{-19} \text{ coulomb}}{1.60 \times 10^{-19} \text{ coulomb}} = 3.00 \quad \text{and}$$

$$\frac{9.60 \times 10^{-19} \text{ coulomb}}{1.60 \times 10^{-19} \text{ coulomb}} = 6.00.$$

Each of the observed charges is an integral multiple of $1.60 \times 10^{-19}$ coulomb, supporting the contention that $1.60 \times 10^{-19}$ coulomb is the charge of an electron.

# THE BOHR ATOM

● **PROBLEM** 668

Determine the radii of the first and second quantum orbits in the hydrogen atom.

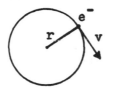

Solution: We will derive the Bohr theoretical expression for the radius of the quantum orbit of an electron.

Consider an electron of mass m and charge e moving at a constant distance r from the nucleus with a tangential velocity v, as shown in the accompanying figure. In order for this orbit to be stable, the Coulombic force attracting the electron to the nucleus ($e^2/r^2$) must be balanced by the centrifugal force on the electron ($mv^2/r$), which, if not compensated for, would compel the electron to leave its orbit. Hence, the magnitude of these two forces must be equal.

$$\frac{e^2}{r^2} = \frac{mv^2}{r}$$

Solving for r, we obtain $r = \frac{e^2}{mv^2}$ .

Bohr introduced quantization of the electron's orbit by assuming that the electron's angular momentum (mvr) is given by:

$$mvr = n\left(\frac{h}{2\pi}\right),$$

where h is Planck's constant and n is an index, called the principal quantum number, which takes on the values n = 1 for the first orbit, n = 2 for the second orbit, and so on. Solving for v,

$$v = \frac{nh}{2\pi mr}$$

We can now obtain an expression for r which includes quantization and is in terms of known quantities. Using the above expressions for v and r,

$$r = \frac{e^2}{mv^2} = \frac{e^2}{m}\left(\frac{2\pi mr}{nh}\right)^2 = \frac{4\pi^2 e^2 mr^2}{n^2 h^2},$$

$$r = \frac{n^2 h^2}{4\pi^2 e^2 m}.$$

The radius of the first orbit (n = 1), is called the Bohr radius. Solving for r: e = $4.80 \times 10^{-10}$ esu and m = $9.11 \times 10^{-28}$ g.

$$r_1 = \frac{1^2 \times h^2}{4\pi^2 e^2 m} = \frac{1 \times (6.626 \times 10^{-27} \text{ erg-sec})^2}{4\pi^2 \times (4.80 \times 10^{-10} \text{ esu})^2 \times 9.11 \times 10^{-28} \text{ g}}$$

$$= 5.29 \times 10^{-9} \text{ cm} = 0.529 \text{ Å}.$$

Let the radius of the nth orbit be denoted $r_n$. Then

$$r_n = \frac{n^2 h^2}{4\pi^2 e^2 m} = n^2\left(\frac{h^2}{4\pi^2 e^2 m}\right) = n^2 r_1 = n^2 \times 0.529 \text{ Å}.$$

Hence, the radius of the second quantum orbit is

$$r_2 = n^2 \times 0.529 \text{ Å} = (2)^2 \times 0.529 \text{ Å}$$

$$= 2.12 \text{ Å}.$$

● **PROBLEM** 669

At what value of Z would the innermost orbit be expected to be pulled inside a nucleus of radius $1.0 \times 10^{-13}$ cm, assuming simple Bohr theory and Bohr radius = $0.5292 \times 10^{-8}$ cm? (Assume N = 1.)

Solution: Niels Bohr stated, for a hydrogen atom, that only certain permitted values for the radius of the electron path exist. He found that the radius, r, of this electron path was equal to a whole number, N, squared, divided by the atomic number, Z, times the Bohr radius, $a_0$, which is constant. That is

$$r = \frac{N^2}{Z} \; a_0.$$

The above question, can be answered by using this equation. N is equal to the quantum number corresponding to the energy level occupied by the electron. It has values of $n = 1, 2, 3, 4, \ldots$ . One is given r and $a_0$ and is asked to compute Z. The electron occupying the innermost orbital is in the ground state, thus N = 1. Thus,

$$Z = \frac{N^2}{r} \; a_0 = \frac{(1)^2}{1 \times 10^{-13} \text{ cm}} \times .5292 \times 10^{-8} \text{ cm}$$

$$= 5.292 \times 10^4.$$

● **PROBLEM** 670

Assuming the Bohr radius is $0.5292 \times 10^{-8}$ cm, what frequency of a light wave has the same wavelength as the Bohr radius?

Solution: The question asks one to determine the frequency of a light wave with a wavelength equal to the Bohr radius, $.5292 \times 10^{-8}$ cm. In answering this question, it should be remembered that there is a relation between frequency and wavelength, $c = \lambda \nu$, where c = the speed of light ($2.998 \times 10^{10}$ cm/sec), $\lambda$ = wavelength, and $\nu$ = frequency. One can solve $\nu$ in this equation by substituting in the values of c and $\lambda$. Thus

$$\nu = \frac{c}{\lambda} = \frac{2.998 \times 10^{10} \text{ cm/sec}}{.5292 \times 10^{-8} \text{ cm}} = 5.665 \times 10^{18} \text{ sec}^{-1}.$$

● **PROBLEM** 671

The atomic radius of hydrogen is 0.037 nm. Compare this figure with the length of the first Bohr radius. Explain any differences.

Solution: The first Bohr radius, r. is equal to $\frac{N^2}{Z}$ $a_0$, where $a_0$ is .05292 nm, N is an integer, and Z the atomic number. For hydrogen, Z = 1. For the first Bohr radius N = 1. Thus

$$r = \frac{(1)^2}{1} \; a_0 = .05292 \text{ nm.}$$

Thus, the atomic radius of hydrogen (.037 nm) is considerably smaller than the first Bohr radius. This can be explained by the fact that the atomic radius is calculated from the bond length of a $H_2$ molecule. There is considerable orbital overlap in bonded atoms, which means the bond length is shorter than the sum of the radii of lone atoms.

# THE DE BROGLIE EQUATION

● PROBLEM 672

It has been suggested that the time required for light to travel 1 cm be used as a unit of time called the "jiffy". How many seconds correspond to one jiffy? The speed of light is about $3.0 \times 10^8$ m/sec.

Solution: The relationship between distance, speed, and time is

$$\text{distance} = \text{speed} \times \text{time, or, time} = \frac{\text{distance}}{\text{speed}} .$$

The time it takes light to travel 1 cm is:

$$\text{time} = \frac{\text{distance}}{\text{speed}} = \frac{1 \text{ cm}}{3 \times 10^8 \text{ m/sec}}$$

$$= \frac{1 \text{ cm}}{3 \times 10^8 \text{ m/sec} \times 10^2 \text{ cm/m}} = \frac{1 \text{ cm}}{3 \times 10^{10} \text{ cm/sec}}$$

$$= 3.3 \times 10^{-11} \text{ sec.}$$

Hence, 1 jiffy = $3.3 \times 10^{-11}$ sec.

● PROBLEM 673

Consider a molecule of hydrogen at 478.15 K and moving with a velocity of $2 \times 10^5$ cm/sec. What is the de Broglie wavelength of this hydrogen atom?

Solution: The de Broglie wavelength ($\lambda$) is related to the momentum of a particle by the de Broglie equation:

$$\lambda = h/mv,$$

where h is Planck's constant, m is the mass of the particle, and v is its velocity (momentum = mv). Calculate the de Broglie wavelength of the hydrogen atom by directly substituting into this equation.

The mass of a hydrogen atom is m = $3.34 \times 10^{-24}$ g. Planck's constant is h = $6.626 \times 10^{-27}$ erg-sec = $6.626 \times$

$10^{-27}$ g-cm$^2$/sec. Hence, the de Broglie wavelength is:

$$\lambda = \frac{h}{mv} = \frac{6.626 \times 10^{-27} \text{ g-cm}^2/\text{sec}}{3.34 \times 10^{-24} \text{ g} \times 2 \times 10^5 \text{ cm/sec}} = 9.9 \times 10^{-9} \text{ cm}$$

$$= 9.9 \times 10^{-9} \text{ cm} \times 10^8 \text{ Å/cm} = 0.99 \text{ Å}.$$

● **PROBLEM** 674

Determine the de Broglie wavelength of an electron (mass = $9.11 \times 10^{-28}$ g) having a kinetic energy of 100 eV.

Solution: Using the value for the kinetic energy of the electron, we will calculate its momentum and substitute this into the de Broglie equation to get the wavelength of the corresponding matter wave.

The kinetic energy, k, of a particle of mass, m, moving with velocity v is k = ½ mv$^2$ and its momentum, p, is p = mv. Kinetic energy and momentum are related as follows:

$$k = \tfrac{1}{2} mv^2 = \frac{1}{2m} m^2 v^2 = \frac{1}{2m} (mv)^2 = \frac{p^2}{2m} = k \text{ or } p = (2mk)^{\frac{1}{2}}.$$

The kinetic energy of our electron is k = 100 eV = 100 eV $\times$ 1.602 $\times$ 10$^{-12}$ erg/eV = 1.602 $\times$ 10$^{-10}$ erg. Hence, its momentum is:

$$p = (2mk)^{\frac{1}{2}} = (2 \times 9.11 \times 10^{-28} \text{ g} \times 1.602 \times 10^{-10} \text{ erg})^{\frac{1}{2}}$$

$$= (.2919 \times 10^{-36})^{\frac{1}{2}} \text{ erg}^{\frac{1}{2}} \text{ g}^{\frac{1}{2}}$$

$$= (.2919 \times 10^{-36})^{\frac{1}{2}} (g - \text{cm}^2/\text{sec}^2)^{\frac{1}{2}} \text{ g}^{\frac{1}{2}}$$

$$= 5.403 \times 10^{-19} \text{ g-cm/sec}.$$

The de Broglie equation is:

$$\lambda = h/mv = h/p,$$

where h is Planck's constant (h = 6.626 $\times$ 10$^{-27}$ erg/sec) = 6.626 $\times$ 10$^{-27}$ g-cm$^2$/sec. Hence, the de Broglie wavelength of a 100 eV electron is:

$$\lambda = \frac{h}{p} = \frac{6.626 \times 10^{-27} \text{ g-cm}^2/\text{sec}}{5.403 \times 10^{-19} \text{ g-cm/sec}} = 1.226 \times 10^{-8} \text{ cm}$$

$$= 1.226 \times 10^{-8} \text{ cm} \times 10^8 \text{ Å/cm} = 1.226 \text{ Å}.$$

What is the de Broglie wavelength of a radium alpha particle (a helium nucleus, $He^{2+}$) having an energy of 4.8 MeV (million electron volts)?

Solution:  The de Broglie wavelength, $\lambda$, of a particle of mass, m, moving at speed, v, is given by

$$\lambda = \frac{h}{mv} ,$$

where h is Planck's constant (h = $6.6 \times 10^{-27}$ erg-sec). The mass of an alpha particle is $6.6 \times 10^{-24}$ g.

The energy of an alpha particle is given by E = $\frac{1}{2} mv^2$ = 4.8 MeV. Since 1 MeV is equal to $1.6 \times 10^{-6}$ erg, we have:

$$\frac{1}{2} mv^2 = 4.8 \times 1.6 \times 10^{-6} \text{ erg.}$$

We will employ this value of the energy in order to obtain the value of mv without having to find the velocity of the particle.

$$(mv)^2 = 2(\tfrac{1}{2} mv^2) \ m$$

$$= 2(4.8 \times 1.6 \times 10^{-6} \text{ erg})(6.6 \times 10^{-24} \text{ g})$$

$$= 1.0 \times 10^{-28} \text{ erg-g} = 1.0 \times 10^{-28} \text{ g}^2 \text{ cm}^2/\text{sec}^2,$$

where we have used 1 erg = $1 g - cm^2/sec^2$. Taking the square root of both sides gives:

$$(mv)^2 = 1.0 \times 10^{-28} \text{ g}^2 \text{ cm}^2/\text{sec}^2$$

$$mv = 1.0 \times 10^{-14} \text{ g cm/sec.}$$

Substituting this value into the expression for the de Broglie wavelength gives:

$$\lambda \ = \frac{h}{mv} = \frac{6.6 \times 10^{-27} \text{ erg-sec}}{1.0 \times 10^{-14} \text{ g-cm/sec}} = \frac{6.6 \times 10^{-27} \text{ g-cm}^2/\text{sec}}{1.0 \times 10^{-14} \text{ g-cm/sec}}$$

$$= 6.6 \times 10^{-13} \text{ cm.}$$

A chemistry student observed each of the following objects: (a) a 10,000 kg truck moving at 27.8 m/sec, (b) a 50 mg flea flying at 1 m/sec, and (c) a water molecule moving at 500 m/sec. The student proceeded to calculate the wavelength in centimeters of each object. What were these wavelength? h = $6.626 \times 10^{-34}$ J sec.

Solution: The answer to this question requires the use of de Broglie's equation to solve for the wavelength of a moving particle. It states that $\lambda = h/mv$, where $\lambda$ = wavelength, m = mass, v = velocity, and h = Planck's constant. One is given h, m, and v. Substitute these values into the equation and solve for $\lambda$. 1 cm = $10^{-2}$ m.

(a) $\lambda = \dfrac{h}{mv} = \dfrac{6.626 \times 10^{-34} \text{ J sec}}{10,000 \text{ kg} \cdot 27.8 \text{ m/sec}} = 2.4 \times 10^{-39}$ meters

$= 2.4 \times 10^{-37}$ cm

(b) $\lambda = \dfrac{h}{mv} = \dfrac{6.626 \times 10^{-34} \text{ J sec}}{5 \times 10^{-5} \text{ kg} \cdot 1 \text{ m/sec}} = 1.3 \times 10^{-29}$ meters

$= 1.3 \times 10^{-27}$ cm

(c) A mole of water has a mass of 18 g. Thus,

$\lambda = \dfrac{h}{mv} = \dfrac{6.626 \times 10^{-34} \text{ J sec}}{\dfrac{.018 \text{ kg/mole}}{6.02 \times 10^{23} \text{ molecules/mole}}\ 500 \text{ m/sec}}$

$= 4.43 \times 10^{-11}$ m $= 4.43 \times 10^{-9}$ cm

# WAVE FUNCTIONS

● PROBLEM 677

Calculate the uncertainty in the location of a 100 g turtle crossing a road at a velocity of 1 mm/sec, the uncertainty of the latter being $10^{-4}$ mm/sec. (h = 6.626 × $10^{-27}$ erg-sec and 10 mm = 1 cm.)

Solution: This problem demands the use of the Heisenberg Uncertainty Principle. This principle states that one cannot simultaneously locate both the position of a particle and its momentum with absolute precision. Whenever one quantity, either position or momentum, is known precisely, the other is known less precisely. Heisenberg showed that the uncertainty in the determination of momentum of a particle, $\Delta p$, and the simultaneous determination of its location, $\Delta x$ is related by the equation $\Delta p \Delta x \geq h$, where h is Planck's constant. Momentum p is equal to mv, where m is mass and v the velocity of the mass, thus $\Delta p = m\Delta v$. One substitutes this into the Heisenberg inequality attaining $m\Delta v \Delta x \geq h$. One can now solve for $\Delta x$. 1 erg = g cm$^2$/sec$^2$, 1 mm = .1 cm.

$\Delta x = \dfrac{h}{m\Delta v} = \dfrac{6.626 \times 10^{-27} \text{ erg-sec}}{(100 \text{ g})(10^{-4} \text{ mm/sec})} \cdot \dfrac{1 \text{ mm}}{.1 \text{ cm}}$

$= 6.626 \times 10^{-24}$ cm.

Consider a kinetic experiment in which the chemical reaction is accompanied by a temperature change. Using the
Heisenberg uncertainty principle, calculate the uncertainty
in a simultaneous measurement of time and temperature.

Solution: The Heisenberg uncertainty principle relates
the uncertainty in energy, $\Delta E$, and the uncertainty in
time, $\Delta t$, arising from a simultaneous measurement of the
two, by the inequality $\Delta E \Delta t \geq h$, where h is Planck's
constant. Thus, the product of the two uncertainties cannot be less than Planck's constant. In order to apply the
Heisenberg uncertainty principle to our problem, we must
relate temperature and energy.

The internal energy of a substance is approximately
proportional to its absolute temperature, T. In fact,
for ideal gases, the two are exactly proportional. Let
the constant of proportionality for our particular system
be c. Then, $E = cT$ and $\Delta E = c\Delta T$, where $\Delta T$ is the uncertainty in the temperature measurement. Then the uncertainty relation becomes:

$$\Delta E \Delta t = c\Delta T \ \Delta t \geq h \quad \text{or,} \quad \Delta T \ \Delta t \geq h/c.$$

At what value or values of r can nodes in the wave
function for a 3s electron bound to a nucleus of charge
+11 be predicted? Bohr's radius = 0.0529 nm.

Solution: One must know two important things to answer
this question: (1) The wave function expression for a
one-electron atom for 3s, and (2) The fact that at a node,
$\Psi$ or the wave function is zero. If $\Psi^2$ is the probability
of finding an electron in a given area and a node is an
area of zero probability, then $\Psi$ must also be zero.

The wave function expression relates $\Psi$ to radius r.

The wave function for a 3s electron is

$$\Psi_{300} = \frac{1}{81\sqrt{\pi}} \left(\frac{Z}{a_0}\right)^{3/2} \left(27 - \frac{18\,Zr}{a_0} + \frac{2Z^2\,r^2}{a_0^2}\right) e^{-Zr/3a_0}$$

For $\Psi_{300}$ to be equal to zero the factor

$\left(27 - \frac{18\,Zr}{a_0} + \frac{2\,Z^2\,r^2}{a_0^2}\right)$ must be equal to zero because the

other factors in the equation can never be equal to zero.

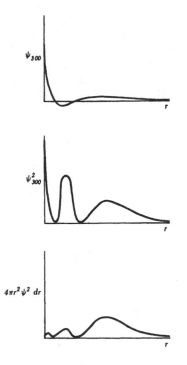

Graphs of $\psi$, $\psi^2$, and
$4\pi r^2 \psi^2$ dr for a 3s electron.

Using $0 = 27 - \dfrac{18Zr}{a_0} + \dfrac{2Z^2 r^2}{a_0^2}$  solve for r.

$$0 = 27 - \frac{18\,(11)\,(r)}{(.0529)} + \frac{2\,(11)^2\,(r)^2}{(.0529)^2}$$

$$= 27 - 3.74 \times 10^3 \; r + 8.67 \times 10^4 \; r^2$$

Using the quadratic equation one finds that r is equal to 0.0092 or 0.034 nm.

● **PROBLEM** 680

The mass of an electron is $9.11 \times 10^{-31}$ and Planck's constant $= 6.63 \times 10^{-34}$ J sec. How much energy does an electron possess in its five lowest states, if it is placed in a cube of dimension 50 nm?

<u>Solution</u>:  Wave function theory states that the permitted values of the energy of a particle in a box is given as

$$E = \frac{N^2 \, h^2}{8 \, m \, L^2} ,$$

where E = energy, N = principal quantum number, h = Planck's constant, m = mass of particle, and L = length of box. Substitute in the known values and solve for E. The principal quantum number N corresponds to the energy levels of the electrons. The five lowest energy states are N = 1, N = 2, N = 3, N = 4, and N = 5, where N = 1 is the lowest and N = 5 the highest energy level.

Substituting:

$$E = \frac{N^2 h^2}{8m L^2} = \frac{N^2 (6.63 \times 10^{-34} \text{ J sec})^2}{8(9.11 \times 10^{-31} \text{ kg})(5 \times 10^{-8} \text{M})^2}$$

$$= N^2 (2.41 \times 10^{-23})$$

(Note: 50 nm = $5.0 \times 10^{-8}$ M)

1st energy level (lowest): $N^2 = 1$, $E = 2.41 \times 10^{-23}$ J

2nd energy level        : $N^2 = 4$, $E = 9.64 \times 10^{-23}$ J

3rd energy level        : $N^2 = 9$, $E = 2.16 \times 10^{-22}$ J

4th energy level        : $N^2 = 16$, $E = 3.85 \times 10^{-22}$ J

5th energy level        : $N^2 = 25$, $E = 6.025 \times 10^{-22}$ J.

● **PROBLEM** 681

If one increases the angle away from the Z axis, how does the wave function for a $2P_Z$ electron change?

Solution: The wave function of a $2P_Z$ electron is

$$\Psi = \frac{1}{4\sqrt{2\pi}} \left(\frac{Z}{a_0}\right)^{5/2} e^{-Zr/2a_0} \, r \cos \Theta,$$

where $\Psi$ is the wave function and $\Theta$ = the polar angle. The polar angle measures the angular deviation away from the vertical, or Z, axis. $\Psi^2$ is proportional to the electron density at any given radius.

One needs to consider only $\Psi$ and $\Theta$ to answer the question. To see how the wave function, $\Psi$, changes when the angle from the Z axis is increased, take two values for $\Theta$, and see what happens to $\Psi$. For example, when $\Theta = 0$ (that is, looking along the Z axis), cos $\Theta$ is one, the maximum value a cosine can assume. The wave function is dependent upon $\Theta$. The cosine $\Theta$ is at a maximum, then $\Psi$ is at a maximum. Thus, $\Psi^2$ is also at a maximum and the electron density is greatest when looking along the Z axis. When $\Theta = 90°$ (that is, when one is looking 90° away from the Z axis, i.e. along the X or Y axis), cos θ is zero, the minimum value a cosine can assume. This means, therefore, that $\Psi^2$ is not at a minimum, which means that the probability of finding the electron at this location is zero. In summary, when the angle was increased from 0° to 90°, the wave function decreased. Thus, when one increases the angle away from the Z axis, the wave function and the electron density for a $2P_z$ electron decreases.

● **PROBLEM** 682

Does the electron probability distribution for 1s and 2s support or refute the Bohr picture of the shell?

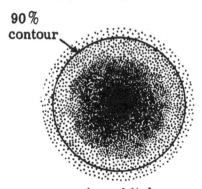

90%
contour

1s orbital

Schematic representations of the 1s orbital of hydrogen: a cross section of the probable density of the electron cloud; this cross section represents a slice through the middle of the cloud.

spherical node

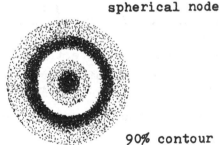

90% contour

2s orbital

Solution: This question involves the comparison of the Bohr concepts and the electron probability distribution for the 1s and 2s orbitals. Bohr believed that the shells have spherical symmetry with the average distance from the nucleus of the 2s electron being greater than that for the 1s. The electron probability distribution (orbitals) for 1s and 2s shells are, spherical in symmetry with 2s having a greater radius than the 1s. The Bohr picture does agree very well with the electron distribution for 1s. However, the Bohr picture does not describe as accurately the spread-out nature of the true distribution or the node separating the two regions which the 2s electron occupies. This is

true because the Bohr picture describes a continuous
density of the electron clouds which is not present in the
2s shell. This is seen in the illustration above.

● PROBLEM 683

Give qualitative explanation for the observed splitting of
the ground state energy of the hydrogen atom in the presence
of a magnetic field.

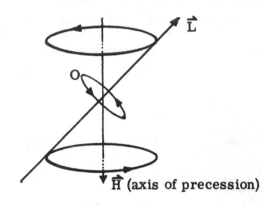

$\vec{H}$ (axis of precession)

Solution:  This phenomenon is called Zeeman splitting,
after the Dutch physicist Pieter Zeeman who first discovered
the effect in 1896.

     Consider an electron moving in the ground state orbital
of a hydrogen atom and describing a circular orbit (labelled
"O" in the diagram). This motion will give rise to an
angular momentum vector $\vec{L}$ directed along the axis of
rotation. In the presence of a magnetic field $\vec{H}$, the angular
momentum vector will precess about the magnetic field axis.

     Now, since the angular momentum of an electron must be
quantized, the projection of $\vec{L}$ onto $\vec{H}$ (that is, the component of the vector $\vec{L}$ along axis $\vec{H}$) must also be quantized.
From quantum theory, we know that this projection must take
on values which are integral multiples (positive, negative,
or zero) of $h/2\pi$, where h is Planck's constant. These integers are the possible values of the magnetic quantum
number, m. For the hydrogen atom, the magnetic quantum
number can take on the values m = - 1, 0, + 1.

     Thus, we observe three lines for hydrogen, in a
magnetic field. The high energy line corresponds to the
case where the projection of $\vec{L}$ onto $\vec{H}$ has the same direction as $\vec{H}$ (m = 1), the intermediate line occurs when

$\overline{L}$ and $\overline{H}$ are perpendicular and the projection of $\overline{L}$ onto $\overline{H}$ is thus zero (m = 0), and the low energy line corresponds to the case where the projection of $\overline{L}$ onto $\overline{H}$ has the opposite direction of $\overline{H}$ (m = - 1).

## ATOMIC SPECTROSCOPY

● PROBLEM 684

The yellow light of a sodium lamp has an average wave-length of 5890 Å. Calculate the energy in (a) electron volts and (b) kilocalories per mole.

Solution: Experiments involving light suggest that it is either made up of waves or particles. This problem deals with the particle nature of light. The following equations are used.

$$E = h\nu \qquad \text{and} \qquad c = \lambda\nu,$$

where E is the energy of the particle of light, h Planck's constant, $6.626 \times 10^{-34}$ Js, $\nu$ the frequency of the light, c the speed of light and $\lambda$ is the wavelength of the light. One can derive a suitable expression to solve this problem from these equations. This is done by substituting $c/\lambda$ for $\nu$. Thus,

$$E = \frac{hc}{\lambda}.$$

$\lambda = 5890$ Å

$h = 6.626 \times 10^{-34}$ Js

$c = 3.0 \times 10^{8}$ m/s

Before placing these values into the equation, one must convert the wavelength of light from angstroms, Å, to meters, m. 1 Å = $1 \times 10^{-10}$ m; thus, the wavelength is $5890 \times 10^{-10}$ m. Hence,

$$E = \frac{(6.626 \times 10^{-34} \text{ Js})(3.0 \times 10^{8} \text{ m/s})}{(5890 \times 10^{-10} \text{ m})} = 3.4 \times 10^{-19} \text{ J.}$$

Once the energy is known, one must convert joules to electron volts and then to Kcal/mole. The conversion factors are: 1 eV = $1.6 \times 10^{-19}$ J

1 eV = 23.06 Kcal/mole.  Thus,

(a)  $E = \dfrac{(3.4 \times 10^{-19} \text{ J})}{(1.6 \times 10^{-19} \text{ J/eV})} = 2.1 \text{ eV}$        and

(b)  $E = (2.1 \text{ eV})(23.06 \text{ Kcal/mole-eV}) = 48.4 \text{ Kcal/mole}$.

● **PROBLEM** 685

Calculate the energy (in ergs) of a single quantum of each of the following: (a) 0.5 Å X-ray, (b) a radio wave of 900 kilocycles/sec. $10^{-8}$ cm = 1 Å, h = 6.626 × $10^{-27}$ erg-sec, c = 2.99 × $10^{10}$ cm/sec and 900 kilocycles/sec = 9 × $10^{5}$ sec$^{-1}$.

Solution:  The energy (E) of a single quantum can be calculated using the formula E = hν, where h = Planck's constant and ν = frequency. Sometimes the frequency is not known, but the wavelength, λ, is. λ and ν are related via the expression λν = c, where c = speed of light. Thus, if one solves for ν and then for E.

Proceed as follows:

(a) It is given that the wavelength of the X-ray = .5 Å or

$.5 \text{ Å} \left( \dfrac{10^{-8}}{\text{Å}} \text{ cm} \right) = .5 \times 10^{-8} \text{ cm}$, thus λ = .5 × $10^{-8}$ cm.

and $\nu = \dfrac{c}{\lambda} = \dfrac{2.99 \times 10^{10} \text{ cm/s}}{.5 \times 10^{-8} \text{ cm}} = 5.98 \times 10^{18} \text{ sec}^{-1}$.

Solving for E:

$E = h\nu = (6.626 \times 10^{-27} \text{erg-sec})(5.98 \times 10^{18} \text{ sec}^{-1})$

$= 4.0 \times 10^{-8} \text{ erg}$.

(b) Use E = hν. One is given the frequency as 900 kilocycles/sec = 9 × $10^{5}$ sec$^{-1}$. Thus,

$E = (6.626 \times 10^{-27} \text{ erg-sec})(9 \times 10^{5} \text{ sec}^{-1})$

$= 5.96 \times 10^{-21} \text{ erg}$.

How much energy is emitted by Avogadro number of atoms, if they each emit a light wave of 400 nm wavelength?

Solution: Energy is related to wavelength in the following equation, $E = hc/\lambda$, where E is energy, h Planck's constant $(6.626 \times 10^{-34}$ Js), c the speed of light $(3.0 \times 10^{8}$ m/sec) and $\lambda$ the wavelength. Using this equation, one calculates the energy emitted by each atom. To find the total amount of energy produced by Avogrado's number of atoms, multiply the amount of energy one atom emits by $6.02 \times 10^{23}$.

Solving for E:   1 nm = $10^{-9}$ m.

$$E = \frac{hc}{\lambda} = \frac{(6.626 \times 10^{-34} \text{ Js})(3.0 \times 10^{8} \text{ m/s})}{(400 \times 10^{-9} \text{ m})}$$

$$= 4.97 \times 10^{-19} \text{ J/atom}$$

Thus, the total amount of energy emitted is

$$E = (4.97 \times 10^{-19} \text{ J/atom})(6.02 \times 10^{23} \text{ atoms/mole})$$

$$= 2.98 \times 10^{5} \text{ J}.$$

The flame test for barium involves placing a barium-containing compound in a Bunsen burner flame. The barium compound then decomposes to give barium atoms which subsequently undergo an electronic transition of energy $3.62 \times 10^{-12}$ erg. What color flame would this give rise to?

Solution: To solve this problem we use the relationship $E = hc/\lambda$, where E is the energy of the transition, h is Planck's constant, c is the speed of light, and $\lambda$ is the wavelength of the emitted light. Solving for $\lambda$,

$$\lambda = \frac{hc}{E} = \frac{6.62 \times 10^{-27} \text{ erg-sec} \times 3.0 \times 10^{10} \text{ cm/sec}}{3.62 \times 10^{-12} \text{ erg}}$$

$$= 5.490 \times 10^{-5} \text{ cm}.$$

Converting to Å,

$$\lambda = 5.490 \times 10^{-5} \text{ cm} = 5.490 \times 10^{-5} \text{ cm} \times 10^{8} \text{ Å/cm} = 5490 \text{ Å},$$

which corresponds to a green flame.

To measure the wavelength of lines in atomic emission and absorption spectra, one uses a spectroscope. Two lines in the yellow emission spectra of hot Na metal, the so-called sodium-D lines, are used to calibrate this instrument. If the wavelength of one of these lines is 5890 Å, find the energy of the electronic transition associated with this line. $h = 6.62 \times 10^{-27}$ erg-sec.

Solution: The energy absorbed or released in electron transition is given by the formula $E = h\nu$, where $E$ = energy, $h$ = Planck's constant and $\nu$ = frequency.

$\nu$ can be found using the sodium wavelength of 5890 Å. Wavelength ($\lambda$), frequency ($\nu$), and speed of light ($c$) are related by the equation $\lambda\nu = c$ ($c = 3.00 \times 10^{10}$ cm sec$^{-1}$). To find $\nu$, substitute these values into the equation and solve.

There are $10^{-8}$ cm per Å. Consequently, 5890 Å = $5.89 \times 10^{-5}$ cm. Substituting, we obtain

$(5.89 \times 10^{-5} \text{ cm})\nu = 3 \times 10^{10}$ cm/sec    which yields

$\nu = 5.093 \times 10^{14}$ sec$^{-1}$.

To find $E$, note that $E = h\nu$. Thus,

$E = h(5.093 \times 10^{14})$

$= (6.62 \times 10^{-27} \text{ erg-sec})(5.093 \times 10^{14} \text{ sec}^{-1})$

$= 3.37 \times 10^{-12}$ erg.

The sodium D-lines, two very distinct lines of yellow light emitted by heated sodium metal, are often used in the calibration of spectrometers. The wavelength of one of these lines is 5890 Å. How much energy does an electron emit (or absorb) in undergoing the electronic transition associated with this line?

Solution: To solve this problem we will combine two relationships: $\nu = c/\lambda$, where $\nu$ is the frequency, $\lambda$ the wavelength, and $c$ the speed of light; and Planck's relationship, $E = h\nu$, where $E$ is the energy and $h$ is Planck's constant, $6.62 \times 10^{-27}$ erg-sec. Then, $E = h\nu = hc/\lambda$. Since 1 Å = $10^{-8}$ cm, 5890 Å = 5890 Å $\times 10^{-8}$ cm/Å = $5.890 \times 10^{-5}$ cm. The energy associated with the

5890 $\overset{\circ}{A}$ sodium D-line is then

$$E = \frac{hc}{\lambda} = \frac{6.62 \times 10^{-27} \text{ erg-sec} \times 3.0 \times 10^{10} \text{ cm/sec}}{5.89 \times 10^{-5} \text{ cm}}$$

$$= 3.37 \times 10^{-12} \text{ erg.}$$

● **PROBLEM** 690

In order for a photon of light incident upon metallic potassium to eliminate an electron from it, the photon must have a minimum energy of 4.33 eV (photoelectric work function for potassium). What is the wavelength of a photon of this energy? Assume h = 6.626 × 10$^{-27}$ erg-sec and c = 2.998 × 10$^{10}$ cm/sec.

Solution:  This problem illustrates two fundamental relationships. The first is that the product of wavelength $\lambda$ and frequency $\nu$ is the speed of light, c = $\lambda\nu$ (or $\nu$ = c/$\lambda$). The second is the relationship between frequency and energy E, E = h$\nu$, where h is Planck's constant. Combining these equations we obtain

E = h$\nu$ = h (c/$\lambda$) = hc/$\lambda$,  or $\lambda$ = hc/E.

Before applying this equation, we must convert the energy E from electron volts (eV) to the more convenient unit of energy, the erg. Since 1 eV = 1.602 × 10$^{-12}$ erg,

E = 4.33 eV = 4.33 eV × 1.602 × 10$^{-12}$ erg/eV

$$= 6.94 \times 10^{-12} \text{ erg.}$$

The wavelength of light corresponding to 4.33 eV is then

$$\lambda = \frac{hc}{E} = \frac{6.626 \times 10^{-27} \text{erg-sec} \times 2.998 \times 10^{10} \text{ cm/sec}}{6.94 \times 10^{-12} \text{ erg}}$$

$$= 2.864 \times 10^{-5} \text{ cm} = 2.864 \times 10^{-5} \text{ cm} \times 10^8 \overset{\circ}{A}/\text{cm}$$

$$= 2864 \overset{\circ}{A}.$$

● **PROBLEM** 691

Calculate the number of electrons that can be removed from metallic cesium with the energy required to remove one electron from an isolated cesium atom. Assume the following: the ionization potential of Cs = 3.89 eV, h = 6.63 × 10$^{-34}$ J sec, and Cs metal is sensitive to red light of 700 nm.

Solution:  This problem entails the understanding of the photoelectric effect. The photoelectric effect states that when one shines light of a given frequency on a metallic material, electrons will be emitted. The equation: E = hv (where E = energy, v = frequency, and h = Planck's constant) determines the amount of energy produced by a given frequency of light. The ionization potential is the amount of energy needed to cause the electrons to be ejected from the metal. To solve this problem, therefore, one must calculate how much energy is present in light with a wavelength of 700 nm and then compare this to the ionization potential of a Cs atom.

It is given that the wavelength of the light is 700 nm. The frequency of this light is equal to speed of light/wavelength = $(3.00 \times 10^8$ m/s)/$700 \times 10^{-9}$ m. Thus,

$$E = (6.63 \times 10^{-34} \text{ J sec}) \frac{(3.00 \times 10^8 \text{ m/s})}{700 \times 10^{-9} \text{ m}}$$

$$= 2.84 \times 10^{-19} \text{ J/photon}$$

1 eV = $1.60 \times 10^{-19}$ J. The ionization energy is 3.89 eV or $(3.89 \text{ eV})(1.60 \times 10^{-19} \text{ J/eV}) = 6.22 \times 10^{-19}$ J.

Because Cs is sensitive to light with a wavelength of 700 nm, one photon of this light flashed on the metal Cs will cause 1 electron to be emitted from the Cs. Therefore, the number of electrons emitted when $6.22 \times 10^{-19}$ J, (the ionization potential of an isolated Cs atom), is added to the metallic system is found by dividing $6.22 \times 10^{-19}$ J by $2.84 \times 10^{-19}$ J

$$\text{no. of electrons emitted} = \frac{6.22 \times 10^{-19} \text{ J}}{2.84 \times 10^{-19} \text{ J}} = 2.2 \text{ electrons}$$

Therefore, the maximum number of electrons that can be emitted is 2.

● **PROBLEM** 692

The accompanying figure shows an energy-level diagram. Make a comparison of the energy of the n = 5 and n = 4 transition for an electron with a nucleus of Z = 3 and the energy of the transition from n = 2 to n = 1 for an electron with a nucleus of Z = 2.

Solution:  When an atom is in the ground state, it is at its lowest energy level, n = 1. If energy is added, the electrons become "excited" and move to a higher energy level; n = 2, 3, 4 ... . After a time, the electron falls back to lower energy states and releases the energy needed to move it to the higher energy levels. The equation

$$E_b - E_a = Z^2 \frac{e^2}{2a_0} \left( \frac{1}{n_a^2} - \frac{1}{n_b^2} \right),$$

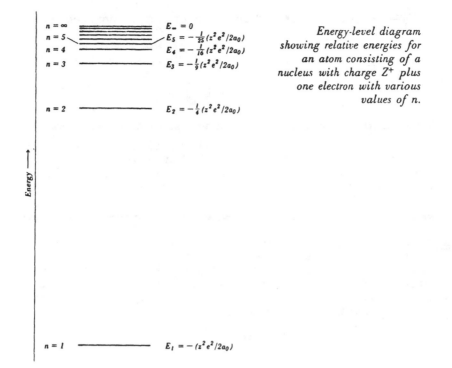

*Energy-level diagram showing relative energies for an atom consisting of a nucleus with charge $Z^+$ plus one electron with various values of n.*

where n = the (quantum) energy level, E = energy, e = charge on electron, $a_0$ = Bohr radius, and Z = atomic number, measures this energy release between energy levels. You are asked to compare the energy releases between two different atoms and energy levels. You do not need to know the actual energy released. Because $e^2/2a_0$ is a constant, it need not be evaluated. (It will cancel out when you compare).

Therefore, you have for n = 5 to n = 4

$$E_b - E_a = \frac{e^2}{2a_0}\left[\frac{1}{n_a^2} - \frac{1}{n_b^2}\right] Z^2 = \frac{e^2}{2a_0}\left[\frac{1}{16} - \frac{1}{25}\right] \times 9$$

For n = 2 to n = 1, you have

$$E_b - E_a = \frac{e^2}{2a_0}\left[\frac{1}{n_a^2} - \frac{1}{n_b^2}\right] Z^2 = \frac{e^2}{2a_0}\left[\frac{1}{1} - \frac{1}{4}\right] \times 4$$

By comparison, in the form of a ratio, one has

$$\frac{\frac{e^2}{2a_0}\left[\frac{1}{16} - \frac{1}{25}\right] \times 9}{\frac{e^2}{2a_0}\left[\frac{1}{1} - \frac{1}{4}\right] \times 4} = \frac{\left[\frac{1}{16} - \frac{1}{25}\right] \times 9}{\left[\frac{1}{1} - \frac{1}{4}\right] \times 4} = \frac{.2025}{3}$$

$\dfrac{.2025}{3} \times 100\% = 6.8\%$. Thus, the energy involved in

the n = 5 to n = 4 transition is 6.8% as large as compared to the energy emitted in the n = 2 to n = 1 transition.

● **PROBLEM** 693

What wavelength of light is needed to excite an electron in a 0.2 nanometer (1 nm = $10^{-9}$ m) box from the ground state to the second excited state? What wavelength of light is emitted when the same electron falls from the second excited state to the first excited state?

Solution: The solution to this problem requires the application of two equations, one to determine the energies of the various states, from which may be obtained the energy of a transition between states, and one to relate energy to wavelength. The first equation is the energy of a particle-in-a-box,

$$E_n = \dfrac{n^2 h^2}{8mL^2} \, ,$$

where n is the principal quantum number (n = 1, 2, 3, ...), h is Planck's constant ($6.626 \times 10^{-27}$ erg-sec = $6.626 \times 10^{-34}$ J-sec), m is the mass of the particle, and L is the length of the box. For an electron in a 0.2 nm

$$= 0.2 \times 10^{-9} \text{ m} = 2 \times 10^{-10} \text{ m box,}$$

$$E_n = \dfrac{n^2 h^2}{8mL^2} = \dfrac{n^2 \, (6.626 \times 10^{-34} \text{ J-sec})^2}{8 \times 9.11 \times 10^{-31} \text{ kg} \times (2 \times 10^{-10} \text{ m})^2}$$

$$= 1.506 \times 10^{-18} \text{ J} \times n^2.$$

The second equation relates the energy difference, $\Delta E$, between two states and the wavelength $\lambda$: $\Delta E = hc/\lambda$, or $\lambda = hc/\Delta E$, where c is the speed of light.

In the following table, we calculate the energies corresponding to the ground state (n = 1), the first excited state (n = 2), and the second excited state (n = 3).

| n | $E_n = n^2 \times 1.506 \times 10^{-18}$ J |
|---|---|
| 1 | $E_1 = (1)^2 \times 1.506 \times 10^{-18}$ J $= 1.506 \times 10^{-18}$ J |
| 2 | $E_2 = (2)^2 \times 1.506 \times 10^{-18}$ J $= 6.024 \times 10^{-18}$ J |
| 3 | $E_3 = (3)^2 \times 1.506 \times 10^{-18}$ J $= 1.355 \times 10^{-17}$ J |

In going from the ground state to the second excited state, the electron must absorb a quantum of light of energy

$\Delta E = E_3 - E_1 = 1.355 \times 10^{-17}$ J $- 1.506 \times 10^{-18}$ J $= 1.204 \times 10^{-17}$ J. The corresponding wavelength is

$$\lambda = \frac{hc}{\Delta E} = \frac{6.626 \times 10^{-34} \text{ J} \times 3 \times 10^{8} \text{ m/sec}}{1.204 \times 10^{-17} \text{ J}}$$

$$= 16.510 \times 10^{-9} \text{ m} = 16.510 \text{ nm}.$$

In falling from the second excited state to the first excited state the electron will emit a quantum of light of energy $\Delta E = E_3 - E_2 = 1.355 \times 10^{-17}$ J $- 6.024 \times 10^{-18}$ J $= 7.526 \times 10^{-18}$ J. This corresponds to a wavelength of

$$\lambda = \frac{hc}{\Delta E} = \frac{6.626 \times 10^{-34} \text{ J} \times 3 \times 10^{8} \text{ m/sec}}{7.526 \times 10^{-18} \text{ J}}$$

$$= 26.412 \times 10^{-9} \text{ m} = 26.412 \text{ nm}.$$

● **PROBLEM** 694

Consider an atom containing one electron and having an atomic number Z. It is observed that light of a wavelength of 0.1 nm will excite this atom from the ground state to infinite ionization. Determine the value of Z.

Solution: We know the wavelength $\lambda$ of the exciting light and the quantum numbers of the ground state ($n_g = 1$) and the ionized state ($n_i = \infty$). We must determine Z, the number of protons in the nucleus of the atom.

There is a relationship to find Z using $n_g$ and $n_i$. It is provided by the Bohr theory of the atom. It is:

$$\frac{c}{\lambda} = \frac{Z^2 e^2}{2ha_0}\left(\frac{1}{n_g^2} - \frac{1}{n_i^2}\right),$$

where c is the speed of light ($c = 2.9969 \times 10^{10}$ cm/sec), e is the electronic charge ($e = 4.80 \times 10^{-10}$ esu), h is Planck's constant ($h = 6.626 \times 10^{-27}$ erg-sec), and $a_0$ is the Bohr radius ($a_0 = 0.05292 \times 10^{-7}$ cm). If we combine some of these constants, we arrive at the value $e^2/2ha_0 = 3.290 \times 10^{15}$ sec$^{-1}$. Hence, our expression becomes

$$\frac{c}{\lambda} = Z^2 \left(\frac{e^2}{2ha_0}\right)\left(\frac{1}{n_g^2} - \frac{1}{n_i^2}\right)$$

$$= Z^2 \times 3.290 \times 10^{15} \text{ sec}^{-1} \left(\frac{1}{n_g^2} - \frac{1}{n_i^2}\right).$$

657

To solve this problem, we must solve this equation for Z and substitute $c$, $n_g$, $n_i$, and $\lambda$ ($\lambda = 0.1$ nm $= 0.1 \times 10^{-9}$ m $= 10^{-10}$ m $= 10^{-8}$ cm) into the resulting equation.

Solving for Z, we obtain

$$Z^2 = \frac{c}{\lambda} \left[ 3.290 \times 10^{15} \text{sec}^{-1} \left( \frac{1}{n_g^{\,2}} - \frac{1}{n_i^{\,2}} \right) \right]^{-1} , \quad \text{or}$$

$$Z = \left( \frac{c}{\lambda \times 3.290 \times 10^{15} \text{ sec}^{-1} \left( \dfrac{1}{n_g^{\,2}} - \dfrac{1}{n_i^{\,2}} \right)} \right)^{\frac{1}{2}}$$

Substituting the values of $c$, $\lambda$, $n_g$, and $n_i$ into this expression yields

$$Z = \left( \frac{2.9969 \times 10^{10} \text{ cm/sec}}{10^{-8} \text{ cm} \times 3.290 \times 10^{15} \text{ sec}^{-1} \left( \frac{1}{1} - \frac{1}{\infty} \right)} \right)^{\frac{1}{2}} \cong 30 ,$$

where we used the equality $1/\infty = 0$. Hence, this atom must have 30 protons.

● **PROBLEM** 695

The Rydberg-Ritz equation governing the spectral lines of hydrogen is $\frac{1}{\lambda} = R \left( \dfrac{1}{n_1^{\,2}} - \dfrac{1}{n_2^{\,2}} \right)$ , where R is the Rydberg constant, $n_1$ indexes the series under consideration ($n_1 =$ 1 for the Lyman series, $n_1 = 2$ for the Balmer series, $n_1 =$ 3 for the Paschen series), $n_2 = n_1 + 1$, $n_1 + 2$, $n_1 + 3$, ... indexes the successive lines in a series, and $\lambda$ is the wavelength of the line corresponding to index $n_2$. Thus, for the Lyman series, $n_1 = 1$ and the first two lines are 1215.56 Å ($n_2 = n_1 + 1 = 2$) and 1025.83 Å ($n_2 = n_1 + 2 =$ 3). Using these two lines, calculate two separate values of the Rydberg constant. The actual value of this constant is $R = 109678$ cm$^{-1}$.

Solution: The first thing to do is to convert the wavelengths from Å to more manageable units, i.e. centimeters. Usin g the relationship 1 Å $= 10^{-8}$ cm, the first two Lyman lines are 1215.56 Å $= 1215.56 \times 10^{-8}$ cm for $n_2 = 2$, and 1025.83 Å $= 1025.83 \times 10^{-8}$ cm for $n_2 = 3$. Solving the Rydberg-Ritz equation for R, one obtains·

$$R = \left( \lambda \left[ \frac{1}{n_1^{\,2}} - \frac{1}{n_2^{\,2}} \right] \right)^{-1} ,$$

For the first line,

$$R = \left[ \lambda \left( \frac{1}{n_1{}^2} - \frac{1}{n_2{}^2} \right) \right]^{-1} = \left[ 1215.56 \times 10^{-8} \text{ cm} \left( \frac{1}{1^2} - \frac{1}{2^2} \right) \right]^{-1}$$

$$= 109689 \text{ cm}^{-1}$$

and for the second line,

$$R = \left[ \lambda \left( \frac{1}{n_1{}^2} - \frac{1}{n_2{}^2} \right) \right]^{-1} = \left[ 1025.83 \times 10^{-8} \text{ cm} \left( \frac{1}{1^2} - \frac{1}{3^2} \right) \right]^{-1}$$

$$= 109667 \text{ cm}^{-1}.$$

The first of these is 0.0100% greater than the true value, and the second is 0.0100% less than the true value.

● **PROBLEM** 696

Wave number ($\bar{\nu}$) are the reciprocals of wavelengths, $\lambda$, and are given by the expression $\bar{\nu} = 1/\lambda$. For the hydrogen atom, the Bohr theory predicts that the wave number for the emission line associated with an electronic transition from the energy level having principal quantum number $n_2$ to that with principal quantum number $n_1$ is

$$\bar{\nu} = R_H \left( \frac{1}{n_1{}^2} - \frac{1}{n_2{}^2} \right) ,$$

where $R_H$ is the Rydberg constant. In what region of the electromagnetic spectrum would there appear a spectral line resulting from the transition from the tenth to the fifth electronic level in hydrogen?

Solution: This problem can be solved by applying the Rydberg formula used for determining the wavenumbers of the spectral lines.

The Rydberg constant is $1.10 \times 10^5 \text{ cm}^{-1}$. $n_2 = 10$ and $n_1 = 5$. Substituting:

$$\bar{\nu} = R_H \left( \frac{1}{n_1{}^2} - \frac{1}{n_2{}^2} \right) = 1.10 \times 10^5 \text{ cm}^{-1} \left( \frac{1}{5^2} - \frac{1}{10^2} \right)$$

$$= 1.10 \times 10^5 \text{ cm}^{-1} \left( \frac{1}{25} - \frac{1}{100} \right) = 1.10 \times 10^5 \text{ cm}^{-1} (0.04-0.01)$$

$$= 1.10 \times 10^5 \text{ cm}^{-1} \times 0.03$$

$$= 3.3 \times 10^3 \text{ cm}^{-1}.$$

This line appears in the infrared region of the spectrum (approximately 20 $cm^{-1}$ to $10^4$ $cm^{-1}$).

● **PROBLEM** 697

What is the maximum number of emission lines one would expect to see in a molecule containing only the six electronic energy levels depicted in diagram A?

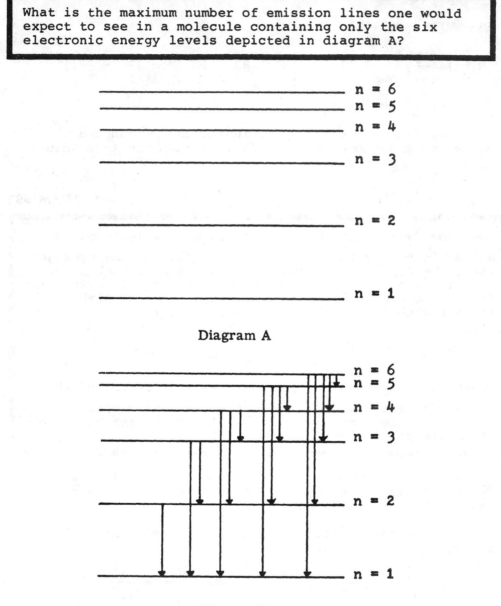

Diagram A

Diagram B

Solution: Emission is observed when an electron falls from an electronic energy level of higher energy to one of lower energy.

From the energy level n = 6, electrons can fall to
energy levels of n = 5, 4, 3, 2, and 1. Thus, five emission
lines are drawn from the energy level n = 6. By using a
similar process, one determines that the maximum number of
emission lines from energy levels n = 5, 4, 3, 2, and 1
are four, three, two, one, and zero, respectively. Because
there are no energy levels lower than n = 1, no emission
lines are obtained from it. The total number of emission
lines is found by taking the sum of all the emission lines
from the various energy levels. Thus, 5 + 4 + 3 + 2 + 1 +
0 = 15 emission lines can be expected. The possible
emission processes are depicted in diagram B.

● **PROBLEM** 698

Find the $\lambda$'s (wavelengths) for the first four hydrogen
spectral lines in the Paschen series. Find the wavelength
of the series limit.

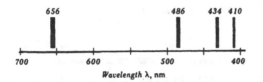

Hydrogen line spectrum.
(Numbers shown are
wavelengths in nanometers,
or tens of angstrom units.)

Solution: The Paschen series illustrates the spectrum
hydrogen exhibits in the infrared region. An example of a
hydrogen line spectrum is shown in the accompanying
figure. The series of lines of the Paschen series fit the
equation

$$\nu = 3.29 \times 10^{15} \left( \frac{1}{9} - \frac{1}{b^2} \right) ,$$

where $\nu$ = frequency and b = 4, 5, 6, 7, 8, ... . Thus,
if you want the first four hydrogen spectral lines, you
use b = 4, 5, 6, and 7. Your answer would be in terms of
frequency, however, you want the wavelengths. To obtain
this, employ the relation between frequency and wavelength,
c = $\lambda\nu$, where c = speed of light, $\lambda$ = wavelength and $\nu$ =
frequency. (c = $2.998 \times 10^{10}$ cm/sec.)

Once you know $\nu$, you can find $\lambda$. Therefore,

1st line: b = 4, $\nu = 3.290 \times 10^{15} \left( \frac{1}{9} - \frac{1}{16} \right)$

$$= 1.599 \times 10^{14} \text{ sec}^{-1}.$$

2nd line: b = 5, $\nu = 3.290 \times 10^{15} \left( \frac{1}{9} - \frac{1}{25} \right)$

$$= 2.340 \times 10^{14} \text{ sec}^{-1}.$$

3rd line: $b = 6$, $\nu = 3.290 \times 10^{15}\left(\dfrac{1}{9} - \dfrac{1}{36}\right)$

$$= 2.742 \times 10^{14} \text{ sec}^{-1}.$$

4th line: $b = 7$, $\nu = 3.290 \times 10^{15}\left(\dfrac{1}{9} - \dfrac{1}{49}\right)$

$$= 2.984 \times 10^{14} \text{ sec}^{-1}.$$

The series limit is $b = $ infinity. If $B \to \infty$, then $\dfrac{1}{B^2} \to 0$. Series limit: $\nu = 3.290 \times 10^{15}\left(\dfrac{1}{9} - 0\right) = 3.656 \times 10^{14}$

The wavelengths $(\lambda) = \dfrac{\text{speed of light (c)}}{\text{frequency } (\nu)}$

1st line: $\lambda = 1.87 \times 10^{-4}$ cm   2nd line: $\lambda = 1.28 \times 10^{-4}$ cm

3rd line: $\lambda = 1.09 \times 10^{-4}$ cm   4th line· $\lambda = 1.004 \times 10^{-4}$ cm

Series limit: $.820 \times 10^{-4}$ cm.

● **PROBLEM** 699

The first two lines in the Lyman series for hydrogen are 1215.56 Å and 1025.83 Å. These lines lie in the ultraviolet region of the spectrum. For each of these lines calculate the following: (a) the corresponding energy in ergs; (b) the corresponding energy in Kcal/mole; (c) the frequency in sec$^{-1}$.

Solution:  The problem involes the application of two basic formulas. The first is the relationship between wavelength ($\lambda$) and frequency ($\nu$). The product of the two is the speed of light c, $\lambda\nu = c$ (or $\nu = c/\lambda$). The second is the relationship between frequency and energy (E), $E = h\nu$, where h is Planck's constant.

First calculate the frequencies for the two lines. Distinguish $\lambda$, $\nu$, and E for these lines with the subscripts "1" and "2". Thus,

$\lambda_1 = 1215.56$ Å          and   $\lambda_2 = 1025.83$ Å.

To express these in the more convenient units of centimeters, use the relationship 1 Å $= 10^{-8}$ cm. Then,

$\lambda_1 = 1215.56 \times 10^{-8}$ cm    and $\lambda_2 = 1025.83 \times 10^{-8}$ cm.

Using the first of the two formulas, one obtains:

$$\nu_1 = \frac{c}{\lambda_1} = \frac{2.9979 \times 10^{10} \text{ cm/sec}}{1215.56 \times 10^{-8} \text{ cm}} = 2.4663 \times 10^{15} \text{ sec}^{-1}$$

$$\nu_2 = \frac{c}{\lambda_2} = \frac{2.9979 \times 10^{10} \text{ cm/sec}}{1025.83 \times 10^{-8} \text{ cm}} = 2.9224 \times 10^{15} \text{ sec}^{-1}.$$

The energies for these two lines are calculated using the second of the two relationships. Planck's constant is $h = 6.626 \times 10^{-27}$ erg-sec. The units of frequency are $\text{sec}^{-1}$, the units of energy (found by using $E = h\nu$), will then be erg-sec $\times$ $\text{sec}^{-1}$ = erg. To convert from ergs to Kcal/mole, first convert to joules (1 joule = $10^7$ ergs), then from joules to Kcal (1 Kcal = $4.184 \times 10^3$ joules), and finally multiply by Avogadro's number ($6.022 \times 10^{23}$/mole) in order to put this on a per mole basis. Hence, the conversion factor from ergs to Kcal/mole is

1 erg = 1 erg $\times$ 1 joule/$10^7$ ergs $\times$ 1 Kcal/$4.184 \times 10^3$ joules

$$\times \ 6 \ 022 \times 10^{23}/\text{mole}$$

$$= 1.439 \times 10^{13} \text{ Kcal/mole.}$$

Applying the second equation one obtains:

$E_1 = h\nu_1 = 6.626 \times 10^{-27}$ erg-sec $\times$ $2.4663 \times 10^{15}$ $\text{sec}^{-1}$

$$= 1.6342 \times 10^{-11} \text{ erg}$$

$$= 1.6342 \times 10^{-11} \text{ erg} \times 1.439 \times 10^{13} \text{ Kcal/mole-erg}$$

$$= 235.2 \text{ Kcal/mole.}$$

$E_2 = h\nu_2 = 6.626 \times 10^{-27}$ erg-sec $\times$ $2.9224 \times 10^{15}$ $\text{sec}^{-1}$

$$= 1.9364 \times 10^{-11} \text{ erg}$$

$$= 1.9364 \times 10^{-11} \text{ erg} \times 1.439 \times 10^{13} \text{ Kcal/mole-erg}$$

$$= 278.6 \text{ Kcal/mole.}$$

● PROBLEM 700

Show that in the Balmer series that the frequencies of successive lines tend toward a limiting value of cR/4, where R is the Rydberg constant and c the speed of light.

Solution: Here, one must try to relate frequency $\nu$ with c and R. The Rydberg constant, R, comes from the Rydberg-Ritz equation,

$$\frac{1}{\lambda} = R \left[ \frac{1}{(n_1)^2} - \frac{1}{(n_2)^2} \right],$$

where $\lambda$ is wavelength. The electron falls from an energy level of $n_2$ to $n_1$. The speed of light is related to the frequency in the equation, $\nu = c/\lambda$.

These equations are combined.

(c) $$\left( \frac{1}{\lambda} \right) = cR \left[ \frac{1}{(n_1)^2} - \frac{1}{(n_2)^2} \right] = \frac{c}{\lambda} = \nu$$

Thus, $$\nu = cR \left[ \frac{1}{(n_1)^2} - \frac{1}{(n_2)^2} \right]$$

In the Balmer series $n_1$ is defined as being equal to 2. For successive lines, $n_2 \to \infty$ and therefore $1/(n_2)^2$ approaches 0. The above equation becomes

$$\nu = cR \left[ \frac{1}{(2)^2} - \frac{1}{(\infty)^2} \right] = \frac{cR}{4} .$$

● **PROBLEM** 701

Find the difference in characteristic vibrational frequency of a diatomic molecule of mass 20 amu if (a) the mass is equally distributed between the two ends, and (b) one atom of 1 amu is at one end and an atom of 19 amu is at the other.

Solution: The vibrational frequency is frequency of the oscillation between the two atoms of the molecule. It is expressed quantitatively by

$$\nu = \tfrac{1}{2} \pi \sqrt{K/M_{eff}}$$

where $\nu$ is the vibrational frequency, K a force constant, and $M_{eff}$ the effective mass of the molecule.

$$M_{eff} = \frac{M_A M_B}{M_A + M_B}$$

where $M_A$ is the mass of one atom and $M_B$ is the other.

(a) If the mass is equally distributed between both atoms, each weighs 10 amu.

$$M_{eff} = \frac{10 \times 10}{10 + 10} = 5$$

Assume that K = 1.

$$\nu = \frac{1}{2\pi} \sqrt{\frac{1}{5}} = 0.0711$$

(b) $M_{eff} = \frac{1 \times 19}{20} = \frac{19}{20}$

$$\nu = \frac{1}{2\pi} \sqrt{\frac{19}{20}} = 0.1632$$

Therefore, the atoms in the molecule described in b are vibrating more quickly than those in molecule a.

● **PROBLEM** 702

Calculate the characteristic vibrational frequency of the lowest vibration state for $H_2$. Assume k = 5.1 newtons/cm and h = $6.63 \times 10^{-34}$ J-sec.

Solution:  Picture a diatomic molecule, such as $H_2$, as two balls connected by a spring. The spring represents the chemical bond. When the two balls (atoms) are pulled apart the string acts as a restoring force to bring them back together to the equilibrium separation. When they are compressed, the string tends to force them apart. Such a system, in which the restoring force is proportional to the amount of distortion, is called a harmonic oscillator. The vibrational frequency of the spring going back and forth is given by the formula

$$\nu = \frac{1}{2\pi} \sqrt{k/M_{eff}}$$

where $\nu$ = characteristic vibrational frequency of the particular bond, k = force constant and $M_{eff}$ = effective mass for the vibrating atoms:

$$M_{eff} = \frac{M_A M_B}{M_A + M_B}$$

If the masses of the two atoms are equal, $M_{eff}$ becomes

$\frac{M^2}{2M} = \frac{1}{2} M$.

To find the characteristic vibration frequency, use this formula.

To find the $M_{eff}$, find the mass of one hydrogen atom. The number of grams in one mole (molecular weight) is one. There are $6.02 \times 10^{23}$ atoms of H in one mole. Thus, the mass of one atom is

$$\frac{1 \text{ g/mole}}{6.02 \times 10^{23}/\text{mole}} = 1.66 \times 10^{-24} \text{ g.} \quad \text{Thus,}$$

$$M_{eff} = \frac{(1.66 \times 10^{-24} \text{ g})}{2} = 8.31 \times 10^{-25} \text{ g.}$$

Solving for $\nu$:

$$k = 5.1 \frac{\text{newtons}}{\text{cm}} = 5.1 \frac{\text{kg M}}{\text{cm} \cdot \text{sec}^2}$$

$$\nu = \frac{1}{2\pi} \sqrt{\frac{5.1 \dfrac{\text{kg M}}{\text{cm sec}^2} \times 1000 \dfrac{\text{g}}{\text{kg}} \times 100 \dfrac{\text{cm}}{\text{m}}}{8.31 \times 10^{-25} \text{ g}}}$$

$$= 1.25 \times 10^{14} \text{ sec}^{-1}.$$

Note: 1000 g/kg and 100 cm/m are conversion factors used to obtain the correct units.

● **PROBLEM** 703

Find the wavelength of light required for the ionization of sodium atoms to occur. The ionization potential of sodium is $8.17 \times 10^{-19}$ J.

Solution: The energy necessary to ionize a sodium atom is equal to the ionization potential. Here, one is looking for the wavelength of light, which, when flashed upon sodium, will possess an energy equal to the ionization potential. Wavelength is related to energy by the equation $E = hc/\lambda$, where E is the energy, h Planck's constant ($6.626 \times 10^{-34}$ J sec), c the speed of light ($3.0 \times 10^8$ m/sec) and $\lambda$ the wavelength. One can rewrite this equation as $\lambda = hc/E$ and then substitute the given to solve for $\lambda$.

$$\lambda = \frac{(6.626 \times 10^{-34} \text{ J sec})(3.0 \times 10^8 \text{ m/s})}{(8.17 \times 10^{-19} \text{ J})}$$

$$= 2.43 \times 10^{-7} \text{ m} = 2.43 \times 10^{-7} \text{ m} \times \frac{1 \text{ Å}}{10^{-10} \text{ m}} = 2430 \text{ Å.}$$

● **PROBLEM** 704

Find the wavelength ($\lambda$) of the first transition, from the first excited state to the ground state in the Lyman and Paschen series. These series show the lines of the emission spectra of $H_2$ gas. The Lyman series defines its lowest energy level $n_1$ as being equal to 1, the Paschen series defines $n_1$ as 3.

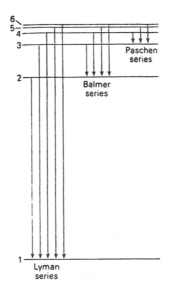

Schematic energy
level diagram for an excited hydrogen atom. Jumps of excited electrons from higher to lower levels account for the different lines in the
hydrogen emission spectrum.

Solution: When an atom is in its ground state, its e-
lectrons are in the lowest energy state possible. When
energy is added to the system, its electrons are promoted
to a higher energy level. These electrons are said to be
in the excited state. After a short time, the electrons
fall back to lower energy levels and eventually to their
ground states. The same amount of energy that was needed
to excite the electron is released when the electron falls
back to its ground state. The Paschen and Lyman series
differ in the value of their lowest energy states.

The Paschen series is the spectra of electrons falling
from excited states to the third energy level ($n_1 = 3$),
whereas the Lyman series is the spectra that is exhibited
when the electrons fall from excited states to the first
energy level ($n_1 = 1$). This is illustrated in the
accompanying figure.

Wavelength and the energy levels of the transition
are related in the Rydberg-Ritz equation,

$$\frac{1}{\lambda} = R \left( \frac{1}{n_1{}^2} - \frac{1}{n_2{}^2} \right),$$

where $\lambda$ is the wavelength, R the Rydberg constant, and
$n_1$ and $n_2$ are the energy levels of the transition. The
electron falls from $n_2$ to $n_1$, thus, $n_2$ is greater than $n_1$.

Solving for $\lambda$:

Lyman series: In the Lyman series $n_1 = 1$. The first

transition is from $n_2 = 2$ to $n_1 = 1$. One solved for $\lambda$ by using the Rydberg-Ritz equation.

$$\frac{1}{\lambda} = (109,678\ cm^{-1}) \left[ \frac{1}{(1)^2} - \frac{1}{(2)^2} \right]$$

$$\frac{1}{\lambda} = 82258\ cm^{-1}$$

$$\lambda = \frac{1}{82258\ cm^{-1}} = 1.21 \times 10^{-5}\ cm.$$

Wavelengths are usually expressed in angstroms. $1\ \overset{\circ}{A} = 10^{-8}\ cm$.

$$(1.21 \times 10^{-5}\ cm)\ \frac{1\ \overset{\circ}{A}}{10^{-8}\ cm} = 1210\ \overset{\circ}{A}$$

Paschen series: The first transition in the Paschen series is from $n_2 = 4$ to $n_1 = 3$ because the base energy level in this series is 3. Here again, one uses the Rydberg-Ritz equation to solve for $\lambda$.

$$\frac{1}{\lambda} = 109678\ cm^{-1}\ \left[ \frac{1}{(3)^2} - \frac{1}{(4)^2} \right] = 5332\ cm^{-1}$$

$$\lambda = \frac{1}{5332\ cm^{-1}} = 1.87 \times 10^{-4}\ cm.$$

Converting to angstroms:

$$1.87 \times 10^{-4}\ cm \times \frac{1\ \overset{\circ}{A}}{10^{-8}\ cm} = 18700\ \overset{\circ}{A}.$$

● **PROBLEM** 705

A gaseous substance exhibits a line spectrum, whereas a liquid or a solid substance exhibits a continuous spectrum. Give a qualitative explanation for these phenomena.

Solution: The basic difference between gases and the condensed phases of matter, which gives rise to these different phenomena, is that in gases the atoms or molecules are sufficiently far apart to have a negligible interaction, whereas in liquids and solids, the interaction between the constituent atoms and molecules is appreciable.

In the gas phase, the molecules are so far apart that they may be considered to be isolated from one another and therefore behave independently of one another. Thus, the spectrum arises only from electronic, vibrational, or rotational transitions of individual atoms or molecules, and not from any collective behavior. But since these transitions are well defined energetically, the spectrum exhibits sharp lines.

In a liquid or solid, the atoms or molecules are sufficiently close together to affect each other's energy levels. That is, in addition to the individual energy levels of each atom or molecule, there exist energy levels for the collective excitation of groups of atoms or molecules. In general, the interaction of one atom or molecule with another decreases with internuclear distance or intermolecular distance in a continuous fashion, so that a continuous spectrum is observed.

# CHAPTER 19

# NUCLEAR CHEMISTRY

Basic Attacks and Strategies for Solving Problems in this Chapter. See pages 671 to 685 for step-by-step solutions to problems.

Nuclear chemistry is the chemistry of reactions that involve the nuclei of atoms rather than the electrons in the outer orbitals. They are, in general, much more energetic than ordinary (non-nuclear) reactions and often result in conversion of appreciable mass to energy according to Einstein's well known equation

$$E = mc^2 \qquad\qquad 19\text{-}1$$

The key to solving problems involving nuclear reactions lies in understanding that, instead of balancing equations to obtain the same number of elements on both sides of the equation, you must balance the sum of the atomic numbers and the sum of atomic weights, usually in amu (atomic mass units), on each side of the equation. Elements are, in general, not conserved. An illustration involving a typical nuclear reaction follows.

$$_{7}^{14}N + {}_{2}^{4}He \rightarrow {}_{8}^{17}O + {}_{1}^{1}H \qquad\qquad 19\text{-}2$$

The subscripted leading number is the atomic number and is redundant since the chemical symbol specifies the atomic number. Nitrogen **always** has an atomic number of 7; H **always** has an atomic number of 1; etc. However, it is convenient to carry the atomic number specifically in nuclear reactions since they must balance. The superscripted number preceding the chemical symbol is the atomic weight in amu and differs for different isotopes of the same element. The atomic weight may be thought of as the sum of neutrons and protons in the nucleus. Note that the precise atomic weight will, in general, differ from the approximate integral value used to balance equations. This difference is, in fact, a reflection of mass converted to energy in nuclear reactions.

In Equation 19-2, the sum of atomic numbers (9) is equal on both sides of the equation as is the sum of atomic weights (18). This is a requirement for balancing nuclear reactions.

There are several subatomic species that commonly enter into nuclear reactions. The most common ones are defined below with their corresponding atomic number and approximate (integral) atomic weight. These atomic numbers and atomic weights of subatomic species can be used to balance nuclear reactions in the manner identical to that used with atomic nuclei.

alpha particle ($^4$He nucleus)    $_2^4\alpha$    (also $_2^4$He)

beta particle    $_{-1}^0\beta$    (an electron)

positron    $_1^0\beta$

gamma ray    $_0^0\gamma$    (a gamma ray has no mass, only energy)

proton (H$^1$ nucleus)    $_1^1p$    (also $_1^1$H)

neutron    $_0^1n$

For balancing reactions, it is often convenient, though not precisely correct, to consider a neutron as a proton plus an electron and a proton as a neutron plus a positron.

Many elements have isotopes that are unstable and spontaneously decompose. Such isotopes are called radioactive isotopes or radioisotopes because their spontaneous decomposition is typically accompanied by the emission of high energy radiation. Radioactive decay is a first-order reaction. As discussed in Chapter 13, the half-life in a first-order reaction is independent of the amount of reactant initially present, so the rate constant for radioactive decay is often expressed as a half-life for a particular species. Equation 13-10 gives the rate constant in terms of the half-life for first-order reactions. It applies for radioactive decay reactions.

$k = \ln(2)/t_{1/2}$ (for first-order reactions)          13-10

$A/[A_0] = e^{(-kt)} = e^{\{-t\,(\ln 2)/t_{1/2}\}}$          19-3

671-B

The energy associated with nuclear reactions is often sufficiently large that the accompanying change in mass is significant. When precise weights for each species are known, the energy of a nuclear reaction can be calculated as follows.

$$\Delta E = (\Delta m)c^2 \qquad \qquad 19\text{-}4$$

where

$$\Delta m = \Sigma\,(m)_{products} - \Sigma\,(m)_{reactants} \qquad \qquad 19\text{-}5$$

and is the mass change associated with the reaction.

Step-by-Step Solutions to
Problems in this Chapter,
"Nuclear Chemistry"

● **PROBLEM** 706

The mass of an electron is $9.109 \times 10^{-28}$ g. What is the atomic weight of electrons?

Solution: Atomic weight is defined as the weight of one mole of particles. There are $6.02 \times 10^{23}$, or Avogadro's number, of particles in 1 mole. Therefore, the atomic weight of electrons is equal to $6.02 \times 10^{23}$ times the mass of one electron.

atomic weight of electrons = $9.109 \times 10^{-28}$ g $\times 6.02 \times 10^{23}$/mole

= $5.48 \times 10^{-4}$ g/mole.

● **PROBLEM** 707

A chemist is given an unknown element X. He finds that the element X has an atomic weight of 210.197 amu and consists of only two isotopes, $^{210}X$ and $^{212}X$. If the masses of these isotopes are, respectively, 209.64 and 211.66 amu, what is the relative abundance of the two isotopes?

Solution: The relative abundance of the two isotopes is equal to their fraction in the element. The sum of the fractions of the isotopes times their respective masses is equal to the total atomic weight of X. The element X is composed of only two isotopes. The sum of their fractions must be equal to one.

Solving: Let y = the fraction of the 210 isotope. Since, the sum of the fractions is one, 1 - y = the fraction of the 212 isotope. The sum of the fractions times their masses, equals the atomic weight of X.

209.64 y + 211.66(1 - y) = 210.197;

y = 0.7242 = fraction of $^{210}$X. 1 − y = 0.2758 = fraction of $^{212}$X.

By percentage the relative abundance of $^{210}$X is 72.42% and $^{212}$X is 27.58%.

Natural chlorine is a mixture of isotopes. Determine its atomic weight if 75.53% of the naturally occurring element is chlorine 35, which has a mass of 34.968, and 24.47% is chlorine 37, which has a mass of 36.956.

Solution: If naturally occurring chlorine contains 75.53% chlorine 35 and 24.47% chlorine 37, one mole of chlorine contains .7553 mole chlorine 35 and .2447 mole chlorine 37. The weight of one mole of chlorine is, then, equal to the sum of the weight of .7553 mole $^{35}$Cl and .2447 mole $^{37}$Cl.

atomic weight of Cl = (.7553)(34.968) + (.2447)(36.956)

$$= 26.34 + 9.11 = 35.45.$$

Chlorine is found in nature in two isotopic forms, one of atomic mass 35 amu and one of atomic mass 37 amu. The average atomic mass of chlorine is 35.453 amu. What is the percent with which each of these isotopes occurs in nature?

Solution: This is a mathematical problem, the solution to which centers on defining the average value properly. Consider a set of N observations or measurements, $M_1$, $M_2$, ..., $M_N$, Let the probability that the observation $M_1$ is made be $p_1$, the probability that the observation $M_2$ is made be $p_2$, and so on. Then, the average A is defined as:

$$A = p_1 M_1 + p_2 M_2 + \ldots + p_N M_N.$$

Since, by definition, the sum of the probabilities must be one, then:

$$1 = p_1 + p_2 + \ldots + p_N.$$

For this problem there are two observations, hence N = 2. Let the first observation be $M_1$ = 35 amu and the second observation be $M_2$ = 37 amu. The average A = 35.453 amu. Hence,

$$A = p_1M_1 + p_2M_2$$

$$35.453 \text{ amu} = p_1 \times 35 \text{ amu} + p_2 \times 37 \text{ amu}$$

where      $1 = p_1 + p_2$.

One must determine $p_1$ and $p_2$ (the probabilities of occurrence of isotope 35 and 37, respectively).

From the second of these relationships, $1 = p_1 + p_2$ one obtains $p_2 = 1 - p_1$. Then

$$A = p_1M_1 + p_2M_2 = p_1M_1 + (1 - p_1)M_2$$

$$= p_1M_1 + M_2 - p_1M_2$$

$$= M_2 + (M_1 - M_2)p_1$$

or,  $p_1 = \dfrac{A - M_2}{M_1 - M_2} = \dfrac{35.453 \text{ amu} - 37 \text{ amu}}{35 \text{ amu} - 37 \text{ amu}} = 0.7735$.

Then, $p_2 = 1 - p_1 = 1 - 0.7735 = 0.2265$.

Thus, the isotope of mass 35 amu occurs with probability of 0.7735 or 0.7735 × 100 % = 77.35 % and the isotope of mass 37 amu occurs with a probability of 0.2265 or 0.2265 × 100 % = 22.65 %.

In reality, the two isotopes do not have integral atomic masses, and the percent occurrences calculated above are not exactly correct.

● **PROBLEM** 710

Chromium exists in four isotopic forms. The atomic masses and percent occurrences of these isotopes are listed in the following table:

| Isotopic mass (amu) | Percent occurrence |
|---|---|
| 50 | 4.31% |
| 52 | 83.76% |
| 53 | 9.55% |
| 54 | 2.38% |

Calculate the average atomic mass of chromium.

Solution:  We will make use of the definition of average:

$$A = p_1M_1 + p_2M_2 + \ldots + p_N M_N$$

where A is the average value, $M_i$ is the atomic mass of isotope "i" and $p_i$ is the corresponding probability of occurrence. For the four isotopes of chromium. we have:

$M_1$ = 50 amu     $p_1$ =   4.31% = 0.0431

$M_2$ = 52 amu     $p_2$ = 83.76% = 0.8376

$M_3$ = 53 amu     $p_3$ =   9.55% = 0.0955

$M_4$ = 54 amu     $p_4$ =   2.38% = 0.0238

Hence, the average atomic mass of chromium is

$A = p_1M_1 + p_2M_2 + p_3M_3 + p_4M_4$

= 0.0431 × 50 amu + 0.8376 × 52 amu + 0.0955 × 53 amu

   + 0.0238 × 54 amu

= 2.155 amu + 43.555 amu + 5.062 amu + 1.285 amu

= 52.057 amu.

Given that the masses of a proton, neutron, and electron are 1.00728, 1.00867, and .000549 amu. respectively, how much missing mass is there in $^{19}_{9}F$ (atomic weight = 18.9984)?

Solution: The total number of particles in $^{19}_{9}F$ and their total weight can be calculated. The amount of missing mass in $^{19}_{9}F$ will be the difference of this calculated weight and the given atomic weight of $^{19}_{9}F$. The subscript number 9, in $^{19}_{9}F$, indicates the atomic number of fluorine (F). Because the atomic number equals the number of protons, there are 9 protons in F. The superscript, 19, indicates the total number of particles in the nucleus. Since the nucleus is composed of protons and neutrons, and there are 9 protons, there are 10 neutrons present. In a neutral atom, the number of electrons equals the number of protons. Thus, there are 9 electrons. The total number of particles is, thus, 28. The mass and quantity of each particle is now known. Calculating the total weight contribution of each type of particle:

Protons:  9 × 1.00728   =   9.06552

Neutrons: 10 × 1.00867   = 10.0867

Electrons:  9 ×  .000549  =   .004941

Total mass = 19.1572.

It is given that the mass of the fluorine atom is 18.9984. Therefore, the missing mass is 19.1572 - 18.9984 = 0.1588 amu.

Calculate $\Delta E$ for the proposed basis of an absolutely clean source of nuclear energy

$$^{11}_{5}B + ^{1}_{1}H \rightarrow ^{12}_{6}C \rightarrow 3\ ^{4}_{2}He$$

Atomic masses: $^{11}B = 11.00931$, $^{4}He = 4.00260$, $^{1}H = 1.00783$.

Solution: The energy of nuclear reactions, $\Delta E$, is calculated from the difference between the masses of products and reactants in accordance with the Einstein Law. Einstein's Law can be stated $\Delta E = \Delta mc^2$, where $\Delta m$ is the difference in the masses of the products and the reactants, and c is the speed of light ($3 \times 10^{10}$ cm/sec). The total reaction here can be written

$$^{11}_{5}B + ^{1}_{1}H \rightarrow 3\ ^{4}_{2}He.$$

Thus, $\Delta m$ is equal to the mass of $^{11}_{5}B$ and $^{1}_{1}H$ subtracted from the mass of $3\ ^{4}_{2}He$. The mass of $3\ ^{4}_{2}He$ is equal to 3 times the mass of $^{4}_{2}He$.

$\Delta m = (3 \times$ m of $^{4}_{2}He) - ($m of $^{11}_{5}B +$ m of $^{1}_{1}H)$

$\quad = (3 \times 4.00260) - (11.00931 + 1.00783)$

$\quad = 12.0078 - 12.01714 = -9.34 \times 10^{-3}$ g/mole.

One can now solve for $\Delta E$ by using $\Delta m$.

$\Delta E = \Delta mc^2 = -9.34 \times 10^{-3}$ g/mole $\times (3.0 \times 10^{10}$ cm/sec$)^2$

$\quad = -9.34 \times 10^{-3}$ g/mole $\times 9.0 \times 10^{20}$ cm$^2$/sec$^2$

$\quad = -8.406 \times 10^{18}$ g cm$^2$/mole sec$^2$

$\quad = -8.406 \times 10^{18}$ ergs/mole.

There are $4.18 \times 10^{10}$ ergs in 1 Kcal, thus ergs can be converted to Kcal by dividing the number of ergs by the conversion factor $4.18 \times 10^{10}$ ergs/Kcal.

$$\text{no. of Kcal} = \frac{-8.406 \times 10^{18} \text{ ergs/mole}}{4.18 \times 10^{10} \text{ ergs/Kcal}}$$

$$= -2.01 \times 10^8 \text{ Kcal/mole.}$$

$\Delta E$ for this reaction is $-2.01 \times 10^8$ Kcal per mole.

Calculate $\Delta m$, the difference in mass between the final and initial nuclei in grams per mole for the emission of a γ ray, $^{19}_{8}O^*$ (excited state) → $^{19}_{8}O$ (ground state) + γ ($\Delta E = 1.06 \times 10^8$ Kcal/mole).

Solution:  One solves for Δm by using Einstein's Law, which relates Δm and ΔE.

$$\Delta E = \Delta m c^2,$$

where ΔE is the energy, Δm is the change in mass, and c is the speed of light, $3 \times 10^{10}$ cm/sec. When using Einstein's Law, the ΔE, in Kcal/mole, must be first converted to ergs/mole. There are $4.18 \times 10^{10}$ ergs in one Kcal, therefore one converts Kcal to ergs by multiplying the number of Kcal by $4.18 \times 10^{10}$ ergs/Kcal.

no. of ergs = $4.18 \times 10^{10}$ ergs/Kcal × $1.06 \times 10^8$ Kcal/mole

$\qquad$ = $4.43 \times 10^{18}$ ergs/mole.

One can now solve for Δm.

$\Delta E = \Delta m c^2$ $\qquad\qquad$ $\Delta E = 4.43 \times 10^{18}$ ergs/mole

$\qquad\qquad\qquad\qquad$ $c = 3 \times 10^{10}$ cm/sec

$$\text{ergs} = \frac{g \ cm^2}{sec^2}$$

$$4.43 \times 10^{18} \ \frac{g \ cm^2}{sec^2 \ mole} = \Delta m \left(3 \times 10^{10} \ \frac{cm}{sec}\right)^2$$

$$\frac{4.43 \times 10^{18}}{(3 \times 10^{10})^2} \ \frac{g}{mole} = \Delta m$$

$4.92 \times 10^{-3}$ g/mole = Δm

When a γ particle is emitted, one knows that the final state weighs less than the initial state, therefore,

$\Delta m = -\ 4.92 \times 10^{-3}$ g/mole.

● **PROBLEM** 714

The first step in the radioactive decay of $^{238}_{92}U$ is $^{238}_{92}U = ^{234}_{90}Th + ^4_2He$. Calculate the energy released in this reaction. The exact masses of $^{238}_{92}U$, $^{234}_{90}Th$, and $^4_2He$ are 238.0508, 234.0437 and 4.0026 amu, respectively. 1.0073 amu = $1.673 \times 10^{-24}$ g.

Solution:  The energy released in this process can be determined from the change in mass that occurs. Energy and mass are related in the following equation,

$$\Delta E = \Delta mc^2$$

where, $\Delta E$ is the change in energy, $\Delta m$ the change in mass and c the speed of light ($3.0 \times 10^{10}$ cm/sec). $\Delta m$ is found by subtracting the mass of $_{92}^{238}U$ from the sum of the masses of $_{90}^{234}Th$ and $_2^4He$.

$$\Delta m = (234.0437 \text{ amu} + 4.0026 \text{ amu}) - 238.0508 \text{ amu}$$

$$= - .0045 \text{ amu}.$$

Energy is expressed in ergs ($g\text{-}cm^2/sec^2$), therefore, $\Delta m$ must be converted to grams before solving for $\Delta E$.

$$\Delta m = (- .0045 \text{ amu}) \frac{(1.673 \times 10^{-24} \text{ g})}{(1.0073 \text{ amu})}$$

$$= - 7.47 \times 10^{-27} \text{ g}.$$

Solving for $\Delta E$:

$$\Delta E = (- 7.47 \times 10^{-27} \text{ g})(3.0 \times 10^{10} \text{ cm/sec})^2$$

$$= - 6.72 \times 10^{-6} \text{ g cm}^2/sec^2 = - 6.72 \times 10^{-6} \text{ erg}$$

Therefore, $6.72 \times 10^{-6}$ ergs are released.

● **PROBLEM** 715

Complete the following nuclear equations.
(a) $_7N^{14} + _2He^4 \rightarrow _8O^{17} + \ldots$
(b) $_4Be^9 + _2He^4 \rightarrow _6C^{12} + \ldots$
(c) $_{15}P^{30} \rightarrow _{14}Si^{30} + \ldots$
(d) $_1H^3 \rightarrow _2He^3 + \ldots$

Solution: The rules for balancing nuclear equations are: (1) the superscript assigned to each particle is equal to its mass number and the subscript is equal to its atomic number or nuclear charge; (2) a free proton is the nucleus of a hydrogen atom, and is therefore written as $_1H^1$; (3) a free neutron has no charge and is therefore assigned zero atomic number. Its mass number is one and its notation is $_0n^1$; (4) an electron, $\beta^-$, has zero mass and its atomic number is - 1, hence the notation $_{-1}e^0$; (5) a positron has zero mass and its atomic number is + 1, hence the notation $_{+1}e^0$; (6) an alpha particle ($\alpha$-particle) is a helium nucleus, and is represented by $_2He^4$ or $\alpha$; (7) Gamma radiation ($\gamma$) is a form of light, and has no mass and no charge; (8) in a balanced equation, the sum of the subscripts must be the same on both sides of the equation; the sum of the superscripts must also be the same on both sides of the equation.

In equation (a), $_7N^{14} + _2He^4 \rightarrow {}_8O^{17} + \ldots$, the sum of the subscripts on the left is $(7 + 2)=9$. The subscript of one of the products is 8, thus, the other product must have a subscript or net charge of 1. The sum of the superscripts on the left is $(14 + 4)= 18$. The superscript of one of the products is 17, thus the other product on the right must have a superscript or mass number of 1. The particle with a + 1 nuclear charge and a mass number of 1 is the proton, $_1H^1$.

In equation (b), $_4Be^9 + _2He^4 \rightarrow {}_6C^{12} + \ldots$, the nuclear charge of the second product particle (that is, its subscript) is $(4 + 2) - 6 = 0$. The mass number of the particle (its superscript) is $(9 + 4) - 12 = 1$. Thus, the particle must be the neutron, $_0n^1$.

In equation (c), $_{15}P^{30} \rightarrow {}_{14}Si^{30} + \ldots$, the nuclear charge of the second particle is $15 - 14 = + 1$. Its mass number is $30 - 30 = 0$. Thus, the particle must be the positron, $_{+1}e^0$.

In equation (d), $_1H^3 \rightarrow {}_2He^3 + \ldots$, the nuclear charge of the second product is $1 - 2 = - 1$. Its mass number is $3 - 3 = 0$. Thus, the particle must be a $\beta^{-1}$ or an electron, $_{-1}e^0$.

● **PROBLEM** 716

What is the total binding energy of $_6C^{12}$ and the average binding energy per nucleon?

Solution: The mass of an atom, in general, is not equal to the sum of its component masses. The mass of the component parts (protons, neutrons, and electrons) is slightly greater than the mass of the atom. This difference in mass has an energy equivalent $(E = mc^2)$, which is called the binding energy of the nucleus.

Although binding energy refers to the nucleus, it is more convenient to use the mass of the whole atom in calculations. Then, $M_n = M - ZM_e$, where $M_n$, $M$, and $M_e$ are the nuclear, atomic, and electron masses, respectively, and Z is the atomic number. Since a carbon 6 atom, $C_6^{12}$, is made up of 6 protons and 6 electrons (or 6 $H^1$ atoms) plus 6 neutrons, then the binding energy, (b.e.) can be represented as follows:

b.e. $= M_n + ZM_e - M$

b.e. for $_6C^{12} = 6$[mass of electron and proton ($H^1$) + mass of neutron] - atomic mass of $_6C^{12}$.

In other words, a mass difference equation can be written in terms of whole atom masses.

678

Mass of 6 $H^1$ atoms = 6 × 1.0078 =  6.0468

Mass of 6 neutrons = 6 × 1.0087 =  6.0522

Total mass of component particles= 12.0990

Atomic mass of $_6C^{12}$                    = 12.0000

Loss in mass of formation of $_6C^{12}$ =  0.0990

Binding energy (931.5 MeV/Δmass)(.0990 Δmass) = 92.22 MeV

Since there are 12 nucleons (protons plus neutrons),
the average binding energy per nucleon is

$$\frac{92.22 \text{ MeV}}{12 \text{ nucleons}} = 7.68 \text{ MeV.}$$

● **PROBLEM** 717

Calculate the maximum kinetic energy of the β⁻ emitted
in the radioactive decay of $He^6$.

Solution: To solve this problem, one must first know that
β⁻ is an electron of negative unit charge and zero mass.
Thus, the equation for this particular decay process is

$$_2^6He \quad \rightarrow \quad _3^6Li + \beta^-.$$

The kinetic energy of the β⁻ particle comes from the
difference in mass (or energy, $E = mc^2$), between products
and reactants. To compute the mass change during this
process, only the whole atomic mass of $He^6$ and $Li^6$ need
be considered. Once the difference in mass is known, a conversion factor is used (931. MeV = 1 amu.) Thus

Mass of $He^6$     =   6.01890  amu

Mass of $Li^6$     =   6.01513  amu

Loss in mass        0.00377  amu

Energy equivalent = (931 MeV/ 1 amu)(0.00377 amu) = 3.51 MeV.

Therefore, the maximum kinetic energy of the β⁻ particle
is 3.51 MeV.

● **PROBLEM** 718

$N^{13}$ decays by β⁺ emission. The maximum kinetic energy of
the β⁺ is 1.19 MeV. What is the nuclidic mass of $N^{13}$?

Solution:   In the emission of a β⁺ particle, the daughter nuclide has a Z value one unit less than the parent with no change in A (mass). Thus,

$$_Z P^A \rightarrow {}_{Z-1} D^A + \beta^+,$$

where P and D are the parent and daughter nuclides, respectively. The emitted positron (β⁺) is unstable and is usually consumed after being slowed down by collisions. This is shown in the following reaction

$$\beta^+ + \beta^- \rightarrow 2\gamma$$

In this reaction, 2 photons of light are produced.

The reaction in this problem is

$$_7 N^{13} \rightarrow {}_6 C^{13} + \beta^+$$

In this type of process, a simple difference of whole atom masses is not desired. Whole atom masses can be used for mass difference calculations in all nuclear reactions, except in β⁺ processes where there is a resulting annihilation of two electron masses (one β⁺ and one β⁻). Thus,

mass difference = $(M_n$ for $_7 N^{13}) - (M_n$ for $_6 C^{13}) - M_e$

$= [(M$ for $_7 N^{13}) - 7 M_e] - [(M$ for $_6 C^{13}) - 6 M_e] - M_e$

$= (M$ for $_7 N^{13}) - (M$ for $_6 C^{13}) - 2 M_e$

$= (M$ for $_7 N^{13}) - 13.00335 - 2(0.00055)$

$= (M$ for $_7 N^{13}) - 13.00445$

Here, $M_n$ is the nuclear mass (mass of neutrons and protons), M is the atomic mass, and $M_e$ is the mass of an electron. This expression is equal to the mass equivalent of the maximum kinetic energy of the  β⁺.

$$\frac{1.19 \text{ MeV}}{931.5 \text{ MeV/amu}} = 0.00128 \text{ amu}$$

Then,    0.00128 amu = $(M$ for $_7 N^{13}) - 13.00445$    or,

M for $_7 N^{13}$ = 13.00445 + 0.00128 = 13.00573 amu.

● PROBLEM 719

2.000 picogram (pg) of $^{33}P$ decays by $_{-1}^{0}\beta$ emission to 0.250 pg in 75.9 days. Find the half-life of $^{33}P$.

Solution:  The half-life is defined as the time it takes for ½ of the amount of a certain compound present to decompose. For example, if a substance has a half-life of 1 day, after one day there will only be ½ of the original amount left. When given the original and final amount of a substance, after a given time has elapsed, the number of half-lives that have passed can be found. This is done by dividing the original amount by the final amount and determining how many factors of 2 are present in the quotient. The half-life is then found by dividing the time elapsed by the number of half-lives.

Solving for the half-life of $^{33}P$:

$$\frac{original\ amount}{final\ amount} = \frac{2.000\ pg}{0.250\ pg} = 8$$

8 = 2 × 2 × 2, therefore 3 half-lives have passed. One is given that these half-lives elapse in 75.9 days.

$$half\text{-}life = \frac{75.9\ days}{3} = 25.3\ days.$$

● PROBLEM 720

The radioactive decay constant for radium is $1.36 \times 10^{-11}$. How many disintegrations per second occur in 100 g of radium?

Solution:  The number of disintegrations per second of a given amount of a particular element can be determined by using the following equation.

$$D = \lambda N$$

where D is the number of disintegrations per second, $\lambda$ the decay constant and N the number of atoms present.

In this problem one is given $\lambda$ and must determine N before solving for D. One is told that 100 g of radium is present, the number of moles present is determined by dividing 100 g by the molecular weight of radium, 226 g/mole.

$$no.\ of\ moles = \frac{100\ g}{226\ g/mole} = 0.442\ moles$$

There are $6.02 \times 10^{23}$ particles per mole, thus, one can calculate the number of atoms (N) in 0.442 moles.

$$N = (0.442\ moles)(6.02 \times 10^{23}\ atoms/mole)$$

$$= 2.66 \times 10^{23}\ atoms$$

Solving for D:

$$D = \lambda N = (1.36 \times 10^{-11})(2.66 \times 10^{23}) = 3.62 \times 10^{12}\ dis/sec.$$

One curie is defined as the amount of material, which is equivalent to one gram of radium, that gives $3.7 \times 10^{10}$ nuclear disintegrations per second. The half-life of $^{210}$At is 8.3 hr. How many grams of astatine (At) would equal one curie?

Solution: The half life is defined as the time it takes for one half of the amount of a radioactive substance to disintegrate. To find the number of grams of $^{210}$At that would be equal to 1 curie, one should first find the number of disintegrations of $^{210}$At per second. One can then solve for the number of moles necessary and from that solve for the number of grams.

Solving:

(1) determine the half-life in seconds. This unit is used because curies are measured in disintegrations per second.

$t_{\frac{1}{2}}$ = 8.3 hr × 60 min/hr × 60 sec/min = 29880 sec

(2) Solving for the number of disintegrations of $^{210}$At in 1 curie.

Let x = the total number of nuclei of $^{210}$At in 1 curie, x/2 the number of nuclei left after 1 half-life, 29880 sec. The following ratio can be set up:

$$\frac{(^{210}\text{At}) \text{ number of nuclei after 1 half life}}{\text{half-life}}$$

$$= 3.7 \times 10^{10} \text{ dis/sec.}$$

One can solve this ratio for x, the original number of nuclei present.

$$\frac{x/2}{29880 \text{ sec}} = 3.7 \times 10^{10} \text{ dis/sec}$$

$$\frac{x}{2} = 29880 \text{ sec} \times 3.7 \times 10^{10} \text{ dis/sec}$$

$$x = 2 \times 29880 \text{ sec} \times 3.7 \times 10^{10} \text{ dis/sec} = 2.21 \times 10^{15} \text{ dis.}$$

Thus, $2.21 \times 10^{15}$ dis of $^{210}$At are equal to 1 curie.

(3) Solving for the number of moles. There are $6.02 \times 10^{23}$ nuclei per mole. Therefore, the number of moles present is equal to the number of disintegrations divided by $6.02 \times 10^{23}$.

$$\text{no. of moles} = \frac{2.21 \times 10^{15} \text{ nuclei}}{6.02 \times 10^{23} \text{ nuclei/mole}} = 3.67 \times 10^{-9} \text{ moles}$$

(4) One solves for the number of grams necessary by multiplying the number of moles by the molecular weight of $^{210}$At (MW = 210)

no. of grams = 210 g/mole × 3.67 × $10^{-9}$ moles

$$= 7.71 \times 10^{-7} \text{ g.}$$

This would be the number of grams necessary at $t = t_{\frac{1}{2}}$.

● **PROBLEM** 722

How many alpha particles per second would be emitted from $4 \times 10^{-12}$ g of $^{210}$Po ($t_{\frac{1}{2}}$ = 138 days)?

<u>Solution</u>: When given the mass and the half-life of a substance, one can find the number of disintegrations per second ($dN/dt$) by using the following equation:

$$\frac{dN}{dt} = - \frac{\ell n\ 2}{t_{\frac{1}{2}}} \times N,$$

where $t_{\frac{1}{2}}$ is the half-life in second, N is the original number of nuclei.

(1) Solving for the half-life in seconds:

$t_{\frac{1}{2}}$ = 138 day × 24 hr/day × 60 min/hr × 60 sec/min

$= 1.192 \times 10^{7}$ sec

(2) Determining the number of nuclei present. One is given that there was originally $4 \times 10^{-12}$ g of $^{210}$Po present. The number of nuclei present is found by multiplying the number of moles by Avogrado's number ($6.02 \times 10^{23}$), the number of nuclei in one mole. The number of moles is found by dividing $4 \times 10^{-12}$ g by the molecular weight of $^{210}$Po. (MW = 210).

no. of moles $= \dfrac{4 \times 10^{-12} \text{ g}}{210 \text{ g/mole}} = 1.905 \times 10^{-14}$ moles

no. of nuclei = 1.905 × $10^{-14}$ moles × 6.022

× $10^{23}$ nuclei/mole

$= 1.147 \times 10^{10}$ nuclei

(3) Solving for dN/dt.

$$\frac{dN}{dt} = - \frac{\ell n\ 2}{t_{\frac{1}{2}}} \times N \qquad\qquad N = 1.147 \times 10^{10} \text{ nuclei}$$

$$t_{\frac{1}{2}} = 1.192 \times 10^{7} \text{ sec}$$

$$\ell n \ 2 = 0.693$$

$$\frac{dN}{dt} = - \frac{.693}{1.192 \times 10^7 \ sec} \times 1.147 \times 10^{10} \ nuclei$$

$$= - \ 666.94 \ nuclei/sec$$

Because dN/dt is negative, it means that the change in the number of nuclei per unit time is caused by disintegrations of the nuclei. This is the disintegration rate at t = 0.

● **PROBLEM** 723

A sample of bristle cone pine wood of age 7,000 ± 100 years, known by counting growth rings, possesses an activity of 6.6 d(disintegrations)/minute-g of carbon. Calculate the exact age of the wood sample from radiochemical evidence.

Solution: Radioactive decay is related to the rate of disintegration and the half-life by the following equation:

$$q_t = q_0 \ (0.5)^{t/t_{\frac{1}{2}}}$$

where $q_t$ is the rate of disintegration of the substance containing carbon, $q_0$ is the rate of disintegration of carbon (14 d/min-g), t is the time elapsed during the reaction, and $t_{\frac{1}{2}}$ is the half-life of the element concerned. For carbon, $t_{\frac{1}{2}}$ = 5730 yrs. Here one needs to solve for t.

$$q_t = q_0 \ (0.5)^{t/t_{\frac{1}{2}}} \qquad\qquad q_t = 6.6 \ d/min-g$$

$$q_0 = 14 \ d/min-g$$

$$t = ?$$

$$t_{\frac{1}{2}} = 5730 \ yrs$$

$$6.6 \ d/min-g = 14 \ d/min-g \times (0.5)^{t/5730 \ yrs}$$

$$\frac{6.6 \ d/min-g}{14 \ d/min-g} = 0.5^{t/5730 \ yrs}$$

$$.471 = 0.5^{t/5730 \ yrs}$$

$$\log .471 = t/5730 \ yrs \times \log 0.5$$

$$- \ .327 = t/5730 \times (- \ .301)$$

$$\frac{- \ .327}{- \ .301} = \frac{t}{5730 \ yrs}$$

$$t = \frac{.327 \times 5730 \text{ yrs}}{.301} = 6225.0 \text{ yrs.}$$

Therefore, the tree is 6,225 years old.

# CHAPTER 20

# ORGANIC CHEMISTRY I: NOMENCLATURE AND STRUCTURE

> **Basic Attacks and Strategies for Solving Problems in this Chapter. See pages 687 to 716 for step-by-step solutions to problems.**

Organic chemistry is the chemistry of substances that contain carbon. The number of catalogued organic compounds is in the millions and is expanding rapidly. In this chapter, the rules for the structure of some simple organic compounds and the associated rules for naming them are outlined.

Hydrocarbons contain only carbon and hydrogen, and are the simplest organic compounds. Their classical names form the basis of naming other organic compounds. Those compounds for which all possible carbon bonds contain hydrogen are called alkanes or saturated hydrocarbons. Unsaturated hydrocarbons have carbon-carbon double bonds, triple bonds, or aromatic (benzene-like) bonds.

The structure of a sample of normal or straight-chain alkanes is shown below. The chemical formula for the alkanes is $C_nH_{(2n+2)}$ and a sample of the names follow in Table 1.

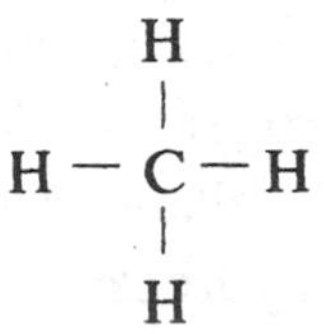

```
        H
        |
  H  —  C  —  H     methane
        |
        H
```

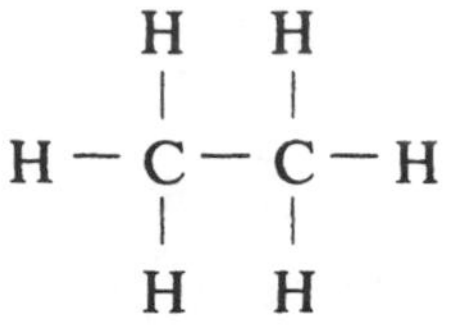

```
    H   H
    |   |
H — C — C — H     ethane
    |   |
    H   H



    H   H   H
    |   |   |
H — C — C — C — H     propane
    |   |   |
    H   H   H



    H       H
    |       |
H — C — ... — C — H     general
    |       |           alkane
    H       H
```

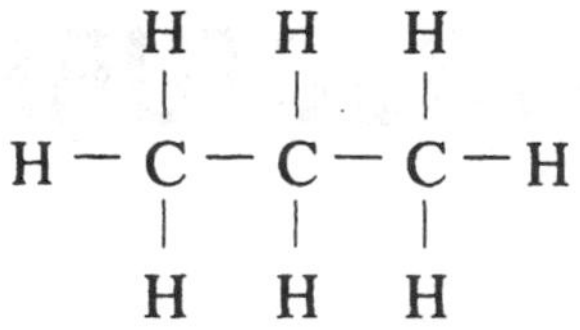

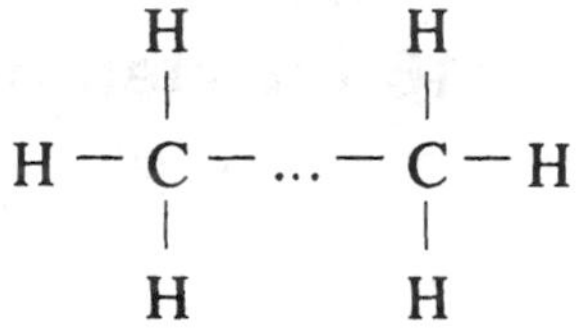

| | Table 1: Names of Some Common Alkanes | |
|---|---|---|
| $n$ | Formula $[C_nH_{(2n+2)}]$ | Name |
| 1 | $CH_4$ | methane |
| 2 | $C_2H_6$ | ethane |
| 3 | $C_3H_8$ | propane |
| 4 | $C_4H_{10}$ | butane |
| 5 | $C_5H_{12}$ | pentane |
| 8 | $C_8H_{18}$ | octane |
| 10 | $C_{10}H_{22}$ | decane |
| 12 | $C_{12}H_{26}$ | dodecane |
| 16 | $C_{16}H_{34}$ | hexadecane |
| 20 | $C_{20}H_{42}$ | eicosane |

The structure of alkanes need not be in a straight chain. Branched chains can form numerous isomers with the same chemical formula, but with different structure and different physical and chemical properties. The $C_{20}$ alkane, for example, has over 366,000 isomers and the $C_{25}$ alkane has over 37,000,000. The straight-chain isomers are called normal hydrocarbons and their name is preceded by a small $n$ as in $n$-butane, $n$-octane, etc.

The analogous series of hydrocarbons with one double bond $(C_nH_{2n})$ are called the alkenes; and with one triple bond $[C_nH_{(2n-2)}]$ are called the alkynes.

Double and triple bonds refer to the C — C bonds. Carbon has four valence or bonding electrons and it is possible for one, two, or three to form a bond with an adjacent carbon. An example of a simple alkene and alkyne follow. They are named with the same prefix as the alkanes but with the *-ene* or *-yne* suffix respectively.

$$H_2C = CH_2 \qquad \text{ethene}$$

$$HC \equiv CH \qquad \text{ethyne}$$

Many of these compounds have common names which were assigned before a formal naming procedure was established. For the two shown above, $C_2H_4$ is marketed as ethylene, and $C_2H_2$ is marketed as acetylene.

The other common hydrocarbon structure is the benzene ($C_6H_6$) structure and its derivatives. Benzene derivatives are called aromatic compounds and are formed by substitution for hydrogen on a benzene ring.

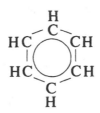

Now we are in a position to outline the IUPAC (International Union of Pure and Applied Chemists) naming system for hydrocarbons. The following rules apply.

1) Select and name the longest continuous chain of carbon atoms.

2) Number the carbon atoms in the chain beginning with the end nearest a branch.

3) Assign the names and (position) numbers that indicate the branches or substituents. If more than one of the same substituent appears, prefixes (*di-*, *tri-*, etc.) are used.

4) The longest continuous chain is numbered so that the substituents have the lowest possible numbers.

The method is illustrated by naming the isomers of pentane.

1) *n*-pentane

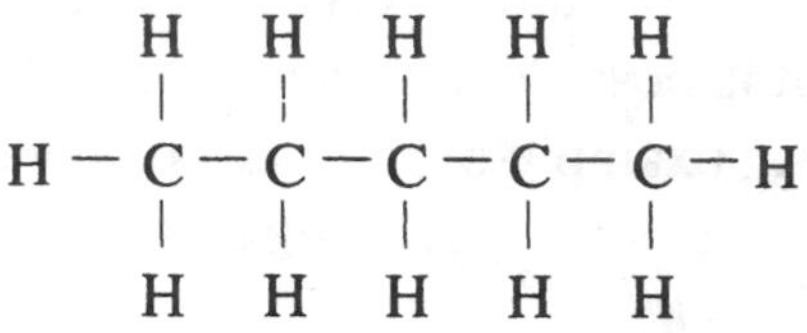

$$H - \underset{\underset{H}{|}}{\overset{\overset{H}{|}}{C}} - \underset{\underset{H}{|}}{\overset{\overset{H}{|}}{C}} - \underset{\underset{H}{|}}{\overset{\overset{H}{|}}{C}} - \underset{\underset{H}{|}}{\overset{\overset{H}{|}}{C}} - \underset{\underset{H}{|}}{\overset{\overset{H}{|}}{C}} - H$$

2) 2-methylbutane

$$\overset{\overset{C}{|}}{C} \\ C - C - C - C$$

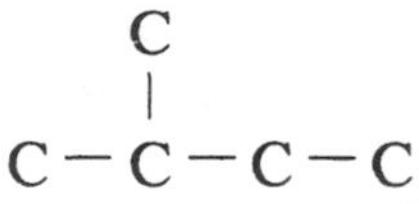

(hydrogens not shown)

3) 2,2 dimethylpropane

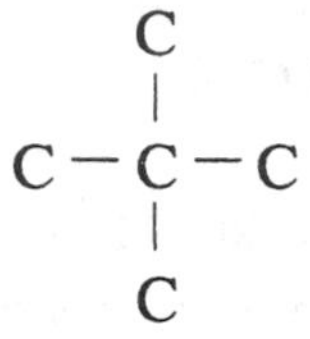

$$\overset{\overset{C}{|}}{\underset{\underset{C}{|}}{C - C - C}}$$

(hydrogens not shown)

Other, nonhydrocarbon, functional groups can also replace hydrogen on any of the hydrocarbons molecules discussed above. Table 2 shows the names used when other functional groups are substituted.

## Table 2: IUPAC Names for Substituents on Organic Compounds

| FAMILY | FUNCTIONAL GROUP | SUBSTITUENT NAME |
|---|---|---|
| Alcohols | $-OH$ | add suffix -*ol* (e.g., 1-propanol) |
| Halogens | Fl, Cl, Br, I, | add prefix: *fluoro, chloro, bromo, iodo* |
| Ethers | $C - O - C$ | add internal -*oxy* (e.g., methoxy, ethoxy) |
| Carboxylic acids | 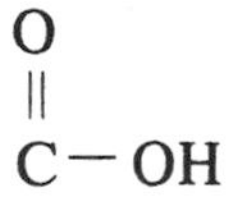 $\overset{\overset{O}{\|\|}}{C} - OH$ | add suffix -*oic acid* (e.g., ethanoic acid) also carboxy |
| Aldehydes | 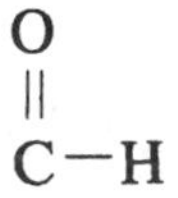 $\overset{\overset{O}{\|\|}}{C} - H$ | add suffix -*al* (e.g., pentanal) |
| Ketones | 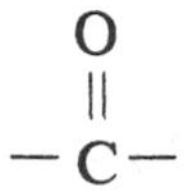 $- \overset{\overset{O}{\|\|}}{C} -$ | add suffix -*one* (e.g., pentanone) |
| Amides | 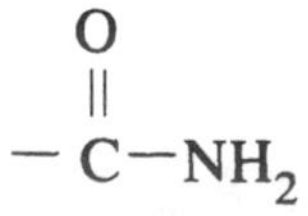 $- \overset{\overset{O}{\|\|}}{C} - NH_2$ | add suffix -*amide* (e.g., ethanamide) |

In all cases the position of the substituent is numbered in the usual fashion.

It must be emphasized that there are many common names of organic compounds that were used before the IUPAC formal naming rules were established. The common names remain in use — particularly in commerce — resulting in many compounds being known by more than one name.

Step-by-Step Solutions to
Problems in this Chapter,
"Organic Chemistry I:
Nomenclature and Structure"

## ALKANES

● PROBLEM 724

Explain the terms primary, secondary, and tertiary in
regards to covalent bonding in organic compounds.

Solution: If a carbon atom is bound to only one other
carbon atom, then the former carbon atom is called
primary. If a carbon atom is bonded to two other carbon
atoms, then that carbon atom is called secondary. If
a carbon atom is bonded to three other carbon atoms,
then that carbon atom is called tertiary.

Any group that is attached to a primary, secondary,
or tertiary carbon is called a primary, secondary, or
tertiary group. For example,

$$
\begin{array}{c}
\text{H} \\
| \\
\text{H} \underset{}{\overset{}{\text{H}}}\!\!-\!\!^1\text{C}\!-\!\text{H} \\
| \quad | \quad \text{H} \quad \text{H} \\
\text{H}\!-\!^2\text{C}\!-\!\!-\!^3\text{C}\!-\!\!-\!^4\text{C}\!-\!^5\text{C}\!-\!\text{H} \\
| \quad | \quad | \quad | \\
\text{H} \quad \text{H} \quad \text{H} \quad \text{H}
\end{array}
$$

Carbons 1, 2, and 5 are primary, carbon 4 is
secondary, and carbon 3 is tertiary. The hydrogen atoms
attached to carbons 1, 2, and 5 are called primary
hydrogens, those attached to carbon 4 are called secondary hydrogens, and those attached to carbon 3 are called
tertiary hydrogens. This same principle applies to
alcohols, and depending upon where the hydroxyl group
(- OH) is attached (that is, primary, secondary, or
tertiary carbon), the alcohol is called primary,
secondary, or tertiary, respectively.

Name each of the following alkanes. Indicate which, if any, are isomers.

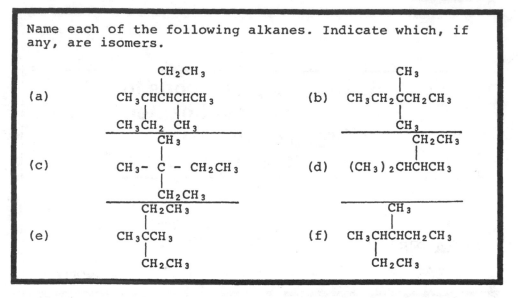

(a)

$$CH_2CH_3$$
$$|$$
$$CH_3CHCHCHCH_3$$
$$|\quad|$$
$$CH_3CH_2 \quad CH_3$$

(b)

$$CH_3$$
$$|$$
$$CH_3CH_2CCH_2CH_3$$
$$|$$
$$CH_3$$

(c)

$$CH_3$$
$$|$$
$$CH_3-\;C\;-\;CH_2CH_3$$
$$|$$
$$CH_2CH_3$$

(d)

$$CH_2CH_3$$
$$|$$
$$(CH_3)_2CHCHCH_3$$

(e)

$$CH_2CH_3$$
$$|$$
$$CH_3CCH_3$$
$$|$$
$$CH_2CH_3$$

(f)

$$CH_3$$
$$|$$
$$CH_3CHCHCH_2CH_3$$
$$|$$
$$CH_2CH_3$$

Solution: Isomers are related compounds that have the same molecular formula but different structural formulas.

Isomerism is not possible among the alkanes until there are enough carbon atoms to permit more than one arrangement of the carbon chain.

To name the above compounds, one uses a set rules to provide each compound with a clear name. These rules for nomenclature are the IUPAC rules (International Union of Pure and Applied Chemistry), and are referred to as systematic nomenclature.

One of the rules of the IUPAC system is to choose the largest chain of carbon atoms in the molecule and call it parent compound. Thus,

(a) has a 6 carbon parent chain with methyl groups bonded to the second and fourth carbon atoms of the parent chain. There exists, also, an ethyl group bonded to the third carbon atom.

As such, the name of this organic molecule is 3-ethyl-2, 4-dimethylhexane.

(b) has 5 carbon parent molecule and 2 methyl groups on carbon number 3. Therefore, the name is 3,3-dimethylpentane.

(c) has a 5 carbon parent molecule with 2 methyl groups bonded to the third carbon of the parent chain. Therefore, the name of this structure becomes 3,3 di-methylpentane.

(d) has a 5 carbon parent molecule, and 2 methyl groups attached to carbon numbers 2 and 3.

Therefore, the name is 2,3-dimethylpentane.


(e) has a 5 carbon parent molecule, and 2 methyl groups attached to carbon number 3.

Therefore, the name is 3,3-dimethylpentane.


(f) has a 6 carbon parent molecule, and 2 methyl groups attached to carbon numbers 3 and 4.

Therefore, the name is 3,4-dimethylhexane.


To find which compounds are isomers, one counts the number of carbon and hydrogen atoms contained in the molecule. If the total number of both carbon and hydrogen atoms in one molecule is the same as in another molecule, but they have different structural formulas, they are isomers.

If the calculation of total carbon and hydrogen molecules is made, b, c, d, and e become isomers, with b and e being same compound.

● **PROBLEM** 726

Each of the following is an incorrect name for an alkane. Write a structural formula for each and provide the correct systematic name: (a) 2-ethylbutane; (b) 2-isopropylpentane; (c) 1,1-dimethylpentane; (d) 2,2-dimethyl-4-ethylpentane.

Solution: First, one writes the formula starting with the parent molecule. Next, one locates the number of the carbon on the parent molecule that is attached to a particular group by the number in front of that particular group in the compound's name.

Thus, (a) 2-ethylbutane becomes

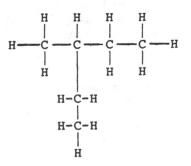

(b) 2-isopropylpentane becomes

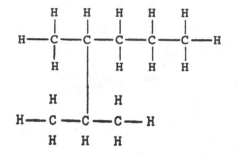

(c) 1,1-dimethylpentane becomes

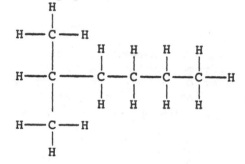

(d) 2,2-dimethyl-4-ethylpentane

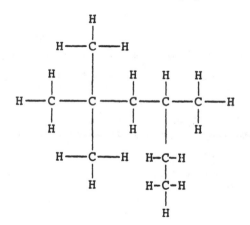

Each name is incorrect in the same respect; the named parent molecule is not the longest chain of the compound. If a compound twists or bends in free space it must be taken into account when naming the compound.

Thus, the correct names are

(a) 3-methylpentane, since the longest chain has 5 carbons and a methyl group attached to carbon number 3.

(b) 2,3-dimethylhexane since the longest chain has 6 carbons and methyl groups are attached to carbon numbers 2 and 3.

(c) 2-methylhexane, since the longest chain has 6 carbons and a methyl group is attached to carbon number 2.

(d) 2,2,4-trimethylhexane, since the longest chain has 6 carbons and three methyl groups attached to carbons 2, 2 and 4; 1 methyl group attached to carbon 4 and 2 methyl groups attached to carbon 2.

● **PROBLEM** 727

Write the structural formulas for all chlorine derivatives having molecular formula $C_5H_{11}Cl$.

Solution: In hydrocarbon derivatives, the hydrogen atoms are replaced by other atoms or groups of atoms, such as oxygen, chlorine, hydroxyl (-OH), or nitro ($-NO_2$), which produce chemically active centers in an otherwise less active molecule.

To solve this problem, one first writes the structural isomers with five carbons in a continuous chain, in which each different hydrogen atom of the alkane has been replaced by a chlorine atom:

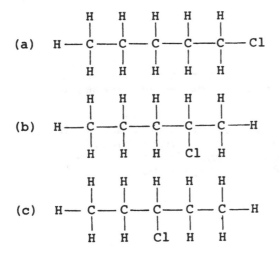

Their names are (a) 1-chloropentane, (b) 2-chloro-pentane, (c) 3-chloropentane.

Second, one writes the structural isomers with four carbons as the parent molecule, in which two hydrogen atoms have been replaced by a chlorine atom and a methyl group:

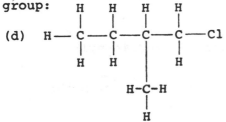

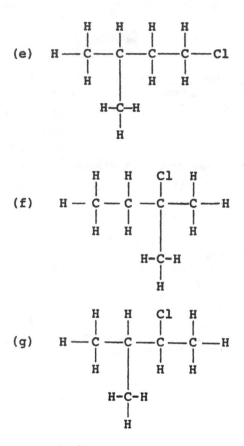

(e)

(f)

(g)

Their names are (d) 1-chloro-2-methylbutane, (e) 1-chloro —3-methylbutane, (f) 2-chloro —2-methylbutane, and (g) 2-chloro —3-methylbutane.

Third, one writes the structure of the isomer with three carbons for a parent molecule, in which three hydrogen atoms of the three carbon chain have been replaced by a chlorine atom and two methyl groups:

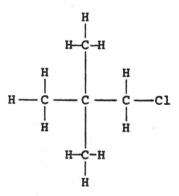

This compound's name is 1-chloro —2,2-dimethyl-propane or 1-chloro neopentane.

Hence, there are eight structural isomers with the molecular formula $C_5H_{11}Cl$.

A chemist has at his disposal the following hydrocarbons:
(a) 3,3-dimethylpentane, (b) n-heptane, (c) 2-methyl-
heptane, (d) n-pentane, and (e) 2-methylhexane. Arrange
these compounds in order of decreasing boiling points
(without referring to tables).

Solution: To rank the boiling points of these substances,
you must consider the rules which govern the property of
boiling points.

Except for very small alkanes, the boiling point
rises 20 to 30 degrees for every carbon that is added to
the chain. (Alkanes are saturated hydrocarbons of the
general formula $C_nH_{2n+2}$, where n = number of carbon atoms.)
Alkanes of the same carbon number but different structures
(isomers) will have different boiling points. A branched-
chain isomer has a lower boiling point than a straight-
chain isomer. This is because the shape of a branched-
chain molecule tends to approach a sphere, decreasing the
surface area, so that the intermolecular forces are
easily overcome at a lower temperature. With this in mind,
you can proceed to solve the problem. Because these rules
pertain only to structures, you must write them out.

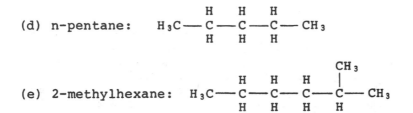

c. has the greatest carbon content, 8 carbon atoms,
which means it has the highest boiling point. a, b and e
all have 7 carbon atoms; thus carbon content is equal.
But only b is _not_ branched, which means it has the
highest boiling point out of this group. e is the next

highest since it is less branched than a. d has the
lowest boiling point, since it has the lowest carbon
content (5 atoms). Thus, you rank them in order of decreasing boiling point as c, b, e, a, d.

It is known that the anti conformation of n-butane is
.8 Kcal/mole more stable than the Gauche conformation.
Explain why.

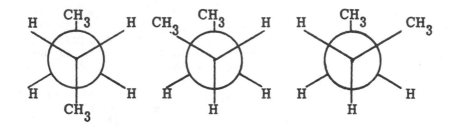

anti conformation          Gauche conformations

Figure A

<u>Solution</u>:  Different arrangements of atoms that can be
converted into one another by rotation about single bonds
are called conformations. In the anti conformation of
n-butane the methyl groups (CH$_3$) are as far apart as possible. The gauche conformations of n-butane are with the
methyl groups only 60° apart. This can be seen more
clearly by using Newman projections for the conformations
of n-butane. See figure A.

Notice that in the gauche conformations, the methyl
groups are closer together than in the anticonformation.
When these groups are close together, as in the gauche
conformations, steric hindrance occurs between them.
This means that there is not really enough room for these
two groups to be so close together. This is due to the
fact that around any nuclei there is a certain volume
which is taken up by the atoms' electrons. This volume
is determined by the van der Waal's radii of a particular
atom. When atoms are too close to permit this volume, a
repulsive force called van der Waals' repulsion occurs.
This force acts to create a greater distance between the
nucleii. This, then, raises the energy (level and lowers
the stability) of the gauche conformation. In the anti
conformation, the methyls are much further apart, so that
you don't have van der Waal's repulsion, thus the
molecule is much more stable.

You are given 3-bromohexane. Draw the stereoisomers of this compound; specify the R and the S enantiomers.

Solution: Non-superimposable mirror-image stereo-isomers are called enantiomers. It is not enough that a molecule has a mirror image, for everything has a mirror image; the mirror images must not be able to be superimposed on each other. The arrangement of atoms that characterize a particular stereoisomer is called its configuration. You want to specify the configuration in terms of R and S. Let us see what this means.

In assigning configuration, you do the following after determining the chiral carbon: (1) Assign a sequence of priority to the four atoms or groups of atoms attached to the chiral center. (A chiral center consists of a carbon atom to which 4 different groups are attached.) (2) Visualize the molecule so that the group of lowest priority is directed away from you. Now, observe the arrangement of the remaining groups. If, in going from the group of highest priority to the group of second priority your eye travels in a clockwise direcion, the configuration is R; if counterclockwise, the configuration is S. Now, how do you determine priority? If the four atoms bonded to chiral center are all different, priority depends on atomic number. The atom of higher atomic number receives the higher priority. If they are isotopes, the higher mass number has priority. Now, if you can't determine priority from this, because you have 2 or more atoms that are the same attached to the chiral center, make a similar comparison of the next atoms (i.e., their atomic numbers) as you move out from the chiral center. With this in mind, you can proceed as follows:

(a) 3-bromohexane. Recall, enantiomers are mirror image isomers so that you can say the enantiomers I and II, where the asterisk designates the chiral carbon.

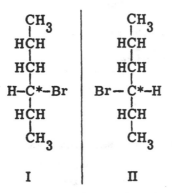

I       II

Figure A

To specify R and S, the lowest priority group must be drawn away from you. H has the lowest priority since it has the smallest atomic number. It can be drawn away from you by placing it in a vertical position. By convention, groups in vertical positions are considered away from you. Thus, you can rewrite these enantiomers as shown in figure B.

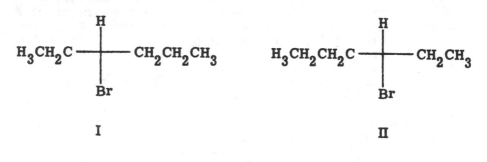

## Figure B

The hydrogens attached to the chiral center are now vertical, and, as such, away from us. The other groups surrounding the chiral center are Br, $CH_2CH_3$, and $CH_2CH_2CH_3$. Of the three, Br has highest priority because its atomic number is higher than carbons. The other two have carbons attached to the chiral center, so that no priority is possible here. You must move out from the chiral center. The next atoms reached are, again, carbon. No priority determination is possible. If you move out to the next atom, you find a carbon atom in $CH_2CH_2CH_3$ and a hydrogen atom in $CH_2CH_3$ (it doesn't have another carbon).

Thus, because carbon has a larger atomic number than hydrogen, $CH_2CH_2CH_3$ has priority over $CH_2CH_3$. In summary, for the groups or atoms attached to the chiral center, the order of increasing priority can be shown by

$$H, \ CH_2CH_3, \ CH_2CH_2CH_3, \ Br$$

$$\longrightarrow$$

increasing priority

Recalling step (2), in going from highest to next lowest, then to next lowest, in the enantiomer on the left, you move in a clockwise direction. For enantiomer I, you move in a counterclockwise direction. Thus it is S and enantiomer II is R.

● **PROBLEM** 731

A solution containing 6.15 g of cholesterol per 100 ml of solution was made. When a portion of this solution was placed in a 5-cm polarimeter tube, the observed rotation is − 1.2°. Determine the specific rotation of cholesterol.

Solution: A polarimeter measures the rotation of plane polarized light, which is light whose vibrations take place in only one plane. An optically active substance has the ability to rotate the plane of polarized light. In essence, then, a polarimeter, measures optical activity. Species that possess optical activity have a specific rotation. It is defined in the following equation:

$$\text{specific rotation} = \frac{\text{observed rotation (degrees)}}{\text{length (dm)} \times g/cc}$$

The length of the polarimeter tube must be converted to decimeters. (1 cm = 0.1 dm) Thus, 5 cm = .5 dm. For water, 100 ml = 100 cc. Substituting, you obtain,

$$\text{specific rotation} = \frac{-1.2°}{.5 \times \frac{6.15}{100}} = -39.02°.$$

# ALKENES AND ALKYNES

● **PROBLEM** 732

Classify each of the following as a member of the methane series, the ethylene series, or the acetylene series: $C_{12}H_{26}$, $C_9H_{16}$, $C_7H_{14}$, $C_{26}H_{54}$.

Solution: Before beginning this problem, one should first know the general formulas for each of the series. For the alkanes (methane series), the general formula is $C_nH_{2n+2}$, where n is the number of carbon atoms and 2n + 2 is the number of hydrogens. Molecules of the ethylene series, also called the alkene series, have two adjacent carbon atoms joined to one another by a double bond. Any member in this series has the general formula $C_nH_{2n}$. The acetylene series, commonly called the alkyne series, has two adjacent carbon atoms joined to one another by a triple bond. The general formula for this series is $C_nH_{2n-2}$. With this in mind one can write:

| | | |
|---|---|---|
| $C_{12}H_{26}$ : | $C_nH_{2n+2}$ | : alkane series |
| $C_9H_{16}$ : | $C_nH_{2n-2}$ | : acetylene series |
| $C_7H_{14}$ : | $C_nH_{2n}$ | : ethylene series |
| $C_{26}H_{54}$ : | $C_nH_{2n+2}$ | : alkane series |

Draw the structure of 4-ethyl-3,4-dimethyl-2-hexene.

Solution:  To draw the structure of more complex compounds, such as this one, certain steps must be followed.

(1) Identify the parent compound that associated with the longest carbon chain that contains the functional group. In 4-ethyl-3,4-dimethyl-2-hexene, the parent compound is hexene.

(2) Draw the parent carbon skeleton, in this case, a six carbon chain. Do not put any hydrogen atoms in yet.

C—C—C—C—C—C

(3) Number the carbon atoms starting at either end. This is important; otherwise it may get confusing when one adds the functionality.

1   2   3   4   5   6

C—C—C—C—C—C

(4) Add the suffix functionality, in this case -2-ene. "Ene" tells one that a double bond is present, while "2" indicates that it is at the second carbon.

1   2   3   4   5   6

C—C≡C—C—C—C

(5) Add the prefix functionality, starting at the beginning of the name and continuing until the parent name is reached. Here, the prefixes are 4-ethyl-3,4-dimethyl-.

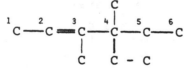

Now, the hydrogen atoms can be added to give a complete structure.

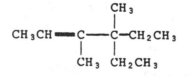

Write condensed structural formulas for all the alkynes, i.e. unsaturated compounds with triple bonds, with a molecular formula of $C_5H_8$.

<u>Solution</u>: A condensed structural formula provides all the information represented by other structural formulas (i.e. Lewis diagrams, bond diagrams), but it is not as cumbersome.

To solve this problem, first write all the structural isomers with five carbons in a continuous chain:

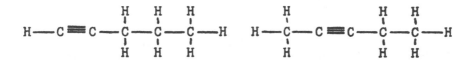

Second, write all the structural isomers with a four carbon parent molecule and one carbon in a branch:

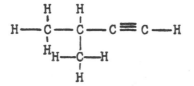

No other structural formulas are possible.

Thus, there are only three structural formulas possible for $C_5H_8$.

Next, one writes these formulas in the condensed form. Namely,

(a) $CH \equiv CCH_2CH_2CH_3$ ; (b) $CH_3C \equiv CCH_2CH_3$ ; (c) $CH_3CHC \equiv CH$.
$\qquad\qquad\qquad\qquad\qquad\qquad\qquad\qquad\qquad\qquad\qquad\qquad | $
$\qquad\qquad\qquad\qquad\qquad\qquad\qquad\qquad\qquad\qquad\qquad\; CH_3$

● **PROBLEM** 735

In the following pair of geometric isomers, designate which is cis and which is trans.

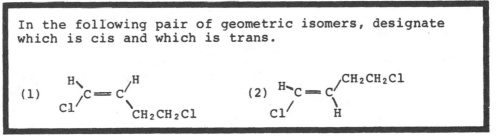

<u>Solution</u>: Isomers, i.e. compounds of the same molecular formula, that differ only in their geometry are termed geometric isomers. Geometric isomerism is one type of stereoisomerism; that is, isomerism due to the arrangement of atoms in a molecule in three dimensions. Geometric isomers are different compounds with different properties. Since the order of attachment of the atoms is the same, they are not structural isomers. The isomer that has the

two groups on the same side of the double bond is called the cis isomer and the isomer that has the two groups on the opposite sides is called the trans isomer. In (1),

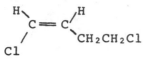

the hydrogens and, as such, the Cl and $CH_2CH_2Cl$ groups are located on the same sides of the double bond. Therefore, this is the cis isomer. In (2),

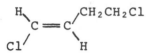

however, the groups are located on the opposite side of the double bond. Consequently, this is the trans isomer.

● PROBLEM 736

Which of the following compounds can exhibit geometric and/or optical isomerism? (a) $H_2C{=\!=}C(Cl)CH_3$; (b) $ClFC{=\!=}CHCl$; (c) $CH_3CH_2CH{=\!=}CHCH(CH_3)_2$; (d) $CH_3CHClCOOH$; (e) $HC{\equiv}CCH{=\!=}CHCl$; (f)$ClCH{=\!=}CHCHClCH_3$; (g) cyclohexene; (h) $H_3CN{=\!=}NCH_3$; (i)$[NH(CH_3)(C_2H_5)(C_3H_5)]^+Cl^-$.

Solution: Optical isomerism refers to the ability of a substance to rotate plane polarized light. Such substances exist as enantiomers (species which are interchangeable only by the breaking and the reforming of bonds and not by only the rotation about bonds). When a mirror is placed between enantiomers, it shows them to be mirror images of each other. If the mirror image of a compound is not equivalent to the original compound, the original compound and its mirror image are said to be non-super-imposable or enantiomeric. If a certain compound possesses an enantiomer, it must possess a chiral center. For a molecule to be chiral, it must have a carbon atom attached to four different groups. Thus, you can identify optical isomerism by detecting whether there exists a chiral carbon atom in the molecule. Only (d), (f), and (i) fit into this general category:

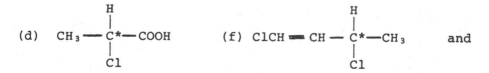

similarly with (i). In (i) the nitrogen atom is the chiral center.

Geometric isomerism refers to cis-trans isomerism. When a compound possesses a double bond, it is possible for geometric isomerism to exist. A geometric isomer is said to exist when two different groups are bonded to each of the two carbon atoms forming the double bond. To determine whether a compound is the cis or the trans isomer, these groups are first assigned priority in order of atomic number. The atom with the highest atomic number is given highest priority. A plane is then imagined along the length of the double bond. If the atoms of greatest priority are both on one side of the plane the isomer is said to be cis. This can be seen in the diagram below. (The numbers indicate priority ratings.)

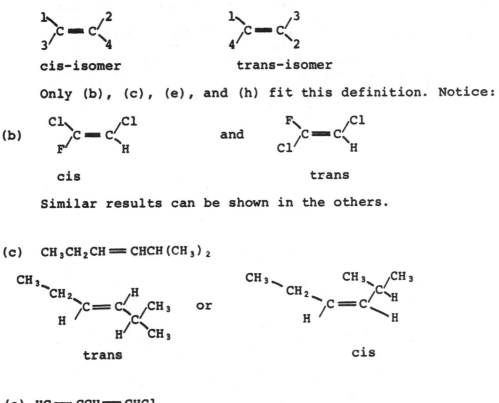

cis-isomer                    trans-isomer

Only (b), (c), (e), and (h) fit this definition. Notice:

(b)

and

cis                              trans

Similar results can be shown in the others.

(c)   $CH_3CH_2CH = CHCH(CH_3)_2$

or

trans                              cis

(e)  $HC \equiv CCH = CHCl$

cis                              trans

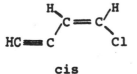

(h)  $H_3CN = NCH_3$

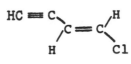

cis                              trans

From the following thermochemical data, determine which is more stable and why:

cis-2-butene $\xrightarrow{H_2}$ n-butane    $\Delta H° = 28.6$ Kcal/mole

trans-2-butene $\xrightarrow{H_2}$ n-butane    $\Delta H° = -27.6$ Kcal/mole

Solution:  One method of finding the relative stabilities between two organic compounds is by a comparison of their heats of hydrogenation. This is exactly what the problem presents to you. Both cis and trans-2-butene, upon hydrogenation, yield the same product. Yet, the trans-2-butene evolves 1 Kcal/mole less heat and therefore the trans-form must be stabilized to that extent relative to the cis-isomer. To find out why trans-isomer is 1 Kcal/mole more stable than the cis -isomer, you must write their structures.

cis-2-butene:

$$CH_3 \diagdown C = C \diagup CH_3$$
$$\diagup H \qquad \diagdown H$$

trans-2-butene:

$$H \diagdown C = C \diagup CH_3$$
$$CH_3 \diagup \qquad \diagdown H$$

Notice, in the cis configuration, the methyl groups ($CH_3$) are located on the same side of the double bond, while in trans, they are found on opposite sides. The cis-isomer will exhibit steric interference, which means that because of the size of the methyl groups, they will tend to "crowd" each other. To avoid this "crowding", there will be repulsion between the methyl groups which, of course, decreases the stability of the molecule. The trans-isomer has no crowding problem as the methyls are located on opposite sides. As such, its stability will not be reduced by steric effects, which explains why it is more stable than the cis-isomer.

# ALCOHOLS

Name the following systematically:

(a) $CH_3CHOHCH_2CH_2CH_2CH_3$

(b) $CH_3CHCH_2CH_2OH$
        |
       $CH_2CH_3$

(c) $CH_3CH_2CH_2COHCH_3$
             |
        $CH_3CHCH_3$

(d) $CH_2OHCHCH_2CHCH_3$
        |     |
      $CH_3$  $CH_3$

Solution: All of these compounds are alcohols; they fit into the general formula R-OH, where R is any alkyl group.

In the systematic naming of any alcohol, the following rules should be followed:

(1) The longest chain that contains the hydroxyl group (OH) is considered the parent compound.

(2) The ⁻e ending of the name of this carbon chain is replaced by ⁻ol.

(3) The locations of the hydroxyl and any other groups are indicated by the smallest possible numbers.

Thus, compound (a)

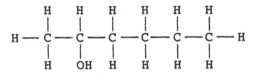

has a 6-carbon chain and an hydroxyl group on carbon number 2. The name of this compound is 2-hexanol.

Compound (b) has a 5-carbon chain,

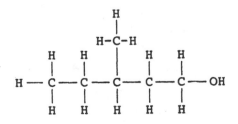

because this is the longest chain that contains the hydroxyl group. There is a methyl group on carbon number 3, and an hydroxyl group on carbon number 1. The name of this compound is, therefore, 3-methyl-1-pentanol.

Compound (c),

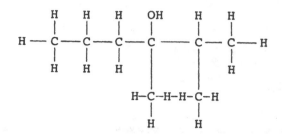

has a 6-carbon chain that contains the OH group, two
methyl groups on carbons 2 and 3, and an hydroxyl group
on carbon number 3. Thus, the name is 2,3-dimethyl-3-
hexanol.

Compound (d)

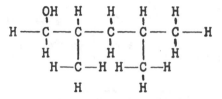

has a 5-carbon chain, two methyl groups on carbons 2 and
4, and an hydroxyl group on carbon 1. Thus, the name is
2,4-dimethyl-1-pentanol.

● **PROBLEM** 739

Name the compound shown in figure A by the IUPAC system.

$$CH_3CH_2 \qquad CH_3$$
$$HC \text{————} CH \text{——} OH$$
$$CH_3CH_2CH_2CH_2 \text{—} C$$
$$\|$$
$$CH_3CH_2 \text{—} C \text{—} CH_2CH_3$$

Figure A

Solution: In naming complex open chain organic compounds,
first find the longest continuous chain containing the
functional groups. Write down the parent name. The parent
for this particular compound is heptene. Number the chain
starting from the end that will give the smallest prefix
numbers for the functional groups. This is shown in figure B.

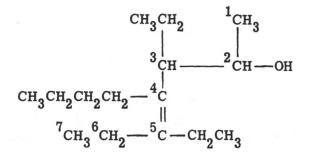

Add the suffix functionality with the appropriate numbering: 4-hepten-2-ol. Add the prefix functionality, remembering to group together like prefixes. Then, double check to make sure a substituent has not been forgotten or one substituent has not been included twice. By following these steps, one arrives at the name for the structure:

4-n-butyl-3,5-diethyl-4-hepten-2-ol.

● **PROBLEM** 740

Which of the following, if any, are not alcohols derived from the methane series of hydrocarbons: $C_6H_5OH$, $C_{17}H_{33}OH$, $C_4H_8OH$, $C_9H_{19}OH$?

Solution: Alcohols are derived from molecules whose hydrogen atoms have been replaced by one or more hydroxyl (-OH) groups. The simplest alcohols are derived from the **alkanes** or the methane series and contain only one hydroxyl group per molecule. These have the general formula ROH, where R is an alkyl group of composition, $C_nH_{2n+1}$. Thus, the alcohols to be derived from the methane series will follow this formula.

(a) $C_6H_5OH$ has 6 carbons. Therefore, its hydrogen content should be $2n + 1 = 2(6) + 1 = 13$. Because there exist only 5 hydrogens, excluding the H from OH, it cannot be derived from the methane series.

Compound (b), $C_{17}H_{33}OH$, also does not fit the general formula ($n = 17$, $2n + 1 = 35$), and thus, is not derived from the methane series. If it did fit, it would possess 35 hydrogens instead of 33.

Compound (c), $C_4H_8OH$, does not fit the general formula ($n = 4$, $2n + 1 = 9$) and, thus, is not derived from the methane series. If it did fit, the alkyl group it would posses 9 H instead of 8.

Compound (d), $C_9H_{19}OH$, does fit the general formula ($n = 9$, $2n + 1 = 19$), and is derived from the methane series.

705

# OTHER FUNCTIONAL GROUPS

● PROBLEM 741

Name each of the organic compounds shown in Figure A below.

Figure A

(a) $CH_3 - CH = CH_2$

(b) $CH_3 - \underset{\underset{C_2H_5}{|}}{CH} - CH_2 - CH = CH - CH_3$

(c) $CH_3CH = CHCH_2CH_2CH_3$

(d)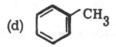

Solution: A system of nomenclature called the IUPAC system has been formulated so that all organic molecules may have their structures defined adequately. In this system for hydrocarbons (i.e. compounds containing only hydrogen and carbon atoms) the longest chain is taken as the basic structure, and the carbon atoms are numbered from the end of the chain closest to a branch chain or other modification of simple alkane structure. The position of substituents in the chain are denoted by the number of carbon atom or atoms to which they are attached. In the problem, you are not given alkanes, you are given alkenes, compounds that contain a double bond between a pair of carbon atoms (i.e. unsaturated), and a ring compound. In unsaturated compounds, the rules are the same, except that the position of the double bond is indicated; the numbering starts at the end of the chain nearest the double bond. For rings, name the ring and any substituent present based on which carbon is located. Thus, you proceed as follows:

(a) There is only one chain, and it possesses three carbon atoms. The double bond is located on the first carbon, not the second, since you want to use the lowest possible number. As such, you form the molecule propene. Three carbons suggest propane. It's an alkene, so that you change ane to ene. Therefore, propane becomes propene. You have no need to name the position of the double bond here; the double bond can only be in two positions and both yield the same molecule. That is, $CH_2 = CH - CH_3$ is equivalent to $CH_3 - CH = CH_2$.

(b) The longest chain containing the double bond has 7 carbon atoms, so that "hep" prefix is suggested. Because it is a double bond, you add the suffix ene to obtain heptene. The double bond is located on the second carbon, not the third, since you want the lowest number. You have, therefore, the 2-heptene. Using this numbering system, you see that CH₃, a methyl group, is located on the fifth carbon. Thus, the name of the molecule is 5-metyl 2 heptene.

(c) Has a six carbon chain, which suggests the prefix "hex". There is a double bond so that the molecule has the suffix ene. Thus, you obtain hexene. The double bond is located on the second carbon, which means the name of the molecule is 2-hexene.

(d): This ring compound has a methyl group (CH₃) positioned on a benzene ring. The compound can be called methyl benzene. It is also given    the special name of toluene.

● PROBLEM 742

Which of the following compounds are saturated? Which are unsaturated? (a) $CH_3CH_2CH_2CH_3$, (b) $CH_2 = CHCH_2CH_3$, (c) cyclohexane, (d) cyclohexene, (e) benzene.

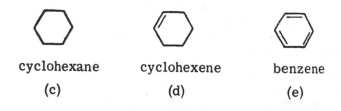

cyclohexane      cyclohexene      benzene

(c)                  (d)                  (e)

Solution: The most fundamental class of organic compounds is the saturated hydrocarbons. A hydrocarbon is a compound that contains only carbon and hydrogen. Saturated is used to describe the absence of double bonds and/or triple bonds. The term comes from the fact that these compounds do not react with hydrogen because they are saturated with hydrogen. When a double and/or triple bond is present, the compound is termed unsaturated.

Thus, compounds a and c are saturated and compounds b, d, and e are unsaturated.

● PROBLEM 743

Explain the use of the terms ortho, meta, and para in systematic nomenclature of benzene hydrocarbons.

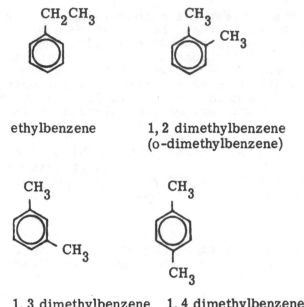

ethylbenzene

1, 2 dimethylbenzene
(o-dimethylbenzene)

1, 3 dimethylbenzene
(m-dimethylbenzene)

1, 4 dimethylbenzene
(p-dimethylbenzene)

Solution:  In establishing a consistent systematic scheme
for the naming of benzene hydrocarbons several problems
arise. The prefixes ortho, meta, and para (abbreviated
o, m, and p) are commonly used to designate the 1,2-,
1,3-, and 1,4-, relationship of substituents on the
benzene ring. The problem arises when, for example, the
compound named systematically as 1,2-dimethylbenzene is
also known as ortho-xylene or ortho-methyltoluene or
ortho-dimethylbenzene.

Four isomeric compounds are possible with benzene
hydrocarbons having the molecular formula $C_8H_{10}$. One is
obtained by substituting an ethyl group for a hydrogen
of benzene, whereas the other three are obtained by substituting methyl groups for two hydrogens.

● **PROBLEM** 744

Write structural formulas for the following: (2) 2-hexanone;
(b) 2-methylbutanal; (c) O-methylbenzaldehyde; (d) methyl
phenyl ketone.

Figure A                Figure B

<u>Solution</u>:  The aldehydes are derivatives of hydrocarbons whose molecules have a double bond to oxygen in place of two hydrogens at the end of a chain. Those derived from alkane hydrocarbons have the general formula RCHO, where R is the hydrogen for the first member of the series and an alkyl group for higher homologs.

The longest chain containing the -CHO group is considered the parent compound and is named by changing the -e ending of the corresponding alkane to -al.

The ketones are derivatives of hydrocarbons whose molecules have a double bond to oxygen in place of two hydrogens at a position other than the end of the carbon chain. Those derived from alkane hydrocarbons have the general formula $RR'C = O$, where R and R' are alkyl groups.

The longest chain containing the carbonyl group is considered the parent compound and is named by changing the -e ending of the corresponding alkane to -one. Where it is necessary, the locations of the carbonyl carbon and attached groups are indicated by numbers.

Thus, compound (a) is a ketone whose longest chain has 6 carbons to which oxygen is attached on carbon number 2. The formula is

$$CH_3CCH_2CH_2CH_2CH_3$$
$$\overset{\|}{O}$$

Compound (b) is an aldehyde whose longest chain has 4 carbons to which a methyl group is attached to carbon 2 and whose end carbon atom is - CHO. The formula is

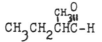

Compound (c) is an aldehyde derived from benzene by replacing a hydrogen with the $- C = O$  group. This
$$\underset{H}{|}$$
compound also has a methyl group attached to the ortho position on the benzaldehyde ($C_6H_5CHO$) ring. The structure can be written as shown in Figure A.

Compound (d) is an aromatic ketone (any compound that contains benzene derivatives is called aromatic), derived from phenols. Phenols are hydroxy derivatives of aromatic compounds whose general formula is ArOH (Ar is an aromatic group. Phenyl is $C_6H_5-$ and is attached to
$C = O$ in this case. The formula can be written as

$C_6H_5CCH_3$  or as shown in Figure B.
$\overset{\|}{O}$

Write a structural formula for each of the following molecular formulas. Name each structure that you draw. (a) $CH_2O$; (b) $C_2H_6O_2$; (c) $C_2H_6O$; (d) $C_2H_4O_2$; (e) $C_2H_4O$.

<u>Solution</u>:  In order to do this problem, one must know how carbon, hydrogen, and oxygen bond. Carbon can share its 4 valence electrons to form 4 covalent bonds, hydrogen can share its 1 valence electron to form 1 covalent bond, and oxygen can share its 6 covalent electrons to form only 2 covalent bonds. Thus, with this knowledge in mind, one can construct several types of molecular structures. A few of these structures have the following functional groups: Alcohols contain the hydroxyl group, - OH, acids contain the carboxyl group,

 (represented as - COOH), aldehydes and ketones

contain the carbonyl group, $\diagdown C = O$.

Structure (a), $CH_2O$, is an aldehyde,

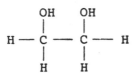

and has the name formaldehyde.

Structure (b), $C_2H_6O_2$, is an alcohol

and has the name 1,2 dihydroxyethane or ethylene glycol.

Structure (c), $C_2H_6O$, is either an alcohol or an ether

and is named ethanol or dimethyl ether, respectively.

Structure (d) $C_2H_4O_2$, is an acid

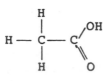

and has the name acetic acid.

Structure (e), $C_2H_4O$, is an aldehyde

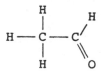

and has the name acetaldehyde.

● PROBLEM 746

Ethanol and methyl ether are isomers, having the molecular formula $C_2H_6O$. Based on structure, which one would be expected to have the higher vapor pressure? The higher boiling point? The greater solubility in water? Explain why in each case.

Solution: The ethers are a class of compounds that contain two hydrocarbon groups bound to an atom of oxygen. The general formula for ethers is ROR´ where R and R´ are alkyl groups and may or may not be the same

Both alcohols and ethers derived from alkanes have the same general formula $C_nH_{2n+2}O$ (ethanol is isomeric with methyl ether). Unlike alcohols, ethers do not form associated molecules by means of hydrogen bonds. Thus, the boiling point of an ether is considerably below the boiling point of its isomeric alcohol. Methyl ether is a gas, bp - 23°C, whereas ethanol is a liquid, bp + 78°C. Therefore, it

$$
\begin{array}{cc}
\text{H} & \text{H} \\
| & | \\
\text{H—C—C—OH} & \quad \text{H—C—O—C—H} \\
| & | \\
\text{H} & \text{H}
\end{array}
$$

        ethanol              methyl ether

is reasonable to expect that methyl ether has a higher vapor pressure than ethanol.

Weak hydrogen bonding between water molecules and ether molecules leads to slight water solubility. The functional group that is characteristic of ethers, the C O C group, is chemically inactive compared to the hydroxyl group of alcohols. Thus, ethanol has the greater solubility.

● PROBLEM 747

Name and classify each of the following: $CH_3COC_4H_9$; $C_2H_5OH$; $CH_3NH_2$; $(C_2H_5)_2O$; $CH_3COOC_3H_7$; $CH_3COOH$.

<u>Solution</u>:  All of these compounds are derivatives of hydrocarbons, i.e. compounds that contain only hydrogen and carbon atoms. The derivative of a hydrocarbon is the substitution of elements or radicals for one or more hydrogen atoms. These derivatives may be classified into functional groups, which, to a great extent, determine the properties of the compound. This is the key to naming these compounds, identify the functional groups and name it accordingly. First one must describe these functional groups.

   <u>Alcohols</u>: They have the general formula R-OH, where R is any alkyl group.

   <u>Amines</u>:  They are derivatives of ammonia ($NH_3$), where one or more H's have been replaced by an alkyl group. In other words, $R-NH_2$, $R-\overset{\displaystyle R}{\underset{\displaystyle |}{N}H}$, or $R-\overset{\displaystyle R}{\underset{\displaystyle |}{N}}-R$.

   <u>Ethers</u>: General formula is R-O-R´, where R may be the same as R´.

   <u>Aldehydes</u>: General formula is $R-\overset{\displaystyle O}{\overset{\displaystyle \|}{C}}-H$

   <u>Ketones</u>:  General formula is $R-\overset{\displaystyle O}{\overset{\displaystyle \|}{C}}-R´$, where R can equal R´. In both, ketones are aldehydes, $\overset{\diagdown}{\underset{\diagup}{C}}=O$ is called the carbonyl group.

   <u>Carboxylic acids</u>: General formula is $R-\overset{\displaystyle O}{\overset{\displaystyle \|}{C}}-OH$, where $\overset{\displaystyle O}{\overset{\displaystyle \|}{\underset{\diagup}{C}}}-OH$ is called the carboxyl group.

   <u>Esters</u>: General formula is $R-\overset{\displaystyle O}{\overset{\displaystyle \|}{C}}-OR´$,  where R can equal R´. With this in mind, proceed as follows:

   $CH_3COC_4H_9$ can be written as $CH_3-\overset{\displaystyle O}{\overset{\displaystyle \|}{C}}-C_4H_9$. This, then fits into the general formula of a ketone and is called  methyl n-butyl ketone.

   $C_2H_5OH$ fits into the alcohol classification group. It is named ethanol or ethyl alcohol.

   $CH_3NH_2$ fits the description of an amine and is named methylamine.


   $(C_2H_5)_2O$ can be written as $H_5C_2-O-C_2H_5$, which fits the description of an ether. It is named ethyl ether.

$CH_3COOC_3H_7$ indicates that an ester is present, once it is rewritten as $CH_3 — \overset{\overset{\displaystyle O}{\|}}{C} — OC_3H_7$. It is named as n-propyl acetate.

$CH_3COOH$ can be written as $CH_3 — \overset{\overset{\displaystyle O}{\|}}{C} — OH$. The carboxyl group is present, which tells you that it is an organic acid. The specific name is acetic acid.

● **PROBLEM** 748

Write the structure for methyl-6-chloro-5,8-diethyl-8-iodo-3,4,4-trimethyl-2,6-decadienoate.

<u>Solution</u>: To draw this structure one should follow certain rules.

(1) Identify and draw the parent name of the compound. The parent is decane,

```
10   9   8   7   6   5   4   3   2   1
 C — C — C — C — C — C — C — C — C — C .
```

(2) Add the suffix functionality. Note that in this compound two different types of functional groups are designated in the suffix endings, double bonds and an ester. Start with the double bonds: -2,6-decadienoate.

```
10   9   8   7   6   5   4   3   2   1
 C — C — C — C = C — C — C — C = C — C
```

Now consider the ester group $\left( - \overset{\overset{\displaystyle O}{\|}}{C} - O - \right)$. The carbonyl (C = O) of the ester must be at carbon number 1.

The alkyl portion of the ester precedes the name (and the prefixes) as a separate word. In this case it is methyl ($CH_3$) so that the structure becomes

```
10    9    8    7    6    5    4    3    2       1
 C — C — C — C = C — C — C — C = C —  C — OCH₃
                                              ‖
                                              O
```

(3) Add the prefix functionality in the order given, 6-chloro-5,8-diethyl-8-iodo-3,4,4-trimethyl-.

```
         I          Cl        C              O
         |          |         |              ‖
C — C — C — C = C — C — C — C = C — C —OCH₃
         |              |   |
         C              C   C
         |              |
         C              C
```

713

Next, fill in the hydrogens, and the structure is complete.

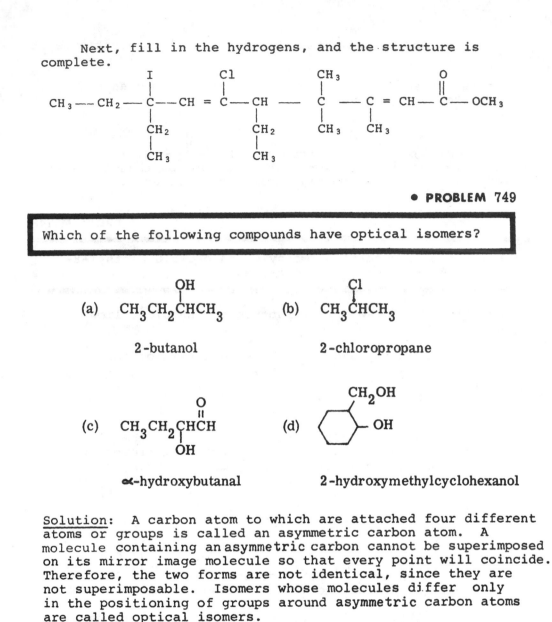

Which of the following compounds have optical isomers?

(a)   
$$\overset{\text{OH}}{\underset{|}{\text{CH}_3\text{CH}_2\text{CHCH}_3}}$$

2-butanol

(b)   
$$\overset{\text{Cl}}{\underset{|}{\text{CH}_3\overset{\bullet}{\text{C}}\text{HCH}_3}}$$

2-chloropropane

(c)   
$$\text{CH}_3\text{CH}_2\underset{\underset{\text{OH}}{|}}{\overset{\overset{\text{O}}{\|}}{\text{CHCH}}}$$

∝-hydroxybutanal

(d)   CH$_2$OH — OH

2-hydroxymethylcyclohexanol

Solution:  A carbon atom to which are attached four different atoms or groups is called an asymmetric carbon atom. A molecule containing an asymmetric carbon cannot be superimposed on its mirror image molecule so that every point will coincide. Therefore, the two forms are not identical, since they are not superimposable. Isomers whose molecules differ only in the positioning of groups around asymmetric carbon atoms are called optical isomers.

To solve this problem, one must find which carbon is asymmetric, if any. If there is a carbon that has four different groups attached to it, then it must be asymmetric and there will be a possible geometric isomer.

In (a), 2-butanol, there exists an optical isomer, since there are four different groups attached to a central carbon atom.

In (b), 2-chloropropane, the compound has no optical isomer since two of the four groups are the same.

In (c), α-hydroxybutanal, the compund has an optical isomer, since it has four different groups attached to a central carbon atom.

In (d), 2-hydroxymethylcyclohexanol, the compound has no optical isomer, since two of the four groups attached to the central carbon atom are the same.

Which of the following organic compounds would you predict to be associated liquids: (a) $CH_3OH$; (b) $CH_3OCH_3$; (c) $CH_3Cl$; (d) $CH_3NH_2$.

Solution: To answer this question, one should first understand the term, associated liquids. Liquids whose molecules are held together by hydrogen bonding are associated liquids. Hydrogen bonding is an especially strong kind of attraction between a hydrogen atom (that is bound to a highly electronegative atom) and another highly electronegative atom. This type of bond is formed by purely electrostatic forces between the positive end of one polar molecule and the negative end of another polar molecule. For hydrogen bonding to be important, the electronegative atoms must be F, O, or N.

Thus, once the structures of the species are known, one can tell if they are associated liquids.

(a) $CH_3OH$, (methanol);

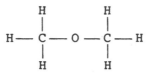

It is an associated liquid because oxygen is bonded to the hydrogen atom. And the hydrogen atom thus forms a bond with another oxygen molecule of methane.

Liquid (b), $CH_3OCH_3$, (dimethyl ether);

is not an associated liquid; the oxygen is not bonded directly to any hydrogen atom and therefore cannot form any hydrogen bonds.

Liquid (c), $CH_3Cl$, (methyl chloride);

is not an associated liquid because Cl is not electro-negative enough to form a stable hydrogen bond.

Liquid (d), $CH_3NH_2$, (methyl amine);

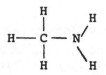

is an associated liquid, N is highly electronegative and the hydrogen can form hydrogen bonds with nitrogen atoms of other $CH_3NH_2$ molecules.

# CHAPTER 21

# ORGANIC CHEMISTRY II: REACTIONS

Basic Attacks and Strategies for Solving Problems
in this Chapter. See pages 717 to 753 for step-by-
step solutions to problems.

Any extensive treatment of organic reactions would require many volumes. The summary here treats only the following simple reactions involving
alkanes, alkenes, alkynes, and alcohols.

a) combustion

b) substitutions for hydrogen

c) substitution reactions across double bonds

d) formation of esters

All hydrocarbons can be oxidized (burned) to form carbon dioxide and
water. In fact this reaction is the basis of the energy derived from gasoline in
motor vehicles.

$$C_nH_m + (n + m/4)O_2 \rightarrow nCO_2 + m/2\ H_2O \qquad\qquad 21\text{-}1$$

Except for the combustion reaction above, alkanes are relatively inert
organic molecules. It is possible, however, under appropriate conditions, to
substitute other atoms or radicals for the hydrogen in saturated hydrocarbons.
A typical example is the substitution of halogens catalyzed by uv light.

$$CH_4 + Cl_2 \rightarrow CH_3Cl + HCl \qquad\qquad 21\text{-}2$$

This is shown structurally for the chlorination of ethane by the following
reaction.

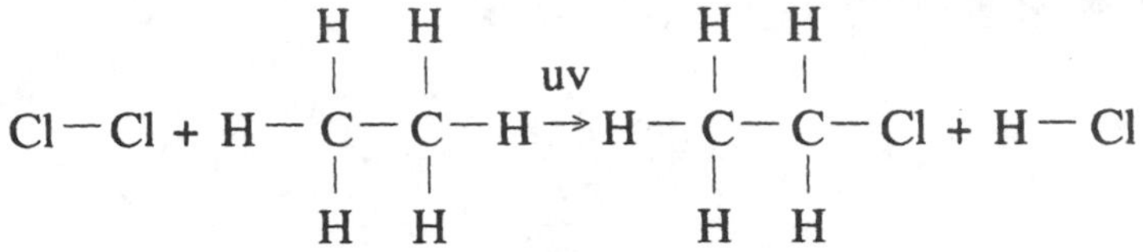

$$\text{Cl-Cl} + \text{H}-\underset{\underset{\text{H}}{|}}{\overset{\overset{\text{H}}{|}}{\text{C}}}-\underset{\underset{\text{H}}{|}}{\overset{\overset{\text{H}}{|}}{\text{C}}}-\text{H} \xrightarrow{uv} \text{H}-\underset{\underset{\text{H}}{|}}{\overset{\overset{\text{H}}{|}}{\text{C}}}-\underset{\underset{\text{H}}{|}}{\overset{\overset{\text{H}}{|}}{\text{C}}}-\text{Cl} + \text{H}-\text{Cl}$$

21-3

Alkenes and alkynes, by contrast, are much more reactive than alkanes — particularly at the site of the double or triple bond. Substitutions across the double (or triple) bond occur much more readily than do substitution reactions in alkanes. An example of the chlorination of ethene is shown below.

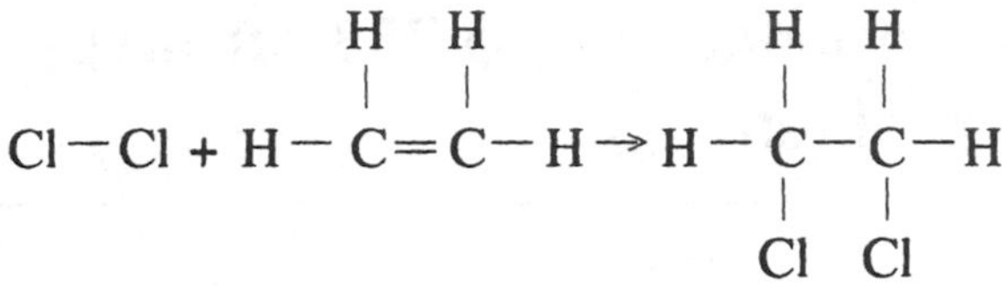

$$\text{Cl-Cl} + \text{H}-\overset{\overset{\text{H}}{|}}{\text{C}}=\overset{\overset{\text{H}}{|}}{\text{C}}-\text{H} \rightarrow \text{H}-\underset{\underset{\text{Cl}}{|}}{\overset{\overset{\text{H}}{|}}{\text{C}}}-\underset{\underset{\text{Cl}}{|}}{\overset{\overset{\text{H}}{|}}{\text{C}}}-\text{H}$$

21-4

Double and triple bonds can also be hydrogenated easily to form saturated hydrocarbons. The reaction is analogous to Equation 21-4 except that hydrogen, rather than chlorine, is added to the double bond.

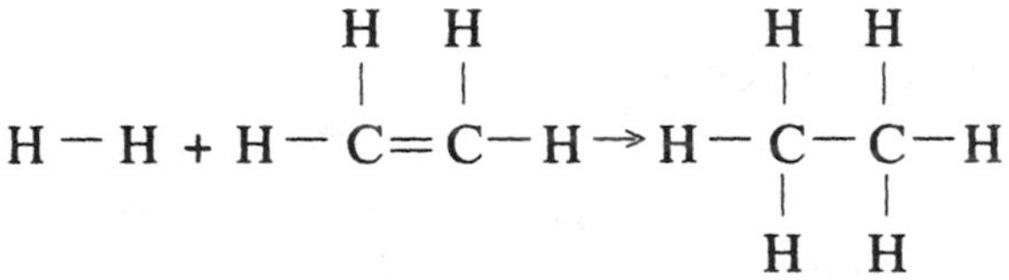

$$\text{H}-\text{H} + \text{H}-\overset{\overset{\text{H}}{|}}{\text{C}}=\overset{\overset{\text{H}}{|}}{\text{C}}-\text{H} \rightarrow \text{H}-\underset{\underset{\text{H}}{|}}{\overset{\overset{\text{H}}{|}}{\text{C}}}-\underset{\underset{\text{H}}{|}}{\overset{\overset{\text{H}}{|}}{\text{C}}}-\text{H}$$

21-5

The esterification of alcohols and organic acids is mentioned here because of the analogy with acid-base reactions. The hydroxyl ($OH^-$) group in an alcohol can react with the hydrogen ($H^+$) in a carboxylic acid to form water and an ester — the organic analogue of a salt. Organic esters are very fragrant and many common fruit flavorings (e.g., banana, pineapple, orange, etc.) are esters. A simple reaction of an alcohol and carboxylic acid to form an ester follows.

ethanol + acetic acid → water + ethyl acetate

$$C_2H_5OH + CH_3CO_2H \rightarrow H_2O + CH_3CO_2 \, C_2H_5$$

21-6

The myriad of other organic reactions which can occur are treated in a variety of organic chemistry texts.

Step-by-Step Solutions to
Problems in this Chapter,
"Organic Chemistry II:
Reactions"

## ALKANES

● **PROBLEM** 751

4 liters of octane    gasoline weigh 3.19 kg. Calculate what
volume of air is required for its complete combustion at
S.T.P.

Solution:  To answer this problem, you need to write the
balanced equation for the combustion of octane gasoline.
This means knowing what the molecular formula of octane
gasoline is and what is meant by combustion. Octane is a
saturated hydrocarbon, i.e., it is an alkane. Now, by
saturated hydrocarbon, you mean a compound that contains
only single bonds between the carbon to carbon and carbon
to hydrogen bonds. Alkanes have the general formula $C_N H_{2N+2}$,
where N = number of carbon atoms. Since the prefix "oct"
means eight, you know there are 8 carbon atoms, which
indicates that 18 hydrogen atoms are present. Thus, gasoline
octane has the formula $C_8 H_{18}$. Now, by combustion you mean
the reaction of an organic compound with oxygen to produce
$CO_2$ and $H_2O$. With this in mind, you can write the balanced
equation for the reaction as

$$2C_8 H_{18} + 25O_2 \rightarrow 16CO_2 + 18H_2O.$$

To find out what volume of air is required for
combustion, you need the volume of $O_2$ required, since 20%
of air is oxygen ($O_2$). To find the amount of $O_2$ involved,
use the fact that at S.T.P. (standard temperature and
pressure)  1 mole of any gas occupies 22.4 liters. Thus,
if you knew how many moles of $O_2$ were required, you would
know its volume. You can find the number of moles by using
stoichiometry. You are told that there exists 3.19 kg
or 3190 g (1000 g = 1 kg). The molecular weight (MW) of
octane is 114 grams/mole. Thus, since

mole = $\frac{\text{grams (weight)}}{\text{M.W.}}$ , you have $\frac{3190}{114}$ = 27.98 moles of

gasoline. From the equation's coefficients, you see that for every 2 moles of gasoline, 25 moles of $O_2$ are required. Thus, for this number of moles of gasoline, you need (27.98) $\frac{25}{2}$ = 349.78 moles of $O_2$.

Recalling that 1 mole of gas occupies 22.4 liters at S.T.P., 349.78 moles of $O_2$ occupies (349.78)(22.4) = 7835.08 liters. Oxygen is 20% of the air. Thus, the amount of air required is

$\frac{5 \text{ liters air}}{1 \text{ liter } O_2}$ × 7835.08 liters $O_2$ = 39,175.40 liters air.

● **PROBLEM** 752

What is your understanding of a substitution reaction mechanism? Illustrate your answer with a specific example.

Solution:  A substitution reaction is one in which one atom or group of atoms replaces another. In this type of reaction, the overall change taking place in the alkane is that one sigma bond is broken and a new sigma bond is formed. At elevated temperatures, or in the presence of ultraviolet light, one or more hydrogen atoms in an alkane molecule may be replaced by atoms of chlorine, bromine, or other similar atoms.

The substitution of a chlorine atom for hydrogen in alkane molecules proceeds by a free-radical mechanism and is a chain reaction. In the initiation step of chlorination of methane, for example, high temperature or light furnishes energy for dissociating chlorine molecules into chlorine radicals as shown (unpaired electrons are shown with a dot):

(1)      $Cl_2 \xrightarrow{\text{light}}$      $\cdot\, Cl + \cdot\, Cl$

The chlorine atom collides with a methane molecule to form a methyl radical and a molecule of hydrogen chloride.

(2) $CH_4 + \cdot\, Cl \longrightarrow \cdot\, CH_3 + HCl$

Subsequent collision of a methyl radical with a chlorine molecule produces the desired compound, $CH_3Cl$, and another chlorine atom

(3)  $\cdot\, CH_3 + Cl_2 \longrightarrow CH_3Cl + \cdot\, Cl$,

so that the whole chain may be repeated.

Equations 2 and 3 are called propagation steps in the chain reaction. Termination steps of this chain reaction are 4 and 5:

$$\cdot \text{Cl} + \cdot \text{Cl} \longrightarrow \text{Cl}_2$$

(4) $\quad \cdot \text{Cl} + \cdot \text{CH}_3 \longrightarrow \text{CH}_3\text{Cl}$

(5) $\quad \cdot \text{CH}_3 + \cdot \text{CH}_3 \longrightarrow \text{CH}_3\text{CH}_3$

These chain-terminating reactions tend to stop the reaction between methane and chlorine by consuming radicals necessary for the propagation steps.

● **PROBLEM** 753

A chemist has tert-butane. If he adds $Br_2$ and light, what are the monobromo substitution products expected? Which one will predominate and why? How would you prepare the predominating compound by an addition reaction?

Solution: The first question concerns substitution in alkanes. You are performing a halogenation substitution reaction and are asked to predict the monohalogenation products expected for tert-butane. This necessitates the writing of its structure and analyzing its hydrogens, since they will be replaced by the bromine atoms. Tert butane can be written

$$\begin{array}{c} \text{CH}_3 \\ | \\ \text{CH}_3 - \text{C} - \text{CH}_3 . \\ | \\ \text{H} \end{array}$$

All the hydrogens on the methyl groups are equivalent. This means that if you substitute a bromine atom for one hydrogen atom from a methyl group, it is the same as substituting for any other hydrogen on one of the other methyl groups. The other hydrogen (the one attached to the 3° carbon) is not equivalent to the hydrogens bound to the 1° carbons. Thus you expect two monobromo products. They are

and You

could have placed the Br on the other two methyls and the same molecule would be obtained, for Example,

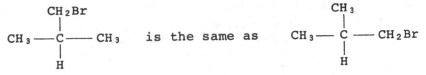

is the same as

When the Br is attached to the 3° carbon tert-butyl bromide is obtained. This product predominates greatly. To see why, consider the mechanism of this reaction. The mechanism is as follows (let x = a halogen).

$$x_2 \longrightarrow 2x \cdot \quad \leftarrow (\text{radicals denoted by dots})$$

$$x \cdot + \text{alkane} \longrightarrow Hx + \text{alkane} \cdot$$

$$\text{alkane} \cdot + x_2 \longrightarrow \text{alkyl halogen} + x \cdot$$

You see that the reacton proceeds by radical formation. To form isobutyl bromide, a 1° radical.

CH₃
|
CH₃—— C ——CH₂ · **must be formed first.**
|
H

T-butyl bromide goes through a 3° radical, i.e.,

CH₃
|
CH₃——C ——CH₃
·

A 3° radical is more stable and, thus, has a longer lifetime than the 1° radical. Since the life time of the 3° radicals is longer, there will be more 3° radicals to react with the bromine radicals, and the amount of t-butyl bromide formed exceeds the amount of iso-butyl bromide formed.

You can prepare t-butyl bromide, by an addition reaction. HBr can be added across the double bond of iso-butylene as shown in the following reaction.

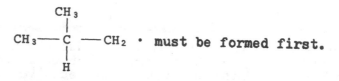

It has been shown that when an inorganic acid, such as HBr, adds across a double bond in an olefin the hydrogen ion will add to the carbon with the most hydrogen atoms already bound to it. This is called Markovnikov's rule.

● **PROBLEM** 754

Ethyl chloride can be chlorinated to give two isomers, A and B. A can be chlorinated to give two isomers of $C_2H_3Cl_3$, while B gives only one isomer of $C_2H_3Cl_3$ upon further chlorination. Indicate the equations for the chlorination of ethyl chloride and identify A and B structurally.

Solution: You are told about isomerism in the compounds A and B, and their chlorinated products. When two compounds are isomers they have the same molecular formula, but differ in the sequence of linkages of atoms. These are structural isomers. Ethyl chloride has the structural formula

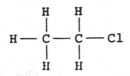

The equation for chlorination is its reaction with $Cl_2$ in the presence of light. Note that there exist two possible carbons on which the chlorines can bond. These are the two isomers A and B. Thus you can write

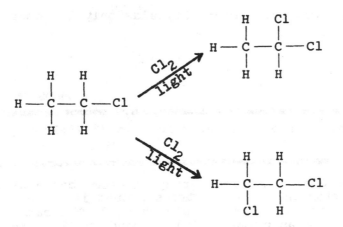

The question is which one is A, as the other must be B. You are told that A gives 2 isomers, while B yields 1, if they are chlorinated again. Let's start with

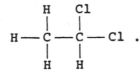

If you chlorinate it, you would obtain

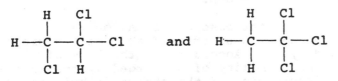

These are two different compounds in the arrangement of atoms. They are structural isomers. Thus, isomer A must be

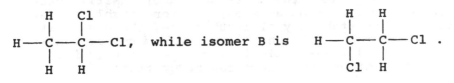

721

If this is correct, then the chlorination of B should yield only one product. Let's see if this is true.

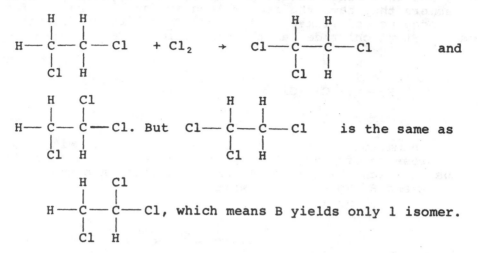

$$H-\underset{\underset{Cl}{|}}{\overset{\overset{H}{|}}{C}}-\underset{\underset{H}{|}}{\overset{\overset{H}{|}}{C}}-Cl \quad + Cl_2 \quad \rightarrow \quad Cl-\underset{\underset{Cl}{|}}{\overset{\overset{H}{|}}{C}}-\underset{\underset{H}{|}}{\overset{\overset{H}{|}}{C}}-Cl \qquad \text{and}$$

$H-\underset{\underset{Cl}{|}}{\overset{\overset{H}{|}}{C}}-\underset{\underset{H}{|}}{\overset{\overset{Cl}{|}}{C}}-Cl$. But $Cl-\underset{\underset{Cl}{|}}{\overset{\overset{H}{|}}{C}}-\underset{\underset{H}{|}}{\overset{\overset{H}{|}}{C}}-Cl$ is the same as

$H-\underset{\underset{Cl}{|}}{\overset{\overset{H}{|}}{C}}-\underset{\underset{H}{|}}{\overset{\overset{Cl}{|}}{C}}-Cl$, which means B yields only 1 isomer.

● **PROBLEM** 755

The ease of formation of carbonium ions is $3°>2°>1°>Ch_3^+$. Explain why.

Solution:  A carbonium ion is a group of atoms containing a carbon atom **with only six electrons around it.**
Thus the carbonium ion is positively charged. The carbonium ions are classified as primary ($1°$), secondary ($2°$), and tertiary ($3°$) after the carbon bearing the positive charge. For example,

$$H \underset{+}{\overset{\overset{\ddot{} }{H}}{:C:R}} \qquad\qquad R \underset{+}{\overset{\overset{\ddot{} }{H}}{:C:R}} \qquad\qquad R \underset{+}{\overset{\overset{\ddot{} }{R}}{:C:R}}$$

   primary           secondary          tertiary

where R can be any alkyl group.

To understand why the ease of formation is $3°>2°>1°>CH_3^+$, you need to understand some of the properties of a charged system. According to the laws of electrostatics, the stability of a charged system is increased by the distribution of the charge. Alkyl groups, for example, tend to donate electrons, so that any positive charge is dispersed. Thus, the more alkyl groups donating electrons, the greater the distribution of positive charges in a carbonium ion, and, as such, the more stable it is. The greater the stability of a substance, the easier it is to form. In our example, the $3°$ carbonium ion has 3 alkyl groups, the $2°$ ion has 2 alkyl groups, and the $1°$ ion has one alkyl group. Note $CH_3^+$ can be written

H
|
H——C——H, which shows that no alkyl groups are present.
+

Therefore, $3° > 2° > 1° > CH_3^+$ in ease of formation, since this sequence is also the sequence for the number of alkyls that distribute the positive charge, which results in increased stability.


# ALKENES AND ALKYNES

What is your understanding of an addition reaction mechanism? Illustrate your answer with a specific example.

Solution: An important difference between saturated and unsaturated hydrocarbons is the general type of reaction that each will undergo. Alkanes react by substitution, while alkenes react by addition reactions (as well as substitution). In an addition reaction, the pi bond of the double bond is destroyed and two sigma bonds are formed, one to each of two carbon atoms.

A mixture consisting of only hydrogen and an alkene undergoes no detectable addition reaction even at high temperature and pressure. However, if a small amount of finely divided metal, such as platinum, palladium, or nickel is added as a catalyst, addition of hydrogen to the alkene occurs quite readily even at room temperature. (The same is true for alkynes, i.e. unsaturated compounds containing a triple bond.)

According to the commonly accepted view, alkenes and alkynes, with relatively loosely held electrons in their pi bonds, act as electron-pair donors, that is, Lewis bases, and protons from addition reagents, HX (where X is H, Cl, Br, ...), act as Lewis acids, that is, electron-pair acceptors. The first step, then, in the addition of HX to an alkene is the addition of a proton in a Lewis acid-base reaction. The process involves the formation of alkyl cations and chlorine anions as reaction intermediates. Such alkyl cations are called carbonium ions; they are groups of atoms that contain a carbon atom having only six electrons.

An example of an addition reaction would be

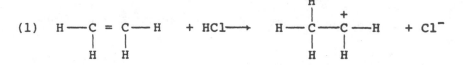

(1)  H—C = C—H  + HCl ⟶  H—C—C—H  + Cl⁻

ethene                     a carbonium ion    anion

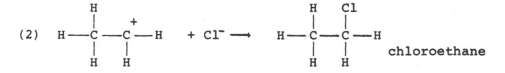

(2)  H—C—C—H  + Cl⁻ ⟶  H—C—C—H

chloroethane

● PROBLEM 757

Write the structural formulas of the hydrogenation products
of the following olefins: (a) propylene; (b) 2-methyl-2-
pentene; (c) 1-methyl cyclohexene.

<u>Solution</u>:  The most important reactions of alkenes
(olefins) involve addition of reagents to the double bonds.
An alkene can be converted to an alkane by addition of
hydrogen to the double bond. These reactions are usually
carried out by using a high pressure of hydrogen gas in
the presence of a catalyst such as finely divided platinum,
palladium, or nickel.

To solve this problem, draw the structural formulas
of each alkene. The addition of hydrogen breaks the double
bond, and the $H_2$ molecule attaches itself to the carbon
atoms involved. Thus,

(a)

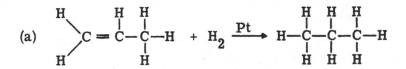

(b)

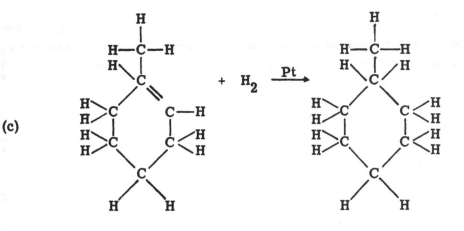

(c)

• PROBLEM 758

Give the name of the compound formed by the addition of HCl to (a) isobutene; (b) 3-methyl-2-pentene.

Solution: When a halogen acid, such as HCl, is added to alkenes, an addition reaction occurs. Reactions with the addition of an acid to unsymmetrical molecules (where the two doubly bonded carbon atoms have different numbers of hydrogen atoms attached to them) usually follow a predictable course: the hydrogen atom of the acid adds to that carbon atom which has attached to it the greater number of hydrogen atoms. The acid anion then adds to that carbon atom which has the lesser number of hydrogen atoms. This is known as Markovnikov's rule.

To solve this problem, write the structural formulas, set up the reaction equation, and then follow the rules just laid out to find the product. Once the product is known in terms of its structure, the next step is to name the compound using the IUPAC system. Thus,

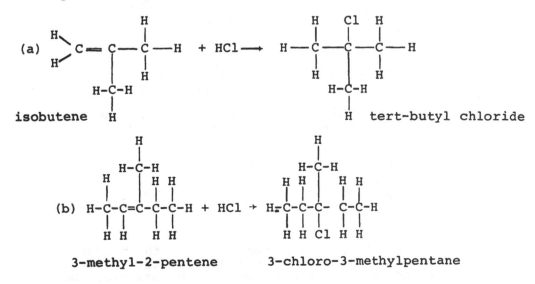

725

Predict the products of the following reactions:

(a) $CH_3 \overset{\underset{\displaystyle CH_3}{|}}{— C} = CH_2 + Br \longrightarrow$

(b) $CH_3 \overset{\underset{\displaystyle CH_3}{|}}{— C} = CH_2 + HCl \longrightarrow$

(c) $H_3C \overset{\underset{\displaystyle Br}{|}}{\overset{\overset{\displaystyle H}{|}}{— C —}} CH_2 — CH_3 \quad + KOH \longrightarrow$

<u>Solution</u>:  To predict the products, you need to identify the type of reaction that will take place and its mechanism. The first two reactions, (a) and (b), involve addition across a double bond, and the last reaction, (c) concerns the dehalogenation of an alkyl halide using the base, KOH. In other words, it is an elimination reaction.

The general mechanism of addition to a double bond is as follows:

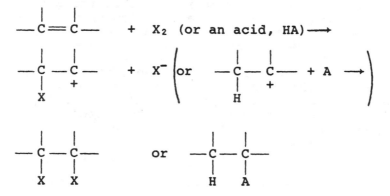

Note: From the mechanism, observe that the reaction goes through a species called a carbonium ion.

The stability of a carbonium ion follows the order: $3° > 2° > 1°$. This is shown schematically as

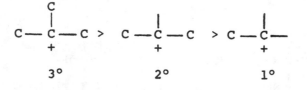

A reaction will proceed so that the product formed in greatest amounts has gone through the mechanism that gives the most stable carbonium ion possible. Thus, you have

(a)

$$CH_3 - \overset{\overset{\displaystyle CH_3}{|}}{C} = CH_2 + Br_2 \rightarrow \quad CH_3 - \overset{\overset{\displaystyle CH_3}{|}}{\underset{\underset{\displaystyle Br}{|}}{C}} - CH_2Br$$

You do not have to consider carbonium ion formation here, since the final product would be the same in either case, however, for (b), the carbonium ion must be considered. There are two possible reactions:

(1) $CH_3 - \overset{\overset{\displaystyle CH_3}{|}}{C} = CH_2 + HCl \rightarrow CH_3 - \overset{\overset{\displaystyle CH_3}{|}}{\underset{\underset{\displaystyle +}{}}{C}} - CH_3 + Cl^- \rightarrow CH_3 - \overset{\overset{\displaystyle CH_3}{|}}{\underset{\underset{\displaystyle Cl}{|}}{C}} - CH_3$

(2) $CH_3 - \overset{\overset{\displaystyle CH_3}{|}}{C} = CH_2 + HCl \rightarrow CH_3 - \overset{\overset{\displaystyle CH_3}{|}}{\underset{\underset{\displaystyle H}{|}}{C}} - CH_2^+ + Cl^- \rightarrow CH_3 - \overset{\overset{\displaystyle CH_3}{|}}{\underset{\underset{\displaystyle H}{|}}{C}} - CH_2Cl$

Two different products are obtained. Notice that in (1), the reaction proceeds through a 3° carbonium, whereas in (2) the reaction proceeds through a 1° carbonium ion. Because 3° is more stable than 1°,

$CH_3 - \overset{\overset{\displaystyle CH_3}{|}}{\underset{\underset{\displaystyle Cl}{|}}{C}} - CH_3$ greatly predominates over $CH_3 - \overset{\overset{\displaystyle CH_3}{|}}{\underset{\underset{\displaystyle H}{|}}{C}} - CH_2Cl.$

(c) In considering this part, remember that potassium hydroxide (KOH) eliminates the halogen from an alkyl halide to give alkenes, compounds containing double bonds. With this in mind, two possible reactions and products are possible:

$H_3C - \overset{\overset{\displaystyle H}{|}}{\underset{\underset{\displaystyle Br}{|}}{C}} - CH_2CH_3 + KOH \rightarrow H_3C - \overset{\overset{\displaystyle H}{|}}{C} = CHCH_3$

or

$H_3C - \overset{\overset{\displaystyle H}{|}}{\underset{\underset{\displaystyle Br}{|}}{C}} - CH_2CH_3 + KOH \rightarrow H_2C = \overset{\overset{\displaystyle H}{|}}{C} - CH_2CH_3$

The first product, $H_3C - \overset{\overset{\displaystyle H}{|}}{C} = CHCH_3$, greatly predominates over the second. The reason stems from the fact that alkenes are more stable when the double bond is surrounded by the maximum number of alkyl groups.

Write equations for the following (a) cracking of a hydrocarbon - use $C_{14}H_{30}$, (b) incomplete combustion of acetylene, (c) addition of $Br_2$ to a mixture of ethane, ethylene,
and ethyne.

Solution:   (a) By cracking, you mean the process where
petroleum fractions containing a high number of carbon
atoms are subjected to high heat and pressure, under which
they break down to gasoline and lower carbon number hydro-
carbons.Octane-gasoline has the formula $C_8H_{18}$. Thus, you
can write the balanced equation for cracking of $C_{14}H_{30}$ as
$C_{14}H_{30} \longrightarrow C_8H_{18} + C_6H_{12}$.

     (b) By complete combustion, you mean the reaction of
an organic compound with $O_2$ to produce water and carbon
dioxide. When combustion is incomplete, as in car engines,
some carbon monoxide ($CO$) or free carbon is formed in
addition to $H_2O$ and $CO_2$. An example of incomplete combustion with acetylene could have the equation

   $HC \equiv CH$   + $2O_2$ $\longrightarrow$   $H_2O$  +  $CO_2$  + CO.
   acetylene

     (c) To write equations for this mixture, you need
to consider some structural characteristics of the components. All are organic compounds that belong to different
classes. Ethane ($C_2H_6$) is an alkane (a saturated hydrocarbon). This means it contains only single bonds between
the  carbons, and between carbon and hydrogen.  Ethylene
($C_2H_4$) is an alkene (an unsaturated hydrocarbon). It
possesses a double bond between the carbon atoms. Ethyne
($C_2H_2$) or acetylene is an alkyne (an unsaturated hydrocarbon). It possesses a triple bond between the carbon
atoms. Unsaturated hydrocarbons undergo addition reactions to the double or triple bond, saturated ones do
not. Thus, when $Br_2$ is added to the mixture, ethene and
ethyne will undergo an addition reaction. Ethane will not
undergo any reaction.Therefore,

   $C_2H_6$ (ethane) + $Br_2$ $\longrightarrow$    no reaction

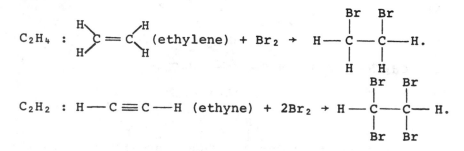

You have an unknown liquid hydrocarbon that may be
hexane, 1-hexene or cyclohexene.
What simple chemical tests could you use to distinguish
which it is?

Hexane: $CH_3CH_2CH_2CH_2CH_2CH_3$

1-Hexene: $CH_3CH_2CH_2CH_2CH = CH_2$

Cyclohexene:

**Figure A**

Solution: Organic compounds can be distinguished by
their functional groups. A functional group is a group
of atoms that characterize a compound's reactions. Thus,
to solve this problem, you want to use reactions that
clearly distinguish between functional groups. Proceed
as follows: Hexane is a saturated hydrocarbon, which
means all of its carbons have single bonds and the
molecule is made up of carbon and hydrogen only. 1-
hexene and cyclohexene both possess a double bond, which
means they are unsaturated. A characteristic feature of
unsaturated compounds is the fact that they undergo
addition reactions. Thus, if you add $Cl_2$ in $CCl_4$, you get

addition to the double bond in both                 and

$CH_3CH_2CH_2CH_2CH = CH_2$, but not in $CH_3CH_2CH_2CH_2CH_2CH_3$. To
distinguish between unsaturated compounds, use an
ozonolysis reaction. When you add ozone to a double bond
it cleaves so that carbonyl compounds are formed, i.e.,

compounds of the form $\underset{R'}{\overset{R}{>}}C = O$, where R is H and/or carbon

groups, likewise with R . If you perform this reaction on
1-hexene, you obtain $CH_3CH_2CH_2CH_2CHO$ (**pentanal**) and formaldehyde ($CH_2O$), the latter is a gas recognizable by its
odor, while the reaction on cyclohexene yields only
$OCHCH_2CH_2CH_2CH_2CHO$ (hexanedial).

What are the products when the following hydrocarbons are oxidized with acidic permanganate solutions? (a) 2-butene; (b) 3,4-dimethyl-3-hexene; (c) 2-methyl-2-butene.

Solution: Alkenes readily react with a number of oxidizing agents. A test for the presence of an olefin is the reaction with an acidic solution of permanganate ion ($MnO_4^-$). The purple color of permanganate ion disappears  as the olefin is oxidized. The course of such a reaction is that the olefin cleaves into two oxidized fragments. A carbon atom with two alkyl groups attached is converted to the carbonyl group of ketone, while a carbon atom with one attached hydrogen becomes the carboxyl group of an acid. This is summarized by the following

$$\underset{R}{\overset{H}{\diagdown}} C = C \underset{R'}{\overset{R''}{\diagup}} \quad \xrightarrow{MnO_4^-} \quad R-C\underset{O}{\overset{OH}{\diagup}} \quad + \quad O=C\underset{R'}{\overset{R''}{\diagup}}$$

To solve this problem, write the structural formula of each hydrocarbon, find which alkyl groups correspond to R, R', and R" in the general formula, and then substitute into the general formula.

(a) 
$$H-\underset{\underset{H}{|}}{\overset{\overset{H}{|}}{C}}-\underset{\underset{H}{|}}{\overset{}{C}}=\underset{\underset{H}{|}}{\overset{}{C}}-\underset{\underset{H}{|}}{\overset{\overset{H}{|}}{C}}-H \quad \xrightarrow{MnO_4^-} \quad 2CH_3-C\underset{O}{\overset{OH}{\diagup}}$$

The product of this reaction is acetic acid.

(b)
$$H-\underset{\underset{H}{|}}{\overset{\overset{H}{|}}{C}}-\underset{\underset{H}{|}}{\overset{\overset{H}{|}}{C}}-\underset{|}{\overset{}{C}}=\underset{|}{\overset{}{C}}-\underset{\underset{H}{|}}{\overset{\overset{H}{|}}{C}}-\underset{\underset{H}{|}}{\overset{\overset{H}{|}}{C}}-H \quad \xrightarrow{MnO_4^-} \quad 2 \ O=C\underset{CH_2CH_3}{\overset{CH_3}{\diagup}}$$

$$H-C-H \quad H-C-H$$
$$| \qquad \qquad |$$
$$H \qquad \qquad H$$

The product of this reaction is methyl ethyl ketone.

(c)
$$H-\underset{\underset{H}{|}}{\overset{\overset{H}{|}}{C}}-\underset{\underset{H}{|}}{\overset{}{C}}=\underset{|}{\overset{\overset{H-C-H}{|}}{C}}-\underset{\underset{H}{|}}{\overset{\overset{H}{|}}{C}}-H \quad \xrightarrow{MnO_4^-} \quad CH_3-C\underset{O}{\overset{OH}{\diagup}} \quad + \quad CH_3\underset{\overset{||}{O}}{C}CH_3$$

The products of this reaction are acetic acid and acetone.

# ALCOHOLS

Perform the following conversions (synthesis) using any
inorganic reagents you require:

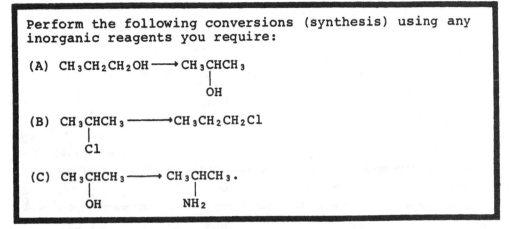

(A) $CH_3CH_2CH_2OH \longrightarrow CH_3\underset{\underset{OH}{|}}{C}HCH_3$

(B) $CH_3\underset{\underset{Cl}{|}}{C}HCH_3 \longrightarrow CH_3CH_2CH_2Cl$

(C) $CH_3\underset{\underset{OH}{|}}{C}HCH_3 \longrightarrow CH_3\underset{\underset{NH_2}{|}}{C}HCH_3$.

<u>Solution</u>: In any synthesis problem you can use a step by
step approach. Try to find reactions that lead to intermediates from which the product can be obtained. In (A)
you want to shift the hydroxyl group from the 1° carbon
to the 2° carbon. You can do this through the corresponding alkene. Once the alkene is obtained, the addition of
acid and water will generate the desired product. To
obtain the intermediate, add acid to the starting material.
Because it is an alcohol, it will dehydrate to the alkene.
You have, then,

$$CH_3CH_2CH_2OH + H^+ \longrightarrow CH_3CH = CH_2 + H_2O$$

$$CH_3CH = CH_2 + H^+/H_2O \longrightarrow CH_3\underset{\underset{OH}{|}}{C}HCH_3$$

In (B) you want to switch the Cl from the 2° carbon
to the 1° carbon. Convert the starting material to an
alkene and from this the product can be obtained. You
have, therefore, the following process:

$$CH_3\underset{\underset{Cl}{|}}{C}HCH_3 \xrightarrow{\text{KOH}} CH_3CH = CH_2 + KCl + H_2O$$

Potassium hydroxide has the ability to dehalogenate
an alkyl halide to an alkene.

$$CH_3CH = CH_2 \xrightarrow{\text{HCl, peroxides}} CH_3CH_2CH_2Cl.$$

With peroxides, the hydrogen from the HCl adds to
the carbon with the **least** number of carbons attached. You

are performing an addition reaction across a double bond. In (C), you are converting the OH group to a $NH_2$ group. You can convert the OH to Cl, first, since upon addition of $NH_3$, the $NH_2$ group is obtained. To obtain the chlorine, dehydrate the alcohol with acid to form the alkene, then add HCl. You have the following sequence:

$$CH_3CHCH_3 \xrightarrow[H_2O]{H^+} CH_3CH = CH_2 \xrightarrow{HCl}$$
$$\qquad | \qquad\qquad\qquad$$
$$\qquad OH \qquad\qquad\qquad$$

$$CH_3CH - CH_3 \xrightarrow{NH_3} CH_3 - CH - CH_3 + HCl$$
$$\qquad | \qquad\qquad\qquad\qquad\qquad |$$
$$\qquad Cl \qquad\qquad\qquad\qquad NH_2$$

● **PROBLEM 764**

Synthesize $CH_2BrCH_2Br$ from $CH_3CH_2OH$.

Solution: You are given the alcohol, ethanol, and are asked to synthesize $CH_2BrCH_2Br$, 1,2 dibromoethane. The fact that it is a dibromo compound suggests that the bromine added across a double bond. Therefore, if you could convert the alcohol to an olefin, namely, ethylene, you could add $Br_2$ in the solvent $CCl_4$, and via an addition reaction, obtain the desired product. To obtain ethylene, remember that the alcohol can be dehydrated to the corresponding olefin by the addition of acid. This well-documented mechanism is as follows:

$$CH_3CH_2OH + H^+(acid) \xrightarrow[\substack{160°C\ or \\ greater}]{heat} CH_3CH_2\overset{\overset{H^+}{\cdot\cdot}}{O}H \longrightarrow CH_3CH_2^+ + HOH$$
$$\xrightarrow{\qquad} CH_2 = CH_2 + H^+$$
$$\qquad ethylene$$

When you add the acid, you protonate the alcohol. Water is expelled and you obtain a carbonium ion, $CH_3CH_2^+$. The carbonium expels a proton ($H^+$), so that ethylene is obtained. Now that you have $CH_2 = CH_2$, add $Br_2$ and $CCl_4$ and you obtain $CH_2BrCH_2Br$, your desired product. The $Br_2$ is added across the double bond in an addition reaction.

● **PROBLEM 765**

Ethanol, $C_2H_5OH$, reacts with HBr, hydrogen bromide, to give ethyl bromide, $C_2H_5Br$, and water. Ethanol, however, does not react with NaBr. Explain. Suggest a mechanism for the reaction with HBr.

Solution: This problem entails an understanding of nucleophilic displacement. In such reactions, a stronger Brönsted base, i.e. one that is a proton acceptor, displaces the weaker Brönsted base. This is why the leaving groups are usually extremely weak bases, such as $I^-$, $Br^-$, and $Cl^-$. With this in mind, notice that NaBr is a salt. This means that in solution, NaBr has dissociated into $Na^+$ and $Br^-$ ions. When NaBr is present, then, you are asked why $Br^-$ cannot replace $OH^-$ to form ethyl bromide. Recall, displacement only occurs if the base which you displace is a weaker base. $Br^-$ is a weak base, as mentioned. It is a weaker base than $OH^-$. Thus, $Br^-$ cannot replace $OH^-$. You can suggest a mechanism that occurs for the reaction with HBr by realizing that HBr is an acid. Thus, in solution, you have $H^+$ and $Br^-$ ions. The $H^+$ protonates the OH group. $Br^-$ then proceeds to displace the water. Why? While $Br^-$ is a weaker base than $OH^-$, it is a stronger base than $H_2O$ (water), a very weak base. In other words, a nucleophilic displacement occurs. Overall, it can be written as

$$C_2H_5OH + HBr \longrightarrow C_2H_5OH_2^+ + Br^- \longrightarrow C_2H_5Br + H_2O.$$

● **PROBLEM** 766

An excellent method of making alcohols is the reaction of an aldehyde on a Grignard reagent. If you perform this reaction, you must scrupulously dry the aldehyde before adding it to the Grignard reagent. Why?

Solution: To answer this question, you must know what a Grignard reagent is and some of its properties. The general formula of a grignard reagent is RMgX, where R is an alkyl or aryl group, Mg is magnesium, and X is a halogen atom (usually bromine or chlorine). Grignard reagents are extremely reactive. You can consider it to be the magnesium salt of the weak acid R-H. Thus, if there exists in solution a stronger acid, R-H will be displaced from its salt. Water is a stronger acid than R-H. Thus, if $H_2O$(water) is present, the following reaction occurs:

$$RMgX + H_2O \longrightarrow RH + Mg(OH)X.$$

Thus, if the aldehyde is not dried, water will be present, and it will then react with the grignard reagent producing RH. The presence of water reduces or eliminates the grignard reagent from solution, so that little or none is present to react with the aldehyde.

A chemist dissolves methyl alcohol in n-octane. He adds this to excess methylmagnesium iodide, which is dissolved in the high-boiling solvent, n-butyl ether. A gas is evolved, which is collected at STP. It occupies a volume of 1.04 cc. What is the gas and how many grams of methyl alcohol were present?

**Solution**: Grignard reagents, which have the general formula of RMgX (R is an alkyl or aryl group, Mg is magnesium, and X is a halogen atom), are extremely reactive. In the presence of any acid stronger than an alkane, the grignard reagent is displaced to form RH. This happens because the grignard reagent may be considered the magnesium salt of a weak acid RH. Thus, a stronger acid displaces RH from its salt. Alcohols are stronger acids than alkanes. Therefore, when methyl alcohol ($CH_3OH$) is added to methyl magnesium iodide ($CH_3MgI$), the following reaction occurs:

$$CH_3OH + CH_3MgI \longrightarrow CH_4 + H_3COMgI.$$

At S.T.P. ($0°C$ and 1 atm), $CH_4$ exists as a gas. To find the amount of $CH_3OH$ present, remember that at STP a mole of gas occupies 22.4 liters. The gas produced here occupies only 1.04 cc or 0.00104 liters (1000 cc = 1 liter). Let x = moles of methane gas produced. Thus, you can set up the proportion

$$\frac{22.4 \text{ liters}}{1 \text{ mole}} = \frac{.00104 \text{ liters}}{x \text{ mole}}$$

Solving, $x = 4.64 \times 10^{-5}$ moles. From the stoichiometry of the equation, 1 mole of $CH_3OH$ produces 1 mole of $CH_4$. Thus, you have $4.64 \times 10^{-5}$ moles of $CH_3OH$. The molecular weight (MW) of $CH_3OH$ is 32 g/mole. By definition, 1 mole equals the weight in grams divided by MW. Substituting and solving for weight in grams,

$$4.64 \times 10^{-5} \text{ moles} = \frac{\text{grams of } CH_3OH}{32 \text{ g/mole}} ;$$

grams of $CH_3OH$ present $= 32 (4.64 \times 10^{-5})$

$$= 1.48 \times 10^{-3} \text{ grams.}$$

How would you distinguish between the following pairs: (a) isobutane and isobutylene (b) sec-butyl alcohol and n-heptane.

**Solution**: To distinguish between two compounds, try to

select for a characterization test. Begin by writing the structures of the compounds involved.

(a) isobutane 

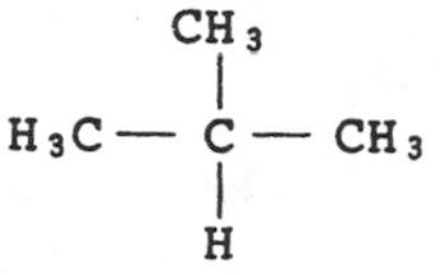

$$H_3C - \underset{\underset{H}{|}}{\overset{\overset{CH_3}{|}}{C}} - CH_3$$

isobutylene 

$$CH_3 - \underset{\underset{}{\overset{\overset{CH_3}{|}}{C}}}{} = CH_2$$

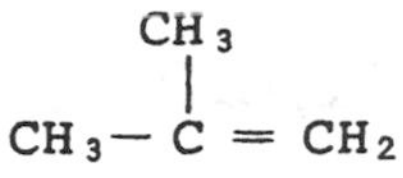

You see that isobutylene is an alkene, and that isobutane is an alkane. Thus, you should be able to distinguish isobutylene from isobutane by an addition reaction. Addition reactions are characteristic of alkenes and other unsaturated compounds. When $Br_2/CCl_4$, a reddish brown solution, is in contact with an alkene, the following reaction occurs:

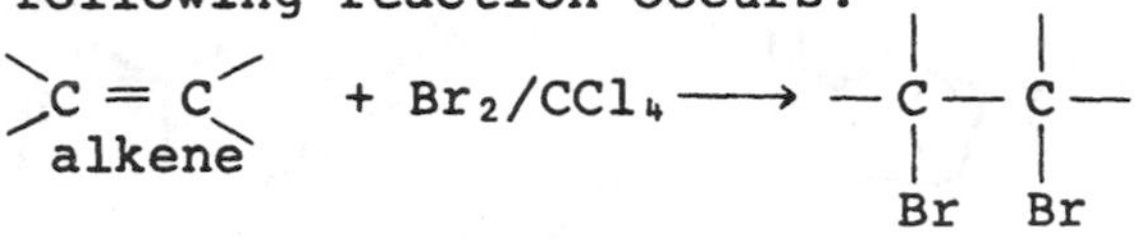

$$\underset{alkene}{\overset{}{>}C = C<} + Br_2/CCl_4 \longrightarrow -\underset{\underset{Br}{|}}{C} - \underset{\underset{Br}{|}}{C} -$$

When the alkene consumes the $Br_2$, the red color disappears. Thus, alkenes can be detected by their ability to decolorize the red solution of $Br_2/CCl_4$.

You can also use a cold, dilute, neutral permanganate solution, the Baeyer test, instead of $Br_2/CCl_4$ to detect the presence of an alkene. Permanganate solution is purple colored, upon reaction with an alkene, the purple disappears and is replaced by a brown manganese dioxide precipitate. The reaction can be written

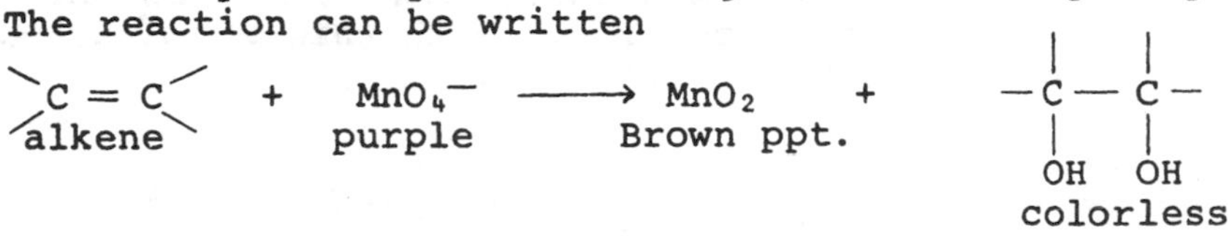

$$\underset{alkene}{\overset{}{>}C = C<} + \underset{purple}{MnO_4^-} \longrightarrow \underset{Brown\ ppt.}{MnO_2} + \underset{colorless}{-\underset{\underset{OH}{|}}{C} - \underset{\underset{OH}{|}}{C} -}$$

Alkenes do give this reaction. Thus, you now have two simple tests to distinguish isobutane from isobutylene.

(b) sec-butyl alcohol 

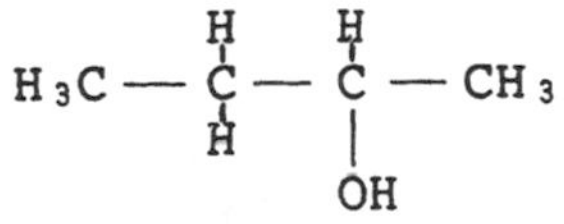

$$H_3C - \underset{\underset{H}{|}}{\overset{\overset{H}{|}}{C}} - \underset{\underset{OH}{|}}{\overset{\overset{H}{|}}{C}} - CH_3$$

n-heptane 

$$H_3C - \underset{\underset{H}{|}}{\overset{\overset{H}{|}}{C}} - \underset{\underset{H}{|}}{\overset{\overset{H}{|}}{C}} - \underset{\underset{H}{|}}{\overset{\overset{H}{|}}{C}} - \underset{\underset{H}{|}}{\overset{\overset{H}{|}}{C}} - \underset{\underset{H}{|}}{\overset{\overset{H}{|}}{C}} - CH_3$$

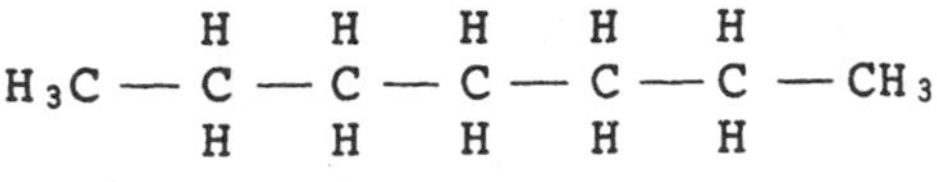

These two can be distinguished by using a chromic anhydride reagent. Secondary alcohols, such as sec-butyl alcohol, are oxidized by chromic anhydride, $CrO_3$, in aqueous sulfuric acid. The clear orange solution turns opaque, blue-green when the alcohol is added. In other words,

$$\underset{1°\ or\ 2°}{ROH} + \underset{\underset{orange}{clear,}}{HCrO_4^-} \longrightarrow \underset{solution}{Opaque,\ blue-green}$$

N-heptane is an alkane, and, as such, does not turn the solution blue green. Thus, you can distinguish sec-butyl alcohol from n-heptane by adding chromic anhydride.

# OTHER FUNCTIONAL GROUPS

Predict the final hydrogenation products in each of the
compounds in figure A.

(a) $CH_3C\equiv CH$,     (b) 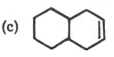,     (c) 

Figure A

Solution: Hydrogenation refers to the reduction of
unsaturated hydrocarbons, by addition of the $H_2$ across the
bond. Frequently a catalyst is used to increase the rate
of the reaction. After complete hydrogenation, the product
contains only single bonds. A triple bond requires two
moles of hydrogen gas, while a double bond requires only
1 mole for complete hydrogenation. With this in mind, you
have the reactions shown in figure B.

(a) $CH_3C\equiv CH$ + $2H_2$ $\xrightarrow[\text{catalyst}]{\text{Pt}}$ $CH_3CH_2CH_3$
(propane)

(b)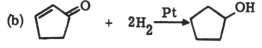
( cyclopentanol)

(c)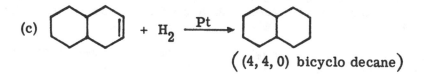
$\Big((4,4,0)$ bicyclo decane$\Big)$

Figure B

Explain why benzene is a planar molecule, while cyclohexane, $C_6H_{14}$, contains a puckered ring of six carbon atoms.

Solution: The best way to approach this problem is to consider the geometry of the carbon-carbon bonding. To do this, consider its hybridization. Benzene can be written as shown in figure A.

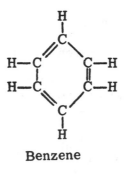

Benzene

Figure A

Notice that each carbon has 4 bonds, and is bonded to three other elements (2 other carbons and 1 hydrogen). This suggests $sp^2$ hybridization. $sp^2$ hybridization allows bonding to three other elements, and leaves a "free" p orbital. This p orbital can overlap with another p orbital to give a pi bond, which serves to form the double bond. With $sp^2$ hybridization, the molecule must assume a geometry where the orbitals are 120° apart. This is the only way that the repulsion which occurs when like charges are brought together can be avoided. See figure B.

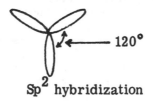

Sp² hybridization

Figure B

An angle of 120° creates a planar geometry. This is true because a planar molecule with bond angles of 120° is a configuration where the 3 atoms involved are the farthest apart. Consider cyclohexane. It can be written as shown in figure C.

Each carbon atom is bound to four other atoms (2 other carbons and 2 hydrogens). This suggests $sp^3$ hybrodization. With $sp^3$ hybridization, a molecule must assume a geometry that permits 109°28' bond angles. The atoms of the bond are arranged in a tetrahedral formation, this allows

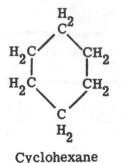

Cyclohexane

Figure C

for the greatest distance between the atoms.

The way cyclohexane can accomplish this is to assume a puckered ring of six carbon atoms, as can be seen in figure D.

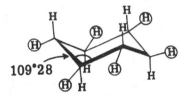

Cyclohexane (puckered ring formation)

Figure D

● PROBLEM 771

Predict the reaction of propene with $Br_2/CCl_4$, $H_2O/H_2SO_4$, and HCl. Predict the reaction of toluene with these substances.

Solution: The procedure for solving this problem is as follows: write the structures of toluene and propene; classify these compounds from their structures; once these compounds are classified, you can predict the reaction that will occur with the given reagents $Br_2/CCl_4$, $H_2O/H_2SO_4$, and HCl.

The structure of propene is $H_2C = CH \longrightarrow CH_3$. Notice, it has a double bond which indicates it is an alkene. Alkenes undergo the characteristic reaction of addition to the double bond. Thus,

$$CH_2 = CH \longrightarrow CH_3 \ + \ Br_2/CCl_4 \longrightarrow \underset{Br}{CH_2} \longrightarrow \underset{Br}{CH} \longrightarrow CH_3$$

Note: Adding $Br_2$ breaks the double bond in a $CCl_4$ solvent.

$$CH_2 = CH \!-\! CH_3 + H_2O/H_2SO_4 \longrightarrow CH_3 \!-\! \underset{\underset{OH}{|}}{CH} \!-\! CH_3$$

$$CH_2 = CH \!-\! CH_3 + HCl \longrightarrow CH_3 \!-\! \underset{\underset{Cl}{|}}{CH} \!-\! CH_3$$

The structural formula of toluene is shown in figure A.

Figure A

The circle inside the ring indicates resonance. Its true structure is a hybrid (figure B),

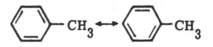

Figure B

which means it is stable to addition reactions. Thus, toluene will not undergo any reaction with these reactants.

● **PROBLEM** 772

Define, illustrate, or explain each of the following: esterification, saponification, elastomer and nitration.

Nitrobenzene

Figure A

Solution: Esterification is the process that occurs during the reaction of an alcohol (R'-OH) with a carboxylic

acid (R-COOH) to produce an ester (R-$\overset{\overset{\displaystyle O}{\|}}{C}$-OR'). It can be illustrated as follows in the formation of the ester ethyl acetate

$$CH_3COOH \quad + \quad HOC_2H_5 \quad \xrightarrow{H_2SO_4} \quad CH_3COOC_2H_5 \quad + \quad H_2O$$

acetic acid       ethyl                 ethyl    acetate       water
               alcohol

Saponification is the term used to describe those reactions of an ester ($R-\overset{\displaystyle O}{\overset{\displaystyle \|}{C}}-OR$) with a base to produce soft and hard soaps. The process can be illustrated by the formation of sodium stearate

$$(C_{17}H_{35}COO)_3C_3H_5 + 3NaOH \longrightarrow 3C_{17}H_{35}COONa + C_3H_5(OH)_3$$

Ester                   Base      Soap (sodium     Glycerol
  (tristearin)                       stearate)

Elastomer is the general name of all rubber sub-stitutes. They tend to be better than natural rubber. For example, most elastomers are more oil resistant and some are less permeable than rubber.

Nitration: This is the name of the reaction in which nitric acid ($HNO_3$) is mixed with sulfuric acid ($H_2SO_4$) and added to a benzene ring, such that an $NO_2$ is substituted for a hydrogen atom attached to the ring to form nitro-benzene (fig. A).

● **PROBLEM** 773

Draw the structure of the isomer of trinitrobenzene that would be easiest to synthesize from benzene, nitric acid, and sulfuric acid.

Solution: The nitration of benzene proceeds quickly if a mixture of concentrated nitric and sulfuric acids is used. For each mole of nitric acid dissolved in sulfuric acid, four moles of particles are formed:

$$2H_2SO_4 + HNO_3 \longrightarrow H_3O^+ + 2HSO_4^- + NO_2^+$$

The nitronium ion ($NO_2^+$) is the reagent responsible for the nitration of aromatic compounds.

When benzene derivatives undergo substitution, a "directing effect" on the position of the new substituent is observed. The first phase of the nitration reaction is the attack of the nitronium ion on the electrons of the benzene ring (seen in figure A). The loss of a hydrogen atom, or proton, to $HSO_4^-$ (a proton acceptor) leads to the restoration of the original $\pi$-bonding system.

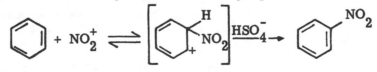

Figure A

The attack on the benzene ring is generally accomplished by an electron-deficient species such as a positive ion, or the positive end of a polar molecule. The electron withdrawing nitro group makes it more difficult for an electrophilic species to add onto the benzene ring; vigorous conditions are required for further nitration of nitrobenzene. The nitro group will "direct" the substitution of additional nitronium ions in a meta orientation.

With this in mind, the structure of the isomer of trinitrobenzene is at once known. Each nitronium ion is directed to the meta position of the other ion. Thus, the structure is that of figure B

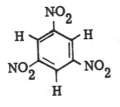

Figure B

and has the formula 1,3,5-trinitrobenzene.

● PROBLEM 774

Explain why the hydrolysis of an ester to yield alcohol and acid is best carried out in basic rather than acidic solution. Use equations to illustrate your answer.

Solution: The best way to solve this problem is to consider what products will be formed with acid and base.

The general formula of an ester is $R' - \overset{\overset{O}{\|}}{C} - OR$.

Acid: $R' - \overset{\overset{O}{\|}}{C} - OR \xrightarrow[H_2O]{H^+} R' - \overset{\overset{O}{\|}}{C} - OH + HOR$

The carboxylic acid and alcohol are generated. But esters are generated from carboxylic acid and alcohols by the use of $H^+$. In other words, there exists nothing to prevent the reaction from going back in the other direction. That is, you have the following equilibrium:

ester $\underset{H_2O}{\overset{H^+}{\rightleftarrows}}$ carboxylic + alcohol

Let us consider the reaction with base.

base: $R' - \overset{\overset{O}{\|}}{C} - OR \xrightarrow[H_2O]{OH^-} R' - \overset{\overset{O}{\|}}{C} - O^- + HOR$

With the basic reaction, you generate an alcohol and a carboxylate anion, not an acid. This is the key. If the base is KOH,

$$R - \overset{\overset{\displaystyle O}{\|}}{C} - O^-K^+$$ is produced. If NaOH you have

$$R - \overset{\overset{\displaystyle O}{\|}}{C} - O^-Na^+.$$ Due to the fact that the carboxylate anion, which forms a salt, is generated, the reaction is prevented from going in the opposite direction. As such, your yields upon hydrolysis are greater with the base than with acid.

● **PROBLEM** 775

Explain why aldehydes but not ketones can be oxidized to carboyxlic acids.

Solution: An examination of the structures of aldehydes, ketones, and carboxylic acids will reveal the solution to this problem. In answering this problem, keep in mind that a carbon atom can participate in only four bonds.

Aldehydes and ketones have the carbonyl functional group; i.e., $\overset{\diagdown}{\diagup}C = O$. The aldehyde has the general formula

$$R \overset{\overset{\displaystyle O}{\|}}{\underset{}{C}} H$$

where R is any carbon group. The ketone's formula is

where R and R' are different or the same carbon groups. A carboxylic acid has the formula

$$R - \overset{\overset{\displaystyle O}{\|}}{C} - OH,$$

where R is any carbon group. Now, let us look at the oxidation:

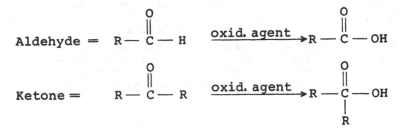

But, notice, in the oxidation from the ketone, the resulting compound has a carbon atom with 5 bonds, which is not permissible. Such a situation does not occur with oxidation in aldehydes. This explains why aldehydes, but not ketones, can be oxidized to carboxylic acid.

● **PROBLEM** 776

Write the reaction equation for the formation of the corresponding ether from methanol. How many ethers are possible when a mixture of $CH_3OH$ and $CH_3CH_2OH$ reacts with the acid $H_2SO_4$?

<u>Solution</u>: The formation of ethers from alcohols requires acid. In acid solution, the hydroxyl group of an alcohol is protonated to form the oxonium ion:

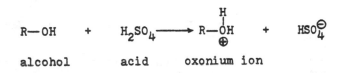

A free electron pair on the oxygen of another molecule of alcohol reacts with the oxonium ion to form an ether. This is accompanied by water loss:

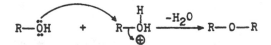

If you had a mixture of $CH_3OH$ and $CH_3CH_2OH$, three ethers would result. Upon the addition of acid, methanol molecules can react with other methanol molecules to form methyl ether. It can also react with ethanol ($CH_3CH_2OH$) to form methyl ethyl ether. The ethanol molecules can react with other ethanol molecules to form ethyl ether. (Ethanol can react with methanol, but this yields methyl ethyl ether.) Thus, there are 3 possible ethers: methyl ether, ethyl ether and methyl ethyl ether.

● **PROBLEM** 777

Complete the equations in Figure A. Name the reactants and the products.

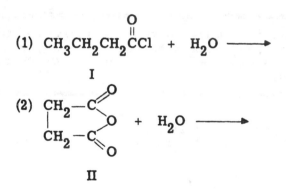

Figure A

Solution:  Compound I is an acid chloride and compound II is an anhydride.  Both of these substances are derivatives of carboxylic acids. Such species undergo nucleophilic substitution reactions. In these reactions, a stronger base displaces a weaker base. Thus, to complete these reactions, investigate whether $H_2O$ is a stronger base than the $Cl^-$ ion or the carboxylate anion of the acid halide and the anhydride, respectively. A stronger base will replace a weaker base in a compound. $Cl^-$ ion and carboxylate anion are weak bases, so that the following occurs

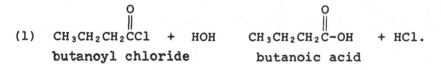

(1)    butanoyl chloride            butanoic acid

The stronger base $OH^-$ ion, replaced the weak $Cl^-$ ion base. Reacton 2 is shown in Figure B.

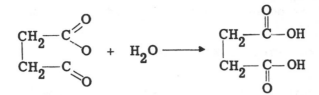

succinic anhydride          succinic acid

Figure B

Again, OH⁻ replaced the weaker carboxylate anion base.

● **PROBLEM** 778

Predict the products expected for the following reactions:
(1) concentrated $H_2SO_4$ on ethyl alcohol (2) citric acid
plus KOH.

Solution:   (1) Here, you have an acid, $H_2SO_4$, reacting with
the alcohol, ethanol. There exist two reactions that can
occur. Which ever one predominates depends on the conditions.
If the temperature is raised to 140°C, you have the formation
of ethyl ether, $C_2H_5OC_2H_5$. Let us investigate why. When
the $H_2SO_4$ is added to ethanol, it protonates the OH, i.e.,

$$C_2H_5OH \ + H_2SO_4 \longrightarrow C_2H_5\overset{H}{\underset{+}{O}H} + HSO_4^-$$

The protonated alcohol undergoes the following reaction:

$$C_2H_5\overset{H}{\underset{+}{O}H} \longrightarrow C_2H_5^+ \ + HOH.$$

A water molecule is expelled. In the solution you have
other ethanol molecules, thus, the following can occur;

$$C_2H_5\overset{..}{\underset{..}{O}}H \ + C_2H_5^+ \longrightarrow C_2H_5\overset{\overset{C_2H_5}{|}}{\underset{..+}{\overset{..}{O}}H}$$

The $C_2H_5^+$ needs electrons, which can be supplied by
the oxygen atom in another alcohol molecule. Once this
occurs, a proton is expelled to obtain the ether. The
proton reacts with the existing $HSO_4^-$, so that $H_2SO_4$ is
not diminished.  You have the following:

$$C_2H_5\overset{\overset{C_2H_5}{|}}{\underset{..+}{\overset{..}{O}}}H \longrightarrow C_2H_5OC_2H_5 + H^+$$

$$H^+ + HSO_4^- \longrightarrow H_2SO_4$$

If the temperature is raised above 160°C, a different set of reactions occur. Again, the acid protonates the

alcohol to form $C_2H_5\overset{\overset{H}{\cdot\cdot}}{O}H$ . Water departs to yield $C_2H_5^+$.

Let us write this as $H-\overset{\overset{\displaystyle H}{|}}{\underset{\underset{\displaystyle H}{|}}{C}}-\overset{\overset{\displaystyle H}{|}}{\underset{\displaystyle +}{C}}-H$. This time, however,

it does not react with another alcohol molecule but expels a proton immediately to form the alkene, ethylene or ethene.

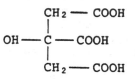

This is the final product, ethylene, at a temperature greater than 160°C.

(2) Here you have citric acid reacting with a base, KOH. This suggests the formation of a salt and water. To predict which salt, write out the structure of citric acid. The structure is,

$$OH-\overset{\overset{\displaystyle CH_2-COOH}{|}}{\underset{\underset{\displaystyle CH_2-COOH}{|}}{C}}-COOH$$

To understand how it reacts, you need to know the origin of its acidic nature. The acidic nature is determined by the H of the carboxylic group, COOH. Thus, you can anticipate the following reaction:

$$OH-\overset{\overset{\displaystyle CH_2-COOH}{|}}{\underset{\underset{\displaystyle CH_2-COOH}{|}}{C}}-COOH \quad + \quad 3KOH \quad \to \quad OH-\overset{\overset{\displaystyle CH_2-COO^-K^+}{|}}{\underset{\underset{\displaystyle CH_2-COO^-K^+}{|}}{C}}-COO^-K^+ \quad + \quad 3HOH$$

● **PROBLEM** 779

Indicate how you would prepare (1) acetone from acetic acid and (2) acetic acid from ethyl alcohol.

Solution:  This problem is concerned with synthesis of organic compounds. In such problems you want to write out the structures of the initial and final products involved, so that you can identify functional groups. From this, you employ reactions that change the functional groups until the desired product is obtained.

(1)   Acetone : $CH_3 — \overset{\overset{\textstyle O}{\|}}{C} — CH_3$; Acetic Acid : $CH_3 — \overset{\overset{\textstyle O}{\|}}{C} — OH$

If you had calcium acetate, $\left( CH_3 — \overset{\overset{\textstyle O}{\|}}{C} — O \right)_2 Ca$, you could obtain acetone by heating it, since calcium acetate forms calcium carbonate and acetone upon the addition of heat. To obtain calcium acetate, add $Ca(OH)_2$, which is a base. Because it is a base, the acid reacts with it to produce the calcium acetate, which is the salt, and water. Thus, the sequence of reactions becomes

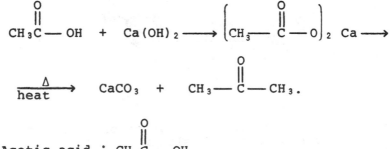

$$CH_3 \overset{\overset{\textstyle O}{\|}}{C} — OH \ + \ Ca(OH)_2 \longrightarrow \left( CH_3 — \overset{\overset{\textstyle O}{\|}}{C} — O \right)_2 Ca \longrightarrow$$

$$\xrightarrow[\text{heat}]{\Delta} \ CaCO_3 \ + \ CH_3 — \overset{\overset{\textstyle O}{\|}}{C} — CH_3 .$$

(2) Acetic acid : $CH_3 \overset{\overset{\textstyle O}{\|}}{C} — OH$

Ethyl alcohol : $CH_3CH_2OH$.  To prepare acetic acid from ethyl alcohol, you must know one important fact: Ethyl alcohol is a primary alcohol, i.e., one where the OH group is located on a carbon atom bonded to only one other carbon atom. Primary alcohols can be directly oxidized to the corresponding carboxylic acid by the addition of $KMnO_4$, potassium permanganate. Thus, you perform the following reaction:

$$CH_3CH_2OH \xrightarrow{\ KMnO_4\ } CH_3\overset{\overset{\textstyle O}{\|}}{C} — OH$$

ethyl alcohol                              acetic acid

● PROBLEM 780

Explain how you would prepare benzoic acid from benzene.
Use any inorganic reagents.

Benzene       Benzoic acid

Figure A

**Solution:** One of the most common ways of preparing benzoic acid from benzene is through an intermediate grignard. To plan this synthesis, first write your starting material and desired product. Then, think of inorganic reagents that lead you step by step through intermediates until you obtain the product.

A grignard reagent can be represented as R-MgX, where R is any aryl or alkyl group and X is a halogen. To obtain this grignard intermediate, you must halogenate the benzene ring. This can be accomplished by using a solution of halogen and $FeX_3$, where X is the halogen. Let the halogen be Br. Thus, you have the reaction in Figure B.

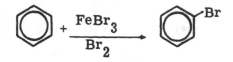

**Figure B**

This is electrophilic substitution of a hydrogen in the ring by a bromine atom. By electrophilic, you mean electron seeking. By addition of Mg, in dry ether, you obtain the grignard reagent. See figure C.

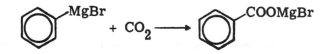

**Figure C**

Now that you have the grignard, add carbon dioxide. The reaction proceeds as shown in figure D.

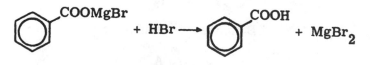

**Figure D**

With the addition of the acid, HBr, the reaction in figure E proceeds to the desired product benzoic acid.

**Figure E**

Indicate how you would perform the following syntheses:
(1) aniline from benzene, (2) ethyl amine from ethane.

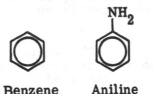

Benzene    Aniline

Figure A

Solution: To solve this problem you need to write out the structures of the initial and final compounds. In any synthesis, think of reactions that lead you step by step to the desired result.

If you had nitrobenzene, you could go directly to aniline because nitrobenzene can be reduced with metal and acid to aniline. To obtain nitrobenzene, perform a nitration reaction, as shown below in figure B.

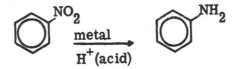

Benzene              Nitrobenzene

Figure B

(2) Ethyl amine ≡ $C_2H_5NH_2$, Ethane ≡ $C_2H_6$. Now, if you had ethyl chloride, $C_2H_5Cl$, you could obtain ethyl amine directly, since upon the addition of ammonia ($NH_3$) ethyl chloride is converted to the corresponding amine. To obtain ethyl chloride, halogenate ethane with $Cl_2$ and light.

$$C_2H_6 + Cl_2 \xrightarrow{\text{light}} C_2H_5Cl + HCl$$

$$C_2H_5Cl + NH_3 \rightarrow C_2H_5NH_2 + HCl.$$

Show the mechanism for the transesterification of methyl benzoate with ethanol in an acidic solution.

Solution: Begin this problem by defining transesterification. An ester is an organic compound

$$\text{of general formula } R-\overset{\overset{\displaystyle O}{\|}}{C}-OR' \text{, where R and R' are alkyl}$$

groups that can be the same or different. In transesterification, the alcoholic portion of the ester is exchanged for another alcoholic group. You are asked to give the mechanism for the reaction shown in figure A.

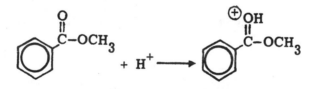

methyl benzoate    CH₃CH₂OH, H⁺ (ethanol)    ethyl benzoate + methanol

$$\text{methyl benzoate} \xrightarrow[\text{(ethanol)}]{CH_3CH_2OH, \; H^+} \text{ethyl benzoate} + CH_3OH$$

Figure A

You will note that you exchange the methyl alcohol portion with ethyl alcohol. From radioactive isotopes, the following mechanism shown in figure B has been established.

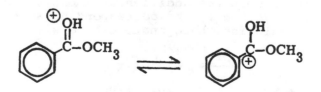

Figure B

Due to the electronegativity of oxygen, the oxygen of

the carbonyl group, $\overset{\diagdown}{\diagup}C=O$, has a slightly negative charge and the carbon has a slightly positive charge. The proton from the acid is attracted to the oxygen as in figure C. These two are in equilibrium with each other.

Figure C

The carbon atom is positively charged and seeks electrons. They can be obtained from the unpaired electrons on the oxygen atom in the ethanol molecule. Thus, the oxygen bonds to that carbon; you still have the positive charge, though.

A proton comes off of the oxygen from ethyl and is transferred to the oxygen of methyl. See figure D.

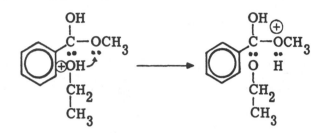

Figure D

Methanol (CH₃OH) leaves, allowing the compound shown in figure E to expel a proton, which leads to the formation of ethylbenzoate. The acid expelled acts as a catalyst for the transesterification of another compound in the solution.

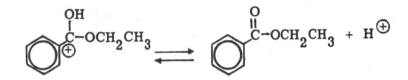

Figure E

● **PROBLEM** 783

Explain how you would prepare aspirin from phenol and acetic acid. You may use any other inorganic reagents.

Solution: The best or the most direct method of preparing aspirin from phenol is by use of the Kolbe reaction, which requires a knowledge of the chemistry of phenols. To plan this synthesis, first write your starting material, phenol, and end product, aspirin, and then think of reactions that direct you in a step by step manner to the desired end. These structures are shown in figure A.

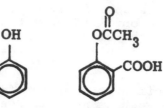

phenol        aspirin (Acetylsalicylic acid)

Figure A

As mentioned, you want to employ the Kolbe reaction; it produces the carboxylate salt shown in figure B from phenol, which begins to look like aspirin. The procedure is shown in figure C.

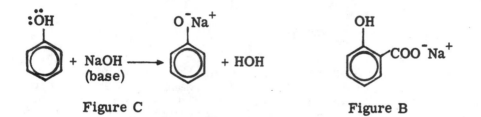

Figure C                                Figure B

The hydrogen of the OH group in phenol is acidic, due to the fact that the ring delocalizes the unpaired electrons of oxygen in resonance structures. Thus, when base is present, you have a neutralization reaction to produce sodium phenoxide,

$O^-Na^+$

a salt, and water. Now, that you have this material,

perform the Kolbe reaction. You have the reaction shown in figure D.

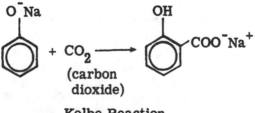

Kolbe Reaction

Figure D

If you acidify (by adding a mineral acid like HCl), you obtain the reaction in figure E. Salicylic acid is produced. You are told that acetic acid is available. The only difference between salicylic acid and aspirin is an ester linkage. Thus, if you perform an esterification reaction with acetic acid on the alcohol portion of salicylic acid, you obtain aspirin.

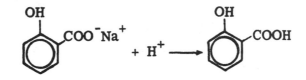

Figure E

By esterification you mean the reaction of a carboxylic acid (R-COOH) and an alcohol (R-OH) to produce an ester and water. Thus, you perform the reaction in figure F.

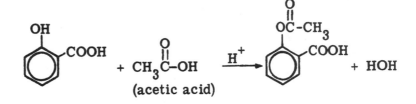

Figure F

The acid ($H^+$) is a catalyst, which speeds up the process.

# CHAPTER 22

# BIOCHEMISTRY I

> **Basic Attacks and Strategies for Solving Problems in this Chapter. See pages 755 to 788 for step-by-step solutions to problems.**

Biochemistry is that branch of chemistry which deals with the chemistry in living cells and organisms. Because living systems are organic, biochemistry is heavily dependent on organic chemistry. However, it is a very specialized part of organic chemistry, and the two should not be confused as being equivalent. Biochemistry, like organic chemistry, is a huge area with many very complicated reactions and mechanisms of importance to living systems. Only the simplest of the ideas can be introduced in this chapter.

Amino acids — molecules that contain both an amino ($NH_2$) group and a carboxylic acid ($CO_2H$) group — are the building blocks of proteins in living systems. These acids have two acid-base sites per molecule — one is the carboxylic acid hydrogen, the other the basic amino site. These sites can ionize independently or together according to the overall reaction.

$$\underset{H_2N-\underset{\underset{R}{|}}{C}H-\underset{\underset{O}{\|}}{C}-OH}{} \leftrightarrow \underset{{}^+H_3N-\underset{\underset{R}{|}}{C}H-\underset{\underset{O}{\|}}{C}-O}{} \qquad 22\text{-}1$$

where R group is an arbitrary organic group that will be different for different amino acids. The ion on the right side of Equation 22-1, with both a positive and negative charge, is called a zwitterion. There are two ionization constants for this ion — one for the ionization of the $CO_2H$ group and one for the $NH_3+$ $NH_2$ group. The pK$_1$ and pK$_2$ constants, shown in Problem 792 for several amino acids, are the negative logarithms of the ionization constant for these reactions. The pH that exists when there is no net charge on the ion is called the isoelectric pH and is written as pI.

The production of energy in biochemical systems is essential to cell growth and metabolism. The energy is typically associated with oxidation of carbohydrates, sugars, fats, or amino acid (protein) molecules. These reactions can be treated like any other reactions. Heats of reaction follow the same rules, and the solution of problems involving energy production for biochemical reactions needs no special techniques or procedures different from other reactions.

Many biochemical molecules are made up of repeating structures of smaller molecular units. For example, proteins are formed by the selective ordering of different amino acids. Because the molecules are huge, molecular weights are difficult to measure precisely. However, the molecular weight of the building block can be determined; it is often referred to as the minimum molecular weight and is typically several thousand grams/mole. The mole ratios in the structure represented by the minimum molecular weight or minimum-sized building block remain constant throughout the molecule.

Molecules like amino acids form isomers that are molecular mirror images of each other—as a right and left hand are mirror images. Such molecules exhibit optical activity (i.e., they rotate the plane of polarized light). If this rotation is clockwise, the isomer is called the dextro (abbreviated as d) isomer; if the rotation is counterclockwise, the isomer is call the levo (abbreviated as l) isomer. The amino acids that form proteins in living systems are all levo isomers.

Step-by-Step Solutions to
Problems in this Chapter,
"Biochemistry I"

## CELLULAR CONSTRUCTION AND DIMENSIONS

● **PROBLEM** 784

A chemist has an E. coli bacterium cell of cylindrical shape. It is $2\mu$ (microns) long and $1\mu$ in diameter and weighs $2 \times 10^{-12}$ g. (a) How many lipid molecules are present, assuming their average molecular weight to be 700 and lipid content to be 2%? (b) If the cell contains 15,000 ribosomes, what percent of the volume do they occupy? You may assume them to be spherical with a diameter of 180 $\overset{\circ}{A}$; also $10^4 \overset{\circ}{A}$ = 1 micron.

Solution: (a) Because you are told the molecular weight of an average lipid molecule, you need only compute their total weight in the E. coli cell; and then find the number of moles present. A mole is defined as weight in grams/molecular weight (MW). From this, you use the fact that in one mole of any substance, $6.02 \times 10^{23}$ molecules (Avogadro's number) exist.

The lipid content is given as 2%. If the total weight of the cell is $2 \times 10^{-12}$ g, the lipid molecules must have a total weight of $(.02)(2 \times 10^{-12})$ or $4 \times 10^{-14}$ grams. The average molecular weight of a lipid molecule is given as 700 g/mole. Thus, you have $4 \times 10^{-14}/700 =$ $5.71 \times 10^{-17}$ moles of lipid molecules. If $6.02 \times 10^{23}$ molecules are in a mole, then in $5.71 \times 10^{-17}$ moles, there are $(6.02 \times 10^{23})(5.71 \times 10^{-17}) = 3.44 \times 10^7$ lipid molecules.

(b) Calculate the volume of both the E. coli cell and the total volume occupied by the ribosomes.

Volume of the E. coli cell: It is of cylindrical shape. The volume of a cylinder can be found from the product of its length and the area of its base. The area of the base, a circle. is $\pi r^2$, where r = radius. Because the diamater = 2r,the radius of the cylindrical E. coli cell is 1μ divided by 2 or 0.5 μ. This is converted to $5 \times 10^3$ Å, by the conversion factor $10^4$ Å = 1 micron. Similarly, $2\mu = 2 \times 10^4$ Å = length of E. coli cell. Its total volume = (area of circle)(length) = [$\pi(5 \times 10^3)^2$ $2 \times 10^4$] = $1.57 \times 10^{12}$.

Volume of ribosomes: Each ribosome has a diameter of 180 Å or a radius of 90 Å. It is given that it is a sphere. The volume of a sphere is given by the formula $4/3 \ \pi r^3$, where r = radius. Substituting the volume of one ribosome = $4/3 \ \pi \ (90)^3 = 3.05 \times 10^6$. If there exists 15,000 ribosomes, their total volume must be $(15,000)(3.05 \times 10^6) = 4.58 \times 10^{10}$. Therefore, the percentage volume it occupies =

$$\frac{\text{total volume of ribosomes}}{\text{volume of E. coli cell}} \times 100 = \frac{4.58 \times 10^{10}}{1.57 \times 10^{12}} \times 100 = 2.9\%.$$

● **PROBLEM** 785

An E. coli culture will grow to a limiting concentration of about $10^9$ cell per $cm^3$. At this concentration what percentage of the total volume of the culture medium is occupied by the cells? Assume that an E. coli cell is cylindrical and that it is 2.0μ long and has a 1.0μ diameter.

Solution:  This percentage can be found by first determining the volume that one E. coli cell occupies and then the volume that $10^9$ cells occupy. This volume divided by 1 $cm^3$ multiplied by 100 equals the percent of the total volume that the cells occupy.

The volume of a cylinder is equal to its length multiplied by the area of its cross-section. Area of a circle = $\pi r^2$, where r = radius of the circle. 1μ = $10^{-4}$ cm

Volume of 1 E. coli cell = $(2.0 \times 10^{-4}$ cm)$(\pi)$

$$\times (0.5 \times 10^{-4} \text{ cm})^2$$

$$= 1.57 \times 10^{-12} \text{ cm}^3$$

There are $10^9$ cells per $cm^3$ of solution, therefore the volume that the cells occupy in 1 $cm^3$ of solution is $10^9 \times 1.57 \times 10^{-12}$ $cm^3$  or  $1.57 \times 10^{-3}$ $cm^3$.

Percent of total volume that the E. coli cells occupy $= \dfrac{1.57 \times 10^{-3} \text{ cm}^3}{1.0 \text{ cm}^3} \times 100 = .157\%.$

a) Calculate the ratio of the volume of a hepatocyte
to that of an E. coli cell. Assume the hepatocyte to be
a cube 20μ on an edge. b) Calculate the ratio of their
surface areas. c) Calculate the surface/volume ratios
for each.

Solution: A hepatocyte is an eucaryote, a large complex
cell, and an E. coli cell is a procaryote, a small
simple cell. An E. coli cell is cylindrical with a length
of 2.0μ and a cross-sectional diameter of 1.0 μ. The
volume of a cylinder is equal to the product of its length
and the area of its cross-section.

Volume of an E. coli cell = $2.0μ × π × (0.5μ)^2 = 1.57 μ^3$

Volume of a hepatocyte = $(20μ)^3 = 8.0 × 10^3 μ^3$

The volume ratio is

$$\frac{\text{hepatocyte}}{\text{E. coli}} = \frac{8.0 × 10^3 μ^3}{1.57 μ^3} = 5.1 × 10^3 : 1.$$

b) The surface area of a solid cylinder is equal to
$2πrℓ + 2πr^2$, where r is the cross-sectional radius and ℓ
is the length.

Surface area of an E. coli = $2π(0.5μ)(2.0μ) + 2π(0.5μ)^2$

$$= 6.28μ^2 + 1.57μ^2 = 7.85μ^2$$

The surface area of a cube is $6s^2$, where s is the
length of the edge.

Surface area of a hepatocyte = $6(20μ)^2 = 2.40 × 10^3 μ^2$

The ratio of their surface areas is

$$\frac{\text{hepatocyte}}{\text{E. coli}} = \frac{2.40 \times 10^3 \ \mu^2}{7.85 \ \mu^2} = 306 : 1.$$

c) Surface to volume ratios:

E. coli; $\dfrac{\text{surface}}{\text{volume}} = \dfrac{7.85\mu^2}{1.57\mu^3} = 5\mu^{-1}$

Hepatocyte; $\dfrac{\text{surface}}{\text{volume}} = \dfrac{2.40 \times 10^3\mu^2}{8.0 \times 10^3\mu^3} = 0.3\mu^{-1}.$

● **PROBLEM** 787

The hydrated $K^+$ and $Cl^-$ ions are approximately spherical in shape, with a diameter of 6.0 Å. What fraction of the total volume of the water phase of an E. coli cell is occupied by these ions?

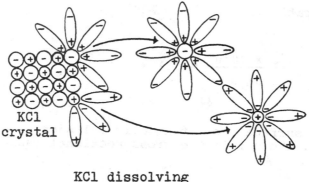

KCl crystal

KCl dissolving
in water.

Solution: A hydrated ion is an ion that is completely surrounded by water molecules.Water gathers around ions because of the strong electrostatic attraction between the water dipoles and the ions. The O atom of a water molecule is slightly negative and is attracted to the positive $K^+$ ion and the H atom of the water is slightly positive and is attracted to the negative $Cl^-$ ion.

The concentration of KCl in an E. coli cell is $150 \times 10^{-3}$ M. Because KCl dissociates into $K^+$ and $Cl^-$ ions, the ion concentration in the cell is $2(150 \times 10^{-3}$ M). If one knows the volume of the aqueous phase of an E.coli cell, one can determine the number of moles of the ions present. An E. coli is cylindrical; it is 2.0$\mu$ long and has a diameter of 1.0$\mu$. The volume of a cylinder is equal to $\ell\pi r^2$, where $\ell$ is the length of the cylinder and r is the radius of its cross-section. $1\mu = 10^{-4}$ cm.

Volume of a single E. coli cell = $(2.0 \times 10^{-4}$ cm$)(\pi)$

$$\times (0.5 \times 10^{-4}$ cm$)^2$$

$$= 1.57 \times 10^{-12} \text{ cm}^3$$

$$= 1.57 \times 10^{-15} \ \ell$$

80% of the E. coli cell is in the aqueous phase. and the $K^+$ and $Cl^-$ ions will only exist in this aqueous phase.

Volume of the aqueous phase = .80 $\times$ $(1.57 \times 10^{-15} \ \ell)$

$$= 1.26 \times 10^{-15} \ \ell$$

The number of moles of ions can now be found.

Moles of ions = $(2)(150 \times 10^{-3}$ moles/liter$)(1.26 \times 10^{-15} \ell)$

$$= 3.78 \times 10^{-16} \text{ moles}$$

There are $6.02 \times 10^{23}$ ions per mole.

No. of ions = $(3.78 \times 10^{-16}$ moles$)(6.02 \times 10^{23}$ ions/mole$)$
$$= 2.28 \times 10^8 \text{ ions.}$$

Each ion is taken to be spherical with a diameter of 6.0 Å. The volume of a sphere is equal to $4/3 \ \pi r^3$, where r is the radius. 1 Å = $10^{-8}$ cm.

Volume of 1 ion = $4/3 \ \pi (3.0 \times 10^{-8}$ cm$)^3$

$$= 1.131 \times 10^{-22} \text{ cm}^3$$

The total volume occupied by the ions is $2.28 \times 10^8$ ions times the volume of 1 ion.

Total volume of ions = $(2.28 \times 10^8$ ions$) \times$
$$(1.131 \times 10^{-22} \text{ cm}^3/\text{ion})$$
$$= 2.58 \times 10^{-14} \text{ cm}^3$$

The fraction of the total volume of the cell occupied by the ions is equal to the volume of the ions divided by the volume of the cell.

fraction of volume occupied by the ions = $\dfrac{2.58 \times 10^{-14} \text{ cm}^3}{1.57 \times 10^{-12} \text{ cm}^3}$

$$= 1.65 \times 10^{-2} \text{ or about 2%.}$$

## pH AND pKa

● PROBLEM 788

Calculate the $H^+$-ion concentrations of the following fluids: (a) blood plasma, (b) intracellular fluid of muscle, (c) gastric juice (pH = 1.4), (d) tomato juice, (e) grapefruit juice, (f) sea water. Use the accompanying table.

| pH of some fluids | |
|---|---|
| | pH |
| Seawater | 7.0–7.5 |
| Blood plasma | 7.4 |
| Interstitial fluid | 7.4 |
| Intracellular fluids | |
| Muscle | 6.1 |
| Liver | 6.9 |
| Gastric juice | 1.2–3.0 |
| Pancreatic juice | 7.8–8.0 |
| Saliva | 6.35–6.85 |
| Cow's milk | 6.6 |
| Urine | 5–8 |
| Tomato juice | 4.3 |
| Grapefruit juice | 3.2 |
| Soft drink (cola) | 2.8 |
| Lemon juice | 2.3 |

Solution: The pH is defined as $-\log [H^+]$. Therefore, $[H^+]$ can be found by calculating $10^{-pH}$.

(a) blood plasma - pH = 7.4

$7.4 = -\log [H^+]$; $[H^+] = 10^{-7.4} = 3.98 \times 10^{-8}$ M

(b) intracellular fluid - pH = 6.1

$6.1 = -\log [H^+]$; $[H^+] = 10^{-6.1} = 7.94 \times 10^{-7}$ M

(c) gastric juice - pH = 1.4

$1.4 = -\log [H^+]$; $[H^+] = 10^{-1.4} = 3.98 \times 10^{-2}$ M

(d) tomato juice - pH = 4.3

$4.3 = -\log [H^+]$; $[H^+] = 10^{-4.3} = 5.01 \times 10^{-5}$ M

(e) grapefruit juice - pH = 3.2

$3.2 = -\log [H^+]$; $[H^+] = 6.31 \times 10^{-4}$ M

(f) sea water - pH = 7.0

$7.0 = -\log [H^+]$; $[H^+] = 1.0 \times 10^{-7}$ M

● **PROBLEM** 789

An E. coli cell is 1.0µ in diameter and 2.0µ long and may be assumed to be cylindrical. It contains 80% water. If the intracellular pH is 6.4, calculate the number of $H^+$ ions in a single cell.

Solution: One can find the number of $H^+$ ions in one

E. coli cell from the dimensions of the cell and the pH, because pH = - log [H$^+$]. [H$^+$] is found in moles per liter and the volume of the cell can be calculated in liters. Solving for [H$^+$]:

6.4 = - log [H$^+$]

[H$^+$] = $10^{-6.4}$ = 3.98 × $10^{-7}$ moles/liter

To determine the number of    ions in the cell, the number of moles present must be determined first. This can be found from the volume. The volume of a cylinder is equal to $\ell \pi r^2$, where $\ell$ is the length and r is the radius of the cross-section. $1\mu$ = $10^{-4}$ cm.

Volume of the cell = (2 × $10^{-4}$ cm)($\pi$)(0.5 × $10^{-4}$ cm)$^2$

= 1.57 × $10^{-12}$ cm$^3$ = 1.57 × $10^{-15}$ $\ell$

If 80% of the cell is H$_2$O, 80% of the volume equals the volume of H$_2$O present. The H$^+$ ions can only exist in the aqueous phase of the cell.

Volume of H$_2$O = .80 × 1.57 × $10^{-15}$ $\ell$

= 1.26 × $10^{-15}$ $\ell$

To solve for the number of moles of H$^+$, multiply [H$^+$] by the volume of H$_2$O present.

No. of moles of H$^+$ = 1.26 × $10^{-15}$ $\ell$ × 3.98 × $10^{-7}$ moles/$\ell$

= 5.01 × $10^{-22}$ moles

There are 6.02 × $10^{23}$ ions per mole.

No. of H$^+$ ions = 6.02 × $10^{23}$ ions/mole × 5.01 × $10^{-22}$ mole

= 3.02 × $10^2$ ions = 302 H$^+$ ions.

● **PROBLEM** 790

Calculate the number of H$^+$ ions in the inner compartment of a single rat liver mitochondrion, assuming it is a cylinder 1.8$\mu$ long and 1.0$\mu$ in diameter at pH 6.0.

Solution: One can calculate the number of moles of H$^+$ ions per liter from the pH because pH = - log [H$^+$]. After determining the volume of the mitochondrion, one can solve for the number of H$^+$ ions present.

pH = - log [H$^+$]

6.0 = - log [H$^+$]

[H$^+$] = $10^{-6}$ M.

The volume of a cylinder is equal to its length times the area of its cross-section.

$$\ell \times \pi r^2,$$

where $\ell$ is the length of the cylinder and r is the radius of the cross-section. $1\mu = 10^{-4}$ cm.

Volume $= 1.8 \times 10^{-4}$ cm $\times \pi \times (0.5 \times 10^{-4}$ cm$)^2$

$$= 1.4 \times 10^{-12} \text{ cm}^3 = 1.4 \times 10^{-15} \ \ell$$

Solving for the number of moles of $H^+$ ion,

No. of moles $= 10^{-6}$ moles/liter $\times 1.4 \times 10^{-15} \ \ell$

$$= 1.4 \times 10^{-21} \text{ moles}$$

There are $6.02 \times 10^{23}$ ions per mole.

No. of $H^+$ ions $= 1.4 \times 10^{-21}$ moles $\times 6.02 \times 10^{23} \ \dfrac{\text{ions}}{\text{mole}}$

$$= 843 \ \ H^+ \text{ ions.}$$

● **PROBLEM** 791

At pH values of 4 to 9, natural amino acids exist as polar or zwitter ions: $H_2\overset{+}{N}CHZC\bar{O}_2$. At a pH of 12, what would be the predominant ionic type? What would it be at a pH of 2? Can the uncharged molecular form, $H_2NCHZCO_2H$ ever predominate at any pH?

<u>Solution</u>: To answer these questions, you need to consider the chemical composition of amino acids.

Amino acids are compounds whose molecules possess both the amino (- $NH_2$) and the carboxy (- $CO_2H$) functional groups. An amino group is basic - if it is an electron donor. The carboxylic group is acidic; it releases protons ($H^+$). In a zwitter ion, as indicated, the carboxylic group has donated its proton to become $COO^- \equiv C\bar{O}_2$ and $NH_2$ has received a proton to become $NH_3^+$. The presence of such an ion is dependent on pH, which is a measure of the $H^+$ or hydronium ion ($H_3O^+$) concentration. Thus, pH gives an indication of the acidity or basicity of a solution. The lower the pH, the more acidic the solution, and, as such, the greater the concentration of $H^+$ ions. The higher the pH, the more basic the solution, and, the lower the concentration of $H^+$ ions. One can now proceed as follows:

pH = 12. At this pH, which is high, the $H^+$ concentration is low; the solution is basic. Thus, the amino group ($NH_2$) will probably not be protonated. The carboxylic group, which is acidic, can and does react in the basic solution to go to COO⁻. Thus, at pH = 12, $H_2NCHZCO_2^-$ predominates. At pH = 2, the reverse is true. The solution is very acidic, so that $NH_3$ will be protonated to $NH_3^+$. $CO_2H$, the acid portion, has no base to react with. Thus, it does not ionize to $CO_2^-$. Thus, $H_2NCHZCO_2H$ can never predominate for the molecule contains both acid and base groups; they interact with each other to form zwitter ions.

● **PROBLEM** 792

Calculate the $pH_I$ values of glycine, alanine, serine and threonine from their $pK^1$ values.

The pK' values for the ionizing groups of some amino acids (25°C)

|  | $pK_1'$ α-COOH | $pK_2'$ α-NH$_3^+$ | $pK_R'$ R group |
|---|---|---|---|
| Glycine | 2.34 | 9.6 | |
| Alanine | 2.34 | 9.69 | |
| Leucine | 2.36 | 9.60 | |
| Serine | 2.21 | 9.15 | |
| Threonine | 2.63 | 10.43 | |
| Glutamine | 2.17 | 9.13 | |
| Aspartic acid | 2.09 | 9.82 | 3.86 |
| Glutamic acid | 2.19 | 9.67 | 4.25 |
| Histidine | 1.82 | 9.17 | 6.0 |
| Cysteine | 1.71 | 10.78 | 8.33 |
| Tyrosine | 2.20 | 9.11 | 10.07 |
| Lysine | 2.18 | 8.95 | 10.53 |
| Arginine | 2.17 | 9.04 | 12.48 |

Solution: $pH_I$ is defined as the isoelectric pH. At the isoelectric point, there is no net electrical charge on the molecule and the molecule will not move in an electric field. $pH_I$ can be determined using the following equation.

$$pH_I = \tfrac{1}{2} (pK_1' + pK_2')$$

One can now solve for the $pH_I$ values of the various amino acids mentioned.

glycine: $pH_I = \tfrac{1}{2}(2.34 + 9.6) = 5.97$

alanine: $pH_I = \tfrac{1}{2}(2.34 + 9.69) = 6.02$

serine: $pH_I = \tfrac{1}{2}(2.21 + 9.15) = 5.68$

threonine: $pH_I = \tfrac{1}{2}(2.63 + 10.43) = 6.53.$

Paper electrophoresis at pH 6.0 was carried out on a mixture of glycine, alanine, glutamic acid, lysine, arginine and serine. (a) Which compound moved toward the anode? (b) Which moved toward the cathode? (c) Which remained at the origin?

Solution: A method of separating amino acids is paper electrophoresis. A drop of the solution containing the amino acids is placed on a filter paper sheet, which is then moistened with a buffer of a given pH. The ends of the sheet dip into electrode vessels, and a high-voltage electrical field is applied while cooling. Because of their different $pK^1$ values, the amino acids migrate in different directions and at different rates, depending on the pH of the system and the emf (electromotive force) applied. To determine whether an amino acid will move towards the anode (positive end) or the cathode (negative end), one must determine the charge on the particular amino acid at pH = 6. This is done by calculating the isoelectric $pH_I$ (the pH, where the amino acid is electrically neutral) of each amino acid. $pH_I = \frac{1}{2}(pK_1' + pK_2')$. If $pH_I$ is less than pH = 6, then the amino acid will be negatively charged at pH = 6 and will move towards the cathode. When the $pH_I$ of an amino acid is near 6, the amino acid will not move.

glycine: $pH_I = \frac{1}{2}(2.34 + 9.6) = 5.97$. This is very near pH = 6, and glycine will remain in the center.

alanine: $pH_I = \frac{1}{2}(2.34 + 9.69) = 6.02$. This is also very close to 6 and alanine will not move from the center.

glutamic acid: As glutamic acid is titrated with base, the following species come about:

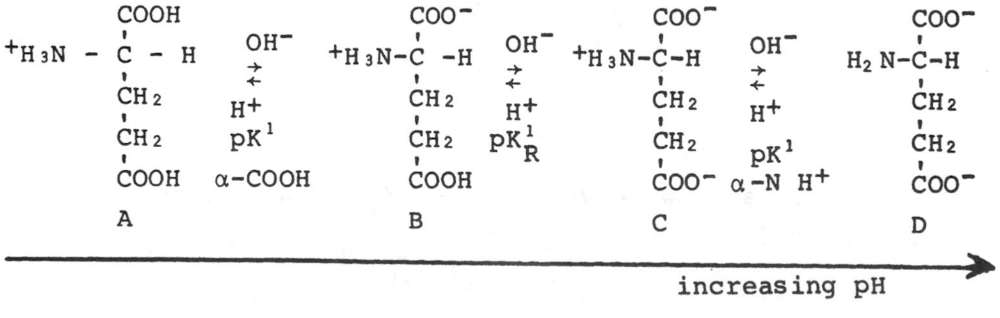

Species B is the neutral species, here. Therefore, $pH_I = \frac{1}{2}(pK^1 + pK_R^1) = \frac{1}{2}(2.19 + 4.25) = 3.22$. 3.22 is much less than 6 and therefore glutamic acid will be negatively charged at pH = 6 and will move towards the anode.

lysine: Using a method similar to the one used above for glutamic acid, one can determine that $pH_I = \frac{1}{2}(pK_2^1 + pK_R^1)$; $pH_I = \frac{1}{2}(8.95 + 10.53) = 9.74$. Lysine will be positively charged at pH = 6 and will be attracted towards the cathode. Arginine: $pH_I = \frac{1}{2}(pK_2' + pK_R') = \frac{1}{2}(9.04 + 12.48) = 10.76$. Arginine will be positively charged and will move towards the cathode.

serine: $pH_I = \frac{1}{2}(pK_1^1 + pK_2^1) = \frac{1}{2}(2.21 + 9.15) = 5.68$. 5.68 is close to 6.0 and serine will remain close to the center of the strip.

● **PROBLEM** 794

To 1.0 liter of a 1.0M solution of glycine at the isoelectric pH is added 0.3 mole of HCl. What will be the pH of the resultant solution? What would be the pH if 0.3 mole of NaOH were added instead?

Solution: When a species is held at its isoelectric pH, the molecules are all neutral. If there is 1.0 liter of 1.0M glycine present, there are 1.0 liter × 1.0 moles/liter =1.0 moles of glycine present. When 0.3 moles of HCl are added to this solution, 0.3 moles of the glycine become protonated and 0.7 moles are left unprotonated. The $K_a^1$ for this reaction is 2.34. One can solve for the resulting pH by using the Henderson-Hasselbach equation,

$$pH = pK_a^1 + \log \frac{[HA]}{[H_2A^+]}$$

$$pH = 2.34 + \log \frac{0.7}{0.3} = 2.34 + .37 = 2.71.$$

Similar logic is used in solving for the pH when 0.3 moles of NaOH are added. Here 0.3 moles of glycine will be negatively charged and 0.7 moles will remain neutral. The $pK^1$ for this reaction is 9.6 and the Henderson-Hasselbach equation becomes

$$pH = pK_a^1 + \log \frac{[A^-]}{[HA]}$$

$$pH = 9.6 + \log \frac{0.3}{0.7} = 9.23.$$

# ENERGY CONVERSION

● **PROBLEM** 795

It is known that energy from proteins is derived by breaking off the amino group as $NH_3$. How would you compare the energy derived from oxidation of a protein with energy obtained from oxidation of carbohydrates?

Solution:  The amount of energy derived from oxidation of an organic compound depends to a great extent on the number of bonds broken. You can compare the energy released upon oxidation of a carbohydrate and protein by noting how many bonds are broken in the process.

    You are told that the amino group comes off as $NH_3$. The general formula for amino acids, of which proteins are composed is

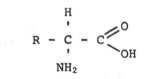

    When the amino group is broken off, only the C-N bond is broken. The bonds between the nitrogen and hydrogen are not broken when the proteins are oxidized. In carbohydrates, which are simple sugars or materials that can be hydrolyzed to simple sugars, all bonds are broken. Thus, because only one bond is broken in the protein's component, the energy released from a protein molecule is not as great.

● **PROBLEM** 796

A spherical cell of radius $10\mu$ and pH 7.0 originally contained ATP at 6.0 mM (millimolar) concentration. All the ATP was then hydrolyzed to ADP and $P_i$. If the free energy lost during the phosphate transfer is used to heat the water in the cell (assume that it is 100% $H_2O$), what is the increase in temperature?

Solution:  For the reaction,

    $ATP + HOH \rightarrow ADP + P_i$

at 37° and pH 7.0, $\Delta G' = - 7.30$ Kcal/mole. If one

determines the number of moles ATP present, one can calculate the number of Kcal produced in this reaction. Because 1 Kcal will raise the temperature of 1000 g of $H_2O$ $1°$, one must find the volume of the cell to determine the amount the temperature is raised due to this reaction.

The volume of a sphere is equal to $4/3 \ \pi r^3$, where r is the radius. $1\mu = 10^{-4}$ cm.

Volume of the cell = $4/3 \ \pi (10^{-3} \ cm)^3$

$$= 4.19 \times 10^{-9} \ cm^3 = 4.19 \times 10^{-12} \ \ell$$

The concentration of ATP in the cell is 6.0 mM. The number of moles present is found by multiplying the molarity by the volume.

No. of moles of ATP = $6.0 \times 10^{-3} \ \dfrac{moles}{liter} \times 4.19 \times 10^{-12} \ \ell$

$$= 2.51 \times 10^{-14} \ moles$$

7.30 Kcal of heat are produced per mole of ATP hydrolyzed.

Heat evolved = 7.30 Kcal/mole $\times$ $2.51 \times 10^{-14}$ mole

$$= 1.83 \times 10^{-13} \ Kcal = 1.83 \times 10^{-10} \ cal.$$

Each cal will raise 1 g of $H_2O$ 1 degree. There are $4.19 \times 10^{-9}$ g of $H_2O$ present (since the density of $H_2O$ is 1000 g/$\ell$ and its volume is $4.19 \times 10^{-12}$) and $1.83 \times 10^{-10}$ cal produced.

Rise in temp = $\dfrac{1°C \ g}{cal} \times 1.83 \times 10^{-10} \ cal \times \dfrac{1}{4.19 \times 10^{-9} \ g}$

$$= 4.37 \times 10^{-2} \ °C.$$

● PROBLEM 797

Skeletal muscle contains about $5 \times 10^{-6}$ mole of ATP and $30 \times 10^{-6}$ mole of phosphocreatine per gram wet weight. Calculate how much work (in calories) a 400-gram muscle can theoretically carry out at the expense of its high-energy phosphate bonds alone, assuming that both glycolysis and respiration are inhibited.

Solution: When they are hydrolyzed, the high-energy phosphate bonds of ATP and phosphocreatine contribute $7.3 \times 10^3$ and $10.3 \times 10^3$ cal/mole, respectively. With this information, one can calculate the amount of work that can be carried out by this muscle.

total work = work done by ATP + work done by phosphocreatine

The number of moles of each compound is found by multiplying the number of moles per gram by the weight of the muscle. This product is then multiplied by the number of cal/mole that are produced when the phosphate bonds of the particular compound are broken.

No. of cal produced by ATP = $5.0 \times 10^{-6}$ moles/g $\times$ 400 g

$$\times\ 7.3 \times 10^3 \text{ cal/mole}$$

$$= 14.6 \text{ cal.}$$

No. of cal produced
  by phosphocreatine = $30 \times 10^{-6}$ mole/g $\times$ 400 g

$$\times\ 10.3 \times 10^3 \text{ cal/g}$$

$$= 123.6.$$

Total no. of cal produced = 14.6 + 123.6 = 138.2 cal.

● **PROBLEM** 798

The hump on a camel's back is mostly depot fat. Assuming complete hydrolysis of the fat, and its subsequent oxidation to $CO_2$ and water, calculate the number of kilocalories that would be released from 25 kg of camel fat.

Solution: Depot fat is usually made up of a mixture of three triglycerides - triolein, tristearin and tripalmitin. These fats are produced by a reaction of three moles of their respective fatty acids (oleic, stearic or palmitic) and one mole of glycerol. The structure of a fatty acid is R - $\overset{\overset{O}{\|}}{C}$ - OH, where R is a long chain saturated or unsaturated alkyl group. The equation for this reaction is

It has been found that 9 Kcal of energy is released by each gram of fat hydrolyzed and then oxidized. In this problem, one is asked to find the amount of energy released when 25 kg of fat is oxidized. 25 kg is equal to 25000 g. Therefore, the number of kilocalories produced in this reaction is

$25000 \text{ g} \times 9 \text{ Kcal/g} = 2.25 \times 10^5 \text{ Kcal.}$

● **PROBLEM** 799

A chemistry student wrote on an exam that "Complete oxidation of a mole of glucose releases more energy than oxidation of a mole of lactose." Why did this student not receive full credit for his answer?

Solution: Oxidation is the addition of oxygen to organic compounds to produce water and carbon dioxide plus heat. The more carbon present in the molecule, the more energy released. Also, more energy will be released if the percentage of oxygen in the compound is low. Consider the structures of glucose and lactose, which are carbohydrates or simple sugars. Hydrolysis is the reaction of a compound with dilute acid or base and water to break down into its simple components.

Notice that if you hydrolyze lactose by adding dilute acid, you can cleave the disaccharide bond to form a glucose molecule and a galactose molecule. Now, if the lactose molecule already contains a glucose molecule plus galactose, it would yield more energy upon oxidation. Thus, this student's statement is false.

● **PROBLEM** 800

In the Citric Acid Cycle (also known as the Krebs Cycle), citrate is converted to isocitrate. From the algebraic sign of $\Delta G^\circ$ for the isomerization, what is the favored or spontaneous direction of the reaction? Calculate $\Delta G^\circ$ for the following reaction and explain what bearing this reaction will have on the isomerization of citrate and the operation of the Krebs Cycle:

$$\text{Isocitrate}^{3-} + \tfrac{1}{2}O_2(g) + H^f \rightleftarrows \alpha\text{-ketoglutarate}^{2-} + H_2O(\ell) + CO_2(g)$$

$\Delta G_f$ in Kcal/mole are for citrate$^{3-}$ = $-279.24$; isocitrate$^{3-}$ = $-277.65$; $H^f = 0$, $O_2 = 0$; $\alpha$ - ketoglutarate$^{2-}$ = $-190.62$; $H_2O(\ell)$ = $-56.69$; $CO_2(g) = -94.26$.

Solution: The Krebs Cycle is a series of oxidations aided by enzymes that leads to citric acid and eventually to $CO_2$, $H_2O$ and ATP.

To find $\Delta G$ of the isomerization, set up the equation

$$\text{citrate}^{3-} \rightleftarrows \text{isocitrate}^{3-}$$

and use the Gibbs free energy of formation values given.

$$\Delta G^\circ_{\text{reaction}} = \sum G^\circ_{f,\text{ products}} - \sum G^\circ_{f,\text{ reactants}}$$

$$= (-277.65 \text{ Kcal/mole}) - (-279.24 \text{ Kcal/mole})$$

$$= +1.59 \text{ Kcal/mole}$$

Thus, isomerization of citrate to isocitrate is not a spontaneous process, since $\Delta G^\circ$ for the reaction is positive.

To find $\Delta G$ of the following reaction:

$$\text{Isocitrate}^{3-} + \tfrac{1}{2}O_2(g) + H^f \rightleftharpoons \alpha\text{-ketoglutarate}^{2-} + H_2O(\ell) + CO_2(g)$$

use $\Delta G^\circ_{\text{reaction}} = \sum G^\circ_{f,\text{ products}} - \sum G^\circ_{f,\text{ reactants}}$

$$= \left( \Delta G^\circ_{f,CO_2} + \Delta G^\circ_{f,H_2O} + \Delta G^\circ_{f,\alpha\text{-ketoglutarate}} \right)$$

$$- \left( \Delta G^\circ_{f,\ H}{}^f + \Delta G^\circ_{f,\frac{1}{2}O_2} + \Delta G^\circ_{f,\text{ Isocitrate}} \right)$$

$$= (-94.26 - 56.69 - 190.62) - (0 + 0 - 277.65)$$

$$= -63.92 \text{ Kcal/mole.}$$

$\Delta G^\circ$ for the oxidation and decarboxylation of isocitrate is $-63.92$ Kcal/mole; this large free energy decrease will favor the constant removal of isocitrate and permit the cycle to move forward despite the non-spontaneity of the isomerization.

# MOLECULAR WEIGHTS, MOLE RATIOS, DENSITIES AND MONOMER UNITS

● **PROBLEM** 801

Ribonuclease is 1.65% leucine and 2.48% isoleucine by weight. Calculate its minimum molecular weight.

Solution: If Ribonuclease is 1.65% by weight leucine, .0165 times the minimum molecular weight of ribonuclease is equal to the molecular weight of leucine. (MW of leucine = 130)

MW of Leu = MW of Ribonuclease × .0165

$$\text{MW of Ribonuclease} = \frac{130}{.0165} = 7.89 \times 10^3$$

If Ribonuclease is 2.48% by weight isoleucine (MW = 131), one can solve for the molecular weight of ribonuclease using a similar equation as the one used for leucine.

$$\text{MW of Ribonuclease} = \frac{\text{MW of Ile}}{.0248} = \frac{131}{.0248} = 5.29 \times 10^3$$

Therefore, to cover both conditions, the minimum molecular weight must be $7.89 \times 10^3$.

● **PROBLEM** 802

A solution contains 1 mg per ml of myosin and $10^{14}$ latex particles per ml. When a given volume of this solution is dried on a grid and viewed under the electron microscope, a typical field contains 122 protein molecules and 10 latex particles. Calculate the molecular weight of myosin.

Solution: If 1 ml of solution contains $10^{14}$ latex particles and 10 latex particles are present in a certain volume, the following ratio can be set up to solve for this unknown volume.

$$\frac{10^{14} \text{ particles}}{1 \text{ ml}} = \frac{10 \text{ particles}}{x \text{ ml}}$$

$$x \text{ ml} = \frac{10 \text{ particles} \times 1 \text{ ml}}{10^{14} \text{ particles}} = 10^{-13} \text{ ml}$$

There is 1 mg of protein in ml of solution. Thus in $10^{-13}$ ml, there is $10^{-13}$ mg of protein. It is given that in $10^{-13}$ mg of protein there are 122 molecules.

It is known that there are $6.02 \times 10^{23}$ molecules per mole (Avogrado's Number), so that the molecular weight of myosin can be found by the following proportion:

$$\frac{10^{-16} \text{ g}}{122 \text{ molecules}} = \frac{MW}{6.02 \times 10^{23} \text{ molecules/mole}}$$

$$MW = \frac{10^{-16} \text{ g} \times 6.02 \times 10^{23} \text{ molecules/mole}}{122 \text{ molecules}}$$

$$= 4.93 \times 10^5 \text{ g/mole}.$$

● **PROBLEM** 803

Cytochrome C is a molecule with an iron-porphyrin head connected to a protein tail. Analysis shows the molecule is 0.45 percent iron by weight. Calculate the molecular weight of cytochrome C.

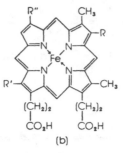

A cytochrome.
R' and R" are
different alkyl
groups.

Solution: From the accompanying figure, one can see that
there is one iron atom in each cytochrome molecule. Thus,
there is 1 mole of Fe atoms in 1 mole of cytochrome C.
If each molecule is 0.45 percent iron by weight, then in
1 mole of cytochrome C, 0.45 percent of the total molecular weight of cytochrome C is equal to the weight of
one mole of Fe. MW of Fe =

$$(4.5 \times 10^{-3})(\text{MW of cytochrome C}) = 55.8 \text{ g/mole}$$

$$\text{MW of cytochrome C} = \frac{55.8 \text{ g/mole}}{4.5 \times 10^{-3}} = 12,400 \text{ g/mole}.$$

● **PROBLEM** 804

The hard shell of crustaceans (lobsters, etc.) and insects
(roaches, etc.) is a polysaccharide called chitin. On
enzymatic hydrolysis of chitin, N-acetylglucosamine is
obtained. This molecule resembles glucose except that at

$$\text{C-2 a } - \overset{\overset{\displaystyle O}{\|}}{N} - C - CH_3 \text{ is attached instead of } - OH.$$ (a)
Write an open chain formula for N-acetylglucosamine.
(b) The structure of chitin is analogous to that of
cellulose. Draw a formula containing two joined N-acetylglucosamine units. (c) If the molecular weight of chitin
is 150,000, how many units are in the polymer?

Solution: A polysaccharide is a chain made up of many
simple sugars. Chitin is a long unbranched chain of many
molecules of the sugar, N-acetylglucosamine.

(a) One is given that the only difference between

glucose and N-acetylglucosamine is that a $- N - \overset{\overset{\displaystyle O}{\|}}{C} - CH_3$
group replaces the - OH on C-2. The open-chain formula
of N-acetylglucosamine can therefore be written as shown
in Figure A.

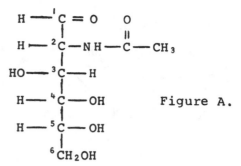

Figure A.

(b)  From the open chain formula, one can see that the ring structure can be written as shown in Figure B for N-acetylglucosamine.

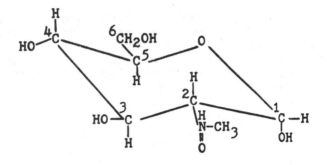

**Figure B**

For the ring structure to form, the O in the - OH on C-5 joins C-1.

From the figure of cellulose, Figure C, one can see that the structure of two N-acetylglucosamine joined together can be written as shown in Figure D.

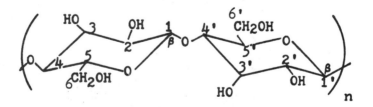

**cellulose: the repeating cellobiose unit**

**Figure C**

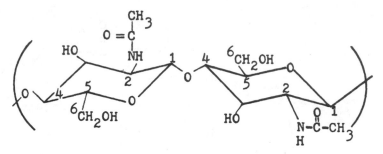

## Figure D

Because the O in the - OH on C-1 joins to a C-4 atom on the adjacent molecule and the ring structures are called glycosides, this is termed a 1,4-glycosidic linkage.

(c)  To determine the number of units in one strand of chitin, one can divide the total weight of the molecule by the weight of one glycoside. From the ring structure of a glycoside joined to another glycoside, one sees that there are 8C's, 5 O's, 13 H's and 1 N. The weight of one mole of this glycoside is $(8 \times 12) + (5 \times 16) + (13 \times 1) + (14) = 203$ g/mole.

$$\text{No. of units} = \frac{150,000 \text{ g}}{203 \text{ g/mole(unit)}} = 739 \text{ units}.$$

● **PROBLEM** 805

Calculate the density of the tightly coiled tropocollagen molecule, which may be considered to be a cylinder 2,800 Å long and 14 Å in diameter. It contains three polypeptide chains of 1,000 amino acid residues each.

## Figure A

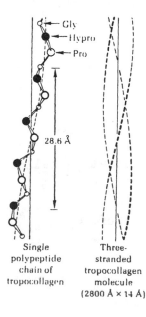

← Gly
← Hypro
← Pro

28.6 Å

Single polypeptide chain of tropocollagen

Three-stranded tropocollagen molecule (2800 Å × 14 Å)

*Conformation of polypeptide chains in triple-stranded tropocollagen molecule. Each chain is a coil with many repeating sequences of Gly-Pro-Hypro.*

Solution: Tropocollagen is a triple helix made up of 3 polypeptide strands as shown in the figure. It is found in tendons. Density is defined as mass divided by volume and is usually in the form of $g/cm^3$. Because the tropocollagen is considered to be a cylinder, its volume is equal to the height times the area of the cross-section.

The diameter of the cross-section is 14 $\overset{o}{A}$ or $1.4 \times 10^{-7}$ cm. Its radius is $7.0 \times 10^{-8}$ cm and its area ($\pi r^2$, where r = radius) equals $(7.0 \times 10^{-8}$ cm$)^2$ $\pi$ or $1.54 \times 10^{-14}$ cm$^2$. Solving for its volume:

volume = $1.54 \times 10^{-14}$ cm$^2$ × $2.800 \times 10^{-5}$ cm (height)

= $4.31 \times 10^{-19}$ cm$^3$

The mass of the tropocollagen is found by first determining the number of moles of amino acid residues present. The average molecular weight of amino acid residues is 120. Therefore, the mass of the collagen is equal to 120 g/mole × no. of moles of amino acid residues. Collagen is made up of 3 polypeptide strands each containing 1000 amino acid residues. Therefore 3000 amino acid residues are present. 3000 amino acid residues is equal to

$$\frac{3000 \text{ residues}}{6.02 \times 10^{23} \text{ residues/mole}} = 4.98 \times 10^{-21} \text{ moles}$$

The mass of the collagen molecule is, then,

mass = $4.98 \times 10^{-21}$ moles × 120 g/mole = $5.98 \times 10^{-19}$ g

The density is equal to mass/volume.

$$\text{density} = \frac{5.98 \times 10^{-19}}{4.31 \times 10^{-19} \text{ cm}^3} = 1.39 \text{ g/ cm}^3.$$

● **PROBLEM** 806

If a membrane contains 40% by weight of lipids and 60% protein, calculate the molar ratio of lipids to proteins. Assume that the lipid molecules have an average molecular weight of 800 and the proteins an average molecular weight of 50,000.

Solution: Assume that one has 100 g of membrane. From the information given, 40 g of this will be composed of lipids and 60 g will be protein. Determine the number of moles present in 40 g of lipids and 60 g of protein. The molar ratio will be equal to the number of moles of lipids divided by the number of moles of protein.

$$\text{number of moles} = \frac{\text{number of grams}}{\text{molecular weight}}$$

$$\text{no. of moles of lipids} = \frac{40 \text{ g}}{800 \text{ g/mole}} = 0.05 \text{ moles}$$

$$\text{no. of moles of protein} = \frac{60 \text{ g}}{50,000 \text{ g/mole}}$$

$$= 1.2 \times 10^{-3} \text{ moles}$$

$$\text{molar ratio} = \frac{\text{lipids}}{\text{protein}} = \frac{0.05 \text{ moles}}{1.2 \times 10^{-3} \text{ moles}} = 41.7.$$

● **PROBLEM** 807

Most membranes of animal cells contain about 60% protein and 40% phosphoglycerides. (a) Calculate the average density of a membrane, assuming that protein has a density of 1.2 g/cm³ and phosphoglyceride a density of 0.92 g/cm³. (b) If a sample of membrane material were centrifuged in NaCl solution of 1.05 specific gravity, would it sediment or float?

Solution: The density of any substance is its mass divided by its volume. It is usually written in terms of g/cm³.

(a) The average density of the membrane is made up of 60% protein and 40% phosphoglycerides. From the data given, one knows that 1 cm³ of protein weighs 1.2 g and that 1 cm³ of phosphoglyceride weighs 0.92 g. One cm³ of membrane is made up of .60 × 1.2 g + .40 × 0.92 g or 1.09 g. The average density of the membrane is 1.09 g/cm³.

(b) If a compound is placed in a solution, the compound will sink if its density is greater than that of the solution. The density of the NaCl solution is 1.05 and that of the membrane is 1.09, therefore the membrane will sediment.

● **PROBLEM** 808

A sample of polymer contains 0.50 mole fraction with molecular weight 100,000 and 0.50 mole fraction with molecular weight 200,000. Calculate (a) the number average molecular weight, $M_n$ and (b) the weight average molecular weight, $M_w$.

Solution: Not all particles (or molecules) in polymers have the same weight. Therefore, one seeks some average value and information regarding the distribution of the individual values.

(a) The ordinary (unweighted) arithmetic mean is the number average; each item is counted once.

The number average molecular weight is equal to the weight of the whole sample divided by the number of molecules in it.

$$M_n = \frac{\sum\limits_{i} n_i M_i}{\sum\limits_{i} n_i} = \sum\limits_{i} X_i M_i$$

Here, $n_i$ is the number of molecules of molecular weight $M_i$ per gram of dry polymer and $X_i$ is the mole fraction of each component.

Solving for $M_n$:

$$M_n = 0.5(100,000) + 0.5(200,000)$$

$$= 50,000 + 100,000 = 150,000 \text{ g/mole.}$$

(b) In the weight-average $M_w$, each item counts not as a single unit but in proportion to its weight. The molecular weight $M_i$ is multiplied by the weight $n_i M_i$ of material of that molecular weight rather than by the number of molecules. The weight average molecular weight $M_w$ is defined

$$M_w = \frac{\sum\limits_{i} n_i M_i^2}{\sum n_i M_i}$$

Because $\dfrac{\sum\limits_{i} n_i M_i}{\sum\limits_{i} n_i} = \sum\limits_{i} X_i M_i; \quad M_w = \dfrac{\sum\limits_{i} X_i M_i^2}{\sum\limits_{i} X_i M_i}$

Solving for $M_w$:

$$M_w = \frac{0.5(100,000)^2 + 0.5(200,000)^2}{0.5(100,000) + 0.5(200,000)}$$

$$= \frac{(1.0 \times 10^5)^2 + (2.0 \times 10^5)^2}{(1.0 \times 10^5) + (2.0 \times 10^5)} = \frac{5.0 \times 10^{10}}{3.0 \times 10^5}$$

$$= 1.67 \times 10^5 \text{ g/mole.}$$

● **PROBLEM** 809

For a condensation polymerization of a hydroxyacid with a residue weight of 200, it is found that 99% of the acid groups are used up.  Calculate a) the number average molecular weight and b) the weight average molecular weights.

Solution:  a) The number average molecular weight is the average weight of all particles present, where each unit is given equal weight. Thus, the number average molecular weight is equal to the individual residue weight times the average number of residues. This number can be calculated from the fact that 99% of the acid groups have reacted to form the polymer chain. When monomers join to form a polymer, the last monomer is left unreacted at the end of the chain. If 99% of the acid reacts then 1% is left unreacted, so that 1% of the average number of monomers is equal to 1. Let the number average of the monomers equal $\bar{X}_n$. Then,

$$.01 \; \bar{X}_n = 1$$

$$\bar{X}_n = 100$$

One can now solve for the number-average molecular weight, $M_n$.

number-average mol.wt = mol.wt of a monomer $\times \bar{X}_n$

$$= 200 \; g/mole \times 100$$

$$= 2.0 \times 10^4 \; g/mole.$$

b) The weight average molecular weight, $M_w$, weights molecules proportionally to their molecular weight in the averaging process. It is found from some complex derivations that

$$M_w = M_n(1 + p), \quad \text{where p is the extent of the re-}$$

action. In this problem, p is .99. From this one can solve for $M_w$

$$M_w = (2.0 \times 10^4 \; g/mole)(1 + 0.99)$$

$$= 3.98 \times 10^4 \; g/mole.$$

● **PROBLEM** 810

For a condensation polymerization of a hydroxyacid in which 99% of the acid groups are used up, calculate (a) the average number of monomer units in the polymer molecules, (b) the probability that a given molecule will have the number of residues given by this value, and (c) the weight fraction having this particular number of monomer units.

Solution: (a) The average number of monomer units is equal to

$$\bar{X}_n = \frac{1}{1 - p} \; ,$$

where $\bar{X}_n$ is the average number of units and p is the extent of the reaction.

$$\bar{X}_n = \frac{1}{1 - 0.99} = 100$$

(b) To determine the number of molecules that have a particular number of monomers examine the number of molecules of various molecular weights. To obtain the distribution of molecular weights, consider the possibilities $\pi_i$ of finding molecules of various degrees of polymerization (those containing i number of monomers) in a partially polymerized sample. The probability $\pi_1$ of finding an unreacted A (or B) group in the mixture is 1 - p. The probability $\pi_2$ that an AB has reacted with another AB to form ABAB is p(1 - p), since the probability of two independent events is the product of the two independent probabilities for having a bond p and not having a bond 1 - p. The probability $\pi_3$, ABABAB, that is, two successive bonds, is $p^2(1 - p)$. To generalize, consider

$$\pi_1 = 1 - p$$

$$\pi_2 = p(1 - p)$$

$$\pi_3 = p^2(1 - p)$$

$$\pi_i = p^{i-1}(1 - p)$$

Here, p = 0.99 and i = 100 as shown in the previous section.

$$\pi_{100} = 0.99^{100-1}(1 - 0.99).$$

$$= 0.99^{99}(0.01) = 3.7 \times 10^{-3}$$

(c) The weight fraction, $W_i$, of polymers containing this particular number of monomer units in an i-mer (a polymer containing i units) is equal to

$$W_i = i\, p^{i-1}(1 - p)^2$$

$$W_i = 100(0.99^{99})(1 - 0.99)^2 = 3.70 \times 10^{-3} \text{ monomers.}$$

# PHYSICAL ASPECTS OF BIOCHEMISTRY

● **PROBLEM** 811

Calculate the percentage of the total wet weight of a hepatocyte that is contributed by the cell membrane. Assume the hepatocyte to be a cube 20μ on an edge and that the membrane is 90 Å thick and that its specific gravity, as well as that of the entire cell, is 1.0.

Solution: Because the specific gravity is the same for the membrane and the rest of the cell, the weight percentage is equal to the volume percentage. If the hepatocyte is assumed to be a cube, its volume is equal to $s^3$, where s is length of the side. The membrane is composed of six rectangular solids. Each solid is 90 Å by 20μ × 20μ and one is found on each face of the cube. The volume of the membrane is equal to the length times the product of the height and width times 6. The percentage of the total weight of the cell that is taken up by the membrane is the quotient of the volume of the membrane divided by the total volume of the cell multiplied by 100. Since 1 Å = $10^{-4}$ μ

Volume of cell = $(20μ)^3$ = $8.0 \times 10^3$ $μ^3$

Volume of the membrane = $(9.0 \times 10^{-3}μ)(20μ)(20μ) \times 6$

$$= 21.6 \ μ^3$$

Percent of weight taken up by the membrane =

$$= \frac{21.6 \ μ^3}{8.0 \times 10^3 \ μ^3} \times 100 = .27\%.$$

● **PROBLEM** 812

Calculate the molecular weight of a pure isoelectric protein if a 1% solution gives an osmotic pressure of 46 mm of $H_2O$ at 0°C. Assume that it yields an ideal solution.

Solution: The relationship between molecular weight and osmotic pressure is

$$M = \frac{c}{\pi} RT,$$

where M is the molecular weight, c the concentration in grams per liter, R the gas constant (0.082 liter atm $deg^{-1}$ $mole^{-2}$), T the absolute temperature, and $\pi$ the osmotic pressure in atmospheres. One is given these quantities in the problem, but they are expressed in different units than those needed to solve for molecular weight. Converting these quantities:

Concentration: The solution is 1% by weight protein. One liter of water weighs 1000 g; 1% of 1000 is 10 so that c = 10.

Osmotic pressure: The osmotic pressure is 46 mm $H_2O$. To convert to atmospheres, 46 mm $H_2O$ must be converted to mm Hg.

$$mm\ Hg = 46\ mm\ H_2O \times \frac{.9970\ density\ of\ H_2O}{13.534\ density\ of\ Hg} = 3.39.$$

There are 760 mm Hg in 1 atm. Therefore, the number of atmospheres is equal to

$$\frac{3.39\ mm\ Hg}{760\ mm\ Hg/atm} = 4.45 \times 10^{-3}\ atm.$$

Temperature: T = 0°C + 273 = 273°K.

Solving for M:

$$M = \frac{10\ g/liter}{4.45 \times 10^{-3}\ atm} \times .082\ \frac{liter\text{-}atm}{°K\text{-}mole} \times 273°K = 50,300\ g/mole$$

● **PROBLEM** 813

The protein human plasma, albumin, has a molecular weight of 69,000. Calculate the osmotic pressure of a solution of this protein containing 2 g per 100 $cm^3$ at 25°C in (a) Torr and (b) millimeters of water. The experiment is carried out using a salt solution for solvent and a membrane permeable to salt.

Solution: When a semipermeable membrane separates a solution of a protein in water from pure water, osmosis will occur. The concentration - or the thermodynamic activity - of water molecules in the protein solution is less than in pure water, and the system will compensate for this difference by net movement of the water from the pure water compartment into that containing the protein solution, until the concentration of the water is the same on both sides of the membrane. Osmotic pressure is the force that must be applied to just prevent such osmotic flow. Osmotic pressure can be determined by using the following equation.

$$\pi = \frac{c\ R\ T}{M},$$

where $\pi$ is the osmotic pressure, c is the concentration in g/liter, R is the gas constant (0.82 liter atm/mole °K), T is the absolute temperature and M is the molecular weight. Solving for $\pi$.

$$\pi = \frac{(20 \text{ g/liter})(.082 \text{ liter atm/mole °K})(298°K)}{69,000 \text{ g/mole}}$$

$$= 7.08 \times 10^{-3} \text{ atm.}$$

Note: $2g/100 \text{ cm}^3$ corresponds to 20 g/liter.

(a) There are 760 Torr in 1 atm.

$\pi$ in Torr = $7.08 \times 10^{-3}$ atm $\times \frac{760 \text{ Torr}}{\text{atm}}$ = 5.38 Torr

(b) 5.38 Torr is equivalent to 5.38 mmHg. This can be converted to mm $H_2O$ by multiplying 5.38 mmHg by the density of Hg divided by the density of $H_2O$, i.e.,

5.38 mmHg $\times \frac{13.534 \text{ g/ml}}{0.997 \text{ g/ml}}$ = 73.0 mm $H_2O$.

● **PROBLEM** 814

The specific rotation $\left( [\alpha]_D^{25°} \right)$ of L-alanine is + 1.8. Calculate the observed rotation of a 1.10 M solution in a 2.5-dm polarimeter tube at 25°C.

Solution: All amino acids with the exception of glycine show optical activity, i.e., they can rotate the plane of plane-polarized light when examined in a polarimeter. Optical activity is shown by compounds capable of existing in two forms, whose structures are nonsuperimposable mirror images of each other. Optical activity is expressed quantitatively as the specific rotation $[\alpha]_D^{25°}$:

$$[\alpha]_D^{25°} = \frac{\text{observed rotation} \times 100}{\text{optical path length (dm)} \times \text{concentration (g/100 ml)}}$$

With this information, one can find the observed rotation of L-alanine after converting the concentration from moles/liter to g/100 ml. The molecular weight of alanine is 89.

$$\frac{1.10 \text{ moles}}{1 \text{ liter } (= 1000 \text{ ml})} \times 89 \text{ g/mole} = \frac{97.9 \text{ g}}{1000 \text{ ml}} = \frac{9.79 \text{ g}}{100 \text{ ml}}$$

Solving for the observed rotation:

observed rotation =

$$\frac{[\alpha]_D^{25°} \times \text{ optical path length (dm)} \times \text{concentration (g/ml)}}{100}$$

$$\text{observed rotation} = \frac{+1.8 \times 2.5 \text{ dm} \times 9.79}{100} = 0.44°.$$

● **PROBLEM** 815

A student wanted to produce a sample of lactic acid. He carried out the following synthesis: $CH_3CH_2CO_2H$ → $CH_3CHClCO_2H$ → $CH_3CHOHCO_2H$. He obtained a product that appeared to be lactic acid, and yet, it was optically inactive. Does this mean the product was not truly lactic acid?

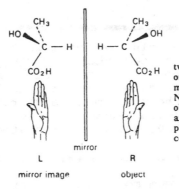

Models showing the two possible arrangements in space of the four groups around the asymmetric carbon atom of lactic acid. Note that the two forms have an object-mirror image relationship and that one cannot be superimposed on the other for all points to coincide.

Solution: A carbon atom that has four different atoms or groups attached to it is termed an asymmetric carbon atom or a chiral center. These four groups can give two spatial arrangements, whose relationship to each other is that of an object and its mirror image. For example, see the accompanying figure. The two arrangements are not identical, because they are not superimposable. In such cases, these molecules, i.e. those which differ only in the positioning of groups around asymmetric carbon atoms, are called optical isomers.

Such isomers can rotate plane polarized light. Plane polarized light vibrates in only one plane. When substances can rotate plane polarized light, they are optically active. Each of the two optical isomers rotates the light in a different direction. The dextro-rotary isomer rotates it to the right, while the levo-rotatory rotates it to the left. If both isomers are present in equal amounts, the light will not be rotated at all. For one rotates it to the left and the other rotates it to the right, which means there is a cancellation of the rotations. When plane-polarized light is not rotated, the solution is optically inactive. With this information, one can now answer the question.

The student formed $CH_3CHOHCO_2H$, which he believed was lactic acid. The solution was optically inactive, and yet, it is known that lactic acid is optically active. The only possible explanation for this is that the student formed equal amounts of the two possible optical isomers. As mentioned, this results in optical inactivity. There is every reason to believe that

$$
CH_3 - \overset{\overset{\displaystyle H}{|}}{\underset{\underset{\displaystyle OH}{|}}{C^*}} - CO_2H \quad \text{and} \quad CH_3 - \overset{\overset{\displaystyle OH}{|}}{\underset{\underset{\displaystyle H}{|}}{C^*}} - CO_2H \quad \text{were}
$$

formed in equal amounts. The star (*) indicates the asymmetric carbon. Thus, the student did form lactic acid. It is just that a mixture of the two optical isomers in equal proportions is present.

● **PROBLEM** 816

Insulin controls the glucose level in the blood. In 1969, Hodgkin announced the secondary structure of insulin, the first protein hormone to be resolved by X-ray diffraction. (a) Chemial analysis shows insulin is 2.15% zinc by weight. Calculate the minimum molecular weight. (b) Osmotic pressure studies show the molecular weight of insulin is about 6000. How many zinc atoms does the molecule contain?

Solution: The minimum molecular weight of insulin can be determined by assuming that there is one zinc atom per molecule. This means that in 1 mole of insulin there exists 1 mole of zinc. From the knowledge that zinc is 2.15% by weight of insulin, and that the MW of zinc is 65.3, one can determine the minimum MW of insulin by using the following relationship.

$$(.0215)(\text{MW of insulin}) = (\text{MW of zinc})$$

$$\text{MW of insulin} = \frac{\text{MW of Zn}}{(.0215)} = \frac{65.3 \text{ g/mole}}{(.0215)} = 3037 \text{ g/mole}$$

(b) If the true molecular weight of insulin is about 6000, then two of the units weighing 3037 g/mole make up each molecule of insulin. Thus, there are 2 atoms of Zn per molecule.

● **PROBLEM** 817

The molar absorptivity of phytochome (a light-sensitive pigment thought to control the process of flowering in plants) is 76,000. What will be the absorbance of a $5.0 \times 10^{-6}$ molar aqueous solution of phytochome if the pathlength of the cell is 1.0 cm?

Solution: The fraction of the incident light absorbed by a solution at a given wavelength is related to the thickness of the absorbing layer and to the concentration of the absorbing species. These two relationships are combined into the Lambert-Beer Law, given in integrated form as

$$\log \frac{I_0}{I} = ac\ell,$$

where $I_0$ is the intensity of the incident light. I is the intensity of the transmitted light, a is the molar absorbing index (the molar extinction coefficient), c is the concentration of the absorbing species in moles per liter, and $\ell$ is the thickness of the light absorbing sample.

$\log \frac{I_0}{I}$ is also known as the absorbance A. Solving for A,

$$A = ac\ell$$

$$A = (76,000)(5.0 \times 10^{-6} \text{ moles/liter})(1 \text{ cm}) = 0.38.$$

● **PROBLEM** 818

Calculate the sedimentation coefficient of tobacco mosaic virus from the fact that the boundary moves with a velocity of 0.454 cm/hr in an ultracentrifuge at a speed of 10,000 rpm at a distance of 6.5 cm from the axis of the centrifuge rotor.

Solution: When a sedimenting boundary of protein moves at a constant rate, the centrifugal force just counterbalances the frictional resistance of the solvent. The rate of sedimentation is expressed as the sedimentation coefficient, s

$$s = \frac{dx/dt}{\omega^2 x},$$

where x is the distance from the center of rotation, $\omega$ is the angular velocity in radians per second, and t is the time in seconds. One is given dx/dt, $\omega$ and x in the problem, but they are not in units that are used in determining s. Converting to proper units:

dx/dt; This should be in cm/sec but is given in cm/hr.

$$0.454 \text{ cm/hr} = 0.454 \text{ cm/hr} \times 1 \text{ hr/3600 sec}$$

$$= 1.26 \times 10^{-4} \text{ cm/sec}.$$

$\omega$; $\omega$ is given as 10,000 rpm (revolutions per minute) but must be expressed in radians per second for use. There are $2\pi$ radians in one revolution.

$$\frac{10,000 \text{ rmp} \times 2\pi}{60 \text{ sec}} = 1.05 \times 10^3 \text{ radians/sec}$$

Solving for s:

$$s = \frac{dx/dt}{\omega^2 \ x} = \frac{1.26 \times 10^{-4} \text{ cm/sec}}{(1.05 \times 10^3/\text{sec})^2 \ 6.5 \text{ cm}} = 1.76 \times 10^{-11} \text{ sec.}$$

● **PROBLEM** 819

Ribonuclease has a partial specific volume of $0.707 \text{ cm}^3 \text{ g}^{-1}$ and a diffusion coefficient of $13.1 \times 10^{-7} \text{ cm}^2 \text{ sec}^{-1}$ corrected to water at 20°C. The density of water at 20°C is $0.998 \text{ g/cm}^3$. Calculate the molecular weight of ribonuclease using the Svedberg equation. Assume that the sedimentation coefficient for ribonuclease is $2.02 \times 10^{-13}$ sec.

Solution: The molecular weight M of a protein can be calculated from the sedimentation coefficient by means of the Svedberg equation, which is derived by equating the centrifugal force with the opposing frictional force, the condition existing when the rate of sedimentation is constant:

$$M = \frac{RT_s}{D(1 - \bar{v}\rho)},$$

where R is the gas constant in ergs per mole per degree ($8.314 \times 10^7 \text{ ergs/}°K \text{ mole}$), T the absolute temperature, s the sedimentation coefficient, $\bar{v}$ the partial specific volume of the protein, $\rho$ the density of the solvent, and D the diffusion coefficient. All of the values necessary to solve for M are given. 20°C = 293°K.

Solving for M:

$$M = \frac{(8.314 \times 10^7 \text{ ergs/}°K \text{ mole})(293°K)(2.02 \times 10^{-13} \text{ sec})}{(13.1 \times 10^{-7} \text{ cm}^2 \text{ sec}^{-1})(1-(0.707 \text{ cm}^3 \text{ g}^{-2})(0.998 \text{ g/cm}^3))}$$

$$= 13,000 \text{ g/mole.}$$

● **PROBLEM** 820

The diffusion coefficient for serum globulin at 20° in a dilute aqueous salt solution is $4.0 \times 10^{-11} \text{ m}^2 \text{ s}^{-1}$. If the molecules are assumed to be spherical, calculate their molecular weight. Given: $\eta_{H_2O} = 0.001005$ Pascal-sec at 20° and $\bar{v} = 0.75 \text{ cm}^3/\text{g}$ for the protein.

Solution: For spherical molecules, the molecular weight M is related to the diffusion coefficient D by the following equality.

$$D = \frac{RT}{N_A 6 \pi \eta} \left[ \frac{4 \pi N_A}{3M \bar{\nu}} \right]^{1/3} ,$$

where R is the gas constant in joules (8.314 J/mole °K), $N_A$ is Avogradro's Number ($6.02 \times 10^{23}$), $\eta$ is the viscosity of the solvent, and $\bar{\nu}$ the partial specific volume of the protein.

Solving for M;

$D = 4.0 \times 10^{-11} \ m^2/s$

$$= \frac{(8.314 \ J/mole \ °K)(293°K)}{(6.02 \times 10^{23}/mole)6\pi(.001005 \ Js/m^2)} \times$$

$$\left[ \frac{4\pi(6.02 \times 10^{23} \ mole)}{3 \times M \times 0.75 \times 10^{-6} \ m^3/s} \right]^{1/3}$$

$4.0 \times 10^{-11} \ m^2/s = (2.14 \times 10^{-19} \ m^2/s) \times$

$$(3.36 \times 10^{30}/M)^{1/3}$$

$1.87 \times 10^8 = (3.36 \times 10^{30}/M)^{1/3}$

$6.53 \times 10^{24} = 3.36 \times 10^{30}/M$

$M = 5.15 \times 10^5 \ g/mole.$

● **PROBLEM** 821

The sedimentation and diffusion coefficients for hemoglobin corrected to 20° in water are $4.41 \times 10^{-13}$ sec and $6.3 \times 10^{-11} \ m^2/s$, respectively. If $\bar{\nu} = .749 \ cm^3/g$ and $\rho_{H_2O} = 0.998 \ g/cm^3$ at this temperature, calculate the molecular weight of the protein. If there is 1 g-atom of iron per 17,000 g of protein, how many atoms of iron are there per hemoglobin molecule?

Solution: One can calculate molecular weight by using the Svedberg equation. This equation is written

$$M = \frac{RTs}{D (1 - \bar{\nu}\rho)} ,$$

where M is the molecular weight, R is the gas constant in ergs ($8.314 \times 10^7$ ergs/mole °K), T is the absolute temperature, s is the sedimentation coefficient, D is the diffusion coefficient, $\rho$ is the density of the solvent, and $\bar{v}$ is the partial specific volume of the protein. Solving for M:

$$M = \frac{(8.314 \times 10^7 \text{ ergs/mole } ^\circ K)(293^\circ K)(4.41 \times 10^{-13} \text{ sec})}{(6.3 \times 10^{-11} \text{ m}^2/\text{s})[1-(0.749 \text{ cm}^3/\text{g})(.998 \text{ g/cm}^3)]}$$

$$= 6.75 \times 10^8 \text{ ergs } \frac{s^2}{m^2 \text{ mole}}$$

$$= 6.75 \times 10^8 \left(\frac{g \text{ cm}^2}{\text{sec}^2}\right)\left(\frac{\text{sec}^2}{m^2 \text{ mole}}\right)\left(\frac{1 \text{ m}^2}{10^4 \text{ cm}^2}\right)$$

$$= 6.75 \times 10^4 \text{ g/mole}$$

$$1 \text{ erg} = \frac{g \text{ cm}^2}{s^2} ; \qquad 1 \text{ m}^2 = 10^9 \text{ cm}^2$$

To determine the number of Fe atoms per molecule of hemoglobin, calculate the number of moles of hemoglobin in 17000 g. It is given that there is 1 mole of Fe in 17,000 g of Hgb. The number of Fe atoms per molecule is then equal to the number of moles of Fe divided by the number of moles of hemoglobin.

$$\text{No. of moles of Hgb} = \frac{17,000 \text{ g}}{6.75 \times 10^4 \text{ g/mole}} = 2.52 \times 10^{-1} \text{ moles}$$

$$\text{No. of Fe atoms per Hgb molecule} = \frac{1 \text{ mole}}{2.52 \times 10^{-1}} = 4.0 \text{ atoms.}$$

# CHAPTER 23

# BIOCHEMISTRY II

Basic Attacks and Strategies for Solving Problems in this Chapter. See pages 789 to 807 for step-by-step solutions to problems.

Proteins are among the most important, if not the single most important, class of biochemical molecules. They are formed by the selective ordering of amino acids into huge molecules with molecular weights in the tens and hundreds of thousands. The peptide linkage, shown in Equation 23-1, is the basis of the bond that is used to form proteins or polypeptide chains.

$$\begin{array}{cc} H & O \\ | & \| \\ -N- & C- \end{array} \quad \text{(peptide bond)} \qquad\qquad 23\text{-}1$$

This bond is hydrolyzed in acid to give back the amino and carboxylic acid groups.

$$\begin{array}{cc} H & O \\ | & \| \\ -N-C- \end{array} + H_2O \rightarrow \begin{array}{c} H \\ | \\ -H-N \end{array} + \begin{array}{c} O \\ \| \\ HO-C- \end{array}$$

Proteins are formed by polypeptide chains containing hundreds of amino acids, some of which can be found in an d-helix structure. To work those problems in this chapter that deal with the helical structure, the principles of stoichiometry can be applied directly as in problems in previous chapters.

Enzymes act as incredibly selective catalysts for biochemical reactions. They have the ability to promote one reaction while completely blocking competing reactions. The "lock and key" theory has been proposed as the mechanism for these very selective catalysts. Enzymes have a chemical structure that mates with the structure of the reactant. When the reactant and the substrate adsorb one on the other, the single reaction is catalyzed by that configuration, and no others can occur. This kind of mechanism is illustrated in Problem 829.

Carbohydrates are the most abundant biochemical molecules in plants and animals. They form support structures such as cell walls, and provide chemical energy. Carbohydrates are formed from sugars, and are often referred to as saccharides. Monosaccharides are the monomers from which complicated carbohydrates are constructed; they are principally sugars including amino sugars or sugars which contain $NH_2$ substitutions.

Carbohydrates, such as starch, can be hydrolyzed to simple sugars or saccharides. That is why bread, which is principally starch, will begin to taste sweet if held in the mouth for an extended period. The starch molecules hydrolyze to form glucose or other simple sugars which are sweet to the taste.

Nucleic acids are the third broad class of biochemical polymers. Like proteins and carbohydrates (polysaccharides), they form an essential component of living materials. The best known, deoxyribonucleic acid (DNA), are huge molecules with molecular weights of $10^9$ or greater. These molecules form the genetic blueprint for reproduction of cells; hence they contain the basic genetic information responsible for heredity. These molecules are extremely complicated and only very elementary treatment, using the stoichiometric principles already discussed, is covered in this text.

Step-by-Step Solutions to
Problems in this Chapter,
"Biochemistry II"

## PROTEINS

● PROBLEM 822

If you hydrolyze a peptide bond in a protein molecule,
what type of compounds are formed?

Solution: When proteins are boiled in dilute acids or
bases they are hydrolyzed, i.e. degraded or broken down,
to amino acids. Amino acids are compounds whose molecules
possess both the amino (- $NH_2$) and the carboxyl (- $CO_2H$)
functional groups. The general formula may be written as

```
          H
          |
H - N - C - C - OH,      where Z = side group
    |   |   ||
    H   Z   O
```

The union of amino acids, which result in the protein,
is due to bonds that come about as a result of the elimination of a hydrogen from the - $NH_2$ group and an OH from
the - $CO_2H$ group. This linkage is termed a peptide bond.
It is depicted as

```
              O
      H       ||
    - N - C -
```

peptide bond

When this substance is hydrolyzed, the following reaction
occurs. In acid:

```
      O                          O
      ||                         ||
  H   ||             H           ||
- N - C -  + H O+  →  - NH + HO - C -.  You obtain
```

(back) the amino and carboyxlic group.

Note: $NH_2$ is basic, so that in acid, you actually get
- $NH_3^+$. Thus, you obtain a compound that possesses an amino
and a carboxylic group, which is, as you recall, an amino
acid. With base, the same result is obtained, but, you
immediately obtain $NH_2$ and not $NH_3^+$. The carboxylic portion,
which is acidic, reacts with the $OH^-$ from base to yield $COO^-$.

● **PROBLEM** 823

Calculate the length (in Å) of a polypeptide chain containing 105 amino acid residues if (a) it exists entirely
in α-helical form, or (b) it is fully extended.

The average dimenions of the
α-helix. The pitch and the
rise per residue correspond
to the major and minor peri-
odicities of 5.4 and 1.5 Å,
respectively. This drawing
shows a left-handed α-helix.

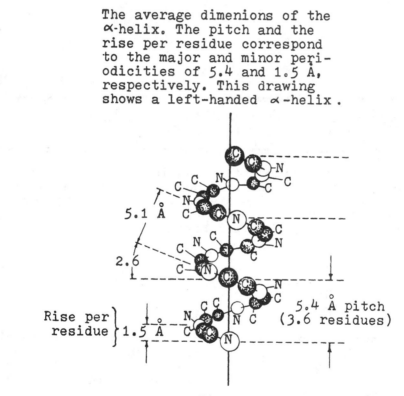

Fig.A

Solution:    (a) Peptide chains may assume the α-helix configuration spontaneously, because this form is stable and has
the least free energy, providing there are no opposing inter-
actions of the R groups or of the solvent. The α-helix is
pictured in Figure A. As shown, when the peptide chain assume
this configuration each amino acid residue contributes 1.5 Å
to the length of the chain. If a chain in the α-helical configuration contains 105 amino acid residues, its length is
1.5 Å x 105 = 157.5 A.

(b) If the chain is fully extended, each amino acid residue contributes its full length to the chain. Figure B shows that the length of an average amino acid is 3.6 Å. The length of the chain in this conformation is 3.6 Å × 105 = 378.0 Å.

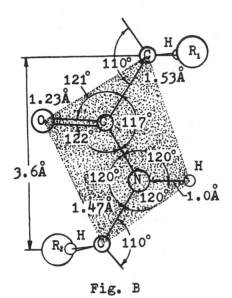

Dimensions of the peptide linkage from x-ray data. The six atoms in the shaded zone lie in a plane. Because the central C-N bond has some double-bond character, this plane tends to be rigid.

Fig. B

● **PROBLEM** 824

On a given diet yielding 3000 Kcal per day, a 70 kg man excretes 27.0 g of urea daily. What percentage of his daily energy requirement is met by protein? Assume that 1.0 gram of protein yields 4 Kcal and 0.16 g of nitrogen as urea.

Solution: If 1.0 g of protein yields 0.16 g of urea, and 27.0 g of urea is excreted daily, then the man eats 27.0 g/0.16 g or 168.75 g of protein a day. Each gram of protein contributes 4 Kcal towards the energy requirement. Therefore, 168.75 g of protein produce 4 × 168.75 or 675 Kcal. 675 Kcal is 675/3000 × 100 or 22.5% of the daily energy requirement.

● **PROBLEM** 825

A turn of α-helix (3.6 residues) is 5.41 Å in length measured parallel to the helix axis. If your hair grows 6 in. per year, how many amino acid residues must be added to each α-helix in the keratin fiber per second?

Solution: Human hair is an example of a keratin fiber, and is found in the α-helix conformation. One is given that in each 5.41 $\overset{o}{A}$ of length, there are 3.6 residues. After converting 6 in to $\overset{o}{A}$, one can determine the number of residues that are added per year to each strand. From this, the number of residues added per second can be calculated.

Solving for length:

1 in = 2.54 cm,     1 $\overset{o}{A}$ = $10^{-8}$ cm

length in $\overset{o}{A}$ = 6 in $\times$ 2.54 cm/in $\times$ 1 $\overset{o}{A}/10^{-8}$ cm =

$$= 1.52 \times 10^9 \ \overset{o}{A}$$

Calculating the number of residues:

number of residues = length in $\overset{o}{A}$ $\times$ $\dfrac{3.6 \text{ residues}}{5.41 \ \overset{o}{A}}$

$$= 1.52 \times 10^9 \ \overset{o}{A} \times \dfrac{3.6 \text{ residues}}{5.41 \ \overset{o}{A}}$$

$$= 1.01 \times 10^9 \text{ residues}$$

There are $1.01 \times 10^9$ residues produced per year.

Solving for the number of residues produced per second

residues produced per second =

$1.01 \times 10^9 \ \dfrac{\text{res}}{\text{year}} \times \dfrac{1 \text{ year}}{365 \text{ days}} \times \dfrac{1 \text{ day}}{24 \text{ hrs}} \times \dfrac{1 \text{ hr}}{60 \text{ min}} \times \dfrac{1 \text{ min}}{60 \text{ sec}}$

$$= 32 \text{ res/sec.}$$

● **PROBLEM** 826

There are some 25,000 ribosomes in an E. coli cell. If the structucal proteins of these ribosomes were stretched out end to end as fully extended polypeptide chains, how many times could they encircle the E. coli cell? Assume that the ribosomes are 180 $\overset{o}{A}$ in diameter, with a specific gravity of 1.0, and that they contain 40% protein. Assume that the E. coli cell is a sphere 1μ in diameter.

Solution: Ribosomes are complexes of nucleic acids and protein, and are found in the cytoplasm of the cell. This problem is solved by using the following series of steps.

(a) the volume of a ribosome

(b) the total weight of all the ribosomes present and the weight of the protein

(c) the number of amino acid residues in the protein

(d) the length of the extended protein chain and the number of times it will encircle an E. coli cell.

Solving:

(a) A ribosome is assumed to be spherical and, thus, its volume can be found by using the formula $4/3 \pi r^3$, where r is the radius of the sphere. If the diameter of a ribosome is 180 Å its radius is 90 Å or $90 \times 10^{-8}$ cm.

$$\text{volume of one ribosome} = \frac{4}{3} \pi (90 \times 10^{-8} \text{ cm})^3$$

$$= 3.06 \times 10^{-18} \text{ cm}^3$$

(b) Because the specific gravity of the ribosome is 1.0, one ribosome weighs $3.06 \times 10^{-18}$ g. There are 25,000 ribosomes present.

$$\text{total weight} = 3.06 \times 10^{-18} \text{ g/ribosome} \times 25,000 \text{ ribosomes}$$

$$= 7.65 \times 10^{-14} \text{ g}$$

The protein comprises 40% of the ribosome.

$$\text{weight of the protein} = (.4)(7.65 \times 10^{-14} \text{ g})$$

$$= 3.06 \times 10^{-14} \text{ g}$$

(c) The molecular weight of an average amino acid residue is 120. One can find the number of moles of amino acid residues present by dividing the weight of the protein by 120. The number of residues is found by multiplying the number of moles by $6.02 \times 10^{23}$ residues/mole

$$\text{no. of amino acid residues} = \frac{3.06 \times 10^{-14} \text{ g}}{120 \text{ g/mole}}$$

$$\times 6.02 \times 10^{23} \frac{\text{residues}}{\text{mole}}$$

$$= 1.54 \times 10^8 \text{ residues}$$

(d) When a protein is fully extended, each amino acid residue contributes 3.6 Å or $3.6 \times 10^{-4}$ μ to the length of the chain.

$$\text{length of the chain} = 3.6 \times 10^{-4} \mu \times 1.54 \times 10^8$$

$$= 5.54 \times 10^4 \mu$$

The diameter of the E. coli is 1μ, which means its radius $(\frac{1}{2})1\mu = 0.5\mu$. As such the circumference of the E. coli equals $2\pi r$, where r is radius, or $2\pi(0.5\mu) =$

$\pi\mu = 3.14\mu$. Therefore, the protein, when fully extended, can be wrapped around the E. coli

$$\frac{5.54 \times 10^4}{3.14} = 1.76 \times 10^4 \text{ times.}$$

If an E. coli cell contains $10^6$ protein molecules, each of mol. wt. 40,000, calculate the total length of all the polypeptide chains of a single E. coli cell, assuming they are entirely $\alpha$-helical.

Solution: When a protein exists in the $\alpha$-helical configuration, each amino acid contributes 1.5 Å to its length. To find the total length of all the polypeptide chains present in one E. coli cell, first calculate the the number of amino acids present and then multiply this number by 1.5 Å. If there are $10^6$ protein molecules, there are $\frac{10^6 \text{ molecules}}{6.02 \times 10^{23} \text{ molecules/mole}}$ or $1.66 \times 10^{-18}$ moles present. The weight of the proteins becomes $1.66 \times 10^{-18}$ moles $\times$ 40,000 g/mole or $6.64 \times 10^{-14}$ g. The average molecular weight of an amino acid is 120, therefore the average weight of one amino acid residue is

$$\frac{120 \text{ g/mole}}{6.02 \times 10^{23} \text{residues/mole}} \text{ or } 1.99 \times 10^{-22} \text{ g/residue.}$$

One can solve for the number of residues present by dividing the total weight of the protein chains by the weight of an amino acid residue.

$$\text{no. of amino acid residue} = \frac{6.64 \times 10^{-14} \text{ g}}{1.99 \times 10^{-22} \text{g/residue}}$$

$$= 3.33 \times 10^8 \text{ residues.}$$

It is now possible to solve for the length of the protein chain.

$$\text{length of chain} = \text{no. of residues} \times 1.5 \text{ Å}$$

$$= 3.33 \times 10^8 \text{ residues} \times 1.5 \text{ Å/residue}$$

$$= 5.0 \times 10^8 \text{ Å}$$

$$= 5.0 \times 10^8 \text{ Å} \times 1 \text{ cm}/10^8 \text{ Å} = 5.0 \text{ cm.}$$

If a man is inhaling air composed of 19% $O_2$ and 0.003%

CO by volume, what is the ratio of the concentrations of carboxyhemoglobin (HbCO) and oxyhemoglobin (HbO$_2$)? Explain why [CO] should be kept to a minimum.

Solution: The concentration of carbon monoxide [CO] must be kept to a minimum due to hemoglobins' (Hb) greater affinity for CO than O$_2$ (oxygen). This occurs by the reaction HbO$_2$ + CO $\rightleftarrows$ HbCO + O$_2$. Because of this reaction in the presence of CO, less O$_2$ is available for body cells. It also results in the reduction of oxy-hemoglobin dissociation into hemoglobin and oxygen so that anoxia (oxygen starvation) occurs.

The ratio of [HbCO] and [HbO$_2$] can be calculated by the Haldane equation:

$$\frac{[HbCO]}{[HbO_2]} = M \frac{p(CO)}{p(O_2)} \quad , \qquad \text{where } p(CO) \text{ and } p(O_2)$$

are the partial pressures (or volume concentrations) of the CO and O$_2$ gases and M is a constant that depends on the species. Given p(CO), p(O$_2$) and M = 250, by substituting one obtains

$$\frac{[HbCO]}{[HbO_2]} = 250 \left( \frac{0.00003}{0.19} \right) = \frac{0.03947}{1} = 0.03947.$$

# ENZYMES

• PROBLEM 829

Enzyme behavior may be described by the lock-and-key theory. Explain and describe this theory.

Solution: Millions of chemical reactions occurring in the body require catalysts to make them go to completion. Enzymes serve as these catalysts. Their behavior and/or function is described by the lock-and-key theory.

According to this theory, enzymes have definite three-dimensional structures. They are arranged so that the substrate (the substance upon which the enzyme acts) fits into the enzyme's structure. Only a specific kind of substrate can fit into a given enzyme. When the substrate and enzyme unite, i.e., form an aggregate, the substrate is exposed for a reaction to occur. In Fig. A, the three-dimensional shape of the enzyme is arranged to accommodate the substrate. The active site is that portion of the enzyme that catalyzes the reaction. In Fig. B, the substrate is held in position by the enzyme while the reaction occurs. In Fig. C, the product(s) of the reaction leaves the enzyme, thus, freeing the enzyme to catalyze another reaction.

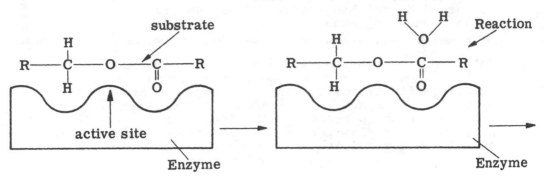

Figure A                              Figure B

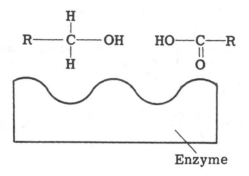

Figure C

● **PROBLEM** 830

The rate of a certain biochemical reaction at body temperature in the absence of enzyme has been measured in
the laboratory. The rate of the same reaction, when
enzyme-catalyzed in the human body, is $10^6$ times faster.
Explain this phenomenon.

Solution: For any reaction to proceed, a certain amount
of energy is needed. This energy is called the energy of
activation. When a catalyst, such as an enzyme, is added
to the system, the energy of activation is lowered.
Because the energy requirement for the reaction is less,
the reaction proceeds more quickly.

● **PROBLEM** 831

Starch (amylose), $(C_6H_{10}O_5)x$, is a polymer of the sugar
glucose, $C_6H_{12}O_6$, linked by - O - bonds; each starch molecule may contain 200 to 1000 glucose units. A mixture
of amino acids added to a solution of amylose produces

no detectable reaction. One molecule of the enzyme amylase, containing the same amino acids, however breaks (hydrolyzes) 4000 -O- bonds per second. (a) Estimate the mass in picograms of glucose produced by an amylase molecule in 1 day. (b) The heat in calories liberated to a cell by the oxidation of the glucose: $C_6H_{12}O_6(aq) + 6O_2(g) \rightarrow 6CO_2(g) + 6H_2O(\ell)$. $\Delta H = - 686$ Kcal. How many calories does the cell obtain from the mass of glucose calculated in part (a)?

Solution:    (a) Five steps can be used to solve for the number of picograms of glucose liberated from the amylose by 1 amylase in a day.

(1) find the number of seconds in one day

(2) calculate the number of glucose molecules formed per day

(3) determine the weight of one molecule of glucose

(4) solve for the number of grams of glucose formed per day

(5) convert from grams to picograms.

Solving:

(1) There are 60 seconds in a minute, 60 minutes in an hour and 24 hours in a day. One can solve for the number of seconds in a day by multiplying these factors together  The units will cancel out.

60 sec/min $\times$ 60 min/hour $\times$ 24 hours/day = $8.64 \times 10^4$ sec/day

(2) The problem states that one amylase molecule liberates 4000 molecules per second. Therefore the number of molecules liberated per day can be found by multiplying the number of seconds in a day by the number of molecules formed per second.

no. of glucose molecules formed per day =

= $8.64 \times 10^4$ sec/day $\times$ 4000 molecules/sec

= $3.456 \times 10^8$ molecules/day

(3) The weight of one molecule of glucose is found by dividing the molecular weight of glucose by the number of molecules per mole. (MW of $C_6H_{12}O_6$ = 180 g/mole, Avogadro's Number = $6.02 \times 10^{23}$ molecules/mole)

weight of one molecule =

$$= \frac{180 \text{ g/mole}}{6.02 \times 10^{23} \text{ molecule/mole}} = 2.99 \times 10^{-22} \text{ g/molecule}$$

(4) After one knows the number of molecules formed per day and the weight of one molecule, one can find the weight

in grams of the glucose formed in one day by one amylase.

no. of grams of glucose formed per day =

$= 2.99 \times 10^{-22}$ g/molecule $\times 3.456 \times 10^8$ molecules/day

$= 1.03 \times 10^{-13}$ g/day.

(5) There are $10^{12}$ picograms in one gram. Grams can be converted to picograms by multiplying the number of grams by the conversion factor $10^{12}$ pg/g.

no. of pg produced $= 1.03 \times 10^{-13}$ g/day $\times 10^{12}$ pg/g

$= 1.03 \times 10^{-1}$ pg/day.

(6) From the equation given one knows that 686 Kcal are liberated when 1 mole of glucose is oxidized. One can find the number of calories liberated by

1) determining the number of moles of glucose present

2) solving for the number of Kcal produced

3) converting Kcal to calories.

Solving·

1) One has solved for the number of molecules produced per day previously (part (a)(2)). Therefore one can find the number of moles produced by dividing the number of molecules produced by the number of molecules in one mole.

$$\frac{3.456 \times 10^8 \text{ molecules/day}}{6.022 \times 10^{23} \text{ molecules/mole}} = 5.74 \times 10^{-16} \text{ moles/day.}$$

2) There are 686 Kcal liberated per mole of glucose oxidized, therefore the number of Kcal formed when $5.74 \times 10^{-16}$ moles are oxidized is equal to

$5.74 \times 10^{-16}$ moles/day $\times 686$ Kcal/mole =

$= 3.94 \times 10^{-13}$ Kcal/day.

3) There are 1000 cal in 1 Kcal, thus Kcal can be converted to cal by multiplying by the conversion factor 1000 cal/Kcal

no. of cal produced $= 3.94 \times 10^{-13}$ Kcal/day $\times 1000$ cal/Kcal

$= 3.94 \times 10^{-10}$ cal/day.

● **PROBLEM** 832

Catalase, the enzyme that assists the breakdown of hydrogen

peroxide into oxygen and water, has a turnover number of $2 \times 10^8$. In one hour, what weight of hydrogen peroxide could be decomposed by one molecule of catalase?

Solution: Enzymes may be defined as catalysts. A substrate is the compound that the enzyme acts upon to form the product. One aspect of enzymatic reactions is the rate at which the enzyme-substrate complex is formed and shifted to product. One can express this rate as the turnover number. It is defined as the number of molecules of substrate converted to product by one molecule of enzyme in one minute at optimum conditions.

In this problem, you are given one molecule of enzyme, and excess substrate. The turnover rate is given as $2 \times 10^8$. You are asked for the weight of the substrate decomposed in one hour. To find the weight, you need the number of molecules decomposed. According to the definition of turnover number, $2 \times 10^8$ molecules, decompose in one minute. Thus, in 1 hour, 60 minutes $\times 2 \times 10^8 = 1.2 \times 10^{10}$ molecules of hydrogen peroxide $(H_2O_2)$ is decomposed. To calculate the total weight of these molecules, one uses Avogadro's number. It states that there are $6.02 \times 10^{23}$ molecules in one mole. Thus, $1.2 \times 10^{10}$ molecules represents $1.2 \times 10^{10}/6.02 \times 10^{23} = 1.99 \times 10^{-14}$ moles of $H_2O_2$ decomposed. A mole is defined as weight in grams/molecular weight. The molecular weight of $H_2O_2 = 34$. Thus, weight of $H_2O_2$ decomposed =

$$(1.99 \times 10^{-14} \text{ moles})(34 \text{ g/mole}) = 6.766 \times 10^{-13} \text{ grams.}$$

● **PROBLEM** 833

Meats possess large amount of connective tissue that can be tenderized by the action of proteolytic enzymes. Can such enzymes harm the stomach lining when the food is eaten?

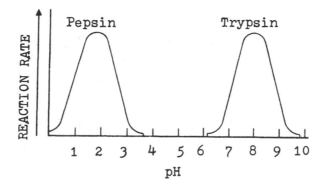

799

Solution: This question deals with proteolytic enzymes, which may be defined as protein degradation catalysts. Two factors that affect enzymic activity are pH, and temperature. Whether these enzymes damage the stomach or not will depend, to a great extent, on the conditions in the stomach. The accompanying figure shows how pH can alter enzyme activity of pepsin and trypsin. From the diagram, you see that these enzymes do not function well at extremely low pH. Due to the presence of HCl in the stomach as a hydrolysis agent, the acidity of the stomach is high, which means an environment of low pH exists. It becomes doubtful that these proteolytic enzymes would have much activity at the low pH of the stomach. This means that little, if any, damage would be done to the stomach. Heating alters enzymic activity, also. It is likely that most of the enzyme would be inactivated by the cooking of the meat.

# CARBOHYDRATES

● **PROBLEM** 834

Explain why it has been necessary to postulate the existence of a cyclic form of glucose?

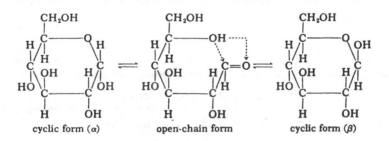

cyclic form ($\alpha$)          open-chain form          cyclic form ($\beta$)

Solution: Glucose ($C_6H_{12}O_6$), is one of the isomeric aldohexoses - a carbohydrate or sugar. Glucose exists in two forms, one of which is the mirror image of the other. Its configurational formulas may be written

```
    H - C = O              H - C = O
        |                      |
    H - C - OH            HO - C - H
        |                      |
   HO - C - H             H - C - OH
        |                      |
    H - C - OH           HO - C - H
        |                      |
    H - C - OH           HO - C - H
        |                      |
      CH2OH                  CH2OH

  (+) - glucose          (-) - glucose
```

At one time, it was believed that these glucose isomers existed in this linear form. Now, however, it is believed that cyclic forms exist, as indicated in the accompanying diagram. This has been postulated for the following reasons: It has been found that when you take a solution of pure (+) - glucose or a solution of pure (-) - glucose form, a pair of optical isomers are obtained, i.e., ones with the ability to rotate plane-polarized light. For this to occur, in a solution of pure (+) or (-) glucose, another asymmetric carbon must be present.

An asymmetric carbon is one with 4 different atoms or molecules attached. Asymmetric carbons can generate optical isomers. In the formation of cyclic structures,

the carbon atom of the aldehyde group ( C = O) must be

the additional asymmetric carbon. This would account for the two optical isomers formed. In other words, that glucose exists in cyclic forms was deduced from the fact that only in such a structure is an additional asymmetric carbon generated, which would account for the optical isomers produced.

● **PROBLEM** 835

There exists a carbohydrate that is slightly soluble in water, does not have a sweet taste, but does dissolve slowly when heated with dilute HCl. To what class of carbohydrates does it probably belong?

Solution:  Carbohydrates may be defined as simple sugars and the substances that hydrolyze to yield simple sugars. Carbohydrates are placed in three classes: (1) Monosaccharides - those that do not undergo hydrolysis; (2) Disaccharides - those that may be hydrolyzed to two monosaccharide molecules; and (3) Polysaccharides - those which form many monosaccharide molecules after hydrolysis.

It is given that the carbohydrate dissolves in dilute HCl. Thus, one can immediately eliminate monosaccharides as a possible case. The fact that it dissolves when HCl is present suggests that a hydrolysis reaction has taken place. Only disaccharides and polysaccharides have the ability to undergo such a reaction. Thus, the carbohydrate belongs to either the class of disaccharides or polysaccharides. To determine which one it is, use the information that it is not sweet or very soluble in water. The more carbons in a molecule the more unlikely it will be soluble in water. Carbon-carbon bonds are nonpolar, while water is a polar molecule. Thus, a polysaccharide, which possesses many more non-polar bonds than a disaccharide, should have a tendency to be less soluble

in water. Note also that among the disaccharides exist
the very sweet sugars of sucrose (table sugar), maltose
(malt sugar), and lactose (milk sugar). The classic
examples of polysaccharides are, however, starch and
cellulose, which are not sweet at all. Thus, one can conclude that this carbohydrate probably belongs to the class
of polysaccharides.

● **PROBLEM** 836

Explain how the absorption of a digested fat differs
from the absorption of digested starch.

Solution: After fat is digested, you have a mixture of
glycerol and fatty acids as well as mono- and diglycerides.
These products adhere to bile salts and pass into the
epithelial cells that line the intestines. At this point,
they are reassembled into triglycerides. They leave and enter
the lymphatic fluid as components of fat droplets or chylomicrons. The lymphatic system is connected to the venous
circulation, so that the chylomicrons enter the blood,
giving it a milky appearance.

Starch absorption - The end products of carbohydrates,
such as starch, after digestion are glucose, fructose, and
galactose and other six carbon sugars. These are absorbed
through the villi of the small intestine and enter the
venous circulation, where an enzyme present in red blood
cells converts some of them to glucose. Others are converted to glucose by different enzymes, in the liver.

In summary, the differences are as follows: Starch
must be hydrolyzed before it is absorbed, while fats do
not have to be. The other major difference in absorption
is that digested starch is absorbed into the bloodstream
directly, whereas fats are absorbed into the lymphatic
system.

● **PROBLEM** 837

Fats are not digested in the stomach even though there
exists a lipase in the gastric fluid. Explain why.

Solution: The general formula of a fat or lipid may be
written $(RCO_2)_3C_3H_5$, where the R's can be the same or
different. It may be defined as an ester of glycerol
and three long-chain carboyxlic acids. These ester
linkages may be hydrolyzed (degraded) by digestive
enzymes called lipases. Such enzymes can function only
under certain specific conditions. This is the reason
why the lipase in gastric juice cannot digest lipids

in the stomach. This lipase has an optimum pH of 7, a neutral solution.

In the stomach, however, due to the presence of HCl for other digestive purposes, there is an acidic environment. Thus, the lipase is virtually inactive under these conditions. As such, the fat will not be digested.

But note, even if the lipase were active, lipid would still be undigested. To understand why, consider the following: Fats are insoluble in water, and, as such, tend to form globules in the body. Digestive enzymes cannot penetrate such globules. Only those fat molecules at the surface can be acted upon. The surface area can be increased, however, by breaking the globule into smaller droplets - a process called emulsification. This process occurs in the small intestine, not the stomach, when the fat is mixed with bile, an akaline fluid secreted by the liver and stored in the gall bladder.

● **PROBLEM** 838

Given that 1 gram of sugar, starch, or other dry carbo-hydrate requires one liter of oxygen for complete oxi-dation and one gram of vegetable oil or other dry fat requires about 2 liters, answer the following: (a) Why is there a difference in the volume of oxygen required. (b) If mineral oil, paraffin wax, and hydrocarbons are foods would they have fewer calories per unit weight than vegetable oils? (c) Can hydrocarbons be a food source for man?

Solution:  Consider what is meant by oxidation and the general chemical composition of the substances mentioned.

Complete oxidation reactions may be defined as the addition of oxygen to an organic compound to produce water ($H_2O$) and carbon  dioxide ($CO_2$) plus heat. The general formula of a fat may be written $(RCO_2)_3C_3H_5$, where the R's can be the same or different molecules. In other words, a fat may be defined as an ester of glycerol and three long-chain carboxylic acids. Carbo-hydrates are simple sugars or the substances that hydrolyze to yield simple sugars of general formula $C(H_2O)x$, where x may range from 3 to many thousands. Thus, you can proceed as follows:

(a) In fats, from the general formula, you see the percentage of C and H in fats is greater than that in carbohydrates. The percentage of O is less in fats than in carbohydrates. The amount of oxygen required in combustion is a direct function of the C, H, and O percentages in the organic compound. The greater the C, H percentage, the more oxygen is required. Thus, fats should need much more oxygen for complete combustion than carbohydrates.

(b) Mineral oil, paraffin wax, or hydrocarbons should yield more calories per unit weight than vegetable oil. From (a), you know that the amount of oxygen required is proportional to the amount of heat produced upon oxidation; this is a direct function of C, H, and O percentage in the organic compound. Mineral oil, paraffin wax, and hydrocarbons have a higher C and H content than fats. (Hydrocarbons are 100% H and C; whereas fats are part oxygen.) Therefore, mineral oil, paraffin and hydrocarbons yield more calories (or heat).

(c) Hydrocarbons can not be used as a food source. Food is not useful to the body until the food particles have been degraded and oxidized by enzymes to yield heat. The human body possesses no enzyme systems that can perform the degradation of hydrocarbons.

● **PROBLEM** 839

If you administer too much insulin to a diabetic, it may cause a condition called "insulin shock". Explain this condition. Why does it come about?

Solution: To solve this problem, you must know how insulin, a hormone, (chemicals that stimulate or depress certain functions of the body) works in the body.

The addition of insulin in the body stimulates glycogenesis, a process which removed glucose from the blood and converts it to glycogen. Insulin also facilitates the transport of glucose across the membranes of muscle cells and adipose (fatty) tissue.

If too much insulin is added, too much glucose is removed from the blood. The concentration of glucose in the blood becomes inadequate, a condition called hypoglycemia results. This "insulin shock" or hypoglycemia may be characterized by all or any of the following symptoms: weakness, trembling, profuse perspiration, rapid heart beat, delirium, and loss of consciousness.

NUCLEIC ACIDS

● **PROBLEM** 840

The single DNA molecule in an E. coli chromosome (MW = $2.80 \times 10^9$) contains about 4.5 million mononucleotide units which are spaced about 3.4 Å apart. Calculate the total length of this DNA molecule and compare it with the length of the E. coli cell.

Solution: There are 4.5 million mononucleotide units spaced 3.4 Å apart in a single E. coli DNA molecule. To solve the for length of this molecule, multiply the number of units by the length of the space.

length = $(4.5 \times 10^6$ units)$(3.4$ Å/unit) = $1.53 \times 10^7$ Å

The length of the E. coli cell is 20,000 Å. Therefore, the DNA molecule is $\frac{1.53 \times 10^7}{20,000}$ or 765 times as long as the cell.

● PROBLEM 841

A single DNA strand begins with the sequence ATGGACGTATTC. Write the complementary DNA strand sequence. (A is adenine, T thymine, G guanine and C cytosine.)

Solution: DNA, deoxyribonucleic acid, may be described as the genetic blueprint of cells and is responsible for heredity.

DNA strands consist of a sugar (deoxyribose) – phosphate "backbone" and four important bases. The bases are divided into two main classes: purines and pyrimidines. There exists two bases under each class. The purines, are adenine and guanine; the pyrimidines are cytosine and thymine. In a particular DNA strand, these bases alternate with each other in a specific way - the sequence. DNA structures consist of 2 strands, each the complement of the other. Each complement refers to the fact that the bases on one strand pair up with the bases on the other strand in a particular fashion through hydrogen bonding. This bonding holds the two strands together. Adenine (A) will pair with thymine (T). Guanine (G) will pair with cytosine (C). Only these pairings produce effective hydrogen bonding.

Thus, to determine the complementary DNA strand sequence of ATGGACGTATTC, determine the bases that correspond to the bases present for effective hydrogen bonding to occur. In the complementary strand thymine will occupy the same spaces as adenine, adenine will occupy same positions as thymine, quanine the same positions as cytosine and cytosiné the same as guanine. Thus the complementary strand is TACCTGCATAAG. This is seen in the accompanying figure.

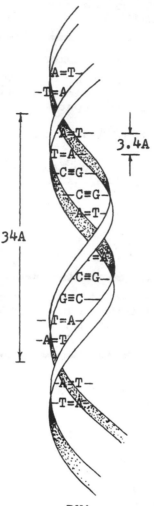

DNA

It is known that DNA (Deoxyribonucleic acid) exists as
a two stranded molecule. Evidence for this stems from
the fact that when $^{15}N$ isotope is added to bacteria
that are reproducing, 50% of the nitrogen in the DNA of
the first new generation of bacteria is $^{15}N$. (a) How
does this observation support the double helix (2
strands) structure of DNA; (b) If the first new generation of bacteria were grown and reproduced on only
$^{14}N$, what isotopic makeup would you expect to find in
the DNA of the second generation?

Solution:  (a) You are told that the new generation had
DNA of 50% $^{15}N$ isotope.

The parents of this generation possess DNA with
only $^{14}N$. For their progeny, the new generation, to have
50% $^{15}N$, DNA had to be synthesized. If all the DNA of
the progeny were synthesized, its DNA content would be

100% $^{15}$N. But only 50% of the nitrogen is $^{15}$N. This means, therefore, that half of the DNA from the parents, which is not newly synthesized, is used in the new generation. Half of the DNA is newly synthesized ($^{15}$N) and half already present ($^{14}$N). This implies the existence of two strands. This can be better seen if you write

←DNA strands

$^{15}$N        $^{14}$N        .        If each strand is 100%

of that type of nitrogen, and there are 2 strands, each type makes up only 50% of the total nitrogen.

(b) The new generation is reproducing on $^{14}$N. The progeny will have one half of its DNA newly synthesized on $^{14}$N and half of its DNA already present. You can picture it as the following:

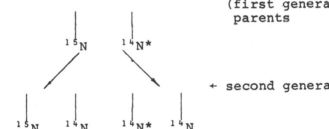

(first generation progeny)
parents

$^{15}$N        $^{14}$N*

← second generation progeny

$^{15}$N        $^{14}$N        $^{14}$N*        $^{14}$N

The starred nitrogen 14 is that already present. The unstarred was that newly synthesized. From this, you see that half of the second new DNA generation would be expected to contain only $^{14}$N strands (starred and unstarred), and the other half would contain 50% $^{14}$N and 50% $^{15}$N.

# CHAPTER 24

# POLLUTION PROBLEMS

● PROBLEM 843

One of the major atmospheric pollutants emitted by fuel-combustion power stations is a sulfur oxide mixture generally designated as $SO_x$. It consists mainly of $SO_2$ but may contain anywhere from 1 to 10 % $SO_3$ computed as percent by weight of the total mass. What would be the approximate range to assign to x?

Solution:  One finds x by determining the number of O atoms present for each S present. For purposes of calculation assume that one always has 100 g of the mixture of $SO_2$ and $SO_3$. Thus when 1 % of the mixture is $SO_3$, there will be 1 g of $SO_3$ present and 99 g of $SO_2$ present. One then finds the number of moles present of each element. The ratio of S : O is then determined.

Solving for x when 1 % of the mixture is $SO_3$: (MW of $SO_2$ = 64, MW of $SO_3$ = 80). The number of moles present is equal to the number of grams present divided by the molecular weight.

For $SO_3$: $\dfrac{1 \text{ g}}{80 \text{ g/mole}} = 1.25 \times 10^{-2}$ moles

For $SO_2$: $\dfrac{99 \text{ g}}{64 \text{ g/mole}} = 1.55$ moles.

Because there is one S present in each compound the total amount of S is 1.55 moles + $1.25 \times 10^{-2}$ moles or 1.56 moles. In 1 mole of $SO_3$ there are 3 moles of O, therefore in $1.25 \times 10^{-2}$ moles of $SO_3$ there is three times that amount.

no. of moles of O in $SO_3$ = $3 \times 1.25 \times 10^{-2} = 3.75 \times 10^{-2}$

There are 2 moles of O in 1 mole of $SO_2$, thus in 1.55 moles of $SO_2$ there are 3.10 moles of O. The total number of moles of O present is equal to the sum of the moles contributed by the $SO_2$ and the $SO_3$.

total no. of moles of O = $3.75 \times 10^{-2} + 3.10 = 3.14$ moles

The formula for this compound is $S_{1.56}O_{3.14}$. The simplest formula is $SO_{2.01}$. When there is 1 % of $SO_2$ present, x = 2.01. x = 3.14/1.56.

When there is 10 % of $SO_3$ present, a similar method is used to find x.

1) Find the number of moles present of $SO_2$ and $SO_3$.

2) Determine the number of moles of S and O contributed by the $SO_2$ and $SO_3$

3) write the formula and simplify.

Solving for x:

1) For a 100 g sample of gas where 10 % of the total weight is $SO_3$, 10 g of the gas is $SO_3$ and 90 g is $SO_2$.

no. of moles of $SO_3$ = $\dfrac{10 \text{ g}}{80 \text{ g/mole}}$ = $1.25 \times 10^{-1}$ moles.

no. of moles of $SO_2$ = $\dfrac{90 \text{ g}}{64 \text{ g/mole}}$ = $1.41$ moles

2) The total number of moles of S present is equal to the sum of the number of moles of $SO_2$ and $SO_3$ present because each compound contributes 1 S per mole of compound present.

no. of moles of S = $1.25 \times 10^{-1} + 1.41 = 1.54$ moles

no. of moles of O from $SO_3$ = $3 \times 1.25 \times 10^{-1}$

$$= 3.75 \times 10^{-1} \text{ moles.}$$

no. of moles of O from $SO_2$ = $2 \times 1.41 = 2.82$ moles

total no. of moles of O = $3.75 \times 10^{-1} + 2.82 = 3.20$ moles.

3) The formula for this compound can be written $S_{1.54}O_{3.20}$. x = 3.20/1.54= 2.08, thus the simplest formula is $SO_{2.08}$.

The range of x then is from 2.01 to 2.08.

● **PROBLEM** 844

Analysis of the exhaust composition of the supersonic transport for one hour of flight yielded the following information:

| Compound | Molecular Weight (g/mole) | Mass (g) |
|---|---|---|
| $H_2O$ | 18 | $8.0 \times 10^7$ |
| $CO_2$ | 44 | $6.6 \times 10^7$ |
| CO | 28 | $3.6 \times 10^6$ |
| NO | 30 | $3.6 \times 10^6$ |

Determine the mole fraction of $CO_2$ in this mixture.

Solution: To determine the mole fraction of $CO_2$ in the mixture, we divide the number of moles of $CO_2$ by the total number of moles present. The number of moles of each component is found by dividing the mass of that component by its molecular weight. Thus, for each component:

moles $H_2O$ = $8.0 \times 10^7$ g/18 g/mole = $4.4 \times 10^6$ moles

moles $CO_2$ = $6.6 \times 10^7$ g/44 g/mole = $1.5 \times 10^6$ moles

moles CO = $3.6 \times 10^6$ g/28 g/mole = $1.3 \times 10^5$ moles

moles NO = $3.6 \times 10^6$ g/30 g/mole = $1.2 \times 10^5$ moles

The total number of moles is

$$4.4 \times 10^6 + 1.5 \times 10^6 + 1.3 \times 10^5 + 1.2 \times 10^5$$

$$= 6.2 \times 10^6 \text{ moles.}$$

The mole fraction of $CO_2$ is then

$$\text{mole fraction } CO_2 = \frac{\text{moles } CO_2}{\text{total moles}} = \frac{1.5 \times 10^6 \text{ moles}}{6.2 \times 10^6 \text{ moles}} = 0.24.$$

● **PROBLEM** 845

Phosphate can be removed from sewage effluent by adding magnesium oxide and ammonium ion according to the reaction

$$5H_2O + H_2PO_4^- + MgO + NH_4^+ \rightarrow Mg(NH_4)PO_4 \cdot 6H_2O.$$

The magnesium ammonium phosphate hexahydrate $(Mg(NH_4)PO_4 \cdot 6H_2O)$ precipitates out. In theory, how much of this could be produced if a city processes 5 million gal. (41.7 million lb.) daily of wastewater containing 30 ppm $H_2PO_4^-$? (1 pound = 453.59 g.)

Solution: The best way to approach this problem is to calculate how much $H_2PO_4^-$ is present in grams. Once the weight in grams is known one can calculate from the stoichiometry of the equation, the number of moles of $Mg(NH_4)PO_4 \cdot 6H_2O$ that will precipitate.

The wastewater has 30 ppm (parts per million) of $H_2PO_4^-$. In other words, 30 lbs. of $H_2PO_4^-$ are present in one million lbs. of wastewater. The city produces 41.7 million lbs. of wastewater. Thus, 30(41.7) = 1,251 lbs. of $H_2PO_4^-$ should be present. Because 1 pound = 453.59 grams; 1,251 lbs = 567,454 grams.

Compute how many moles of $H_2PO_4^-$ are present. A mole is defined as grams (weight)/molecular weight (M.W.). The M.W. of $H_2PO_4^- = 96.97$ g/mole. The number of moles of $H_2PO_4^-$ is

$$= \frac{567,454 \text{ g}}{96.97 \text{ g/mole}} = 5852 \text{ moles.}$$

According to the reaction equation, one mole of $H_2PO_4^-$ generates one mole of precipitate. The M.W. of the precipitate = 245 g/mole. Recalling the definition of a mole, the amount of $Mg(NH_4)PO_4 \cdot H_2O = (5852)(245) = 1.433 \times 10^6$ g $= 1.433 \times 10^3$ kg.

● PROBLEM 846

The $SO_2$ content in air may be determined by passing the air through water to produce a solution of sulfurous acid, $H_2SO_3$. The subsequent reaction of this solution with potassium permanganate, $KMnO_4$, establishes the amount of $SO_2$ absorbed from the air. (a) Write the balanced equation for the reaction of $SO_2$ with water to produce $H_2SO_3$. (b) Write the balanced equation for the reaction of $H_2SO_3$ (or $SO_3^{2-}$) with $KMnO_4$ (or $MnO_4^-$) to form $MnSO_4$, $K^+$ and $H_2SO_4$, among other products. (c) If 1000 liters of air are passed through water and the resulting solution requires $2.5 \times 10^{-5}$ moles of $KMnO_4$ for complete reaction (part b), what is the concentration (in ppm) of $SO_2$ in the air. Assume STP conditions.

Solution: (a) An $SO_2$ molecule reacts with a water molecule to form an $H_2SO_3$ molecule according to the following equation:

$$SO_2 + H_2O \rightarrow H_2SO_3$$

However, $H_2SO_3$ is a diprotic acid and dissociates into 2 protons and a sulfate ion

$$H_2SO_3 \rightarrow 2H^+ + SO_3^=$$

Thus, for every mole of $SO_2$ consumed 1 mole of $SO_3^=$ are produced.

(b) This reaction is a typical redox reaction and the equation must be balanced using the standard rules.

From the problem, one knows initially that

$$H_2SO_3 + KMnO_4 \rightarrow MnSO_4 + H_2SO_4 + K^+$$

This reaction is not balanced in terms of charge and elemental constituents. The first rule is to separate the above reaction into a reduction and an oxidation reaction.

red:          $KMnO_4 \rightarrow MnSO_4 + K^+$

Mn is reduced from +7 to +2 state.

ox:                    $H_2SO_3 \rightarrow H_2SO_4$

S is oxidized from +4 to +6 state.

Next one balances the oxygen and hydrogen atoms in these reactions using $OH^-$, $H^+$, and $H_2O$ molecules on either side of the equation depending upon whether the reaction is acidic or basic. Once the masses are balanced, electrons are used to balance charges and are placed on the appropriate side of the equation. Thus,

red:                    $KMnO_4 \rightarrow MnSO_4 + K^+$

The potassium ion and the $SO_4$ ion do not play significant roles in the balancing of this equation and are therefore unnecessary.

$$MnO_4^- \rightarrow Mn^{+2}$$

This reaction is acidic and therefore will be balanced using $H^+$ and $H_2O$.

$$8H^+ + MnO_4^- \rightarrow Mn^{+2} + 4H_2O$$

The $H_2O$ is placed on the right to balance the oxygen in $MnO_4^-$; the H in $H_2O$ is balanced on the left with $H^+$. The equation will be balanced in terms of elements and charge when electrons are added. Thus,

$$8H^+ + 5e^- + MnO_4^- \rightarrow Mn^{++} + 4H_2O$$

ox:                    $H_2SO_3 \rightarrow H_2SO_4$

the hydrogen ions are unnecessary and play insignificant roles in balancing this equation; therefore,

$$SO_3^= \rightarrow SO_4^=$$

Next place water on the left and balance using a sufficient number of hydrogen ions on the right.

$$H_2O + SO_3^= \rightarrow SO_4^= + 2H^+$$

Now balance off the charge using electrons:

$$H_2O + SO_3^= \rightarrow SO_4^= + 2H^+ + 2e^-$$

Once both oxidation and reduction processes are balanced, they are added together in the following manner.

red:    $[8H^+ + 5e^- + MnO_4^- \rightarrow Mn^{++} + 4H_2O] \times 2$

ox:     $[H_2O + SO_3^= \rightarrow SO_4^= + 2H^+ + 2e^-] \times 5$

These reactions are solved simultaneously to eliminate the electrons in both reactions. Thus,

$$16H^+ + 2MnO_4^- \rightarrow 2Mn^{++} + 8H_2O$$

$$\underline{5H_2O + 5SO_3^= \rightarrow 5SO_4^= + 10H^+}$$

Total: $6H^+ + 2MnO_4^- + 5SO_3^= \rightarrow 2Mn^{++} + 5SO_4^= + 3H_2O$

(c) To solve this part of the problem, one must know the number of moles of $SO_2$ in air is equal to the number of moles of $SO_3^=$ that are represented in the equation for the reaction with $KMnO_4$. This amount is determined from the stoichiometry.

For every mole of $SO_2$ consumed, 1 mole of $SO_3^=$ is formed, and for every 1 mole of $SO_3^=$ consumed, 2/5 mole of $MnO_4^-$ is also consumed. Thus,

number moles of $SO_2 = 5/2$ $(2.5 \times 10^{-5}) = 6.25 \times 10^{-5}$ moles

To obtain a concentration in units of ppm (parts per million), one needs to know the volume of $SO_2$ consumed and the total volume of air. The total volume of air is $1 \times 10^3$ liters. The volume of $SO_2$ consumed is

volume of $SO_2 = (6.25 \times 10^{-5}$ moles$)(22.4$ liter/mole$)$

$$= 1.4 \times 10^{-3} \; \ell,$$

since, at STP, one mole of gas will occupy 22.4 liters.

Thus, the ratio of $SO_2$ volume to total volume of air is

$$\frac{1.4 \times 10^{-3} \text{ liter}}{1 \times 10^3 \text{ liter}} = 1.4 \times 10^{-6}.$$

To convert this ratio into ppm, we make the denominator $10^6$ liters,

$$\frac{1.4 \times 10^{-3} \text{ liter}}{1 \times 10^3 \text{ liter}} \times \frac{10^3}{10^3} = \frac{1.4}{1 \times 10^6}$$

The concentration of $SO_2$ in the air is 1.4 ppm.

● **PROBLEM** 847

An approach to controlling industrial pollution is to recover waste products in usable form. In the reaction,

$$2H_2S + SO_2 \rightarrow 3S + 2H_2O$$

waste $H_2S$ and $SO_2$ are recovered as elemental sulfur liquid. How much sulfur can be recovered if 650,000 kg of $SO_2$ are deposited into the air?

<u>Solution</u>: From the stoichiometry of this equation, 2 moles of $H_2S$ react with 1 mole of $SO_2$ to produce 3 moles of S and

2 moles of $H_2O$.

To solve this problem, calculate how many moles of $SO_2$ are polluted into the air by dividing its weight by its molecular weight (MW $SO_2$ = 64 g/mole).

From the stoichiometry of the equation, determine how many moles of S would be produced, and then calculate the weight of S by multiplying the number of moles of S times its molecular weight (MW S = 32 g/mole). Thus,

$$\text{number moles of } SO_2 = \frac{650,000,000 \text{ g } SO_2}{64 \text{ g/mole}} = 1.02 \times 10^7 \text{ moles}$$

From the stoichiometry of the equation 1 mole $SO_2 \rightarrow$ 3 mole S or $1.02 \times 10^7$ mole $SO_2 \rightarrow 3.06 \times 10^7$ mole S.

$$\text{weight of S} = (3.06 \times 10^7 \text{ mole S})(32 \text{ g/mole})$$

$$= 97.9 \times 10^7 \text{ g} = 979,000 \text{ kg.}$$

● **PROBLEM** 848

Iodine pentoxide ($I_2O_5$) is a very important reagent because it can oxidize carbon monoxide (CO), which is a major pollutant of the atmosphere, to carbon dioxide ($CO_2$) in the absence of water. The reaction involved is

$$I_2O_5 + 5CO \rightarrow I_2 + 5CO_2$$

How many kg of $I_2O_5$ is needed to completely clear 140,000 kg of CO from the air?

Solution: From the stoichiometry of this equation, 1 mole of $I_2O_5$ completely reacts with 5 moles of CO. To solve this problem, calculate the number of moles of CO that will react by dividing its weight by its molecular weight of 28 g/mole. Determine, also, how many moles of $I_2O_5$ will completely react with this, and then calculate the weight of $I_2O_5$ that will completely react by multiplying the number of moles of $I_2O_5$ by its molecular weight (MW of $I_2O_5$ = 334 g/mole). Thus,

$$\text{number moles CO} = \frac{140,000,000 \text{ g CO}}{28 \text{ g/mole}} = 5.00 \times 10^6 \text{ mole.}$$

This means that one fifth of this number of moles will be the moles of $I_2O_5$ required. Recall, there will exist 5 moles of CO for every mole of $I_2O_5$. Therefore,

$$\text{moles } I_2O_5 = \frac{1}{5}(5 \times 10^6) = 1 \times 10^6 \text{ moles.}$$

This means that

$$(1 \times 10^6 \text{ moles})(334 \text{ g/mole}) = 3.34 \times 10^8 \text{ grams of}$$

$I_2O_5$ is needed to completely clear 140,000 kg of CO from the air.

● **PROBLEM** 849

The average concentration of particulate lead in a city with highly polluted air is 5 μg/m$^3$. If an adult respires 8500 liters of air daily and if 50 per cent of the particles below 1 μm in size are retained in the lungs, calculate how much lead is absorbed in the lungs each day. Of the particulate lead in urban air, 75 per cent is less than 1 μm in size.

Solution: To solve this problem (1) determine the total percentage of particulate lead remaining in the lungs, (2) determine the amount of lead absorbed into the lungs.

The problem states 75 % of particulate lead is less than 1 μm in size, therefore, 25 % of the lead is greater than 1 μm in size and does not remain inside the lungs. Of the 75 % particulate lead less than 1 μm in size, only 50 % of this is retained in the lungs. The total percentage of lead retained is 37.5 %.

The total volume of air-intake is 8500 liters, so that

total volume of air-intake = (8500)(10$^{-3}$ m$^3$/liter) = 8.5 m$^3$

The total amount of lead-intake is

amount of lead = (8.5 m$^3$)(5 μg/m$^3$) = 42.5 μg lead.

The total amount of lead absorbed is

amount of lead absorbed = (42.5 μg)(0.375) = 15.9 μg.

● **PROBLEM** 850

In 1932, many cities showed higher sunlight levels than in the previous year. Explain this observation.

Solution: To answer this problem, consider the effect of atmospheric particles on the amount of sunlight received and what happened in 1932.

By 1932 three years of slackened industrial activity had passed. Then as now, industries released massive amounts of particulate matter, i.e. solid and liquid aerosols, into the atmosphere. These particles have the ability to scatter and absorb light. Because industrial activity was drastically reduced, fewer particles were released that could scatter or absorb the sunlight. The cities, therefore, received more sunlight in 1932 than in years when industry was thriving.

It is interesting to note that because of particulate matter, cities receive about 15 to 20 % less solar radiation than do rural areas.

● PROBLEM 851

The air pollutant, sulfur dioxide, is considered particularly foul. Explain why there is so much concern with this compound.

Solution: To solve this problem, consider the reactions that can occur between sulfur dioxide ($SO_2$) and other elements, and/or compounds, in the natural environment.

$SO_2$ can enter the atmosphere from both natural and man-made sources. Besides causing eye irritation and plant damage, it can cause other health problems because of its ability to form acids in the atmosphere. When $SO_2$ comes in contact with water ($H_2O$), the following reaction occurs.

$$SO_2 + H_2O \rightarrow H_2SO_3$$
$$\text{sulfurous acid}$$

$SO_2$ can also be oxidized in air to sulfur trioxide ($SO_3$) by the reaction

$$2SO_2 + O_2 \rightarrow 2SO_3.$$

Sulfur trioxide, when exposed to water, produces sulfuric acid by the reaction

$$SO_3 + H_2O \rightarrow H_2SO_4$$
$$\text{sulfuric acid}$$

These two acids, sulfuric and sulfurous, account for the corrosion of metal parts exposed to the atmosphere. They are also responsible for the decay of stone statues and buildings. This occurs because most stones consist of $CaCO_3$ (calcium carbonate), which, upon addition of $H_2SO_4$, undergoes the following reaction:

$$CaCO_3 + H_2SO_4 \rightarrow CaSO_4 + H_2O + CO_2.$$

$CaSO_4$ is soluble in water and will wash away with the rain causing the wearing away of the stone.

● PROBLEM 852

If the carbon dioxide content of the air doubled, what

would happen to the temperature of the earth. Explain your answer.

Solution: This problem requires a knowledge of the greenhouse effect.

The light energy from the sun tends to be of short wavelengths that are not absorbed by the atmosphere. Upon striking the earth's surface much of this energy is converted to thermal or infrared energy. Infrared energy can be absorbed by carbon dioxide and thus is not permitted to escape from the earth. As the concentration of $CO_2$ (carbon dioxide) is increased, the efficiency of this light energy-trapping is increased, which causes a rise in temperature on the earth. This phenomenon is termed the greenhouse effect.

If the $CO_2$ concentration were doubled a rise in the earth's temperature would be expected. It has been estimated that the temperature increase would be 2-4°C.

● **PROBLEM** 853

The following observations have been made on a typical day in Los Angeles: The hydrocarbon content in the atmosphere starts to rise at about 6 A.M. It peaks at 8 A.M. and dies off at 10 A.M. A similar peaked curve occurs at 6 P.M. About 2 hours after each peak, the ozone concentration rises. How can you explain these observations?

Solution: The peaks in hydrocarbon content occur during the height of the rush hour - 8 A.M. and 6 P.M. This suggests that busses and automobiles may be responsible. Both machines emit large quantities of hydrocarbons into the atmosphere from their exhaust systems. The hydrocarbon content rises at 6 A.M. since, at this time, the rush is just beginning.

To understand why the ozone concentration rises after 2 hours, while the hydrocarbon concentration decreases after 2 hours (10 A.M.), consider the chemical reactions of hydrocarbons with elements that make up the atmosphere. Before doing this, consider the following: It is found that high-temperature combustion generates NO which, on exposure to oxygen, is converted to $NO_2$. $NO_2$ undergoes the following chain reactions:

$$NO_2 + light \rightarrow NO + O$$

$$O + O_2 \rightarrow O_3$$

$$O_3 + NO \rightarrow O_2 + NO_2$$

The $NO_2$ undergoes a reaction with light to produce NO and monoatomic oxygen (O), which is highly reactive. This, then reacts with oxygen (diatomic - $O_2$) to produce $O_3$ (ozone), a toxic element in the lower atmosphere. However, the ozone reacts with the NO that was previously produced to again form $NO_2$, which allows the cycle to repeat.

Now, based on the above information, consider what happens at rush hour. The cars and busses release hydrocarbons. Among these fragments of hydrocarbons are free radicals which are <u>extremely</u> reactive. They proceed to react with NO, reducing its concentration. When the NO is depleted, there is little of it available to react with ozone to produce $NO_2$ and $O_2$. Thus, $O_3$ builds up. It is found that this process takes 2 hours, which explains why the $O_3$ concentration increases after the peak hydrocarbon concentration.

● **PROBLEM** 854

Given that the earth's mean radius = $6.37 \times 10^6$ m, normal atmosphere pressure = $1.013 \times 10^5$ N/m$^2$, and the gravitational acceleration = 9.8 m/sec$^2$, what is the mass of the homosphere?

<u>Solution</u>: The homosphere is that part of the atmosphere that extends from the surface of the earth up to 80 kilometers. The constituent gases in the homosphere are well mixed. The proportions of the principal constituents at high altitudes do not differ to a great extent from those found at sea level. To find the mass of the homosphere, calculate the total mass of the atmosphere.

The homosphere is equal to 99.99% of the atmosphere and, hence, their masses are approximately equal.

The mass of the atmosphere can be derived from the atmospheric pressure. Since

$$\text{Pressure} = \frac{\text{Force}}{\text{Area}} \quad \text{and} \quad \text{Force} = \text{mass} \times \text{gravity}$$

Combining these two equations one obtains

$$\text{Pressure} = \frac{\text{mass} \times \text{gravity}}{\text{Area}}$$

Since one is solving for mass the equation can be rearranged to

$$\text{mass} = \frac{\text{Pressure} \times \text{Area}}{\text{gravity}}$$

One is given the atmospheric pressure and the force of gravity. The surface area of a sphere is equal to $4\pi r^2$. Substituting and solving for the mass of the homosphere: (N = kg-m/s$^2$)

$$mass = \frac{(1.013 \times 10^5 \text{ kg-m/m}^2\text{s}^2)(4\pi)(6.37 \times 10^6 \text{ m})^2}{(9.8 \text{ m/sec}^2)}$$

$$= 5.27 \times 10^{18} \text{ kg}.$$

● **PROBLEM** 855

Explain what is meant by primary, secondary, and tertiary treatment of sewage.

Solution: Sewage treatment is carried out in three stages known as primary, secondary, and tertiary. Each uses a different process to eliminate different substances from the sewage.

Primary treatment uses mechanical means to separate out solid objects such as sticks and rags. These solid objects are removed by means of a coarse screen. The liquid that passes through this screen flows into settling tanks, where insoluble material settles and forms sludge. This removes approximately one third of the pollutants.

Whereas primary treatment is mechanical, secondary treatment is biological. The liquid or effluent from the primary facility filters slowly through a bed of rock. Bacteria present consume most of the organic material. In many plants, oxygen is bubbled in to accelerate the growth of bacteria and, as such, the decomposition of the organic waste. This process removes about 90 % of the biodegradable materials in the water.

The last type of treatment, tertiary, is the most expensive. It is used to remove organic chemicals, nutrients, and excessive salts. To accomplish this, the treatment may include such processes as chemical coagulation, distillation and reverse osmosis.

● **PROBLEM** 856

What is eutrophication? What problems are associated with it? Why would a reduction of phosphate ions from waterways alleviate eutrophication in lakes and streams?

Solution: The enrichment of water with nutrients is a naturally occurring biological process called eutrophication. It is caused by excess nutrients in water that enable algae to grow to great abundance. The problem of massive algae growth occurs when it dies. As it dies, it sinks to the bottom of lakes and rivers and begins to decay. This decaying process consumes oxygen thereby deleting the oxygen content of the water. Fish and other aquatic life dependent on oxygen suffocate. Therefore, the decaying of excessive

amounts of algae can lead to the death of marine life.
This is one major problem associated with eutrophication.

A second problem also results from the depletion of
oxygen. Remember that anaerobic bacteria, i.e. bacteria
that can live without oxygen, still function effectively
under the conditions described above. These bacteria feed
on decaying algae. As a result of their metabolic process,
they produce compounds such as hydrogen sulfide, which
cause the putrid odor associated with decaying organic
matter. The water becomes foul smelling and foul tasting.

A reduction of phosphate would alleviate eutrophication because it is one of the major nutrients of algae. A
smaller amount of phosphate in the water would limit the
growth of algae.

● **PROBLEM** 857

The combined effects of pollution have been known to bring
about the phenomena of synergism and antagonism. Explain
each of these terms. Provide specific examples of each.

Solution:  The effects of individual pollutants can be
**disastrous**  to health, while the combined action of two
or more pollutants can be completely different on health
and/or the environment.

When the combined effects of two or more **pollutants**
are more severe **or** qualitatively different from the individual effects, a condition called synergism exists. For
example, in 1966 and 1967, there was injury to peanut crops
due to the synergistic action of ozone and sulfur dioxide.

A condition termed antagonism exists when the combined
effects are less severe. For example, cyanides in industrial
wastes are quite poisonous to aquatic life, however, in the
presence of nickel, a nickel-cyanide complex is formed. This
nickel-cyanide complex is not as toxic as the cyanides
originally produced.

● **PROBLEM** 858

(A) Explain the purpose of a coliform bacteria count in
natural water. (B) What is meant by hard or refractory
organic compounds?

Solution:  (a) One of the most important problems
associated with polluted water is the possible presence
of disease-causing bacteria. A test for the presence of
bacteria is made by measuring the amount of common coliform bacteria in water. (Coliform bacteria are harmless

and reside in the large intentines of humans.) Coliform bacteria are present in plentiful quantities in feces, which become part of sewage wastes. Since coliform bacteria do not thrive in an aquatic environment, they give an indication of how recently and to what extent sewage pollution has occurred by their count (which is proportional to the amount of disease-causing bacteria present). If the count is high, human sewage contamination has taken place, possibly accompanied by the presence of dangerous bacteria.

(b) Hard or refractory organic compounds are nondegradable substances that decay very slowly in water. Degradation refers to the breaking down of substances into simpler forms. Examples of nondegradable substances are certain pesticides and detergents.

● PROBLEM 859

A chemist has a quantity of steam. He wants to determine whether or not it has been polluted with acidic mine drainage (AMD). How should he go about determining this?

Solution: AMD occurs yearly when millions of tons of sulfuric acid seep out from coal and copper mines to pollute waterways.

It begins by the oxidation of iron pyrite ($FeS_2$) to sulfate:

$$FeS_2 \xrightarrow[H_2O]{O_2} Fe^{2+} + SO_4^{2-}.$$

The ferrous ions are then oxidized to ferric ions:

$$Fe^{2+} \xrightarrow[H_2O]{O_2} Fe^{3+}.$$

This process can be sped up by microorganisms. The ferric ions can react with water to give hydrogen ions and a precipitate of hydrated ferric oxide:

$$Fe^{3+} \xrightarrow{H_2O} Fe_2O_3 \cdot H_2O + H^+.$$

Thus, acidic mine drainage is a diluted sulfuric acid solution carrying iron. If the steam has AMD, it must carry iron. Addition of thiocyanate to ferric (iron) solutions will yield an intensely red-colored complex ion. The reaction is as follows:

$$Fe^{3+} + SCN^- \rightarrow FeSCN^{2+}$$

thio-      red-colored
cyanate    complex ion

Thus, the chemist can detect for AMD pollution in the steam by adding thiocyanate. If a red-colored material is formed, he knows it is contaminated.

A chemist dilutes two 10-ml samples of waste water to 300 ml with aerated water. One sample is analyzed for dissolved oxygen at 20°C immediately, but the second sample is incubated for 5 days before being analyzed for dissolved oxygen. The results are

$$(DO)^{20°C}_{0 \text{ days}} = 7.9 \text{ mg/liter and } (DO)^{20°C}_{5 \text{ days}} = 1.0 \text{ mg/liter}$$

What is $BOD_5$ of the waste water?

**Solution:** BOD stands for Biochemical Oxygen Demand. It measures the potential oxygen requirement of sewage and of

sewage polluted waters. $(DO)^{20°C}$ stands for the concentration of dissolved oxygen at 20°C.

The numerical value of BOD can be found by measuring the dissolved oxygen content (DO) of the sample before and after the 5 day incubation period. BOD is the difference between these dissolved oxygen concentrations (after accounting for corrections of dilution and any additional organic material that might be present in the diluting water).

Proceed as follows:

The 10 ml sample was diluted to 300 ml or 0.3 ℓ. The **DO** concentrations at day 0 and day 5 are multiplied by 0.3 ℓ to determine the amount of oxygen used on those days

$(7.9 \text{ mg/ℓ})(0.3 \text{ ℓ}) = 2.37 \text{ mg}$        Day 0

$(1.0 \text{ mg/ℓ})(0.3 \text{ ℓ}) = 0.3 \text{ mg}$        Day 5

These amounts are subtracted to determine the BOD of the sample.

$$2.37 - 0.3 = 2.07 \text{ mg}$$

The original sample was 10 ml; the density of water is 1 g/ml. Thus, 10 ml weighs 10 g. The BOD of this sample is 2.07 mg/10 g. BOD's are usually expressed in ppm. Converting 2.07 mg/10 g to ppm one ontains 207 ppm.

For a diversified warm-water biota, including game fish, the DO concentrations should be at least 5 mg/ℓ or 5 ppm. A body of water is considered polluted when the DO concentration drops below the level necessary to sustain a normal biota for that water. The BOD range for untreated municipal sewage is 100-400 ppm.

# CHAPTER 25

# APPLIED GAS PROBLEMS

● **PROBLEM** 861

The pressure in an automobile tire is measured by a tire gauge in winter at 0°C as 30 pounds per square inch (psi) (2.0415 atm). The gauge measures the difference between the actual tire pressure and the atmospheric pressure (15 psi). Assume that the tire does not leak air and does not change its volume. What will the new gauge pressure be if the same tire is measured during the summer at 50°C?

Solution: We will solve this problem by employing a modified form of the ideal gas law. The ideal gas equation is PV = nRT, where P is the total pressure (P = atmospheric pressure + gauge pressure), V is the tire volume, n the number of moles of air, R the gas constant, and T the absolute temperature. Bearing in mind the fact that n, V, and R are constant, we may write this equation for the two separate sets of conditions, as

$$P_{summer} V = n\, RT_{summer} \quad \text{and} \quad P_{winter} V = n\, RT_{winter}.$$

Dividing the first by the second gives

$$\frac{P_{summer} V}{P_{winter} V} = \frac{n\, RT_{summer}}{n\, RT_{winter}} \quad \text{or} \quad \frac{P_{summer}}{P_{winter}} = \frac{T_{summer}}{T_{winter}}.$$

Solving for $P_{summer}$, we have

$$P_{summer} = \frac{T_{summer}}{T_{winter}} \times P_{winter}.$$

$$P_{winter} = \text{atmospheric pressure} + \text{gauge pressure}$$

$$= 15 \text{ psi} + 30 \text{ psi} = 45 \text{ psi } (3.06 \text{ atm}).$$

$T_{summer}$ = 50°C = 323°K, and $T_{winter}$ = 0°C = 273°K. Hence,

$$P_{summer} = \frac{T_{summer}}{T_{winter}} \times P_{winter} = \frac{323°K}{273°K} \times 45 \text{ psi} = 53 \text{ psi}.$$
$$(3.606 \text{ atm})$$

The gauge pressure will be $P_{summer}$ - atmospheric pressure, or 53 psi - 15 psi = 38 psi (2.586 atm)

● **PROBLEM** 862

The pressure in interplanetary space is estimated to be of the order of $10^{-16}$ torr. This corresponds to a density of about 4 molecules per cubic centimeter. What temperature must be assigned to the molecules in interplanetary space for this to be true?

Solution:  To solve this problem, one assumes that interplanetary space is an ideal gas at a low concentration and at a very low pressure. This means that the ideal gas law can be used, i.e.,

$$PV = n\,RT$$

The pressure, P, is known $10^{-16}$ torr = $10^{-16}$ **torr x** (1 atm/760 torr) = $1.32 \times 10^{-19}$ atm and R, the gas constant = 0.082 atm-liter/mole-°K, is also known. The volume, V, and the number of moles, n, are unknown, but if n is divided by V, then this value becomes a density (moles/liter) and can be determined from the given density of 4 molecules/cm. To do this, (1) convert the number of molecules to moles using Avogadro's number and (2) convert $cm^3$ to units of $\ell$ using the conversion 1 $cm^3$ = 1 m$\ell$ = 0.001 $\ell$. Thus, the density is

$$\left( \frac{4 \text{ molecules}}{1 \text{ cm}^3} \right) \left( \frac{1000 \text{ cm}^3/\ell}{6.02 \times 10^{23} \frac{\text{molecules}}{\text{mole}}} \right)$$

$$= 6.64 \times 10^{-21} \text{ moles}/\ell$$

Rearranging the ideal gas low from PV = n RT to $\frac{P}{R} \cdot \frac{V}{n} = T$, and substituting the known values, yields

$$T = \frac{(1.32 \times 10^{-19} \text{ atm})}{\left[ 0.082 \frac{\text{atm-liter}}{°K\text{-mole}} \right]} \times \left( \frac{1}{6.64 \times 10^{-21} \frac{\text{moles}}{\ell}} \right)$$

$$= 2.42 \times 10^2 \text{ °K}$$

T = 242°K = 242°K - 273° = - 31°C.

Assume that a thermonuclear explosion 1000 ft above ground quickly heats 10,000 ft.$^3$ of air at standard pressure from 273°K to 1,092,273°K. The air would attempt to expand to what volume as an outgoing pressure-blast wave develops?

Solution: To solve this problem, Charles' Law is used,

$$\frac{V_1}{V_2} = \frac{T_1}{T_2} \ ,$$

where V = volume and T = absolute temperature. The pressure, (since the expansion is done in open space) remains at standard pressure of 1 atm. Thus,

$V_1 = 10,000$ ft.$^3$, $T_1 = 273°K$,  $T_2 = 1,092,273°K$
and $V_2$ is solved for

$$V_2 = \frac{V_1 T_2}{T_1} = \frac{(10,000 \text{ ft.}^3)(1,092,273°K)}{(273°K)} = 40,010,000 \text{ ft.}^3 .$$

The air expands by a factor of

$$\frac{40,010,000 \text{ ft.}^3}{10,000 \text{ ft.}^3} = 4,001.$$

An aerosol spray at room temperature, 25°C, is under a pressure of 10 atm at a volume of 0.5 ℓ. If someone releases all the contents to fill an empty 4.6 ℓ box at standard pressure, what is the temperature of the contents inside the box?

Solution: To solve this problem the general gas law, assuming ideal gas conditions, for two different states is used.

$$\frac{P_1 V_1}{T_1} = \frac{P_2 V_2}{T_2} \ ,$$

where $P_1 = 10$ atm, $V_1 = 0.5$ ℓ, $T_1 = 298°K$, $P_2 = 1$ atm, (standard pressure), $V_2 = 4.6$ ℓ, and $T_2$ is unknown. Solving for $T_2$ and substituting the known values into this equation gives

$$T_2 = \frac{P_2 V_2 T_1}{P_1 V_1} = \frac{(1 \text{ atm})(4.6 \text{ ℓ})(298°K)}{(10 \text{ atm})(0.5 \text{ ℓ})} = 274°K = 1°C.$$

The decrease in temperature is significant; the change is from 25°C to near freezing.

In the Van Slyke method for determining $CO_2$ capacity in blood, you place the sample over mercury in a closed flask. $CO_2$ is released from the sample by the addition of acid. The volume and pressure of the released is then measured. In a 0.2 ml sample of blood, the $CO_2$ released exerts a pressure of 162 mm Hg at a temperature of 27°C and occupies a volume of 0.5 cc. What is the corresponding volume of the $CO_2$ at standard temperature and pressure (STP)?

Solution: STP refers to conditions where the temperature = 0°C or 273°K and the pressure = 760 mm Hg or 1 atm. To find the volume of the $CO_2$ released under these conditions, apply the Combined Gas Law.

$$\frac{P_1 V_1}{T_1} = \frac{P_2 V_2}{T_2} = \text{constant}$$

where P = pressure, V = volume, and T = absolute temperature and the subscripts refer to different states.

To solve this problem substitute the known values into this equation:

$P_1 = 162$ mm $\qquad\qquad$ $P_2 = 760$ mm

$V_1 = 0.5$ cc $= 0.5$ ml $\qquad$ $V_2 = ?$

$T_1 = 27°C = 300°K$ $\qquad$ $T_2 = 273°K$

where $P_2$ and $T_2$ are the standard conditions at STP. Thus,

$$\frac{(162 \text{ mm})(0.5 \text{ ml})}{(300°K)} = \frac{(760 \text{ mm})V_2}{(273°K)}$$

Solve for $V_2$, the volume of $CO_2$ at STP,

$$V_2 = \frac{(162 \text{ mm})(0.5 \text{ ml})(273°K)}{(760 \text{ mm})(300°K)} = 0.0969 \text{ ml} = 0.0969 \text{ cc.}$$

Carbon monoxide poisons humans by irreversibly binding to hemoglobin in the bloodstream. Although not dangerous in minute concentrations over long exposure times, a CO concentration of 0.4 % by volume is quickly lethal. If a defective automobile in a sealed garage having a volume of $4.1 \times 10^4$ liters produces 0.60 mole of CO per minute, how long will it take to reach this lethal concentration of CO? Assume that the volume remains constant at 27°C and the pressure at 1 atm.

Solution: This problem is solved by first converting the

concentration, 0.4 % by volume, to a partial pressure and determining the number of moles that will give rise to this pressure. The time it takes to reach this pressure is then determined by dividing the number of moles of CO corresponding to this pressure by the rate of CO formation.

When the concentration of CO is lethal, it occupies 0.4 % of the atmosphere by volume. Therefore, its partial pressure, p, is 0.4 % of the total pressure, or p = 0.4 % × 1 atm = 0.004 atm.

The ideal gas equation reads $pV = nRT$, or $n = \frac{pV}{RT}$ where n is the number of moles, V is the volume, R the gas constant, and T the absolute temperature. Since V = $4.1 \times 10^4$ liters, R = 0.082 liter-atm/mole-deg, and T = 27°C = 300°K, the number of moles of CO corresponding to a partial pressure of 0.004 atm is

$$n = \frac{pV}{RT} = \frac{0.004 \text{ atm} \times 4.1 \times 10^4 \text{ liters}}{0.082 \text{ liter-atm/mole-degree} \times 300°K} = 6.7 \text{ moles.}$$

Since the automobile produces CO at a rate of 0.60 mole/min, it will take 6.7 moles/0.60 mole/min $\cong$ 11 minutes for CO to reach the lethal concentration.

● **PROBLEM** 867

The average density of the Universe is very low, various estimates of the average density being between 1.0 × $10^{-30}$ g/cm³ and 3.0 × $10^{-28}$ g/cm³. Using an intermediate value for the average density, 1.5 × $10^{-29}$ g/cm³, and assuming that the Universe consists solely of hydrogen atoms, what is the average volume of space that contains a single hydrogen atom?

Solution: To solve this problem, we will convert the density (in g/cm³) to a molar density (moles/cm³) and then express this as the number of atoms per unit volume (number density). Inverting the number density will give us the volume occupied by a single atom.

To convert from density to molar density, we divide by the atomic weight of hydrogen (1.0 g/mole). This stems from the fact that grams/atomic weight = mole. Hence,

molar density = density ÷ atomic weight

$$= 1.5 \times 10^{-29} \text{ g/cm}^3 \div 1.0 \text{ g/mole}$$

$$= 1.5 \times 10^{-29} \text{ mole/cm}^3.$$

To convert from molar density to number density, we must multiply by the number of atoms per mole (Avogadro's number). Hence number density = molar density × Avogadro's number

$$= 1.5 \times 10^{-29} \text{ mole/cm}^3 \times 6 \times 10^{23} \text{ atoms/mole}$$

$$= 9.0 \times 10^{-6} \text{ atom/cm}^3.$$

The average volume occupied by a single H atom is the reciprocal of the number density, or

$$\text{volume per atom} = \frac{1}{\text{number density}} = \frac{1}{9.0 \times 10^{-6} \text{ atom/cm}^3}$$

$$= 1.1 \times 10^5 \text{ cm}^3/\text{atom}.$$

Hence, on the average, each $1.1 \times 10^5$ cm³ (110 liters) is occupied by a single hydrogen atom.

● **PROBLEM** 868

An anaesthetic can be prepared by mixing gaseous cyclo-propane ($C_3H_6$, molecular weight = 42 g/mole), and oxygen ($O_2$, molecular weight = 32 g/mole). If a gas cylinder is prepared with cyclopropane at a partial pressure of 170 torr (0.223 atm) and oxygen at a partial pressure of 570 torr(0.75 atm), calculate the ratio of the number of moles of cyclopropane to that of oxygen $\left[ n_{C_3H_6}/n_{H_2O} \right]$.

Solution: We solve this problem by relating the ratio of moles, $n_{C_3H_6}/n_{H_2O}$, to the ratio of partial pressures, $P_{C_3H_6}/P_{H_2O}$. For a given volume V and fixed absolute temperature T, the ideal gas law reads:

$$P_{C_3H_6} = n_{C_3H_6} \frac{RT}{V}$$

for cyclopropane, and

$$P_{O_2} = n_{O_2} \frac{RT}{V}$$

for oxygen, where R is the gas constant. Dividing the first of these equations by the second gives

$$\frac{P_{C_3H_6}}{P_{O_2}} = \frac{n_{C_3H_6}}{n_{O_2}} \frac{RT/V}{RT/V} = \frac{n_{C_3H_6}}{n_{O_2}}$$

Since $P_{C_3H_6}$ = 170 torr and $P_{O_2}$ = 570 torr,

$$\frac{P_{C_3H_6}}{P_{O_2}} = \frac{170 \text{ torr}}{570 \text{ torr}} = 0.30 = \frac{n_{C_3H_6}}{n_{O_2}}$$

or, $n_{C_3H_6}/n_{O_2}$ = 0.30.

A 7600 liter compartment in a space capsule, maintained at an internal temperature of 27°C, is designed to hold one astronaut. The human body discharges 960 g of carbon dioxide gas ($CO_2$, molecular weight = 44 g/mole) each day. If the initial partial pressure of carbon dioxide in the compartment is zero, how much $CO_2$ must be pumped out the first day to maintain a partial pressure of no more than 4.1 torr (0.0054 atm) ?

Solution: To solve this problem we will calculate the mass of $CO_2$ that corresponds to a partial pressure of 4.1 torr and subtract it from the mass of $CO_2$ discharged by the astronaut.

The ideal gas law reads

$$p_{CO_2} V = n_{CO_2} RT$$

where $p_{CO_2}$ is the partial pressure of $CO_2$, V is the volume of the cabin, $n_{CO_2}$ is the number of moles of $CO_2$, R is the gas constant, and T is the absolute temperature of the cabin. Since $n_{CO_2} = m/MW$, where m is the mass of $CO_2$ and MW is the molecular weight, we can write this equation as

$$p_{CO_2} V = \frac{m}{MW} RT$$

$$m = \frac{p_{CO_2} V (MW)}{RT}$$

Now, $p_{CO_2}$ = 4.1 torr = 4.1 torr × 1 atm/760 torr = 4.1/760 atm, V = 7600 liters, MW = 44 g/mole, R = 0.082 liter-atm/mole-degree, and T = 27°C = 300°K. Hence,

$$m = \frac{p_{CO_2} V (M.W.)}{RT}$$

$$= \frac{4.1/760 \text{ atm} \times 7600 \text{ liters} \times 44 \text{ g/mole}}{0.082 \text{ liter-atm/mole-degree} \times 300°K} \cong 73 \text{ g.}$$

The astronaut discharges 960 g of $CO_2$ each day, and we wish to maintain the amount at 73 g. Hence, the first day, 960g- 73 g = 887 g of $CO_2$ must be pumped out.

If the partial pressure of water vapor in air at 20°C is 10 mm, answer the following questions: (A) What is the relative humidity? (B) What is the approximate dew point?

The vapor pressure of water is 17.54 mm at 20°C and 9.21 mm at 10°C.

Solution: (A) Relative humidity is defined as the ratio of the partial pressure of water vapor in the air to the vapor pressure of water at that temperature of the air. Given that the partial pressure of water vapor at 20°C is 10 mm and that its vapor pressure is 17.54 mm at 20°C:

$$\text{relative humidity} = \frac{10 \text{ mm}}{17.54 \text{ mm}} = 0.57.$$

Expressed as a percent; relative humidity = 57 %.

(B) Most of the time, air is not saturated with water vapor. However, it can be saturated if it is cooled to a temperature where the vapor pressure of water becomes equal to the partial pressure of the water vapor in the air. This is the temperature of the dew point. To calculate the dew point in this problem, change the temperature until the vapor pressure equals 10 mm (the partial pressure). Given that the vapor pressure is 17.54 at 20°C and 9.21 at 10°C, the dew point must be between these two temperatures. To find the exact temperature (the dew point) the following proportion can be set up where 10 mm is the vapor pressure: Let x = dew point.

$$\frac{10 \text{ mm}}{17.54 \text{ mm}} = \frac{x}{20°C}$$

Solving for x; x = 11.4°C.

● **PROBLEM** 871

A controlled environment for guinea pigs is made by constructing a 294 liter box to be maintained at 21°C and 40 % relative humidity, and in which the entire air volume is **changed every minute.** In order to maintain this relative humidity, what weight of water must be added to the dry air flow every minute. The equilibrium vapor pressure for water at 21°C is about 19 torr.

Solution: To solve this problem, we use the relative humidity and the equilibrium vapor pressure of water to obtain the partial pressure of water. The ideal gas law is then used to calculate the amount of water corresponding to this partial pressure.

The relative humidity is defined as the percent ratio of the partial pressure p to the equilibrium vapor pressure, $p_e$. Hence,

**relative humidity** $= \dfrac{p}{p_e} \times 100$ %, or

$$p = p_e \times \frac{\text{relative humidity}}{100\ \%} = 19 \text{ torr} \times \frac{40\ \%}{100\ \%}$$

$$= 19 \times \frac{40}{100} \text{ torr} = 19 \times 0.4 \text{ torr}.$$

Converting to atmospheres, $p = 19 \times 0.4$ torr $\times$ 1 atm/760 torr = 0.01 atm.

The ideal gas equation may be written as

$$pV = nRT = \frac{m}{MW} RT,$$

where V is the volume (V = 294 liters), R is the gas constant (R = 0.082 liter-atm/mole-degree), T is the absolute temperature (T = 21°C = 294°K), and n = numbers of moles = m/MW, where m is the mass, and MW is the molecular weight of $H_2O$ (MW = 18 g/mole). Solving for m we obtain

$$m = \frac{PV(MW)}{RT} = \frac{0.01 \text{ atm} \times 294 \text{ liters} \times 18 \text{ g/mole}}{0.082 \text{ liter-atm/mole-degree} \times 294°K} \cong 2.2 \text{ g}.$$

Thus, 2.2 g of $H_2O$ must be supplied to the dry air flow every minute to maintain a relative humidity of 40 %.

● PROBLEM 872

The pistons in an automobile engine are driven by the following reaction between octane ($C_8H_{18}$, molecular weight = 114 g/mole), and oxygen ($O_2$, molecular weight = 32 g/mole), which takes place in the cylinders.

$$2C_8H_{18} + 25O_2 \rightarrow 16CO_2 + 18H_2O$$

If the cylinders have a total volume of 6.15 liters, and one-fifth of the air filling this volume at 1 atm and 27°C is oxygen, what weight of octane is necessary to combine exactly with the oxygen?

Solution: To solve this problem, we first determine the partial pressure of oxygen. Using this value and the ideal gas law, we then calculate the number of moles of oxygen present and finally, the number of moles of octane required to react with it.

One-fifth the total volume is occuped by oxygen. This corresponds to 20 % oxygen by volume (1/5 × 100 % = 20 %). The partial pressure of oxygen, $p_{O_2}$, is equal to the percent oxygen multiplied by the total pressure, or $p_{O_2}$ = 20 % × 1 atm = 0.20 atm.

The number of moles, $n_{O_2}$, of oxygen present is calculated from the ideal gas equation,

$$n_{O_2} = \frac{P_{O_2} V}{RT} \quad ,$$

where V is the volume (V = 6.15 liters), R is the gas constant (R = 0.082 liter-atm/mole-degree), and T is the absolute temperature (T = 27°C = 300°K). Then,

$$n_{O_2} = \frac{P_{O_2} V}{RT} = \frac{0.20 \text{ atm} \times 6.15 \text{ liters}}{0.082 \text{ liter-atm/mole-deg} \times 300°K} = 0.05 \text{ mole.}$$

From the equation for the reaction between octane and $O_2$, we see that 2 moles of octane react with 25 moles of $O_2$. Setting up the following proportion

$$\frac{2 \text{ moles } C_8H_{18}}{25 \text{ moles } O_2} = \frac{n_{C_8H_{18}}}{n_{O_2}}$$

and solving for the number of moles of octane, $n_{C_8H_{18}}$, we obtain

$$n_{C_8H_{18}} = \frac{2 \text{ moles}}{25 \text{ moles}} \times n_{O_2} = \frac{2 \text{ moles}}{25 \text{ moles}} \times 0.05 = 0.004 \text{ mole.}$$

The required mass of octane is obtained by multiplying the number of moles of octane by the molecular weight of octane, or

0.004 mole × 114 g/mole = 0.46 g.

● **PROBLEM** 873

$SO_2$ (sulfur dioxide) can be used as a preservative for meats, wine, and beer; as a bleaching agent for foods; in pulp and paper industry; and in tanning. Express the equilibrium constant, $k_2$, in terms of the equilibrium constant $k_1$, in the two gaseous equilibria involving $SO_2$. Namely,

$$SO_2(g) + \tfrac{1}{2} O_2(g) \; \rightleftharpoons \; SO_3(g) \qquad\qquad k_1$$

$$2SO_3 \; \rightleftharpoons \; 2SO_2(g) + O_2(g) \qquad\qquad k_2$$

Solution: An equilibrium expression is an equation where the equilibrium constant is equated with the ratio of concentrations of products to reactants, each raised to the power of its coefficients in the chemical reaction. For $k_1$,

$$k_1 = \frac{[SO_3]}{[SO_2][O_2]^{\frac{1}{2}}}$$

For $k_2$, $\quad k_2 = \frac{[O_2][SO_2]^2}{[SO_3]^2}$ .

Now that this is written, $k_2$ can now be expressed in terms of $k_1$ by solving these equations in terms of some common concentration. For example, solve the $k_1$ equation for $[SO_3]$.

$$[SO_3] = k_1 \ [SO_2] \ [O_2]^{\frac{1}{2}}.$$

$$[SO_3]^2 = k_1^2 \times [SO_2]^2 \ [O_2].$$

Substitute this into the $k_2$ equation,

$$k_2 = \frac{[O_2][SO_2]^2}{[SO_3]^2}$$

The result is $k_2 = \dfrac{[O_2][SO_2]^2}{k_1^2 \ [SO_2]^2[O_2]}$ . This reduces to

$$k_2 = \frac{1}{k_1^2} \ .$$

# CHAPTER 26

# APPLIED LIQUID AND SOLUTION PROBLEMS

● **PROBLEM** 874

The maximum safe concentration of fluoride ion in drinking water is 2.0 ppm (parts per million). Express this concentration in milligram %.

Solution: Milligram percent is the number of milligrams of solute in 100 ml. of liquid. It is given that there are 2 g. of fluoride in $1.0 \times 10^6$ g $H_2O$. Because 1 ml of $H_2O$ weighs 1 g, there are 2 g of fluoride in $1 \times 10^6$ ml of $H_2O$. Solving for mg %:

$$\text{milligram } \% = \frac{100 \text{ ml} \times 2g \times 1000 \text{ mg/g}}{1 \times 10^6 \text{ ml}} = .20 \text{ mg } \%.$$

The concentrations of substances in body fluids such as urine or blood are rather low, and for this reason they are generally expressed in milligram %.

● **PROBLEM** 875

Suppose a rain cloud has a water volume of $10^5$ liters. If the cloud forms over a chemical factory which puts $SO_3$ into the air and each $SO_3$ molecule bonds with one out of $10^4$ $H_2O$ molecules (to form $H_2SO_4$), what is the molarity of the $H_2SO_4$ in the cloud?

Solution: To solve this problem, the number of moles of $H_2SO_4$ must be determined. This is done by applying the stoichiometric rules to the equation of this particular reaction:

$$H_2O + SO_3 \rightarrow H_2SO_4$$

According to this equation for every mole of $SO_3$ consumed, one mole of $H_2SO_4$ is produced or, in other words, if the number of moles of $SO_3$ consumed is known, then the number of moles of $H_2SO_4$ produced is also known. The number of moles of $SO_3$ can be found since it is the same number as $H_2O$ (from the stoichiometry of the equation). The number of moles of $H_2O$ consumed can be found by dividing its weight by its molecular weight (MW = 18.02) or

$$\text{number of moles of } H_2O = \frac{(10^5 \text{ liter})(1000 \text{ g/liter})}{18.02 \text{ g/mole}}$$

$$= 5.5 \times 10^6 \text{ moles}$$

Note: 1000 g/liter is the density of water (rain).

However, the number of water molecules is needed and thus,

number of $H_2O$ molecules

$$= (5.5 \times 10^6 \text{ moles})(6.02 \times 10^{23} \text{ molecules/mole})$$

$$= 3.3 \times 10^{30} \text{ molecules.}$$

The problem states that $SO_3$ molecules bond with one out of $10^4$ $H_2O$ molecules. From this, the number of $SO_3$ molecules is determined,

$$\frac{3.3 \times 10^{30} \ H_2O \text{ molecules}}{10^4 \ \frac{H_2O \text{ molecules}}{SO_3 \text{ molecules}}} = 3.3 \times 10^{26} \ SO_3 \text{ molecules.}$$

The number of moles of $SO_3$ can now be obtained.

$$\text{number of moles of } SO_3 = \frac{(3.3 \times 10^{26} \ SO_3 \text{ molecules})}{6.02 \times 10^{23} \ \frac{\text{molecules}}{\text{mole}}}$$

$$= 5.5 \times 10^2 \text{ moles}$$

or $5.5 \times 10^2$ moles of $H_2SO_4$ are formed.

The molarity of the $H_2SO_4$ produced is thus

$$\frac{5.5 \times 10^2 \text{ moles}}{10^5 \text{ liter}} = 5.5 \times 10^{-3} \text{ M.}$$

● **PROBLEM** 876

What is the molarity of ethanol in 90-proof whiskey? Density of alcohol = 0.8 g/ml.

Solution: Molarity is defined as the number of moles of solute per liter of solution. By definition, 90-proof whiskey is 45 % ethanol by volume. Thus in 1 ℓ (= 1000 ml)

of whiskey there is (45 %)(1000 ml) = 450 ml of ethanol.
To determine the number of moles of ethanol in 1 ℓ of
whiskey first calculate the weight of ethanol in 1 ℓ
by multiplying the density of ethanol by the volume of
ethanol in 1 ℓ. Then, divide the weight of ethanol in
1 ℓ by its molecular weight (MW ethanol = 46 g/mole).
Therefore,

$$(450 \text{ ml ethanol})(0.8 \text{ g/ml}) = 360 \text{ g ethanol}$$

$$\text{molarity} = \frac{\text{number moles ethanol}}{1 \text{ ℓ whiskey}} = \frac{360 \text{ g}/46 \text{ g/mole}}{1 \text{ ℓ}} = 7.83 \text{ M.}$$

● **PROBLEM** 877

A nurse mistakenly administers a dilute $Ba(NO_3)_2$ (a
soluble, strong electrolyte) solution to a patient for
radiographic investigation. What treatment would you
provide to prevent the absorption of soluble barium
and subsequent barium poisoning?

Solution: An electrolyte is a compound that conducts
electricity when dissolved in water. Electrolytes are
able to do this by dissociating into ions in water.
Because $Ba(NO_3)_2$ is a strong electrolyte, it will completely dissociate into ions according to the reaction

$$Ba(NO_3)_2 (s) \rightarrow Ba^{2+} (aq) + 2NO_3^- (aq)$$

To prevent barium poisoning, the $Ba^{2+}$ ions, must
be removed from the solution as an insoluble compound
so that they cannot be absorbed. One way to precipitate
out the barium ion ($Ba^{2+}$) is by adding a dilute solution
of $Na_2SO_4$. The reason is that $Na_2SO_4$ reacts with $Ba^{2+}$
and $NO_3^-$ ions to produce $BaSO_4$ and $NaNO_3$. The barium
sulfate is extremely insoluble as indicated by a solubility
product constant of $1.6 \times 10^{-9}$ mole$^2$ liter$^{-2}$. (A solubility
product constant is indicative of a salt's solubility in
solution. If this value is low, the salt is insoluble or
barely soluble.)

Therefore, by adding the $Na_2SO_4$, the $Ba^{2+}$ precipitates out and is not absorbed.

● **PROBLEM** 878

A household cleaning solution has a hydronium ion
concentration of $10^{-11}$ M. What is the pH of the solution?
Is the solution acidic or basic?

Solution: The pH of any solution by definition is

839

determined by using the following equation.

$$pH = - \log [H^+]$$

Thus, if hydronium concentration $= 10^{-11}$ M,

$$pH = - \log 10^{-11} \quad \text{or pH} = 11.$$

The solution is basic because it has a pH greater than 7.

10 cc of battery acid (density 1.21) is diluted to 100 cc. 30 cc of the dilute solution is neutralized by 45 cc of 0.5 N base. What is the normality of the diluted acid? What is its strength in grams per liter? What is the strength of the undiluted 1.21 density acid in g/ℓ?

Solution: To solve this problem, use ie following relationship for the neutralization of acids and bases

$$N_a V_a = N_b V_b$$

where $N_a$ is the normality of the acid at volume $V_a$ and $N_b$ is the normality of the base at volume $V_b$. Thus, the normality of the diluted acid is

$$N_a = \frac{N_b V_b}{V_a} = \frac{(0.5 \text{ N})(0.045 \text{ } \ell)}{(0.030 \text{ } \ell)} = 0.75 \text{ N}.$$

Note: 1 cc = .001 ℓ.

The undiluted acid has a density of 1.21 g/cc; thus 10 cc of acid, weighs 12.1 g. Thus, when 90 cc of $H_2O$ is added to form 100 cc of the diluted acid the weight of the entire solution becomes 90 + 12.1 = 102.1 g, but the weight of the acid is still 12.1 g. In a liter of diluted acid, there are

$$\frac{12.1 \text{ g}}{100 \text{ cc}} \times \frac{10}{10} = \frac{121.0 \text{ g}}{1000 \text{ cc}} = 121 \text{ g/}\ell \text{ of acid.}$$

For the undiluted acid, the density is:

$$\frac{1.21 \text{ g}}{1 \text{ cc}} \times \frac{1000}{1000} = \frac{1210 \text{ g}}{1000 \text{ cc}} = 1210 \text{ g/}\ell.$$

A sample of blood has freezing point of - 0.560°C. What is the effective molal concentration of the solute in the serum? According to this concentration, what is the osmotic pressure in the blood serum at body temperature (37°C)?

Solution: The molal freezing point depression and boiling point elevation for the ideal 1 molal solution are also called the cryoscopic and ebullioscopic constants.

The mathematical expressions for the freezing point depression is

$$\Delta T_f = m K_f$$

where $K_f$ is the freezing point constant for water, 1.86°C, m is the effective molality of the solute, and $\Delta T_f$ is the change in the freezing point. For this particular problem $\Delta T_f = 0.560$°C and $K_f = 1.86$°C, so that

$$m = \frac{\Delta T_f}{K_f} = \frac{0.560°C}{1.86°C} = 0.30 \text{ m.}$$

Hence, 0.30 m is the effective molal concentration of solute in serum.

Osmosis is the diffusion of a solvent through a semipermeable membrane into a more concentrated solution. The osmotic pressure of a solution is the minimum pressure that must be applied to the solution, in excess of the pressure on the solvent, to prevent the flow of solvent from pure solvent into the solution. An empirical osmotic pressure equation for solutions is

$$\pi = CRT,$$

where $\pi$ is the osmotic pressure, C is the concentration in molality or molarity, R is the gas constant (0.082 atm-liter/°K-mole), and T is the temperature (in °K).

Thus, at 37°C = 310°K and at 0.30 m, blood serum has an osmotic pressure of

$\pi$ = CRT = (0.30 m)(0.082 atm-liter/°K mole)(310°K)

= 7.6 atm.

Calculate the osmotic work done by the kidneys in secreting 0.158 moles of $Cl^-$ in a liter of urine water at 37°C when the concentration of $Cl^-$ in plasma is 0.104 M, and in urine 0.158 M.

Solution: Work done on a system is measured by the change in free energy of the system.

To calculate the osmotic work done by living systems in transferring solutes from regions of lower concentration to those of higher, the following equation is used:

$$\Delta G = G_2 - G_1 = n \; RT \; \ln \frac{c_2}{c_1}$$

where R = gas constant, T = temperature, n = number of moles, G = free energy, and c = concentration.

This equation gives the free energy change ($\Delta G$) corresponding to a change in the concentration of one component (solute). For n moles, if $c_2$ is the final concentration and $c_1$ the initial concentration, then, from the equation, an increase in concentration will cause the solution to gain free energy with respect to that particular component ($\Delta G$ will be positive), and that dilution with respect to the same component will cause the solution to lose free energy ($\Delta G$ will be negative). A loss in free energy represents a loss in the capacity to do work.

For this problem,

R = 1.987 cal/mole °K,   T = 273° + 37°C = 310°K

n = 0.158 moles $Cl^-$, $c_2$ = 0.158 M, $c_1$ = 0.104 M.

Thus,

osmotic work = $\Delta G$ = (0.158)(1.987)(310) $\ln \dfrac{0.158 \text{ M}}{0.104 \text{ M}}$

$= (2.303)(0.158)(1.987)(310) \log 1.52$

$= 40.7$ cal.

$CF_4$ is used as a low-temperature refrigerant under the name Freon-14. Its normal boiling point is - 127.8°C and its heat of vaporization is 12.62 kJ/mole.  At what pressure should you pump liquid Freon-14 to set a temperature of - 135°C?

Solution: Assuming that $CF_4$ vapor obeys the ideal gas law, and that the heat of vaporization $\left(\Delta H_{vap}\right)$ is independent of temperature, then the following equation can be used to calculate the vapor **pressures at two different** temperatures.

$$\log \frac{p_2}{p_1} = \frac{\Delta H_{vap} (T_2 - T_1)}{2.303 \ R \ T_1 T_2}, \text{ where } T_1 \text{ and } T_2 \text{ are absolute}$$

temperatures, R = gas constant, and $p_1$ and $p_2$ are pressures. Any units of pressure may be **chosen** as long as the same units are used for both pressures, and any units of heat may be used as long as $\Delta H_{vap}$ and RT have the same units of energy.

Thus, substituting into the above equation the known data

$$p_1 = 1 \text{ atm}, \quad T_1 = -127.8° + 273° = 145.2°K$$

$$T_2 = -135° + 273° = 138°K, \ \Delta H_{vap} = 12620 \text{ J/mole}$$

$$R = 8.314 \text{ J/mole }°K$$

$$\log \frac{p_2}{(1 \text{ atm})} = \frac{(12620 \text{ J/mole})(138°K - 145.2°K)}{2.303(8.314 \text{ J/mole }°K)(145.2°K)(138°K)}$$

$$= \frac{(12620)(-7.2)}{383662.77} = -\frac{90864}{383662.77} = -.2368$$

$$p_2 = (1 \text{ atm}) \ 10^{-.2368} = 0.58 \text{ atm}.$$

● **PROBLEM** 883

A recipe for artificial sea water is NaCl, 26.518 g/kg; $MgCl_2$, 2.447 g/kg; $MgSO_4$, 3.305 g/kg; $CaCl_2$, 1.141 g/kg; KCl, .725 g/kg; $NaHCO_3$, .202 g/kg; NaBr, .083 g/kg. All of these salts are dissolved in water to make a total weight of 1000 g. What are the chlorinity and salinity of the solution?

Solution: The chlorinity of a solution is defined as the percent of $Cl^-$ ions in the solution by weight. The salinity is a measure of the salt content of seawater.

To solve this problem, the number of moles of $Cl^-$ must be determined. Once this value is known the weight of $Cl^-$ can be determined and thus the percentage of $Cl^-$ can be found.

The molecular weights of the salts containing $Cl^-$ are 58.5 g/mole for NaCl, 95.3 g/mole for $MgCl_2$, 111 g/mole for $CaCl_2$, and 74.5 g/mole for KCl.

Since there are 1000 g in 1 kg of sea water, the weights of the salts are 265.18 g for NaCl, 24.47 g for MgCl$_2$, 11.41 g for CaCl$_2$, and 7.25 g for KCl. Thus,

$$\text{moles of Cl}^- \text{ from NaCl } = \frac{26.518 \text{ g}}{58.5 \text{ g/mole}} = .453 \text{ moles}$$

$$\text{moles of Cl}^- \text{ from MgCl}_2 = \frac{2(2.447 \text{ g})}{95.3 \text{ g/mole}} = .0514 \text{ moles}$$

$$\text{moles of Cl}^- \text{ from CaCl}_2 = \frac{2(1.141 \text{ g})}{111 \text{ g/mole}} = .0206 \text{ moles}$$

$$\text{moles of Cl}^- \text{ from  KCl } = \frac{.725 \text{ g}}{74.5 \text{ g/mole}} = .00973 \text{ moles}$$

Total number of moles of Cl$^-$ = .535 moles

The total weight of Cl$^-$ in the sea water is

(.535 moles)(35.5 g/mole) = 19.0 g.

Thus the chlorinity of the sea water is

$$\frac{19.0 \text{ g Cl}^-}{1000 \text{ g sea water}} \times 100 = 1.9\% \text{ chlorinity.}$$

This value of chlorinity is average for sea water.

To find the salinity of the sea water use the formula:

% salinity = 1.805 (% Cl) + 0.03 %

= (1.805)(1.9%) + 0.03 % = 3.433 %.

● **PROBLEM** 884

A 25.00-ml sample of sea water is titrated with 2.50 M AgNO$_3$ solution, using K$_2$CrO$_4$ as an indicator, until the presence of excess Ag$^+$ is shown by the reddish color of Ag$_2$CrO$_4$ at the end point. The density of the sea water is 1.028 g/cc . Find the chlorinity and salinity of the sea water if 53.50 ml of AgNO$_3$ solution were used to complete the precipitation of silver halides.

Solution: To solve this problem use the following equation:

$$M_1 V_1 = M_2 V_2 ,$$

where M = molarity and V = volume. This equation will determine the number of moles of Cl$^-$ consumed in the titration. The equation of the reaction is

$$AgNO_3 + Cl^- (aq) \rightarrow AgCl \downarrow + NO_3^- (aq)$$

The problem states that 0.0535 liter of $AgNO_3$ solution at 2.5 M concentration reacts with an unknown $Cl^-$ concentration. What actually needs to be known is the total weight of $Cl^-$ in 25.00 ml of sea water.

$$(2.5 \text{ moles/liter})(0.0535 \text{ liter}) = 0.13 \text{ moles } AgNO_3$$

However, according to the stoichiometry of the titration reaction, 0.13 moles of $Cl^-$ also is consumed. Therefore, the weight of $Cl^-$ in the sea water is

$$(0.13 \text{ moles } Cl^-)(35.5 \text{ g/mole}) = 4.62 \text{ g}.$$

The total weight of sea water is

$$(25 \text{ ml})(1.028 \text{ g/ml or cc}) = 25.7 \text{ g sea water}.$$

Thus, the percentage of $Cl^-$ in sea water is

$$\frac{4.62 \text{ g } Cl^-}{25.7 \text{ g sea water}} \times 100 = 18.0 \text{ \% } Cl^-$$

The chlorinity of sea water is 18 %.

The salinity is determined from the following equation

$$\text{\% salinity} = 1.805\left(Cl \text{ \%}\right) + 0.03 \text{ \%}.$$

Thus, $\text{\% salinity} = (1.805)(18 \text{ \%}) + 0.03 \text{ \%} = 32.52 \text{ \%}.$

● **PROBLEM** 885

Hard water is often softened by a commercial water softener. Explain, in general, its operation.

Solution: Hard water possesses dissolved calcium or magnesium ions. Softening the water means the removal of these ions. Commercial water softeners perform this task by a process known as ion exchange. In this process, water is passed through a bed of cation-exchange resin or zeolite. A cation-exchange resin is an insoluble polymer of polystyrene or other material that possesses sulfonic acid groups attached to it. The resin is activated by NaOH or $NaCO_3$ that produces the corresponding sodium salt.

Calcium and magnesium ions are held more tightly by the resin than are $Na^+$ ions. Consequently, when the hard water is passed through this resin, the hard-water ions are held by the resin, and the sodium ions are released. This is the operation of a commercial water softener.

Hydrogen chloride (HCl) is used industrially in the manufacture of fabric dyes. The acidic properties of HCl arise because the H-Cl bond is polar. The observed dipole moment, $u_{obs}$, of HCl is 1.03 D and the H-Cl bond distance is 1.28 Å. Estimate the percent ionic character of the HCl bond. Electron charge, $e = 4.8 \times 10^{-18}$ esu.

Solution: A dipole moment is an indication of bond polarity. The magnitude of the dipole moment is the product of the electronic charge and the distance between the charge centers.

To determine the ionic character of HCl, the magnitude of the dipole moment is needed.

$$\text{Percent ionic character} = \frac{u_{obs}}{u_{ionic}} \times 100, \text{ where}$$

$u_{ionic}$ = the magnitude of dipole moment for total ionic character. If HCl were ionic, one electron would be transferred from H to Cl to give the ionic structure $H^+Cl^-$. Thus, the electron charge would be the charge on one electron, $4.8 \times 10^{-18}$ esu. The distance between the two atoms is given as 1.28 Å or $1.28 \times 10^{-8}$ cm. (There are $10^{-8}$ cm per Å.) Recalling the definition of dipole moment,

$$u_{ionic} = 4.80 \times 10^{-10} \text{ esu} \times 1.28 \times 10^{-8} \text{ cm}$$

$$= 6.14 \times 10^{-18} \text{ esu-cm.}$$

Dipole moment is usually expressed, however, in Debyes. There are $10^{18}$ Debyes per esu-cm. Thus,

$$6.14 \times 10^{-18} \text{ esu-cm} = 6.14 \text{ Debyes} = u_{ionic}.$$

Therefore, percent ionic character =

$$= \frac{1.03}{6.14} \times 100\% = 16.8 \text{ \%.}$$

# CHAPTER 27

# APPLIED STOICHIOMETRY PROBLEMS

● **PROBLEM** 887

The compound calcium cyanamide, $CaCN_2$, is prepared in considerable amounts for use as a solid fertilizer. The solid mixed with the water in the soil slowly adds ammonia and $CaCO_3$ to the soil.

$$CaCN_2 + 3H_2O \rightarrow CaCO_3 + 2NH_3$$

What weight of ammonia is produced as 200 kg of $CaCN_2$ reacts?

**Solution:** From the stoichiometry of this equation, 1 mole of $CaCN_2$ produces 2 moles of ammonia. The number of moles of $CaCN_2$ (MW = 80 g/mole) in 200 kg is

$$\frac{200,000 \text{ g } CaCN_2}{80 \text{ g/mole}} = 2,500 \text{ moles } CaCN_2$$

Thus, (2,500 × 2=) 5,000 moles of ammonia is produced and the weight of ammonia is its molecular weight (MW = 17 g/mole) times the number of moles produced:

$$(17 \text{ g/mole})(5000 \text{ moles}) = 85,000 \text{ g} = 85 \text{ kg}$$

of ammonia produced.

● **PROBLEM** 888

A tomato weighing 60.0 g is analyzed for sulfur content by digestion in concentrated $HNO_3$ followed by precipitation with Ba to form $BaSO_4$. 0.466 g of $BaSO_4$ is obtained. If all of the sulfur is converted to $BaSO_4$, what is the percentage of sulfur in the tomato?

**Solution:** The percentage of sulfur in the tomato is the weight of sulfur in the tomato divided by the total weight

of the tomato multiplied by a factor of 100. To find the
percentage, the weight of sulfur must be calculated. This
is done by calculating the amount of sulfur precipitated
out or, in other words, the amount of sulfur present in
0.466 g of BaSO₄.

Using the Law of Definite Proportions, 32 g (1 mole)
of sulfur are consumed to form 233 g (1 mole) of $BaSO_4$.
Thus, once the percentage by weight of S in $BaSO_4$ is
determined, then the weight of sulfur can be calculated.

The fraction of S in $BaSO_4$ is

$$\frac{\text{MW of S}}{\text{MW of BaSO}_4} = \frac{32 \text{ g/mole}}{233 \text{ g/mole}} = 0.14.$$

The weight of sulfur in 0.466 g of $BaSO_4$ is thus,

$$(0.14)(0.466 \text{ g}) = 0.065 \text{ g}.$$

Now, that the weight of sulfur is known, the percentage
of sulfur in the tomato is,

$$\frac{0.065 \text{ g of S}}{60.0 \text{ g of tomato}} \times 100 = 0.11 \text{ \%}.$$

● **PROBLEM** 889

Solder is a lead-tin alloy. 1.00 gram of the alloy treated
with warm 30 % $HNO_3$ gave a lead nitrate solution and some
insoluble white residue. After drying, the white residue
weighed 0.600 g and was found to be $SnO_2$. What was the
percentage tin in the sample?

$$Sn + 4HNO_3 \rightarrow SnO_2 + 4NO_2 + 2H_2O$$

Solution:  From the stoichiometry of this equation, 1 mole
of Sn produces 1 **mole of SnO₂.** Thus, if the number of moles
of $SnO_2$ is known, the number of moles of Sn is also known
(it is equivalent to the number of moles of $SnO_2$). Once the
number of moles of Sn is known, the weight of Sn can be
calculated and thus the percentage of Sn by weight in
1.00 gram of solder alloy can be determined.

The number of moles of $SnO_2$ (MW = 151 g/mole) is

$$\frac{0.600 \text{ g of SnO}_2}{151 \text{ g/mole}} = 0.00397 \text{ moles.}$$

The weight of 0.00397 moles of Sn (at. wt. 119 g/mole)
is (0.00397 moles)(119 g/mole) = 0.472 g. Thus, the percentage of Sn in the alloy is

$$\frac{0.472 \text{ g Sn}}{1.00 \text{ g alloy}} \times 100 = 47.2 \text{ \%}.$$

0.20 g of cuttings from a silver coin (containing Cu) was completely dissolved in $HNO_3$. Then a NaCl solution was added to precipitate AgCl, which, when collected and dried, weighed 0.2152 g. What is the indicated percent of Ag in the coin?

Solution: To solve this problem, we must (1) set up the reactions (2) determine the number of moles AgCl precipitated (3) determine the number of moles of Ag used from the stoichiometry of the equations (4) calculate the weight of Ag used (5) calculate the percent of Ag in the coin.

The equations are

(i)     $Ag + Cu + HNO_3 \rightarrow AgNO_3 + Cu(NO_3)_2 \downarrow$

(ii)    $AgNO_3 + NaCl \rightarrow AgCl \downarrow + NaNO_3$.

The stoichiometry of these equations says that 1 mole of AgCl is produced from 1 mole of Ag metal. From the data given, the AgCl weighed 0.2152 g. Therefore, the number of moles of AgCl precipitated can be calculated, since the molecular weight of AgCl is 143.4 g/mole. Therefore,

$$\frac{.2152 \text{ g}}{143.4 \text{ g/mole}} = 1.50 \times 10^{-3} \text{ moles of AgCl precipitated out.}$$

Thus, $1.50 \times 10^{-3}$ moles of Ag metal are consumed. The weight of Ag metal is the number of moles of Ag times its atomic weight (At. wt. = 107.868 g/mole), which is equal to

$(1.50 \times 10^{-3} \text{ moles})(107.868 \text{ g/mole}) = 0.162 \text{ g}.$

The percent of Ag metal in the coin is

$$\frac{0.162 \text{ g Ag}}{0.20 \text{ g}} \times 100 = 81 \text{ \%}.$$

The molecular weight of nicotine, a colorless oil, is 162.1 and it contains 74.0 % carbon, 8.7 % hydrogen, and 17.3 % nitrogen. Using three significant figures for the atomic weights, calculate the molecular formula of nicotine.

Solution: A molecular formula gives the actual composition in number of atoms per molecule. It also summarizes the weight composition of the substance. Hence, to calculate such a formula, the molecular weight, the weight composition of the substance, and the atomic weights of the constituent elements must be known.

The weight composition of the substances in nicotine, assuming there are 100 g of it, are

C = 74.0 g, H = 8.7 g, N = (100 - 74.0 - 8.7) = 17.3 g.

The atomic weights of these substances are

C = 12.0 g/mole, H = 1.01 g/mole, N = 14.0 g/mole.

The number of moles of each constituent is, therefore,

$$\text{moles C} = \frac{74.0 \text{ g}}{12.0 \text{ g/mole}} = 6.17 \text{ moles}$$

$$\text{moles H} = \frac{8.7 \text{ g}}{1.01 \text{ g/mole}} = 8.61 \text{ moles}$$

$$\text{moles N} = \frac{17.3 \text{ g}}{14.0 \text{ g/mole}} = 1.24 \text{ moles.}$$

Thus, the mole ratios of C : H : N are 6.17 : 8.61 : 1.24. The empirical formula is

$$C_{6.17}H_{8.61}N_{1.24} \quad \text{or} \quad C_{\frac{6.17}{1.24}}H_{\frac{8.61}{1.24}}N_{\frac{1.24}{1.24}} = C_5H_7N_1.$$

To determine the molecular formula, which is related to the formula weight of the empirical formula by a whole number, divide the molecular weight (MW Nicotine = 162.1) by the formula weight and multiply this whole number times the ratio of atoms in the empirical formula. Thus,

the formula weight of $C_5H_7N_1$ = 81.05

the molecular weight of nicotine = 162.1

the whole number = $\frac{162.1}{81.05}$ = 2.

The molecular formula is

$(C_5H_7N_1)_2 = C_{10}H_{14}N_2.$

● **PROBLEM** 892

The analysis of a sample of soil weighing 0.4210 g yields a mixture of KCl and NaCl weighing 0.0699 g. From the KCl, 0.060 g AgCl is precipitated out. Calculate the percent of KCl in the soil.

Solution:  To solve this problem

(1) set up the equation for the reaction of KCl to AgCl

(2) determine the number of moles of AgCl precipitated

(3) from the stoichiometry of equation determine how many moles of KCl were consumed

(4) calculate the weight of KCl consumed

(5) calculate the percentage of KCl in the soil by dividing the weight of KCl by the total weight of soil and multiplying by 100.

The equation for the reaction is

$$KCl + Ag^+ \rightarrow AgCl + K^+$$

The number of moles of AgCl is

$$\frac{\text{weight of AgCl}}{\text{MW of AgCl}} = \frac{0.060 \text{ g}}{143.5 \text{ g/mole}} = 4.18 \times 10^{-4} \text{ moles}$$

From the stoichiometry of the equation, $4.18 \times 10^{-4}$ moles of KCl is consumed.

The weight of KCl consumed is the number of moles of KCl times its molecular weight. Thus,

weight of KCl $= (4.18 \times 10^{-4}$ moles$)(74.6$ g/mole$) = 0.0312$ g

The percentage of KCl in the soil is

$$\frac{\text{weight of KCl}}{\text{weight of soil}} \times 100 = \frac{0.0312 \text{ g KCl}}{0.4210 \text{ g Soil}} = 7.41 \text{ \%}.$$

● **PROBLEM** 893

When ammonium nitrate ($NH_4NO_3$) is heated, laughing gas ($N_2O$) and water are produced. If the laughing gas is collected at STP (Standard Temperature and Pressure) and one starts with 10 grams of $NH_4NO_3$, how many liters of laughing gas could be collected?

Solution: When any gas is collected at STP (0°C and 1 atm), one mole of it will occupy 22.4 liters. The balanced equation for this chemical reaction is

$$NH_4NO_3 \rightarrow N_2O + 2H_2O.$$

One mole of $NH_4NO_3$ yields one mole of $N_2O$. If the number of moles of $NH_4NO_3$ that reacted is known, the number of moles of laughing gas produced is also known.

The molecular weight of ammonium nitrate = 80 g/mole. 10 g of ammonium nitrate (10 g/80 g/mole) = 0.125 moles of $NH_4NO_3$ produce 0.125 moles of laughing gas.

Recall that 1 mole of any gas (at STP) occupies 22.4 liters. To find out what volume 0.125 moles occupy, set up the following proportion:

$$\frac{1 \text{ mole}}{22.4 \text{ liters}} = \frac{0.125}{x}$$ , where x = the volume. Solving,

x = 2.8 liters = volume of laughing gas collected.

● PROBLEM 894

Solid ammonium nitrate, amitol, is much used as an explosive; it requires a detonator explosive to cause its decomposition to hot gases.

$$NH_4NO_3 \rightarrow N_2O + 2H_2O$$

Assume 900 g $NH_4NO_3$ almost instantaneously decomposed. What total volume of gas in liters forms at 1092°C?

Solution: From the stoichiometry of the equation, 1 mole of $NH_4NO_3$ produces a total of 3 moles of products (1 mole of $N_2O$ + 2 moles of $H_2O$).

The number of moles of $NH_4NO_3$ 900 g (MW = 80 g/mole) is

$$\frac{900 \text{ g}}{80 \text{ g/mole}} = 11.25 \text{ moles of } NH_4NO_3.$$

There are, thus, a total of 3 × 11.25 moles = 33.75 moles of products. Using the relationship that 1 mole of any ideal gas has a volume of 22.4 ℓ at STP (Standard Temperature and Pressure), the total volume of gaseous products is

(22.4 ℓ/mole)(33.75 moles) = 756 ℓ at STP.

To obtain the volume at 1092°C, Charles' Law is used

$$\frac{V_1}{V_2} = \frac{T_1}{T_2}$$

where $V_1$ = 756 ℓ, $T_1$ = 273°K (the STP temperature), and $T_2$ = 1092° + 273° = 1365°K. Solving for $V_2$,

$$V_2 = \frac{V_1 T_2}{T_1} = \frac{(756 \text{ ℓ})(1365°K)}{(273°K)} = 3780 \text{ ℓ}.$$

Lunar soil contains ilmenite, a mineral with the composition $FeO \cdot TiO_2$. It has been suggested that moon explorers might obtain water and oxygen from lunar soil by reducing it with hydrogen (brought from earth) according to the following reactions. Beginning with soil heated to 1300°C in a solar furnace:

$$FeTiO_3 + H_2 \longrightarrow H_2O + Fe + TiO_2$$

$$2H_2O \xrightarrow{\text{electrolysis}} 2H_2 + O_2$$

How much water and oxygen could be obtained from 50 kg of soil if the soil is taken to be 5 per cent ilmenite?

Solution: From the stoichiometry of these equations, 1 mole of $FeO \cdot TiO_2$ is consumed to form 1 mole of $H_2O$ and 2 moles of $H_2O$ is consumed to form 1 mole of $O_2$.

If 5 % (0.05) of the soil is $FeO \cdot TiO_2$ then only

(50,000 g)(0.05) = 2,500 grams are ilmenite.

The molecular weight of ilmenite is 152 g/mole. Thus, the number of moles of ilmenite is

$$\frac{2,500 \text{ g ilmenite}}{152 \text{ g/mole ilmenite}} = 16.4 \text{ moles.}$$

Thus, there must be 16.4 moles $H_2O$ and the weight of $H_2O$ is

(16.4 moles)(18.02 g/mole) = 296 g $H_2O$.

If there are 16.4 moles of $H_2O$, then there are 8.2 (or half 16.4 moles) moles of $O_2$. Thus,

(8.2 moles)(32 g/mole) = 262 g $O_2$.

The equation for cream of tartar and baking powder, in action with hot water is:

$$NaHCO_3 + KHC_4H_6O_6 \rightarrow NaKC_4H_4O_6 + H_2O + CO_2 \text{(g)}$$

From the use of 4.2 g of $NaHCO_3$ (baking powder), what volume of $CO_2$ should be liberated through the dough at 273°C?

Solution: From the stoichiometry of the equation, 1 mole of $NaHCO_3$ produces 1 mole of $CO_2$. The molecular weight of $NaHCO_3$ is 84 g/mole. Thus, 4.2 g is (4.2 g/84 g/mole ) 0.05 moles. 1 mole of any ideal gas at STP (Standard Temperature and Pressure) occupies 22.4 ℓ; thus the volume of $CO_2$ gas liberated at STP would be

$(22.4 \ \ell/mole)(0.05 \ mole) = 1.12 \ \ell$.

However, this must be converted from 1.12 $\ell$ at STP ($0°C = 273°K$, 1 atm) to a temperature of $273°C = 546°K$.

To this conversion, Charles' Law is used

$$\frac{T_1}{T_2} = \frac{V_1}{V_2} \ ,$$

where T is temperature in $°K$, and V is the volume at that temperature. Thus,

$$\frac{546°K}{273°K} = \frac{V_1}{1.12 \ \ell}$$

Volume of $CO_2 = V_1 = 2.24 \ \ell$.

● **PROBLEM** 897

Sea shells are mostly calcium carbonate. When heated, calcium carbonate decomposes into calcium oxide (lime) and carbon dioxide. (a) Write the balanced equation for the reaction. (b) Calculate the number of kilograms of calcium carbonate needed to produce 10 kg of lime.

<u>Solution</u>:  Calcium carbonate decomposes as follows:

$$CaCO_3 \rightarrow CaO + CO_2$$

From the stoichiometry of this equation, 1 mole of $CaCO_3$ decomposes into 1 mole of CaO and 1 mole of $CO_2$. If the number of moles of lime is known, the number of moles of $CaCO_3$ can be determined. The number of moles of lime is its weight divided by its molecular weight (56 g/mole) or

number of moles of lime = $\frac{10,000 \ g}{56 \ g/mole}$ = 179 moles.

Since 179 moles of lime are produced, 179 moles of $CaCO_3$ are consumed. The weight of 179 moles of $CaCO_3$ is the weight of one mole of $CaCO_3$ (or the molecular weight). Thus,

the weight of $CaCO_3$ = (179 moles)(100 g/mole) = 17,900 g

or 17.9 kg of $CaCO_3$ consumed.

To obtain 500 kg of glass composed of equimolar proportions of $Na_2SiO_3$ and $CaSiO_3$, what weights of $Na_2CO_3$, $CaCO_3$, and $SiO_2$ should be used?

**Solution:** When a mixture of limestone, sodium carbonate, and sand is melted together, a clear, homogeneous mixture of sodium and calcium silicates is produced.When the fused liquid is cooled, it becomes more and more viscous and finally hardens to a transparent rigid mass called glass.

$$CaCO_3 \quad + \quad Na_2CO_3 \quad + 2SiO_2 \rightarrow CaO \cdot SiO_2 + Na_2O \cdot SiO_2$$
(limestone)   (sodium carbonate)   (sand)    (glass)

$$+ \ CO_2$$

From the stoichiometry of this equation, an equal number of moles of $CaCO_3$ and $Na_2CO_3$ plus twice as many moles of $SiO_2$ are consumed to produce an equal number of moles of $CaSiO_3$ and $Na_2SiO_3$. To solve this problem, determine the number of moles of glass substituents produced; this number will be equal to the number of moles of $CaCO_3$ and $Na_2CO_3$ and half the number of moles of $SiO_2$. From the number of moles and each constituent's molecular weight, the actual weights can be determined.

The molecular weights are

MW of $Na_2SiO_3$ = 122 g/mole; MW of $CaSiO_3$ = 116 g/mole;

MW of $CaCO_3$ = 100 g/mole; MW of $Na_2CO_3$ = 106 g/mole;

MW of $SiO_2$ = 60 g/mole.

Since the number of moles of each product is unknown (but equal), the number of moles is set equal to x and

(x)(122 g/mole) + (x)(116 g/mole) = 500,000 g of glass

238 x = 500,000 g

the number of moles = 2101 moles.

Thus, the number of moles of the reactants needed are $CaCO_3$ = 2101 moles; $Na_2CO_3$ = 2101 moles; $SiO_2$ = 4202 moles; and the weights of each are

$CaCO_3$ : (2101 moles)(100 g/mole) = 210,100 g = 210.1 kg

$Na_2CO_3$ + (2101 moles)(106 g/mole) = 222,706 g = 222.7 kg

$SiO_2$ : (4202 moles)(60 g/mole) = 252,120 g = 252.1 kg.

Write equations to show how $Ca(HCO_3)_2$ is involved in
(a) stalagmite formation in caves, (b) formation of scale
in tubes or pipes carrying hot water, (c) the reaction with
soap ($C_{17}H_{35}COONa$), and (d) reaction with $Na_2CO_3$ in water
softening.

Solution: (a) To answer this, one must know the composition
of stalagmites, which are rocks composed of $CaCO_3$ and are
formed in underground caverns. Thus, one knows that somehow
the reaction $Ca(HCO_3)_2 \rightarrow CaCO_3$ occurs.

Since $Ca^{+2}$, $CO_3^{-2}$ are not reduced or oxidized, no
electrons are exchanged.

One needs to balance the reaction with respect to mass.

$$Ca(HCO_3)_2 \rightarrow CaCO_3 + H_2O + CO_2$$

This equation has both charge and mass balanced.

(b) The formation of scales in pipes is the result of
$CaCO_3$ build up from the decomposition of bicarbonates. The
reaction for this process is

$$Ca(HCO_3)_2 \rightarrow CaCO_3 + H_2O + CO_2$$

(c) The reaction of $Ca(HCO_3)_2$ with soap ($C_{17}H_{35}COONa$)
occurs when the $Na^+$ on the soap molecule is replaced by the
$Ca^{++}$.

$$Ca(HCO_3)_2 + 2C_{17}H_{35}COONa \rightarrow (C_{17}H_{35}COO)_2 Ca + H_2O + CO_2$$

(d) Permanent hardness in water results from chlorides
or sulfates of metals, such as calcium or magnesium. Boiling
does not remove these salts; a substance called    water
softener must be added to the water to form a precipitate
with the salts. Water containing $Ca(HCO_3)_2$ can be softened
by adding the proper amount of $Na_2CO_3$. The reaction is

$$Ca(HCO_3)_2 + Na_2CO_3 \rightarrow CaCO_3(s) + 2NaHCO_3 .$$

A red signal flare for use on railroads is made by mixing
strontium nitrate, carbon, and sulfur in the proportions
as shown by the equation

$$2Sr(NO_3)_2 + 3C + 2S \rightarrow$$

Complete and balance the equation (no atmospheric oxygen
is necessary).

Solution: Until the beginning of the twentieth century, large amounts of $KNO_3$ (saltpeter) were used in the preparation of black gunpowder. This powder was prepared by mixing powdered saltpeter, carbon, and sulfur. When ignited, sulfur and carbon are very rapidly oxidized by $KNO_3$, producing several gaseous products.

$$4KNO_3 + S + 4C \rightarrow 2K_2O + 2N_2 + SO_2 + 4CO_2 + heat$$

This equation can be applied to solve this problem. The only difference is that a red flare uses $Sr(NO_3)_2$ to give it a distinctive color and is mixed in different proportions with carbon and sulfur. $Sr(NO_3)_2$ also oxidizes carbon and sulfur to form the same gaseous products as black gunpowder.

Thus, the products for the oxidation of carbon and sulfur are $CO_2$ and $SO_2$ and the product of reduction is $N_2$ (going from $N^{+5}$ in $NO_3^-$ to $N^0$ in $N_2$). The complete reaction is:

$$2Sr(NO_3)_2 + 3C + 2S \rightarrow 2SrO + 2N_2 + 3CO_2 + 2SO_2.$$

● **PROBLEM** 901

Ethylene glycol, $C_2H_4(OH)_2$, which is widely used as antifreeze, may be converted into an explosive in a manner similar to the manufacture of nitroglycerine. (a) Write an equation for the preparation of $C_2H_4(NO_3)_2$. (b) Write an equation for the detonation of the latter compound into $CO_2$, $N_2$, and $H_2O$.

Solution: The manufacture of most explosives involves the use of concentrated sulfuric acid. The action of concentrated nitric acid on such compounds as glycerin seems to be markedly hastened by concentrated sulfuric acid because water is removed from the reaction.

$$\underset{glycerine}{C_3H_5(OH)_3} \quad + \quad 3HNO_3 \quad \xrightarrow{H_2SO_4} \quad \underset{nitroglycerin}{C_3H_5(NO_3)_3} \quad + 3H_2O$$

Most of the compounds formed by the action of nitric acid on organic substances are explosive. Thus, for part (a),

$$C_2H_4(OH)_2 + 2HNO_3 \xrightarrow{H_2SO_4} C_2H_4(NO_3)_2 + 2H_2O$$

An explosion occurs as a result of a rapid chemical reaction attended by the formation of a large volume of gas. The equation for the detonation of nitroglycerine is

$$4C_3H_5(NO_3)_3 \rightarrow 12CO_2 + 10H_2O + 6N_2 + O_2$$

For the detonation of $C_2H_4(NO_3)_2$, the products $CO_2$, $N_2$, and $H_2O$ are known, thus only balancing need be done.

$$C_2H_4(NO_3)_2 \xrightarrow{\text{detonation}} CO_2 + N_2 + H_2O$$

after balancing:

$$C_2H_4(NO_3)_2 \longrightarrow 2CO_2 + N_2 + 2H_2O.$$

● **PROBLEM** 902

One problem concerning space flight is the removal of carbon dioxide ($CO_2$, molecular weight = 44.0 g/mole) emitted by the human body (about 924 g of $CO_2$ per day). One solution is to react $CO_2$ with sodium hydroxide (NaOH, molecular weight = 40.0 g/mole) to form water and sodium carbonate ($Na_2CO_3$, molecular weight = 94 g/mole) according to the following reation:

$$2NaOH + CO_2 \rightarrow Na_2CO_3 + H_2O.$$

How much NaOH must be carried on board a space capsule to remove the $CO_2$ produced by an astronaut on a 10 day flight?

**Solution:** Two moles of NaOH are required to remove one mole of $CO_2$, as indicated by the coefficients in the reaction. The number of moles of $CO_2$ produced is calculated from the mass emitted by the astronaut during the flight. From this, the number of moles of NaOH required for the removal of the $CO_2$ is calculated, and this number of moles is then converted to grams.

At the rate of 924 g per day, in 10 days the astronaut will produce 924 g/day × 10 days = 9240 g of $CO_2$. Dividing this mass by the molecular weight of $CO_2$, it is calculated that the astronaut will produce a total of

$$\frac{9240 \text{ g}}{44.0 \text{ g/mole}} = 210 \text{ moles of } CO_2.$$

The number of moles of NaOH required to remove 210 moles of $CO_2$ is 2 × 210 moles = 420 moles. Multiplying this by the molecular weight, the minimum amount of NaOH that must be carried on board is

420 moles × 40 g/mole = 16,800 g = 16.8 kg.

● **PROBLEM** 903

Paper pulp (almost 100 % cellulose) is prepared commercially by digesting wood chips in a hot aqueous solution of calcium bisulfite, $Ca(HSO_3)_2$. The latter dissolves lignin and resins in the wood, leaving nearly pure cellulose. The sulfite solution is prepared by the following reactions:

$$S + O_2 \rightarrow SO_2$$

$$SO_2 + H_2O \rightarrow H_2SO_3$$

$$CaCO_3 + 2H_2SO_3 \rightarrow Ca(HSO_3)_2 + CO_2 + H_2O$$

(a) For every 250 kg of limestone ($CaCO_3$) used in the process, what weight of sulfur would be required? (b) What weight of $Ca(HSO_3)_2$ is produced from the limestone and sulfur in part (a)?

Solution: From the stoichiometry of the above equations, 1 mole of sulfur yields 1 mole of $H_2SO_3$ and 1 mole $CaCO_3$ uses 2 moles of $H_2SO_3$. Thus, 1 mole of $CaCO_3$ uses 2 moles of sulfur. The molecular weights of $CaCO_3$ and S are 100 g/mole and 32 g/mole, respectively. If 250 kg =250,000 g of $CaCO_3$ are used then, the number of moles of $CaCO_3$ is

$$\frac{250,000 \text{ g } CaCO_3}{100 \text{ g/mole}} = 2500 \text{ moles.}$$

For every 2500 moles $CaCO_3$ used in the process, twice that amount, or 5000 moles, of sulfur is needed. The weight of sulfur required is (5000 moles)(32 g/mole) = 160,000 g = 160 kg.

For every mole of $CaCO_3$ consumed, one mole of $Ca(HSO_3)_2$ is produced. Thus, if 2500 moles of $CaCO_3$ is consumed then, 2500 moles of $Ca(HSO_3)_2$ are produced. To obtain the weight of $Ca(HSO_3)_2$ produced multiply its molecular weight (MW of $Ca(HSO_3)_2$ = 202 g/mole) by the number of moles produced. The weight of $Ca(HSO_3)_2$ produced is

(2500 moles)(202 g/mole) = 505,000 g = 505 kg.

● **PROBLEM** 904

Lithopone white paint is made by heating a mixture of BaS and $ZnSO_4$ according to the equation

$$BaS + ZnSO_4 \rightarrow BaSO_4 + ZnS$$

What is the weight ratio of products $BaSO_4$ and ZnS in the paint?

Solution: From the stoichiometry of this equation, 1 mole of BaS + 1 mole of $ZnSO_4$ → 1 mole of $BaSO_4$ + 1 mole of ZnS. Assuming that 1 mole of each product is contained in this paint, their weights would be equivalent to their molecular weights. The molecular weight of $BaSO_4$ is 233 g/mole; for ZnS it is 97 g/mole. The weight ratio is the ratio of weight of one substance compared to the weight of another substance. Thus, the weight ratio of $BaSO_4$ to ZnS in lithopone is

$$BaSO_4 : ZnS = 233 : 97 \quad \text{or} \quad 2.4 : 1.$$

The initial assumption that only 1 mole of each product is contained in the paint is therefore not necessary. The weight ratio would be the same no matter how many moles of products were formed, since the products are formed in equimolar amounts.

● **PROBLEM** 905

Manufacturers use soda-lime glass for containers. This type of glass can be prepared by melting sodium carbonate ($Na_2CO_3$), limestone ($CaCO_3$), and sand ($SiO_2$) together. The equation for this reaction is

$$Na_2CO_3 + CaCO_3 + 6SiO_2 \rightarrow Na_2O \cdot CaO \cdot 6SiO_2 + 2CO_2$$
$$\text{soda-lime glass}$$

If a manufacturer wanted to make 5000 wine bottles, each weighing 400 g, how many grams of sand would be required?

Solution: From the reaction equation, it can be seen that 6 moles of $SiO_2$ are consumed for every mole of soda-lime glass formed. Thus, one can determine the amount of sand needed to form the 5000 bottles once the number of moles of glass that makes up these bottles is found. To solve:

(1) determine the weight of the bottles and the number of moles of soda-lime glass that this weight is equivalent to. (2) Calculate the number of moles of sand needed and from this its weight.

Solving:

(1) weight of bottles = (400 g/bottle)(5000 bottles)

$$= 2.0 \times 10^6 \text{ g.}$$

The molecular weight of soda-lime glass is 478 g/mole.

number of moles = $\dfrac{2.0 \times 10^6 \text{ g}}{478 \text{ g/mole}}$ = $4.18 \times 10^3$ moles of glass.

(2) For each mole of glass formed, 6 moles of sand must react. Thus,

no. of moles of sand = $6 \times 4.18 \times 10^3 = 2.51 \times 10^4$ moles.

The molecular weight of sand is 60.1 g/mole. Hence,

the amount of sand needed = $(2.51 \times 10^4 \text{ moles})(60.1 \text{ g/mole})$

$$= 1.51 \times 10^6 \text{ g.}$$

An explorer, lost in the desert, stumbled on an abandoned automobile whose gas tank still held 5 liters of fuel. Gasoline has a density of 0.67 g/cc. Assuming gasoline to be composed of $C_7H_{16}$ molecules, how many liters of water could he produce by burning the fuel and trapping the products? The reaction is

$$C_7H_{16}(\ell) + O_2(g) \rightarrow CO_2(g) + H_2O(g)$$

Solution: There is one mole of $H_2O$ produced by each mole of $C_7H_{16}$ burned. To find the amount of water formed by burning the gasoline determine the number of moles of fuel present. Once the number of moles formed is determined, the number of grams of $H_2O$ can be found.

There are 5 liters of gasoline present. It is given that 1 cc of $C_7H_{16}$ weighs 0.67 g. In 1 liter there are 1,000 cc, therefore the gasoline weighs

$$5000 \text{ cc} \times 0.67 \text{ g/cc} = 3350 \text{ g}$$

One finds the number of moles present by dividing the total weight by the molecular weight. (MW of $C_7H_{16}$ = 100).

$$\text{moles} = \frac{3350 \text{ g}}{100 \text{ g/mole}} = 33.50 \text{ moles.}$$

Because 33.50 moles of $C_7H_{16}$ are present, 33.50 moles of $H_2O$ can be formed. The number of grams of $H_2O$ formed is found by multiplying the molecular weight of $H_2O$ by the number of moles present (MW of $H_2O$ = 18).

$$\text{grams} = 33.50 \text{ moles} \times 18.0 \text{ g/mole} = 603.0 \text{ g.}$$

All the silicon and most of the oxygen in the earth's crust exists in the form of silicate minerals, such as granite. If the basic chemical unit of all silicates is the $SiO_4^{4-}$ ion, which consists of an $Si^{4+}$ ion and four $O^{2-}$ ions, write the electronic configurations of the $Si^{4+}$ and $O^{2-}$ ions in silicate materials.

Solution: An electronic configuration of an element (or ion) gives a picture of the arrangement of electrons that surround a nucleus.

The atomic number gives the total number of electrons of the atom (not the ion). With this in mind, write the electron configuration of the oxygen and silicon atoms. The electronic configuration of oxygen with atomic number 8 is $1s^2 2s^2 2p^4$ and for silicon with atomic number 14 it is $1s^2 2s^2 2p^6 3s^2 3p^2$.

To find the electronic configurations of the ions $Si^{4+}$ and $O^{2-}$ note each ion's charge. The charge of 4+ on Si indicates that it lost 4 electrons (electrons are negatively charged). Thus, if the silicon atom had a configuration of $1s^2 2s^2 2p^6 3s^2 3p^2$, and it loses 4 electrons, the configuration becomes $1s^2 2s^2 2p^6$ (ions will form by losing only high energy electrons).

The oxygen atom configuration was $1s^2 2s^2 2p^4$. The fact that the charge was 2- indicates that the atom gained two electrons. Consequently, the configuration becomes $1s^2 2s^2 2p^6$.

# CHAPTER 28

# APPLIED THERMOCHEMICAL PROBLEMS

● PROBLEM 908

A snowball at - 15°C is thrown at a large tree trunk. If
we disregard all frictional heat loss, at what speed must
the snowball travel so as to melt on impact? The specific
heat of fusion of ice is 79.7 cal/gm.

Solution:  To solve this problem, one must know that the
translational energy or kinetic energy of the snowball
is completely converted into heat energy. The kinetic energy
is given by the equation

$$k.e. = \tfrac{1}{2} mv^2,$$

where m = mass and v = velocity. The total heat energy for
this particular problem is

$$Q = m\Delta Tc + m\Delta H_{fus},$$

where m = mass, $\Delta T$ = change in temperature, c = heat
capacity (for $H_2O$, 0.5 cal/°C gm), and $\Delta H_{fus}$ = 79.9 cal/gm.

These two equations are set equal to each other. The
kinetic energy must be large enough so that the snowball
can increase its temperature to 0°C and then be completely
converted to liquid water at 0°C. Thus,

$$m\Delta Tc + m\Delta H_{fus} = \tfrac{1}{2} mv^2$$

$$\Delta Tc + \Delta H_{fus} = \tfrac{1}{2} v^2$$

$$(15°C) (0.5 \text{ cal/°C gm}) + 79.7 \text{ cal/gm} = \tfrac{1}{2} v^2$$

$$v^2 = 174 \text{ cal/gm} = (174 \text{ cal/mg})(4.18 \text{ J/cal}) = 727 \text{ J/gm}$$

$$= 727 \text{ kgm}^2/\text{s}^2/\text{gm} = 7.27 \times 10^5 \text{ m}^2/\text{s}^2$$

[Note: 1 kg = 1000 gm; $727 \times 1000 \, m^2/s^2 = 7.27 \times 10^5 \, m^2/s^2$]

Thus,     $v = \sqrt{7.27 \times 10^5 \, m^2/s^2} = 853 \, m/s$.

● **PROBLEM** 909

Two possible metabolic fates of α-D-glucose are given below. Compare the standard free energy changes at 25°C for these processes. Which process represents the more efficient utilization of α-D-glucose?

(a) α-D-glucose (1 M) → 2 ethanol (1 M) + $2CO_2$(g)

(b) $6O_2$(g) + α-D-glucose (1 M) → $6CO_2$(g) + 6 $H_2O$ (ℓ)

$\Delta G_f^o$ in Kcal/mole are $CO_2$(g) = - 94.26, $H_2O$ (ℓ) = - 56.69, ethanol (1 M) = - 43.39, α-D-glucose = - 219.22.

Solution:  For the standard free energy changes,

$$\Delta G_{reaction} = \sum \Delta G_{formation\ of\ products} - \sum \Delta G_{formation\ of\ reactants}$$

The free energy of any element in its standard state is taken as zero.

Thus, for reaction (a), $\Delta G_{f,\ reaction\ (a)}^o$ is:

$$2\Delta G_{f,CO_2}^o + 2\Delta G_{f,\ ethanol}^o - \Delta G_{f,\ \alpha-D-glucose}^o.$$

Substituting the known values into this equation yields

2(- 94.26) + 2(- 43.39) - (- 219.22) = - 56.08 Kcal/mole

For reaction (b), $\Delta G_{f,\ reaction\ (b)}^o$ is

$$6\Delta G_{f,\ H_2O}^o + 6\Delta G_{f,\ CO_2}^o - \Delta G_{f,\ \alpha-D-glucose}^o - \Delta G_{f,\ O_2}^o$$

Substituting in the known values yields

6(- 56.69) + 6(- 94.26) - (- 219.22) - (0) = - 685.5 Kcal/mole

Reaction (b) is more efficient since its standard free energy change of formation is more negative or, in other words, more exothermic and releases more energy per mole of α-D-glucose.

One of the methods for the synthesis of dilute acetic acid from ethanol is the use of acetobacter. A dilute solution of ethanol is allowed to trickle down over beechwood shavings that have been inoculated with a culture of the bacteria. Air is forced through the vat countercurrent to the alcohol flow. The reaction can be written as

$$C_2H_5OH \text{ (aq)} + O_2 \rightarrow CH_3COOH \text{ (aq)} + H_2O \text{ (}\ell\text{)}$$
$$\text{acetic acid}$$

Calculate the Gibbs Free Energy, $\Delta G°$, for this reaction, given the following Gibbs Free Energies of Formation: Ethanol = $-43.39$ Kcal/mole, $H_2O$ $(\ell)$ = $-56.69$ Kcal/mole, and Acetic acid = $-95.38$ Kcal/mole.

Solution: The $\Delta G°$ of a reaction is a measure of the free energy, i.e., the energy available to do work. It can be calculated from the free energy of formation, $\Delta G°_f$, of the reactants and products. $\Delta G°_f$ is a measure of the free energy needed to form a compound from its elements in their standard states. By definition, the $\Delta G°_f$ of an element is equal to 0.

The $\Delta G°$ of the reaction

$$C_2H_5OH + O_2 \rightarrow CH_3COOH + H_2O$$

can be found by taking the sum of the $\Delta G°_f$'s of the products minus the sum of the $\Delta G°_f$'s of the reactants, where products and reactants are each multiplied by their molar amount, indicated by the coefficients in the chemical equation. In other words,

$$\Delta G_{reaction} = \sum \Delta G_{formation\ of\ products} - \sum \Delta G_{formation\ of\ reactants}$$

For the reaction in this problem,

$$\Delta G_{reaction} = \left[ \left( \Delta G°_{f\ CH_3COOH} + \Delta G°_{f\ H_2O} \right) - \left( \Delta G°_{f\ C_2H_5OH} + \Delta G°_{f\ O_2} \right) \right]$$

$\Delta G°_{f\ O_2}$ is zero, because $O_2$ is an element in its standard state. Substituting the $\Delta G°_f$'s given and solving, one obtains:

$$\Delta G = [(-95.38 - 56.69) - (-43.39 + 0)]$$

$$= -108.7 \text{ Kcal/mole.}$$

One method being developed to efficiently use solar energy for the heating of homes is the use of sodium sulfate decahydrate $(Na_2SO_4 \cdot 10H_2O)$. It is placed in a bin on the roof of a house. During the day, the bin stores heat collected from the sun via the endothermic reaction:

$$Na_2SO_4 \cdot 10H_2O \rightarrow NaSO_4(s) + 10H_2O\ (\ell).$$

At temperatures above 32.4°C, the reaction goes completely to the right. At night, when the temperature falls below 32.4°C, the reaction goes completely to the left. If the efficiency of the system is 100 %, how much heat could 322 kg of $Na_2SO_4 \cdot 10H_2O$ provide to a home at night? Use the following information:

| Compound | Standard Heat of Formation at 298°K |
|---|---|
| $Na_2SO_4 \cdot 10H_2O$ (s) | - 1033.5 |
| $Na_2SO_4$ (s) | - 330.9 |
| $H_2O$ ($\ell$) | - 68.3 |

Solution: This problem can be solved once the amount of heat one mole of $Na_2SO_4 \cdot 10H_2O$ absorbs during the day is known.

An endothermic reaction is one that absorbs heat from the surroundings. The amount of heat absorbed can be calculated from the Standard Heats of formation given. The heat of the reaction (i.e. the amount of heat absorbed or released) is equal to the $\Delta H^\circ_{products}$ minus $\Delta H^\circ_{reactants}$, where $\Delta H^\circ$ = Standard Heats of Formation.

During the day (when the reaction absorbs heat and proceeds to the right),

$$\text{Heat of reaction} = \Delta H^\circ_{products} - \Delta H^\circ_{reactants}$$

$$= \left( \Delta H^\circ_{Na_2SO_4(s)} + 10\Delta H^\circ_{H_2O(\ell)} \right) - \left( \Delta H^\circ_{Na_2SO_4 \cdot H_2O} \right)$$

$$= (- 330.9 - 683) - (- 1033.5)$$

$$= + 19.6 \text{ Kcal mole}^{-1}.$$

Thus, 1 mole of $Na_2SO_4 \cdot 10H_2O$ absorbs 19.6 Kcal mole$^{-1}$ of heat from the sun during the day. At night, the reaction is reversed, which means it is exothermic, or released heat.

If the reaction absorbs   19.6 Kcal mole$^{-1}$ of heat when it proceeds to the right, then it must release 19.6 Kcal/mole of heat to the house at night, when it proceeds to the left.

The number of moles of $Na_2SO_4 \cdot 10H_2O$ present in 322 kg can be calculated.

In 1 kg there are 1,000 g, thus 322 kg = $322 \times 10^3$ g. A mole is defined as grams(mass)/molecular weight. The molecular weight of $Na_2SO_4 \cdot 10H_2O$ = 322 g/mole. This means that

$$\frac{322 \times 10^3 \text{ g}}{322 \text{ g/mole}} = 1 \times 10^3 \text{ moles of } Na_2SO_4 \cdot 10H_2O \text{ is present.}$$

Therefore, this amount of $Na_2SO_4 \cdot 10H_2O$ should give

$(1 \times 10^3 \text{ moles})(19.6 \text{ Kcal/mole}) = 1.96 \times 10^4$ Kcal

of heat to the house at night.

● **PROBLEM** 912

When an inch of rain falls on New York, it results in a rainfall of $19.8 \times 10^9$ liters. If the temperature drops to 0°C after this rainfall and the density of rain ($H_2O$) is 1.00 g/cm$^3$, how much heat is released when this quantity of water freezes? You may assume $\Delta H_{fus}$ = 1.4 Kcal mole$^{-1}$.

Solution:  $\Delta H_{fus}$ is the heat of fusion. It is defined as the amount of heat necessary to melt 1 g of the solid to a liquid at the same temperature. In this problem, water is freezing to ice.

To find out the exact amount of heat released, the number of moles of rain that fell must be known.

To determine the number of moles of rain that fell, use the fact that its volume is $19.8 \times 10^9$ liters and density = 1.00 g/cm$^3$.

Mass = (density)(volume)

= (1.00 g/cm$^3$)($1.98 \times 10^{13}$ cm$^3$) = $1.98 \times 10^{13}$ g.

(note: $1.98 \times 10^9$ liters is equivalent to $1.98 \times 10^{13}$ cm$^3$ by the conversion factor 1000 cm$^3$/$\ell$).

From the mass of the water, the number of moles of $H_2O$ can be computed.

$$\text{number of moles} = \frac{\text{grams (mass)}}{\text{molecular weight (M.W.)}}$$

The M.W. of water = 18 g/mole. Therefore, in $1.98 \times 10^{13}$ grams of water, there are

$$\frac{1.98 \times 10^{13} \text{ g}}{18 \text{ g/mole}} = 1.1 \times 10^{12} \text{ moles.}$$

Recalling that $\Delta H_{fus} = 1.4$ Kcal/mole, $1.1 \times 10^{12}$ moles of water loses

$(1.4) \left(1.1 \times 10^{12}\right) = 1.54 \times 10^{12}$ Kcal of heat when it freezes.

# CHAPTER 29

# APPLIED ENERGY PROBLEMS

● **PROBLEM** 913

During its lifetime of about $10^{10}$ years, a normal star radiates an energy of about $1.0 \times 10^{52}$ ergs. What is the energy equivalent in kilowatthours (kwh)?

Solution: This problem illustrates the conversion between watts (w) and ergs/sec (1 w = $10^7$ ergs/sec). One kilowatt (kw) is equal to $10^3$ watts. Converting kw to erg/sec, we have 1 kw = $10^3$ w × $10^7$ ergs/sec - w= $10^{10}$ ergs/sec. Also, converting seconds to hours (1 hour = 3600 sec), we have 1 kw = $10^{10}$ ergs/sec = $10^{10}$ ergs/sec × 3600 sec/hour = $3.6 \times 10^{13}$ ergs/hour. Therefore, 1 kw = $3.6 \times 10^{13}$ ergs/hour or 1 kwh = $3.6 \times 10^{13}$ ergs. The total energy radiated by the star is $1.0 \times 10^{52}$ ergs. Converting to kwh we get

$$\frac{1.0 \times 10^{52} \text{ ergs}}{3.6 \times 10^{13} \text{ ergs/kwh}} = 7.8 \times 10^{38} \text{ kwh.}$$

Therefore the total energy radiated by the star is $7.8 \times 10^{38}$ kwh.

● **PROBLEM** 914

The world's total reserve of coal and natural gas is estimated to be equivalent to $3.0 \times 10^{19}$ Kcal. The world's present consumption of these fuels is $4.2 \times 10^{27}$ ergs/year. At this rate of consumption, how long will the fuel reserves last?

Solution: This problem is an example of the conversion between calories and ergs (1 cal = $4.2 \times 10^7$ ergs). The total reserve is

$$3.0 \times 10^{19} \text{ Kcal} = 3.0 \times 10^{19} \text{ Kcal} \times 10^3 \text{ cal/Kcal}$$

$$= 3.0 \times 10^{22} \text{ cal},$$

or, $3.0 \times 10^{22}$ cal $\times 4.2 \times 10^7$ ergs/cal $= 1.26 \times 10^{30}$ ergs.

Dividing the total reserve by the rate of consumption gives

$$\frac{1.26 \times 10^{30} \text{ ergs}}{4.2 \times 10^{27} \text{ ergs/year}} = 300 \text{ years.}$$

Hence, at the present rate of consumption, the world's supply of gas and coal will last 300 years.

● **PROBLEM** 915

Calculate the kinetic energy of an automobile weighing 2000 pounds and moving at a speed of 50 miles per hour. Record your answer in (a) Joules and (b) ergs.

Solution:  Kinetic energy is equal to $\frac{1}{2}mV^2$, where m is the mass of the object and V is its velocity. One is given that the mass of the car is 2000 lbs and that its velocity is 50 miles/hr. Solving for the kinetic energy in lbs miles²/hr²:

$$\text{kinetic energy} = \frac{1}{2}(2000 \text{ lb})(50 \text{ mile/hr})^2$$

$$= \frac{1}{2}(2000 \text{ lb})2500 \text{ mile}^2/\text{hr}^2)$$

$$= 2.5 \times 10^6 \text{ lb mile}^2/\text{hr}^2$$

(a) 1 erg = 1 g cm²/sec². Thus, the units of the kinetic energy found must be converted. Pounds must be converted to grams, miles² to cm² and hours² to sec². This is done by the use of conversion factors.

There are 454 g in 1 lb, thus to convert pounds to grams one must multiply the number pounds by 454 g/lb. There are $1.0 \times 10^5$ cm in 0.6214 mi, squaring these two quantities, one finds that there are $1.0 \times 10^{10}$ cm² in .386 mi². The conversion factor from mi² to cm² is then $1.0 \times 10^{10}$ cm²/.386 mi².

There are 3600 seconds in an hour, therefore there are (3600 sec)² in 1 hour² or $1.296 \times 10^7$ sec² per 1 hr². The appropriate conversion factor from hr² to sec² is

$$\frac{1 \text{ hr}^2}{1.296 \times 10^7 \text{ sec}^2}$$

Solving for ergs:

$$\left(2.5 \times 10^6 \frac{\text{lb mi}^2}{\text{hr}^2}\right) \left(454 \frac{\text{g}}{\text{lb}}\right) \left(\frac{1.0 \times 10^{10} \text{cm}^2}{.386 \text{ mi}^2}\right) \left(\frac{1 \text{ hr}^2}{1.296 \times 10^7 \text{ sec}^2}\right)$$

$$= 2.27 \times 10^{12} \frac{\text{g cm}^2}{\text{sec}^2} = 2.27 \times 10^{12} \text{ ergs.}$$

(b) 1 Joule = $10^7$ ergs, thus the number of Joules is equal to the number of ergs divided by $10^7$. The conversion factor used when converting ergs to Joules is 1 Joule/$10^7$ ergs.

$$\text{no. of Joules} = 2.27 \times 10^{12} \text{ ergs} \times \frac{1 \text{ Joule}}{10^7 \text{ ergs}}$$

$$= 2.27 \times 10^5 \text{ Joules}$$

● **PROBLEM** 916

How much work is done when a man weighing 75 kg (165 lb) climbs the Washington monument, 555 ft high? How many kilocalories must be supplied to do this **muscular work,** assuming that 25 % of the energy produced by the oxidation of food in the body can be converted into muscular mechanical work?

Solution: The energy (work) needed is equal to the potential energy the man attains once he is 555 ft. above the ground. The formula for the potential energy is

$$\text{P.E.} = \omega = mgh,$$

where P.E. = $\omega$, which is the work done, m = mass, g is the gravitational acceleration constant = 9.8 m/s$^2$, and h is height. Substituting the given values

$$\omega = (75 \text{ kg})(9.8 \text{ m/s}^2)(555 \text{ ft})(12 \text{ in/ft})\left(\frac{1 \text{ m}}{39.37 \text{ in}}\right)$$

$$= 1.24 \times 10^5 \frac{\text{kg m}^2}{\text{s}^2} = 1.24 \times 10^5 \text{ J}$$

The energy needed is four times greater than the work done.

$$E = 4(1.24 \times 10^5 \text{ J})/(4.18 \text{ J/cal})(10^3 \text{ cal/Kcal})$$

$$= 119 \text{ Kcal.}$$

When air is let out of an automobile tire, the air feels cold to the touch. Explain why you feel this cooling sensation.

Solution: To solve this problem, consider what is happening when the air is being expelled.

For the air to be released, it must displace the atmosphere that immediately surrounds the tire. To displace the atmosphere, it must do work on the atmosphere. In other words, the air must expend energy. There exists a correlation between the energy of a substance and its temperature. As the air is being released, it is doing work. Since it consumes energy, its temperature must fall. As such, you feel a cooling sensation.

# CHAPTER 30

# APPLIED WAVE PHENOMENA PROBLEMS

● PROBLEM 918

A surface probe on Mars transmits radiowaves at a frequency of $6.0 \times 10^5$ sec$^{-1}$ to an earth station $8.0 \times 10^7$ km away. How long does it take radiowaves to traverse this distance?

Solution: Since radiowaves are a form of electromagnetic radiation, they move at the speed of light and we can use the relationship

$$\text{time} = \frac{\text{distance}}{\text{speed}}$$

where the speed is that of light. We do not need to know the frequency. Since 1 km = $10^5$ cm, $8.0 \times 10^7$ km = $8.0 \times 10^7$ km $\times 10^5$ cm/km = $8.0 \times 10^{12}$ cm and we have

$$\text{time} = \frac{\text{distance}}{\text{speed}} = \frac{8.0 \times 10^{12} \text{ cm}}{3.0 \times 10^{10} \text{ cm/sec}} = 2.7 \times 10^2 \text{ sec,}$$

or 4 ½ minutes.

● PROBLEM 919

There is an appreciative sigh throughout the audience in the "mad" scene of "Lucia di Lammermoor", when the soprano hits high C. On the musical scale, high C has a frequency of 1,024 sec$^{-1}$. Assuming the velocity of sound in air is $3.317 \times 10^4$ cmsec$^{-1}$, compute the wavelength of the high C sound wave.

Solution: The relationship between the wavelength ($\lambda$), frequency ($\nu$), and velocity (v) of a sound wave is:

$$\lambda = \frac{v}{\nu} .$$

Given both v and ν, substitute these values into this expression and evaluate.  The wavelength of high C =

$$= \lambda = \frac{3.317 \times 10^4 \text{ cmsec}^{-1}}{1024 \text{ sec}^{-1}} = 32.39 \text{ cm.}$$

● **PROBLEM** 920

What is the de Broglie wavelength of an 88 kg man skiing down Mt. Tremblant in Québec at $5.0 \times 10^5$ cm/sec?

Solution:  The de Broglie wavelength, $\lambda$, of a body of mass moving at speed v is

$$\lambda = h/mv,$$

where h is Planck's constant, h = $6.6 \times 10^{-27}$ erg-sec = $6.6 \times 10^{-27}$ g-cm$^2$/sec. Hence,

$$\lambda = \frac{h}{mv} = \frac{6.6 \times 10^{-27} \text{ g-cm}^2/\text{sec}}{88 \text{ kg} \times 5.0 \times 10^5 \text{ cm/sec}}$$

$$= \frac{6.6 \times 10^{-27} \text{ g-cm}^2/\text{sec}}{88000 \text{ g} \times 5.0 \times 10^5 \text{ cm/sec}} = 1.5 \times 10^{-37} \text{ cm.}$$

The de Broglie wavelength of the skier, $1.5 \times 10^{-37}$ cm, is in the microwave region of the spectrum.

● **PROBLEM** 921

Cobalt-60 ($^{60}_{72}$ CO) is an artificially produced radio-isotope that is important in cancer therapy and in the manufacture and sterilization of certain plastics. If gamma radiation from cobalt-60 has a wavelength of $1.0 \times 10^{-8}$ Å, what is its frequency?

Solution:  Since gamma radiation is a form of electro-magnetic radiation, we can solve this problem by applying the relationship $c = \lambda\nu$, or $\nu = c/\lambda$, where $\nu$ is the frequency, $\lambda$ the wavelength, and c the speed of light. Since 1 Å = $10^{-8}$ cm, $\lambda = 1.0 \times 10^{-8}$ Å = $1.0 \times 10^{-8}$ Å $\times 10^{-8}$ cm/Å = $1.0 \times 10^{-16}$ cm and

$$\nu = \frac{c}{\lambda} = \frac{3.0 \times 10^{10} \text{ cm/sec}}{1.0 \times 10^{-16} \text{ cm}} = 3.0 \times 10^{26} \text{ sec}^{-1}.$$

The accompanying figure shows the wavelength ranges for the different colors in the visible spectrum. If potassium-containing materials emit light of frequency $7.41 \times 10^{14}$ sec$^{-1}$, what color flame would you expect to see when you heat a potassium compound such as potassium chloride in a Bunsen burner flame?

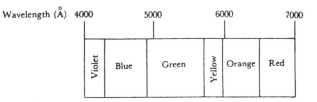

**Solution:** The color that you can expect will depend on the wavelength of the light emitted, as indicated by the diagram. Given the frequency of the light emitted, the wavelength can be determined. The frequency ($\nu$), wavelength ($\lambda$), and speed of light (c) are related by the equation $\nu\lambda = c$. $c = 3.00 \times 10^{10}$ cm/sec. Solving for $\lambda$:

$$(7.41 \times 10^{14})\ \lambda = 3.00 \times 10^{10}.$$

$$\lambda = \frac{3.00 \times 10^{10}}{7.41 \times 10^{14}} = 4.05 \times 10^{-5} \text{ cm}.$$

To convert this into $\overset{\circ}{A}$; remember that there are $10^8\ \overset{\circ}{A}$ per centimeter. Consequently,

$$4.05 \times 10^{-5} \text{ cm} = (4.05 \times 10^{-5})(10^8) = 4050\ \overset{\circ}{A}.$$

According to the diagram, this $\lambda$ corresponds to the violet region.

Given $\Delta H = 2800$ kJ for the production of one mole of $C_6H_{12}O_6$ from $CO_2$ and $H_2O$, calculate what wavelength of light would be needed per molecule of $CO_2$ to achieve this photosynthesis in a one-step process.

**Solution:** To solve this problem:

(1) set up the balanced equation for the production of 1 mole of $C_6H_{12}O_6$ from $CO_2$ and $H_2O$.

(2) From the stoichiometry of this equation determine the number of moles of $CO_2$ necessary to make 1 mole of $C_6H_{12}O_6$.

(3) Calculate the number of molecules of $CO_2$ necessary to make 1 mole of $C_6H_{12}O_6$ using Avogadro's number ($6.022 \times 10^{23}$ molecules/mole).

(4) calculate the energy necessary per molecule of $CO_2$ by dividing $\Delta H$ by the total number of $CO_2$ molecules.

(5) Use the expression $E = h\nu$ where E is the energy of light, h is Planck's constant, and $\nu$ is the frequency of light.

Set the energy per molecule of $CO_2$ equal to $h\nu$ and solve for the frequency, $\nu$.

(6) Use the expression $c = \lambda\nu$, (where c is the speed of light, $\nu$ is the frequency of light (from step 5) and $\lambda$ is the wavelength of light) to find $\lambda$.

To begin, the balanced equation for the production of 1 mole of $C_6H_{12}O_6$ is

$$6CO_2 + 6H_2O \rightarrow C_6H_{12}O_6 + 6O_2$$

Note that this process is photosynthesis.

From the stoichiometry of this equation, 1 mole of $C_6H_{12}O_6$ is produced for every 6 moles of $CO_2$ consumed. And, therefore, the number of molecules of $CO_2$ necessary to make 1 mole of $C_6H_{12}O_6$ is

(6 moles $CO_2$)(6.02 $\times 10^{23}$ molecules/mole)

$$= 3.61 \times 10^{24} \text{ molecules.}$$

The energy necessary per molecule of $CO_2$ is

$$\frac{2,800,000 \text{ J}}{3.61 \times 10^{24} \text{ molecules } CO_2} = 7.75 \times 10^{-19} \text{ J/molecule.}$$

Steps 5 and 6 are combined to give the expression $E = h\frac{c}{\lambda}$, where $h = 6.63 \times 10^{-34}$ J-s, and $c = 3.00 \times 10^{10}$ cm/s. Setting this equal to $7.75 \times 10^{-19}$ J and solving for $\lambda$ gives

$$\lambda = \frac{hc}{E} = \frac{(6.63 \times 10^{-34} \text{ J-s})(3.00 \times 10^{10} \text{ cm/s})}{(7.75 \times 10^{-19} \text{ J})}$$

$$= 2.57 \times 10^{-5} \text{ cm} = 2.57 \times 10^{-5} \text{ cm} \times \frac{10^8 \text{ Å}}{\text{cm}} = 2570 \text{ Å.}$$

The sensitivity of the silver halides to light is the basis of photography. The change from white silver chloride to a gray purple solid is the result of photochemical decomposition,

$$AgCl(s) \xrightarrow{\text{light}} Ag(s) + \tfrac{1}{2} Cl_2(g)$$

The $\Delta H$ of this reaction is 30.362 Kcal/mole. From this, determine the frequency of light needed to decompose 1 molecule of AgCl.

**Solution:** To solve this problem (1) determine the energy consumed per molecule AgCl by dividing the $\Delta H$ given by Avogadro's number. (2) Convert the energy per molecule of AgCl from units of Kcal/molecule to units of J/molecule using the factors 1 Kcal = 1000 cal and 1 cal = 4.18 J. (3) Use the expression $E = h\nu$ where E is the energy in J/molecule, h is Planck's constant, and $\nu$ is the frequency of light. Thus,

(1) $\quad E = \dfrac{30.362 \frac{\text{Kcal}}{\text{mole}}}{6.02 \times 10^{23} \frac{\text{molecules}}{\text{mole}}} = 5.04 \times 10^{-23} \dfrac{\text{Kcal}}{\text{molecule AgCl}}$

(2) $\quad E = \left(5.04 \times 10^{-23} \dfrac{\text{Kcal}}{\text{molecule}}\right)\left(1000 \dfrac{\text{cal}}{\text{Kcal}}\right)\left(4.18 \dfrac{\text{J}}{\text{cal}}\right)$

$\qquad = 2.11 \times 10^{-19}$ J/molecule AgCl

(3) $\quad E = h\nu \qquad$ or $\quad \nu = E/h$

$\qquad \nu = \dfrac{2.11 \times 10^{-19} \text{ J/molecule}}{6.63 \times 10^{-34} \text{ J-s}}$

$\qquad = 3.18 \times 10^{14}$ cycles/sec per molecule AgCl.

Explain how a solar cell is constructed and why it can generate electricity.

**Solution:** When certain materials are exposed to light, a current is induced in the material. The effect is called the photo-electric effect. These certain materials are called semiconductors.

Using the photo-electric effect, solar cells convert sunlight directly into electricity.

A solar cell consists of a solid electrode, attached to the back of a semiconducting material. A semitransparent metal-film electrode is attached to the front of the semi-conductor.

In general, cadmium sulfide crystals serve as a semi-conductor in solar cells. A silver electrode is attached to one side of the cadmium sulfide and a thin layer of indium is placed on the opposite side. (See Figure 1.)

When light strikes the cadmium sulfide through the indium electrode, electrons are released from cadmium sulfide. The released electrons flow through the external circuit and back to the silver electrode, which creates an electrical current. As long as the light hits the cadmium sulfide, electricity will flow.

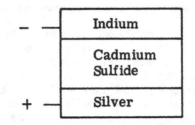

Figure 1

# CHAPTER 31

# APPLIED ORGANIC AND POLYMER PROBLEMS

● **PROBLEM** 926

What would be the simplest and safest reliable method of
distinguishing between the following pairs of materials?
(a) Water and rubbing alchol. (b) Gasoline and kerosene.
(c) Baking soda and baking powder.

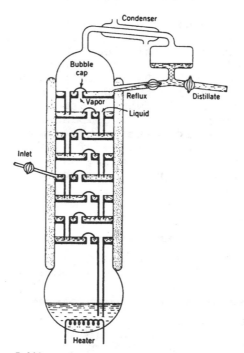

Bubble-cap fractionating column.

Solution:    (a) Pour to a depth of about one-half inch of
each compound into two separate test tubes. Cover each of
the test tubes with your thumb and shake them vigorously.

There will be an outward pressure on your thumb from the contents of one of the test tubes. This tube contains the rubbing alcohol. This pressure is called vapor pressure. It is proportional to the ability of a liquid to vaporize or evaporate. Vapor pressure is a measure of the physical property called volatility. Rubbing alcohol is more volatile than water and will have a greater vapor pressure and lower boiling point.

(b) Gasoline and kerosene are both products of the distillation of crude oil. Distillation is a process used to separate mixtures of liquids. When a liquid is distilled it is boiled in a round bottomed container at the end of a glass enclosed tube. The various compounds in the mixture burn at their own distinctive boiling points. Thus, first the lowest boiling substance will boil and evaporate, then as the mixture becomes hotter the next lowest and so on until no compounds are left in the container. As each compound boils its vapors enter the tube or column where they are condensed and eventually collected. A different vessel is used to collect each fraction with a distinct boiling point and hence the mixture becomes separated into its various components. The more times a fraction is distilled the more pure a sample made. This is the best method for separating gasoline and kerosene. Gasoline burns over a range from 40° to 200°C and kerosene from 175° to 275°C. They can be easily separated by distillation. The accompanying figure shows a fractional distillation of crude oil.

(c) The difference between baking soda and baking powder is that baking soda is a compound and baking powder is a mixture. A compound is a homogeneous substance composed of two or more elements, the proportions of which are fixed and invariable by weight. A mixture is a substance made of two or more elements or compounds in a nonhomogeneous combination. Under a magnifying glass or microscope one will not be able to distinguish the components of a compound whereas one can readily identify the components of a mixture. Baking powder is a mixture of baking soda, an acid substance and starch, while baking soda is made up of a single compound, sodium hydrogen carbonate ($NaHCO_3$).

● **PROBLEM** 927

Distinguish between a soap and a detergent. How does a soap clean? Explain the function of phosphate in a detergent.

Solution: Soaps are usually potassium or sodium salts of fatty acids (long carbon chain carboxylic acids of formula R-COOH, where R is a long alkyl group). They are produced commercially by heating a fat with an aqueous sodium or potassium hydroxide solution.

The structure of soap explains its cleansing action. Soap is a long hydrocarbon (non-polar) group with an ionic group at one end. (See figure A.)

$$CH_3 \; CH_2 \; CH_2 \; CH_2 \; CH_2 \; CH_2 \; CH_2 \; (CH_2)_{10} \; \overset{\overset{\text{O}}{\|}}{C}-O^-Na^+$$

Long hydrocarbon chain                    ionic group

Sodium Stearate

Figure A

$$C_{9-15} \; H_{19-31} \langle\!\!\!\!\!\bigcirc\!\!\!\!\!\rangle SO_3{}^{\ominus} \, Na^{\oplus}$$

Alkyl-Benzene Sulfonate

Figure B

The ionic group is water-soluble and the hydrocarbon group is oil-soluble. Dirt particles are oil-containing. The hydrocarbon end of the soap is attached to and dissolves in dirt particles, leaving the ionic end exposed to water. The dirt particle is dispersed in the water as a stable emulsion and removed from the article being cleaned.

It is known that the cleansing action of soap is reduced by hard water, which contains calcium or magnesium ions. To avoid this problems, soaps have been replaced by detergents; they are more resistant to precipitation by hard water. The active ingredients, or surfactants, of detergent possess the same principal features as soaps. Detergents have a long hydrocarbon chain and an ionic group, usually a sulfonyl group. One example is the sodium salt of alkylbenzene sulfonate, (shown in Fig. B).

In addition to surfactants, detergents possess builders. The most widely used builder is $Na_5P_3O_8$. The phosphate in the detergent functions as a water softener by tying up hard-water ions. It aids the surfactant by dispersing the suspended dirt. It can also act as a buffer, thereby maintaining the pH of the solution at the optimum point for effective washing action.

● PROBLEM 928

Explain why tetraethyl lead is added to gasoline.

Solution: To explain the function of this additive in gasoline, consider how an automobile uses gasoline. The composition of gasoline varies according to the time of year, the refinery at which it was processed and the climate. Basically, gasoline is composed of hydrocarbon isomers containing between seven and ten carbon atoms. In cold climates, hydrocarbons such as butanes are used, which have a lower boiling point, so that the volatility of the gasoline is increased. The increased volatility allows the engine to start more easily. In warmer climates, higher-boiling components are used so that the gasoline's volatility is decreased; this minimizes such problems as vapor lock.

These changes must not alter the octane rating of the gasoline.

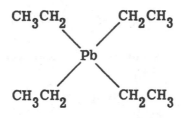

Tetraethyl lead

The octane rating is a measure of the manner in which gasoline burns in the engine. The fuel should be consumed at the end of the piston cycle and the gasoline should burn such that there is steady pressure on the piston. If these conditions are not met as closely as possible, the automobile engine will knock or ping. This can eventually damage some engine parts. The quantitative measure of a fuel's ability to knock resulted in the formation of octane rating. By definition, an octane rating of 100 is the highest premium gasoline, while an octane of zero is the lowest. The octane rating of a part-icular gasoline can be raised by blending in compounds of relatively high octane rating. The only problem is that such compounds are extremely expensive.

The discovery of additives solved the problem of improving octane rating. Tetraethyl lead (as pictured above) is one of these additives that can improve the octane rating without using the expensive high-octane hydrocarbons.

● **PROBLEM** 929

For which of the following two reactions should ΔS be more positive? Why?

Solution: If the heat gained by a system is equal to that lost by the surroundings, then the entropy change for the surroundings is the negative of the entropy change for the system; for both the system and surroundings taken together, ΔS is zero if the transfer of heat is carried out reversibly.

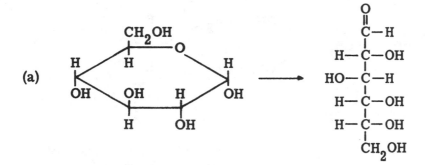

(a)

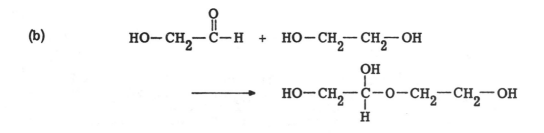

(b)

    The entropy of a system increases with the increasing randomness of the molecules. Thus, for reaction (a), ΔS is positive owing to the loss of rigidity or increasing randomness of reactants. For reaction (b), ΔS is negative owing to the loss of independent translational motion of the reactants. In other words, the product is more stable or more ordered than the reactants and, thus, the entropy decreases for the reaction.

● **PROBLEM** 930

Kodel is a polyester fiber. The monomers are terephthalic acid and 1,4-cyclohexanedimethanol. Write the structure of a segment of Kodel containing at least one of each monomer unit.

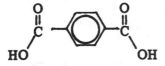

terephthalic acid

Figure A

$CH_2-CH_2$

$HOCH_2-CH$      $HC-CH_2OH$

$CH_2-CH_2$

1,4 - cyclohexanedimethanol

Figure B

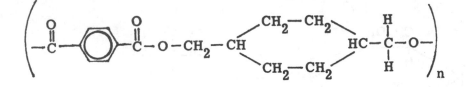

Figure C

Solution: Kodel is a polymer of significantly different properties than its constituent monomers. This particular polymer segment is made by the combination of a carboxylic acid and an alcohol; this addition reaction is called esterification.

Structures of the monomers are shown above in Figure A and B.

The esterification reaction occurs when the -OH group on the acid breaks off and forms a water molecule with one of the hydrogen atoms of the alcohol. Thus, a monomer of Kodel looks like Figure C (above) and forms a continuous chain.

● **PROBLEM** 931

Explain each of the following terms: (a) synthetic polymer. (b) Thermosetting. (c) Thermoplastic. (d) Elastomers.

Solution: A synthetic polymer is a large molecule with a repeating structure. It may be synthesized by chemically combining many monomer or single units. Polymers which are used for molding are often referred to as resins or plastics.

The polymers may be divided into three general types:

Thermoplastic polymers - These substances will melt on being heated and can be molded easily into new shapes. Examples include polyvinylchloride, polyethylene,and nylon.

Thermosetting polymers - These substances slowly char rather than melt. They are originally molded in their final shape or are machined into shape. They cannot be molded again after cooling. The best known example of this is Bakelite; it is used in the electrical industry as an insulating material.

Elastomers - This type of material may be characterized as rubbery or elastic. Examples include natural rubber and the various synthetic rubbers such as neoprene.

# CHAPTER 32

# APPLIED BIOLOGICAL PROBLEMS

● PROBLEM 932

Per day, an average adult may lose liquid water in the
following liter amounts: as urine 1.0; as sweat, 0.6;
as fecal matter, 0.1; in expired air, 0.4. How many
gallons of water pass through a human being, assuming
these data hold for 60 years?

Solution:  From this data, one can see that 1.0 + 0.6
+ 0.1 + 0.4 = 2.10 liters of water are lost per day. In
60 years, there are 365 × 60 = 21,900 days. In 60 years,
21,900 days × 2.10 $\ell$/day = 4.60 × 10$^4$ $\ell$ of water pass
through one person. 1 gallon = 4.224 liters. Converting
liters to gallons,

$$4.60 \times 10^4 \ \ell \times \frac{1 \text{ gallon}}{4.224 \text{ liters}} = 1.089 \times 10^4 \text{ gallon.}$$

● PROBLEM 933

The gypsy moth produces a natural attractant, $C_{18}H_{34}O_3$.
If a female moth is trapped behind a cellophane screen
containing a pinhole and the carbon dioxide she produces
diffuses through the pinhole at the rate of 1 milli
micromole per 90 seconds, what quantity of attractant
will diffuse through the orifice in the same amount of
time (90 seconds)?

Solution:  The rate of escape of gases through an orifice
is inversely proportional to the square root of the density
of the gas. This principle is known as Graham's Law. If
two gases are compared at the same temperature and pressure
(as in this problem), then

$$\frac{u_1}{u_2} = \sqrt{\frac{d_2}{d_1}}$$

where u is the velocity (or the rate at which the gas diffuses out of the orifice) and d is the density of the particular gas.

To solve this problem, this equation is modified somewhat to avoid using densities in the expression (since the densities are unknown). However, the density is defined as mass per unit volume, and since the volume per mole of both gases, $C_{18}H_{34}O_3$ and $CO_2$, are the same, Graham's Law becomes

$$\frac{u_1}{u_2} = \sqrt{\frac{d_2}{d_1}} = \sqrt{\frac{m_2/V}{m_1/V}} = \sqrt{\frac{m_2}{m_1}} \, ,$$

where V is volume and $m_1$ and $m_2$ are the masses of the gases. Another modification is necessary before solving this problem: the mass of any substance is its molecular weight

times the number of moles of that substance. However, assume that 1 mole is involved of both substances and thus molecular weights of the gases (which can be determined) is substituted. Graham's Law then becomes

$$\frac{u_1}{u_2} = \sqrt{\frac{M_2}{M_1}}$$

$u_1$ for $CO_2$ = $10^{-9}$ mole/90 sec = $1.1 \times 10^{-11}$ moles/sec. Note: 1 milli micromole = $10^{-9}$ mole.

$M_1$ for $CO_2$ = 44 g/mole

$M_2$ for $C_{18}H_{34}O_3$ = 298 g/mole.

Thus, solve for $u_2$.

$$\frac{1.1 \times 10^{-11} \text{moles/sec}}{u_2} = \sqrt{\frac{298 \text{ g/mole}}{44 \text{ g/mole}}} = 2.6$$

$$u_2 = 4.2 \times 10^{-12} \text{ moles/sec.}$$

1 picomole = $10^{-12}$ moles.

$u_2$ = $4.2 \times 10^{-12}$ moles/sec = 4.2 picomoles per sec

$u_1$ = $1.1 \times 10^{-11}$ moles/sec = 11.0 picomoles per sec.

The $CO_2$ diffuses 1 milli micromole within 90 sec

([ 0.011 milli micromole/sec][90 sec] = 1 milli micromole) and $C_{18}H_{34}O_3$ diffuses

(0.0042 milli micromoles/sec)(90 sec) = 0.38 milli
micromoles

within the same time as the $CO_2$.

In order to attract males for mating, females of many insect species secrete chemical compounds called pheromones. One compound of formula $C_{19}H_{38}O$ (282 g/mole) is among these pheromones, $10^{-12}$ g of it must be present to be effective. How many molecules must be present to be effective?

**Solution:** Avogadro's number ($6.02 \times 10^{23}$) is the number of particles per mole of any substance. To find the number of molecules to be effective, determine the number of moles present, and multiply this quantity by Avogadro's number.

Solving: $10^{-12}$ g of pheromones must be present to attract males. The number of moles is equal to number of grams/molecular weight.

number of moles = $\dfrac{10^{-12} \text{ g}}{282 \text{ g/mole}}$ = $3.55 \times 10^{-15}$ moles

number of molecules = $3.55 \times 10^{-15}$ moles $\times$ $6.02 \times$

$$10^{23}/\text{mole}$$

$$= 2.12 \times 10^{9} \text{ molecules.}$$

A laboratory analyst puts Fehling solution in a test tube and then adds the urine of a diabetic which contains some sugar. On heating, what red copper compound precipitates?

**Solution:** Copper forms two series of compounds: Copper (I) (cuprous), and copper II (cupric) compounds. Copper (I) oxide ($Cu_2O$) occurs naturally as the mineral cuprite. This red oxide may also be obtained by careful oxidation of copper in air or by precipitation of $Cu^{+}$ ions from solutions with sodium or potassium hydroxide. Copper (I) hydroxide is not known. Copper (I) oxide is also obtained as a reduction product of Fehling's solution, a reagent used in the diagnosis of diabetes to test for the presence of reducing sugars in the urine.

Fehling's solution consists of an aqueous solution of copper (II) sulfate, sodium hydroxide, and sodium potassium tartrate. The tartrate forms a complex with $Cu^{2+}$. As a result, so few free $Cu^{2+}$ are left in solution that no precipitate of copper (I) hydroxide is obtained when the sodium hydroxide is added. Reducing sugars reduce $Cu^{2+}$ to $Cu^{+}$, and $Cu_2O$ separates as a reddish precipitate.

Most bread is made from flour, water, fat, sugar and salt. Yeast is added as a leavening or "raising agent". Besides providing flavor, explain the function of salt in bread.

Solution:  The small portion of salt in bread (0.88 – 1.18 % by weight) helps to regulate the rate of fermentation of the yeast. Fermentation is the breakdown of complex molecules in organic compounds. Without salt, fermentation proceeds too rapidly and the bread becomes too porous or "light". With excess amounts of salt, fermentation is too slow, which results in the bread becoming too compact or heavy. Thus, the amount of salt present is crucial to the formation of quality bread.

Define antibiotics.  Where are they derived from?  What is meant by broad-spectrum antibiotics?

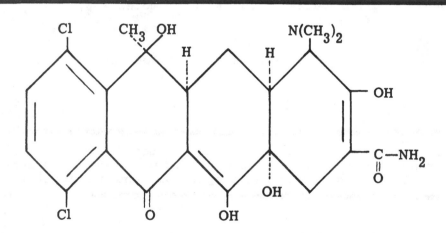

Aureomycin

Solution:  Compounds that can inhibit and/or destroy bacteria may be defined as antibiotics.  These substances may be bacteriostatic (growth-inhibiting) or bactericidal (bacteria killing) in their action.  Currently, antibiotics are derived from microorganisms and are, in effect, metabolic products of these organisms.  Common classes include the penicillins and the tetracyclines.

Tetracyclines have been isolated from "streptomyces" bacteria strains and all have a basic structure of four fused rings.  These compounds are also called broad-spectrum antibiotics because of their effectiveness against a wide variety of microorganisms.  Aureomycin, the first member of the tetracycline family, has its structure written above:

What is an antipyretic? Explain how it functions in humans.

Solution: Antipyretics are fever-reducing substances that relieve headaches and minor pains. They are also important in the treatment of arthritis and rheumatic fever. They belong to the class of medicinal compounds called salicylates, which, in turn, are termed analgesics or pain-killers.

The antipyretic action of these drugs appears to be accomplished by the influence of the drugs on a center in the hypothalamus of the brain. The hypothalamus causes small blood vessels in the skin to dilate (to swell or expand) thus enabling the body to lose heat.

Associate a medicinal property for each of the following families of compounds: (a) Phenothiazines, (b) Barbiturates, and (c) Amphetamines.

Solution: (a) Phenothiazines are one major type of tranquilizers used to treat psychoses. In general, tranquilizers are compounds which act as depressants of the central nervous system with highly selective action on brain function.

(b) There exist drugs that induce relaxation and sleep, which are sometimes used to treat patients with organic and emotional disorders. The barbiturate family is the most common and widely used for this sedative and hypnotic purpose.

(c) Amphetamines belong to the class of anti-depressant drugs. These drugs, as their name implies, are used to treat depression disorders, one of the conditions associated with mental illness. They stimulate the central nervous system, and are used to bring about an increase in alertness, elevation of blood pressure and increase in heart action.

The daily protein intake of a typical American diet consists of meat, dairy products and eggs. Its average net protein utilization (NPU) is 70. If the daily protein requirement of the body is 0.3 g per pound of body weight, what is the actual protein requirement for an average American?

<u>Solution</u>:  The biological value of a food is the ratio of the nitrogen retained in the body for growth and maintenance to the nitrogen absorbed in digestion. This expresses the fraction of the protein in the food which is used by the body. Another factor is how much of the protein of a food is digested. This factor is called the digestibility. The NPU of a food is the product of the biological value and the digestibility. The reason that nitrogen content is used as a measure of the amount of protein taken in by the body is because proteins are long chains of amino acids. On one end of each amino acid there is an amino group which contains one nitrogen atom and two hydrogen atoms. Therefore, the amount of nitrogen present is proportional to the amount of protein present.

The actual protein requirement is defined by the following equation.

$$\begin{pmatrix} \text{basic protein} \\ \text{requirement, grams} \\ \text{per pound} \end{pmatrix} \left( \frac{100}{\text{NPU}} \right) = \text{actual protein requirement.}$$

Substituting and solving for the actual protein requirement of an average American:

$$\left( \frac{0.3 \text{ g}}{\text{pound}} \right) \left( \frac{100}{70} \right) = \frac{0.43 \text{ g}}{\text{pound}}$$

Hence, a 150 lb man needs to eat

$$\left( \frac{0.43 \text{ g}}{\text{lb}} \right) \times (150 \text{ lb}) = 64.5 \text{ g of protein per day to}$$

avoid degradation of his own body protein.

# MATHEMATICAL REVIEW

## SCIENTIFIC NOTATION

● **PROBLEM** 941

Use scientific notation to express each number.
(a) 4,375  (b) 186,000  (c) 0.00012  (d) 4,005

Solution: A number expressed in scientific notation is written as a product of a number between 1 and 10 and a power of 10. The number between 1 and 10 is obtained by moving the decimal point of the number (actual or implied) the required number of digits. The power of 10, for a number greater than 1, is positive and is one less than the number of digits before the decimal point in the original number. The power of 10, for a number less than 1, is negative and is one more than the number of zeros immediately following the decimal point in the original number. Hence,

(a)  $4,375 = 4.375 \times 10^3$    (b)  $186,000 = 1.86 \times 10^5$

(c)  $0.00012 = 1.2 \times 10^{-4}$    (d)  $4,005 = 4.005 \times 10^3$

● **PROBLEM** 942

Express  $\dfrac{6,400,000}{400}$  in scientific notation.

Solution: In order to solve this problem, we express the numerator and denominator as the product of a number between 1 and 10 and a power of 10. This is known as scientific notation. Thus

$$6,400,000 = 6.4 \times 1,000,000 = 6.4 \times 10^6$$
$$400 = 4 \times 100 = 4 \times 10^2$$

Thus,

$$\frac{6,400,000}{400} = \frac{6.4 \times 10^6}{4.0 \times 10^2}$$

Since  $\dfrac{ab}{cd} = \dfrac{a}{c} \cdot \dfrac{b}{c}$ : $= \dfrac{6.4}{4.0} \times \dfrac{10^6}{10^2}$,

Since  $\dfrac{a^x}{a^y} = a^{x-y}$ : $= 1.6 \times 10^4$

## MIXTURES

● **PROBLEM** 943

How many liters of a liquid that is 74 percent alcohol must be combined with 5 liters of one that is 90 percent alcohol in order to obtain a mixture that is 84 percent alcohol?

<u>Solution</u>: If we let x represent the number of liters needed of the first liquid and remember that 74 percent of x is 0.74x, then the table (see table) shows all the data given in this problem.

|  | Number of liters | Percentage of alcohol | Number of liters of alcohol |
|---|---|---|---|
| First liquid | x | 74 | 0.74x |
| Second liquid | 5 | 90 | 0.90(5) = 4.5 |
| Mixture | x + 5 | 84 | 0.84(x + 5) |

We are told that we are combining the number of liters of alcohol in the 74 percent alcohol (0.74x) with the number of liters of alcohol in the 90 percent alcohol (4.5) to obtain the number of liters of alcohol in the 84 percent alcohol [0.84(x + 5)]. Thus

$$.74x + 4.5 = .84(x + 5)$$

Multiplying both sides by 100,

$$74x + 450 = 84 (x + 5)$$
$$74x + 450 = 84x + 420$$
$$30 = 10x$$
$$x = 3$$

Therefore, 3 liters of liquid that is 74 percent alcohol must be combined with 5 liters of one that is 90 percent alcohol to obtain a mixture of 84 percent alcohol.

● **PROBLEM** 944

How much water must be added to 5 liters of 90% ammonia solution to reduce the solution to a 60% solution?

<u>Solution</u>: The volume of the original solution is 5 liters. If we let x represent the number of liters of water added to the original solution, then the final solution will have a volume of (5 + x) liters. No ammonia is added so the only ammonia in the final solution

is the ammonia in the original solution. We can now record our given information in tabular form:

|  | Total Volume (liters) | Percent Concentration | liters of 100% Ammonia |
|---|---|---|---|
| Original solution | 5 | 90 | 4.5 |
| Final solution | 5 + x | 60 | 4.5 |

Note: The liters of 100% ammonia for the original solution was obtained in the following way: multiply the total volume of the original solution by the percent concentration of the original solution. Therefore,

$$5 \times 90\% = 5 \times \frac{90}{\underset{20}{100}} = \frac{90}{20} = \frac{9}{2} = 4.5 \text{ liters of } 100\%$$

ammonia. Also, the liters of 100% ammonia for the final solution is the same as the liters of 100% ammonia for the original solution; that is, 4.5 liters, since no ammonia was added.

The formula for the amount of substance in solution for the final solution is (total volume) X (percent concentration) = liters of 100% ammonia. Therefore: (5 + x) X 60% = 4.5
or
$$(5 + x) \frac{60}{100} = 4.5$$

or $$(5 + x) \cdot .60 = 4.5$$

or $$.60(5 + x) = 4.5$$

$$3 + .60x = 4.5$$

$$.6x = 4.5 - 3$$

$$.6x = 1.5$$

$$x = \frac{1.5}{.6} = 2.5 \text{ liters}$$

Therefore, 2.5 liters of water must be added to reduce the concentration to 60%.

Check: If 2.5 liters of water are added to the original solution, we will have 7.5 liters of solution. The amount of 100% ammonia remains the same 4.5 liters. The concentration equals

$$\frac{4.5}{7.5} \times 100 = \frac{450}{7.5} = 60\%.$$

How many grams of silver alloy which is 28% silver must be mixed with 24 grams of silver which is 8% silver to produce a new alloy which is 20% silver?

Solution: Let  x = number of grams of 28% silver to be used.  The relationship used to set up the equation is

Weight of 28% silver + Weight of 8% silver = Weight of silver in mixture.

$$.28x + .08(24) = .20(x + 24)$$
$$28x + 8(24) = 20(x + 24)$$
$$8x = 288$$
$$x = 36 \text{ grams of silver}$$

Check:   Weight of 28% silver = (.28)(36) = 10.08
Weight of 8% silver  = (.08)(24) = 1.92
Total amount of silver = 12 grams
The total mixture contains 24 + 36 = 60 grams, and 12 grams is 20% of 60 grams.

A chemist has an 18% solution and a 45% solution of a disinfectant. How many liters of each should be used to make 12 liters of a 36% solution?

Solution:   Let  x = Number of liters from the 18% solution
And  y = Number of liters from the 45% solution

(1)    $x + y = 12$

(2)    $.18x + .45y = .36(12) = 4.32$

Note that .18 of the first solution is pure disinfectant and that .45 of the second solution is pure disinfectant.  When the proper quantities are drawn from each mixture the result is 12 liters of mixture which is .36 pure disinfectant, i.e., the resulting mixture contains 4.32 liters of pure disinfectant.

When the equations are solved, it is found that

$$x = 4 \quad \text{and} \quad y = 8.$$

## EXPONENTIAL NOTATION

Simplify:

(a)  $2^3 \cdot 2^2$     (b)  $a^3 \cdot a^5$     (c)  $x^6 \cdot x^4$

Solution:  If  a  is any number and  n  is any positive integer, the product of the  n  factors  a·a·a ... a  is denoted by  $a^n$ .  a  is called the base and  n  is called the exponent.  Also, for base  a  and exponents  m  and  n,  m  and  n  being positive integers, we have the law:

$$a^m \cdot a^n = a^{m+n} .$$

Therefore,

(a) $\quad 2^3 \cdot 2^2 = (2 \cdot 2 \cdot 2)(2 \cdot 2) = 8 \cdot 4 = 32$

or $\quad\quad 2^3 \cdot 2^2 = 2^{3+2} = 2^5 = 32$

(b) $\quad a^3 \cdot a^5 = (a \cdot a \cdot a)(a \cdot a \cdot a \cdot a \cdot a)$

$$= (a \cdot a \cdot a \cdot a \cdot a \cdot a \cdot a \cdot a) = a^8$$

or $\quad\quad a^3 \cdot a^5 = a^{3+5} = a^8$

(c) $\quad x^6 \cdot x^4 = x^{6+4} = x^{10} .$

● **PROBLEM** 948

Simplify the following expressions:

(a) $\quad -3^{-2}$ $\quad\quad$ (b) $\quad (-3)^{-2}$ $\quad\quad$ (c) $\quad \dfrac{-3}{4^{-1}}$

<u>Solution</u>:

(a) Here the exponent applies only to 3.

Since $x^{-y} = \dfrac{1}{x^y}$, $\quad -3^{-2} = -(3^{-2}) = -\dfrac{1}{3^2} = -\dfrac{1}{9}.$

(b) In this case the exponent applies to the negative base. Thus, $\quad (-3)^{-2} = \dfrac{1}{(-3)^2} = \dfrac{1}{(-3)(-3)} = \dfrac{1}{9}.$

(c) $\quad \dfrac{-3}{4^{-1}} = \dfrac{-3}{(\frac{1}{4})^1} = \dfrac{-3}{\frac{1^1}{4^1}} = \dfrac{-3}{\frac{1}{4}}.$

Division by a fraction is equivalent to multiplication by that fraction's reciprocal, thus

$$\dfrac{-3}{\frac{1}{4}} = -3 \cdot \dfrac{4}{1} = -12,$$

and $\quad\quad \dfrac{-3}{4^{-1}} = -12.$

● **PROBLEM** 949

Simplify the quotient $\dfrac{2x^0}{(2x)^0} .$

<u>Solution</u>: The following two laws of exponents can be used to simplify the given quotient:

1) $a^0 = 1$ where a is any non-zero real number, and

2) $(ab)^n = a^n b^n$ where a and b are any two numbers.

In the given quotient, notice that the exponent in the numerator applies only to the letter x. However, the exponent in the denominator applies to both the number 2 and the letter x; that is, the exponent in the denominator applies to the entire term (2x). Using the first law, the numerator can be rewritten as:

$$2x^0 = 2(1) = 2$$

Using the second law with n = 0, the denominator can be rewritten as:

$$(2x)^0 = 2^0x^0$$

Using the first law again to further simplify the denominator:

$$(2x)^0 = 2^0x^0$$
$$= (1)(1)$$
$$= 1$$

Therefore,

$$\frac{2x^0}{(2x)^0} = \frac{2}{1} = 2$$

● **PROBLEM** 950

Determine the value of $(0.0081)^{-3/4}$.

Solution: $(0.0081) = .3 \times .3 \times .3 \times .3 = (.3)^4$,

therefor $(0.0081)^{-3/4} = \left(.3^4\right)^{-3/4}$

Recalling the property of exponents,

$$\left(a^x\right)^y = a^{x \cdot y}$$

we have,

$$\left(.3^4\right)^{-3/4} = .3^{(4)(-3/4)} = .3^{-3}.$$

Since $a^{-x} = \frac{1}{a^x}$, $.3^{-3} = \frac{1}{.3^3} = \frac{1}{0.027} = \frac{1}{\frac{27}{1000}}$

Division by a fraction is equivalent to multiplication by its reciprocal, thus,

$$\frac{1}{\frac{27}{1000}} = \frac{1000}{27}.$$

Hence,

$$(0.0081)^{-3/4} = \frac{1000}{27}.$$

Simplify $\left[\dfrac{1600 \times 10{,}000}{2000}\right]^{1/3}$ .

**Solution:** Observe $1600 = 16 \times 100 = 16 \times 10^2$

$$10{,}000 = 10^4$$
$$2{,}000 = 2 \times 10^3 .$$

Thus, $\left[\dfrac{1600 \times 10{,}000}{2000}\right]^{1/3} = \left[\dfrac{(16 \times 10^2)(10^4)}{2 \times 10^3}\right]^{1/3}$ .

Using the associative property,

$$= \left[\dfrac{16 \times (10^2 \times 10^4)}{2 \times 10^3}\right]^{1/3}$$

Recall: $a^x \cdot a^y = a^{x+y}$ ,

$$= \left[\dfrac{16 \times 10^6}{2 \times 10^3}\right]^{1/3}$$

Since $\dfrac{ab}{cd} = \dfrac{a}{c} \cdot \dfrac{b}{d}$ ,

$$= \left[\dfrac{16}{2} \times \dfrac{10^6}{10^3}\right]^{1/3}$$

Recall $\dfrac{a^x}{a^y} = a^{x-y}$ ,

$$= \left(8 \times 10^3\right)^{1/3}$$

Since $(ab)^x = a^x b^x$ ,

$$= 8^{1/3} \times 10^{3(1/3)}$$
$$= 2 \times 10^1$$
$$= 20$$

## SQUARE ROOTS

Evaluate $\sqrt{400}$ .

**Solution:** $400 = 4 \times 100$

Thus, $\sqrt{400} = \sqrt{4 \times 100}$

Since $\sqrt{ab} = \sqrt{a}\,\sqrt{b}$,

$$\sqrt{400} = \sqrt{4}\,\sqrt{100}$$
$$= 2 \cdot 10$$
$$= 20$$

Check: If $\sqrt{400}$ is 20, then $20^2$ must equal 400, which is true. Hence, 20 is the solution.

Evaluate $16^{-\frac{3}{4}}$ .

Solution:
$$16^{-\frac{3}{4}} = \frac{1}{16^{\frac{3}{4}}}$$

$$= \frac{1}{\left(\sqrt[4]{16}\right)^{3}} \quad .$$

Note that $2^{4} = 2 \cdot 2 \cdot 2 \cdot 2 = 16$, hence $\sqrt[4]{16} = 2$. Thus, $16^{-\frac{3}{4}}$

$$= \frac{1}{2^{3}} = \frac{1}{2 \cdot 2 \cdot 2} = \frac{1}{8} \quad .$$

Find the indicated roots.

(a) $\sqrt[5]{32}$  (b) $\pm \sqrt[4]{625}$  (c) $\sqrt[3]{-125}$  (d) $\sqrt[4]{-16}$.

Solution:  The following two laws of exponents can be used to solve these problems:  1) $\left(\sqrt[n]{a}\right)^{n} = \left(a^{1/n}\right)^{n} = a^{1} = a$, and 2) $\left(\sqrt[n]{a}\right)^{n} = \sqrt[n]{a^{n}}$.

(a) $\sqrt[5]{32} = \sqrt[5]{2^{5}} = \left(\sqrt[5]{2}\right)^{5} = 2$.  This result is true because $(2)^{5} = 32$, that is, $2 \cdot 2 \cdot 2 \cdot 2 \cdot 2 = 32$.

(b) $\sqrt[4]{625} = - \sqrt[4]{5^{4}} = \left(\sqrt[4]{5}\right)^{4} = 5$. This result is true because $\left(5^{4}\right) = 625$, that is, $5 \cdot 5 \cdot 5 \cdot 5 = 625$.

$-\sqrt[4]{625} = -\left(\sqrt[4]{5^{4}}\right) = -\left[\left(\sqrt[4]{5}\right)^{4}\right] = -\left[5\right] = -5$.  This result is true because $(-5)^{4} = 625$, that is, $(-5) \cdot (-5) \cdot (-5) \cdot (-5) = 625$.

(c) $\sqrt[3]{-125} = \sqrt[3]{(-5)^{3}} = (\sqrt[3]{-5})^{3} = -5$.  This result is true because $(-5)^{3} = -125$, that is, $(-5) \cdot (-5) \cdot (-5) = -125$.

(d)  There is no solution to $\sqrt[4]{-16}$ because any number raised to the fourth power is a positive number, that is, $N^{4} = (N) \cdot (N) \cdot (N) \cdot (N) = $ a positive number $\neq$ a negative number, $-16$.

Simplify:  (a) $\sqrt[3]{-512}$  (b) $\sqrt[4]{\frac{81}{16}}$  (c) $\sqrt[3]{-16} \div \sqrt[3]{-2}$.

Solution: (a) By the law of radicals which states that $\sqrt[n]{ab} = \sqrt[n]{a}\,\sqrt[n]{b}$ where a and b are any two numbers, $\sqrt[3]{-512} = \sqrt[3]{8(-64)} = \sqrt[3]{8}\sqrt[3]{-64}$. Therefore, $\sqrt[3]{-512} = \sqrt[3]{8}\sqrt[3]{-64} = (2)(-4) = -8$. The last result is true because $(2)^3 = 8$ and $(-4)^3 = -64$.

(b) By another law of radicals which states that $\sqrt[n]{\dfrac{a}{b}} = \dfrac{\sqrt[n]{a}}{\sqrt[n]{b}}$ where a and b are any two numbers, $\sqrt[4]{\dfrac{81}{16}} = \dfrac{\sqrt[4]{81}}{\sqrt[4]{16}}$ . Therefore, $\sqrt[4]{\dfrac{81}{16}} = \dfrac{\sqrt[4]{81}}{\sqrt[4]{16}} = \dfrac{3}{2}$. The last result is true because $(3)^4 = 81$ and $(2)^4 = 16$.

(c) By the law of radicals stated in example (b), $\sqrt[3]{-16} \div \sqrt[3]{-2} = \dfrac{\sqrt[3]{-16}}{\sqrt[3]{-2}} = \sqrt[3]{\dfrac{-16}{-2}} = \sqrt[3]{8} = 2$. The last result is true because $(2)^3 = 8$.

## LOGARITHMS

• PROBLEM 956

Find $\log_{10} 100$.

Solution: The following solution presents 2 methods for solving the given problem.

Method I. The statement $\log_{10} x = y$ is equivalent to $10^y = x$, hence $\log_{10} 100 = x$ is equivalent to $10^x = 100$. Since $10^2 = 100$, $\log_{10} 100 = 2$.

Method II. Note that $100 = 10 \times 10$; thus $\log_{10} 100 = \log_{10} (10 \times 10)$. Recall: $\log_x (a \times b) = \log_x a + \log_x b$, therefore

$$\log_{10} (10 \times 10) = \log_{10} 10 + \log_{10} 10$$

$$= \quad 1 \quad + \quad 1$$

$$= 2.$$

• PROBLEM 957

Find the values of the following logarithims:

a) $\log_{10} 10$     b) $\log_{10} 100$     c) $\log_{10} 1$

d) $\log_{10} 0.1$     e) $\log_{10} 0.01$

__Solution__:  The logarithmic expression $N = \log_b x$ is equivalent to $b^N = x$.  Hence,

a) Let $N_1 = \log_{10} 10$.  Then the logarithmic expression $N_1 = \log_{10} 10$ is equivalent to $10^{N_1} = 10$.  Since $10^1 = 10$, $N_1 = 1$.  Therefore, $N_1 = 1 = \log_{10} 10$.

b) Let $N_2 = \log_{10} 100$.  Then the logarithmic expression $N_2 = \log_{10} 100$ is equivalent to $10^{N_2} = 100$.  Since $10^2 = 100$, $N_2 = 2$.  Therefore, $N_2 = 2 = \log_{10} 100$.

c) Let $N_3 = \log_{10} 1$.  Then the logarithmic expression $N_3 = \log_{10} 1$ is equivalent to $10^{N_3} = 1$.  Since $10^0 = 1$, $N_3 = 0$.  Therefore, $N_3 = 0 = \log_{10} 1$.

d) Let $N_4 = \log_{10} 0.1 = \log_{10} \frac{1}{10}$.  Then the logarithmic expression $N_4 = \log_1 0.1 = \log_{10} \frac{1}{10}$ is equivalent to $10^{N_4} = \frac{1}{10}$.  Since $10^{-1} = \frac{1}{10^1} = \frac{1}{10}$, $N_4 = -1$.  Therefore, $N_4 = -1 = \log_{10} 0.1$.

e) Let $N_5 = \log_{10} 0.01 = \log_{10} \frac{1}{100}$.  Then the logarithmic expression $N_5 = \log_{10} 0.01 = \log_{10} \frac{1}{100}$ is equivalent to $10^{N_5} = \frac{1}{100}$.  Since $10^{-2} = \frac{1}{10^2} = \frac{1}{100}$, $N_5 = -2$.  Therefore, $N_5 = -2 = \log_{10} 0.01$.

● **PROBLEM** 958

Find  $\log_{10}\left(10^2 \cdot 10^{-3} \cdot 10^5\right)$.

__Solution__:  Recall  $\log_x (a \cdot b \cdot c) = \log_x a + \log_x b + \log_x c$ .  Thus
$$\log_{10}\left(10^2 \cdot 10^{-3} \cdot 10^5\right) = \log_{10} 10^2 + \log_{10} 10^{-3} + \log_{10} 10^5$$
Recall  $\log_b b^x = x$, since  $b^x = b^x$; therefore, $\log_{10} 10^2 + \log_{10} 10^{-3} + \log_{10} 10^5 = 2 + (-3) + 5 = 4$.  Thus  $\log_{10}\left(10^2 \cdot 10^{-3} \cdot 10^5\right) = 4$.
Another method of finding  $\log_{10}\left(10^2 \cdot 10^{-3} \cdot 10^5\right)$  is to note
$10^2 \cdot 10^{-3} \cdot 10^5 = 10^{2+(-3)+5} = 10^4$  $\left(\text{because } a^x \cdot a^y \cdot a^z = a^{x+y+z}\right)$.
Thus  $\log_{10}\left(10^2 \cdot 10^{-3} \cdot 10^5\right) = \log_{10} 10^4 = 4$.

**• PROBLEM 959**

Find the logarithm of $3^2$.

Solution: Recall that $\log_b x^y = y \log_b x$; thus

$$\log_{10} 3^2 = 2 \log_{10} 3$$

Referring to a table of common logarithms we find:

$$\log_{10} 3 = .4771; \text{ hence,}$$

$$= 2(.4771)$$

$$= .9542 .$$

Thus, $\log_{10} 3^2 = .9542$.

**• PROBLEM 960**

Given $\log_{10}2 = 0.3010$, find $\log_{10}32$.

Solution: Note that,

$$32 = 2 \cdot 2 \cdot 2 \cdot 2 \cdot 2 = 2^5.$$

Thus,

$$\log_{10}32 = \log_{10}2^5.$$

Recall the logarithmic property,

$$\log_b x^y = y \log_b x.$$

Hence,

$$\log_{10}32 = \log_{10}2^5 = 5 \log_{10}2$$

$$= 5 (0.3010)$$

$$= 1.5050$$

**• PROBLEM 961**

If $\log_{10} 3 = .4771$ and $\log_{10} 4 = .6021$, find $\log_{10} 12$.

Solution: Since $12 = 3 \times 4$,

$$\log_{10} 12 = \log_{10}(3)(4).$$

Since $\log_b(xy) = \log_b x + \log_b y$

$$\log_{10}(3 \times 4) = \log_{10} 3 + \log_{10} 4$$

$$= .4771 + .6021$$

● **PROBLEM** 962

Given that $\log_{10} 2 = 0.3010$ and $\log_{10} 3 = 0.4771$, find $\log_{10}\sqrt{6}$ .

Solution: $\sqrt{6} = 6^{\frac{1}{2}}$, thus $\log_{10}\sqrt{6} = \log_{10}6^{\frac{1}{2}}$ . Since $\log_b x^y = y \log_b x$, $\log_{10}6^{\frac{1}{2}} = \frac{1}{2} \log_{10}6$ . Therefore $\log_{10}\sqrt{6} = \frac{1}{2} \log_{10}6$ . $6 = 3 \cdot 2$, hence $\frac{1}{2} \log_{10}6 = \frac{1}{2} \log_{10}(3 \cdot 2)$ . Recall $\log_{10}(a \cdot b) = \log_{10}a + \log_{10}b$. Thus $\frac{1}{2} \log_{10}(3 \cdot 2) = \frac{1}{2}\left(\log_{10}3 + \log_{10}2\right)$ . Replace our values for $\log_{10}3$ and $\log_{10}2$ ,

● **PROBLEM** 963

Find the logarithm of 30,700.

Solution: First express 30,700 in scientific notation. $30,700 = 3.07 \times 10^4$. 4 is the characteristic. To find the mantissa, see a table of common logarithms of numbers The mantissa is 4871. Thus log 30,700 = 4 + .4871 = 4.4871.

## ANTILOGARITHMS

● **PROBLEM** 964

Find log 0.0364 .

Solution: $0.0364 = 3.64 \times 10^{-2}$ . Therefore, the characteristic, the power of 10, is -2. From a table of logarithms, the mantissa for 3.64 is 0.5611. Therefore, log 0.0364 = -2 + 0.5611 = -1.4389.

● **PROBLEM** 965

What is the value of log 0.0148?

Solution: $0.0148 = 1.48 \times 10^{-2}$. The characteristic is the exponent of 10. Hence, the characteristic is -2. The mantissa for 148 can be found in a table of logarithms. The mantissa is 0.1703. Therefore, log 0.0148 = -2 + 0.1703 = -2.0000 + 0.1703 = -1.8297. Notice that the number 0.0148 is less than 1. Therefore, the value of log 0.0148 must be negative, as it was found to be.

Find Antilog$_{10}$ 0.8762 - 2 .

**Solution**: Let N = Antilog$_{10}$ 0.8762 - 2. The following relationship between log and antilog exists: $\log_{10} x = a$ is the equivalent of x = antilog$_{10}$a . Therefore,

$$\log_{10} N = 0.8762 - 2.$$

The characteristic is -2. The mantissa is 0.8762. The number that corresponds to this mantissa is 7.52. This number is found from a table of common logarithms, base 10. Therefore,

$$N = 7.52 \times 10^{-2}$$

$$= 7.52 \times \left(\frac{1}{10^2}\right)$$

$$= 7.52 \times \left(\frac{1}{100}\right)$$

$$= 7.52(.01)$$

$$N = 0.0752 .$$

Therefore, N = Antilog$_{10}$ 0.8762 - 2 = 0.0752.

Find Antilog$_{10}$ 1.4850.

**Solution:** By definition, Antilog$_{10}$ a = N is equivalent to $\log_{10} N = a$. Let Antilog$_{10}$1.4850 = N. Hence, Antilog$_{10}$1.4850 = N is equivalent to $\log_{10} N = 1.4850$. The characteristic is 1. The mantissa is 0.4850. Therefore, the number that corresponds to this mantissa will be multiplied by $10^1$ or 10. The mantissas which appear in a table of common logarithms and are closest to the mantissa 0.4850 are 0.4843 and 0.4857. The number that corresponds to the mantissa 0.4850 will be found by interpolation.

|  | Number | | Logarithms | |
|---|---|---|---|---|
| | 3.05 | | 0.4843 | |
| d | x | | 0.4850 | .0007 |
| .01 | 3.06 | | 0.4857 | .0014 |

Set up the following proportion.

$$\frac{d}{.01} = \frac{.0007}{.0014}$$

cross-multiplying, .0014d = (.01)(.0007), or $d = .01\left(\frac{.0007}{.0014}\right)$

$$= \left(1 \times 10^{-2}\right)\left(\frac{7 \times 10^{-4}}{1.4 \times 10^{-3}}\right)$$

$$= \frac{7 \times 10^{-6}}{1.4 \times 10^{-3}} = \frac{7}{1.4} \times \frac{10^{-6}}{10^{-3}} = 5 \times 10^{-6-(-3)}$$

$$= 5 \times 10^{-3} = 5 \times .001 = .005$$

Hence, d = 0.005

$$x = d + 3.05$$

$$= 0.005 + 3.050$$

$$= 3.055$$

Hence, N = Antilog$_{10}$1.4850 = 3.055 $\times$ 10

$$= 30.550$$

$$= 30.55$$

Therefore Antilog$_{10}$1.4850 = 30.55.

● PROBLEM 968

Find Antilog 2.3625.

Solution:  By definition, b = **Antilog** a, is equivalent to  log b = a.
Let  N = **Antilog** 2.3625.  Therefore, log N = 2.3625.  The characteristic
is 2.  Hence, the number that corresponds to the mantissa 0.3625 will
be multiplied by $10^2$  or  100.  In a table of four-place common logarithms,
the mantissas 0.3617 and 0.3636 are those given that are closest to
0.3625.  Therefore, since the mantissa 0.3625 does not appear in the table,
the number that corresponds to this mantissa will be found through interpolation.

| | | Number | | Log | | |
|---|---|---|---|---|---|---|
| .01 | d | 2.30 | | 0.3617 | .0008 | .0019 |
| | | x | | 0.3625 | | |
| | | 2.31 | | 0.3636 | | |

The following proportion is now established:

$$\frac{d}{.01} = \frac{.0008}{.0019}$$

Cross multiplying,

$$.0019d = (.01)(.0008)$$

$$d = .01\left(\frac{.0008}{.0019}\right)$$

$$= 1 \times 10^{-2}\left(\frac{8 \times 10^{-4}}{1.9 \times 10^{-3}}\right)$$

$$= \frac{8 \times 10^{-2+(-4)}}{1.9 \times 10^{-3}} = \frac{8 \times 10^{-6}}{1.9 \times 10^{-3}}$$

$$= \frac{8}{1.9} \times \frac{10^{-6}}{10^{-3}}$$

$$= 4.2 \times 10^{-6-(-3)}$$

$$= 4.2 \times 10^{-3}$$

$$d = .0042$$

Hence, $x = 2.30 + 0.0042 = 2.3042 \approx 2.304$. Therefore, **Antilog** 2.3625 =

$$N = 2.304 \times 10^2$$

$$= 2.304 \times 100$$

$$= 230.4$$

● **PROBLEM** 969

Determine the value of x such that $10^x = 3.142$.

Solution: The statement $10^x = 3.142$ is equivalent by definition to $\log_{10} 3.142 = x$. Thus we must find log 3.142, using the following interpolation:

$$.01 \left( .002 \left( \begin{array}{c|c} \text{Number} & \text{Log} \\ \hline 3.140 & .4969 \\ 3.142 & \\ \hline 3.150 & .4983 \end{array} \right) x \right) .0014$$

We set up the proportion,

$$\frac{.002}{.01} = \frac{x}{.0014}$$

Cross multiply to obtain,

$$.01x = .0000028$$

$$x = .00028$$

$$x \simeq .0003$$

Thus log 3.142 = .4969 + .0003 = 0.4972
Therefore $x = \log_{10} 3.142 = 0.4972$

## QUADRATIC EQUATIONS

● **PROBLEM** 970

Find the roots of the equation $x^2 + 12 - 85 = 0$.

Solution: The roots of this equation may be found using the quadratic formula

$$x = \frac{-B \pm \sqrt{B^2 - 4AC}}{2A}$$

In this equation $A = 1$, $B = 12$, and $C = -85$. Hence, by the quadratic formula,

$$x = \frac{-12 + \sqrt{144 + 340}}{2} \qquad \text{or} \qquad x = \frac{-12 - \sqrt{144 + 340}}{2}$$

$$x = \frac{-12 + 22}{2} \qquad \text{or} \qquad x = \frac{-12 - 22}{2}$$

Therefore $x = 5$ or $x = -17$. This is equivalent to the statement that the solution set is $\{-17, 5\}$.

● **PROBLEM** 971

Solve for $x$: $4x^2 - 7 = 0$.

<u>Solution</u>: This quadratic equation can be solved for $x$ using the quadratic formula, which applies to equations in the form $ax^2 + bx + c = 0$ (in our equation $b = 0$). There is, however, an easier method that we can use:

Adding 7 to both members, $\qquad 4x^2 = 7$

dividing both sides by 4, $\qquad x^2 = \dfrac{7}{4}$

taking the square root of both sides, $\quad x = \pm\sqrt{\dfrac{7}{4}} = \pm\dfrac{\sqrt{7}}{2}$.

The double sign $\pm$ (read "plus or minus") indicates that the two roots of the equation are

$$+\frac{\sqrt{7}}{2} \quad \text{and} \quad -\frac{\sqrt{7}}{2}.$$

● **PROBLEM** 972

Use the quadratic formula to solve for x in the equation $x^2 - 5x + 6 = 0$.

<u>Solution</u>: The quadratic formula, $x = \dfrac{-b \pm \sqrt{b^2 - 4ac}}{2a}$, is used to solve equations in the form $ax^2 + bx + c = 0$. Here $a = 1$, $b = -5$, and $c = 6$. Hence

$$x = \frac{-(-5) \pm \sqrt{(-5)^2 - 4 \cdot 1 \cdot 6}}{2 \cdot 1} = \frac{5 \pm \sqrt{25 - 24}}{2}$$

$$= \frac{5 \pm \sqrt{1}}{2}$$

$$= \frac{5 \pm 1}{2}$$

$$= \frac{5 + 1}{2} \quad \text{or} \quad \frac{5 - 1}{2}$$

$$= \frac{6}{2} \quad \text{or} \quad \frac{4}{2}$$

$$= 3 \quad \text{or} \quad 2$$

Thus the roots of the equation $x^2 - 5x + 6 = 0$ are $x = 3$ and $x = 2$.

● **PROBLEM** 973

Solve $6x^2 - 7x - 20 = 0$.

_Solution_: $6x^2 - 7x - 20 = 0$ is not factorable. Therefore, find the roots of the quadratic equation $ax^2 + bx + c$ using:

$$x = \frac{-b \pm \sqrt{b^2 - 4ac}}{2a},$$

where $a = 6$, $b = -7$, $c = -20$.

$$x = \frac{7 \pm \sqrt{49 - 4(6)(-20)}}{12}$$

$$x = \frac{7 \pm \sqrt{529}}{12} = \frac{7 \pm 23}{12}.$$

Therefore,

$$x_1 = \frac{7 + 23}{12} = \frac{30}{12} = \frac{5}{2}$$

$$x_2 = \frac{7 - 23}{12} = -\frac{16}{12} = -\frac{4}{3}.$$

# INDEX

Numbers on this page refer to <u>PROBLEM NUMBERS</u>, not page numbers

Numbers on this page refer to **PROBLEM NUMBERS**, not page numbers

# THE PERIODIC TABLE

**KEY**

| | |
|---|---|
| Group Classification → | IVA / IVB |
| Atomic Number → | 22 |
| Symbol → | Ti |
| Atomic Weight → | 47.88 |

( ) indicates most stable or best known isotope

TRANSITIONAL METALS

| 1 IA/IA | 2 IIA/IIA | 3 IIIA/IIIB | 4 IVA/IVB | 5 VA/VB | 6 VIA/VIB | 7 VIIA/VIIB | 8 VIIIA/VIII | 9 VIIIA/VIII | 10 VIIIA/VIII | 11 IB/IB | 12 IIB/IIB | 13 IIIB/IIIA | 14 IVB/IVA | 15 VB/VA | 16 VIB/VIA | 17 VIIB/VIIA | 18 VIII/0 |
|---|---|---|---|---|---|---|---|---|---|---|---|---|---|---|---|---|---|
| 1 H 1.008 | | | | | | | | | | | | | | | | | 2 He 4.003 |
| 3 Li 6.941 | 4 Be 9.012 | | | | | | | | | | | 5 B 10.811 | 6 C 12.011 | 7 N 14.007 | 8 O 15.999 | 9 F 18.998 | 10 Ne 20.180 |
| 11 Na 22.990 | 12 Mg 24.305 | | | | | | | | | | | 13 Al 26.982 | 14 Si 28.086 | 15 P 30.974 | 16 S 32.066 | 17 Cl 35.453 | 18 Ar 39.948 |
| 19 K 39.098 | 20 Ca 40.078 | 21 Sc 44.956 | 22 Ti 47.88 | 23 V 50.942 | 24 Cr 51.996 | 25 Mn 54.938 | 26 Fe 55.847 | 27 Co 58.933 | 28 Ni 58.693 | 29 Cu 63.546 | 30 Zn 65.39 | 31 Ga 69.723 | 32 Ge 72.61 | 33 As 74.922 | 34 Se 78.96 | 35 Br 79.904 | 36 Kr 83.8 |
| 37 Rb 85.468 | 38 Sr 87.62 | 39 Y 88.906 | 40 Zr 91.224 | 41 Nb 92.906 | 42 Mo 95.94 | 43 Tc (97.907) | 44 Ru 101.07 | 45 Rh 102.906 | 46 Pd 106.4 | 47 Ag 107.868 | 48 Cd 112.411 | 49 In 114.818 | 50 Sn 118.710 | 51 Sb 121.757 | 52 Te 127.60 | 53 I 126.905 | 54 Xe 131.29 |
| 55 Cs 132.905 | 56 Ba 137.327 | 57 La 138.906 | 72 Hf 178.49 | 73 Ta 180.948 | 74 W 183.84 | 75 Re 186.207 | 76 Os 190.23 | 77 Ir 192.22 | 78 Pt 195.08 | 79 Au 196.967 | 80 Hg 200.59 | 81 Tl 204.383 | 82 Pb 207.2 | 83 Bi 208.980 | 84 Po (208.982) | 85 At (209.982) | 86 Rn (222.018) |
| 87 Fr (223.020) | 88 Ra (226.025) | 89 Ac (227.028) | 104 Unq (261.11) | 105 Unp (262.114) | 106 Unh (263.118) | 107 Uns (262.12) | 108 Uno (265) | 109 Une (266) | 110 Uun (272) | | | | | | | | |

Group 1: Alkali Metals — Group 2: Alkaline Earth Metals — Group 17: Halogens — Group 18: Noble Gases

**LANTHANIDE SERIES**

| 58 Ce 140.115 | 59 Pr 140.908 | 60 Nd 144.24 | 61 Pm (144.913) | 62 Sm 150.36 | 63 Eu 151.965 | 64 Gd 157.25 | 65 Tb 158.925 | 66 Dy 162.50 | 67 Ho 164.930 | 68 Er 167.26 | 69 Tm 168.934 | 70 Yb 173.04 | 71 Lu 174.967 |
|---|---|---|---|---|---|---|---|---|---|---|---|---|---|

**ACTINIDE SERIES**

| 90 Th 232.038 | 91 Pa 231.036 | 92 U 238.029 | 93 Np (237.048) | 94 Pu (244.064) | 95 Am (243.061) | 96 Cm (247.070) | 97 Bk (247.070) | 98 Cf (251.080) | 99 Es (252.083) | 100 FM (257.095) | 101 Md (258.1) | 102 No (259.101) | 103 Lr (262.11) |
|---|---|---|---|---|---|---|---|---|---|---|---|---|---|

## Fundamental constants and equivalents

| CONSTANT (SUMBOL) | VALUE |
|---|---|
| Avogadro number (N) | $6.0221367(36) \times 10^{23}$ mole$^{-1}$ |
| atomic mass unit ($\mu$) | $1.6605402(10) \times 10^{-24}$ g |
| Bohr radius ($a_0$) | $5.29177249(24) \times 10^{-9}$ cm |
| electron rest mass ($m_e$) | $5.48579903(13) \times 10^{-4}$ awu |
| | $= 9.1093897(54) \times 10^{-28}$ g |
| electronic charge ($e$) | $1.60217733(49) \times 10^{-19}$ C |
| faraday ($F$) | $9.6485309(29) \times 10^4$ C/mol |
| gas constant ($R$) | $8.20562(35) \times 10^{-2}$ $l$ atm mol$^{-1}$ K$^{-1}$ |
| | $8.314510(70)$J mol$^{-1}$ K$^{-1}$ |
| | $1.98428$ cal K$^{-1}$ mol$^{-1}$ |
| molar volume, ideal gas at STP ($V_0$) | $2.41410(19) \times 10^1$ l mol$^{-1}$ |
| neutron rest mass ($m_n$) | $1.6749286(10) \times 10^{-24}$ g |
| | $1.008664904(14)$ amu |
| Planck's constant ($h$) | $6.6260755(40) \times 10^{-34}$ JS |
| or | $6.6260755(40) \times 10^{-27}$ erg sec |
| proton rest mass ($m_\mu$) | $1.627623\ (10) \times 10^{-24}$ g |
| or | $1.007276470(12)$ amu |
| velocity of light, in vacuum ($c$) | $2.99792458 \times 10^8$ msec$^{-1}$ |

### EQUIVALENTS

| | |
|---|---|
| 1 m = 39.370 in. | 1 kg = 2.2046 lb (avoirdupois) |
| 1 in. = 2.54 cm (exact) | 1 lb (avoirdupois) = 453.59 g |
| 1 angstrom (Å) = 1 × 10–8 cm (exact) | 1 oz (avoirdupois) = 28.350 g |
| 1 km = 0.62137 miles | 1 atm = 760 mm Hg (exact) |
| 1 liter = 1.0567 qt (U.S.) | 1 atm = 14.696 psi |
| 1 gal (U.S.) = 3.7854 liters | 1 joule = $10^7$ ergs (exact) |
| 1 cu ft = 28.317 liters | 1 eV = $1.6022 \times 10^{-12}$ erg |
| 1 ml = 1 cc (exact) | 1 eV = $3.8293 \times 10^{-20}$ cal |
| 1 ml = 0.061025 cu in. | 1 cal = $4.1840 \times 10^7$ ergs |
| 1 cu in. = 16.387 ml | 1 cal = $1.1622 \times 10^{-6}$ kW=hr |
| t°C = (t$_{°F}$ − 32)/1.8 | 1 Btu = 251.996 cal |

## Vapor pressure of water at different temperatures

| TEMPERATURE, °C | VAPOR PRESSURE, mm Hg | TEMPERATURE, °C | VAPOR PRESSURE mm Hg |
|---|---|---|---|
| 0 | 4.59 | 29 | 30.06 |
| 5 | 6.55 | 30 | 31.84 |
| 10 | 9.21 | 31 | 33.72 |
| 11 | 9.85 | 32 | 35.69 |
| 12 | 10.52 | 33 | 37.75 |
| 13 | 11.24 | 34 | 39.93 |
| 14 | 11.99 | 35 | 42.20 |
| 15 | 12.79 | 40 | 55.37 |
| 16 | 13.64 | 45 | 71.93 |
| 17 | 14.54 | 50 | 92.59 |
| 18 | 15.48 | 55 | 118.15 |
| 19 | 16.48 | 60 | 149.50 |
| 20 | 17.54 | 65 | 187.68 |
| 21 | 18.66 | 70 | 233.84 |
| 22 | 19.84 | 75 | 289.25 |
| 23 | 21.08 | 80 | 355.33 |
| 24 | 22.39 | 85 | 433.65 |
| 25 | 23.77 | 90 | 525.92 |
| 26 | 25.22 | 95 | 634.02 |
| 27 | 26.75 | 100 | 759.96 |
| 28 | 28.37 | 150 | 3568.2 |

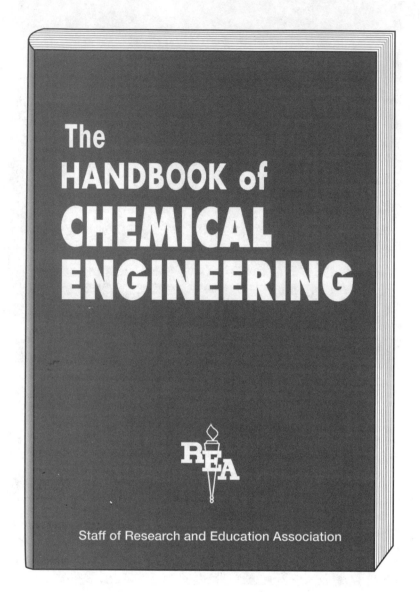

The
HANDBOOK of
CHEMICAL
ENGINEERING

**R**$_{EA}$

Staff of Research and Education Association

*Available at your local bookstore or order directly from us by sending in coupon below.*

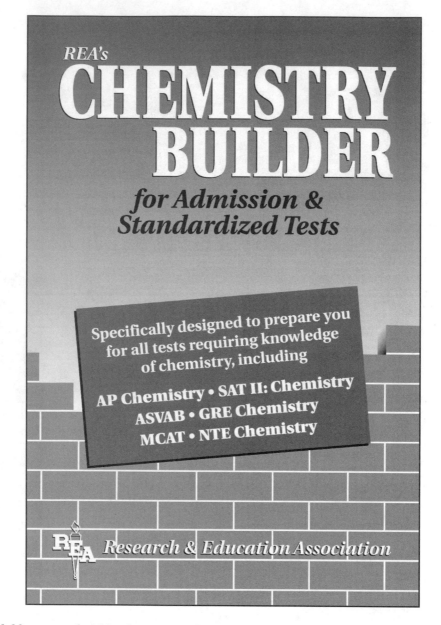

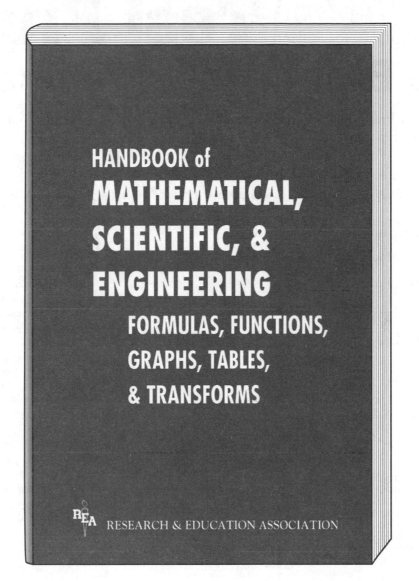

# REA's Test Preps
# The Best in Test Preparation

- REA "Test Preps" are **far more** comprehensive than any other test preparation series
- Each book contains up to **eight** full-length practice tests based on the most recent exams
- **Every** type of question likely to be given on the exams is included
- Answers are accompanied by **full** and **detailed** explanations

*REA publishes over 60 Test Preparation volumes in several series. They include:*

**Advanced Placement Exams (APs)**
Biology
Calculus AB & Calculus BC
Chemistry
Computer Science
Economics
English Language & Composition
English Literature & Composition
European History
Government & Politics
Physics B & C
Psychology
Spanish Language
Statistics
United States History

**College-Level Examination Program (CLEP)**
Analyzing and Interpreting Literature
College Algebra
Freshman College Composition
General Examinations
General Examinations Review
History of the United States I
History of the United States II
Human Growth and Development
Introductory Sociology
Principles of Marketing
Spanish

**SAT II: Subject Tests**
Biology E/M
Chemistry
English Language Proficiency Test
French
German

**SAT II: Subject Tests (cont'd)**
Literature
Mathematics Level IC, IIC
Physics
Spanish
United States History
Writing

**Graduate Record Exams (GREs)**
Biology
Chemistry
Computer Science
General
Literature in English
Mathematics
Physics
Psychology

**ACT** - ACT Assessment

**ASVAB** - Armed Services Vocational Aptitude Battery

**CBEST** - California Basic Educational Skills Test

**CDL** - Commercial Driver License Exam

**CLAST** - College Level Academic Skills Test

**COOP & HSPT** - Catholic High School Admission Tests

**ELM** - California State University Entry Level Mathematics Exam

**FE (EIT)** - Fundamentals of Engineering Exams - For both AM & PM Exams

**FTCE** - Florida Teacher Certification Exam

**GED** - High School Equivalency Diploma Exam (U.S. & Canadian editions)

**GMAT CAT** - Graduate Management Admission Test

**LSAT** - Law School Admission Test

**MAT** - Miller Analogies Test

**MCAT** - Medical College Admission Test

**MTEL** - Massachusetts Tests for Educator Licensure

**MSAT** - Multiple Subjects Assessment for Teachers

**NJ HSPA** - New Jersey High School Proficiency Assessment

**NYSTCE: LAST & ATS-W** - New York State Teacher Certification

**PLT** - Principles of Learning & Teaching Tests

**PPST** - Pre-Professional Skills Tests

**PSAT** - Preliminary Scholastic Assessment Test

**SAT I** - Reasoning Test

**TExES** - Texas Examinations of Educator Standards

**THEA** - Texas Higher Education Assessment

**TOEFL** - Test of English as a Foreign Language

**TOEIC** - Test of English for International Communication

**USMLE Steps 1,2,3** - U.S. Medical Licensing Exams

**U.S. Postal Exams 460 & 470**

---

**RESEARCH & EDUCATION ASSOCIATION**          **website: www.rea.com**
61 Ethel Road W. • Piscataway, New Jersey 08854 • Phone: (732) 819-8880

### Please send me more information about your Test Prep books

Name _____

Address _____

City _____ State _____ Zip _____